WileyPLUS

WileyPLUS is a research-based online environment for effective teaching and learning.

WileyPLUS builds students' confidence because it takes the guesswork out of studying by providing students with a clear roadmap:

- what to do
- how to do it
- if they did it right

It offers interactive resources along with a complete digital textbook that help students learn more. With *WileyPLUS*, students take more initiative so you'll have greater impact on their achievement in the classroom and beyond.

Now available for

Bb

Blackboard

For more information, visit www.wileyplus.com

WileyPLUS

ALL THE HELP, RESOURCES, AND PERSONAL SUPPORT YOU AND YOUR STUDENTS NEED!

www.wileyplus.com/resources

1st DAY OF CLASS ...AND BEYOND!

2-Minute Tutorials and all of the resources you and your students need to get started

WileyPLUS
Student Partner Program

Student support from an experienced student user

Wiley Faculty Network

Collaborate with your colleagues, find a mentor, attend virtual and live events, and view resources
www.WhereFacultyConnect.com

WileyPLUS
Quick Start

Pre-loaded, ready-to-use assignments and presentations created by subject matter experts

Technical Support 24/7
FAQs, online chat, and phone support
www.wileyplus.com/support

Your *WileyPLUS* Account Manager, providing personal training and support

DISCOVERING PHYSICAL GEOGRAPHY

Third Edition

Alan F. Arbogast   Michigan State University

WILEY

VICE PRESIDENT AND PUBLISHER   Petra Recter
EXECUTIVE EDITOR   Ryan Flahive
SENIOR ASSOCIATE EDITOR   Veronica Armour
ASSISTANT EDITOR   Julia Nollen
SENIOR CONTENT MANAGER   Micheline Frederick
SENIOR PRODUCTION EDITOR   Janet Foxman
SENIOR PHOTO EDITOR   Billy Ray
MEDIA SPECIALIST   Anita Castro
CREATIVE DIRECTOR   Harry Nolan
SENIOR DESIGNER   Wendy Lai
MARKETING MANAGER   Suzanne Bochet
PRODUCTION ASSISTANT   John Du Val
FRONT-COVER PHOTO   © Outdoor-Archiv/Alamy
BACK-COVER PHOTO   © Alan Arbogast

Images credited with "USGS" are courtesy of the U.S. Geological Survey. Due to space considerations, the abbreviated version "USGS" appears throughout the text.

This book was set in 10/12 Times Roman by CodeMantra and printed and bound by Quad Graphics/Versailles.

This book is printed on acid-free paper. ∞

Founded in 1807, John Wiley & Sons, Inc., has been a valued source of knowledge and understanding for more than 200 years, helping people around the world meet their needs and fulfill their aspirations. Our company is built on a foundation of principles that include responsibility to the communities we serve and where we live and work. In 2008, we launched a Corporate Citizenship Initiative, a global effort to address the environmental, social, economic, and ethical challenges we face in our business. Among the issues we are addressing are carbon impact, paper specifications and procurement, ethical conduct within our business and among our vendors, and community and charitable support. For more information, please visit our website: *www.wiley.com/go/citizenship*.

Library of Congress Cataloging-in-Publication Data:
Arbogast, Alan F.
  Discovering physical geography/Alan F. Arbogast, Michigan State University. -- Third edition.
    pages cm.

Includes index.
ISBN 978-1-118-52678-1 (pbk.)
  1. Physical geography--Textbooks. I. Title.

GB54.5.A73 2014
910'.02--dc23
                                        2013038705

978-1-118-52678-1 (Main Book ISBN)
978-1-118-70772-2 (Binder-Ready Version ISBN)
Printed in the United States of America
10 9 8 7 6 5 4 3 2

For Jenn, Hannah, and Rosie

ABOUT THE AUTHOR

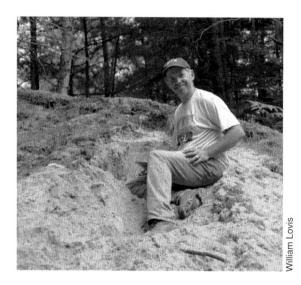

William Lovis

ALAN F. ARBOGAST is Professor and Chairperson in the Department of Geography at Michigan State University in East Lansing, Michigan. He is an award-winning instructor at MSU where he teaches a variety of classes at the undergraduate and graduate levels. Most of these courses are related to physical geography, geomorphology, and human-environment interactions. Alan frequently teaches large lecture classes and is known as an enthusiastic instructor who uses innovative approaches. He has also taught extensively in Australia and New Zealand in association with MSU's study abroad program.

Alan's research focuses on the Holocene landscape evolution of eolian, coastal, and fluvial environments in the Great Lakes region. Most of his research focuses on the age and formation of coastal sand dunes along Michigan's Great Lakes. Alan has published over 30 scientific papers, book chapters, and monographs related to this work. His research has been funded by the National Aeronautics and Space Administration, the National Science Foundation, and the state of Michigan. He is a member of the Association of American Geographers, Geological Society of America, St. George Geographical Society, and American Quaternary Association. Alan is married to Jennifer and has two daughters, Hannah and Rosie.

Follow Alan on Twitter at: @ArbogastDPG, where he posts regularly on topics related to geography and the environment.

Introduction to Physical Geography is a high-enrollment course at most universities. The usual goal of this course is to help students understand the Earth as a natural system and how various processes on the planet operate over time and space. Given the interactions that occur among these natural processes, physical geography requires the integration of many different topics. For example, students are expected to understand how seasonal Earth–Sun relationships affect atmospheric circulation, which in turn influences the distribution of vegetation. In addition, physical geography is an applied discipline that can inform decisions about environmental issues such as global warming, earthquake hazards, coastal erosion in populated areas, soil degradation, and deforestation, to name only a few.

Discovering Physical Geography: A Visually Oriented, Interactive Approach

Everyone associated with geography is fully aware that geographical literacy in the United States is very poor. This poor comprehension exists not only with respect to the fundamental issue of locating places, but also in understanding the age and processes associated with physical landscapes. Because many students enroll in a physical geography course only to fulfill a general education requirement in natural science, they frequently have little enthusiasm for the subject or fail to see the relevance of the class in their lives. In addition, a high percentage of students are simply afraid of science and are thus intimidated by the course. As a result, they believe that their chances of success in the class are low.

Discovering Physical Geography confronts these barriers to learning and classroom success in a number of unique ways. At a fundamental level, the text is written in a conversational style that is easily understandable to the average nonmajor. This writing style was not chosen at the expense of science, as the topics covered are dealt with in the scope and breadth as they are elsewhere. In fact many topics, such as Earth/Sun geometry, reconstructing past climates, ongoing climate change, geologic time, tides, and formation of sand dunes, to name a few, are covered in a more comprehensive way in this text than in others. The discussions are

just presented in a more accessible way to help the nonmajors and those intimidated by science as they grapple with the concepts. My goal was to help them connect with the content a bit more readily.

Another way that this text confronts learning barriers is by offering students rich graphics and striking photos that depict physical processes and the natural variability of the landscape in memorable ways. The quality and breadth of the illustrations are designed to spark students' interest and help them see the relevance of physical geography to their daily lives. The illustrations are accompanied by a dynamic tool, *the Geo Media Library,* which is an interactive, Web-based multimedia resource. *The Geo Media Library* consists of a variety of animations and simulations that allow students to visualize and manipulate many of the factors associated with geographical processes and see the results over time and space. The multimedia will enhance students' learning as they participate more closely in geographical processes and will reinforce the integrative nature of the discipline by showing related variables in motion. This form of active learning will, in turn, help promote long-term retention of the material. The multimedia is fully integrated within the chapter text in distinct sections that direct students to the related modules on the website and explain to students what they should expect to learn by interacting with it.

Each *Geo Media* module also includes a variety of self-assessment questions for students. Students can use these questions to test their understanding of topics, or instructors can assign them as homework. Such questions allow both students and instructors to assess learning. They also provide the foundation for exam questions that are independent of class lectures. The *Geo Media* modules should motivate more students to interact with the textbook and media because they will more readily see their connection with the course. They contain a variety of self-assessment tools that will engage the interest of *all* students enrolled in physical geography, not just those who are scientifically inclined or have a background in Earth science.

Changes for the Third Edition

Students and faculty often wonder why it is necessary to produce new editions of textbooks. The need for new editions in geography is particularly necessary because events frequently occur that change the nature of the cultural and physical landscape. Similarly, scientists learn more about Earth processes between editions that advance the state of knowledge. Since the second edition of this text was written, a number of important events have occurred on Earth that require coverage in a third edition. For example, the extent of the annual Antarctic ozone hole shows signs of shrinking due to international efforts to control the production of chlorofluorocarbons. An intense tsunami hit Japan in March 2011. This tsunami devastated the coast, killed about 16,000 people, and damaged property even on the western seaboard of the United States. Still another major event was Hurricane Sandy, which blasted the northeast coast of the United States (particularly New Jersey and New York) in the fall of 2012. Although this storm was ranked only as a Category 1 hurricane, the geographical area it covered was huge. In addition, it struck the coast at the peak high tide, which magnified beach erosion, property loss, and flooding.

On the climate front, concentrations of atmospheric carbon dioxide on Earth have risen still further, from about 390 ppm in 2010 to 395 ppm in 2013. This continued increase is causing the vast majority of climate scientists to become even more concerned about the effects of climate change and the potential future warming of Earth. As a result, the melting of the Greenland ice cap (as well as other glaciers on Earth) apparently continues to accelerate, including a dramatic increase in the amount of surface melt area in July 2012. Ongoing climate change may have contributed to an intense drought in the central United States from 2011 to the early part of 2013, one that harked back to the Dust Bowl era of the 1930s. Similarly, the occurrence of wildfires appears to be rising in the western United States, often with disastrous results. Although these are far from the only changes in the physical geography of Earth, they illustrate why a new edition is warranted.

Aside from the new coverage of topics in this edition, another reason to create a new one was simply to improve upon the second edition and make it a better resource for students. The first and second editions were well received, and a number of reviewers and users of it offered excellent suggestions to strengthen the text. A key area of emphasis in this edition was to enhance the Human Interactions theme of the text. This theme focuses on some of the many ways that physical geography is relevant to human/environment interactions such as global climate change, agriculture, and energy production. Given the growing relevance of human/environmental issues in the world, and the important role physical geographers play in understanding and solving them, this theme was systematically strengthened in this edition by adding the following *Human Interactions* discussions:

1. *The United States Public Land Survey*—One of the key reference features for geographers is the Earth's grid system, which allows people to determine location based on latitude and longitude. In the United States we also use another type of grid system, one based on the four cardinal directions, to organize property and space in a systematic way. This discussion focuses on how this survey system is organized and used.

2. *Drought in the U.S. Farm Belt*—One of the most indirect ways that physical geography is relevant to people is when weather patterns change and drought conditions evolve. Such a drought enveloped the central United States from 2011 to 2013. The drought was particularly bad in the summer of 2012 when searing temperatures and low rainfall burned crops throughout the region to a crisp. As a result, crop yields fell dramatically and food prices surged. This discussion in Chapter 7 examines the cause of this drought and the impact it had on the regional farm industry.

3. *Coping with Destructive Hurricanes*—The intensity of recent hurricanes, such as Katrina in 2005 and Sandy in 2012, demonstrates why people need to carefully monitor tropical circulatory systems during the fall months. The discussion in Chapter 8 follows the tracks of these two storms and the devastation they caused. It also touches on the political problems associated with dealing with these massive storms.

4. *Human Influence on Vegetation Patterns*—One of the most obvious ways that people impact the environment is by changing the nature of plant cover, such as through deforestation or agriculture. These issues are closely examined in Chapter 10, with an enhanced focus on the role of agriculture in the Great Plains region in North America.

5. *The Anthropocene*—This discussion in Chapter 12 reflects the fact that humans have become the most important variable that influences the shape and patterns of the Earth's landscape. In response to this heavy influence, geologists are strongly considering the establishment of a new period in the geological timescale called the Anthropocene. The discussion weighs the evidence for such a new designation, as well as the most logical time for it to have begun.

6. *Earthquakes as Natural Hazards*—Earthquakes are perhaps the most frightening of natural hazards because they occur without warning. Strong earthquakes can be devastating, with significant loss of human life. The discussion in Chapter 13 examines the concept of earthquake probability and focuses specifically on the tsunamis that blasted Indonesia in 2005 and Japan in 2011. It compares the warning systems in each region at the time the earthquakes and associated tsunamis struck and their impacts in terms of loss of human life.

7. *Developing Unconventional Oil Supplies in North America*—The second edition of the text included a new chapter, Chapter 20 (*Relevance of Physical Geography to Environmental Issues*). This chapter included three case studies that focused on water issues in the arid southwestern United States, soil salinization in California and Australia, and geographers' efforts to stabilize

the panda habitat in China. This chapter now includes a fourth case study that focuses on the challenges we as a nation face to develop unconventional sources of oil, now that the easily mined oil has been consumed. It includes a discussion of fracking, the controversial technique used to extract oil and natural gas from shale deposits where these carbon-based energy supplies are tightly bound.

Special Features of the Text

To help students navigate their way through the book and better appreciate the nature and scope of physical geography, the chapters include a number of special and innovative features:

- **Discover...**—This feature presents an opportunity for students to discover the patterns and causes of particularly interesting geographic phenomena, such as rainbows, wildfires, and unusual clouds (to name a few) on Earth. Each chapter has one such feature, which includes a photo and explanation of how geographers interpret this aspect of the Earth system. The goal of this feature is to make students realize that there is more to the physical landscape than meets the eye, which will hopefully spark their interest in what they see around them. A variety of new such features have been included in this edition, such as the *Mars Rover* (Chapter 2), *What Is the Land Skin Temperature?* (Chapter 5), *Fossil Fuels* (Chapter 12), *Earthquake Prediction and the Italian Legal System* (Chapter 13), *Prehistoric Cliff Dwellings in the Southwestern U.S.* (Chapter 14), and *Coastal Dunes Along the Eastern Shore of Lake Michigan* (Chapter 19).

- **Geo Media Library**—Each chapter contains several multimedia modules that explain to students what they can expect to see and learn as they interact with the simulations and animations on the text's website. The website media also include a variety of self-assessment questions for students.

- **Visual Concept Check**—To provide students with a means of self-testing within the flow of chapter content, this feature offers a scenario with an illustration and questions to test students' understanding of key chapter concepts. Answers to the visual concept checks appear at the end of the chapter.

- **Key Concepts to Remember**—This feature is an interim summary that appears after specific sections of the chapter to help students check their comprehension of the key concepts covered.

- **Locator Maps with Photographs**—Photographs of non-U.S. sites are accompanied by a small map indicating the location of the site shown.

- **Marginal Glossary**—Key terms are set in boldface type in the text and defined at the foot of the page for easy recognition and reference.

- **Summary of Key Concepts**—The main points of the chapter are summarized.

- **Check Your Understanding**—Self-assessment questions at the end of the chapter allow students to test their comprehension.

Acknowledgments

A project of this scope naturally required the help and support of a number of people. I would first like to thank the friendly people at John Wiley & Sons, Inc., for their faith in my vision and for giving me the freedom and resources to see it through. In this context, I have to give special thanks to a group of four who, in reality, are good friends who happened to help me create a great book. Ryan Flahive deserves special kudos for recruiting me, backing me up many times, and making a number of good things happen along the way. I am particularly indebted to Veronica Armour, who has worked with me in the past two editions. She did such a good job managing the second edition that I begged to work with her again on this project. As I expected, she remained an organizational force who effectively guided the process, kept me on track, edited effectively, had great ideas, and just made it a whole lot of fun. Thank you Veronica!. I am also grateful for the awesome job that Janet Foxman did on guiding the production process, and to Billy Ray for the great work on photo research. My thanks also go to Harry Nolan for the beautiful design that supports all our efforts. For the expert proofreading in producing the book, I am extremely grateful to Katrina Avery.

I also want to extend my thanks to the many geographers and scientists who are doing the research that needs to be done in this complex time of Earth history. Their research is crucial to a better understanding of our world and for the development of sustainable land-use practices. Many members of this team served as reviewers for this book, and I want to thank them for taking the time out of their very busy schedules to help out. They include:

John All, *Western Kentucky University*

John Anderton, *Northern Michigan University*

Jake Armour, *University of North Carolina, Charlotte*

Barbara Batterson-Rossi, *Cuyamaca College (California)*

Kevin Baumann, *Indiana State University*

Sheryl Beach, *George Mason University*

Jason Blackburn, *Baton Rouge Community College*

Greg Bohr, *California Polytechnic State University, San Luis Obispo*

Margaret Boorstein, *Long Island University, C.W. Post*

William Budke, *Ventura College (California)*

Michaele Ann Buell, *Northwest Arkansas Community College*

David Cairns, *Texas A&M University*

Tom Carlson, *University of Washington, Tacoma*

Nicole Cerveny, *Mesa Community College*

Philip Chaney, *Auburn University*

Richard Cooker, *Kutztown University*

Ron Crawford, *University of Alaska, Anchorage*

James Davis, *University of Utah*

Lisa DeChano, *Western Michigan University*

Carol DeLong, *Victor Valley College*

Mike DeVivo, *Grand Rapids Community College*

Lynn Fielding, *El Camino College*

Donald Friend, *Minnesota State University, Mankato*

John D. Frye, *University of Wisconsin-Whitewater*

Colleen Garrity, *SUNY Geneseo*

Alan Gaugert, *Glendale Community College*

David Goldblum, *University of Wisconsin, Whitewater*

John Greene, *University of Oklahoma, Norman*

Duane Griffin, *Bucknell University*

Chris Groves, *Western Kentucky University*

Patrick Hesp, *Louisiana State University*

Miriam Helen Hill, *Jacksonville State University*

Curt Holder, *University of Colorado, Colorado Springs*

David Harms Holt, *University of Southern Mississippi*

Michael Holtzclaw, *Central Oregon Community College*

Robert Hordon, *Rutgers University*

Hixiong (Shawn) Hu, *East Stroudsburg University*

Peter Jacobs, *University of Wisconsin*

Scott Jeffrey, *Community College of Baltimore County, Catonsville*

Kris Jones, *Long Beach City College*

Stacy Jorgenson, *University of Hawaii, Manoa*

Theron Josephson, *Brigham Young University, Idaho*

Trudy Kavanagh, *University of Wisconsin, Oshkosh*

Ryan Kelly, *Lexington Community College*

Joseph Kerski, *U.S. Geological Survey*

John Keyantash, *California State University, Dominguez Hills*

Eric Keys, *Arizona State University*

Marti Klein, *Saddleback College*

Dafna Kohn, *Mt San Antonio College*

Jean Kowal, *University of Wisconsin, Parkside*

Jack Kranz, *California State University, Northridge*

Barry Kronenfeld, *George Mason University*

Steve LaDochy, *California State University, Los Angeles*

Charles Lafon, *Texas A&M University*

Phillip Larson, *Arizona State University*

Jeff Lee, *Texas Tech University*

Elena Lioubimtseva, *Grand Valley State University*

James Lowry, *Stephen F Austin State University*

David Lyons, *Century College*

Michael Madsen, *Brigham Young University, Idaho*

Emmanuel Mbobi, *Kent State University, Stark*

Christine McMichael, *Morehead State University*

Beverly Meyer, *Oklahoma Panhandle State University*

Armando V. Mendoza, *Cypress College*

Peter Mires, *Eastern Shore Community College*

Laurie Molina, *Florida State University*

Christopher Murphy, *Community College of Philadelphia*

Steven Namikas, *Louisiana State University*

Andrew Oliphant, *San Francisco State University*

Darren Parnell, *Kutztown University*

Charlie Parson, *Bemidji State University*

Brooks Pearson, *University of West Georgia*

Robert Pinker, *Johnson County Community College*

Rhea Presiado, *Pasadena City College*

Kevin Price, *University of Kansas*

David Privette, *Central Piedmont Community College*

Carl Reese, *University of Southern Mississippi*

David Sallee, *University of North Texas*

Justin Scheidt, *Ferris State University*

Peter Scull, *Colgate University*

Glenn Sebastian, *University of Southern Alabama*

John Sharp, *SUNY New Paltz*

Wendy Shaw, *Southern Illinois University, Edwardsville*

Andrew Shears, *Kent State University*

Binita Sinha, *Diablo Vallege College*

Brent Skeeter, *Salisbury University*

Thomas Small, *Frostburg State University*

Lee Stocks, *Kent State University*

Sam Stutsman, *University of South Alabama*

Aondover Tarhule, *University of Oklahoma*

Roosmarijn Tarhule-Lips, *University of Oklahoma, Norman*

Thomas Terich, *Western Washington University*

Donald Thieme, *Georgia Perimeter College*

U. Sunday Tim, *Iowa State University*

Paul E. Todhunter, *University of North Dakota*

Erika Trigoso, *University of Arizona*

Alice Turkington, *University of Kentucky, Lexington*

Lensyl Urbano, *University of Memphis*

Michael Walegur, *Moorpark College*

David Welk, *State Center Community College District, Clovis Center*

Forrest Wilkerson, *Minnesota State University, Mankato*

Dennis Williams, *Southern Nazarene University*

Joy Wolf, *University of Wisconsin, Parkside*

Lin Wu, *California State Polytechnic University, Pomona*

Ken Yanow, *Southwestern College, Chula Vista*

Brent Yarnal, *Pennsylvania State University*

Hengchun Ye, *California State University, Los Angeles*

Zhongbo Yu, *University of Nevada, Las Vegas*

Guoqing Zhou, *Old Dominion University*

Yu Zhou, *Bowling Green State University*

Tongxin Zhu, *University of Minnesota, Duluth*

Charles Zinser, *SUNY Plattsburgh*

Craig ZumBrunnen, *University of Washington*

A thank you also goes to the following professors and companies for allowing us to use select videos and animations in the Geo Media Library:
AccuWeather, Inc.

Allianz SE

Bill Dietrich, *University of California, Berkeley*

Glacier National Park

Paul Heller, *University of Wyoming*

NOAA Center for Tsunami Research

Andrew Oberhardt

Chris Paola, *University of Minnesota*

Taichiro Sakagami, *Duke University*

Doug Smith, *California State University, Monterey Bay*

The Times-Picayune Publishing Company

University Corporation for Atmospheric Research

U.S. Geological Survey

Last, and most important, I wish to express my heartfelt thanks to my family for their support. My wife, Jennifer, is a wonderful and beautiful person who has enriched my life in ways that I cannot describe. She urged me to write the first edition and has patiently lived with the many ups and downs along the way. I simply could not have done it without her support. My children, Hannah and Rosie, have consistently given me their affection throughout this process and I love them dearly. I hope this effort inspires them someday to stretch the limits of their potential.

Alan F. Arbogast
Professor of Geography
Michigan State University

The Teaching and Learning Package

Discovering Physical Geography, Third Edition, is supported by a comprehensive supplements package that includes an extensive selection of print, visual, and electronic materials.

Additional information, including prices and ISBNs for ordering, can be obtained by contacting John Wiley & Sons. You can find your local representative and their contact information using our "rep locator" website: www.wiley.com/college/rep.

WileyPLUS

WileyPLUS is an innovative, research-based, online environment for effective teaching and learning.

What Do Students Receive with WileyPLUS?

A Research-Based Design

WileyPLUS provides an online environment that integrates relevant resources, including the entire digital textbook, in an easy-to-navigate framework that helps students study more effectively.

- *WileyPLUS* adds structure by organizing textbook content into smaller, more manageable "chunks."

- Related media, examples, and sample practice items reinforce the learning objectives.

One-on-One Engagement

With *WileyPLUS* for *Discovering Physical Geography, Third Edition,* students receive 24/7 access to resources that promote positive learning outcomes. Students engage with related examples (in various media) and sample practice items, including:

- Practice Questions provide immediate feedback to true/false, multiple-choice, and short-answer questions.

- www.ConceptCaching.com: caches linked directly to the e-book give additional examples of the concepts.

- Geo Media modules and questions.

- Interactive Drag-and-Drop Exercises challenge students to correctly label important illustrations from the textbook.

Measurable Outcomes

Throughout each study session, students can assess their progress and gain immediate feedback. *WileyPLUS* provides precise reporting of strengths and weaknesses, as well as individualized quizzes, so that students are confident they are spending their time on the right things. With *WileyPLUS*, students always know the exact outcome of their efforts.

What Do Instructors Receive with *WileyPLUS*?

WileyPLUS provides reliable, customizable resources that reinforce course goals inside and outside of the classroom as well as visibility into individual student progress. Pre-created materials and activities help instructors optimize their time.

Course Materials and Assessment Content

- Lecture Notes PowerPoint Slides
- Classroom Response System (Clicker) Questions
- Image Gallery
- Instructor's Manual
- Gradable Reading Assignment Questions (embedded with online text)
- Geo Media Library
- Testbank

Gradebook

WileyPLUS provides instant access to reports on trends in class performance, student use of course materials, and progress toward learning objectives, helping inform decisions and drive classroom discussions.

WileyPLUS. Learn More. www.wileyplus.com.

Powered by proven technology and built on a foundation of cognitive research, *WileyPLUS* has enriched the education of millions of students in over 20 countries around the world.

Discovering Physical Geography Instructor's Site (www.wiley.com/college/arbogast)

This comprehensive website includes numerous resources to help you enhance your current presentations, create new presentations, and employ our premade PowerPoint presentations. These resources include:

- Image Gallery. We provide online electronic files for the line illustrations and maps in the text, which the instructor can customize for presenting in class (for example, in handouts, overhead transparencies, or PowerPoints).
- A complete collection of PowerPoint presentations, available in beautifully rendered, four-color format, has been resized and edited for maximum effectiveness in large lecture halls.
- A comprehensive Test Bank with multiple-choice, fill-in, matching, and essay questions is distributed via the secure Instructor's website as electronic files, which can be saved into all major word-processing programs.
- Geo Media Library. In addition to the modules from the book, this easy-to-use website offers lecture launchers that help reinforce and illustrate key concepts from the text through the use of animations, videos, and interactive exercises. Students can use the resources for tutorials as well as self-quizzing to complement the textbook and enhance understanding of geography. Easy integration of this content into course management systems and homework assignments gives instructors the opportunity to integrate multimedia with their syllabi and with more traditional reading and writing assignments. Resources include:
  - *Animations:* Key diagrams and drawing from our rich signature art program have been animated to provide a virtual experience of difficult concepts. These animations have proven beneficial to the understanding of this content for visual learners.
  - *Videos:* Brief video clips provide real-world examples of geographic features and put these examples into context with the concepts covered in the text.
  - *Simulations:* Computer-based models of geographic processes allow students to manipulate data and variables to explore and interact with virtual environments.
  - *Interactive Exercises:* Learning activities and games built off our presentation material give students an opportunity to test their understanding of key concepts and explore additional visual resources.
- An online database of photographs, **www.ConceptCaching.com**, allows professors and students to explore what a physical feature looks like. Photographs and GPS coordinates are "cached" and categorized along core concepts of geography. Professors can access the images or submit their own by visiting www.ConceptCaching.com.
- Instructor's Manual. This manual includes chapter overviews, lecture suggestions, and classroom activities.

Wiley Faculty Network

This peer-to-peer network of faculty is ready to support your use of online course management tools and discipline-specific software/learning systems in the classroom. The Wiley Faculty Network will help you apply innovative classroom techniques, implement software packages, tailor the technology experience to the needs of each individual class, and provide you with virtual training sessions led by faculty for faculty.

Student Companion Website (www.wiley.com/college/arbogast)

This easy-to-use and student-focused website helps reinforce and illustrate key concepts from the text. It also provides interactive media content that helps students prepare for tests and improve their grades. This website provides additional resources that complement the textbook and enhance your students' understanding of physical geography:

- Flashcards offer an excellent way to drill and practice key concepts, ideas, and terms from the text.
- The Geo Media Library allows students to explore key concepts in greater depth using videos, animations, and interactive exercises.
- Chapter Review Quizzes provide immediate feedback to true/false, multiple-choice, and short-answer questions.
- Annotated Web Links put useful electronic resources into context.

DISCOVER...—This feature presents a photo and demonstrates how visual clues contained within it can be used to "discover" the character of the landscape or environment. The goal of this feature is to make students realize that there is more to the physical landscape than meets the eye, which will hopefully spark their interest in what they see around them.

• DISCOVER...

FASCINATING CLOUDS ASSOCIATED WITH THUNDERSTORMS

In addition to strong wind, lightning, and thunder, powerful thunderstorms sometimes produce fascinating cloud patterns diagnostic of distinct meteorological processes. One such cloud formation is a *shelf cloud* like the one pictured here. A shelf cloud develops when a cold downdraft within a storm surges outward along the ground after it reaches the surface. As it does so, it lifts the relatively warm and moist air ahead of the storm up to the level of condensation, forming this shelf-like feature. Such a cloud is often accompanied by a cold gust front.

Another interesting and diagnostic cloud formation associated with thunderstorms is *mammatus clouds*. Also known as *mammatocumulus*, which means "mammary cloud," mammatus clouds look like pouches or lobes hanging eerily upside down in the air like those shown here. These clouds are usually found on the base of a spreading anvil downwind of a large thunderstorm. Their formation is poorly understood but may be somehow related to overturning pockets of colder air that descend into warmer air at high altitudes. Although they are usually composed of ice, they also contain liquid water and can seem translucent. Mammatus clouds are more common during warm months and are more often seen in the Midwest and the eastern parts of the country.

GEO MEDIA LIBRARY MULTIMEDIA—Multimedia in every chapter explain to students what they can expect to see and learn as they interact with the simulations, animations, and videos on the text's website. *WileyPLUS* also includes a variety of self-assessment questions for students.

www.wiley.com/college/arbogast   **WileyPLUS**

Formation of a Midlatitude Cyclone

We can now examine how the development of midlatitude cyclones occurs in an animated way by viewing the animation *Formation of a Midlatitude Cyclone*. This module allows you to see the process of cyclogenesis in motion. Figure 8.9 is the foundation of this animation. As you watch the animation, follow how an initial kink in the atmosphere evolves into a mature cyclone that spins counterclockwise in the Northern Hemisphere. This animation will help you better understand how these systems and associated processes cause highly variable weather when they migrate through a region. Once you have completed the animation, be sure to answer the questions at the end of the module to test your understanding of this concept.

KEY CONCEPTS TO REMEMBER—This feature is an interim summary that appears after specific sections of the chapter to help students check their comprehension of the key concepts covered.

KEY CONCEPTS TO REMEMBER ABOUT AIR MASSES AND FRONTS

1. An air mass is a large body of air that has distinctive characteristics and forms in specific geographic regions.

2. Five principal types of air masses affect North America. Continental air masses include continental Polar (cP), continental Arctic (cA), and continental Tropical (cT).

VISUAL CONCEPT CHECK 8.1

This image shows temperature in the United States. Where is the approximate location of the cold front?

a) Western Texas.
b) Florida.
c) It cuts across the states of Washington and Oregon.
d) It extends from western Michigan to central Texas.

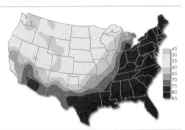

VISUAL CONCEPT CHECK—To provide students with a means of self-testing within the flow of chapter content, this feature offers a scenario with an illustration and questions to test students' understanding of key chapter concepts. Answers to the visual concept checks appear at the end of the chapter.

GLOSSARY—Key terms are set in boldface type in the text and defined at the foot of the page for easy recognition and reference.

rainfall patterns and the actual rotation of a tornado. It does so by making use of the Doppler Effect, described first by the 19th-century physicist Christian Doppler, which states that the frequency of energy waves generated by a moving source changes relative to an observer.

A classic example of this effect is the sound a train makes as it approaches you. Have you ever noticed what happens? The pitch of the train's whistle rises with the approach of the train and then lowers after it passes. Using this effect, Doppler radar can determine that raindrops on one side of a tornado are moving toward the radar detector, while those on

(Figure 8.20a). This rotation is seen by the meteorologist in the weather laboratory as a distinct **hook echo** on the computer screen (Figure 8.20b, c). If such a feature is identified, a tornado warning is quickly issued for the area in the path of the storm and people usually have sufficient time to seek shelter before the storm strikes.

***

Hook echo   *The diagnostic feature in Doppler radar indicating that strong rotation is occurring within a thunderstorm and that tornado development is thus possible.*

SUMMARY OF KEY CONCEPTS

1. An air mass is a large body of air that forms in specific geographic regions and thus has distinctive characteristics. Five principal air masses affect North America. Continental air masses include continental Polar (cP), continental Arctic (cA), and continental Tropical (cT). The maritime air masses are maritime Tropical (mT) and maritime Polar (mP).

2. Air masses have distinct boundaries called fronts. At a stationary front, contrasting air masses are flowing parallel to one another. A warm front is a place where warm air is advancing into relatively cool air. Given that warm air slowly slides over the top of the cooler air along a warm front, rainfall is slow and steady. A cold front is a place where cold air is advancing into relatively warm air. Given the higher density of colder air, rainfall is intense and of short duration along the front because warm air

the polar front jet stream at the 500-mb pressure level. As a cyclone spins, it pulls warm (mT) air up from the south on its eastern side. This warm, moist air encounters cold air as it moves to the north. The cyclone also pulls cold air (cP) down from the north on its western side. This cold, dry air encounters warm air as it moves to the south.

4. In general, the most severe midlatitude storms form along strong cold fronts when warm, moist (mT) air ahead of the front is rapidly forced aloft. Thunderstorms evolve in predictable stages, including the cumulus stage, mature stage, and dissipating stage, that are related to the upward and downward flow of air. The strongest storms associated with midlatitude weather are tornadoes, which are localized bodies of intense low pressure that develop in association with supercell thunderstorms.

SUMMARY OF KEY CONCEPTS—The main points of the chapter are summarized.

CHECK YOUR UNDERSTANDING

1. Define an air mass.

2. What are the specific characteristics of an mT air mass, and how do they differ from the characteristics of a cP air mass?

3. Which air mass is most likely to be associated with precipitation—an mT air mass or a cP air mass? Why?

4. Discuss the evolution and migration of a midlatitude cyclone.

5. Why are midlatitude cyclones a mechanism through which contrasting air masses are mixed?

6. How does the formation of an upper air trough at the 500-mb level result in the development of a midlatitude cyclone?

7. What is the basic difference between a warm front and a cold front? Why is the term *front* used in association with these concepts?

8. Precipitation along a warm front is gradual and long-lasting, whereas it is short-lived and often violent along a cold front. Why does this difference exist?

9. Describe the evolution of a thunderstorm and the various stages it goes through during its life cycle.

10. What is a downdraft, and why is it the first step in the dissipation of a thunderstorm?

11. Discuss the evolution of a hurricane in the Northern Hemisphere, including its various stages, movement, and relationship with ocean temperature.

CHECK YOUR UNDERSTANDING—Self-assessment questions at the end of the chapter allow students to test their comprehension.

BRIEF CONTENTS

1 ▋ Introduction to Physical Geography   2

2 ▋ The Geographer's Tools   12

3 ▋ Earth–Sun Geometry and the Seasons   42

4 ▋ The Global Energy System   58

5 ▋ Global Temperature Patterns   84

6 ▋ Atmospheric Pressure, Wind, and Global Circulation   104

7 ▋ Atmospheric Moisture and Precipitation   136

8 ▋ Air Masses and Cyclonic Weather Systems   168

9 ▋ Global Climates and Global Climate Change   200

10 ▋ Plant Geography   234

11 ▋ The Global Distribution and Character of Soils   264

12 ▋ Earth's Internal Structure, Rock Cycle, and Geologic Time   304

13 ▋ Tectonic Processes and Landforms   332

14 ▋ Weathering and Mass Movement   370

15 ▋ Groundwater and Karst Landscapes   390

16 ▋ Fluvial Systems and Landforms   408

17 ▋ Glacial Geomorphology: Processes and Landforms   446

18 ▋ Arid Landscapes and Eolian Processes   478

19 ▋ Coastal Processes and Landforms   504

20 ▋ Relevance of Physical Geography to Environmental Issues   536

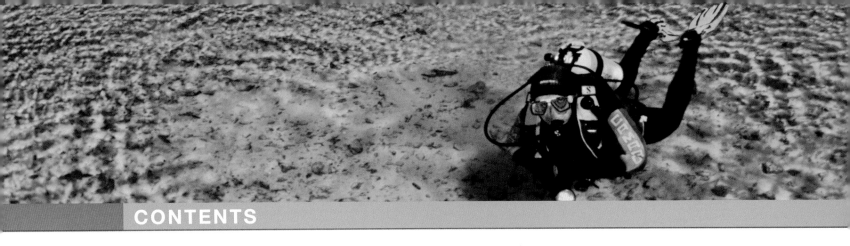

CONTENTS

1 ▮ **Introduction to Physical Geography** 2

The Scope of Geography 4

Defining Physical Geography 5

The Earth's Four Spheres 7

Organization of This Book 8

Exploring Cause-and-Effect Relationships Holistically 9

Emphasis on Human Interactions with the Environment 10

Geo Media Library: An Interactive Tool 10

Focus on Geographical Literacy 10

Physical Geography Is Interesting, Exciting, and Very Relevant to Your Life 11

Geo Media Library: *Stream Meandering* 11

2 ▮ **The Geographer's Tools** 12

The Geographic Grid 14

Latitude 15

Longitude 17

Using the Geographic Grid 19

Human Interactions: The United States Public Land Survey 19

Geo Media Library: *Using the Geographic Grid* 19

Maps—The Basic Tool of Geographers 22

Map Projections 23

Map Scale 27

Isolines 29

Geo Media Library: *Using Maps* 31

Digital Technology in Geography 32

Remote Sensing 32

Global Positioning Systems 37

Geographic Information Systems 39

Geo Media Library: *Using a Geographic Information System* 39

3 ▮ **Earth–Sun Geometry and the Seasons** 42

Our Place in Space 44

The Shape of Earth 44

Earth's Orbit Around the Sun 45

The Earth's Rotation and Axial Tilt 47

The Seasons 50

Solstice and Equinox 50

Human Interactions: How We See Earth–Sun Geometry on Earth 53

Day and Night 53

Geo Media Library: *Orbital View and Earth as Viewed from the Sun* 54

Seasonal Changes in Sun Position (Angle) and Length of Day 54

Geo Media Library: *Celestial Dome* 56

Geo Media Library: *Sun Angle and Length of Day* 56

Geo Media Library: *Earth–Sun Geometry and Ancient Humans* 56

4 ▮ **The Global Energy System** 58

The Electromagnetic Spectrum and Solar Energy 60

The Electromagnetic Spectrum 60

Solar Energy and the Solar Constant 61

Geo Media Library: *The Electromagnetic Spectrum* 62

Composition of the Atmosphere 62

Constant Gases   63

Variable Gases   63

Particulates   69

The Flow of Solar Radiation on Earth   70

Heat Transfer   70

Flow of Solar Radiation in the Atmosphere   71

Interaction of Solar Radiation and the Earth's Surface   74

Geo Media Library: *The Angle of Incidence*   75

The Global Radiation Budget   76

Geo Media Library: *The Global Energy Budget*   77

Human Interactions: Solar Energy Production   80

5   Global Temperature Patterns   84

Layered Structure of the Atmosphere   86

The Troposphere   86

The Stratosphere   88

The Mesosphere   88

The Thermosphere   88

Surface and Air Temperatures   89

Human Interactions: Calculating the Heat Index and Wind Chill   90

Large-Scale Geographic Factors That Influence Air Temperature   90

Local Factors That Influence Air Temperature   93

Geo Media Library: *Surface Temperature*   96

Geo Media Library: *Maritime vs. Continental Effect*   98

The Annual Range of Surface Temperature (A Holistic Assessment)   98

Geo Media Library: *Global Temperature Patterns*   100

Human Interactions: Urban Heat Islands   101

Geo Media Library: Temperature and Location   102

6   Atmospheric Pressure, Wind, and Global Circulation   104

Atmospheric Pressure   106

Factors That Influence Air Pressure   106

Measuring and Mapping Air Pressure   107

Atmospheric Pressure Systems   108

Low-Pressure Systems   108

High-Pressure Systems   108

The Direction of Airflow   110

Unequal Heating of Land Surfaces   110

Pressure Gradient Force   111

Coriolis Force   112

Geo Media Library: *Fluctuations in the Pressure Gradient*   113

Geo Media Library: *The Coriolis Force*   114

Frictional Forces   114

Global Pressure and Atmospheric Circulation   116

Tropical Circulation   116

Midlatitude Circulation   119

Polar Circulation   122

Geo Media Library: *Global Atmospheric Circulation*   123

Seasonal Migration of Pressure Systems   123

Geo Media Library: *The Asian Monsoon*   124

Geo Media Library: *Global Atmospheric Circulation and Water Vapor Movement*   125

Human Interactions: Harnessing Wind Energy   125

Local Wind Systems   128

Land–Sea Breezes   128

Topographic Winds   129

Oceanic Circulation   130

Gyres and Thermohaline Circulation   130

El Niño   133

Geo Media Library: *El Niño*   134

7   Atmospheric Moisture and Precipitation   136

Physical Properties of Water   138

Hydrogen Bonding   138

Thermal Properties of Water and Its Physical States   138

Geo Media Library: *Latent Heat*   140

The Hydrosphere and the Hydrologic Cycle   140

Humidity   142

Maximum, Specific, and Relative Humidity   142

Dew-Point Temperature   145

Geo Media Library: *Atmospheric Humidity*   147

Evaporation   148

Adiabatic Processes   149

The Dry Adiabatic Lapse Rate   149

Geo Media Library: *Adiabatic Processes*   151

The Wet Adiabatic Lapse Rate   151

Cloud Formation and Classification   152

Cloud Classification   153

Fog   153

Precipitation   156

Types of Precipitation   156

Precipitation Processes   156

Geo Media Library: *Orographic Processes*   161

Geo Media Library: *Convectional Precipitation*   164

Human Interactions: Drought in the U.S. Farm Belt   164

8 ▓ **Air Masses and Cyclonic Weather Systems   168**

Air Masses and Fronts   170

Air Masses   170

Fronts   170

Evolution and Character of Midlatitude Cyclones   173

Upper Airflow and the 500-mb Pressure Surface   174

Interaction of Upper Airflow and Surface Airflow   176

Cyclogenesis   177

Geo Media Library: *Formation of a Midlatitude Cyclone*   179

Geo Media Library: *Migration of a Midlatitude Cyclone*   180

Thunderstorms   180

Evolution of Thunderstorms   180

Geo Media Library: *Formation of Thunderstorms*   182

Severe Thunderstorms   182

Tornadoes   184

Geo Media Library: *Tornadoes*   189

Tropical Cyclones   190

Hurricanes   190

Human Interactions: Coping with Destructive Hurricanes   192

Geo Media Library: *Tracks of Hurricane Katrina and Hurricane Sandy*   194

9 ▓ **Global Climates and Global Climate Change   200**

Climate and the Factors That Affect It   202

Köppen Climate Classification   203

Geography of Köppen Climates   207

Tropical (*A*) Climates   207

Arid and Semi-Arid (*B*) Climates   208

Geo Media Library: *Tropical Savanna Climate (Aw)*   208

Mesothermal (*C*) Climates   211

Geo Media Library: *Humid Subtropical Hot-Summer Climate (Cfa, Cwa)*   213

Geo Media Library: *Marine West-Coast Climates (Cfb, Cfc)*   214

Microthermal (*D*) Climates   214

Polar (*E*) Climates   215

Highland (*H*) Climates   216

Geo Media Library: *Global Climates*   217

Geo Media Library: *Remote Sensing and Climate*   217

Reconstructing Past Climates   217

Pollen Records   218

Tree Ring Patterns   219

Ice Core Analysis   221

Geo Media Library: *Reconstructing Past Climates Using Oxygen Isotopes*   224

Causes of Past Climate Change   224

Geo Media Library: *The Milankovitch Theory*   225

Human Interactions and Future Climate Change   226

The Carbon Cycle   227

Is Anthropogenic Climate Change Really Occurring?   229

Predicting Future Climate Change   230

Geo Media Library: *The Greenhouse Effect and Global Climate Change*   231

10 ▓ **Plant Geography   234**

Ecosystems and Biogeography   236

The Process of Photosynthesis   237

Geo Media Library: *Photosynthesis and Respiration* 240

The Relationship of Climate and Vegetation: The Character and Distribution of Global Biomes 241

Forest Biomes 243

Grassland Biomes 247

Desert Biomes 249

Tundra Biome 250

Local and Regional Factors That Influence the Geographic Distribution of Vegetation 251

Slope and Aspect 251

Vertical Zonation 252

Plant Succession 254

Riparian Zones 254

Geo Media Library: *Plant Succession* 255

Human Interactions: Human Influence on Vegetation Patterns 255

Deforestation 256

Agriculture in the Midlatitude Grassland Biome 259

Geo Media Library: *Deforestation* 259

Overgrazing 260

Geo Media Library: *Remote Sensing and the Biosphere* 261

11 ▌ **The Global Distribution and Character of Soils** 264

What Is Soil? 266

Basic Soil Properties 266

Soil-Forming (Pedogenic) Processes 268

Soil-Forming Factors 271

Measurable Soil Characteristics 275

Soil Chemistry 278

Soil pH 278

Colloids and Cation Exchange 279

Geo Media Library: *Soil Colloids and pH* 280

Soil Profiles (Reading the Soil) 280

Time and Soil Evolution 282

Geo Media Library: *Soil Horizon Development* 283

Soil Science and Classification 283

The Twelve Soil Orders 284

Geo Media Library: *African Climate, Vegetation, and Soils* 290

Geo Media Library: *Regional Pedogenic Processes* 296

Geo Media Library: *North American Climate, Vegetation, and Soils* 297

Human Interactions with Soils 299

12 ▌ **Earth's Internal Structure, Rock Cycle, and Geologic Time** 304

Earth's Inner Structure 306

The Major Layers 306

Rocks and Minerals in the Earth's Crust 310

Igneous Rocks 311

Geo Media Library: *Identification of Igneous Environments* 312

Sedimentary Rocks 313

Geo Media Library: *Clastic Rocks* 316

The Rock Cycle 320

Geo Media Library: *The Rock Cycle* 321

Geologic Time 322

Geo Media Library: *Geologic Time* 324

"Telling" Geologic Time 324

Putting Geologic Time in Perspective 324

A Holistic View of Geologic Time and the Rock Cycle: The Grand Canyon and the Spanish Peaks of Colorado 325

Human Interactions with Geologic Time: The Anthropocene 328

13 ▌ **Tectonic Processes and Landforms** 332

Plate Tectonics 334

The Lithospheric Plates 334

Plate Movement 335

Geo Media Library: *Continental Drift* 337

Types of Plate Movements 338

Passive Margins 338

Transform Plate Margins 338

Plate Divergence 339

Plate Convergence 340

Geo Media Library: *Folding* 347

Geo Media Library: *Plate Tectonics* 347

Earthquakes 348

Seismic Processes 348

Human Interactions: Earthquakes as Natural Hazards 353

Geo Media Library: *Types of Faults* 360

Volcanoes 360

Explosive Volcanoes 361

Geo Media Library: *Walk the Pacific Ring of Fire* 363

Fluid Volcanoes 364

Hotspots 365

Geo Media Library: *Volcanoes* 368

14 ▊ **Weathering and Mass Movement** 370

Weathering 372

Mechanical Weathering 373

Chemical Weathering 375

Human Interactions: Acid Rain 379

Geo Media Library: *Weathering* 380

Mass Wasting 381

Rockfall 382

Soil Creep 383

Landslides 384

Flows 385

Avalanches 386

Geo Media Library: *Weathering and Mass Movements* 387

15 ▊ **Groundwater and Karst Landscapes** 390

Movement and Storage of Groundwater 392

Human Interactions with Groundwater 395

The High Plains Aquifer 395

Subsidence 400

Groundwater Contamination 400

Geo Media Library: *Hydrologic Cycle and Groundwater* 402

Karst Landforms and Landscapes 402

Caves and Caverns 402

Karst Topography 404

16 ▊ **Fluvial Systems and Landforms** 408

Overland Flow and Drainage Basins 410

Origin of Streams 410

Drainage Basins 411

Drainage Patterns, Density, and Stream Ordering 413

Geo Media Library: *The Rhine River* 415

Hydraulic Geometry and Channel Flow 416

Fluctuations in Stream Discharge 416

Fluvial Processes and Landforms 420

Erosion and Deposition 420

Stream Gradation 422

Geo Media Library: *Bedload Transport and Braided Streams* 423

Geo Media Library: *The Graded Stream* 425

Evolution of Stream Valleys and Floodplains 427

Geo Media Library: *Stream Meandering* 428

Geo Media Library: *Fluvial Geomorphology and Stream Processes* 429

Human Interactions with Streams 438

Urbanization 438

Artificial Levees 439

Dams and Reservoirs 440

17 ▊ **Glacial Geomorphology: Processes and Landforms** 446

Development of a Glacier 448

The Metamorphosis of Snow to Glacial Ice 448

The Glacial Mass Budget 449

Geo Media Library: *The Glacial Mass Budget* 450

Glacial Movement 450

Types of Glaciers 452

Glaciers in Mountainous Regions 452

Geo Media Library: *Glaciers in the Cascade Mountains* 455

Continental Glaciers 455

Glacial Landforms 456

Landforms Made by Glacial Erosion 457

Deposition of Glacial Drift and Resulting Landforms 461

Geo Media Library: *Depositional Glacial Landforms* 465

History of Glaciation on Earth 465

The Wisconsin Glaciation and Evolution of the Great Lakes 467

Probable Human Impact on Glaciers 470

Geo Media Library: *Glaciers and Climate Change* 472

Periglacial Processes and Landscapes 473

Permafrost 474

Ground Ice and Associated Landforms 475

18 ▊ **Arid Landscapes and Eolian Processes** 478

Arid Landscapes 480

Desert Geomorphology 480

Eolian Erosion and Transport   485

The Fluid Behavior of Wind, Erosion, and
     Sediment Transport   485

Eolian Deposition
and Landforms   488

Airflow and the Formation
     of Sand Dunes   488

Classification of Sand Dunes
     and Related Landforms   491

Deposition of Windblown Dust   493

Geo Media Library: *Eolian Processes and
     Landforms*   497

Human Interactions
with Arid Landscapes
and Eolian Processes   497

Desertification in the African
     Sahel   498

Desertification in the Great Plains of the United
     States: The Dust Bowl   500

19 ▦ **Coastal Processes and
        Landforms**   504

Oceans and Seas on Earth   506

The Nature of Coastlines:
Intersection of Earth's Spheres   507

Processes That Shape the Coastline   507

Geo Media Library: *Tides*   513

Coastal Landforms   515

Erosional Coastlines   515

Geo Media Library: *Waves and Coastal
     Erosion*   519

Depositional Coastlines   519

Geo Media Library: *Longshore Processes and
     Depositional Coastlines*   525

Human Interactions with
Coastlines   529

Coastal Engineering   530

Global Climate Change and the Impact on
     Coastlines   532

Geo Media Library: *Evolution of the Louisiana
     Coastline*   534

20 ▦ **Relevance of Physical Geography to
        Environmental Issues**   536

A Short History of
Human Population   538

Technological Development and Population
     Growth   538

The Impact of Growing
     Human Population on
     the Natural Environment   539

Case 1: Water Issues in the Arid
American Southwest   542

Establishing Water Rights   543

Will Water Supplies Disappear?   544

A Drought-Prone Future?   547

Case 2: Soil Salinization in
Arid and Semi-Arid Lands   548

Ancient Sumeria   550

The Current Global Extent
     of Salinization   551

The San Joaquin Valley in
     California   552

Australia   555

The Challenge of Sustainable Agriculture in Semi-
     Arid and Arid Regions   558

Case 3: Developing Unconventional Oil
Supplies in North America   558

Oil Supply, Demand, and Geography   558

Developing Unconventional
     Sources of Oil   560

Is the Environment Threatened
     by Fracking?   562

Case 4: Giant Panda Conservation in
China   565

The Wolong Nature Reserve   567

▦ **Glossary**   573

▦ **Index**   586

DISCOVERING
PHYSICAL GEOGRAPHY

Third Edition

WILEY

CHAPTER ONE

INTRODUCTION TO PHYSICAL GEOGRAPHY

I want to welcome you to this introductory textbook about physical geography, an exciting scientific discipline that examines the Earth and how it functions. Physical geographers study a wide variety of interesting things, ranging from climate and weather to soils, glaciers, and beaches, to name just a few. Their purpose is to understand the location and character of certain features on the landscape, such as mountain ranges and river valleys, and to explain why they came to be and how they differ from a geographical perspective. This text seeks to encapsulate the essence of physical geography in a way that is appropriate for a beginning student of the discipline such as yourself. As you read through it, your first goal should be to investigate the various physical processes on Earth and how they operate and combine to form distinctive geographical features and patterns. At a secondary level, you will be asked to examine how aspects of physical geography affect human lives and, in turn, how people impact them. This opening chapter outlines the topics discussed in this book and places them in the context of the overall discipline of geography. Then we discuss the various components and features of the book and how they will assist with your learning.

This view of Mt. Hood in Oregon reflects many geographical processes discussed in this text, including the character of the atmosphere, how water is stored and flows within the hydrosphere, the role of climate and its impact on vegetation, and the way landscapes evolve over time.

CHAPTER PREVIEW

The Scope of Geography

Defining Physical Geography

Organization of This Book

 Stream Meandering

LEARNING OBJECTIVES

1. Comprehend the character and scope of geography as a scientific discipline.

2. Discuss the concept of spatial analysis and how it relates to geography.

3. Define the subdisciplines of physical geography.

4. Explain the concept of a natural system.

5. Define the four Earth spheres.

6. Describe how the scientific method is used in physical geography.

7. Discuss why physical geography is relevant to many human/environment issues.

The Scope of Geography

When most people are asked to describe the nature of geography, a common response is that the discipline focuses primarily on the locations of countries, capital cities, rivers, and oceans. They also assume that most of the work geographers do involves maps of some kind. Although such an understanding of the discipline is accurate to some degree, the field actually encompasses *far* more than the average person realizes. In fact, geography is a discipline that is highly relevant to everyday life in a myriad of ways, ranging from analysis of traffic patterns to importation of economic goods, migration of ethnic groups, cost of gasoline, and earthquake hazards, to name a very few. In addition to maps, geographers use a fascinating array of techniques and technologies to conduct their work, such as computer models, field reconnaissance, personal interviews, satellite imagery, global positioning systems, and even shovels.

Geography, a word derived from the Greek words for "Earth description," is an ancient discipline that examines the spatial attributes of the Earth's surface and how they differ from one place to another. The concept of geography has likely been important to the human experience for tens of thousands of years. It is easy to imagine, for example, that prehistoric hunters and gatherers were intimately aware of their surroundings, including the location and character of forests, streams, lakes, berry patches, migrating animal herds, and competing groups of people. In short, this *geographical awareness* would have been absolutely essential for people to sustain themselves and their communities. It would also have been critically important to pass this awareness on to future generations so that they, in turn, could successfully compete for resources.

So, for thousands of years at least, geography was a descriptive discipline that focused on the generalized location and character of places and features on the landscape. As civilization evolved, however, geography slowly became an academic discipline with numerous specialized subfields. Scientists became experts in areas such as geology, meteorology, ecology, and human cultural differences. Interest in geography grew especially between the 15th and 19th centuries when explorers such as Christopher Columbus, Ferdinand Magellan, James Cook, Charles Darwin, and Lewis and Clark began to investigate parts of the world that were previously unknown to people of European descent (including Americans). These explorers, as well as many others, brought detailed descriptions of exotic places and animals to a keenly interested public. The new knowledge and perspectives gained from this time provided a major driving force for the development of the modern world.

The trend toward increased specialization in geography has continued to the present time. Most geographers generally consider themselves primarily either physical or human specialists. Within these two broad fields are a range of geographical subdisciplines, as shown in Figure 1.1a. Although each of these subfields has a unique focus, such as soils or agricultural land use, geographers draw from many of these subfields when they analyze any particular spatial pattern. For example, to fully understand human settlement patterns in Africa, it is important to consider the interaction of subfields such as climatology, soils, and vegetation. In turn, to understand the nature of soils in any given place, you must consider the effects of climatology, vegetation, geomorphology (Figure 1.1b), and perhaps even regional cultural practices.

(a)

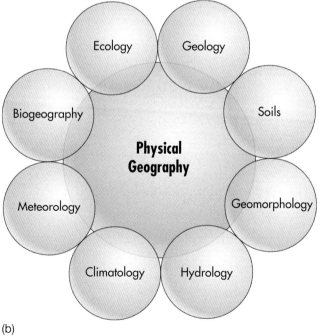

(b)

Figure 1.1 Subdisciplines of geography. (a) The field of geography can be broadly subdivided into physical and human geography. Many subfields occur within these two broad categories, with many that overlap. (b) Physical geography itself overlaps with several other areas of science. Many of these areas require a good understanding of chemistry and physics as well.

Although this book focuses on physical geography, it is useful to know that all subfields of geography are based on the same five themes of location, movement, place, human/environment interactions, and region. Location refers to the exact position where something is found on Earth. As the name implies, movement is related to the way that geographical variables flow from one locality to another. The concept of place is an important theme in geography because it reflects the characteristics that make a certain location distinct. Human/environment interactions acknowledge the complex bond between people and nature. Finally, a region is an area that shares one or more common characteristics.

In the context of these five themes, all geographers share a common methodology that makes them part of the overall discipline. Geographers use a method known as **spatial analysis**, which attempts to explain patterns or distributions of specific variables across physical space. At a fundamental level, most geographers want to know the answer to a pair of important questions about any geographical variable: *where* and *why*. In other words, geographers want to know where a specific environmental or cultural variable is located and why it happens to be there.

To see how these questions are important and work in tandem, let's briefly consider an example involving cultural diversity in the Middle East. As you may know, this region has many different religious sects, dialects, and tribal identities. Although this diversity has produced a rich heritage, it has also resulted in a great deal of conflict between and among various cultural groups. A cultural geographer can examine the spatial distri-

---

Spatial analysis *A method of analyzing data that specifically includes information about the location of places and their defining characteristics.*

bution of the many groups in the area (in other words, *where are they?*), as well as study why people in one place differ in, say, their political or religious views, from people someplace else. The geographer might look for similarities (such as language or religion) among groups across physical space, which might explain why certain people align themselves politically with others. In the course of this study, the geographer would have integrated several variables into one picture, including language, religion, history, and climate. Such a study might contribute to an understanding of why people differ across this region and why sources of conflict remain.

Defining Physical Geography

Now that the general character of geography has been defined, let's focus on a more detailed definition of physical geography. You probably already have an interest in this part of the discipline, whether you know it or not. For example, do you like big storms and wonder why they happen? You may know that tornadoes frequently occur in the central United States, especially in the springtime (Figure 1.2a). Do you know why? Maybe you wonder why large mountains are found in Washington but not in Texas (Figure 1.2b). Perhaps you have heard about the Sahara Desert and wonder why it is so dry there and why much of it is covered with sand dunes (Figure 1.2c). Like many people, you might enjoy the seashore and wonder why nice beaches form in some places (Figure 1.2d) but not in others. If you have asked yourself questions like these, then you are probably interested in physical geography at some level. The fact is that most people have such an interest; they just do not realize it.

(a)

(b)

(d)

(c)

Figure 1.2 Some elements of physical geography. (a) A tornado in the central United States (see Chapter 8). (b) The Olympic Mountains overlooking Puget Sound in Washington (see Chapter 13). (c) Sand dunes in the Sahara Desert (see Chapters 9 and 18). (d) Coastline at Big Sur in northern California (see Chapter 19).

(a)

(c)

Roger Harris/Photo Researchers, Inc.

Alan Arbogast

(b)

Science Source

(d)

© David Gowans/Alamy

Figure 1.3 Examples of energy flows on Earth. (a) Earth receives its energy from the Sun in the form of solar radiation (see Chapters 3 and 4). (b) The atmosphere circulates energy around Earth, as can be seen in this stream of clouds (see Chapter 7). (c) Some energy is transferred when water flows from the atmosphere to Earth as rain (see Chapter 5). (d) Some of the energy on the surface of Earth is transferred by flowing water (see Chapter 16).

Simply stated, **physical geography** involves the spatial analysis of the various physical components and natural processes of the Earth. Some examples of Earth's physical components are air, water, rocks, vegetation, and soil. The term **process** broadly refers to a series of actions that can be measured and that produce a predictable end result. In physical geography, these processes are often products of the energy that flows from the Sun to Earth in the form of solar radiation. Once this energy reaches Earth, it then flows from one place to another on the planet in various forms. Some examples of natural processes directly related to the flow of solar radiation (Figure 1.3) are the circulation of the atmosphere (Chapter 7), the distribution of vegetation (Chapter 10), the formation of soils (Chapter 11), and the movement of water in the air and streams, and collection in lakes (Chapters 5 and 16).

As you will see throughout this book, many processes behave in an interconnected way within *natural systems* where one environmental variable has a direct impact on another.

Given these relationships, physical geographers often invoke **systems theory** in their studies because it is a holistic framework through which they can analyze and/or describe a group of variables that work together to produce some definable result. Another way to look at natural systems is to recognize that they are greater than the sum of their parts. Such systems depend on a set of energy inputs that flow in some way to various kinds of predictable outputs. Physical geographers understand that natural systems are complex and that processes within them

Physical geography *Spatial analysis of the physical components and natural processes that combine to form the environment.*

Process *A naturally occurring series of events or reactions that can be measured and that result in predictable outcomes.*

Systems theory *The examination of interactions involving energy inputs and outputs that result in predictable outcomes.*

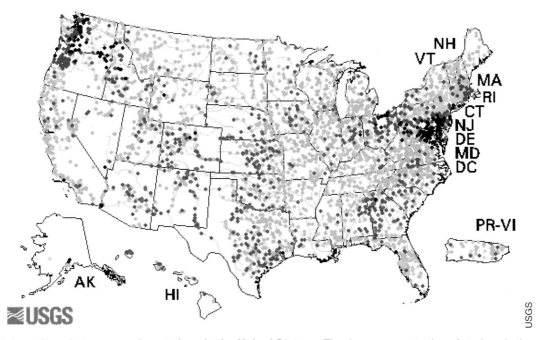

Monday, October 29, 2012 19:30ET

Figure 1.4 Location of stream-gaging stations in the United States. The dense concentration of stations in the eastern half of the country reflects the fact that many more streams occur in the eastern United States than in the interior western states. The map is color-coded to illustrate the amount of water in any given stream on October 29, 2012, relative to its average flow on that day calculated over time. For example, red dots reflect very low water levels, whereas dark blue/purple dots mean that stream flow was relatively high on that day. Black dots represent localities on streams that have very high flow and may be close to flooding. The concentration of black dots along the central Atlantic coast reflects prodigious rains dropped by Hurricane Sandy on this day.

do not occur in isolation from one another. Instead, they are interconnected in often complicated ways that may at first appear chaotic. Despite this apparent complexity, natural systems are, in fact, self-organizing entities that internally adjust toward an equilibrium condition. These adjustments can be readily explained by understanding how dynamic feedbacks, oscillations, and delays in reaction time influence outcomes.

To see an example of how environmental variables relate to one another and can be viewed holistically within a natural system, imagine you want to explain the spatial distribution of rivers in the United States. One way to see the geographical concentration of rivers in the country is with a map showing the location of gaging stations, which are places where the U.S. Geological Survey continuously monitors the flow of water in the streams (Figure 1.4). Observe that the eastern part of the country contains many more gaging stations than the interior west in places such as Nevada and Utah. This pattern reflects the fact that far more streams are found in the eastern part of the United States than in the interior west.

The question a geographer would ask about this pattern is: *Why* do more rivers occur in the eastern United States than in the western part of the country? This question would naturally lead you to holistically examine streams as natural systems that reflect the input of water from some kind of source to the output of water actually flowing in rivers and creeks. Understanding these relationships, in turn, helps explain the geographical patterns observed in Figure 1.4. The simple reason for the geographical

distribution of streams in the United States is that far more precipitation falls in the eastern part of the United States than in the western states. You will later study why this geographical variability occurs, but for now it is sufficient to say it exists because the atmosphere over the eastern United States typically contains more water than in the interior parts of the western states. Thus, more water flows from the atmosphere to the ground as precipitation in the eastern United States than in the western states. Some of this water flows directly across the Earth's surface as input into streams. A great deal of it slowly absorbs into the ground where it is steadily released into streams. As a result of these interconnected processes, the eastern United States contains more streams than the western part of the country.

The Earth's Four Spheres

As you can imagine, a huge number of component/process combinations exist for geographical study in a holistic way. In physical geography, these various combinations can be grouped into the four "spheres" on Earth (Figure 1.5):

1. **Atmosphere**—The atmosphere is the gaseous shell that surrounds Earth. This sphere is composed of many critical components essential to life, such as oxygen, carbon, water, and nitrogen, that flow around Earth.

2. **Lithosphere**—The lithosphere is the solid part of Earth, including soil and minerals. A good example of a natural

Atmosphere *The gaseous shell that surrounds Earth.*

Atmosphere

Hydrosphere

Earth

Lithosphere

Biosphere

Figure 1.5 The four Earth spheres. Each sphere encompasses a major component of the Earth's natural environment.

system in this sphere is the way in which water, minerals, and organic matter flow in the outermost layer of the Earth to form soil. This sphere provides the habitat and nutrients for many life-forms.

3. **Hydrosphere**—The hydrosphere is the part of Earth where water, in all its forms (solid ice, liquid water, and gaseous water vapor), flows and is stored. This sphere is absolutely critical to life and is one with which humans regularly interact—for example, through irrigation and navigation.

4. **Biosphere**—The biosphere is the living portion of Earth and includes all the plants and animals (including humans) on the planet. Various components of this sphere regularly flow from one place to another, both on a seasonal basis and through human intervention. Humans interact with this sphere in a wide variety of ways, with agriculture being an obvious example.

These four spheres overlap to form the natural environment that makes Earth a unique place within our solar system. Physical

---

Lithosphere *A layer of solid, brittle rock that comprises the outer 70 km (44 mi) of Earth.*

Hydrosphere *The part of Earth where water, in all its forms, flows and is stored.*

Biosphere *The portion of Earth and its atmosphere that supports life.*

geography examines the spatial variation within these spheres, how natural systems work within them, the observable outcomes in each, and the manner in which components flow from one sphere to another.

Physical geography can be a descriptive discipline that simply characterizes the nature of the Earth's spheres in specific regions. A simple example of such a descriptive focus would be to acknowledge that the western part of the United States is mountainous, whereas the central part of the country consists mostly of relatively level plains. Physical geography is also a science because research is conducted within the framework of the scientific method, which is the systematic pursuit of knowledge through the recognition of a problem, the formulation of hypotheses, and the testing of hypotheses through the collection of data by measurement, observation, and experiment. The conclusions derived from the systematic application of the scientific method contribute to the formulation of scientific theories and laws that explain how Earth functions.

A simple example of using the scientific method in physical geography would be to test the hypothesis that water in streams within a certain region is acidic. This hypothesis would best be tested by collecting water samples from a number of streams and conducting chemical analyses on them to determine their acid levels. If high acid levels were indeed obtained, then an effort would be made to explain why that pattern occurred. In addition to the understanding produced in these kinds of analyses, physical geographers also test hypotheses about all sorts of natural phenomena by collecting information from the atmosphere, rocks, soils, ice cores, satellite images, the Earth's magnetic field, and even other planets (Figure 1.6). For these investigations to occur, it is essential that scientists understand physical laws and have the ability to mathematically analyze and compare them.

Organization of This Book

The chapters in this book are organized to provide you with a good understanding of the fundamental concepts associated with physical geography. They contain information that ranges in scale from global to local, which will allow you to better grasp your place both in the world and even in your neighborhood. Chapter 2 focuses on the various kinds of tools that geographers use in their work, such as maps, remote sensing, and geographical information systems. Chapters 3 through 5 examine our relationship with the Sun (Chapter 3), the way we receive solar radiation (Chapter 4), and how those interactions relate to temperature (Chapter 5). The processes discussed in these chapters will prepare you for the topics that will be covered in the rest of the book. Chapters 6 through 9 revolve around the atmosphere, including the way that air circulates within it (Chapter 6), precipitation processes (Chapter 7), weather systems (Chapter 8), and global climate patterns (Chapter 9). The text then examines the influence of the atmosphere and how it interacts with Earth's other spheres by focusing on plant geography in Chapter 10 and soils in Chapter 11.

Chapters 12 through 19 deal mainly with the lithosphere and hydrosphere. Chapter 12 describes the Earth's internal structure,

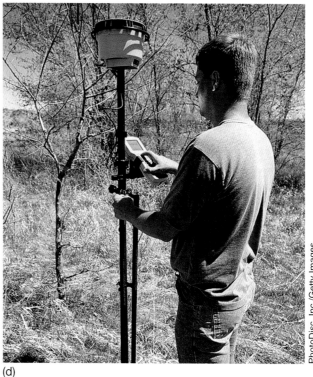

Figure 1.6 Examples of collecting scientific data about Earth. (a) Certain satellites are designed to obtain measurements about the atmosphere, oceans, and the distribution and character of plants, among many other things. (b) To learn about the behavior of streams in the past, scientists study the type of sediment deposited by the stream through time. This picture is from one of my class field trips in the Great Plains. (c) One way to learn about past climate changes on Earth is to obtain samples of ancient ice on the Greenland and Antarctic ice caps. (d) New methods of surveying enable scientists to obtain accurate measurements about elevation and location.

rock cycle, and geologic time. This discussion leads directly into Chapter 13, which focuses on the lithosphere and tectonic landforms. From there, we turn your attention in Chapter 14 to the way that rocks weather and how sediment moves through mass wasting processes. Chapters 15 and 16 discuss the way that water moves on Earth and how it is stored within it. Chapter 15 focuses specifically on groundwater processes and the formation of landforms such as caves. In Chapter 16 we look at how water flows across the surface in stream systems and the landforms that result. Chapters 17 through 19 are devoted to specific geomorphic processes and the resulting landforms, including glaciers (Chapter 17), eolian (wind) processes and arid landscapes (Chapter 18), and coastal regions (Chapter 19).

Exploring Cause-and-Effect Relationships Holistically

As you work through these chapters, you will constantly see how the four Earth spheres interact with each other to produce definable geographical patterns. To understand how such interactions work, let's briefly examine the searing drought in the central United States in the summer of 2012. With this issue in mind, consider the following question, one that encompasses elements of the atmosphere and the hydrosphere: How was the quantity of water in rivers in this part of the country affected by the drought? A testable hypothesis would be that the quantity of water in rivers decreased. You could test this hypothesis by collecting data from gaging stations about the amount of water in the streams during the drought and comparing those values to normal water levels. In all probability, the amount of water dropped significantly during the drought, which, in fact, is reflected in Figure 1.4. The reason for this decreased water level is that a significant drought would result in less water flowing from the atmosphere (as precipitation) to Earth. As a result, less water would then be available as an input to streams across the Earth's surface. In addition, the quantity of water stored in the ground likely decreased, which would also have reduced river levels because a great deal of water in streams is derived from the ground. Potential further impacts of this drought could have been that some forms of vegetation became less common or that the likelihood of fire increased.

This book systematically explores these kinds of cause-and-effect relationships in a variety of ways. One way is through the traditional use of text accompanied by photographs, diagrams, and tables. Each chapter contains detailed discussions that connect important concepts to events that you may have experienced.

Emphasis on Human Interactions with the Environment

Many of the scientific analyses associated with physical geography are driven by the growing impact that human activities have within and among the Earth spheres. Given the nature of human impact on the natural environment, physical geography is at the forefront of research on many environmental issues that face the world today. Among just a very few of these issues are the following.

- **Global climate change**—Human industrial activities are increasing the levels of carbon dioxide in the atmosphere. Abundant scientific evidence suggests that this relationship is contributing significantly to global climate change.
- **Deforestation**—The clearing of the tropical rainforests is occurring at a very rapid rate, leading to soil erosion, loss of wildlife habitat, and species extinctions.
- **Farmland loss**—Due to increasing global population, farmland is being converted to zones of economic development and residential housing. This loss of farmland is resulting in more intensive farming of agricultural soils still in use, which increases the risk of soil erosion and pollution due to the extensive application of fertilizers and pesticides.
- **Natural hazards**—Hazards occur when extreme events result in danger to humans. Examples of natural hazards include hurricanes, tornadoes, flooding, earthquakes, and volcanoes. Natural hazards are a particularly important area of geographical study because as the global population grows, increasing numbers of people are moving into areas that are susceptible to extreme natural events.

A recent natural disaster in the United States vividly illustrates the integrated nature of physical geography and the critical role that geographers play with respect to solving real-world problems. As you probably know, Hurricane Sandy devastated the East Coast in the fall of 2012, causing many deaths and billions of dollars of damage, much of it occurring when the New Jersey shore was extensively flooded. Before the storm reached land, geographers were at the forefront of the effort to monitor the storm's path and predict where the most significant damage would occur. Once the storm passed, geographers began conducting research on the impact that the hurricane had on a variety of issues, including the shape of the coastline, the distribution of wetlands, and the regional economy. These studies have profound implications for future environmental decisions, politics, and economic development in the region. Most chapters of this book contain sections specifically devoted to such human–environment interactions, including a discussion of solar energy production in Chapter 4, wind energy in Chapter 5, and petroleum in Chapter 12. The final chapter (Chapter 20) is devoted entirely to these kinds of issues and demonstrates how physical geography is highly relevant to human–environmental interactions.

Geo Media Library: An Interactive Tool

These holistic discussions in the text are accompanied by graphics and photographs, as well as a more dynamic tool—the Geo Media Library. This consists of a variety of animations and simulations that allow you to visualize and manipulate many of the factors associated with geographical processes and to see the results over time and space. The animations and simulations will enhance your learning as you participate more closely with geographical processes and will reinforce the interactive nature of the discipline by showing related variables in motion. The media are integrated entirely within the chapter text as distinct sections that explain what you should expect to learn by interacting with them.

Here is a good example of how the Geo Media Library will enhance your learning. Let's again use flowing water in rivers as an example because it is easy to visualize. This concept is described in great detail in Chapter 16 where it is accompanied by a variety of diagrams that illustrate how water flows, using flow lines and arrows embedded within them. It is also supplemented by several animations that are accessed through the Geo Media Library or on the text's accompanying website. One of the animations in that chapter shows how streams snake across the river valley in a process called *meandering*. Through this process, the geographical position of streams actually moves through time. The Geo Media Library shows this process in animated form, which will enable you to comprehend it better. Have a look now to see what these animations are like.

Focus on Geographical Literacy

In addition to improving your overall understanding of physical geography and how it relates to human/environmental issues, another goal of this book is to enhance your geographical literacy. It is common knowledge that the overall geographical literacy of most Americans is very poor. How many Americans can identify, for example, the countries of the Middle East, where so much of our national focus is presently centered? Do they know the location and character of Afghanistan, which is where we continue to engage in the longest war in our history? In a similar vein, how many people understand the science associated with the issue of ongoing climate change, or why hurricanes may raise the cost of gasoline at the pump?

Geographical literacy also involves knowing the location of distinct physical regions on Earth, such as the Sahara Desert or the Himalaya Mountains, and why they exist. In this context, you will see that discussions include maps of the places described. I also hope that your *visual* geographical literacy will improve by using this text. In other words, this book is designed to sharpen *your eye* so that when *you* see things in the landscape, which you may have previously ignored, you might better appreciate them and why they occur. Two features—*Discover* and *Visual Concept Check*—are specifically designed to help you improve this aspect of your geographical literacy. The *Discover* feature allows you to see certain parts of the physical landscape and what they mean about environmental conditions at the time or place they formed. The *Visual Concept Check* features are placed after key topic discussions so that you can test your understanding of those topics immediately after encountering them.

Physical Geography Is Interesting, Exciting, and Very Relevant to Your Life

As with any new endeavor, you can expect that improving your understanding of physical geography may be difficult at times. Many students initially avoid this subject because they feel that geography is boring, or they are intimidated by science, or they see no relevance of geography to their lives. If you genuinely give this subject a chance, however, you will see that physical geography is indeed relevant to *your* everyday life and, most of all, is interesting and even exciting. How else can you explain the popularity of weather and nature programs, national and state parks, the

travel industry, mountains, or beautiful coastlines? Why do people go on exotic vacations if not, in part, to enjoy the uniqueness of the physical landscape in new places? With a greater understanding of physical geography, you will appreciate those trips more. You may even appreciate the immediate world around you more.

In addition, an understanding of physical geography will help you make *informed* decisions when you are confronted with important environmental issues in your lifetime. Thus, you will become a better citizen, one who is capable of better protecting the best interests of your family and community. For instance, at some point in the future you may be confronted with a choice of where to place a landfill in your city or town. In order to make the most informed decision, one that perhaps ensures the safety of your drinking water, it will be important to understand the geology of the site, the character of soils, and the way water moves through the ground and is stored within it.

It is also possible that after using this text you may even decide that you want to become a physical geographer, as many people have. A number of excellent, well-paying jobs can be obtained with a specialization in physical geography, including environmental analysts, cultural resource managers, conservation agents, teachers, meteorologists, and landscape architects. If you decide to pursue such a career, you will find that collecting, analyzing, and reporting about geographical data are very rewarding activities. I do that kind of work myself and find it incredibly rewarding. Earth is a beautiful and complex place, and it is fun to understand how it works. Regardless of whether or not you are a geographer, you will understand your home planet better after using this text. Enjoy the ride!

www.wiley.com/college/arbogast

WileyPLUS

Stream Meandering

An excellent way to get a feel for the interactive media in this book is by examining a simple animation showing how streams move across the landscape. To do so, view the animation *Stream Meandering.* This animation allows you to visualize the way that streams migrate horizontally through the process of meandering. Throughout the book, you will frequently encounter modules like this one that will give you a better idea of how geo-

graphical processes function. Some of the modules are simple and require only that you carefully watch them. Others are simulations that allow you to manipulate variables to see how outcomes change. Regardless of the type of media, each module will contain a series of questions that can be used to test your understanding of the concepts after you watch the animation or simulation.

CHECK YOUR UNDERSTANDING

1. Describe the character of geography as a scientific discipline.

2. Explain the concept of spatial analysis and how it is used in geography.

3. What is the nature of physical geography, and why is it a subdiscipline of geography?

4. What is a natural system, and why can it be viewed holistically with distinct inputs and predictable outputs? Provide an example.

5. What is the scientific method, and how is it applied to research problems in physical geography?

6. Compare and contrast the four spheres of Earth and provide an example of how two of them interact with each other.

7. Why can physical geography be used to understand the relationship of humans with the environment?

CHAPTER TWO

THE GEOGRAPHER'S TOOLS

Before you begin to study the various concepts associated with physical geography, it is useful to understand some of the tools that geographers use to gather, manipulate, and display spatial information. Some of these tools, such as maps, are probably familiar because you have used them yourself and because they are generally associated with geography. Other aids, such as remote sensing, Geographic Information Systems (GIS), and Global Positioning Systems (GPS), are more technical and not well understood by most people. This chapter focuses on these tools, with the goal of providing a better understanding of how geographers do their work. The material in this chapter will also help you understand how much of the information presented in this text was obtained and how it should be interpreted.

Geographic technology allows us to view landscapes in unique and insightful ways. This photograph combines a satellite image with a digital elevation model to show the relationship of urban areas (in gray shades) and topography near Los Angeles, California. Of particular interest is the nearby location of the San Andreas Fault, which is the linear feature that angles from the upper center of the image to the lower right.

CHAPTER PREVIEW

The Geographic Grid

 Using the Geographic Grid

Maps—The Basic Tool of Geographers

 Using Maps

Digital Technology in Geography

 Using a Geographic Information System

LEARNING OBJECTIVES

1. Compare and contrast the various components of the Earth's geographic grid.

2. Discuss the concept of map projection and why it is important when designing maps.

3. Explain the concept of map scale and how it is used to represent location.

4. Describe remote sensing and the various ways it is used.

5. Discuss the concept of Global Positioning Systems (GPS) and how location is determined with this method.

6. Understand the nature of Geographical Information Systems (GIS) and how they are used to manage spatial data.

The Geographic Grid

Let's begin the discussion of geographical tools by focusing on the most basic of geographical themes, location. One way that location can be determined is by comparing the location of a place relative to some other place. For example, a geographer might identify Los Angeles as being south of San Francisco or observe that Africa is on the eastern side of the Atlantic Ocean. You have likely made such a comparison yourself, perhaps by indicating that you live on the west side of town or that your favorite park is in the eastern part of a state.

Although a determination of relative location can be useful, such a statement is very general. In most circumstances, geographers desire to know the absolute location of a place, which is most effectively done using the geographic grid. If Earth were flat, then it would be easy to construct a grid similar to a sheet of graph paper using the cardinal directions: north, south, east, and west (Figure 2.1). In this imaginary situation, the grid lines would extend from these four directions and intersect at various places on the grid. With this grid in place, it would be easy to choose a place of interest and then determine the distance from this reference point to your location by measuring along the nearest grid lines, using some unit of measurement such as miles or kilometers.

However, Earth is not flat, but instead has a curved surface. In order to account for Earth's shape when locating places, the grid must consist of a series of intersecting circles, with one set extending north and south and the other set east and west. We can begin by examining the circles that are oriented in a north–south direction. These circles converge at the North and South Poles on their way to the other side of Earth. These circles are called **great circles** because they are the largest circles that can be drawn on a sphere (Figure 2.2a). A good analogy of great

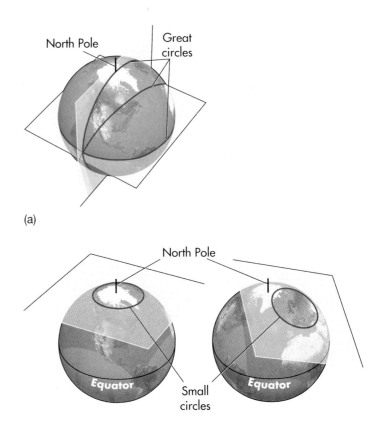

Figure 2.2 Great and small circles. (a) Examples of great circles on Earth. (b) Examples of small circles on Earth.

circles in the Earth's grid system is the set of circles that separate orange segments after the fruit is peeled. If you examine a peeled orange closely, you will see that these great circles converge on the top and bottom of the fruit, and bisect it into many segments.

Great circles can be drawn that bisect Earth in myriad ways (Figure 2.2a). Each of these circles has the center of Earth as its core, and they bisect Earth into two equal halves. Great circles are also important because their outlines reflect the shortest distance between any two points on the planet. This concept is particularly relevant to international travel because pilots plot their courses by great circle routes to reach destinations as quickly as possible and save on fuel costs. If you plan to travel from New York City to Moscow, for example, you would probably think that the shortest route would be to fly due east across the Atlantic Ocean. But, in fact, the great circle route from New York to Moscow passes north over Greenland.

With respect to Earth's grid system, the most important group of great circles is the one that converges at the North and South Poles. Along with this group of circles, another corresponding set of circles has outlines that extend east and west. These outlines are always parallel to one another and thus

Figure 2.1 Hypothetical flat land and grid. If Earth were shaped in this way, you could determine your location by measuring the distance from the nearest grid lines.

Great circles *Circles that pass through the center of the Earth and that divide the planet into equal halves.*

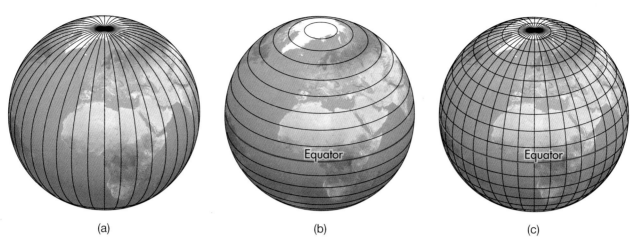

Figure 2.3 Orientation of great and small circles. Together, these circles form Earth's grid system. (a) All the great circles that are oriented north and south intersect at the poles. (b) Except for the great circle at the Equator, all circles oriented east and west on Earth are small circles. (c) Overlay of great and small circles to form the Earth's grid system.

never meet. The largest of these parallel circles (the Equator) lies halfway between the two poles and is a great circle because its outline corresponds with the maximum circumference of Earth. North and south of the Equator, however, the parallel circles have smaller diameters and are known as **small circles** (Figure 2.2b).

To visualize this sequence of small circles, imagine what happens when you slice a lemon or pineapple. First, the slices are generally parallel to one another as you work across the fruit. Second, the circumference of each slice is progressively smaller as you move away from the center of the fruit (the great circle of the fruit). As with great circles, small circles can be aligned in any direction on the globe. Geographers use this network of north–south intersecting great circles and east–west parallel small circles to create a grid on the Earth's surface (Figure 2.3) that allows them to determine any location within about 30 m (100 ft). This system is known as the latitude/longitude coordinate system. Let's see how this system works.

Latitude

Latitude is the portion of the grid system that uses the parallel set of circles with outlines extending east and west. Using these circles, latitude measures location north and south of a geographical reference. The natural reference for latitude is the great circle that bisects the middle of Earth

at its maximum east–west circumference, otherwise known as the **Equator** (Figure 2.4a). The portion of Earth north of the Equator is called the **Northern Hemisphere**, whereas the area south of the Equator is the **Southern Hemisphere**. Latitude is determined using simple geometry by measuring the angle of any point on the Earth's surface north or south of the Equator. This calculation is accomplished by projecting two lines from the center of Earth to its surface. One of these lines always extends from the center of Earth to the surface along the Equator. The other line extends to the north or south from the center of Earth directly to the location in question, depending on whether it is in the Northern or Southern Hemisphere, respectively (Figure 2.4b). We can then calculate the latitude of the location from the geometric arc between the two intersecting lines.

In this system the Equator has a value of 0° latitude. Latitudes close to the Equator have small geometric arcs from the Earth's center and, therefore, have relatively low-latitude designations such as 5° or 10° N or S (see Figure 2.4a). Localities progressively farther north (N) or south (S) of the Equator have progressively greater angles from the Earth's center and, therefore, have progressively larger latitude designations. The maximum angle is perpendicular (90°) to the center of the equatorial great circle. In the Northern Hemisphere, this angle corresponds to the North Pole (90° N), whereas it correlates with the South Pole (90° S) in the Southern Hemisphere.

Small circles *Circles that intersect the Earth's surface and that do not pass through the center of the planet.*

Latitude *The part of the Earth's grid system that determines location north and south of the Equator.*

Equator *The great circle that lies halfway between the North and South Poles.*

Northern Hemisphere *The half of Earth that lies north of the Equator.*

Southern Hemisphere *The half of Earth that lies south of the Equator.*

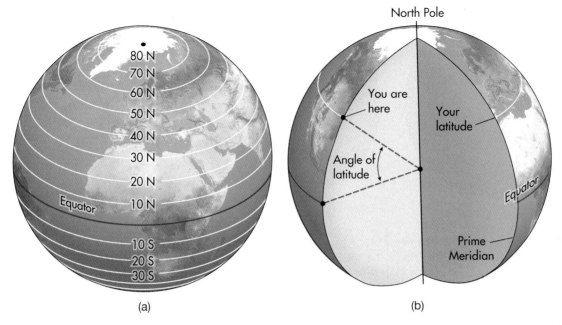

(a)

(b)

Figure 2.4 Calculating latitude. (a) The reference point for latitude is the Equator, which divides Earth into Northern and Southern Hemispheres. Latitude ranges from 0° to 90° N and S. (b) Latitude is calculated by simply measuring the angle of a point relative to the Equator from the center of Earth.

As far as distance is concerned, degrees of latitude are separated by about 110 km (69 mi) from each other, regardless of location. In other words, a location on the Earth's surface can be specified only within 110 km north or south at the level of a degree. It is possible to more closely locate places by subdividing degrees of latitude first into 60 minutes ('), and subsequently each minute into 60 seconds ("). An example of such a latitude designation would be a location in the Northern Hemisphere that has latitude of 39° 52′ 46″ N. At this level of detail, the location of something can be determined within 30.5 m (100 ft) on the Earth's surface.

Another way to improve the level of location precision is simply to convert latitude designations containing minutes and seconds to full decimal notations. This conversion is done with a simple equation that would, for example, convert the previous latitude of 39° 52′ 46″ N to 39.8794° N. The use of decimal notations is very common and most frequently designates such important lines of latitude as the Tropic of Cancer (23.5° N), the Tropic of Capricorn (23.5° S), the Arctic Circle (66.5° N), and the Antarctic Circle (66.5° S) (Figure 2.5). These lines of latitude are related to the Earth/Sun geometric

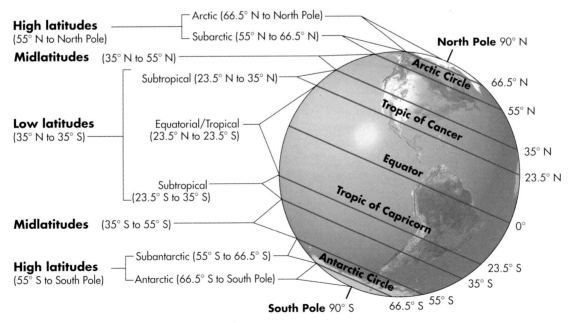

Figure 2.5 Important lines of latitude on Earth and general latitude zones. These zones of latitude often have distinctive climate, vegetation, and soil characteristics that will be explored in this text.

TABLE 2.1   General Zones of Latitude

Latitude Zone	Range	General Characteristics
Low latitudes	About 35° N–35° S	Very warm region of Earth with generally consistent weather over the course of the year
Midlatitudes	About 35°–55° in both hemispheres	Seasonal weather with warm/hot summers and cool/cold winters
High latitudes	From about 55° to 90° in both hemispheres	Very cold region of Earth with very short summers

relationship and will be discussed in much more detail in the next chapter.

Lines of latitude are also called **parallels** because they are parallel to the Equator and each other. The value of any given line of latitude remains the same no matter the location, whether it is on one side of Earth (such as in China) or the other (such as in the United States). A common misconception is that lines of latitude must indicate direction east and west because that is the direction that these lines follow. To remember how the system really works, think of degrees of latitude as being numbered sequentially north and south of the Equator.

Although you can certainly think of lines of latitude in individual terms to locate specific places, it is also useful to think of them collectively by region. In that context, lines of latitude can be lumped into nine geographic zones (see Figure 2.5). These nine regions can be further grouped into three general zones (Table 2.1) in which many distinct physical processes occur. The **low latitudes** range from about 35° N to 35° S of the Equator and are generally the warmest part of Earth. The **midlatitudes** range from about 35° to 55° latitude, in both hemispheres, and are typically regions of Earth that experience highly variable weather over the course of the year. Finally, the **high latitudes** range from 55° to 90° latitude, again in both hemispheres, and are typically the coldest places on Earth.

KEY CONCEPTS TO REMEMBER ABOUT LATITUDE

1. Lines of latitude run exactly east and west; they are also called parallels because they always parallel the Equator.

2. Lines of latitude are determined by measuring the geometric arc between two lines projected from the center of Earth to the surface at the Equator and at the location in question. The reference point for latitude designations is the Equator (or 0°); latitude designations extend to 90° N and S.

3. Lines of latitude never intersect with one another and therefore have the same value over the entire circle.

4. Zones of latitude can be grouped into nine geographic zones. These nine can be further grouped into three broad categories: the low, middle, and high latitudes.

Longitude

The previous discussion demonstrated how latitude is used to identify location north and south of the Equator. Now let's examine how east and west locations are determined with **longitude**. The foundation of the longitude system is the complex of great circles that pass through the poles. Half of such a great circle is a spherically shaped outline that connects each pole. Each of these half circles is called a meridian of longitude, or simply a **meridian**. In contrast to the latitude system, which uses the Equator as the foundation because it is the only great circle of the parallels, the easterly and westerly longitude calculations require an arbitrary reference meridian because all of them are half outlines of the great circles that pass through the poles. The arbitrary reference meridian for longitude is the **Prime Meridian**, which passes through Greenwich, England, at the Old Royal Observatory (Figure 2.6). This meridian was chosen as the reference meridian of longitude at an 1884 conference in Washington, D.C.

Parallels   *Lines of latitude.*

Low latitudes   *The zone of latitude that lies between about 35° N and 35° S.*

Midlatitudes   *The zone of latitude that lies between about 35° and 55° in both hemispheres.*

High latitudes   *The zone of latitude that lies from about 55° to 90° in both hemispheres.*

Longitude   *The part of the Earth's grid system that determines location east and west of the Prime Meridian.*

Meridians   *Lines of longitude.*

Prime Meridian   *The arbitrary reference point for longitude that passes through Greenwich, England.*

Figure 2.6 The Prime Meridian. This line of longitude at the Old Royal Observatory in Greenwich, England, is the reference meridian for the longitude system.

As with the Equator, the Prime Meridian has a value of 0°, and all points north and south along that meridian are at 0° longitude. The geometric basis for determining longitude is that a circle contains 360°. Think of the circle (the Earth) as divided into eastern and western halves by the Prime Meridian. The longitude grid consists of a series of meridians that extend 180° east and west of the Prime Meridian to complete the 360° circle. Using Figure 2.7 as a guide, imagine you are floating

in space and looking at a section of the Earth at the Equator. Longitude is calculated by measuring the arc formed by two lines projected from the Earth's center at the Equator. One line extends to the Prime Meridian, the other to the meridian of the location in question. The angle of the arc formed by the two lines determines the longitude. Calculated this way, smaller geometric arcs from the Prime Meridian have lower longitude designations, and greater arcs have higher designations (see Figure 2.7b). As with latitude, degrees of longitude can be subdivided into minutes and seconds or be given full decimal notations.

KEY CONCEPTS TO REMEMBER ABOUT LONGITUDE

1. Lines of longitude run north and south and are also called meridians.

2. Longitude is determined by measuring the geometric arc between two lines projected from the center of Earth to the surface at the Equator and to the meridian in question. The reference point for longitude is the Prime Meridian (or 0°), which divides Earth into eastern and western halves.

3. Meridians are located east and west of the Prime Meridian and extend 180° E and 180° W to complete a full circle.

4. The distance between meridians is greatest at the Equator but decreases steadily until converging at the poles.

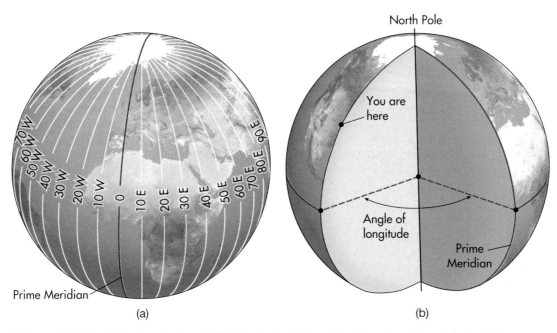

(a)　　　　　　　(b)

Figure 2.7 Calculating longitude. (a) The longitude grid consists of a series of great circles that converge at the poles and that are measured east and west of the Prime Meridian. (b) Longitude is calculated on the angle of arcs measured from the center of Earth and relative to the Prime Meridian.

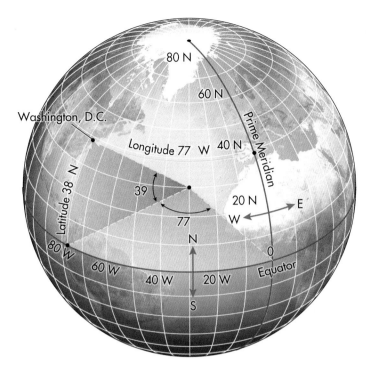

Figure 2.8 Determining latitude and longitude of Washington, D.C. Using the grid network, we can locate Washington, D.C., at approximately 39° N, 77° W.

A primary difference between longitude and latitude is that while lines of latitude are always parallel to one another, lines of longitude converge at the poles (Figure 2.7a). Thus, degrees of longitude are about 110 km (69 mi) apart at the Equator and progressively shorten in distance until they meet at the poles. As with latitude, a common misconception about longitude is that meridians are used to determine location north and south because that is the direction in which the semicircle outlines extend. Instead, remember that longitude determines location east and west of the Prime Meridian.

Using the Geographic Grid

Now that the two grid components of latitude and longitude are in place, they can be overlain to form a complete grid network that encompasses Earth (Figure 2.8). Thus, any location can be described in terms of its position north or south of the Equator and east or west of the Prime Meridian. Imagine, for example, that you wanted to find Washington, D.C., in the grid; Washington is located at 39° N (latitude), 77° W (longitude). To see how you would locate Washington within the grid, examine Figure 2.8 again. Beginning with latitude, find the 38th parallel north of the Equator to identify the location in the Northern Hemisphere. Once this is accomplished, locate the Prime Meridian and arc west to the 77th meridian. The intersection of the 38th parallel and the 77th meridian is the position of Washington, D.C. Notice that you can first find the latitude or you can first find the longitude; the order does not matter, as you wind up at the same place either way. However, remember that by convention the latitude of a location is always specified before the longitude is given.

Human Interactions: The United States Public Land Survey

The preceding discussion about latitude and longitude revolved around the global grid system used to determine location on Earth. In addition to this grid system, the United States also uses another form of grid network that is based on the cardinal directions (north, south, east, west) to organize the landscape in association with property ownership. Have you ever flown over the central part of the United States and noticed that the landscape is systematically organized into hundreds of giant squares? This grid is part of the **United States Public Land Survey** system (USPLS) and is the framework that surveyors used to survey the majority of the country during the era of westward expansion in the 1800s.

The USPLS was originally proposed by Thomas Jefferson and was implemented shortly after the Revolutionary War in the late 1780s to survey the large tracts of land then available for settlement west of the original 13 colonies. Prior to the establishment of the USPLS system, landscape survey in the colonies had followed the *metes and bounds* system, which was based largely on the location of landmarks such as trees,

United States Public Land Survey *The primary way that land is subdivided in the United States by using a grid system based on the four cardinal directions.*

www.wiley.com/college/arbogast

Using the Geographic Grid

In order to help you fully comprehend how the complete geographic grid works, view the animation *Using the Geographic Grid*. In this simulation you will be able to select a number of places on Earth and determine their latitude and longitude. This,

in turn, should help you better understand where locations are in physical space. Once you complete the exercise, be sure to answer the questions at the end to test your understanding of the geographic grid.

Beautiful natural features are scattered around the world on all seven continents. Locate the following places on a globe or map of the world and determine the country in which each one is located.

a) Mont Blanc, 45.50° N, 6.52° E
b) Lake Chad, 13.30° N, 14.00° E
c) Valle de Luna (Valley of the Moon), 23.50° S, 69.00° W

(a)

(b)

(c)

river bends, and rock outcrops. In other words, the survey of individual land plots was determined on the distance and direction of one tree to another or a river bend to a distinct rock outcrop. Although this system made some intuitive sense, it was problematic because landmarks are seldom permanent features, given that trees die and rivers move. As a result, people could not depend on the accuracy of the metes and bounds system over the long term.

After the Revolutionary War ended and settlers began to spill over from the colonies into areas west of the Appalachian Mountains, Jefferson and others realized that a more systematic survey method was required to properly organize the developing territories to avoid the problems caused by the old metes and bounds system. The USPLS was first implemented in association with the *Northwest Ordinance of 1787*, which established a rectangular survey system to facilitate the transfer of federal lands to private citizens in what ultimately became the states of Ohio, Michigan, Indiana, Illinois, and Wisconsin. The system proved successful and was further implemented throughout the rest of the country as settlement progressed. Since the implementation of the system, almost 1.5 billion acres have been surveyed throughout the country through regulation by the U.S. Department of the Interior (Bureau of Land Management).

To understand how the USPLS works, it is useful to think of it as a hierarchical system that is based on an initial set of intersecting lines established within a region of the country, such as the central Great Plains (Figure 2.9a). This first pair of intersecting lines includes a *baseline*, which extends east and west, and a *principal meridian*, which runs north and south. Once this

pair of lines was established, surveyors then created a secondary grid by surveying lines every 6 mi along each baseline and principal meridian (Figure 2.9b). Lines surveyed along the principal meridian (extending east and west) are called township lines, whereas those along the baseline (extending north and south) are called range lines. The intersection of township and range lines—remember all are 6 mi apart—resulted in a grid network that consisted of a series of 36 mi^2 tracts. Each of these 36 mi^2 tracts is called a *township*, with each township identified with a township and range designation. Township designations indicate the location north or south of the baseline, whereas range designations indicate the location east or west of the principal meridian. So, for example, a township might be identified as Township 2 South, Range 3 West, which would mean that it was in the second tier of townships south of a baseline, and in the third column of townships west of a principal meridian.

Once the grid of townships was established, they were further subdivided by survey into thirty-six 1 mi^2 tracts (Figure 2.9b). Each of these tracks is called a *section* and contains 640 acres. They are numbered sequentially within each township, beginning with Section 1 in the upper right corner and Section 36 in the lower right. (Section 14 is identified in Figure 2.9b). It is these sections that you see when you fly over much of the country (Figure 2.9c), with roads marking the edge of each 1 mi^2 tract of land. It is also why roads in rural areas are usually at right angles to each other because they intersect at section corners. Sections can be further subdivided into ½ sections (320 acres) or ¼ sections (160 acres Figure 2.9b). Such ¼ sections were the base land unit of the

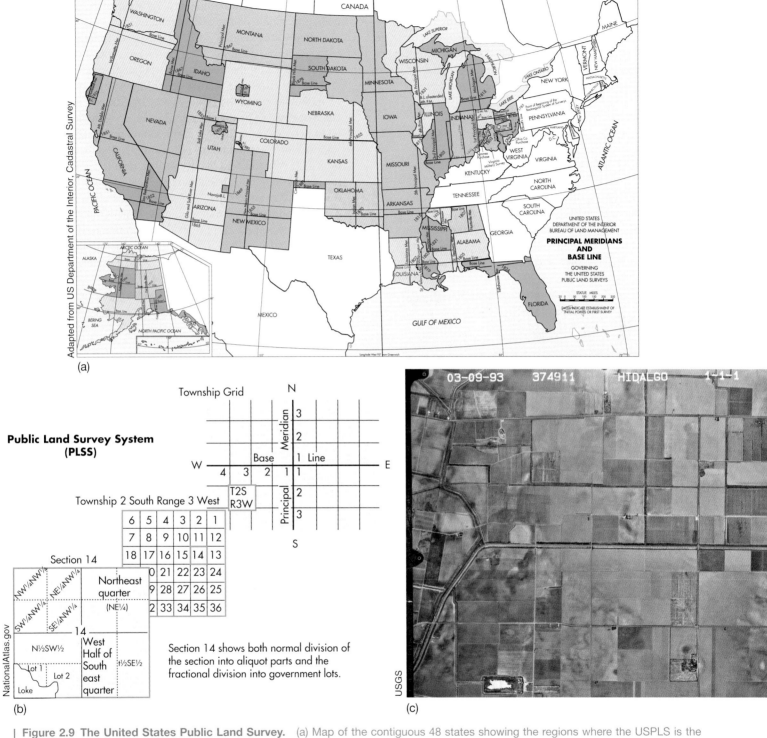

Figure 2.9 The United States Public Land Survey. (a) Map of the contiguous 48 states showing the regions where the USPLS is the basis for land survey. Each one of the color-coded areas has a baseline and principal meridian that is the foundation of the broader grid network. The eastern states are based on the old metes and bounds survey system. Texas is based on a different system. (b) Subdivision of township and range, which is based on the English system of measurement used in the United States. Townships are 36 mi^2 tracts established on the basis of township lines and range lines surveyed every 6 mi on the principal meridian and baseline, respectively. Each township is further subdivided into thirty-six 1 mi^2 sections, which can be, in turn, subdivided into 1/2 sections and 1/4 sections. (c) Aerial photo of a rural landscape in the Great Plains. Each of the visible squares is a section, which contains 640 acres. The section edges are marked by country roads. A highway crosses the center of the image.

famous *Homestead Act of 1862*, which provided for 160 acres of free land for people who settled the undeveloped lands west of the Mississippi River. Taken together, the township and range system establishes the basis for identifying the legal description of land parcels. Such a description includes the state, name of the principal meridian, township and range designations with directions, and the section number, such as: Kansas, Sixth Principal Meridian, T5N, R3W, Section 5.

Maps—The Basic Tool of Geographers

Now that Earth's grid system has been discussed, let's investigate the various tools that geographers use to portray spatial information within the grid. Most everyone knows that the primary tool associated with geography is the map. Figure 2.10 shows a typical thematic map (it focuses on a specific theme) and is one with which you are probably familiar. This map shows the United States and includes such geographic information as state boundaries, location of major cities, the course of major rivers, and vegetation patterns. You probably also know that a map such as this one contains a scale, a compass arrow pointing north, and a legend that defines major categories within the map, such as population density, the area covered by cities, the relative size of highways, or the location of mountain ranges. A wide variety of additional information can be portrayed on maps, including the geographic distribution and thickness of rock units, temperature, precipitation patterns, intensity of solar radiation, vegetation, and soils. Much of the information in this text is presented in map form, so in order to navigate through this book and to be successful in your course, it is helpful to understand the fundamentals of cartography.

Cartography is the subdiscipline of geography that focuses on the many ways to display spatial information so that it can be used and understood efficiently. The essence of cartography is the manner in which Earth is portrayed in a usable fashion. This

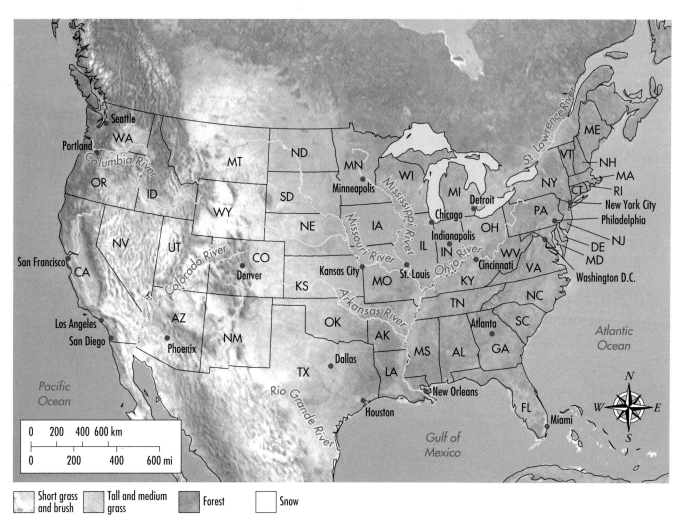

Short grass and brush   Tall and medium grass   Forest   Snow

Figure 2.10 A simple thematic map. This map shows the location of some important features within the United States, including states, major rivers, and cities. Note the legend that reflects general types of vegetation.

Cartography *The design and production of maps.*

(a)

Age Fotostock America, Inc.

(b)

Nasa Jet Propulsion Laboratory

Figure 2.11 The globe and planet Earth. (a) A globe portrays geographic information about Earth in a three-dimensional fashion. (b) Compare the image of the globe with this picture of Earth taken from space. This particular Earth view focuses on Africa.

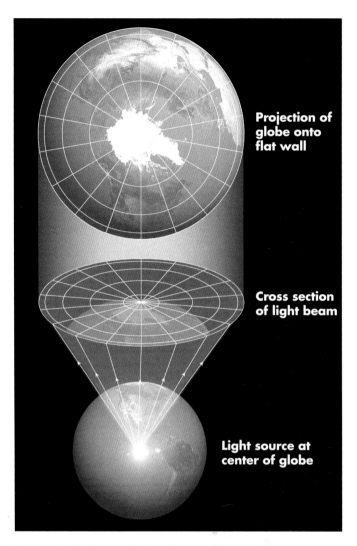

Projection of globe onto flat wall

Cross section of light beam

Light source at center of globe

Figure 2.12 The projection theory. Light from a source projects the information on the curved surface of the Earth onto a plane, forming a flat map.

task can be accomplished in a variety of ways. The most visually accurate and complete way to illustrate the entire Earth is through the use of a globe (Figure 2.11), which depicts locations on the three-dimensional representation of the planet. Globes can be cumbersome to use, however, and often do not illustrate the level of detail desired. Instead, geographers frequently rely on two-dimensional maps to show various geographic locations and attributes. Many factors must be taken into consideration to represent these features most accurately.

Map Projections

The most important step in the cartographic process is the presentation of three-dimensional information of Earth on a two-dimensional map. In order to create a map, locations on the roughly spherical surface of Earth must somehow be transferred to a flat surface, such as paper or a computer screen. This is done through the process of **map projection**.

A simple way to visualize how projection works is to imagine a light source passing through a curved translucent surface that has an image on it, projecting the image from the curved surface onto a flat wall (Figure 2.12). Light from the source (the light bulb) streams out toward the wall in straight lines. Before it reaches the wall, however, the light passes through the spherical surface of Earth. As the light beams move through this spherical surface, imagine that they carry the information from the Earth's surface through which they just passed. When the beams reach the wall, this information is then visibly displayed. Although the image was originally curved on Earth, it becomes flat on the wall. The actual process of map projection theoretically works in the same manner.

Map projection *The representation of the three-dimensional Earth on a two-dimensional surface.*

Figure 2.13 Why map distortion occurs. On the left is the semicircular (undistorted) half of a lime. Once the center of the lime is pressed to the table to make it flat, distortion occurs at the edges. Note that the area covered by the flattened part of the lime is larger than that beneath the uncut part of the lime.

Although projection serves the purpose of creating a flat map that is potentially more usable than a globe, a problem results because it is not possible to flatten a sphere (such as Earth) without some distortion of the image. Look at what happens in Figure 2.13 if a lime is sliced in two so that two hemispheres are created. If either of the two hemispheres is flattened on a table, it covers a larger area than it did when it was intact. Also

notice that the edges of the flattened hemisphere are severely distorted; in this case, they are split. In fact, the only way to make the hemisphere become perfectly flat is to either stretch or cut the edges.

Maps contain the same kind of distortion because spatial information from the curved surface of Earth is essentially spread over a flat surface. Because of this distortion, a variety of map projections have been developed, with each kind presenting location and distortion in a different way that can preserve either shape or size, but not both. Geographers are keenly aware of the strengths and weaknesses of each kind of projection with respect to spatial analysis and select the type of map they want to use accordingly. Although it is not necessary at this introductory level for you to understand maps as thoroughly as a geographer does, you must know something of the properties that result from the various projections that are used to create them. Such a fundamental understanding can be obtained by examining two simple kinds of projections: conformal and equivalent.

Conformal (or True Shape) A **conformal projection** is a map that maintains the correct shape of features on Earth but distorts their size relative to one another. A conformal projection is conceptually constructed by placing a translucent cylinder over the globe and shining a light centered within the globe outward so that an image is transferred to the cylinder (Figure 2.14). If the cylinder is then removed from the globe and cut so that it can be laid flat, a map will result.

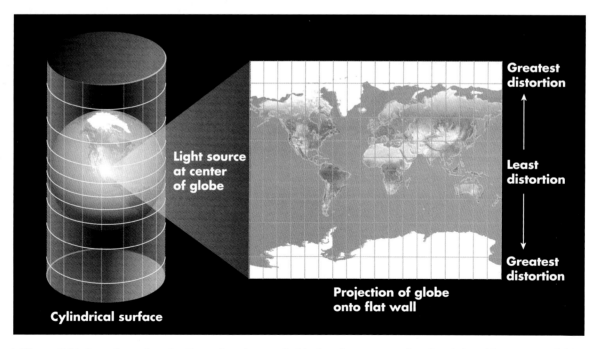

Greatest distortion

Least distortion

Greatest distortion

Light source at center of globe

Cylindrical surface

Projection of globe onto flat wall

Figure 2.14 A conformal projection. Imagine a cylindrical surface surrounding the globe with a source of light at the center of the globe. The surface is unwound to produce a flat map. The regions of greatest distortion occur at the top and bottom of the map.

Conformal projection *A map that maintains the correct shape of features on the Earth but distorts their relative size to one another.*

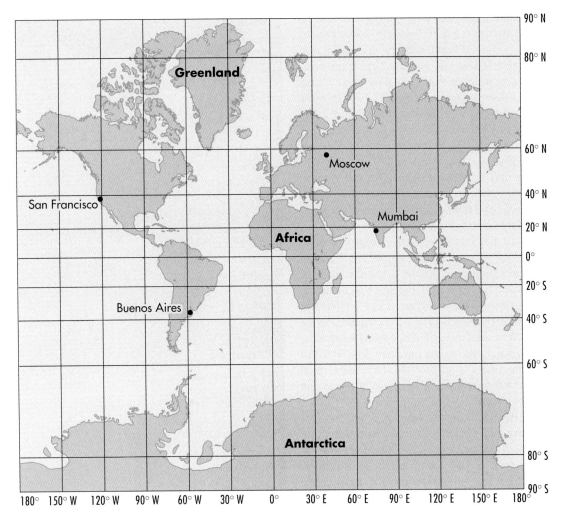

Figure 2.15 A Mercator projection.   This map is a conformal projection that maintains shape at the expense of size at high latitudes. Accordingly, scale increases from low to high latitudes.

One of the most common conformal projections is the Mercator projection (Figure 2.15), which was developed in 1568 by the Flemish geographer and mathematician Gerhardus Mercator. This projection is used frequently because it correctly shows the relative shape of features on the globe. As an example of this accuracy, look at Greenland and the Americas in Figure 2.15 and notice that the actual shapes of these places have been maintained. You might ask yourself, why would this be useful? Imagine that you wanted to compare the shape of coastlines between Greenland and the United States. The only way to accomplish this task accurately would be to use a conformal projection such as the Mercator, which would preserve the correct shape of the coastline and thus make an accurate comparison possible.

Remember, however, that some spatial attribute must be distorted to keep the shape accurate. In the case of the Mercator projection, the distorted feature is size. Notice how lines of latitude and longitude are at right angles to each other because meridians do not converge at the poles; note, too, that the distance between parallels increases with distance from the Equator. This increase occurs because the higher latitudes are

geometrically stretched more during the projection process than the lower latitudes. Another way to look at this type of distortion is to realize that the map in Figure 2.15 was stretched much like the lime was in Figure 2.13.

You can see this distortion particularly well by looking at Greenland and Antarctica in Figure 2.15. Notice that Antarctica covers almost a quarter of the map and is significantly larger than Africa. In reality, however, the Antarctic continent is less than half the size of Africa (8.7 million km² [3.4 million mi²] vs. 18.8 million km² [7.3 million mi²]). The size of Greenland is also distorted, which can be seen through comparison with the United States. Although Greenland appears to be significantly larger than the United States on the map, in reality Greenland is about 2,175,590 km² (840,000 mi²) in size, whereas the United States covers an area of about 9,582,956 km² (3.7 million mi²)—more than four times the size of Greenland.

Equivalent Projection   Once you know how the angular relationships between places compare through use of a conformal projection, you might want to see how their relative size differs. In order to accomplish this visualization, you would need to create an **equivalent projection**, which accurately portrays size features throughout the map. This map can be theoretically constructed by placing a cone over a portion of Earth, with the sharp point over the North Pole (Figure 2.16). In this instance light is projected upward through the cone, which is then laid flat to form the map.

One kind of equivalent projection that results from this projection process is an Albers equal-area projection (Figure 2.16), which was developed in 1805 by the German geographer H. C. Albers. Notice how this map, which focuses on North America, compares with the Mercator projection in Figure 2.15. One primary difference is that the lines of latitude are now curved. Given that these particular lines are actually curved on Earth, this kind of map represents a more accurate presentation of the geographic grid than a Mercator projection.

Another difference lies in the respective shapes of land-masses, which can be nicely seen again at Greenland. Remember that in the Mercator projection (Figure 2.15), the shape of Greenland was accurately portrayed but the size was distorted. How does Greenland now look in Figure 2.16? Notice that Greenland looks squat in the Albers projection compared to the Mercator projection. Why does this difference exist? Remember that a map projection can preserve only one spatial variable at a time. In the case of the Albers projection, the shape of places at high latitudes is distorted so that the relative size of places can be more accurately observed.

You might ask, why does it matter whether or not the sizes of landmasses on Earth on a map are accurate? Remember that the Mercator projection, because it preserves shape, allows you to compare features such as the configuration of coastlines. However, would a Mercator projection provide an accurate comparison of the *length of* coastline between the United States and Greenland? The answer to this question is "no" because the size of continents at high latitudes is severely distorted by a conformal projection. Thus, the only way to compare accurately the length of coastline between Greenland and the United States would be to use an equivalent-area projection.

It is important to emphasize that many different kinds of projections exist, although it is beyond the scope of this text to describe each of them. Depending on the kind of information you want to see, a projection exists for that purpose. Some projections intensively exaggerate the distortion of one property in order to represent another more accurately. Others attempt to show all spatial information as accurately as possible, given the limitations of the technique, and therefore moderate the distortion of both shape and size. A good example of such a projection is the Robinson projection (Figure 2.17), which was developed in 1963 by the American cartographer Arthur H. Robinson. Notice how this particular projection seems to be evenly balanced relative to the Mercator (Figure 2.15) and Albers (Figure 2.16) projections. Greenland and Antarctica are still somewhat enlarged relative to their accurate size, but some semblance of the Earth's curvature exists.

Equivalent projection *A map projection that accurately portrays size features throughout the map.*

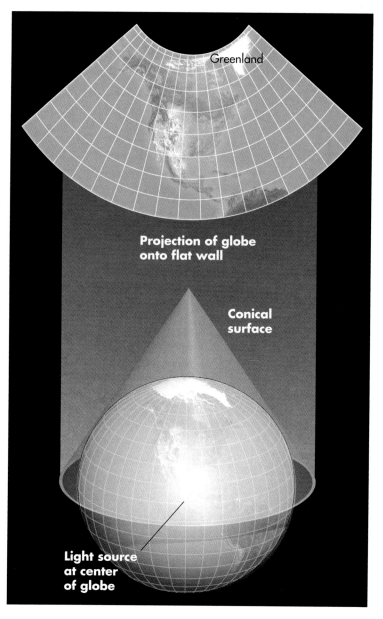

Figure 2.16 Constructing an equivalent projection. The resulting map in this diagram is a specific kind of equivalent projection called an Albers equal-area projection.

KEY CONCEPTS TO REMEMBER ABOUT MAP PROJECTIONS

1. Map projection is a critical part of cartography because it allows the three-dimensional Earth to be represented in two dimensions.

2. A map projection is designed to preserve either the shape or size of geographic features.

3. Conformal projections maintain the angular relationship between geographic features, but distort the relative size.

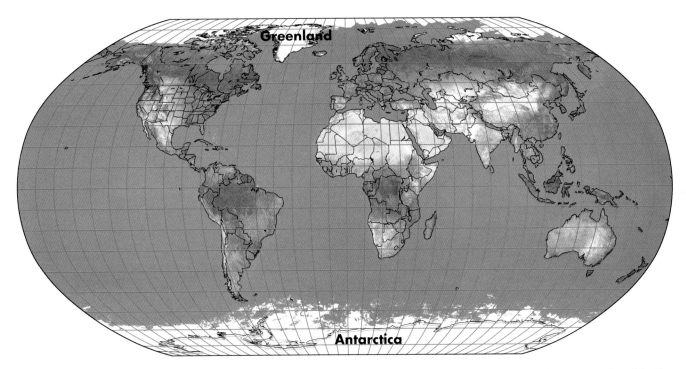

Figure 2.17 The Robinson projection. In an effort to balance the errors associated with projection, this map contains distortion in both shape and size.

4. Equivalent projections maintain the relative size of geographic features at the expense of shape.

5. Several different kinds of projections are used by cartographers, each of which is designed to portray size and shape in a specific way.

Map Scale

Aside from understanding how the concept of map projection works, another important consideration involving map interpretation is the scale. Map scale is a critical part of cartography because it relates the size of features on a map to their actual size in the real world. Similarly, the scale relates the distance between features on the map to the corresponding actual distances on Earth. In other words, **map scale** represents the ratio of the size/distance on the map to the size/distance on the ground.

Map scale can be depicted in several ways.

1. *Written (Verbal) Scale.* Often written as: One Inch = 2000 Feet, which means that 1 in. measured on the map represents 2000 ft on Earth.

2. *Representative Fraction.* An example of a representative fraction is the ratio 1:24,000 or 1/24,000. This scale means that one unit of measurement on the map represents 24,000 of the same unit on the ground. It does not matter what unit of measurement is used—it could be inches, centimeters, etc.—but the units must be the same within the ratio (inches to inches, centimeters to centimeters, etc.).

3. *Graphic Scale or Bar Scale.* These scales show the actual size of units on the map. The advantage of the graphic (or bar) scale is that the scale remains accurate if the map is enlarged or reduced when photocopying.

Maps are made at different scales depending on the geographic qualities they are designed to display, including the area they cover and spatial detail. In general terms, geographers think of maps in terms of being large scale or small scale. Although these designations appear easy to differentiate, the fact is that many nongeographers find them confusing because an inverse relationship exists between map scale and actual Earth features. You might think, for example, that a large-scale map is a map that shows a large part of Earth. Conversely, a small-scale map would be a map that focuses on an area of limited geographic extent, such as a city or park. The fact is that **small-scale maps** are maps that show large areas with relatively limited detail because the representative fraction is a small number; in other words, the size ratio of the map features to the real world is small. **Large-scale maps**, in

---

Map scale *The distance ratio that exists between features on a map and the real world.*

Small-scale map *A map that shows a relatively large geographic area with a relatively low level of detail.*

Large-scale map *A map that shows a relatively small geographic area with a relatively high level of detail.*

contrast, represent much smaller geographic areas with a greater level of detail because the representative fraction is a larger number; that is, the size ratio on the map is closer to the actual world. In order to obtain a better feel for map scale, you should examine the information in Figure 2.18 and Table 2.2.

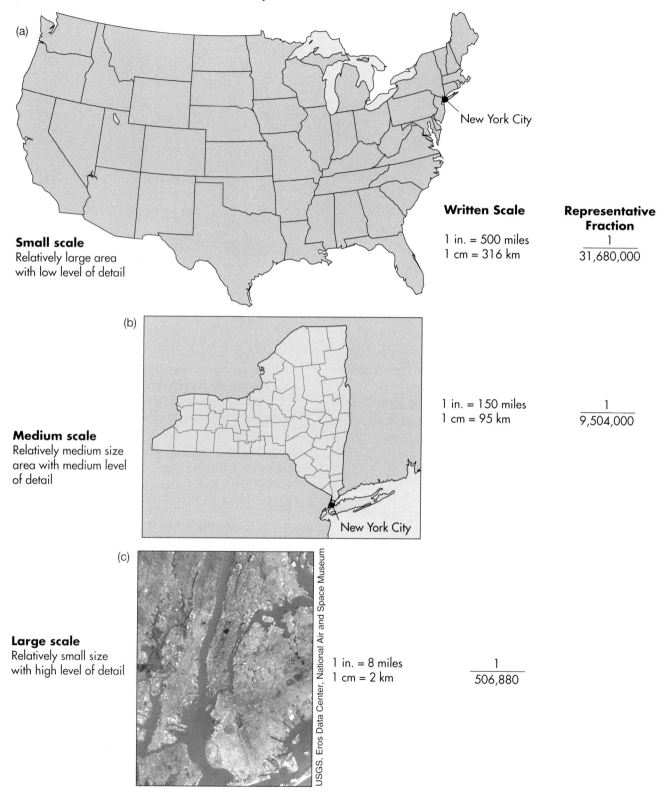

(a)

New York City

Small scale
Relatively large area
with low level of detail

Written Scale

1 in. = 500 miles
1 cm = 316 km

Representative Fraction

$\dfrac{1}{31,680,000}$

(b)

Medium scale
Relatively medium size
area with medium level
of detail

New York City

1 in. = 150 miles
1 cm = 95 km

$\dfrac{1}{9,504,000}$

(c)

Large scale
Relatively small size
with high level of detail

USGS, Eros Data Center, National Air and Space Museum

1 in. = 8 miles
1 cm = 2 km

$\dfrac{1}{506,880}$

Figure 2.18 Examples of map scale. Note that as scale increases, the amount of detail about a place increases. (a) Map of the United States that shows the location of New York City. (b) Map of the state of New York showing the location of New York City. Note that this map shows more detail (county boundaries) of the state of New York than (a). (c) Satellite image of New York City. Note how much more detail of New York City can be seen in this image than in the other maps.

TABLE 2.2   Map Scale

Large-Scale Maps	Small-Scale Maps
Used for maps of small geographic areas.	Used for maps of large geographic areas.
Used to illustrate great detail, such as road networks and locations of parks.	Used to illustrate limited geographic detail, but show spatial relationships of large areas.
Representative fraction is a large number, such as 1:1000.	Representative fraction is a small number, such as 1:25,000.

Isolines

Isolines, another important map component, are lines that connect points of equal value. These lines are particularly useful, for example, when showing regional temperature and precipitation patterns, elevation, or atmospheric pressure. These data can be derived in a variety of ways and then used to illustrate fundamental geographic trends. For example, imagine that you want to create a map showing the pattern of high

Isolines   *Lines on a map that connect data points of equal value.*

temperatures throughout the United States on any given day. The first thing you would do would be to access the temperature measurements taken at all of the National Weather Service sites in the country and show them on a map. This map would contain an outline of the country, state boundaries, the various locations where temperature measurements were taken, and the temperature at each place (Figure 2.19a).

Although this map would contain the information you desire, it would be difficult to discern any geographic pattern because the map contains nothing but a series of numbers. By looking at the numbers closely, you could perhaps determine

(a)

(b)

(c)

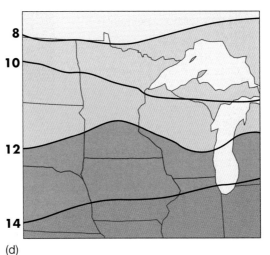

(d)

Figure 2.19 Drawing isolines. (a) Spatial distribution of data points, in this case temperature in degrees centigrade. (b) Interpolating the 14° isotherm. A line is drawn connecting points where 14° measurements were obtained. The position of the 14° isoline is estimated between locations where 13° and 15° temperatures were measured. (c) Isolines drawn at intervals of 2°C. (d) The resulting map shows the spatial distribution of temperature in the upper Midwest.

that it was generally warmer in the southern part of the United States than in the northern part of the country, but the overall geographic patterns would be vague. Geographers typically clarify these patterns by using isolines, which are lines that connect points of equal value at predetermined intervals. In the context of our temperature map, you may decide that you want to show the temperature pattern in the country at 5°C (9°F) intervals, or you may choose another interval. Using a 5°C (9°F) interval as an example, you would draw an isoline connecting all points of equal value at, for example, 5°C (41°F), 10°C (50°F), 15°C (59°F), or 20°C (68°F). In many regions these distinct temperatures might not exist; that is, the temperature at one weather station might be 7°C (44.6°F), whereas at the adjacent station it could be 12°C (53.6°F). Where this kind of pattern exists, you would draw the line between the two weather stations in a process called *interpolation* because you could assume that the 10°C line occurs somewhere midway between 7°C (44.6°F) and 12°C (53.6°F). Figure 2.19 shows this process more closely with some temperature data in the upper Midwest.

In the context of isolines, several different kinds are used when mapping specific types of geographic phenomena. The various kinds of isolines are:

1. **Isobars**—connect points of equal atmospheric pressure.
2. **Isotherms**—connect points of equal air temperature.
3. **Isohyets**—connect points of equal amounts of precipitation.
4. **Isopachs**—connect points of equal sedimentary thickness.
5. **Contours**—connect points of equal elevation.

Contour lines are particularly significant because they illustrate the configuration of the three-dimensional landscape, or the **topography**, on a **topographic map**. Topographic maps contain a great deal of information, especially at a 1:24,000 scale (Figure 2.20). Here are some simple rules to follow when reading these maps:

1. The closer the spacing of contour lines, the steeper the slope. Conversely, contours that have wide spacing represent terrain that is relatively flat.
2. Contours that form closed circles indicate hills. Closed circles with hatch marks indicate closed depressions.
3. Where contours cross a stream, they form V's pointing upstream (which is also uphill).

Isobars *Isolines that connect points of equal atmospheric pressure.*

Isotherms *Isolines that connect points of equal temperature.*

Isohyets *Isolines that connect points of equal precipitation.*

Isopachs *Isolines that connect points of equal sediment or rock thickness.*

Contours *Isolines that connect points of equal elevation.*

Topography *The shape and configuration of the Earth's surface.*

Topographic map *A map that displays elevation data regarding the Earth's surface.*

(a)

USGS

(b)

Richard Price/Taxi/Getty Images

Figure 2.20 A typical topographic map. Topographic maps show elevation patterns of a given geographic area. (a) This topographic map focuses on the Maroon Bells area near Aspen, Colorado. (b) Photograph of the Maroon Bells from Maroon Lake. The arrow in the accompanying topographic map shows the approximate line of sight from Maroon Lake toward the mountain peaks.

Now that you have learned something about cartography, view the animation *Using Maps*. The goal of this animation is to increase your understanding of maps by interacting with them online. This particular animation covers how maps are produced, issues relating to scale, and the various kinds of data that can be presented in a map. Once you finish this exercise, be sure to answer the questions at the end to test your understanding about maps.

VISUAL CONCEPT CHECK 2.2

a) The process of map projection is the method through which three-dimensional features on Earth are presented as a two-dimensional image. Two basic projections, conformal and equivalent, are presented here. Which of these two maps is a conformal projection? Which one is the equivalent projection? How can you tell the difference between the two?

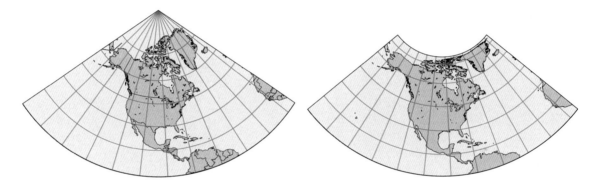

b) The scale of a map depends on the amount and kind of information that a cartographer wishes to present. In this case, two maps are presented that contain information about Seattle, Washington. Which one of the maps is a relatively small-scale map? Which one is the large-scale map? How can you tell the difference between the two?

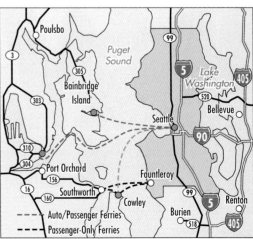

Digital Technology in Geography

As you are no doubt aware, digital technology is an essential part of life in the industrialized world today, with personal computers, cell phones, the Internet, and satellite television common features. These technological advances have also been very important in the recent evolution of geography as a scientific discipline, resulting in expanded work in the field that has increased the need for skilled people at all professional levels at excellent pay. Much current geographic research is focusing on the development and refinement of three digital techniques: Global Positioning Systems (GPS), remote sensing, and Geographic Information Systems (GIS). These techniques, used independently as well as together, enable geographers to use maps and other images in a digital format in order to increase the speed and efficiency of geographic research. In addition, the ability to display geographic information has improved remarkably. The remainder of this chapter focuses on these new and exciting tools that are revolutionizing how people are viewing the planet from a geographic perspective.

Remote Sensing

Until the early 20th century, the only way that a geographer could consistently learn anything about new places firsthand was to personally conduct fieldwork and collect samples or make observations. This limitation began to change somewhat in the early 1900s with the development of airplanes, which people could use to see places without actually setting foot on the ground. With the advent of this new technology, geographers began to use a new approach to spatial analysis called **remote sensing**, which means measuring properties of the environment without direct contact. In this manner, geographers could learn much information about the vegetation patterns of the landscape, the routes that rivers take across the countryside, or even the configuration and distribution of agricultural fields. Remote sensing began to be used in a widespread, systematic fashion in the 1930s when on-board cameras were developed that could take **aerial photographs** (Figures 2.9c, 2.21) at prescribed intervals of time. With these photographs, which could be taken in a succession of linear flight lines, entire counties or even states could be viewed in a systematic way.

Although aerial photography marked a dramatic leap in the geographer's ability to study the landscape, this technology proved limiting because a complete set of photographs of large regions (like a state) was only taken once every decade, or at best, at irregular intervals of time. Although this temporal pattern allowed for some studies of landscape change, perhaps by taking photographs of the same place two or more times over, say, 25 years, geographers were still limited because they could not see

Figure 2.21 A typical aerial photograph. With an aerial photograph you can see a variety of landscape patterns that cannot be viewed from the ground. This aerial photo, for example, shows the pattern of crop fields, trees, roads, and dwellings at Rajasthan, India.

David Zimmerman/Corbis Images

Remote sensing *The method through which information is gathered about the Earth from a distance.*

Aerial photographs *Photographs taken of the Earth's surface from the air.*

THE MARS ROVER

NASA/JPL/Cornell

(a)

Panoramic image of the Martian landscape taken by *Spirit Rover* in 2005. This image was taken with a camera on board the rover and then sent back to Earth via radio waves where it was subsequently processed into this digital file. Given that the distance between Earth and Mars varies between about 78 million km (48 million mi) to 377 million km (230 million mi), it takes between 4 minutes and 20 minutes for the data to reach Earth.

Digital technology allows scientists to examine many aspects of Earth, including soils, the hydrosphere, biosphere, and terrain. One of the most fascinating ways that digital technology is being employed to understand planetary history is on Mars. Mars has always been a source of great interest to scientists because it formed at about the same time as Earth, is similar in size (certainly compared to the giant planets such as Saturn and Jupiter), and is in the same part of the solar system. Most importantly, strong evidence exists that liquid water once flowed on Mars and may even remain there hidden below the surface. If so, it is possible that primitive life once existed on Mars and may still do so in some remote and simplistic way.

In this scientific context, the National Aeronautic and Space Administration (NASA) is currently in the midst of a rigorous program to explore Mars. The most interesting aspect of this program is the deployment of rovers on the surface to investigate the environment and geology. These rovers arrived after a nine-month trip from Earth and have been subsequently directed from scientists on Earth. A pair of rovers, *Spirit* and *Opportunity*, landed on the surface in early 2004 and began exploring the planet, sending digital imagery and data back to scientists on Earth. Although *Spirit* became stuck in sand after six years, *Opportunity* continues to roam the surface with direction from scientists at NASA.

A new rover, *Curiosity*, landed in Gale Crater on Mars in August 2012. *Curiosity* is a much more sophisticated

NASA/JPL-Caltech

(b)

Artist's conception of the *Curiosity Rover* vaporizing a rock on the Martian surface. The rover is about the size of a car and has the ability to drill into rock, excavate shallow trenches, and conduct experiments to search for organic material and various chemicals.

machine than the earlier rovers, having the ability to drill into rock, excavate trenches, collect samples, and conduct complex chemical experiments. This rover's goals are to investigate the Martian climate, geology, and the role of water in shaping the landscape and providing a suitable environment for microbial life. An early finding was that flowing water once existed in the crater. *Curiosity* should run for several years and promises to generate some amazing data that could very well change our view of life in the universe. Pay attention!

the change as it was occurring. The fact is that major changes can occur on a landscape in a very short period of time and geographers usually want to see them quickly.

With the development of satellite technology in the 1950s and 1960s, our ability to remotely sense the landscape increased dramatically. As you know, satellites are mechanical platforms

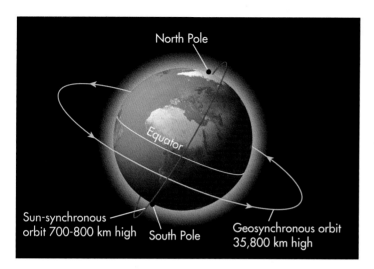

North Pole

Equator

Sun-synchronous
orbit 700-800 km high   South Pole   Geosynchronous orbit
35,800 km high

Figure 2.22 Sun-synchronous and geosynchronous satellite orbits. Sun-synchronous orbits are designed to keep pace with the Sun's westward progress as Earth rotates. The orbit essentially extends between the poles with a small inclination so that the satellite moves slightly westward with each successive day. Geosynchronous orbits hover over the same point on Earth, but to do so, they must stay in a very high orbit over the Equator.

that continuously orbit Earth. These systems are designed for either Sun-synchronous or geostationary orbits (Figure 2.22). A satellite with a **Sun-synchronous orbit** has an orbit that keeps pace with the Sun's westward progress as Earth rotates. Such a position is maintained because the satellite's orbit is basically north–south between the poles with a slight (8°) angular inclination. As a result, a particular satellite may cross the Equator 12 times during the course of a day, with the local time perhaps being 3:00 P.M. at the time of each pass. A Sun-synchronous orbit is usually at an altitude of about 700 km to 800 km (430 mi to 500 mi).

The orbit of Sun-synchronous satellites results in large gaps in image coverage between successive orbits on any given day. Given the slight inclination of the orbit, however, the satellite progresses slightly westward with each new day. In this way, the satellite slightly overshoots the orbital path from the previous day, yielding imagery that overlaps a little so that no information is lost. Because satellites with Sun-synchronous orbits pass over the same part of Earth at about the same time each day, they are particularly good for monitoring landscape change over time.

Several well-known satellite systems have Sun-synchronous orbits. Perhaps the best known of these satellites is the Landsat

system, which is jointly administered by NASA, the National Oceanic and Atmospheric Administration (NOAA), and the U.S. Geological Survey (USGS). This satellite system is the longest continuous Earth-observing project in history, beginning with Landsat 1 in 1972 and continuing through the launch of Landsat 8 in 2013. Landsat 8 orbits Earth at an altitude of about 705 km (438 mi), has a swath width (it can "see") of 185 km (115 mi), and returns to the point of origin every 16 days (233 orbits). Landsat 8 carries two sensors: Operational Land Imager (OLI) and Thermal Infrared Sensor (TIRS). OLI has nine spectral bands, which means it can discriminate emitted energy in several parts of the electromagnetic spectrum, whereas the TIRS sensor has two spectral bands. The **spatial resolution** of the various Landsat sensors ranges from 15 m (49.2 ft) to 60 m (197 ft). Spatial resolution refers to the size of the ground area that the sensor can "see" in detail. A 15-m (49.2-ft) spatial resolution, for example, means that the satellite can discern objects that are at least 15 m × 15 m in size. Each of these 15 m × 15 m (49.2 ft × 49.2 ft) blocks is referred to as a **pixel**.

In addition to the Landsat platform, another satellite with a Sun-synchronous orbit is the Advanced Very High Resolution Radiometer (AVHRR), which is managed by NOAA. This satellite orbits Earth 14 times per day from an altitude of 833 km (517 mi) and has a spatial resolution that ranges between 1.1 km (0.7 mi) and 4 km (2.5 mi), depending on whether the place viewed is directly below the satellite or on the fringe of the area seen.

A **geostationary orbit**, in contrast to a Sun-synchronous orbit, is designed to permanently remain in one place above Earth. This consistent position is accomplished by placing the satellite in an easterly orbit, directly over the Equator, and at a very high altitude. The satellite orbits in the same direction and at the same speed that Earth rotates. With such an orbit, the satellite can view nearly half of Earth at any given time. This kind of orbit is mostly used to observe weather and to facilitate communications. An excellent example of satellites with a geostationary orbit is NOAA's Geostationary Operational Environmental Satellites (GOES), which are used to monitor meteorological conditions (Figure 2.23). This system consists of an array of satellites that are positioned about 35,800 km (22,300 mi) above Earth, high enough to allow very large regions of Earth to be viewed at any given time. The GOES-12 (or GOES East) satellite, for example, is positioned above the Equator at 75° W longitude and views most of North America, all of South America, and most of the Atlantic Ocean. In contrast, the GOES-10 (or GOES West) satellite is located

---

Sun-synchronous orbit *A slightly inclined polar orbit that keeps pace with the Sun's westward progress as Earth rotates, resulting in regular return intervals over every location on Earth.*

Spatial resolution *The area on the ground that can be viewed in detail from the air or space.*

Pixel *The smallest definable area of detail on an image; short for pixel element.*

Geostationary orbit *An orbit where satellites remain over the same place on Earth every day. This orbit is achieved because the satellite is placed very high above Earth and travels at the same speed as the Earth's rotation.*

NASA GSFC

Figure 2.23 A typical GOES image. This example is from the GOES East satellite, which provides daily imagery of the eastern United States. Clouds are bright white, whereas darker areas are zones of clear sky. This image shows that clear skies dominated the eastern part of the country on this particular day.

above the Equator at 135° W longitude and views the western portion of North America and South America, as well as the vast majority of the Pacific Ocean. As with other forms of satellite remote sensing, the GOES-10 and GOES-12 images slightly overlap one another.

Remote sensing operates on the simple principle that objects on Earth emit electromagnetic energy that can be measured. Although the electromagnetic spectrum will be discussed in much greater detail in Chapter 4, it is sufficient for now to say that objects on Earth emit energy that is both visible and invisible to our unaided eyes. This energy can be measured by either satellites or airplane platforms and then stored digitally in a computer for detailed analysis.

Figure 2.24 shows the typical steps through which satellite imagery is processed, with this example focusing on a false-color image of Cape Cod in Massachusetts. The sequence begins with a measurement of **emissivity** by the satellite. These data are then sent to a receiving station on Earth, such as at NASA, where they are archived until distributed to the appropriate users in the scientific community as black-and-white images. The image is then modified by the users through the application of colorizing filters to produce a false-color image that highlights particular geographic features of interest.

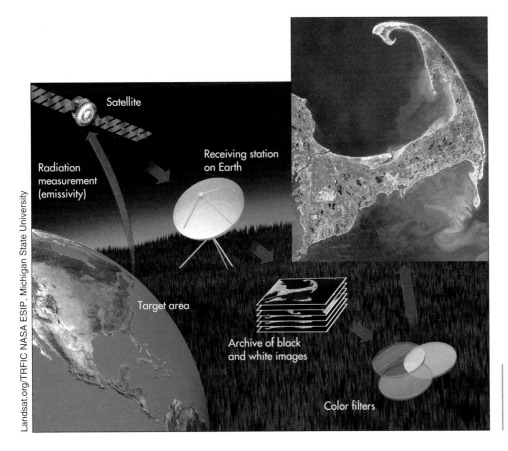

Landsat.org/TRFIC NASA ESIP, Michigan State University

Satellite

Receiving station on Earth

Radiation measurement (emissivity)

Target area

Archive of black and white images

Color filters

Figure 2.24 Processing of remotely sensed images. This sequence of events produces a false-color image of Cape Cod, Massachusetts, that can be geographically analyzed.

Emissivity *The amount of electromagnetic energy released by some aspect of the Earth's surface.*

Digital Technology in Geography • 35

(a)

(b)

Figure 2.25 Infrared satellite images. (a) This image of the region around Baltimore, Maryland, shows areas of vegetation in bright red and urban areas as bluish gray. Areas of dark shading consist of water, specifically part of Chesapeake Bay. (b) Deforestation in the Amazon rainforest as seen from NASA's Terra Satellite. Dark red represents uncut forest, whereas the lighter shades of red indicate the pattern of forest clearing. The dark wavy line in the image is the Jiparaná River. Note how the image path appears to be tilted slightly. This shape reflects the slight inclination of the satellite's Sun-synchronous orbit.

Although much can be learned by viewing Earth as humans normally see it, a great deal more information can be acquired by investigating the landscape in the invisible parts of the spectrum. This analysis often occurs by examining the thermal infrared energy emitted by Earth. All objects emit this kind of energy, but the relative amount varies depending on the temperature—warm objects emit more energy than cold ones—and the overall emissivity of the object. With a thorough understanding of these properties, geographers can see very subtle differences on the landscape and thus determine the spatial distribution of specific features such as vegetation, rock, and water. Areas such as these can be studied frequently in this fashion with infrared aerial photography, while larger regions can be investigated on an annual basis with Sun-synchronous satellites. Figure 2.25 shows a pair of infrared satellite images. This kind of analysis is especially useful, for example, as a means of showing urban patterns (Figure 2.25a) and monitoring deforestation in tropical rainforests (Figure 2.25b).

The methods described so far are passive systems in that they depend on energy emitted by objects on Earth to obtain measurements. Another form of remote sensing relies on active systems that send a beam of wave energy toward Earth. This beam is then partially reflected back to the sensor, which can make measurements about the amount of energy emitted. A good example of this kind of system is *radar*, which uses energy in the microwave part of the electromagnetic spectrum. Radar systems emit short pulses of energy toward Earth and

then measure the time it takes for the pulse to return, plus its strength, to create a picture of the landscape. Such imagery can be obtained either from an airplane (Figure 2.26)

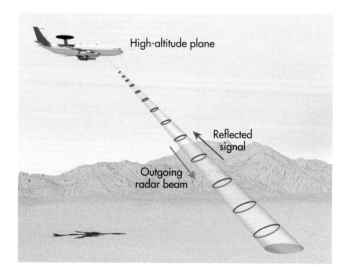

Figure 2.26 Acquisition of Synthetic Aperture Radar (SAR) imagery from an airplane. In this system, a radar unit at the base of the airplane sends radar waves to the ground in a sweeping motion across the landscape. Subsequently, these waves bounce back to the airplane where their timing (from and to the airplane) and strength are calculated and converted into an image of the landscape.

Figure 2.27 Radar image of Shenandoah National Park in the Appalachian Mountains. This image was acquired with C-Band Interferometric Radar during the Shuttle Radar Topography Mission (SRTM) and subsequently color-coded to show elevation. Greens represent topographically lower areas, rising through yellow, red, and magenta, to bluish white at the highest elevations.

or from space. In the latter instance, the most extensive radar imagery was acquired in the Shuttle Radar Topography Mission (SRTM), during which virtually all the surfaces on Earth were mapped with radar. In contrast to passive systems, which can be limited when cloud cover is extensive, radar can "see" through the clouds and can therefore be used all the time. As a result, radar has been extensively used to map the surface of Venus, which is masked by continuous cloud cover. On Earth, radar is especially effective at mapping geological formations in mountainous landscapes such as the Appalachians (Figure 2.27).

Global Positioning Systems

Another way that digital technology has benefited geographers is in determining location within the geographic grid. With the proliferation of satellites in the late 20th century, the U.S. military developed a system that enables people to determine their geographic location quickly and with much greater precision than with the use of a map. This system is called the Global Positioning System, or GPS, and is based on a network of 24 satellites that orbit Earth every 12 hours (Figure 2.28). These satellites continuously transmit a radio signal known as

Pseudo-Random Code (PRC) that is unique to the GPS. This code consists of a complicated sequence of on/off pulses that travel at the speed of light (299,300 km/sec; 186,000 mi/sec) in space and can be acquired by receivers on Earth. Figure 2.29 shows an example of a hand-held GPS receiver. The fundamental principle of GPS is that by knowing the positions of satellites in space, the exact time that a signal is sent from a given satellite and received by a receiver, and the speed of the incoming PRC, you can use this artificial constellation of satellites as locational references through a system of triangulation.

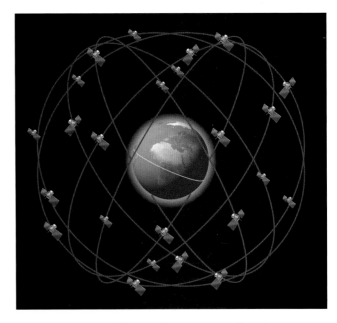

Figure 2.28 The GPS satellite system. Satellites are positioned in six orbital planes, with four satellites in each orbit (24 total), and are about 20,200 km above Earth's surface. This array provides complete coverage of Earth.

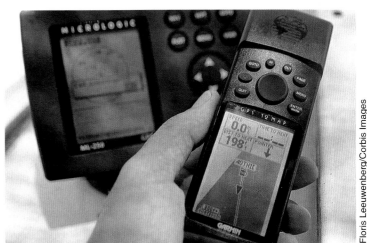

Figure 2.29 A hand-held GPS receiver. This unit receives the PRC signal from satellites in the GPS constellation.

Figure 2.30 GPS triangulation. By determining the distance from three different satellites, the GPS receiver narrows its location to the region where three different spheres intersect. The distance to a fourth satellite can be determined as a check. Locations can be determined to within one meter with expensive receivers.

So how does this system work (Figure 2.30)? Imagine your car or cell phone contains a GPS receiver and receives a signal from a single satellite indicating that it is 7000 km (4350 mi) away from your location. This lone signal indicates that you are anywhere on the surface of a sphere that is 7000 km (4350 mi) from the satellite. If the signal from a second satellite is added, for example, one that is 8000 km (4971 mi) away from the receiver, your location can be narrowed to the region contained within the space where both spheres intersect. Unfortunately, the size of this overlapping region is still considerable and might tell you that your location, for example, is someplace within the eastern United States. However, by adding the distance to a third satellite, say, one that is 9000 km (5592 mi) away, the location is further narrowed to a very small area where all three spheres intersect. To further increase the confidence of the measured location, the distance to a fourth satellite can be measured as a check. In this way, the system is more immune to fluctuations in the speed of the PRC signal as it travels through the atmosphere and inconsistencies that may exist between the time indicated by the incredibly accurate atomic clocks on satellites and less reliable clocks in receivers.

Through this process of triangulation, it is possible to pinpoint geographic location within about 20 m (about 90 ft) with inexpensive (less than $100) systems and to less than 1 m (3.3 ft) with more expensive systems (more than $10,000). This degree of accuracy is especially useful in the military because missiles and smart bombs can be precisely delivered, as you may have seen in television footage from recent conflicts. Within the geographic community, high-resolution GPS has been used in a wide variety of circumstances, including coastal navigation, determining the precise location of study sites, and the monitoring of plate movement along faults in earthquake zones (Figure 2.31). It is now used at an everyday level by millions of ordinary people to find their way around as they travel. You probably have a GPS yourself.

Although GPS is an exciting tool that has great potential for geographic analysis, it has several potential sources of error with which you should be familiar. One potential source of error is that the altitude of individual satellite orbits may vary from time to time. This variation is a problem because the receiver bases its distance calculations on a predetermined satellite altitude. Another source of error is the effect of the atmosphere on the satellite signal as it beams toward Earth. This effect is especially a problem when dust, water vapor, and ionized particles in the atmosphere interfere with the speed of the GPS signal, which leads to errors in distance calculations. A final transmission problem is

Remotely sensed image

Geological faults

Transportation network

Population density

Topography and vegetation

Figure 2.32 GIS layers. Each layer represents a distinctive part of the landscape, such as topography, vegetation, population density, or the local road network. When these layers are integrated, they form a detailed image of the landscape.

Figure 2.31 Using GPS to measure landscape change. Given the high accuracy of some GPS systems, it is possible to monitor changes in ground elevation, such as those that may occur near the Augustine Volcano in Alaska due to renewed volcanic activity.

multipath error, which results when obstructions such as buildings and trees cause the incoming GPS signal to be deflected before it reaches the receiver. This deflection confuses the receiver because two slightly different signals arrive at the same time.

Geographic Information Systems

Another tool that is very important in the way that geographers do their work is Geographic Information

---

Multipath error *Error that results when a satellite signal reaches the GPS receiver directly from the satellite as well as others reflected from nearby buildings or other surfaces.*

Systems, or GIS. A GIS is a system for storing, analyzing, and manipulating spatially referenced data, usually in digital form in a computer. Most GIS databases consist of a series of individual data layers that are considered to be relevant for the study being conducted. A data layer contains measurements obtained with respect to a specific geographic variable, such as vegetation, soils, road networks, municipal boundaries, and the distribution of surface water (hydrology) in lakes and rivers.

Let's choose vegetation as an example. A geographer may desire to map the spatial distribution of trees and grassland in a region, which can be determined by using infrared remote sensing or by digitally tracing old maps. In so doing, a GIS layer is created that can be identified as "vegetation" in the overall database. A variety of attributes can be subsequently assigned to this layer, such as species composition, height, slope of the land, and condition. Subsequently, similar maps can be created for other desired variables, resulting in layers for each of them. Now look at Figure 2.32, which shows a potential series of GIS layers from a theoretical study area. Notice how these layers exist as distinct units that can stand alone. In addition, the layers can be combined to illustrate the overall geographic character of the region. It is beyond the

www.wiley.com/college/arbogast

WileyPLUS

Using a Geographic Information System

To get a feel for GIS, view the animation *Using a Geographic Information System*. This animation focuses on the 2010 oil spill in the Gulf of Mexico and illustrates the way that GIS could be used to assess the character of the spill and its geographical relationship to various aspects of the regional environment and economy. After you complete the exercise, answer the questions at the end to test your understanding of Geographic Information Systems.

A variety of geographic information can be acquired, stored, and analyzed using digital technology. This image is an infrared satellite image of a portion of Washington, D.C. What do the red, black, and bluish-gray areas represent in this image? What are at least four data layers that could be constructed in a GIS?

Landsat/GSFC/NASA

scope of this book to describe GIS in detail, but you should understand that it is an extremely powerful tool that is used in a wide variety of ways, including:

- **Environmental management**—GIS can be used to manage spatial information associated with soils, wetlands, vegetation species, topography, and the location of data collection sites.

- **Municipal planning**—GIS is used in virtually all cities, large towns, and counties to manage spatial information such as road networks, location of sewer and utility lines, and emergency traffic routes.

- **Business needs**—A growing number of companies are incorporating GIS into their operations. GIS can be used, for example, to preselect the most efficient delivery routes that will save fuel costs. In addition, GIS can be used to identify the best location for new shopping centers based on surrounding demographic characteristics.

SUMMARY OF KEY CONCEPTS

1. Lines of latitude are called parallels and determine location north and south relative to the Equator. Latitude designations are calculated using simple geometric principles and extend to points 90° N and S, which are the North and South Poles, respectively. Lines of longitude, in contrast, are called meridians and determine location east and west of the Prime Meridian in Greenwich, England. These designations are also calculated using the geometry of a circle and extend to 180° E and W. Latitude and longitude overlay to form a geographic grid. The United States uses another grid system, the United States Public Land Survey (USPLS), to survey property ownership.

2. Maps are two-dimensional representations of three-dimensional aspects of Earth. They are created through the process of map projection, which creates distortion in some aspect of the map. A conformal projection, for example, maintains the correct shape of features at the expense of their relative size.

3. Map scale refers to the ratio of the size of features on a map relative to the real world. Small-scale maps show relatively large areas but with less detail. A large-scale map, in contrast, shows a relatively small area but with greater detail.

4. "Remote sensing" refers to the ability to observe and monitor aspects of Earth from a distance. The most systematic way to accomplish this task is with satellite platforms such as the Landsat system. These systems measure emitted energy that can be stored as digital data and ultimately accessed and manipulated by scientific users.

5. A Geographic Information System (GIS) is a digital database that contains relevant geographic information such

as the density of roads, location of human structures, and vegetation type, to name but a few. These various data sets are stored as individual layers that can be individually or collectively manipulated and analyzed. In contrast to GIS, GPS stands for Global Positioning System and is the method through which precise locations can be determined through triangulation with a constellation of satellites.

CHECK YOUR UNDERSTANDING

1. What are the two components of the geographic grid, and how are they calculated?

2. Where are the low latitudes on Earth, and how do they differ from the mid- and high latitudes?

3. Discuss the concept of map projection and why it is necessary when constructing maps.

4. What kind of map projection would provide the most accurate comparison of the shape of national boundaries between countries in the low and high latitudes?

5. Which would have the larger scale: a map showing the detail of New York City or a map showing the position of New York City within the state of New York?

6. Define the concept of remote sensing.

7. What is the difference between active and passive remote sensing?

8. Compare and contrast a Sun-synchronous orbit with a geostationary orbit.

9. What is a GPS satellite array, and how is it used to determine location on Earth?

10. What are the three potential sources of error associated with GPS?

11. Describe the character of a geographical information system.

12. Compare and contrast three GIS data layers.

ANSWERS TO VISUAL CONCEPT CHECKS

VISUAL CONCEPT CHECK 2.1

a) Mont Blanc is located in the Alps in France.
b) Lake Chad is located in central Africa.
c) Valle de Luna (Valley of the Moon) is located in eastern Chile.

VISUAL CONCEPT CHECK 2.2

a) The map on the left-hand side of the figure is a conformal projection, whereas the one on the right is an equivalent projection. You can tell the difference because the conformal projection maintains the shape of features on Earth but distorts their relative size, especially at the high latitudes. The equivalent projection, in contrast, maintains relative shape, with size being distorted. To see the difference between the two, note the relative shape and size of Greenland compared to that of North America.

b) The map of the state of Washington has the relatively small scale, whereas the map of just the Seattle area has the larger scale. You can tell the difference because the map of Washington shows a relatively large geographic area that contains very little detail about Seattle. The map of Seattle, in contrast, shows much more detail about the city, including the road pattern and shape of water bodies.

VISUAL CONCEPT CHECK 2.3

This infrared satellite image contains a variety of geographic information. The red, black, and bluish-gray areas represent vegetation, water (mostly the Potomac River), and urban development, respectively. Potential GIS layers include (1) the configuration of water bodies, (2) the distribution of vegetation, (3) the road network, and (4) the pattern of urban development.

CHAPTER THREE | EARTH–SUN GEOMETRY AND THE SEASONS

Of all the spatial relationships that geographers study, the Earth–Sun relationship has the greatest impact on the way Earth functions and the existence and character of life as we know it. In the context of physical geography, this relationship is especially important because most of the physical processes on Earth are powered by incoming radiation received from the Sun. Although Earth receives only about one-half of one-billionth the amount of energy that the Sun radiates, this tiny amount is nonetheless critical to powering physical processes and life on the planet.

At first glance it might appear that the Earth–Sun relationship is simple and easy to comprehend. In fact, this relationship is very complex because it depends on many different geographic variables such as latitude, the position of the Sun in the sky, the time of year, and the nature of the Earth's rotation and axial tilt. It is essential to understand the information in this chapter because it provides the foundation for many geographic patterns you will see in the remainder of the book.

© Paul Souders/Age Fotostock

A multiple exposure such as this one shows the change in the Sun's position during a beautiful arctic sunset in the summer months. This chapter focuses on the character and significance of the Earth/Sun geometric relationship.

CHAPTER PREVIEW

Our Place in Space

The Seasons

Orbital View and Earth as Viewed from the Sun

Human Interactions: How We See Earth–Sun Geometry on Earth

Celestial Dome

Sun Angle and Length of Day

Earth–Sun Geometry and Ancient Humans

LEARNING OBJECTIVES

1. Understand the Earth's position in space.

2. Describe the nature of Earth's axial tilt and orbit around the Sun.

3. Discuss the concept of the subsolar point and its seasonal movement.

4. Explain why the Equator, Tropic of Cancer, and Tropic of Capricorn are important features associated with Earth–Sun geometry.

5. Explain the concept of Sun angle and how it varies both by latitude and seasonally.

6. Understand how we sense the Earth–Sun geometric relationship.

7. Demonstrate understanding of Earth–Sun relationships by accurately predicting outcomes based on variable tilt scenarios.

Our Place in Space

A good way to begin a discussion of Earth–Sun relationships is with a brief discussion of the Universe. The origin of the Universe seems to be best explained by the **Big Bang theory**, which is based on the concept that the Universe emerged from a singular, enormously dense and hot state about 14 billion years ago. The primary evidence for this theory comes in two forms, specifically that (1) the most distant star clusters are moving away from us at greater speeds than those that are relatively close, and (2) the amount of cosmic microwave background radiation is remarkably uniform throughout the Universe, which implies that it represents the leftover energy from an early period of rapid expansion.

According to present estimates, the Universe is approximately 20 billion light-years across, with a light-year being the distance that light travels over the course of a year at the speed of 1,079,252,848.8 km/h (670,616,629.4 mi/h). With respect to the ultimate fate of the Universe, it will either continue to expand for all of eternity or it will one day collapse upon itself, depending on the average density of matter and energy within the entire Universe. Given that the speed of distant star clusters seems to be accelerating, it appears that the Universe may indeed expand forever.

The largest clearly definable unit within the Universe is a galaxy. According to astronomers, the Universe has approximately 50 billion galaxies, with each galaxy containing billions of stars. We happen to live in the Milky Way Galaxy (Figure 3.1), which includes approximately 400 billion stars. The Milky Way is a typical spiral galaxy, consisting of a bright central region with a high density of stars and a flat circular region containing most of the other stars. Younger, brighter stars form in long spiral arms that extend out from the galactic center. Our solar system lies in one of these spiral arms.

The center of our particular solar system is the star that we refer to as the Sun. Eight official planets orbit the Sun (Figure 3.2), with Pluto now considered a dwarf planet. To give you an idea of the size of the solar system, consider that Pluto is approximately 5.9 billion km (3.6 billion mi) from the Sun. A probe launched in January 2006 to study Pluto is currently zooming through space at about 14,500 km/h (9000 mi/h). Even at that remarkable speed, the probe will not reach Pluto until July 2015! Given this great distance, Pluto is much too far away from the Sun to receive sufficient energy to support life. Of the remaining planets, most are either too far away from the Sun (and thus are too cold) or too close (and thus are too hot) for life as we know it to exist. The lone exception is Earth, the third planet from the Sun, whose orbit happens to be a perfect distance from the Sun for supporting life as we know it.

The Shape of Earth

Let's now turn our attention to the shape of Earth. Although Earth appears to be a perfect sphere, it bulges slightly at the Equator and is flattened somewhat at the poles to form a shape known as an *oblate spheroid*. The circumference of Earth measured at the Equator is 40,075 km (24,902 mi), while the circumference measured through the poles is slightly less: 40,008 km (24,860 mi). The bulge at the Equator, which gives Earth a slightly thicker middle, is caused mainly by the centrifugal force of the Earth's rotation and by differences in density of the Earth's crust and gravitational field. This centrifugal force is similar to the sideways push you feel when rounding a curve quickly in your car.

The shape of Earth is an important factor to consider regarding the Earth–Sun relationship. Because Earth presents a curved surface to the arriving rays of the Sun, the intensity of solar radiation received varies by latitude. This difference is a reflection of the **Sun angle**, which is the angle at which the

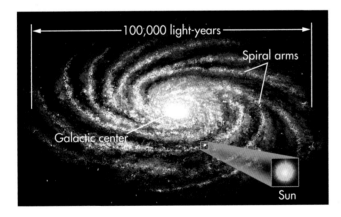

Figure 3.1 The Milky Way Galaxy. Our Sun is but one of approximately 400 billion stars in this galaxy. Note our Sun's location in one of the spiral arms of the galaxy.

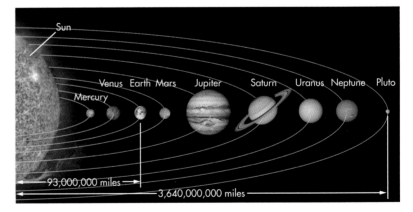

Figure 3.2 Our solar system. The solar system consists of eight official planets (Pluto is now considered a dwarf planet) that orbit the Sun. Earth is the third planet from the Sun. (Not drawn to scale.)

Big Bang theory *The theory that the Universe originated about 14 billion years ago when all matter and energy erupted from a singular mass of extremely high density and temperature.*

Sun angle *The angle at which the Sun's rays strike the Earth's surface at any given point and time. This angle is high at low latitudes and is progressively less at higher latitudes.*

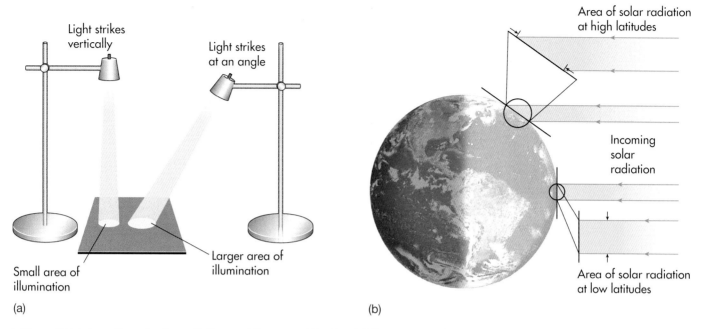

Figure 3.3 Interaction of solar radiation and the curved Earth. (a) You can use a desk lamp to show that light striking a surface at an angle illuminates a larger area (and is more diffuse) than light striking the surface vertically. (b) Solar radiation strikes Earth most directly at low latitudes because the Sun angle is high, resulting in concentrated solar radiation in these regions. In contrast, solar radiation is more diffuse at high latitudes because the surface curves away from the radiation and the Sun angle is relatively low.

Sun's rays strike Earth's surface. In general, the Sun angle is relatively high at low latitudes and becomes progressively less toward the poles because the Earth's surface is curved. The net effect of this variation is that the solar energy received at higher latitudes is spread over a relatively large surface area compared to the same amount of energy received at lower latitudes. As a result, the amount of incoming energy per unit area is relatively low in areas where the Sun angle is low.

Figure 3.3 illustrates this difference. Lower latitudes generally receive more intense solar radiation because the Sun angle is high (between 75° and 90°). The point on Earth where the Sun's rays are perpendicular to the surface (Sun angle = 90°) at any given point in time, and therefore most intense, is known as the **subsolar point**. Higher latitudes, in contrast, receive less intense radiation because the Sun's rays hit the Earth's surface at angles much less than 90° (a more oblique angle) due to the Earth's curvature. You can explore this relationship yourself by shining a desk lamp onto a surface, as in Figure 3.3a, both directly and at an angle. Note that the light is spread more diffusely when shone at a lower angle than when the angle is 90°. The Earth's curvature has an analogous effect on incoming solar radiation, which, in turn, contributes greatly to many geographic patterns on Earth, including atmospheric circulation and the global distribution of climate, vegetation, and even soils.

Earth's Orbit Around the Sun

Let's now turn to the character of Earth's orbit. As Earth revolves around the Sun, it moves in a counterclockwise direction relative to a view above the North Pole on a flat (imaginary) plane referred to as the **plane of the ecliptic** (Figure 3.4). Although it may be difficult to visualize this plane, imagine that the Sun is in the middle of a large, round cake pan. If the planets were marbles on the cake pan, they would roll around the Sun only on the surface of the pan, not above or below it. In a very similar way, Earth orbits the Sun on a flat plane, except that the plane of the ecliptic passes through the center of Earth rather than beneath it, as marbles do on a cake pan.

It takes about 365 days (365.24 days, to be exact) for Earth to make one full revolution around the Sun. In comparison, Pluto requires 248 of our years to complete one orbit. Early astronomers who calculated Earth's orbital time length configured our calendar year to reflect this period of time. You may wonder why February has a variable number of days, depending on the year. Normally, February contains 28 days. Every fourth year (known as *Leap Year*), however, February contains 29 days. This day is added in order to correct for the uneven length of time (365.24 vs. 365 days) it takes for Earth to complete its orbit.

Subsolar point *The point on Earth where the Sun angle is 90° and solar radiation strikes the surface most directly at any given point in time.*

Plane of the ecliptic *The flat plane on which the Earth travels as it revolves around the Sun.*

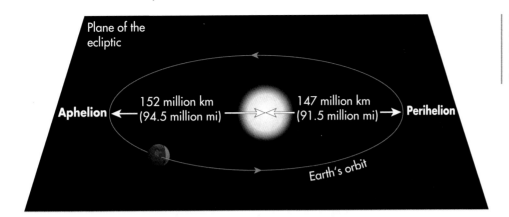

Figure 3.4 The elliptical Earth orbit. As Earth travels on the plane of the eclip-tic, it is closest to the Sun at perihelion on January 3. Earth is farthest away from the Sun at aphelion on July 4.

DISCOVER...

THE SUN

The Sun is easily the most important natural feature when it comes to life on Earth. Although the Sun is about 100 times larger than the Earth, it is quite small compared to the millions of giant stars, multiple stars, pulsating stars, and black holes scattered throughout the Universe. The Sun is incredibly important for us because it provides the power that drives most of the natural processes on Earth. This power is generated because the Sun essentially functions as a gargantuan nuclear reactor that converts hydrogen gas to helium. As a result, the Sun's temperature is incredibly high, ranging from about 15,000,000°C (27,000,000°F) at the center to about 6000°C (10,800°F) at the surface. The heat energy produced by these incredible temperatures is the engine that powers Earth. Traveling at the speed of light, it takes about 8 minutes for the Sun's energy to reach Earth.

In addition to emitting very high temperatures, the Sun also produces strong magnetic fields. These fields cause huge flares of hot gas to erupt from the surface in the shape of loops, as shown here. The Sun also has cooler areas scattered across its surface that appear dark in comparison with the hotter and brighter background. These areas are called sunspots. The number of sunspots at any one time rises and falls in a regular cycle that is closely connected to the Sun's magnetic field.

SOHO Extreme Ultraviolet Imaging Telescope/Photo Researchers

This image illustrates two interesting things about the Sun. First, notice how much bigger the Sun is than Earth, which is the blue dot provided for scale. Second, look at the immense size of the solar flare compared to the Sun. Such flares are loops of gas that follow the pattern of the Sun's magnetic field.

Swedish One-metre Solar Telescope/Photo Researchers

Sunspots are cooler areas on the Sun that appear to be relatively dark compared to the rest of the surface. The largest sunspot in this image is about the same size as Earth.

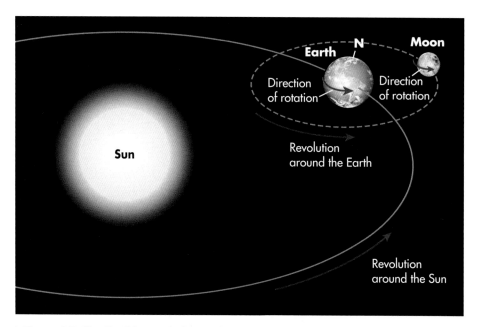

Figure 3.5 The Earth's revolution and rotation. As Earth revolves around the Sun, it rotates in a counterclockwise direction on its axis, when viewed from above the North Pole. Also note the geometrical relationship of Earth and Moon.

The Earth's Rotation and Axial Tilt

As Earth revolves around the Sun, it also rotates (spins) on its **axis**, which is an imaginary line that extends through the center of Earth from pole to pole. When viewed from above the North Pole, Earth rotates in a counterclockwise direction on its axis (Figure 3.5). It takes 24 hours for Earth to make one complete rotation, which is the origin of our day length. During this interval, one-half of Earth is always illuminated, whereas the other is in shadow. The boundary between day and night is known as the **circle of illumination** and is constantly moving across the surface as Earth rotates. You can see a good example of the circle of illumination by looking at a half Moon. Notice that one-half of the Moon is brightly illuminated, whereas the other half is completely dark. The dividing line between those two portions of the Moon is the circle of illumination.

In association with Earth's rotational cycle, humans established 24 time zones that span specific portions of the planet (Figure 3.6). The contiguous United States contains four such areas, including (from east to west) the Eastern, Central, Mountain, and Pacific time zones (Figure 3.7). Each of these time zones covers about 15° of longitude because it takes 1 hour for Earth to rotate that distance (360°/24 h = 15°/h). Given the irregular nature of political boundaries, the edges of time zones are usually asymmetrical because people in individual states or countries want to be entirely within a particular time zone. Notice in Figure 3.7, for example, that the western boundary of the Eastern Time Zone jogs west to encompass most of Michigan's Upper Peninsula even though the far western part of the state nearly reaches the same longitude as St. Louis, which is in the Central Time Zone. This irregular boundary was drawn because most people in Upper Michigan identify with Lower Michigan as far as time is concerned. A sliver of western Upper Michigan lies in the Central Time Zone because it borders Wisconsin. In this border area, it makes good economic sense that people on both sides of the state line be on the same time schedule.

The Prime Meridian in Greenwich, England, was chosen as the standard for the entire time zone system at the 1884 International Prime Meridian Conference in Washington, D.C. This system, referred to as Universal Time Coordinated (UTC), is used by the vast majority of countries on Earth, although some (such as Saudi Arabia) adhere to their own time systems. In the UTC

As Earth orbits the Sun, it does not follow a perfectly circular path, but rather an elliptical one (see Figure 3.4). Note also that the Sun is not located in the exact center of our orbit and that the distance between Earth and Sun varies over the course of the year, with an average distance of approximately 150 million km (93 million mi). The point where Earth is closest to the Sun, approximately 147 million km (91.5 million mi) away, is called **perihelion**. In contrast, the term **aphelion** refers to the point where Earth is farthest from the Sun, about 152 million km (94.5 million mi) away. At the present time, perihelion and aphelion occur in early January and July, respectively.

Given the basic geometric relationship shown in Figure 3.4, what seems odd about the timing of our seasons in the Northern Hemisphere? The answer is that winter in the Northern Hemisphere occurs when we are closest to the Sun, during orbital perihelion. Intuitively, you might think that we would experience our summer when Earth is closest to the Sun. Since that is not the case, we can conclude that the distance of Earth from the Sun does not cause our seasons and that another variable is responsible. This variable is the Earth's axial tilt.

Perihelion *The point of the Earth's orbit where the distance between the Earth and Sun is least (~147 million km or 91.5 million mi).*

Aphelion *The point of the Earth's orbit where the distance between the Earth and Sun is greatest (~152 million km or 94.5 million mi).*

Axis *The line around which the Earth rotates, extending through the poles.*

Circle of illumination *The great circle on Earth that is the border between night and day.*

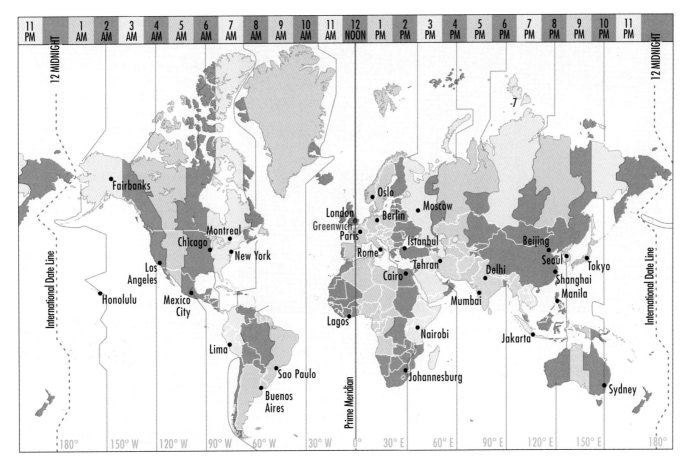

Figure 3.6 World time zones. Earth is subdivided into 24 time zones. The International Date Line, which lies at approximately 180° E and W longitude, is the arbitrary designation used to mark the beginning of each new day.

system, time at any given place is calculated relative to how many hours ahead or behind that particular location is relative to the Greenwich meridian. The beginning and end of each calendar day is the **International Date Line**, which is located at 180° longitude. If you are traveling west, for example, on a flight from San Francisco to Sydney, Australia, you will cross the date line and immediately move into the next day (say, from Monday to Tuesday). Similarly, when you cross the date line while traveling east (from Sydney to San Francisco), you will return to the previous day (such as from Tuesday back to Monday).

Although the axis of Earth is the anchor around which the daily rotation occurs, it is a very important point of reference for another reason as well. On first consideration, you might think that the axis is perpendicular to the plane of the ecliptic, that is, at a 90° angle relative to the plane on which Earth orbits the Sun. In fact, the Earth's axis is not perpendicular to this plane but is instead tilted 23.5° from a line perpendicular to the

---

International Date Line *This line generally occurs at 180° longitude, with some variations due to political boundaries, and marks the transition from one day to another on Earth.*

plane as it revolves around the Sun (Figure 3.8). Remember that Earth maintains this degree of tilt *and* orientation with respect to the Sun throughout the year. This geometric relationship is important to understand because it is the reason for the seasons that we experience, as explained in the next section.

▌**KEY CONCEPTS TO REMEMBER ABOUT OUR PLACE IN SPACE**

1. Earth is the third planet from the Sun.

2. Although Earth is essentially spherical, it is not perfectly round; in fact, it is an oblate spheroid.

3. Earth orbits the Sun on the plane of the ecliptic in an elliptical orbit.

4. Earth rotates on its axis, which results in our day/night cycle.

5. The axis of Earth is tilted at 23.5° from a line perpendicular to the plane of the ecliptic.

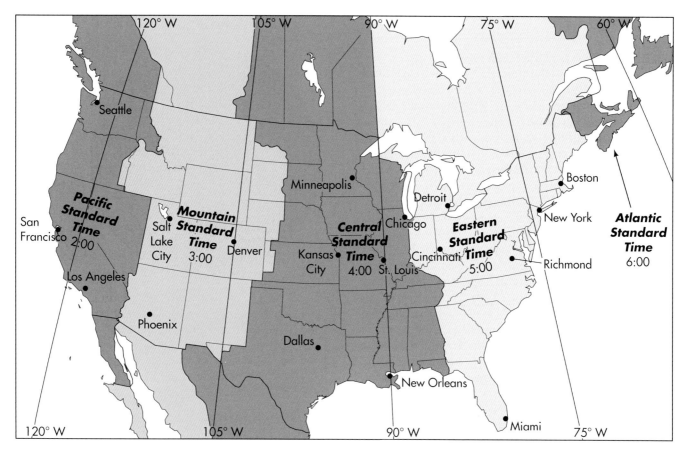

Figure 3.7 Time zones in the contiguous United States and Canada. The most eastern time zone in the contiguous United States is Eastern Standard Time, which is 3 h ahead of the most western time zone, Pacific Standard Time.

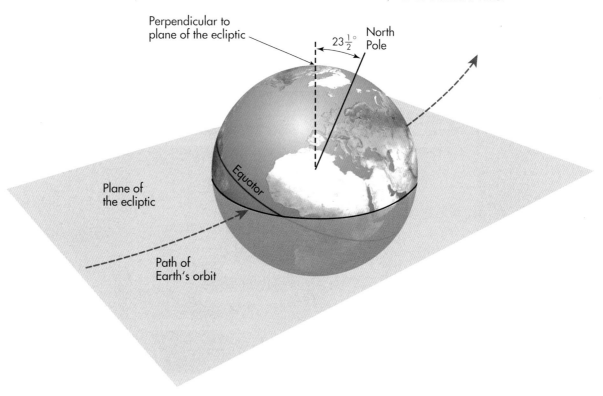

Figure 3.8 The Earth's axial tilt. The Earth's axis is not perpendicular to the plane of the ecliptic. Instead, the axis is tilted 23.5° from a line perpendicular to the plane of the ecliptic.

The Seasons

To understand how the Earth's axial tilt causes the seasons, first imagine what would happen if the Earth's axis were *not* tilted (Figure 3.9). Under these circumstances, as Earth traveled around the Sun on the plane of the ecliptic, the Sun's rays would always strike Earth most directly at the Equator. As a result, the subsolar point would perpetually be at the Equator and the Sun's rays would always strike Earth at progressively lower angles with increasingly higher latitudes (greater than 0°). In this scenario, as Earth revolves around the Sun, all places on Earth would receive a consistent intensity of solar radiation (although this amount would differ with latitude) over the course of the year and distinct seasons would not be experienced anywhere on the planet.

Now picture the Earth's actual axial tilt, which is 23.5° from a line perpendicular to the plane of the ecliptic. Keep in mind that the axis maintains a constant tilt angle *and* orientation as Earth travels around the Sun (Figure 3.10). These geometric relationships—angle and orientation—are the fundamental cause of our seasons because one of the hemispheres is always in the gradual process of tilting more toward the Sun while the orbit progresses, whereas the other one is slowly tilting away from the Sun. You might be thinking, "Wait a minute, you just said that the axial tilt and orientation never change; isn't that contradictory?" To see why not, examine Figure 3.10 and focus on the Northern Hemisphere. Notice that during June, the Northern Hemisphere is tilted *toward* the Sun, whereas in December, it is titled *away*. The opposite is true for the Southern Hemisphere. In both cases the tilt did not change, nor did the orientation of the tilt. Instead, *the position of Earth relative to the Sun changed as its orbit progressed*, which has the effect of moving each hemisphere either toward or away from the Sun's rays. This apparent movement of the hemispheres toward or away from the Sun is the fundamental

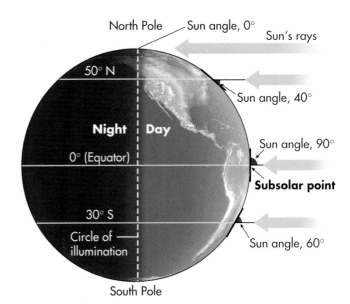

Figure 3.9 Geometric relationship if the Earth's axis were perpendicular to the plane of the ecliptic. If these conditions existed, the subsolar point would always be located at the Equator and the circle of illumination would extend from the North to South Poles. At 50° N the Sun angle would be 40°, whereas it would be 0° at 90° N (the North Pole).

cause of the seasons because it results in the migration of the subsolar point 23.5° N or S of the Equator over the course of the year.

Solstice and Equinox

If you live in the Northern Hemisphere, you probably know that the recognized first day of summer is sometime toward the end of June. Conversely, in the Southern Hemisphere, the first day of summer is toward the end of December. You also probably

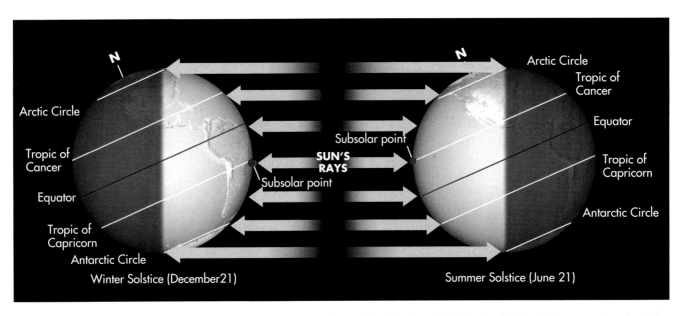

Figure 3.10 Axial geometry of Earth during June and December. Note that the axis is tilted at 23.5°, which causes the orientation of the hemispheres to change with respect to the location of the subsolar point. On December 21, the subsolar point is at the Tropic of Capricorn, whereas it is at the Tropic of Cancer on June 21.

know that certain dates mark the first days of autumn, spring, or winter on the calendar. Why do these dates exist, and what is their significance?

These dates are approximately the same every year because they indicate key periods within the Earth–Sun geometrical relationship. Remember that the Earth's axial tilt, in combination with its orbit, causes the subsolar point to move between 23.5° N and 23.5° S over the course of a year (Figure 3.11). In other words, the Sun angle is 90° at some point on the Earth's surface between these latitudes every day. Although this migration is a seamless process that we barely notice from day to day, early astronomers found it useful to mark the passage of the seasons by noting the date when the Sun's rays are perpendicular to the Equator and to the Tropic of Cancer (23.5° N) and Tropic of Capricorn (23.5° S). These two latitudes represent the highest latitude that the subsolar point reaches in each hemisphere during its seasonal cycle. The term *equinox* refers to the time when the subsolar point is at the Equator and all localities on Earth experience equal hours of daylight and darkness (12 of each). In contrast, the term *solstice* is used to denote when the Sun angle is 90° at either of the tropical boundaries.

In an effort to better understand when these dates occur and how they are related to Earth–Sun geometry, let's examine the passage of the seasons by beginning in March and moving forward in time through the calendar. It will help if you refer to Figures 3.12 and 3.13 to view the process and assume that your viewpoint of time is from the perspective of living in the Northern Hemisphere.

March 20–21   This date is the **Spring Equinox** (also called the *vernal equinox*) and represents the official beginning of the

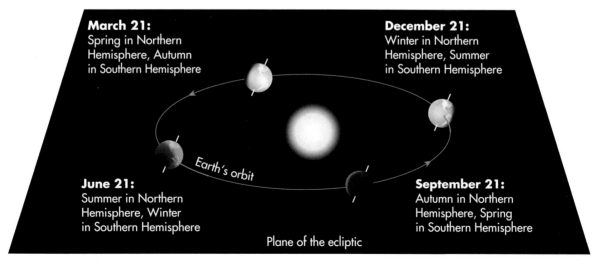

Figure 3.11 Range of seasonal subsolar point movement. The subsolar point migrates between the Tropic of Cancer and Tropic of Capricorn over the course of the year.

spring season in the Northern Hemisphere and the fall season in the Southern Hemisphere. This noteworthy date occurs because the subsolar point is located at the Equator and neither the Northern nor the Southern Hemisphere is tilted toward the Sun on this date. Notice that the circle of illumination extends through each pole in Figures 3.12 and 3.13.

June 20–21   This date is the **Summer Solstice** and represents the first official day of summer in the Northern Hemisphere and winter in the Southern Hemisphere. The Summer Solstice occurs because, following the Spring Equinox, Earth continues to orbit the Sun in a counterclockwise direction when viewed from above. As the orbit progresses, the Northern Hemisphere slowly

Spring Equinox   *Assuming a Northern Hemisphere seasonal reference, the Spring (or vernal) Equinox occurs on March 20 or 21, when the subsolar point is located at the Equator (0°).*

Summer Solstice   *Assuming a Northern Hemisphere seasonal reference, the Summer Solstice occurs on June 20 or 21, when the subsolar point is located at the Tropic of Cancer (23.5° N).*

Figure 3.12 The orbital positions of Earth during the solstices and equinoxes.   Note that the tilt of Earth does not change; instead, the position of Earth relative to the Sun changes. Due to this relationship, the intensity of the solar radiation that strikes any particular place on Earth varies over the course of the year.

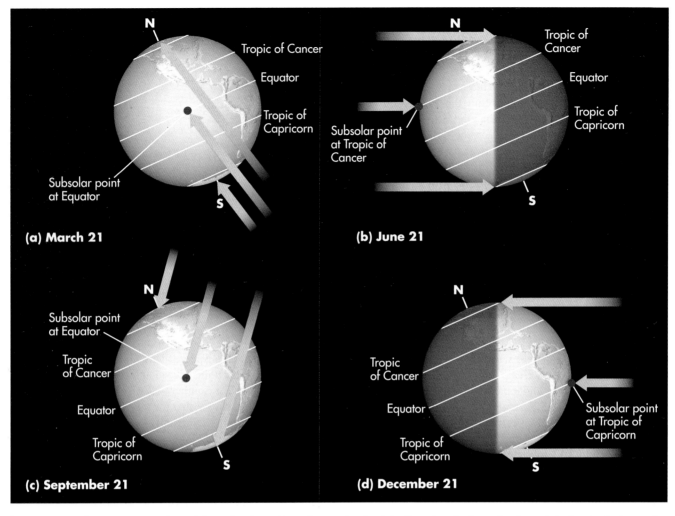

(a) March 21

(b) June 21

(c) September 21

(d) December 21

Figure 3.13 Seasonal migration of the subsolar point. Due to the tilt of the Earth's axis, the subsolar point migrates between the Tropic of Cancer and Tropic of Capricorn over the course of the year. For perspective, imagine that the Sun is in front of Earth in (a) and behind Earth in (c). Further, imagine that it is to the left of Earth in (b) and the right of Earth in (d). Also notice the orientation of the circle of illumination at different times of the year.

VISUAL CONCEPT CHECK   3.1

Suppose the subsolar point is at 10° S latitude. In other words, the Sun angle at that latitude is 90° at solar noon. Which of the following months are logical times of the year when the subsolar point would be at this latitude? Explain.

a) October
b) December
c) February
d) June

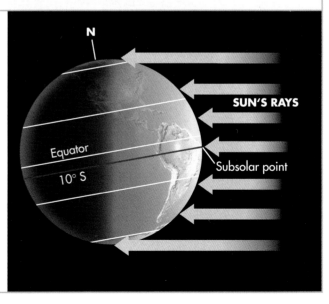

reaches its position of greatest tilt toward the Sun. Meanwhile, the Southern Hemisphere is tilted away from the Sun. During this solstice the subsolar point is located at the **Tropic of Cancer** (23.5° N). In Figure 3.13, notice that all latitudes above 66.5° N (the Arctic Circle) receive 24 hours of daylight during this period of time; this is why these high latitudes are called the *land of the midnight Sun*. At the same time the northern arctic regions are in perpetual sunlight, all locations above 66.5° S (the Antarctic Circle) experience 24 hours of darkness. This continual darkness occurs because the Southern Hemisphere is tilted away from the Sun during this time and those latitudes above the Antarctic Circle never rotate into the illuminated part of Earth.

September 22–23   This date is the **Fall Equinox** (also called the *autumnal equinox*) and represents the first official day of fall in the Northern Hemisphere and spring in the Southern Hemisphere. At this time the Earth's orbit has progressed such that the subsolar point is now once again at the Equator (Figures 3.12 and 3.13). As at the Spring Equinox, all locations on Earth get equal hours of day and night (12 hours daylight, 12 hours darkness) because neither the Northern nor the Southern Hemisphere is tilted toward the Sun and the great circle of illumination extends from pole to pole.

December 21–22   This date is the **Winter Solstice** and represents the first official day of winter in the Northern Hemisphere and summer in the Southern Hemisphere. Given the progression of the orbit since the Fall Equinox, Earth is now positioned such that the Northern Hemisphere is at its greatest tilt away from the Sun, while the Southern Hemisphere is tilted toward it. At this time the subsolar point is located at 23.5° S (Figures 3.12 and 3.13), which is also known as the **Tropic of Capricorn**. As a result of this subsolar position, all latitudes above the Arctic Circle experience 24 hours of darkness, whereas those higher than the Antarctic Circle experience 24 hours of daylight.

KEY CONCEPTS TO REMEMBER ABOUT THE SEASONS

1. If the axis of Earth were not tilted, there would be no seasons.

2. The subsolar point is the place where the Sun's rays strike Earth most directly. This occurs because the Sun angle is 90° at that point and the rays are perpendicular to the surface.

3. Due to the axial tilt, the subsolar point migrates between the Tropic of Cancer and Tropic of Capricorn over the course of the year.

4. A solstice occurs whenever the subsolar point is at either 23.5° N or S latitude. An equinox, in contrast, occurs when the Sun is directly over the Equator.

5. The position of Earth relative to the Sun changes as the orbit progresses.

Human Interactions: How We See Earth–Sun Geometry on Earth

In the previous section you examined Earth–Sun geometry as it looks when viewed from space. Now it is time to explore how this geometry looks from the surface of Earth—in other words, how *we* see and experience it. Let's begin with the basic principle of day and night, which is related to the Earth's rotation on its axis.

Day and Night

Over the course of the day, you can "see" the Earth's rotation by noting the position of the Sun in the sky. Everyone knows that the Sun "rises" in the east and then arcs across the sky toward the west. As the arc progresses over the course of the day, the Sun reaches its highest position at **solar noon,** and then gradually lowers as it "sets" in the west. Solar noon at any given place is not necessarily equivalent to standard clock time because the time of sunrise and sunset varies across a time zone due to the curvature of Earth. Although it appears that the Sun is actually rising in the morning and moving across the sky during the course of the day, the fact is that the apparent motion of the Sun is caused by the rotation of Earth on its axis. In short, the Sun rises, wherever you may live, because your part of Earth has rotated to a position where the Sun begins to illuminate your home.

Tropic of Cancer   *The line of latitude at 23.5° N where the subsolar point is located (and Sun angle is thus 90°) on the Summer Solstice in the Northern Hemisphere.*

Fall Equinox   *Assuming a Northern Hemisphere seasonal reference, the Fall (or autumnal) Equinox occurs on September 22 or 23, when the subsolar point is located at the Equator (0°).*

Winter Solstice   *Assuming a Northern Hemisphere seasonal reference, the Winter Solstice occurs on December 21 or 22, when the subsolar point is at the Tropic of Capricorn (23.5° S).*

Tropic of Capricorn   *The line of latitude at 23.5° S where the subsolar point is located on the Winter Solstice (and Sun angle is thus 90°) in the Northern Hemisphere.*

Solar noon   *The time of day when the Sun angle reaches its highest point as the Sun arcs across the sky.*

Let's reinforce the important concept of axial tilt by interacting with some simulations. First, go to the Geo Media Library and select *Orbital View*. This simulation shows Earth orbiting the Sun, with the Earth's axis indicated. This simulation has close-up views of Earth at the solstices and equinoxes, showing Earth rotating (so lengths of daylight periods are indicated), the Earth's axis (indicating axial tilt), and the portion of Earth that is illuminated at each of the four times. Click on the "Orbit," "Fall Equinox," "Winter Solstice," "Spring Equinox," and "Summer Solstice" buttons to see different perspectives. Pay particularly close attention to the tilt and orientation of Earth, and the fact that these variables do not change during the orbit. In addition, be sure to notice where the subsolar point is at any given point in time and compare this migration to Figure 3.13 in the text. Last, change the axial tilt and explore how the migration of the subsolar point changes. Remember that the subsolar point represents the place where solar radiation is striking Earth most intensely. Notice how,

with increased tilt, the subsolar point migrates a great deal. With less tilt, the subsolar point migrates less.

After you interact with the Orbital View simulation, select *Earth as Viewed from the Sun*. This part of the module provides a different view of the Earth–Sun relationship by simulating Earth as it would appear if viewed from the Sun. As the months progress in time, notice how the *apparent* tilt of Earth changes, relative to your view (on the Sun). Remember that the *actual* tilt does not change; instead, the position of Earth in its orbit around the Sun progresses. In this simulation, you can also adjust the amount of axial tilt on Earth and see how the migration of the subsolar point changes as viewed from the Sun. Again, notice that the subsolar point migrates more with increased tilt and less with decreased tilt. Once you complete this simulation, examine Figure 3.13 to review where the subsolar point strikes Earth, with the tilt at 23.5° on each of the solstices and equinoxes. Be sure to answer the questions at the end of the exercise to make sure you understand this concept.

At the same time that your day is beginning on one side of the circle of illumination, night has begun on the opposite side of Earth because that place has rotated so that the Sun is no longer illuminating it. Remember, the Sun is always illuminating half of Earth at any given time, whereas the other half is in shadow.

Seasonal Changes in Sun Position (Angle) and Length of Day

In addition to the daily or **diurnal cycle** of day and night, which is related to the Earth's rotation, we can also see a seasonal cycle in the Sun's position in the sky, which depends on orbital progression. Whereas the day–night cycle is obvious and can be easily seen because of the observable east–west arc of the Sun, the seasonal migration of the Sun's position is a slow progression in north–south directions relative to the Equator. This migration results in a seasonal variation in the angle of the Sun's rays with respect to any location on Earth. To follow this migration, the best point of reference is the daily position of the Sun at solar noon at any given place because that is when the Sun is at its highest point in the arc at that locality. As we examine the seasonal migration of the noontime Sun, imagine that you are living at 45° N (but remember that what you see is opposite for those who reside in the Southern Hemisphere). Also, keep in mind that the overall seasonal north–south migration of the Sun is occurring in combination with the Sun's daily arc.

If you live in the middle latitudes of the Northern Hemisphere, you will notice that the Sun arcs across the southern part

of the sky over the course of the day. At solar noon, it is directly to the south rather than overhead because the subsolar point never reaches a point higher than 23.5° N at any time in the year. Given that the Sun is in the southern sky at solar noon, shadows from buildings and trees project toward the north. In the context of the seasons, the noon Sun is highest in the sky during summer. In other words, the Sun angle is greatest during summer. This high angle occurs because the Sun is in the Northern Hemisphere during this time of year.

As the season moves from summer to fall, the position of the noontime Sun (as you see it) slides farther south in the sky on a daily basis. Why? This southerly migration occurs because the subsolar point moves toward the Equator (which is to the south) as fall approaches and Sun angle decreases. With the gradual approach of winter, the noontime Sun moves still farther to the south because the subsolar point is migrating toward the Tropic of Capricorn. At this time, Earth is in a position within its orbit where the Northern Hemisphere is tilted away from the Sun. If you happen to live in the Northern Hemisphere, the noontime Sun is lowest in the southern sky (in other words, the Sun angle is the least) on December 21, or the Winter Solstice. From this point on, the noontime Sun slowly migrates back to the north, as Sun angle increases, until the Summer Solstice.

You can see the combined daily and seasonal migrations of the Sun, with respect to the Earth's surface, in a graphical way by examining Figure 3.14. This diagram is called a **celestial dome** and shows the Sun's daily arc and seasonal migration relative to the Earth's surface. In this particular diagram, the dome is orient-

Diurnal cycle *A 24-hour cycle*

Celestial dome *A sphere that shows the Sun's arc, relative to the Earth, in the sky.*

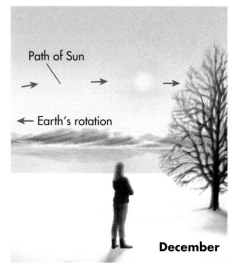

(a)

(b)

Figure 3.14 The celestial dome. (a) The Sun's arc at the Winter Solstice, Equinoxes, and Summer Solstice when viewed from 45° N latitude. (b) The position of the Sun at solar noon varies dramatically between the winter and summer seasons. It is much higher in the sky in summer and lower in the sky in winter. This difference occurs due to the Earth's axial tilt.

ed with respect to a person standing at 45° N latitude. The surface of Earth is presented as a two-dimensional feature that extends from pole to pole, with the arc and position of the Sun shown at three periods: Summer Solstice, Winter Solstice, and Equinox. In other words, this diagram illustrates how you would see the Sun as the day progresses at three different times of year.

One of the most obvious ways in which we notice the seasonal change in Sun angle is with respect to the number of daylight hours, or the length of day. Look again at the celestial dome in Figure 3.14 or the one in the *Celestial Dome* animation and notice how the curvature of Earth influences when the Sun emerges from the horizon in the east or disappears beneath it in the west. You should be able to see that the Sun emerges from the horizon earlier and disappears later in the summer months than during winter. It should make sense, too, that this seasonal variation is more pronounced at higher latitudes than at lower ones. Table 3.1 illustrates this concept by showing the number of daylight hours at five different latitudes in the Northern

Hemisphere at the equinoxes and solstices. Notice how the day length varies among the five sites.

KEY CONCEPTS TO REMEMBER ABOUT HOW WE SEE EARTH–SUN GEOMETRY

1. We experience day and night because Earth rotates on its axis.

2. Due to the Earth's rotation, the Sun arcs from east to west across the sky.

3. As the Sun arcs across the sky, it is at its highest point at solar noon.

4. Sun angle refers to the angle at which solar radiation strikes Earth. Sun angle is generally highest at the low latitudes and progressively lessens toward the poles.

TABLE 3.1	Variations in Day Length on Equinoxes and Solstices across Latitude				
		Day Length on Equinox or Solstice			
Location	Latitude (N)	March 20	June 21	September 22	December 21
Bogotá, Colombia	4°329	12 h, 06 min	12 h, 24 min	12 h, 07 min	11 h, 51 min
Miami, Florida	25°469	12 h, 08 min	13 h, 45 min	12 h, 09 min	10 h, 32 min
Topeka, Kansas	39°079	12 h, 07 min	14 h, 55 min	12 h, 13 min	9 h, 26 min
Calgary, Alberta	45°019	12 h, 12 min	16 h, 33 min	12 h, 15 min	7 h, 56 min
Barrow, Alaska	71°039	12 h, 23 min	24 h, 00 min	12 h, 22 min	00 h, 00 min

Celestial Dome

To enhance your understanding of Sun angle and seasonality, go to the Geo Media Library and select *Celestial Dome*. This animation provides an opportunity to interact with an animated celestial dome. Change the latitude from several options to observe seasonal change in the arc of the Sun, and its height above the horizon at equinox and solstice. Be sure to notice how the Sun arc and angle vary between low- and high-latitude locations, as well as between the hemispheres. Be sure to answer the questions at the end of the exercise to ensure you understand this concept.

Sun Angle and Length of Day

Let's now take a comprehensive look at Earth–Sun geometric relationships and how they influence the length of day. Go once again to the Geo Media Library and select *Sun Angle and Length of Day*. As you watch this animation, notice how and why day length at three different latitudes varies over the course of the year. Relate your observations with the *Celestial Dome* animation and Table 3.1. After you complete the animation, be sure to answer the questions at the end of the exercise to test your understanding of this concept.

Earth–Sun Geometry and Ancient Humans

Earth–Sun geometric relationships have been important to people for a long time. Although ancient humans did not know that Earth revolves around the Sun, they nevertheless understood that Sun angle and day length changed over the course of the year and were somehow associated with the seasons. In many prehistoric human cultures, these changes had important spiritual significance because they were related to agricultural cycles. To learn more about how ancient humans celebrated the Sun, go to the Geo Media Library and select *Earth–Sun Geometry and Ancient Humans*. After you watch the video, be sure to answer the questions at the end of the exercise to test your understanding of this concept.

VISUAL CONCEPT CHECK 3.2

This cross section of the Eastern Time Zone in the United States ranges from Boston, Massachusetts, on the east to Holland, Michigan, on the west. Although both cities occur at the same line of latitude (42° N), the timing of sunrise and sunset varies between the two locations. The Sun rises about an hour earlier in Boston than it does in Holland. Conversely, the Sun sets about an hour later in Holland than it does in Boston. Considering the rotation of Earth and basic Earth–Sun geometry, why does this variation occur?

1. Earth revolves around the Sun within an elliptical orbit in the plane of the ecliptic. As the Earth revolves, it rotates on its axis, which is tilted at 23.5° from a line perpendicular to the plane of the ecliptic. It takes approximately 365 days for one complete revolution to occur.

2. The subsolar point is the line of latitude on Earth where the Sun angle is 90° and thus where solar radiation strikes Earth most directly. Due to the axial tilt, the subsolar point migrates between the Tropic of Cancer and Tropic of Capricorn between June 21–22 (Northern Hemisphere Summer Solstice) and December 21–22 (Northern Hemisphere Winter Solstice), respectively. The Sun is directly over the Equator at solar noon on both the Spring and Fall Equinoxes (March 21–22 and September 21–22, respectively).

3. Day and night on Earth occur because Earth rotates on its axis. It takes 24 hours for a complete rotation to occur.

CHECK YOUR UNDERSTANDING

1. Describe Earth's place in the solar system.

2. What is the shape of Earth's orbit around the Sun, and how do the terms *aphelion* and *perihelion* apply?

3. What is the tilt of Earth's axis relative to the plane of the ecliptic?

4. Define the concept of the subsolar point and how it relates to Sun angle. Describe why it moves on a seasonal basis. What are the four important dates in this movement, and how do they relate to latitude?

5. Chicago, Illinois, and Omaha, Nebraska, are both located in the Central Time Zone, with Chicago being on the east side of the zone and Omaha on the west. With respect to time, which of these two cities is the first to "see" the sunrise? Why does this variation occur?

6. If the axis of Earth were perpendicular to the plane of the ecliptic, where would the subsolar point be on December 21?

7. Imagine that the axis of Earth were tilted at 45°. In this case, where would the Tropic of Cancer be located?

8. If the axis of Earth were tilted at 40°, what would the Sun angle be at 40° S latitude on December 21–22?

9. Which direction, north or south, do shadows project during solar noon on December 21–22 at 40° S latitude?

10. Where does the greatest range in daylight hours occur, at high or low latitudes?

11. Assume that you are standing on the Equator and it is June 21–22. Given your perspective, is the noontime Sun in the northern or southern part of the sky?

ANSWERS TO VISUAL CONCEPT CHECKS

VISUAL CONCEPT CHECK 3.1

Both October and February are logical times of the year when the subsolar point would be at 10° S latitude. In October, the Sun is in the process of moving from the Equator (0°) to the Tropic of Capricorn (23.5° S). After the subsolar point reaches 23.5° S in late December, it begins to move back to the Equator, which it will reach on or about March 21. To do so, however, it must pass by 10° S, which it does in February.

VISUAL CONCEPT CHECK 3.2

The Sun rises and sets earlier in Boston, Massachusetts, than in Holland, Michigan, because the surface of Earth is curved. When the Sun rises in Boston, Holland lies on the part of Earth that is curved away from the Sun's light. As Earth rotates, the Sun ultimately "rises" in Holland, but about an hour later than in Boston. The Sun sets later in Holland than in Boston because Boston rotates out of sunlight about an hour before Holland.

THE GLOBAL ENERGY SYSTEM

You have examined Earth's geometric relationship with the Sun and how it influences the distribution of solar radiation seasonally and by latitude. Now we will look more closely at the way this radiation is intercepted by Earth, how it flows through the atmosphere, how it interacts with the land and ocean surfaces, and how it maintains a balance with outgoing radiation on a global scale. Understanding these relationships is critical to comprehending many Earth processes discussed in later chapters, such as wind, weather, and ocean currents, because the Sun provides the energy required to power them, either directly or indirectly. We begin by investigating the concept of radiation energy and how it is measured.

Science Source/Photo Researchers

Illumination of clouds over the Pacific Ocean by the Sun. The Sun sends a continuous stream of electromagnetic energy to Earth. This energy interacts with Earth in a wide variety of ways that are discussed in this chapter.

CHAPTER PREVIEW

The Electromagnetic Spectrum and Solar Energy

 The Electromagnetic Spectrum

Composition of the Atmosphere

The Flow of Solar Radiation on Earth

 The Angle of Incidence

The Global Radiation Budget

 The Global Energy Budget

Human Interactions: Solar Energy Production

LEARNING OBJECTIVES

1. Describe the fundamental components of the electromagnetic spectrum and how they differ from one another.

2. Discuss the various components of the atmosphere.

3. Compare and contrast the processes of conduction, convection, and radiation as methods of heat transfer.

4. Explain the flow of radiation from the Sun to Earth and how absorption, reflection, scattering, and albedo influence the system.

5. Discuss the various components of the global radiation budget and how the system achieves balance.

6. Understand why solar energy is a renewable energy source and the challenges associated with its production.

The Electromagnetic Spectrum and Solar Energy • 59

The Electromagnetic Spectrum and Solar Energy

The Electromagnetic Spectrum

Radiation is electromagnetic energy that is transmitted in the form of waves. These waves can be measured in terms of their length and amplitude, with **wavelength** being the distance between wave crests, and the term **wave amplitude** referring to half the height between the wave crest and wave trough (Figure 4.1). The entire wavelength range of electromagnetic energy is called the **electromagnetic spectrum**.

Within the electromagnetic spectrum, wavelengths range from *gamma rays* that are less than a nanometer long (one billionth or 10^{-9} of a meter) to *radio waves* that are tens of meters in length (Figure 4.2). Humans can see radiation directly only in the *visible light* part of the spectrum, which ranges from violet with the shortest wavelength of 375 nanometers (nm) to red with the longest wavelength (740 nm).

You should remember two very important principles about electromagnetic radiation. The first principle is that an *inverse* relationship exists between the temperature of an object and the wavelength of the electromagnetic radiation it emits. In other words, hotter objects emit radiation with shorter wavelengths than cooler objects. For example, the Sun has a surface temperature of about 6000°C (11,000°F) and emits energy as **shortwave radiation**, which includes gamma rays, X-rays, ultraviolet (UV) radiation, visible light, and near-infrared radiation. The Earth, in contrast, is a much cooler object (~16°C, 61°F) and thus emits **longwave radiation** in the thermal infrared part of the spectrum (Figure 4.3). Keep in mind, though, that the vast majority of energy released by Earth initially originated at the Sun.

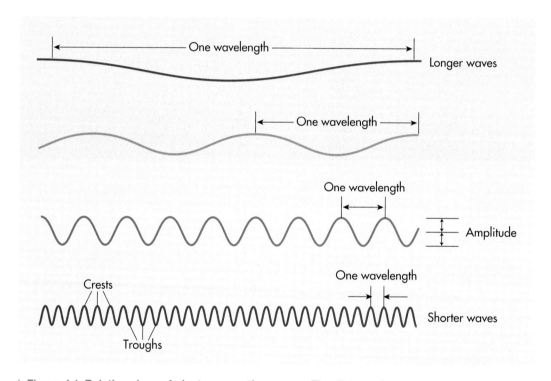

Figure 4.1 Relative sizes of electromagnetic waves. The distance from one crest to the other, or from one trough to another, is called the wavelength. Wave amplitude refers to half the distance from the wave crest to the wave trough.

Wavelength *The distance between adjacent wave crests or wave troughs.*

Wave amplitude *The overall height of any given wave as measured from the wave trough to the wave crest.*

Electromagnetic spectrum *The radiant energy produced by the Sun that is measured in progressive wavelengths.*

Shortwave radiation *The portion of the electromagnetic spectrum that includes gamma rays, X-rays, ultraviolet radiation, visible light, and near-infrared radiation.*

Longwave radiation *The portion of the electromagnetic spectrum that includes thermal infrared radiation.*

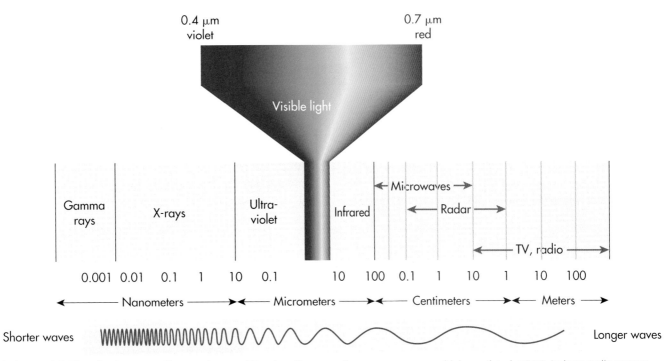

0.4 μm
violet

0.7 μm
red

Visible light

| Gamma rays | X-rays | Ultra-violet | | Infrared | Microwaves | Radar | TV, radio |

0.001 0.01 0.1 1 10   0.1   10 100 0.1 1 10   1 10 100

← Nanometers →◄─ Micrometers ─►◄─ Centimeters ─►◄─ Meters →

Shorter waves   Longer waves

I **Figure 4.2 The electromagnetic spectrum.** Wavelengths range from gamma rays, which are the shortest, to long radio waves.

The second principle of electromagnetic radiation is that a *direct* relationship exists between the absolute temperature of the object and the amount of radiation it emits. This relationship is described by the Stefan–Boltzmann law and means that hotter objects emit more radiation than cooler ones. As a result, Earth emits much less radiation than the Sun. This temperature/emitted radiation relationship is exponential, meaning that a small temperature increase in an object results in very large increases in emitted radiation. For example, you have probably experienced how a blacktop or concrete surface emits much more radiation, and is therefore much hotter, than a plowed field or grassy park that it might surround. This higher release of radiation takes place because the highway or sidewalk is warmer than the field or park.

Solar Energy and the Solar Constant

As we discussed in Chapter 3, solar energy is created in vast quantities by nuclear fusion within the Sun. This energy works its way to the surface of the Sun, where it is emitted as electromagnetic radiation. From this point this energy travels along straight lines (or rays) through space at the speed of light.

Given that the Sun produces energy at a nearly constant rate, the output of solar radiation is also nearly constant. Although some variability in solar output occurs due to the waxing and waning of energy production within an 11-year solar cycle, the overall output of solar energy is considered to be consistent over time. As a result of this generally constant production and emission, the amount of solar radiation received in an area of fixed size in space (outside Earth's atmosphere), and at right angles to the Sun, is also constant. This amount of received energy, which is referred to as the **solar constant**, has a value of about 1370 watts per square meter (W/m^2) at the top of the atmosphere.

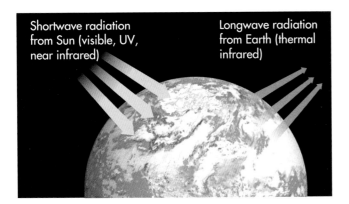

Shortwave radiation from Sun (visible, UV, near infrared)

Longwave radiation from Earth (thermal infrared)

Figure 4.3 Radiation to and from Earth. Earth receives short-wave radiation from the Sun. Some of this radiation reaches the surface of Earth where it is absorbed and then emitted as long-wave energy.

Solar constant *The average amount of solar radiation ($\sim$1370 W/m^2) received at the top of the atmosphere.*

To better visualize the various components of the electromagnetic spectrum, view the animation *The Electromagnetic Spectrum*. You will be able to interact with elements of the electromagnetic spectrum by selecting different wavelengths, such as those in the infrared and X-ray parts of the spectrum. When you select a specific wavelength, an image will appear that is representative of that part of the spectrum. In this way you will develop a better feel for the electromagnetic spectrum, how it is a pervasive part of Earth, and how geographers use it to study the environment. Once you complete the interaction, be sure to answer the questions at the end of the exercise to test your understanding of this concept.

VISUAL CONCEPT CHECK 4.1

Although we take the Sun for granted, it is by far the most important variable with respect to life on Earth. How is solar energy produced? What form does solar energy take as it travels through space? How long does this energy take to reach the Earth?

SOHO's Extreme Ultraviolet Imaging Telescope/Science Photo Library/Photo Researchers

KEY CONCEPTS TO REMEMBER ABOUT THE ELECTROMAGNETIC SPECTRUM AND SOLAR ENERGY

1. The Sun produces energy that radiates to Earth as various sizes of electromagnetic waves.

2. The electromagnetic spectrum is the total wavelength range of electromagnetic energy.

3. An inverse relationship exists between temperature and radiation wavelength.

4. A direct relationship exists between the temperature of an object and the amount of electromagnetic radiation it produces.

5. The Sun produces electromagnetic energy at a generally constant rate over time. The amount of energy that reaches the top of Earth's atmosphere is known as the solar constant.

Composition of the Atmosphere

Before examining how solar radiation interacts with the atmosphere, you first need to understand the fundamental composition of the atmosphere because it is the medium through which radiation flows on its way to Earth's surface. The atmosphere is unique in our solar system because it supports life by providing oxygen and carbon dioxide for animals and photosynthesis, respectively. In addition, the atmosphere serves

NASA Johnson Space Center

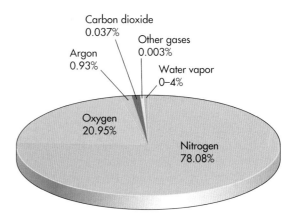

Figure 4.5 Proportion of various gases in the atmosphere. Most of the atmosphere consists of nitrogen and oxygen, which are constant gases. Other gases, such as carbon dioxide and water vapor, are variable gases.

Figure 4.4 The Earth's atmosphere as viewed from space. This image nicely demonstrates the thinness of the atmosphere, which is the faint blue haze that appears on the horizon. Note the infinite blackness of deep space in the background.

as a buffer that shields the planet from the potentially harmful effects of UV radiation from the Sun, allowing mostly visible and infrared wavelengths to reach Earth. The atmosphere behaves as a fluid in much the same manner as water, with flowing currents and eddies. Fluctuations in these currents and eddies shape the course of environmental conditions on the Earth's surface at every moment. Knowing how solar radiation flows in the atmosphere is key to understanding Earth's temperature, atmospheric circulation, and precipitation patterns, which are discussed in later chapters.

In general terms the atmosphere consists of air, an invisible medium that surrounds and protects Earth (Figure 4.4). The fundamental components of the atmosphere can be divided into three categories: (1) constant gases, (2) variable gases, and (3) particulates. Each of these constituents is critical to the way the atmosphere functions because it performs a unique role essential to life on Earth.

Constant Gases

Constant gases maintain more or less the same proportion in the atmosphere. In our atmosphere, nitrogen and oxygen are the primary constant gases and together make up 99% of the atmosphere (Figure 4.5). Argon is the other constant gas, composing

approximately 1% of the atmosphere. It is inert (chemically inactive) and of little importance in natural processes.

Nitrogen occurs in molecular form as two nitrogen atoms bonded together (N_2). This gas makes up 78% of the atmosphere and is derived from the decay and burning of organic material, volcanic eruptions, and the chemical breakdown of specific kinds of rocks. Although nitrogen is largely inert in the atmosphere, it is critical to plant life because it can be transformed, or fixed, into chemical compounds (ammonia or nitrates) in the soil. These compounds are absorbed by plants and incorporated in their tissues as proteins. Nitrogen maintains a constant proportion of the total atmosphere because what is added is balanced by what is removed through precipitation and various biological processes.

Oxygen makes up 21% of the atmosphere and is a by-product of photosynthesis. In contrast to the inert nature of nitrogen, oxygen gas (O_2) is very active and can combine with a variety of other elements through the process of oxidation. Oxygen is essential to animal respiration because it is required to convert foods into energy. Oxygen is a constant gas because the amount produced by plants balances the amount absorbed by various organisms through respiration.

Variable Gases

Variable gases differ in their proportion of the atmosphere over time and space, depending on environmental conditions. Although some of these gases are very important to life, they make up only a tiny portion (less than 1%) of the atmosphere. The most important variable gases are carbon dioxide, water vapor, and ozone.

Constant gases *Atmospheric gases such as nitrogen, oxygen, and argon that maintain relatively consistent levels in space and time.*

Variable gases *Atmospheric gases such as carbon dioxide, water vapor, and ozone that vary in concentration in space and time.*

Water Vapor An important variable gas in the atmosphere is water vapor. Water is found in three physical states: liquid, solid (ice), and gas (water vapor). The amount of water vapor in the atmosphere near Earth's surface is about 2% in most parts of the planet, but can range from just less than 1% over deserts and polar regions to about 4% in tropical zones. Atmospheric water vapor is especially vital because it absorbs and stores heat energy from the Sun and is thus an important component, along with carbon dioxide, of the greenhouse effect. As airflow within the atmosphere moves vapor around, the effect is to moderate temperature and transport energy around Earth.

The amount of water vapor at any given place in the atmosphere depends on many variables, such as the proximity to a large body of water or the air temperature. Remember that a direct relationship exists between the temperature of the air and the amount of water vapor it can hold, with warmer air capable of holding more vapor than cooler air. You have probably experienced this relationship yourself. Try to remember what it feels like on a hot, muggy day in the summer when the air is sticky and uncomfortable. It feels that way in large part because the air holds a lot of water vapor, which is possible because the air is so warm. This vapor/temperature relationship is an important principle in physical geography because it directly influences the process of precipitation. As we will discuss more thoroughly in Chapter 7, precipitation happens when bodies of air cool and water changes from vapor to liquid.

Figure 4.6 shows atmospheric water vapor over the North American continent in February 2009. Notice that on this particular day, the highest concentration of water vapor was in the central and eastern United States. If you happened to be in those areas, you would have likely noticed numerous clouds in the sky. The water vapor in those clouds flowed into the region in association with a body of warm air originating over the tropical Pacific Ocean. Areas of relatively dry air appeared over northwestern Mexico and the upper Midwest.

Carbon Dioxide Another important variable gas is carbon dioxide (CO_2), which currently makes up about 0.040% of the atmosphere (or 400 parts CO_2 for every 1 million parts of the atmosphere). Despite this very small percentage, CO_2 is a critical part of the atmosphere for two very important reasons. First, plants absorb CO_2 and release oxygen as a by-product. Second, atmospheric CO_2 contributes significantly to the **greenhouse effect**, which is the process through which the atmosphere traps longwave radiation. This process warms the atmosphere, which, in turn, warms Earth. Another significant greenhouse gas is methane, which is derived from natural gas and decaying organic matter. Methane currently makes up about 0.00017% of the atmosphere. You may associate the greenhouse effect most closely with the concept of global warming, an environmental issue that greatly

Figure 4.6 Water vapor over the North American continent. This image was taken by the Geostationary Operational Environmental Satellite (GOES) on February 11, 2009. Dark zones represent regions in the atmosphere about 6 km to 10 km (~3.7 mi to 6.2 mi) above the Earth's surface containing very little water vapor, whereas bright areas represent relatively high concentrations of water vapor. In areas other than the tropics and poles, the average water vapor content is about 2% of the air.

concerns many scientists today. Although the greenhouse effect is indeed linked to this issue, it can also be viewed in a positive way because it has helped maintain a constant atmospheric temperature that is hospitable to life. The concern now is that CO_2 levels are rapidly rising beyond recent historical norms due to human industrial activity, specifically the consumption of fossil fuels. As a result, Earth appears to be in the midst of a rapid warming trend that may have major environmental consequences in the future.

You can examine the processes associated with the greenhouse effect more closely in Figure 4.7. Although we will discuss the flow of solar radiation more thoroughly later in the chapter, this diagram shows the significance of CO_2 in the atmosphere and is relevant here. As illustrated in Figure 4.7, Earth receives shortwave radiation from the Sun, which passes through the atmosphere and is absorbed by the surface. Subsequently, the Earth releases energy as longwave radiation. Some of this radiation flows directly back into space, but most is absorbed by the atmosphere, where it is held in large part by CO_2 (water vapor is another greenhouse gas). The atmosphere then redirects some longwave energy back to the surface as **counterradiation**. Some atmospheric longwave energy escapes to space.

The greenhouse effect is the process through which counterradiation is returned to the surface. The really fascinating aspect of the greenhouse effect, as far as life on Earth is con-

---

Greenhouse effect *The process through which the lower part of the atmosphere is warmed because longwave radiation from Earth is trapped by carbon dioxide (CO_2) and other greenhouse gases.*

Counterradiation *Longwave radiation that is emitted toward the Earth's surface from the atmosphere.*

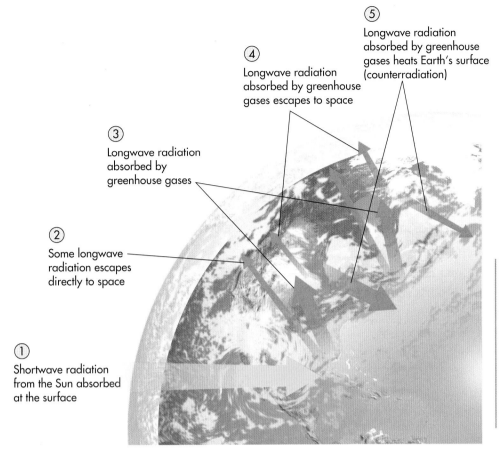

⑤ Longwave radiation absorbed by greenhouse gases heats Earth's surface (counterradiation)

④ Longwave radiation absorbed by greenhouse gases escapes to space

③ Longwave radiation absorbed by greenhouse gases

② Some longwave radiation escapes directly to space

① Shortwave radiation from the Sun absorbed at the surface

Figure 4.7 The greenhouse effect. Earth absorbs shortwave radiation from the Sun and emits it as longwave energy. Some of this emitted longwave radiation flows directly to space, but most is absorbed by carbon dioxide (CO_2) and water vapor in the atmosphere. Subsequently, this longwave radiation either escapes into space or reflects back to the Earth's surface through the process of counterradiation.

cerned, is that the balance is currently just right for the planet to be neither too hot nor too cold. To see the significance of this balance, consider the planets Mars and Venus. In both places, CO_2 is the dominant component of the atmosphere, comprising about 96% of the volume on each planet. On Mars, the overall density (measured by pressure) of the atmosphere at the surface is about 1000 times less than that of Earth; thus, there is comparatively much less CO_2 on Mars than on Earth. As a result, Mars is much colder than Earth, with an average temperature of –63°C (–81°F). With respect to Venus, the atmosphere there is about 90 times denser at the surface than Earth's atmosphere; thus, the concentration of CO_2 on Venus is much higher than on Earth. Consequently, Venus is considered to have a runaway greenhouse effect, with an average temperature of 500°C (932°F)!

Although Venus provides an extreme example of the greenhouse effect, it is a context through which we can view the concerns that many scientists have regarding the ongoing human impact on the greenhouse process through the combustion of fossil fuels for energy. Since the middle of the 19th century, levels of atmospheric CO_2 have steadily increased largely through our consumption of fossil fuels. To see how these levels have changed, examine Figure 4.8, which shows CO_2 measurements obtained from Mauna Loa in Hawaii. These data indicate that atmospheric CO_2 increased approximately 90 parts per million (ppm) between 1958 and 2010. Most climatologists believe this increase is a primary cause of global climate change and is the negative side of the greenhouse effect that many people associate with the relationship. This important issue will be covered in greater detail in Chapter 9.

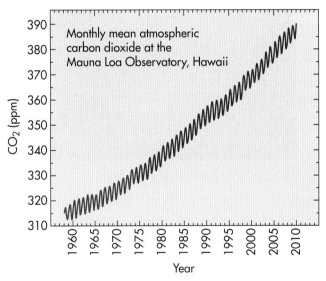

Sources: 1958 –1974 Scripps Institute of Oceanography; NOAA
1974 –2012 National Oceanic and Atmospheric Administration; NOAA

Figure 4.8 Changes in atmospheric carbon dioxide from 1958 to 2010 as measured at Mauna Loa, Hawaii. The vertical axis is the concentration of CO_2 in parts per million (ppm). Note the steady increase during this period of time, which has been attributed to human consumption of fossil fuels. Peaks and valleys in the curve represent seasonal fluctuations, with low points taking place during the Northern Hemisphere summer when plants take up CO_2. (*Sources*: C. D. Keeling and T. P. Whorf. Atmospheric CO_2 records from sites in the SIO Air Sampling Network. In *Trends: A Compendium of Data on Global Change*. U.S. Department of Energy; NOAA; NASA.)

THE FORMATION OF RAINBOWS

In addition to producing clouds, the presence of water vapor in the air can have a very beautiful side effect because it is linked to the formation of rainbows. A rainbow develops when white light from the Sun peeks through the clouds and strikes water droplets in the air from a rainstorm. The water droplets bend the light rays, but by a different amount for each color of light. The result is that the white light is spread out into bands of colors. These colors are reflected back down toward Earth by the water droplets, at an average angle of about 42°. If you are standing in the right place at the right time, you see these colors as a rainbow.

Rainbows require bright sunlight, so you cannot see one during the height of a rainstorm. On the other hand, a rainbow requires the presence of water droplets in the air, so it will disappear after the rain has gone. In other words, the sight of a rainbow usually means that the rain is clearing out and fair weather is ahead.

I A rainbow near Jervis Bay, Australia.

Alan Arbogast

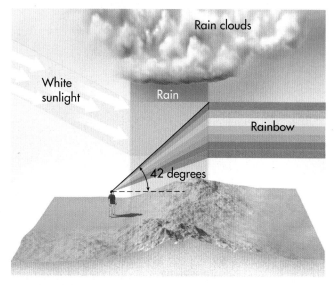

Water droplets in the air bend sunlight by an angle that depends on the wavelength (color) of the light. We see the result as a rainbow. Notice that red appears at the top of the rainbow and blue at the bottom because red light is bent more than blue light.

Ozone   Ozone is a form of oxygen that has three oxygen atoms (O_3), rather than the two found in normal oxygen gas (O_2). Atmospheric ozone forms when gaseous chemicals react in the upper atmosphere (Figure 4.9) with light energy. In the initial stage of this reaction, O_2 absorbs UV energy, which causes the oxygen molecule to split into two oxygen atoms ($O + O$). Subsequently, one of the O atoms combines with an O_2 molecule to form ozone. Ozone is naturally destroyed when it absorbs UV radiation and splits from O_3 into $O_2 + O$. The single oxygen atom can then recombine with another O_2 molecule to form ozone. Ultimately, O_3, O_2, and oxygen atoms (O) are repeatedly formed, destroyed, and re-formed in the ozone layer in a way that absorbs UV radiation each time a transformation takes place.

Ozone primarily occurs in two layers in the atmosphere: at ground level and within the stratosphere. Although diffuse amounts of ozone exist as high as 50 km (32 mi) above Earth, it is most concentrated in the ozone layer between about 15 km and 30 km (9 mi to 19 mi; see Figure 4.10). Ground-level ozone is a form of pollution (Figure 4.11) created when nitrogen oxides and organic gases emitted by automobiles and industrial sources react. This form of ozone has been linked to cardiorespiratory illness and may also cause crop damage.

Human Interactions: The Ozone Hole   The greatest concentrations of ozone are found in the **ozone layer**. This

---

Ozone layer   *The layer of the atmosphere that contains high concentrations of ozone, which protect the Earth from ultraviolet (UV) radiation.*

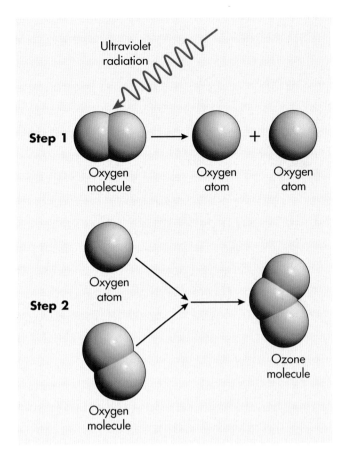

Figure 4.9 Formation of atmospheric ozone. In the first step, UV radiation is absorbed by oxygen molecules, creating free oxygen atoms. In the second step, these free oxygen atoms combine with oxygen molecules to form ozone.

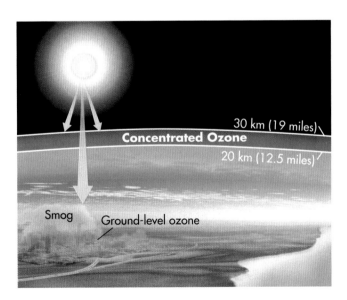

Figure 4.10 Ozone concentrations in the atmosphere. One layer is at high altitudes and absorbs UV radiation from the Sun. Most of this ozone is concentrated in the ozone layer. The second concentration occurs in the lower part of the atmosphere and is associated with pollution, particularly chemical smog in cities.

Figure 4.11 Smog in Singapore. A large component of this haze is ozone created by the interaction of nitrogen oxides and organic gases emitted by automobiles and factories.

layer is critical to life because it absorbs harmful UV radiation from the Sun that would otherwise reach the Earth's surface. A major environmental issue that developed in the last half of the 20th century is the depletion of the ozone layer. This depletion happened on a global basis due, in part, to the release of chlorofluorocarbons (CFCs) associated with air conditioners and other cooling systems. CFC molecules are very stable in the atmosphere and gradually diffuse upward from the surface to the ozone layer. Once they reach this altitude, CFCs absorb UV radiation and break down into chlorine oxide (ClO) molecules that, in turn, attack ozone molecules and convert them into oxygen atoms (Figure 4.12). The net effect of this process is that the ozone layer is depleted and more UV radiation reaches the ground because less is absorbed in the atmosphere.

In the 1980s it was discovered through satellite remote sensing of the atmosphere that the ozone layer was being depleted at an alarming rate. Although some depletion occurs naturally by volcanic aerosols periodically injected into the atmosphere, most of the reduction was attributed to industrial CFC production. The most striking example of ozone depletion is the Antarctic **ozone hole,** which was discovered in the mid-1980s. This feature forms every spring in the Southern Hemisphere (Figure 4.13), because a polar vortex develops during the previous winter that traps the air within it. This vortex is somewhat analogous to a whirlpool in water that does not allow a floating object to get out as it spins.

Over the course of each winter, air becomes very cold and thin clouds of ice form within the atmospheric whirlpool. The floating ice crystals, in turn, facilitate the transformation of

Ozone hole *The decrease in stratospheric ozone observed on a seasonal basis over Antarctica, and to a lesser extent over the Arctic.*

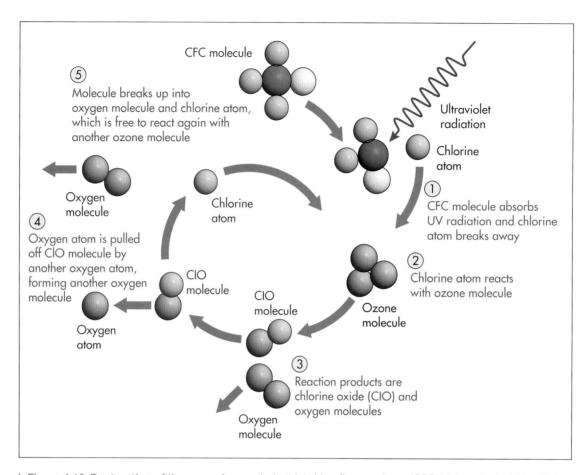

Figure 4.12 Destruction of the ozone layer. Industrial chlorofluorocarbons (CFCs) interact with UV radiation in the ozone layer and release chlorine atoms. Follow the numbered steps to see how this process works. It begins with the breaking away of a chlorine atom from a CFC molecule due to the absorption of UV radiation. The resultant free chlorine atom then begins to interact with an ozone molecule. The chlorine atoms transform ozone molecules into ordinary oxygen molecules without being used up themselves. Thus, one CFC molecule can destroy many ozone molecules.

(a) (b)

Figure 4.13 Antarctic ozone hole. (a) Ozone hole during September 2004. The area of the hole was 24,200,000 km² (9,400,000 mi²), or larger than the combined area of the United States, Canada, and Mexico. Dark blue represents levels of ozone that are about 20% less than normal. (b) Ozone hole in September 2012. The average size of the hole in 2012 was 17,900,000 km² (6,900,000 mi²), making It the second smallest in the past 20 years.

chlorine to chlorine oxide that destroys ozone, with the peak loss taking place in September and October. As a result of this decrease in ozone, the amount of surface exposure to UV radiation at 55° S has increased about 10% per decade since the late 1970s. In addition to the Antarctic ozone hole, a similar but weaker hole develops over the Arctic region in the Northern Hemisphere winter. Significant depletions of ozone have also been measured in the midlatitudes of the Northern Hemisphere. As a result, the average amount of annual exposure to UV radiation at 55° N has increased about 7% per decade since the late 1970s. These increases in UV exposure are of great concern for several reasons, including damage to some forms of aquatic life, reduced crop yields, and increases in the incidence of skin cancer in humans.

Although scientists have legitimate concerns about ozone depletion and its impact on life and human health, the threat may diminish in your lifetime. Shortly after the ozone issue emerged, the world community began working diligently to solve the problem of CFC emissions. These efforts culminated in the 1987 Montreal Protocol, an environmental treaty sponsored by the United Nations and signed by 23 nations. This treaty called for developed countries such as the United States and the nations of the European Union to cut CFC emissions by 50% by 1999 and for developing countries such as various African nations and India to halt CFC use by 2010. The effect of this treaty has been generally positive, with stratospheric CFC concentrations peaking in 1997 and slowly falling since that time. If the current rate of decline continues, the ozone layer could be completely restored by the latter part of this century.

Particulates

The last important variable component of the atmosphere is particulates, which are microscopic bodies carried in the air, existing in both liquid and solid form. The liquid variety comes in the form of clouds and rain, which develop when water vapor changes its physical state due to temperature changes in the atmosphere. Solid particulates take an assortment of forms, including snow, hail, pollutants, wind-blown soil (dust), smoke from wildfires, volcanic ash, pollen grains from plants, and salt spray from breaking waves. Most of the time particulate concentrations are densest near their place of origin, such as during an intense dust storm (Figure 4.14) or near a volcano. Overall, however, they make up less than 1% of the Earth's atmosphere.

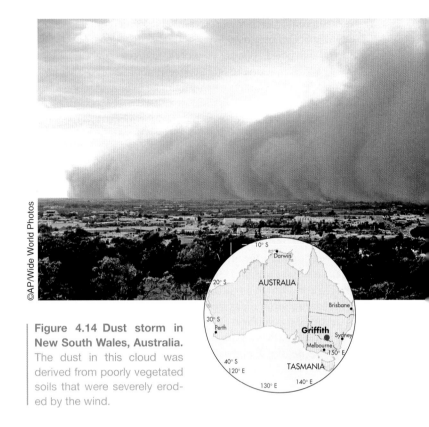

Figure 4.14 Dust storm in New South Wales, Australia. The dust in this cloud was derived from poorly vegetated soils that were severely eroded by the wind.

3. The atmosphere consists largely of nitrogen (78%) and, to a lesser extent, oxygen (21%).

4. Although variable gases and particulates compose less than 1% of the atmosphere, they nevertheless affect atmospheric processes and climate in important ways. A good example of the impact of these gases is the greenhouse effect and the way it moderates temperature on Earth.

Despite their small proportion of the atmosphere, particulates nevertheless play an important role in weather and climate. Precipitation would not occur if dust particles were not present in the atmosphere because they provide a nucleus around which water condenses in the first step of cloud formation. You can see this relationship the next time it rains right after you wash your car, because your car will be coated with fine dust. Other particulates, such as smoke and volcanic ash, are important because they either absorb or reflect solar energy. These combined processes influence local weather and regional climate by moderating the temperature of the atmosphere.

Although most particulates are a positive component of the atmosphere, some can cause negative environmental and health effects, especially the toxic air pollutants associated with human activities. According to the U.S. Environmental Protection Agency, over 100 pollutants are of particular concern. Benzene and perchlorethylene, for example, are pollutants derived from gasoline and dry cleaning facilities, respectively. Other toxic pollutants include dioxin, asbestos, toluene, and metals such

KEY CONCEPTS TO REMEMBER ABOUT THE COMPOSITION OF THE ATMOSPHERE

1. Although the atmosphere is technically about 10,000 km (6000 mi) thick, the vast majority of the air is found in the lowermost 30 km (10 mi).

2. The atmosphere contains constant gases (primarily nitrogen and oxygen), variable gases (mainly carbon dioxide and water vapor), and particulates.

as cadmium, mercury, chromium, and lead compounds. These pollutants can originate from stationary locations, such as factories and power plants, or mobile sources, such as automobiles. When ingested in large concentrations over a long period of time, these pollutants can accumulate in human tissue and ultimately cause cancer, immune disorders, and a variety of neurological, reproductive (e.g., reduced fertility), developmental, respiratory, and other health problems.

The Flow of Solar Radiation on Earth

Once solar radiation reaches Earth, it flows along several pathways within the atmosphere and to the surface. Some radiation flows straight from the Sun to the surface of Earth as **insolation**, which is most accurately defined as the amount of solar radiation measured in watts per square meter (W/m²) that strikes a surface perpendicular to the Sun's incoming rays. Most solar radiation, however, follows a very indirect path. Some of it bounces around like a ping-pong ball, while some of it is directly absorbed by various components of the atmosphere. These various pathways directly influence climate on Earth and will be discussed thoroughly later in the chapter.

Heat Transfer

Before exploring the various pathways that solar radiation follows on Earth, it is useful to first understand how heat is transferred to Earth's atmosphere, land, and oceans. This discussion is based on the first law of thermodynamics, which states that energy cannot be created or destroyed; it can only change forms. Heat is a form of energy, related to the motion of atoms and molecules, which is transferred in several different ways that are easy to imagine (Figure 4.15). This transfer follows the second law of thermodynamics, which can be stated as follows: heat can spontaneously flow from a region of higher temperature to an area of colder temperature, but not the other way around.

The next time that you are around a campfire, consider that heat is transferred by **radiation**. This process involves the creation of electromagnetic waves, either as visible light or

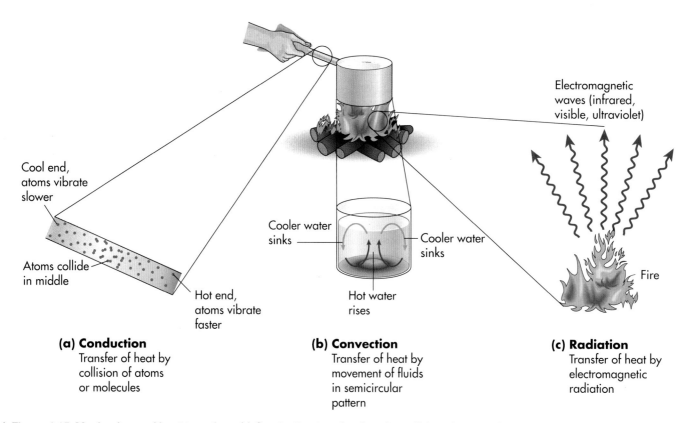

(a) Conduction
Transfer of heat by collision of atoms or molecules

(b) Convection
Transfer of heat by movement of fluids in semicircular pattern

(c) Radiation
Transfer of heat by electromagnetic radiation

Figure 4.15 Mechanisms of heat transfer. (a) Conduction transfers heat by collisions between fast-vibrating atoms or molecules (hot) and slower-vibrating atoms or molecules (cool). (b) Convection is the transfer of heat by the large-scale movement of matter. (c) Radiation is the transfer of heat by electromagnetic radiation, usually in the form of infrared, visible, or UV waves.

Insolation *Amount of solar radiation measured in watts per square meter (W/m²) that strikes a surface perpendicular to the Sun's incoming rays.*

Radiation *Energy that is transmitted in the form of rays or waves.*

The atmosphere contains a wide variety of natural constant gases, variable gases, and particulates. People also contribute to the overall character of the atmosphere. What potential impacts do you think this factory complex has on the atmosphere?

Corbis Images

invisible to the naked eye. Because this radiation exists in wave form, it carries energy and thus can move from one place to another without requiring an intervening medium. When this radiation reaches you, part of the energy of the wave is converted back into heat, which is why you feel warm sitting beside a campfire. Some of the radiation may exist in the form of visible light that we can see, but a great deal of the radiation emitted is infrared waves, whose longer wavelength is usually detectable only with special infrared detectors.

Another form of heat transfer is **conduction**, which involves the diffusion of energy through molecules that are in contact with one another. This diffusion occurs because, as the temperature of molecules increases, they begin to vibrate more rapidly, causing collisions that produce similar motions in adjacent molecules. In this fashion, sensible heat always moves from areas of relatively high temperature to zones of relatively low temperature. You can experience conduction yourself if you pick up a heated pot from a stove or other fire.

The third form of heat transfer is **convection**, which involves the upward movement of heat. The next time you boil water on the stove, notice how a small circular current of water arises within the pot where hot water moves up and is replaced at the bottom by relatively cooler water. As you will see in later chapters, convection is a very important mechanism of heat transfer on Earth and is associated with atmospheric circulation and precipitation.

Flow of Solar Radiation in the Atmosphere

With the mechanisms of heat transfer in mind, it is time to explore the various ways that solar radiation interacts with Earth. As you work your way through this discussion, refer to Figure 4.16, which illustrates the flow of solar radiation on the planet. Of all the incoming radiation that reaches Earth, about 25% flows uninterrupted to the surface as **direct radiation**. However, this amount can vary greatly depending on local geographic variables such as cloud cover or density of atmospheric dust. Figure 4.17 shows, for example, the impact that cloud cover can have on direct solar radiation.

The remaining 75% of incoming insolation is either absorbed or otherwise redirected in the atmosphere (see Figure 4.16). **Absorption** takes place when variable gases and particulates in the atmosphere interrupt the flow of solar radiation by absorbing specific wavelengths. For example, almost all UV wavelengths (those less than 0.3 μm or micrometer in length) are absorbed by oxygen and ozone. Similarly, radiation at the 1.3 μm and 1.9 μm wavelengths is absorbed very strongly by water vapor and CO_2. Overall, approximately 24% of incoming solar radiation is absorbed, with 18% absorbed by atmospheric water vapor and dust, 3% by ozone, and 3% by clouds. Remember, this absorption of solar radiation is important because it helps to moderate temperature in the atmosphere.

Conduction *The transfer of heat energy from one substance to another by direct physical contact.*

Convection *A circular cell of moving matter that contains warm material moving up and cooler matter moving down.*

Direct radiation *Solar radiation that flows directly to the surface of Earth and is absorbed.*

Absorption *The assimilation and conversion of solar radiation into another form of energy by a medium such as water vapor. In this process, the temperature of the absorbing medium is raised.*

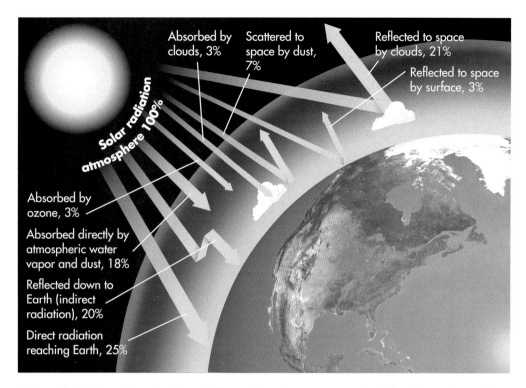

Figure 4.16 Interaction of solar radiation with various components of the atmosphere. Note that solar radiation is absorbed, stored, and reflected in many ways, and that about half the overall amount actually reaches the Earth's surface.

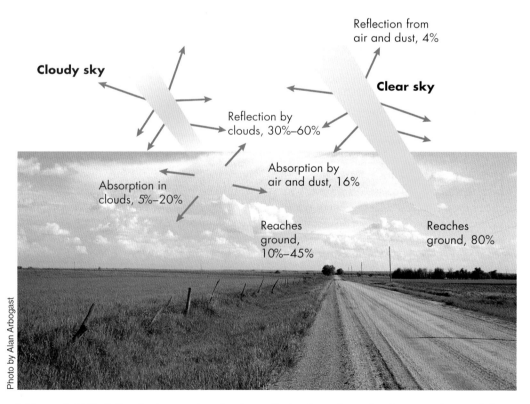

Figure 4.17 Variation in the receipt of solar radiation based on cloud cover. Areas of dense clouds reflect more radiation than areas of clear sky.

White light from Sun

Observer sees white Sun, blue sky at noon

North Pole

Direction of rotation

White light from Sun

Observer sees red sunset

Blue light scattered

(b)

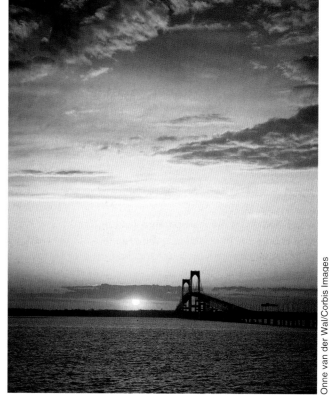

(a)

(c)

Figure 4.18 Colors in the sky. (a) The sky is blue because oxygen and nitrogen molecules scatter light in the blue wavelength. (b) Geometric relationship of insolation, atmospheric thickness, and sky color. (c) Colors at sunset along the Atlantic coast at Rhode Island. Oranges and reds appear because the blue wavelength is completely scattered before it reaches your eye.

Some incoming solar radiation in the atmosphere is also redirected by **reflection**. The amount of radiation reflected depends on the **albedo** of a surface and takes place in the atmosphere when insolation bounces off bright cloud tops. (Albedo refers to the amount of reflection a given surface can cause; a high albedo is typical of bright surfaces such as a mirror. We will come back to this factor shortly.) Approximately 21% of radiation is reflected back to space in this manner, and clouds are by far the most important reflector.

In addition to the amount of radiation reflected, an additional component of incoming radiation is redirected within the atmosphere by a process called **scattering**. Although scattering cannot be directly observed, you can see the indirect effects very clearly in the apparent colors of the sky (Figure 4.18). Have you ever wondered why the sky is blue? The reason is that oxygen and nitrogen molecules cause solar radiation to bounce around in the air in a process called *Rayleigh scattering*. This effect is more pronounced during the middle part of the day when the Sun is high in the sky and solar radiation is streaming more or less directly toward you. The sky appears to be blue because the

Reflection *The process through which solar radiation is returned directly to space without being absorbed by Earth.*

Albedo *The reflectivity of features on the Earth's surface or in the atmosphere.*

Scattering *The redirection and deflection of solar radiation by atmospheric gases or particulates.*

blue wavelength is preferentially scattered, whereas the longer wavelengths pass on through the atmosphere. Approximately 7% of this scattered radiation is redirected back into space.

You may also wonder why sunsets are so colorful, with the orange and red parts of the spectrum visible (Figure 4.18c). This change happens because solar radiation is streaming toward your eye from a much lower angle at dawn or dusk than when the Sun is high in the sky (Figure 4.18b). As a result, the incoming radiation is passing through a thicker slice (when viewed horizontally) of the atmosphere to reach your line of sight. This thickened atmosphere causes the blue wavelengths to be scattered out entirely before they reach you, leaving the longer wavelengths (oranges and red) for you to observe.

As you can see, atmospheric reflection and scattering have a strong influence on the way you perceive Earth. The brightness of cloud tops and the colors of the sky are visible environmental indicators of these processes. As a by-product of these interactions, some reflected and scattered radiation is also redirected downward toward the surface of Earth as **indirect radiation**. Approximately 20% of solar radiation that reaches Earth does so in this indirect fashion.

Interaction of Solar Radiation and the Earth's Surface

Let's now examine the approximately 45% of all solar energy (direct and indirect) that strikes the Earth's surface. What happens to this portion of incoming radiation is important because it influences variables such as temperature, atmospheric circulation, the density and kind of vegetation in a region, soils, and even where glaciers occur. Relationships such as these will be discussed in later chapters. In general, either of two things happens when solar radiation strikes the ground: (1) it is absorbed, or (2) it is reflected.

Absorbed Radiation   Of the amount of solar energy that reaches the ground, 96% is absorbed by the various land and water bodies on the surface, thus heating Earth. One way this heat energy can be stored is in the form of **sensible heat,** which is heat that can be sensed by touching or feeling and can be measured by a thermometer. A second way it can be stored is as latent heat, when water from land and ocean/lake surfaces is transformed into water vapor within the atmosphere through evaporation. **Latent heat** is a form of heat that is hidden and cannot be measured with a thermometer because it goes into breaking bonds between molecules when a substance changes physical state, such as from a liquid to a gas.

Stored energy can be lost from Earth in several ways. Sensible heat can be transferred from the Earth's surface to the atmosphere through the process of convection. In this case, warm air rises upward from Earth and cooler air descends to replace it. Heat can also be removed through the process of **evaporation**, which is the change of liquid water to water vapor by absorption of heat. In addition, absorbed radiation can simply be re-radiated back into space as longwave radiation. When longwave radiation is released, it can either escape directly back into space, be absorbed by atmospheric CO_2 and water vapor, or be backscattered by dust particles.

Reflected Radiation   Reflected radiation is energy that bounces off the Earth's surface in such a way that it does not provide heat. Overall, approximately 3% of incoming radiation is reflected in this way. The proportion of incoming radiation reflected by the surface is, just as for the atmosphere, dependent on the albedo of an object. Although you may not be directly aware of surface albedo, you are probably indirectly aware of its effects. Almost everyone knows how uncomfortable it feels to wear a dark-colored shirt on a very hot day with bright sunshine. The reason for this is that dark objects absorb radiation and therefore increase in temperature. A dark asphalt highway, for example, reflects only about 5% to 10% of incoming solar energy. A snow-capped mountain peak, in contrast, reflects 80% to 95% of incoming solar radiation and thus absorbs little heat. Figures 4.19 and 4.20 show how albedo can influence the amount of energy reflection on different surfaces.

The amount of radiation reflected depends not only on albedo, but also on the Sun angle in the sky. Remember from Chapter 3 that the Sun is always at some angle in the sky relative to any location on the planet. (This concept was illustrated in Figure 3.3, if you wish to review.) With respect to incoming solar radiation, this angle is called the **angle of incidence**. Think of the angle of incidence as being similar to the angle at which you might throw a rock into the water. If you drop the rock straight down into the water, it is immediately absorbed by the water and sinks to the bottom. However, if you throw the rock at a shallower angle (relative to the water), it might skip across the water. The angle of incidence works in much the same fashion in that solar radiation from a high-angle Sun interacts directly with Earth; in addition, the Sun's rays are confined to a relatively small area and are thus intense at that point. When the Sun is lower in the sky, however, more radiation is reflected when it strikes Earth; also, the radiation is spread over a larger area. When this takes place, the radiation is not as intense compared to higher-angle Sun locations.

Indirect radiation   *Radiation that reaches Earth after it has been scattered or reflected.*

Sensible heat   *Heat that can be felt and measured with a thermometer.*

Latent heat   *Heat stored in molecular bonds that cannot be measured.*

Evaporation   *The process by which atoms and molecules of liquid water gain sufficient energy to enter the gaseous phase.*

Angle of incidence   *The angle at which the Sun strikes Earth at any given place and time.*

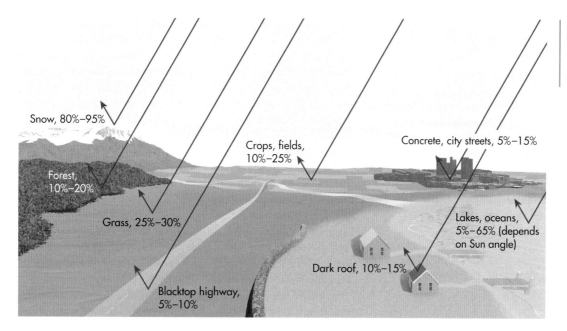

Figure 4.19 Variation of surface albedo. In general, albedo values are higher on brighter surfaces than on darker ones.

Snow, 80%–95%

Forest, 10%–20%

Grass, 25%–30%

Crops, fields, 10%–25%

Concrete, city streets, 5%–15%

Lakes, oceans, 5%–65% (depends on Sun angle)

Dark roof, 10%–15%

Blacktop highway, 5%–10%

When the Sun is directly overhead, water has an albedo value of around 5%, but when the Sun is near the horizon, the albedo value is around 65% because radiation reflects off of the surface more readily. You can see this difference clearly in Figure 4.20, which shows that oceans at higher latitudes have a higher albedo than those at lower latitudes. This variability occurs because the angle of incidence is smaller at higher latitudes and more radiation is reflected off the oceans in those regions.

www.wiley.com/college/arbogast

WileyPLUS

The Angle of Incidence

With the concept of solar radiation and Earth interactions in mind, you can further explore this important concept by viewing the simulation *The Angle of Incidence*. This simulation allows you to interact with the angle of incidence as it relates to latitude. The foundation of this simulation is Figure 3.3, which shows how the angle of incidence influences how much solar radiation is received at any given location. You will be able to choose between several latitude locations in the Northern and Southern Hemispheres to see how the angle of incidence varies between locations. As you work through the various scenarios, notice not only how the flow of solar radiation changes, but also the direction from which it comes. You should observe a big difference between the Northern and Southern Hemispheres. Pay attention to these differences and relate them back to what you have learned about Earth–Sun geometry and seasons.

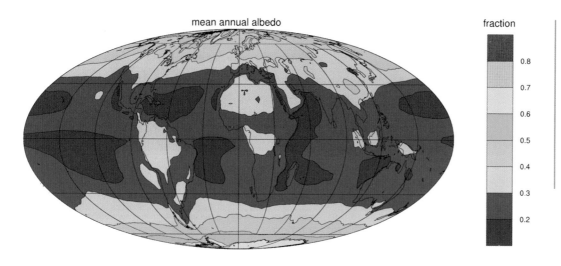

mean annual albedo

fraction

0.8
0.7
0.6
0.5
0.4
0.3
0.2

Figure 4.20 Mean annual albedo across the Earth's surface. The bar code at the right represents percentage albedo. Blue areas indicate zones of relatively low albedo and are concentrated in the oceans where radiation is absorbed. Polar ice caps, in contrast, have relatively high albedo, almost as much as 70%.

1. Once solar radiation reaches the atmosphere, it begins to flow along several different pathways. Some of it flows directly to the surface, whereas other parts are reflected off clouds or scattered by particulates.

2. Insolation that reaches the surface is either absorbed or reflected (a function of albedo).

3. Radiation that is absorbed is re-radiated as longwave radiation.

4. The amount of energy reflected or absorbed depends in large part on the angle of incidence, which, in turn, varies by latitude and season.

The Global Radiation Budget

To fit this discussion into a coherent model, think of Earth's energy flow as being analogous to money that flows in and out of your bank account. Sometimes you have a net surplus of cash because you have made a lot of money, whereas at other times you might even be a little in the hole because you have spent more than you earn. Like your bank account, the Earth's **radiation budget** refers to the balance between incoming (shortwave) radiation and outgoing radiation, which is either re-radiated through reflecting and scattering processes described earlier or released from the Earth in longwave form. The difference between incoming and outgoing values is the **net radiation**. For Earth as a whole, the long-term radiation budget must be balanced; otherwise, Earth would become progressively warmer or cooler. This balance is achieved because incoming radiation flows are matched by outgoing flows.

Figure 4.21 illustrates the fundamental nature of the Earth's radiation budget by incorporating the data from Figure 4.16. To simplify the concept, remember that solar energy is in the form of shortwave radiation and that 69% of it is actually absorbed by some aspect of Earth,

either by clouds, water, or the surface. The remaining 31% of shortwave energy is reflected directly back into space due to albedo or scattering. For the global radiation budget to remain in balance, therefore, Earth must emit all the absorbed energy back to space, which it does in longwave form. This release, plus the 31% of reflected shortwave radiation, keeps the system balanced. Earth's average temperature is moderated because greenhouse gases such as CO_2 and water vapor trap a small amount of longwave radiation.

Aside from the amount of longwave energy that is trapped by the atmosphere, the bulk of it is lost in a way that balances the global radiation budget. This balance is accomplished in a variety of ways (see Figure 4.21). The majority (45%) of longwave energy is lost to space by direct radiation from the surface. This form of loss does not happen evenly across the Earth's surface, but rather has a distinct geographic pattern of its own, with lower latitudes emitting more energy than higher latitudes (Figure 4.22). This pattern makes sense, of course, because low latitudes receive more shortwave radiation than high latitudes. In addition to direct energy loss from the surface, 21% of longwave energy is lost directly by radiation from the atmosphere and another 3% is emitted by ozone.

Although the long-term energy budget for Earth balances on a global scale, it can vary markedly over the short term in different regions, due to the interaction of various environmental factors. Because of this variability, certain locations

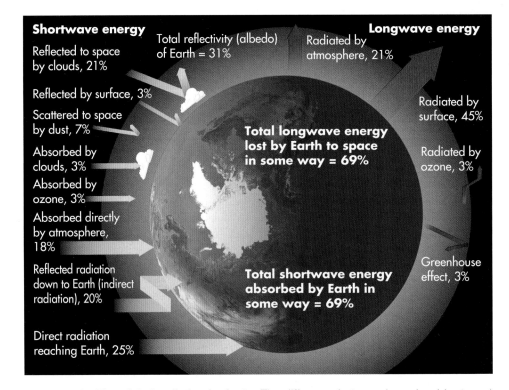

Shortwave energy

Reflected to space by clouds, 21%

Reflected by surface, 3%

Scattered to space by dust, 7%

Absorbed by clouds, 3%

Absorbed by ozone, 3%

Absorbed directly by atmosphere, 18%

Reflected radiation down to Earth (indirect radiation), 20%

Direct radiation reaching Earth, 25%

Total reflectivity (albedo) of Earth = 31%

Longwave energy

Radiated by atmosphere, 21%

Radiated by surface, 45%

Radiated by ozone, 3%

Greenhouse effect, 3%

Total longwave energy lost by Earth to space in some way = 69%

Total shortwave energy absorbed by Earth in some way = 69%

Figure 4.21 The global radiation budget. The difference between incoming (shortwave) radiation and outgoing (longwave) radiation must be balanced; otherwise, the temperature of Earth would either cool or warm dramatically.

Radiation budget *The overall balance between incoming and outgoing radiation on Earth.*

Net radiation *The difference between incoming and outgoing flows of radiation.*

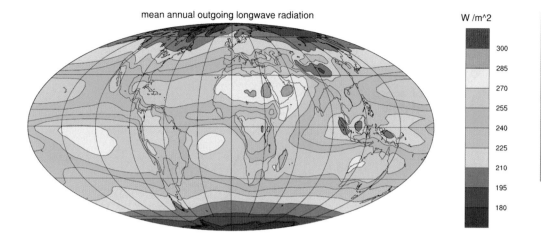

mean annual outgoing longwave radiation

W /m^2

300
285
270
255
240
225
210
195
180

Figure 4.22 Mean annual outgoing longwave radiation on Earth in watts per cubic meter. The original source of this radiation was the Sun. The radiation was then absorbed and stored by the Earth until its release as longwave energy. Note the geography of this process, with most longwave radiation being released in the tropical latitudes.

have a net surplus of radiation, whereas others have a net deficit. In addition, a particular place may have a net surplus of radiation during one time of year but a net deficit at another. These net differences are important because they are the driving mechanism behind climate and many physical processes, such as atmospheric circulation and evaporation. Four primary factors influence net radiation around the globe: (1) the Sun's angle of incidence, (2) latitude, (3) seasonality, and (4) length of day. Secondary factors include the varying output of the Sun due to sunspots and solar flares, the elliptical nature of the Earth's orbit, changes in the thickness and properties

of the atmosphere, and variability in the length of day. The best way to understand the changes in net radiation across the globe is by focusing on the four primary factors because they work in a holistic way.

Remember that the angle at which solar radiation strikes Earth directly affects absorption and reflection. This angle is primarily a function of latitude, with low latitudes having high Sun angles and high latitudes having low Sun angles. Therefore, low latitudes receive high amounts of radiation and have a net annual surplus, whereas the high latitudes have a net deficit because they receive less direct radiation (Figure 4.23).

www.wiley.com/college/arbogast

WileyPLUS

The Global Energy Budget

After you have examined Figure 4.21, view *The Global Energy Budget*. This animation allows you to better visualize the flow of solar radiation from the Sun to Earth. As you watch it, notice the various pathways that radiation takes as it interacts with the atmo-

sphere. In addition, try to integrate all the concepts introduced so far in this chapter to solidify your understanding of the global energy budget. Once you complete the animation, be sure to answer the questions at the end to test your understanding of this concept.

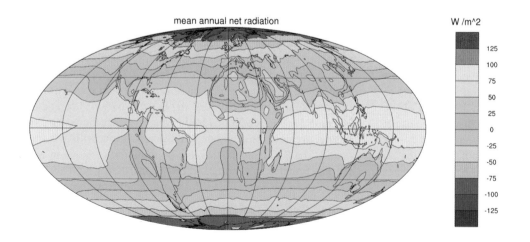

mean annual net radiation

W /m^2

125
100
75
50
25
0
-25
-50
-75
-100
-125

Figure 4.23 Mean annual net radiation at the top of the atmosphere on Earth. Note the relationship between color and value (watts per square meter), which is indicated in the key to the right of the image. Low to middle latitudes have a net surplus of radiation, whereas high latitudes have a net deficit.

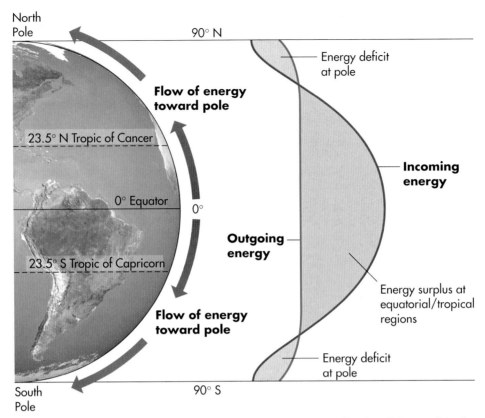

North Pole

90° N

Flow of energy toward pole

23.5° N Tropic of Cancer

0° Equator

0°

23.5° S Tropic of Capricorn

Flow of energy toward pole

South Pole

90° S

Energy deficit at pole

Incoming energy

Outgoing energy

Energy surplus at equatorial/tropical regions

Energy deficit at pole

Figure 4.24 Net radiation and the transfer of heat energy on Earth.   Compare this diagram to Figure 4.23 and note where areas of net surpluses and deficits are found and how energy flows on the planet.

How does this variation in net radiation between latitudes influence natural processes on Earth? We will cover the significance of this spatial relationship in greater detail later in the text, but for now it is sufficient to say that the difference in net radiation is a primary driving force behind many atmospheric circulatory processes, especially large-scale wind patterns. Figure 4.24 shows how this transfer works.

What happens to net radiation patterns when we add the effect of the seasons to the mix of variables? Recall that this seasonal effect takes place because of the geometric relationship between Earth and the Sun, which results in changes in the angle of incidence over the course of the year. Remember, too, that at any given point on Earth, the position of the Sun at solar noon migrates over the course of the year because of axial tilt and orbital position. This seasonal migration of the Sun, in turn, causes the subsolar point (on Earth) to change position with respect to latitude. This shift of the subsolar point north and south of the Equator has a major influence on the seasonal distribution of net radiation, as you can see in Figure 4.25.

Figure 4.26 compares two curves of annual insolation by latitude—one with Earth's actual axial tilt (23.5°) and the other with the axis not tilted. With a tilted Earth, you can see that

annual radiation at low latitudes is about 40% greater than at high latitudes, but that high latitudes still receive a considerable amount. However, if the Earth's axis were perpendicular to the plane of the ecliptic, annual radiation would range from a high value at the Equator, where the Sun angle would consistently be 90°, to *zero* at the poles, where the Sun would always be at the horizon. You can see that axial tilt is an important variable with respect to daily insolation at various latitudes, and that as a result, the poles receive nearly half the amount of radiation as the Equator, even though they are in darkness six months of the year.

A good way to conclude this discussion about the global radiation balance is to examine the effects when all the variables—angle of incidence, latitude, seasonality, and length of day—are considered together. Figure 4.27 graphically illustrates these combined effects on seasonal insolation at the top of the atmosphere at three places: the Equator (0°), 45° N, and the North Pole (90° N). Notice in this diagram that the North Pole has the greatest annual range of daily insolation, with a distinct peak at the Summer Solstice and no radiation received in winter. The Equator, in contrast, receives consistently high amounts of radiation all year. Given its location midway between the

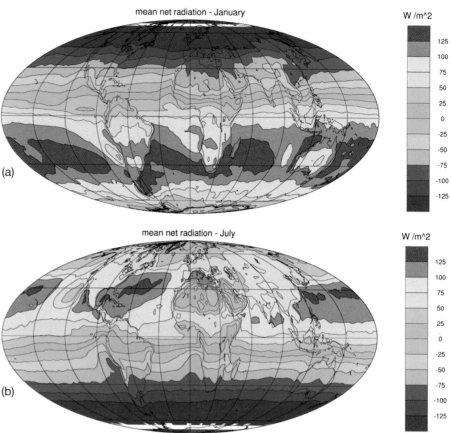

Figure 4.25 Average seasonal changes in net radiation on Earth. (a) Mean January net radiation. (b) Mean July net radiation. In both figures, reds indicate high values of net radiation, measured in watts per square meter, whereas blues equal low or deficit amounts.

mean net radiation - January

W /m^2

mean net radiation - July

W /m^2

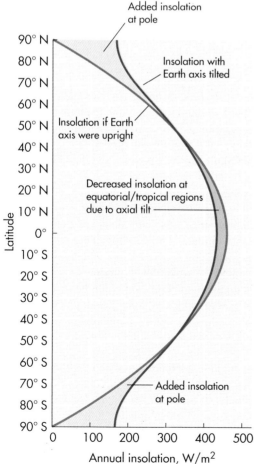

Equator and the North Pole, 45° N has a moderate range of daily insolation, with a peak during the summer months and lower values during winter.

In addition to these patterns, Figure 4.27 reveals two other interesting relationships. The first is that the Equator shows two peaks in daily insolation, rather than one, over the course of the year. The reason for these dual peaks is that the Sun is directly over the Equator two times during the year—specifically, at each Equinox. The second interesting pattern is that the North Pole (90° N) receives more insolation around the Summer Solstice than the Equator, even though the angle of incidence at very high latitudes is still relatively low. Why does this happen? The answer is that latitudes above the Arctic Circle receive radiation for 24 h because the Sun never sets. At the Equator, in contrast, day length is always about 12 h. This pattern shows the significance of day length for the amount of daily insolation received, which, in turn, depends on axial tilt and orbital position.

Figure 4.26 Variation in annual insolation (by latitude) on Earth. The red line shows the variability that exists as a function of the actual axial tilt. The blue line indicates what insolation would be if the axis were not tilted. Note the added insolation at high latitudes due to the tilt of Earth.

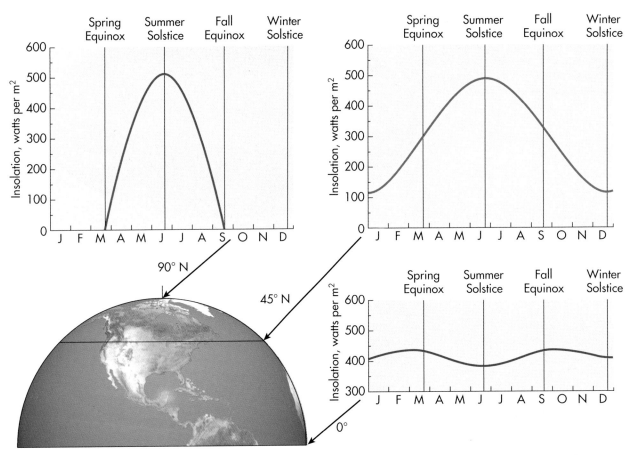

Figure 4.27 Daily insolation through the year at the top of the atmosphere at select latitudes in the Northern Hemisphere. Note that insolation changes at each of these locations on a seasonal basis and that the annual range increases at progressively higher latitudes.

Human Interactions: Solar Energy Production

Humans have attempted to harness solar energy for practical purposes for thousands of years. The Anasazi Indians of the U.S. Southwest, for example, built their cliff dwellings on the south-facing side of canyons so that the Sun would shine into their homes and warm them during the winter. Since the oil price shocks of the 1970s, and to lessen our reliance on fossil fuels, more office buildings and homes today are designed with windows that face the Sun when it is at a low angle during the winter months (Figure 4.28). This angle allows the Sun to shine directly into the building and enables heat to be trapped by the glass. You can experience this kind of *passive solar heating* when you enter your car after you leave the windows rolled up on a sunny day while you shop.

In addition to passive solar systems, efforts are under way to further develop active solar systems that directly capture solar energy and convert it to useful outputs. For example, photovoltaic systems, or *solar cells*, change sunlight directly into electricity; they tend to be used in remote places to power devices such as lighted road signs that are not connected to the electrical grid. On a much larger scale, solar power plants

contain a number of mirrors, or *heliostats*, that indirectly generate electricity because they redirect sunlight into a fluid that warms and produces steam. This steam, in turn, is used to power generators. These reflecting mirror systems are designed to turn to follow the Sun over the course of the day so that they constantly collect energy.

The use of solar power has distinct pros and cons. On the one hand, solar power is a clean and sustainable energy source that does not produce greenhouse emissions and thus does not contribute to global climate change. Also, these systems can produce electricity wherever the Sun shines. On the other hand, the current technology is inefficient in producing an adequate supply. Most photovoltaic modules, for example, convert only about 10% of sunlight to electricity. Solar power plants are inefficient, too, which is why they typically contain dozens (or even hundreds) of collectors that funnel solar energy into a common focal point.

Perhaps the biggest problem associated with capturing solar energy is the simple fact that the Sun does not shine everywhere on a consistent basis (Figure 4.29). The Great Lakes region of North America, for example, is cloudy much of the time and is thus a poor place to produce solar energy on a large scale. In contrast, the best place in the United States for solar energy production is the desert region of the Southwest where

Winter sun angle

Summer sun angle

Seasonal shading to south wall glazing

Louvred vent

Winter blinds

Light colored roof materials

Ventilation

Summer cooling breezes

Bedrooms

Pergola

Living room

Study

House designed to minimize over-shadowing

Courtyard

Deciduous planting

Maximum glazing to south walls

Thermal mass flooring

Screen planting against winter winds

Figure 4.28 A well-designed home with effective passive solar heating. In the Northern Hemisphere, windows are oriented south to allow low-angle winter Sun to warm living areas. Cross-circulation within the home carries warm air upstairs. In the summer, high-angle Sun is blocked by overhang to keep the house cool.

the Sun shines twice as much as in other parts of the country. A number of solar plants have been built here since the early 1980s, including nine Solar Energy Generating Systems (SEGS) that collectively have over 936,000 trough energy collectors (Figure 4.30) covering more than 6.5 km² (2.5 mi²). In contrast to less efficient flat mirrors, trough mirror systems can focus the Sun at 30 to 60 times its normal intensity and contain

synthetic oil that captures heat, reaching temperatures of 390°C (735°F). These plants can collectively produce 354 mW of electricity—enough to meet the power needs of approximately 500,000 people a year. Despite the many advances in solar energy production, it remains a supplementary supply when compared to energy produced by fossil fuels.

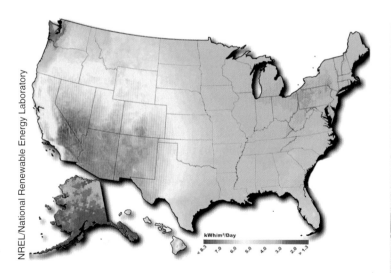

kWh/m²/Day

< 8.3 7.0 6.0 5.0 4.0 3.0 2.0 > 1.3

Figure 4.29 Map of solar energy potential in the United States. The Northeast is a poor place for solar energy production because it is often cloudy, especially in winter. In contrast, the Southwest has a very high potential for producing solar energy because it is a sunny desert environment. Map units are in number of kilowatt hours of production per square meter per day.

Figure 4.30 The Solar Energy Generating System in California's Mojave Desert. This photograph shows a portion of the complex. Over 936,000 trough energy collectors are contained within the entire system, producing enough electricity for 500,000 people per year.

Net radiation refers to the total amount of solar radiation received by Earth at any given place and time. This image shows the geography of mean net radiation during a particular time of year, with yellows and reddish hues representing high values in watts per square meter. Darker colors, in contrast, represent low values of net radiation. By observing this diagram, can you determine what time of year it must be? What must the Earth–Sun geometric relationship be like during this time of year?

Suomi Virtual Museum

KEY CONCEPTS TO REMEMBER ABOUT THE GLOBAL RADIATION BUDGET

1. The global radiation budget refers to the balance between incoming and outgoing radiation on Earth.

2. Over the long term, the global radiation budget is balanced. However, even though the long-term radiation budget is balanced, a great deal of variability occurs across Earth as a whole.

3. In general, low latitudes have a net surplus of radiation, whereas high latitudes have a net deficit. This imbalance is important because it drives atmospheric circulatory processes.

4. Net radiation depends on a complex interaction of several variables, including angle of incidence, latitude, season, day length, and albedo.

5. In an effort to supplement current energy supplies, efforts are under way to increase the production of renewable solar energy. The major limitations are that production of solar energy with current technology is inefficient and the Sun does not shine consistently in many places.

SUMMARY OF KEY CONCEPTS

1. The atmosphere serves as a protective shield that filters the potentially deadly effects of ultraviolet radiation from the Sun, allowing mostly visible and infrared wavelengths to reach Earth.

2. The lower part of the atmosphere is warmed primarily due to the greenhouse effect. This warming occurs because variable gases in the atmosphere, such as carbon dioxide and water vapor (as well as others), trap longwave radiation that is emitted by Earth.

3. Insolation refers to the amount of solar radiation received at the top of the atmosphere. From this point, radiation follows several paths. Approximately 50% of all radiation reaches Earth, with some arriving directly and some indirectly. The remaining radiation is either absorbed in the atmosphere, scattered by dust, or reflected directly back into space.

4. Albedo refers to the reflectivity of surfaces on Earth. Snowy surfaces have the highest albedo, whereas darker surfaces such as roads and oceans have the lowest albedo. Overall, the Earth reflects about 31% of all incoming solar radiation through albedo.

5. The global radiation budget refers to the overall balance of incoming versus outgoing radiation. Earth receives shortwave radiation from the Sun. This radiation is either absorbed by the atmosphere and the various Earth surfaces, or reflected and scattered back to the surface or back into space. The radiation that reaches Earth's surface is absorbed and then re-radiated as longwave radiation. The overall radiation budget must be in balance; otherwise, Earth would become progressively hotter or colder.

6. Efforts are currently under way to more systematically develop solar energy to supplement energy supplies. This supply of energy is renewable and environmentally friendly because its production does not result in CO_2 emissions. Unfortunately, current technology is inefficient, and the production of solar energy is hampered in many places because the Sun shines inconsistently.

CHECK YOUR UNDERSTANDING

1. Discuss the various components of the electromagnetic spectrum.

2. Are wavelengths in the visible part of the electromagnetic spectrum short or long?

3. Why does the Sun emit shortwave radiation, whereas Earth emits longwave radiation?

4. What are the constant gases in the atmosphere, and why are they important?

5. List the variable gases and describe why they change with respect to their proportion of the atmosphere.

6. Why is the sky blue during the day, but red and orange at dawn and dusk?

7. Compare and contrast the methods of heat transfer.

8. Describe some of the variables that influence the amount of radiation that is absorbed on Earth.

9. How does the angle of incidence vary by latitude?

10. If the axis of the Earth were perpendicular to the plane of the ecliptic, how would net radiation change at high latitudes? Would these regions have a greater net surplus or net deficit of radiation?

11. Why does the zone of net radiation surplus migrate (on Earth) over the course of the year?

12. Describe the global radiation budget. What are some of the things that happen to solar radiation when it reaches Earth? Why is there a global balance in the radiation budget?

13. Describe the pros and cons of solar energy production. Why is it considered to be renewable and environmentally friendly? What are the major limitations to the production of solar energy?

ANSWERS TO VISUAL CONCEPT CHECKS

VISUAL CONCEPT CHECK 4.1

Solar energy is produced by the process of nuclear fusion within the Sun. In this fashion, hydrogen is converted to helium at very high temperatures and pressures. Solar radiation travels in the form of electromagnetic energy at 300,000 km (186,000 mi) per second. It reaches Earth in about 8 min.

VISUAL CONCEPT CHECK 4.2

This factory is producing a variety of particulates and variable gases that are streaming into the atmosphere. Some of these industrial by-products will cause solar radiation to be reflected, whereas others will cause absorption or scattering of electromagnetic radiation.

VISUAL CONCEPT CHECK 4.3

This map of net radiation reflects patterns observed at either equinox. Note that the highest values of net radiation are received in the tropical regions, which reflect the location of the subsolar point and a maximum (90°) Sun angle. Such relationships can only occur along the Equator during the Spring and Fall Equinoxes.

CHAPTER FIVE | GLOBAL TEMPERATURE PATTERNS

In the previous chapter we examined the various ways that energy flows in the atmosphere. Now it is time to investigate its primary effects on atmospheric temperature patterns. The word *atmosphere* is derived from the Greek phrase meaning "sphere of air," which implies that the air is uniform in its consistency and character. Actually, the atmosphere consists of several layers, each having different characteristics and interactions with solar radiation. Each of these layers will be examined in this chapter, including how the temperature varies between and within them. Most of the discussion will focus on the environmental variables that influence temperature in the lower part of the atmosphere and how these variables interact to form distinct geographic temperature patterns on Earth.

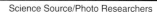

Science Source/Photo Researchers

A profile of the atmosphere can be seen in this sunset image taken from the space shuttle. The deep red band near the horizon is probably caused by smoke from biomass burning at the surface. The purple layer marks the tropopause, which contains a layer of dust at an altitude of 14 km (~8.7 mi). Above the tropopause is the stratosphere, which appears pale in this image.

CHAPTER PREVIEW

Layered Structure of the Atmosphere

Surface and Air Temperatures

 Surface Temperature

 Maritime vs. Continental Effect

 Global Temperature Patterns

Human Interactions: Urban Heat Islands

 Temperature and Location

LEARNING OBJECTIVES

1. Compare and contrast the various layers of the atmosphere and the vertical change in temperature that occurs within them.

2. Describe the way that the various large-scale factors influence surface air temperature on Earth.

3. Explain the character of maritime and continental climates and why they differ from each other.

4. Discuss the urban heat island effect and how human interactions in these environments influence air temperature.

Layered Structure of the Atmosphere

The place to begin a discussion of global temperature patterns is in the atmosphere, which is the medium through which solar radiation flows. The atmosphere technically extends from a very shallow depth within Earth (because soil contains air and water) to a height of about 480 km (300 mi). Most of the atmosphere's mass, however, lies below an altitude of 30 km (18.6 mi). Although we usually think of the atmosphere as a uniform gaseous medium that envelops Earth, it actually contains four distinct layers: the troposphere, stratosphere, mesosphere, and thermosphere. These layers are distinguished by temperature and also by what elements they contain. Figure 5.1 illustrates these major layers and the temperature trends within them.

The Troposphere

The **troposphere** is the lowest layer of the atmosphere, extending from the surface to an average altitude of about 12 km (~7.5 mi). The name of this layer is derived from the Greek word *tropos* for "turning" or "mixing" because it is the most active zone of the atmosphere, with vigorously moving currents of air. Physical geographers and meteorologists are most interested in this atmospheric layer because it contains the vast majority of nonmarine living organisms and is the zone where most weather occurs, such as rain, wind, and snow. Much of this weather occurs because the troposphere contains most of the atmospheric water vapor and particulates.

The troposphere is warmed by longwave radiation emitted from Earth as part of the global energy balance discussed in Chapter 4. Because the source of heat in the troposphere is Earth, temperature decreases with increasing altitude (Figure 5.2) at the **environmental lapse rate**. As altitude (from the Earth) increases in the troposphere, the average temperature drops 6.4°C per

Troposphere *The lowermost layer of the atmosphere, which lies between the Earth's surface and an altitude of about 12 km (~7.5 mi).*

Environmental lapse rate *The decrease in temperature that generally occurs with respect to altitude in the troposphere. This rate is 6.4°C per kilometer or 3.5°F per 1000 ft (a positive lapse rate).*

Figure 5.1 Layers and temperature patterns in the atmosphere. (a) The four major layers in the atmosphere. Note the features such as clouds and auroras that occur within specific layers. (b) Temperature changes with respect to altitude in each of the four major layers of the atmosphere.

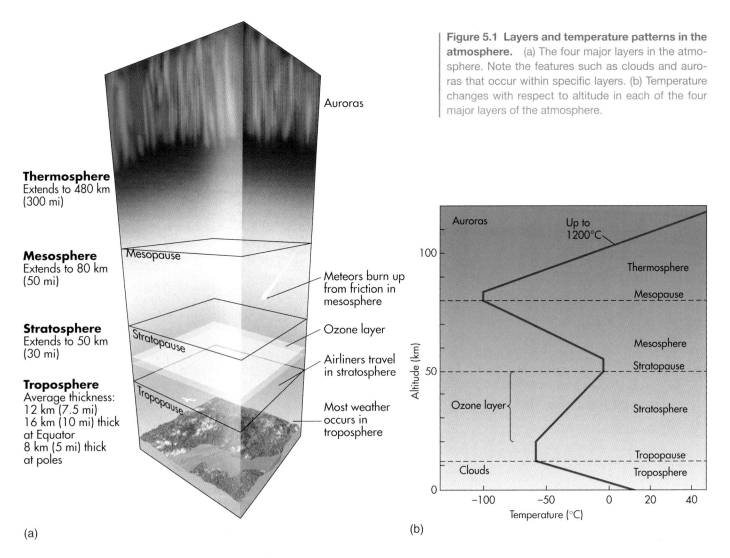

(a)

Thermosphere
Extends to 480 km (300 mi)

Mesosphere
Extends to 80 km (50 mi)

Stratosphere
Extends to 50 km (30 mi)

Troposphere
Average thickness:
12 km (7.5 mi)
16 km (10 mi) thick at Equator
8 km (5 mi) thick at poles

Auroras

Mesopause

Meteors burn up from friction in mesosphere

Stratopause

Ozone layer

Tropopause

Airliners travel in stratosphere

Most weather occurs in troposphere

(b)

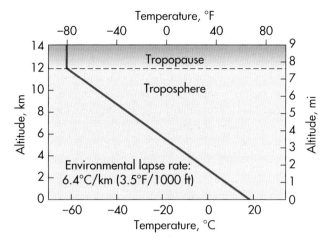

Figure 5.2 Environmental lapse rate in the troposphere. The average temperature decreases with increasing altitude in the troposphere until it reaches −57°C (−70°F).

1000 m or 3.5°F per 1000 ft (a positive lapse rate). At the upper limit of the troposphere, which is called the **tropopause**, the temperature stops decreasing and begins to increase into the next layer of the atmosphere. The tropopause is found at the altitude where the air temperature is −57°C (−70°F), which again varies depending on latitude.

Tropopause *The top part of the troposphere, which is identified by where the air temperature is −57°C (−70°F).*

Temperature inversion *A layer of the troposphere in which the air temperature increases, rather than cools, with altitude.*

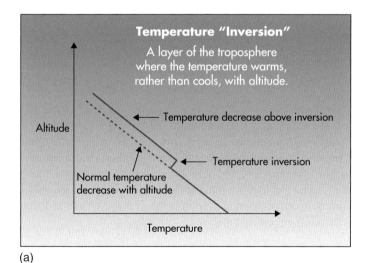

(a)

Figure 5.3 Temperature inversions. (a) A temperature inversion develops when warm air overlies cold air at some altitude within the troposphere. (b) Airflow in the troposphere under normal conditions (top) and when an inversion exists (bottom). Under normal conditions, warm air at the surface easily flows vertically. When a temperature inversion exists, however, the warm layer serves as a cap that keeps air from moving upward (c) Smog in Mexico City. Smoggy conditions can develop in large cities when a temperature inversion bottles up polluted air at low elevations.

The thickness of the troposphere depends largely on temperatures at the Earth's surface; therefore, it varies with season and latitude, ranging in thickness from about 16 km (10 mi) at the Equator to approximately 8 km (5 mi) at the poles. Why is this so? Recall that the low latitudes have a net surplus of radiation because these regions receive more intense solar radiation. Thus, near the Equator, more radiation is transferred back to the atmosphere as longwave radiation, which heats the lower atmosphere. The higher temperature causes the atmospheric gases to expand, making the troposphere extend higher from the Earth's surface at the Equator than at the poles.

Although temperature usually decreases in a fairly predictable way with altitude in the troposphere, conditions sometimes develop when warm air overlies cold air. Such a pattern is called a **temperature inversion** (Figure 5.3a) and can develop

(b)

(c)

at any altitude within the troposphere, from near ground level to thousands of meters up. Temperature inversions are usually associated with local weather phenomena, such as when the ground cools quickly on a clear night or along coastlines when cold ocean water cools the air close to the ground. Inversions are important meteorologically because they act to cap air from freely moving upward (Figure 5.3b, bottom). These conditions can result in the development of fog, as will be discussed in Chapter 7. They also cause smog in many of the world's largest cities, such as Mexico City (Figure 5.3c), Los Angeles, and London, because urban pollutants are trapped under the inversion layer. If the inversion layer persists, smog can build up to become a serious health hazard.

The Stratosphere

The layer of the atmosphere that lies immediately above the troposphere is the **stratosphere** (see Figure 5.1), which ranges in altitude between about 12 km and 50 km (between ~7.5 mi and 30 mi). This portion of the atmosphere derives its name from the Greek word *stratos*, which means "layer." The stratosphere is critically important to life on Earth because it contains the ozone layer, which occurs at an altitude between 20 km and 50 km (12 mi and 31 mi). The concentration of ozone in this portion of the atmosphere is about 10 parts per million by volume (ppmv), whereas it is only 0.04 ppmv in the troposphere.

As discussed in Chapter 4, the ozone layer filters UV radiation from the Sun and re-radiates it as infrared energy. Stratospheric temperature trends reflect this filtering and the overall thickness of the ozone layer. From the top of the tropopause to the base of the ozone layer, temperatures are consistently about −57°C (−70°F) (see Figure 5.1b). Above that altitude, however, temperatures increase with altitude because of the way ozone absorbs UV energy from the Sun. At the lower altitudes in the stratosphere, ozone absorbs UV radiation at wavelengths between 44 nm and 80 nm. It absorbs these wavelengths inefficiently, however. At higher altitudes in the stratosphere, ozone very efficiently absorbs UV at wavelengths between 200 nm and 350 nm. This difference in absorption efficiency explains why temperature increases with altitude in the stratosphere. The top of the stratosphere, or **stratopause**, is marked by the altitude where temperature stops increasing. At this altitude, the average temperature is about −5°C (−23°F).

In addition to its filtering effects, the stratosphere is also important for humans because it is the portion of the atmosphere through which commercial jets fly. The stratosphere is a perfect medium for flight because it contains very little water vapor and few impurities; thus, pilots find few clouds and good visibility. In addition, the air is relatively calm compared to the turbulent troposphere below because air in the stratosphere usually flows parallel to the surface of Earth.

The Mesosphere

The **mesosphere** is a layer of decreasing temperature that occurs from about 50 km to 80 km (30 mi to 50 mi) in altitude (see Figure 5.1); it is the coldest of the atmospheric layers. Its name comes from the Greek word *mesos*, which means "middle." Vertical temperature trends in the mesosphere have a positive lapse rate because temperature decreases with increasing distance from the ozone layer, which is located in the stratosphere below. The altitude at which temperature stops decreasing is known as the **mesopause** and is the upper boundary of the mesosphere. At this altitude, air temperature is about −100°C (−148°F). In the mesosphere, solar radiation reduces gas molecules to individual electrically charged particles called ions. This deep layer of charged particles can disrupt communications between astronauts and ground control and interfere with various satellite communications, such as transmission of television signals.

Another interesting fact about the mesosphere is that it is the layer where most meteors burn up as they fall through the atmosphere. Did you know that most of these *shooting stars* are about the size of one sand grain? They are destroyed because they collide with billions of ions and gas particles as they zip through the mesosphere. These collisions create sufficient heat to burn the tiny rock fragments, creating a short-lived streaking path, long before they strike the ground. Occasionally, the largest rock fragments reach the Earth's surface. If they do, they are called *meteorites*.

The Thermosphere

The **thermosphere** is the upper layer of the atmosphere and occurs between about 80 km and 480 km (50 mi to 300 mi) in altitude. Atmospheric gases in this portion of the atmosphere are sorted into a variety of sublayers based on their molecular mass. Oxygen molecules are few and literally many kilometers (miles) apart from one another. In fact, they are so widely spaced that the boundary with space is very diffuse and thus difficult to precisely determine.

Stratosphere *The layer of the atmosphere, between the troposphere and mesosphere, that ranges between about 12 km and 50 km (between ~7.5 mi and 31 mi) in altitude.*

Stratopause *The upper boundary of the stratosphere where temperature reaches its highest point.*

Mesosphere *A layer of decreasing temperature in the atmosphere that occurs from about 50 km to 80 km (~30 mi to 50 mi) in altitude.*

Mesopause *The upper boundary of the mesosphere where temperature reaches its lowest point.*

Thermosphere *The upper layer of the atmosphere, which occurs between about 80 km and 480 km (~50 mi and 300 mi) in altitude.*

1. An airliner in the midlatitudes takes off and climbs through the troposphere to its cruising altitude. As it rises through the troposphere, does the outside temperature (a) increase, (b) decrease, or (c) remain the same?

2. The pilot informs you that the plane's cruising altitude is 36,000 ft. Is this in the (a) troposphere, (b) stratosphere, or (c) mesosphere?

3. The pilot decides to avoid some turbulence by having the plane climb to 40,000 ft. At this new altitude, is the outside temperature (a) higher, (b) lower, or (c) the same?

George Hall/Corbis Images

The name of this atmospheric layer is derived from the Greek *thermo*, which implies "heat"; this term makes perfect sense if you look at the temperature patterns of the thermosphere in Figure 5.1b. Notice that they increase drastically about 10 km (6 mi) above the mesopause, ultimately reaching 1200°C (2200°F) and higher. These high temperatures occur because intense solar radiation interacts with the upper part of the atmosphere, causing the few oxygen molecules present to vibrate at tremendously high speeds, which creates **kinetic energy**. This form of energy is specifically contained within the molecules by virtue of their spatial relationship to other oxygen molecules. The thermosphere is important for human communications because it enables radiowaves from one location at the surface to bounce off and be received at locations beyond the horizon.

Despite the very high temperatures of the thermosphere, this atmospheric layer does not feel "hot" in the same way you feel heat on the Earth's surface. The reason for this oddity is that individual oxygen molecules in the thermosphere are so far apart from one another. Because of these great distances, the molecules hardly ever come into contact with each other, which means that very little heat is transferred from one to another. At lower levels of the atmosphere, in contrast, temperatures are genuinely higher because trillions of molecules are constantly colliding and a great deal of heat is exchanged.

KEY CONCEPTS TO REMEMBER ABOUT THE STRUCTURE OF THE ATMOSPHERE

1. The atmosphere consists of four major layers, each with distinct temperature characteristics.

Kinetic energy *The energy of motion in a body, measured as temperature, that is derived from movement of molecules within the body.*

2. The troposphere lies closest to the Earth's surface and typically has a positive lapse rate of 6.4°C per 1000 m or 3.5°F per 1000 ft.

3. The next lowest layer is the stratosphere, which contains the ozone layer and thus warms with increasing altitude.

4. Immediately above the stratosphere is the mesosphere, which cools with increasing altitude and is the part of the atmosphere where solar radiation reduces individual molecules to ions.

5. The uppermost layer of the atmosphere is the thermosphere. Although the temperature in this part of the atmosphere is very high, it does not feel hot because the individual air molecules are very far apart.

Surface and Air Temperatures

Most of the atmospheric behavior of interest to geographers occurs near the Earth's surface. Of particular significance are global surface and air temperatures, which measure the amount of sensible heat at the surface or in the atmosphere. Atmospheric temperature is a measure of the kinetic energy contained within a unit of geographical space within the air. Surface temperature, in contrast, is a measure of the kinetic energy contained in a region very close to the Earth's surface. Thus, changes in surface temperature measure the ebb and flow of energy at ground level. As noted in Chapter 4, this variation in energy on Earth depends largely on net radiation. When a net surplus of radiation occurs, temperature increases because the surface absorbs more radiant energy than it is emitting in longwave form. Conversely, temperature decreases with a net deficit of radiation because the surface emits more energy than it absorbs. Recall from Chapter 4 that surface temperature can also change through the processes of conduction, evaporation, and convection.

In contrast to surface temperature, the term *air temperature* refers to the degree of warmth of a portion of the atmosphere. The standard altitude at which air temperature is measured is usually about 1.2 m (4 ft) above the ground surface. Although air temperature usually differs from surface temperature, the amount of heat energy in the ground influences the temperature of the air above it. As you know, air temperature is measured with a thermometer, which is traditionally a hollow glass tube containing a liquid, often mercury, that expands or contracts depending on the amount of energy present. More recently, digital thermometers have become common, relying on small electrical devices that sense temperature.

Three temperature scales are used around the world (Figure 5.4). You are probably most familiar with the Fahrenheit scale, which was named after the German physicist Gabriel Daniel Fahrenheit, who devised the scale in the 18th century. This scale is used by the United States National Weather Service and American news media to report temperature. The common reference temperatures used in the Fahrenheit scale are the freezing and boiling points of water, defined as 32°F and 212°F, respectively.

Another temperature scale that you may be somewhat familiar with is the Celsius scale, named after Anders Celsius, the 18th-century Swedish astronomer who devised it. This scale is part of the International System of Measurement (SI) because it is a decimal scale, with 0°C and 100°C being the freezing and boiling points of water, respectively. The vast majority of the world uses the Celsius scale, including scientists in the United States, which is why we present Celsius measurements in this text first, rather than Fahrenheit. Because most of the world uses the Celsius scale, the stated goal of the U.S. government is to one day convert fully to this form of measurement.

The conversions required to convert from Fahrenheit to Celsius and from Celsius to Fahrenheit are, respectively,

$$F = 9/5C + 32°$$
$$C = 5/9 \ (F + 32°)$$

Thus, 1°C equals 1.8°F and 1°F equals 0.56°C.

The third temperature scale is the Kelvin scale, named for the 19th-century British physicist William Thomson, Lord Kelvin. This scale is used in a great deal of scientific research because it measures absolute temperature, with absolute zero theoretically being the point where an object has no measurable temperature. On the Celsius scale, absolute zero is −273°C. The Kelvin scale has no negative values as the other scales do, but it is similar to the Celsius scale because it maintains a 100° temperature range between the boiling and freezing points of water. In other words, 1 K = 1°C. Given that the Kelvin scale is used by climatologists and meteorologists only for more advanced work, it will not be used further in this book. For reference purposes, however, the conversions between Kelvin and Celsius scales are

$$C° = K + 273$$
$$K = C° + 273$$

Human Interactions: Calculating the Heat Index and Wind Chill

Have you ever noticed that the air occasionally seems warmer or colder than the temperature that is given? This kind of variation occurs when additional environmental factors, such as wind speed and the amount of atmospheric water vapor, come into play. The combined impact of these factors with air temperature has implications for human comfort because they can make the air feel much warmer or colder than it really is.

The measures of human comfort that we use are the *wind chill index* in winter and the *heat index* in summer. The wind chill index is calculated using a variety of parameters, such as actual air temperature, wind speed, average face height, and components of modern heat loss theory. The wind chill index chart in Table 5.1 presents temperature data in the Fahrenheit scale and wind speed in miles per hour. In contrast to wind chill, the heat index measures the apparent temperature based on the combined variables of actual air temperature and relative humidity. (As we will discuss in Chapter 7, relative humidity is the ratio of the specific amount of vapor relative to the amount the air could hold at a given temperature.) The resulting heat index chart, shown in Table 5.2, also uses the Fahrenheit temperature scale.

Large-Scale Geographic Factors That Influence Air Temperature

A common theme in this text is the holistic interaction of geographical variables. These interrelationships are especially important for understanding how temperature varies across Earth, which incorporates some of the concepts covered in Chapters 3 and 4. Some of these factors may be intuitive for you at this stage, but it is nevertheless useful to review them and their relationships. Later in this section we will look at some factors that can influence temperature at a local scale. For now, however, let's focus on three large-scale factors that influence temperature across Earth, no matter the location.

Latitude   Recall from Chapter 4 that the shape of Earth causes rays hitting the surface to differ in the area they cover, according to the latitude at which they strike. In other words, differences in the angle of incidence cause the same amount of energy to be directed at a smaller or larger area on the Earth's surface. When a larger area is covered at a lower incidence angle, as occurs at high latitudes, it results in less energy per unit area on that surface compared to a smaller area struck at a higher incidence angle, similar to what occurs at low latitudes. This difference is most pronounced when comparing high and low latitudes and results in distinct temperature differences between two such regions (see Figure 4.24).

Seasons and Length of Day   As discussed in earlier chapters, Earth's axial tilt causes seasonal migration of the subsolar point because the Northern and Southern Hemispheres are tilted either toward or away from the Sun, depending on the time of

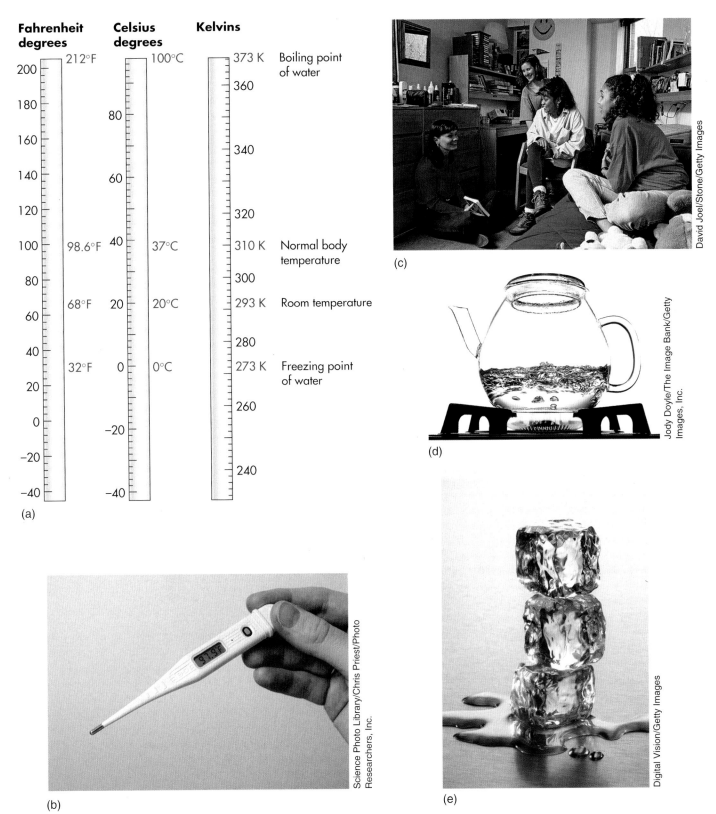

Figure 5.4 Measures of temperature. (a) The Fahrenheit, Celsius, and Kelvin temperature scales. The size of one degree on the Celsius scale is the same as one kelvin; a degree on the Fahrenheit scale is smaller. (Note that units on the Kelvin scale are not called degrees, but rather kelvins.) (b) Temperature is measured with a thermometer. Until recently, most thermometers were filled with alcohol or mercury, which rose or fell within a glass tube if warmed or cooled, respectively. Today, most thermometers contain temperature-sensitive electrodes and display information digitally. (c) For most people a comfortable room temperature is about 21°C to 23°C (69°F to 73°F; 294 K to 296 K). (d) Water boils at a temperature of 100°C (212°F; 373 K). (e) The freezing/melting point of water is 0°C (32°F; 273 K).

TABLE 5.1 The Wind Chill Index

Temperature (°F)

W i n d s p e e d (mph)		40	35	30	25	20	15	10	5	0	−5	−10	−15	−20	−25
	5	36	31	25	19	13	7	1	−5	−11	−16	−22	−28	−34	−40
	10	34	27	21	15	9	3	−4	−10	−16	−22	−28	−35	−41	−47
	15	32	25	19	13	6	0	−7	−13	−19	−26	−32	−39	−45	−53
	20	30	24	17	11	4	−2	−9	−15	−22	−29	−35	−42	−48	−55
	25	29	23	16	9	3	−4	−11	−17	−24	−31	−37	−44	−51	−58
	30	28	22	15	8	1	−5	−12	−19	−26	−33	−39	−46	−53	−60
	35	28	21	14	7	0	−7	−14	−21	−27	−34	−41	−48	−55	−62
	40	27	20	13	6	−1	−8	−15	−22	−29	−36	−43	−50	−57	−64
	45	26	19	12	5	−2	−9	−16	−23	−30	−37	−44	−51	−59	−65
	50	26	19	12	4	−3	−10	−17	−24	−31	−38	−45	−52	−60	−67
	55	25	18	11	4	−3	−11	−18	−25	−32	−39	−46	−54	−61	−68

Frostbite occurs in 15 minutes or less.

Source: NOAA.

TABLE 5.2 The Heat Index

Relative Humidity (%)

Air Temperature (°F)		20	25	30	35	40	45	50	55	60	65	70	75	80	85	90	95	100
	125	141																
	120	130	139	148														
	115	120	127	135	143	151												
	110	112	117	123	130	137	143	150										
	105	105	109	113	118	123	129	135	142	149								
	100	99	101	104	107	110	115	120	126	132	138	144						
	95	93	94	96	98	101	104	107	110	114	119	124	130	136				
	90	87	88	90	91	93	95	96	98	100	102	106	109	113	117	122		
	85	82	83	84	85	86	87	88	89	90	91	93	95	97	99	102	105	106
	80	77	77	78	79	79	80	81	81	82	83	85	86	86	87	88	89	91
	75	72	72	73	73	74	74	75	75	76	76	77	77	78	78	79	79	80
	70	66	66	67	67	68	68	69	69	70	70	70	70	71	71	71	71	72

TABLE 5.2 The Heat Index (*continued*)

Celsius	Fahrenheit	Impact on Humans
27°C–32°C	80°F–90°F	Caution: Fatigue is possible with continued exposure.
32°C–41°C	90°F–105°F	Extreme Caution: Possible sunstroke, heat cramps, and exhaustion.
41°C–54°C	105°F–125°F	Danger: Likely occurrence of sunstroke and heat cramps.
54°C	above 130°F	Extreme Danger: Heat stroke or sunstroke likely.

Source: NOAA.

year. This migration of the subsolar point occurs only in the tropics, but results in dramatic changes in the angle of incidence at all latitudes, which, in turn, influences net radiation. If you need to review this concept again, examine Figures 4.24 and 4.26. A secondary impact of axial tilt and seasons is that length of day can fluctuate a great deal, depending on latitude. Longer days mean that more radiation is received, which, in turn, also influences temperature. Think of your own experience regarding the temperatures where you live during winter and summer.

Time of Day As the day progresses from the morning sunrise, the Sun arcs westward across the sky. Recall from Chapter 3 that the Sun is at its highest point, resulting in the day's most intense radiation, at solar noon. Temperature usually follows the Sun in that it increases as the Sun rises and decreases when it sets. The relationship is not quite that simple, however, because a **temporal lag** occurs between the highest Sun angle and the warmest temperature of the day. In other words, the warmest part of the day occurs later in the day than the peak period of insolation and net radiation.

A good way to see how these geographical variables affect one another is in graphical form. For example, Figure 5.5 shows how season and time of day are interrelated with respect to temperature. These diagrams show three sets of data from a typical observing station at 40° to 45° N latitude in the interior of North America: (a) insolation, (b) net radiation, and (c) air temperature. Parameter values, such as the amount of net radiation, are presented on the vertical (or "y") axis of the respective diagrams, whereas time of day ranges across the horizontal (or "x") axis.

Notice in Figure 5.5 that each of the measured variables at the observation station varies both by season and time of day. With respect to insolation (Figure 5.5a), the amount of insolation is greater on the June solstice than on the December solstice. This seasonal variation in insolation results in a greater

---

Temporal lag *The difference in time between two events, such as when peak insolation and peak temperature occur.*

surplus (above the bold horizontal line) of net radiation during the summer months than in the winter (Figure 5.5b). Notice also that the number of hours of daily insolation and net radiation also vary per day. During the summer, for example, insolation is received for 16 h of the day, whereas during the winter it is only 8 h. This decrease occurs because Earth's orbital position has changed. As a result of this seasonal insolation variation, the time of a net deficit of radiation (below the bold horizontal line) increases from about 8 h in summer to 16 h in winter.

To see how annual insolation/net radiation patterns influence temperature, study Figure 5.5c. As you can imagine, temperature is related to insolation and net radiation in that maxima and minima of each factor occur at about the same time. The pattern is predictable from a seasonal perspective in that temperatures are warmer during the summer and colder during the winter. Examining Figures 5.5a and 5.5b shows a clear correlation among temperature, insolation, and net radiation. Notice, however, that the maximum and minimum temperatures occur at different times than insolation/net radiation fluctuations on a daily basis. In other words, a temporal lag occurs. With respect to maximum temperature, this lag exists because the peak release of longwave energy by Earth occurs sometime after the insolation peak. Similarly, the minimum temperature occurs only after all the stored heat from the day is released.

Local Factors That Influence Air Temperature

In the previous discussion we examined the important factors that influence air temperature on a large scale across Earth. Besides these factors, additional, smaller-scale factors can influence air temperature in specific places. This section of the chapter focuses on some of these kinds of geographic patterns.

Maritime vs. Continental Locations Although the radiation budget is the largest influence on air and surface temperatures, other factors yet to be discussed can play an important role. A good example of a local or regional variable

(a)

(b)

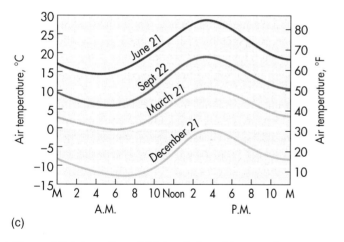

(c)

Figure 5.5 Interaction of (a) insolation, (b) net radiation, and (c) air temperature. These graphs show data taken at a midlatitude location in the Northern Hemisphere. Notice the close relationship among these three variables. Also note that a distinct lag occurs with respect to the temperature response.

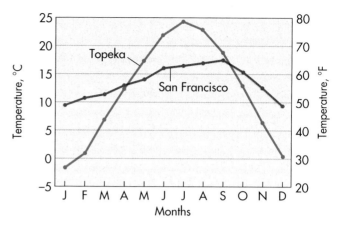

Figure 5.6 Annual temperature data from Topeka, Kansas, and San Francisco, California. A distinct peak in temperature occurs in Topeka during the summer, whereas San Francisco is comparatively moderate all year round.

that influences temperature is the **maritime vs. continental effect**. By definition, **maritime** places are located within or near a very large body of water, such as San Francisco, California, which is on the coast of the Pacific Ocean. In contrast, **continental** localities are surrounded by large landmasses, such as Topeka, Kansas, which is on the eastern edge of the Great Plains.

If you compare Topeka and San Francisco, observe that the annual temperature curves vary significantly (Figure 5.6). Notice that the San Francisco curve does not have a steep peak, but rather changes gradually over the course of the year. In addition, the range between the high and low monthly temperatures is narrow. At Topeka, in contrast, the curve has a much higher peak with a more distinct change between seasons and a greater overall range. This variability exists even though both places are located at about the same latitude (38° N), which would (correctly) lead you to believe that each locality receives about the same amount of insolation.

Maritime vs. continental effect *The difference in annual and daily temperature that exists between coastal locations and those that are surrounded by large bodies of land.*

Maritime *A place that is close to a large body of water that moderates temperature.*

Continental *A place that is surrounded by a large body of land and that experiences a large annual range of temperature.*

WHAT IS THE LAND SKIN TEMPERATURE?

Have you ever heard the old saying that *it's 100 degrees in the shade*? This phrase is intended to reflect the idea that it is really hot outside, so hot that it is even uncomfortable underneath big trees that block the Sun. The old saying makes sense because everyone understands on some level that it is usually hotter in the direct Sun than it is elsewhere. This relationship exists because your skin is heated by both the radiant heat of the Sun and the warm air surrounding you. For similar reasons, official air temperatures are always taken in indirect light because structures can absorb radiant energy, which can affect readings.

Scientists have applied the understanding of temperature variability between sunny and shady areas to a concept called land skin temperature, which reflects the pure heating of the ground by solar radiation, the atmosphere, and other heat flows. The concept of land skin temperature is based on the fact that the many different kinds of surfaces within an area, such as those that are vegetated or consist of exposed rock or sand, have different temperatures based

on factors such as albedo, dryness of the soil, character of sediment, and degree of shade. NASA scientists measure the land skin temperature with the Moderate Resolution Imaging Spectroradiometer (MODIS) satellite. As you can see in the accompanying satellite image of the Taklamakan Desert in China, the surface skin temperature can vary considerably over a small geographical area, ranging from 30°C (86°F) in the crop areas to over 50°C (122°F) in the nearby dune field. This variation occurs because the dark surface of the dune field absorbs more radiation than elsewhere. Why does skin temperature matter to you? Remember that if it really is 100 degrees in the shade where you live on a summer day, it must be *really* hot in the direct Sun.

MODIS image of the Taklamakan Desert in China. Compare the natural color image at the top with the skin temperature image at the bottom. Can you see the geographical variability in temperature and why it occurs?

Surface temperatures can vary dramatically even in conventional suburban landscapes due to degree of direct sunlight, shade, and character of the surface.

Here is a chance to examine the factors that affect temperature more closely. To do so, view the animation *Surface Temperature*. This animation allows you to observe how the variables insolation, net radiation, and time of day influence temperature. The focus of this feature is a hypothetical location at 45° N latitude, one that lies deep within a continent. As you watch this animation, observe how air temperature near the Earth's surface changes over the course of the day on June 21, which is the Summer Solstice in the Northern Hemisphere. In particular, notice the lag that exists with respect to increases and decreases in insolation and temperature over the course of the day. After you complete the animation, be sure to answer the questions at the end to test your understanding of this concept.

Why does this temperature variation occur? The main reason is the maritime vs. continental effect (Figure 5.7). Large bodies of water, such as the Pacific Ocean, store tremendous amounts of thermal energy, primarily because water has a high specific heat; that is, water absorbs a relatively large amount of energy before its temperature rises. In addition, solar radiation can penetrate to great depths in the ocean, heating water below the surface. The ocean currents constantly mix this warm water with the cooler water not exposed to sunlight. As a result, large water bodies maintain a more or less constant temperature for most of the year (Figure 5.8). This effect can be enhanced if the ocean current that flows along the coast is a cold or warm current. Such currents will be described further in Chapter 6.

As water evaporates from the ocean, energy is transferred to the atmosphere in the form of latent heat that, in turn, moderates the air temperature of coastal places such as San Francisco. This moderating effect is magnified on the west coast of North America because the prevailing winds are westerly; that is, they blow from west to east. This circulatory pattern will be discussed in more detail in Chapter 6, but for now it is sufficient to know that these westerly winds blow relatively mild, moist air off the ocean into the continental interior. In this way, San Francisco maintains a moderate air temperature all year round.

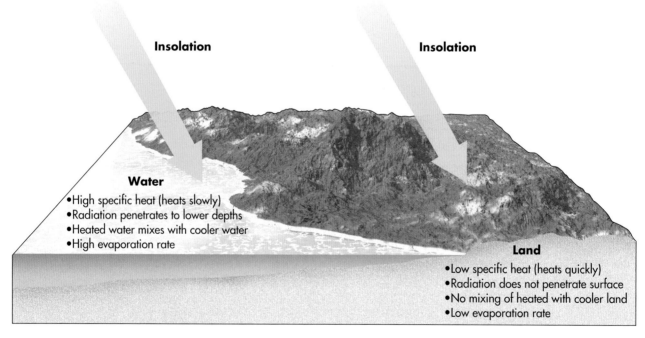

Figure 5.7 Maritime vs. continental contrasts. Several factors explain why temperature differences occur between coastal places and locations deep within continents. Note the different ways in which insolation interacts with land and water.

Figure 5.8 Global sea-surface temperatures during June and December 2000. Although the seasons differ dramatically between the images, notice the similarity in the geographic pattern. For example, the sea-surface temperature in the eastern Pacific shows a small change between the months. Temperatures are in the Celsius scale and are coded by color, as indicated by the horizontal bar between the maps.

PODACC

degC

0 5 10 15 20 25 30

(a)

PODACC

(b)

At continental localities like Topeka, however, the surrounding landmass does not store as much thermal energy as oceans do, largely because of their low specific heat; that is, the land temperature starts to rise after absorbing relatively small amounts of energy. In addition, radiation does not penetrate to great depths in landmasses. As a result, continental localities exhibit dramatic annual temperature variations compared to maritime places. This pattern not only exists over the course of the year, but can also be seen in the daily cycle of temperature (Figure 5.9).

Other Local Factors That Influence Temperature In addition to the maritime vs. continental effect and urban impact on temperature, a variety of other variables can influence temperature. These variables include altitude, the position of topographic barriers, and wind flow patterns. Altitude influences temperature because, as we saw in Figure 5.1b, temperature tends to decrease with increased elevation. As you will see in Chapter 7, topographic barriers such as mountain ranges can have a significant impact on temperature that goes beyond the simple variable of increased altitude. In areas where wind descends a mountain range, temperature can warm considerably due to increased molecular friction as air is compacted.

Figure 5.9 Daily cycle of temperature at San Diego, California, and Las Vegas, Nevada, from July 1 to July 5, 2003. San Diego is a maritime city, whereas Las Vegas is located deep within the desert. Notice how pronounced the daily cycle of temperature is in Las Vegas compared to San Diego.

Surface and Air Temperatures • 97

Maritime vs. Continental Effect

To see how the martime vs. continental effect occurs, view the animation *Maritime vs. Continental Effect*. In this animation, you can observe how temperature varies over the course of the day at Yuma, Arizona, and San Diego, California. Although San Diego and Yuma are both located at about the same latitude (~33° N), and thus receive about the same amount of daily insolation, the range in daily temperature differs dramatically between the two places. Watch how air temperature varies between these two places over the course of the day and year and relate what you see to the previous discussion in this text. After you complete the animation, be sure to answer the questions at the end to test your understanding of this concept.

VISUAL CONCEPT CHECK 5.2

One of the most beautiful cities in North America is Vancouver, British Columbia. This Canadian city is particularly scenic because it lies between the Haro Strait, which is connected to the Pacific Ocean to the west, and the Cascade Mountains to the east. Although Vancouver lies at a fairly high latitude (49° N), it has a very moderate climate with average high temperatures that range from 22°C (71°F) in July to 6°C (43°F) in December. Which one of the following reasons accounts for this temperature pattern?

a) Vancouver is a continental location.
b) Vancouver lies next to a large body of water that has a consistent annual temperature.
c) The wind in Vancouver generally flows from east to west.
d) The Pacific Ocean has a wide range of annual temperature.

North Light Images/Age Fotostock America, Inc.

The Annual Range of Surface Temperature (A Holistic Assessment)

Let's now bring together some of the variables that influence temperature around the world in an effort to see some fundamental global geographic patterns. A good place to begin is by considering Figure 5.10, which shows a hypothetical continent bordered on the east and west by oceans. This theoretical continent straddles the Equator and contains, as a point of reference, the hypothetical position of the 15°C (59°F) isotherm in January and July. This diagram shows the combined effects of the maritime vs. continental relationship and seasonality. Due to seasonality, the isotherm shifts into the Southern Hemisphere during that hemisphere's summer season (January) and back into the Northern Hemisphere during July. This migration occurs because a net surplus of radiation occurs in each hemisphere during the respective summer months due to high Sun angles and increased insolation. The maritime vs. continental effect is evident in the range of latitude that the isotherm shifts in each hemisphere. Notice that, in the oceans, the isotherm does not move much compared to how it migrates on the landmass. This difference is due to the fact that large water bodies heat and cool much more slowly than do continents.

Given the seasonal and geographic pattern observed in Figure 5.10, consider the broad-scale range cycle of temperature on Earth. Once again, a good place to begin is by examining some illustrations that show the fundamental patterns. With this in mind, take a look at Figure 5.11, which shows mean air temperatures on Earth during January and July.

Beginning with the January image, observe that the landmasses in the Northern Hemisphere are quite cold, with temperatures in northeastern Asia (Siberia) of −50°C (−58°F) and −35°C (−31°F) in northern Canada. An interesting pattern exists in North America, where the 0°C (32°F) isotherm extends across the cen-

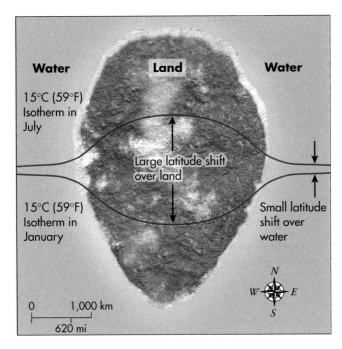

Figure 5.10 Theoretical seasonal migration of the 15°C (59°F) isotherm on a hypothetical continent. This hypothetical continent straddles the Equator, with 15°C (59°F) isotherms in each hemisphere. Notice how far the isotherm migrates over the continent compared to its movement over the oceans.

ter of the United States in the interior of the continent, but crosses into the Pacific Ocean much farther to the north. In other words, January temperatures are much warmer at higher latitudes along the west coast of North America than in the continental interior.

As expected for the Southern Hemisphere summer, mid- to high-latitude temperatures in that part of the world are also relatively warm, ranging from 10°C (50°F) on the southern tip of South America to 30°C (86°F) in the core of Australia. Predictably, temperatures in most of the tropical regions are warm, with an average of about 25°C (77°F) over a large part of the globe. The only significant deviation from this overall tropical pattern exists in western South America, where cool temperatures penetrate into very low latitudes. This pattern exists in part because the Andes Mountains follow much of the west coast of South America and are over 6000 m (19,680 ft) high. In addition, the cold Humboldt Current flows northward along the coast, which contributes to cooler temperatures in this area.

Now turn your attention to the July image in Figure 5.11. What different patterns emerge in this diagram? For one, notice that the Northern Hemisphere landmasses are much warmer than they were in January, with temperatures reaching 10°C (50°F) in northeastern Asia and northern Canada. In other words, a 60°C (108°F) annual range of temperature exists at high latitudes in the Northern Hemisphere. Similarly, the interior of the United States is much warmer as well. Interestingly, the high latitudes in the Southern Hemisphere do not experience such a range of temperature. Winter temperatures on the southern tip of South America, for example, cool to only about 0°C (32°F), which means the annual range there is approxi-

mately 10°C (18°F). Predictably, the tropical regions are warm, with a slight northerly migration of the 25°C (77°F) isotherm. Again, the only variation from this overall tropical pattern is the relatively cool temperatures that follow the Andes Mountains and Humboldt Current in South America.

Why do these annual geographical patterns exist? The answer lies in the holistic interactions of the factors discussed so far, including seasonality, insolation, net radiation, latitude, the maritime vs. continental effect, and the environmental lapse rate in the troposphere. Some patterns are easy to explain. For example, the consistently warm temperatures in the tropics are naturally a function of high Sun angles and associated intense radiation all year long. As already noted, the exception to this tropical pattern exists along the axis of the Andes Mountains. This geography demonstrates that secondary variables—in this case, the troposphere's negative lapse rate and its cooling effect at progressively higher altitudes—also influence temperature patterns.

Another clear pattern is the extreme range of temperature at high latitudes in the Northern Hemisphere, which is easily explained by the highly variable amount of radiation received over the year due to changes in orbital position and the angle of incidence. It should also make sense that temperatures along the

January

July

Figure 5.11 January and July distribution of surface air temperatures on Earth. Notice the geographic patterns and variations that occur between months. Blue isotherms indicate temperatures below 0°C (32°F), whereas red isotherms represent temperatures above 0°C (32°F).

west coast of North America are warmer than in the continental interior in winter. This pattern is represented by the variation in temperature between San Francisco and Topeka, as illustrated in Figure 5.6.

Although the Northern Hemisphere pattern may seem logical, you might also ask yourself the following (and very good) question: Why do the high latitudes in the Southern Hemisphere not experience the same extreme temperature range as those in the Northern Hemisphere? To answer this question, note the size of the Southern Hemisphere continents compared to those in the Northern Hemisphere; they are much smaller than their Northern Hemisphere counterparts. In other words, the Southern Hemisphere is covered more by water than the Northern Hemisphere. Thus, more of a maritime effect occurs in the Southern Hemisphere than in the Northern Hemisphere, where the continental effect dominates. A good example of the continental effect in the Southern Hemisphere, however, is the core of Australia, where the annual range shows more variability due to the relatively large landmass of that country.

KEY CONCEPTS TO REMEMBER ABOUT SURFACE AND AIR TEMPERATURES

1. Air temperature is a measure of the warmth or coolness of a portion of the atmosphere.

2. The surface air temperature is closely related to the temperature of the ground.

3. Air temperature has a close relationship with several factors, including seasonality, angle of incidence, latitude, and proximity to large bodies of water or cities.

www.wiley.com/college/arbogast

WileyPLUS

Global Temperature Patterns

In an effort to better visualize temperature geography on Earth, go to the *Geo Media Library* and select *Global Temperature Patterns*. This animation examines specific regions on Earth, such as the tropics and large landmasses, and how airflow influences seasonal temperature changes. After you interact with the animation, answer the questions at the end to ensure you understand this concept.

VISUAL CONCEPT CHECK 5.3

This image shows global temperature at a particular time of year, with the key showing the range of temperature in degrees Celsius. Given your understanding of the factors that influence temperature, which one of the following choices best explains the pattern you see?

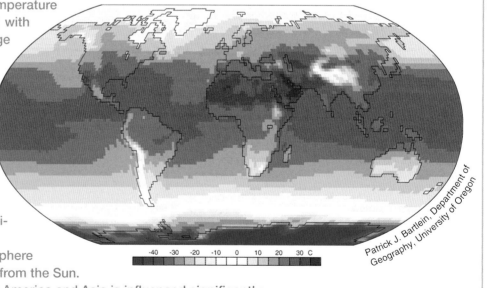

a) It must be summer in the Northern Hemisphere.

b) The Northern Hemisphere must be tilted away from the Sun.

c) The interior of North America and Asia is influenced significantly by the maritime effect on temperature.

d) The Southern Hemisphere is receiving the most insolation.

(a)

Digital Vision/Getty Images

Pete Seaward/Stone/Getty Images

(b)

Figure 5.12 Urban vs. rural landscapes. The different characteristics of city and rural surfaces can significantly influence temperature. (a) Rural surfaces such as this Scottish landscape are cooler because less radiation strikes the ground due to tree cover, and more water is found in soil and on vegetation surfaces. (b) Urban surfaces, in contrast, receive more direct radiation and absorb more radiation. This image shows a portion of the cityscape in Tokyo, Japan.

4. Water stores a large amount of thermal energy and thus moderates the air temperature above it on a seasonal basis. In contrast, landmasses store relatively little thermal energy, which causes a greater annual range of air temperature.

5. Distinct geographic patterns occur with respect to air temperature on Earth. Large landmasses and high latitudes typically exhibit a wide range in temperature. In contrast, locations at low latitudes and near large water bodies have a very narrow temperature range.

Human Interactions: Urban Heat Islands

In addition to the natural variables that influence air temperature, such as latitude, maritime effect, and topography, another important factor is the human influence as it relates to large cities. Urban centers are growing all around the world in response to increased human population and inward migration. Between 1990 and 2010, for example, the collective population in the world's cities grew from two to three billion, so that now approximately half of the people in the world live in cities. As cities continue expanding, more of the natural landscape is transformed into new buildings, homes, and parking lots (Figure 5.12). In this process, vegetation is removed and soils are covered by pavement or other structures. This human-induced change in land cover can cause a measurable temperature increase in and around a city through a variety of ways, resulting in a distinct geographic feature known as an **urban heat island** where temperatures may be as much as 3°C to 4°C (6°F to 8°F) warmer (or more) than the surrounding countryside.

Urban heat island *The relatively warm temperatures associated with cities that occur because paved surfaces and urban structures absorb and release radiation differently from the surrounding countryside.*

Urban environments tend to be warmer than rural areas for several reasons. One of the most important reasons for this effect is that urban surfaces consist of metal, glass, asphalt, and concrete, whereas rural areas are covered with soil in which forest and grass grow. As a result, urban surfaces tend to be darker than rural surfaces, which leads to greater net radiation in cities because albedo is lower. In the context of this higher net radiation, up to 70% of net radiation in cities is converted to sensible heat, much more so than in the countryside. Urban surfaces also conduct more energy than rural soils and are thus warmer. This effect is magnified by the fact that forest canopies in rural areas shade the ground beneath them, keeping them relatively cool. In urban environments, on the other hand, solar radiation strikes the surface directly and is absorbed, later to be released as warming longwave radiation. This re-radiation tends to be trapped for longer periods of time in cities, further warming them, because the irregular geometry of city buildings limits wind speed by about 20% to 30% over the course of the year. This decreased flow of air reduces the loss of heat that would otherwise occur.

Another reason why cities are warmer than rural places is related to the way water is stored and moves in the two areas. In rural areas where soils cover the landscape, rainfall and snowmelt gradually soak into the ground. As described earlier, water has a modifying effect on temperature, which results in soils being cooler than urban surfaces. The presence of water in rural soils also reduces the conductivity of those surfaces, keeping them relatively cool. In contrast, urban surfaces tend to be sealed because they are paved or covered with buildings. The effect of this change in land cover on the way water is stored and moves is significant because water does not absorb into the ground in cities in the same way as it does in the countryside. Instead, precipitation rapidly runs across these surfaces and into storm drains that carry it away. Thus, the modifying effect that water has on the temperature of soils in rural areas does not exist in cities, causing them to be warmer.

Yet another reason cities are warmer than rural areas is the impact that human activities have on urban environments. In hot summer months, the production of electricity to cool homes and businesses releases a great deal of energy, perhaps as much as 25%

(a)

(b)

Figure 5.13 Urban heat island in Atlanta, Georgia. (a) This image from September 2008 shows natural colors where dense urban development is gray, bare ground brownish, and vegetation green. (b) A land surface temperature map from the same day where cooler temperatures are yellow and hotter temperatures are red. Notice the distinct spatial relationship between the two images, with warmer temperatures clearly associated with the dense urban areas.

to 50% of insolation. During the winter months, in contrast, a great deal of sensible heat is generated through artificial heating activities. Humans also alter the heating characteristics of cities through the production of pollutants such as ground-level ozone and other aerosols. In general, about 10 times more human-produced particulates are present in urban environments than in rural areas. Although these pollutants increase the reflectivity of the atmosphere above cities, thus reducing insolation, they cause an increase in the amount of infrared energy re-radiated downward to the surface.

An excellent example of an urban heat island is Atlanta, Georgia. This southeastern city has grown rapidly since the middle of the 20th century, becoming the leading commercial, industrial, and transportation area of the region. During this time, Atlanta has been one of the fastest-growing metropolitan areas in the United States, with population increasing from about 2,300,000 in 1980 to a projected ~6,000,000 in 2015. Associated with this explosive growth has been a dramatic expansion of the urban environment at the expense of agricultural land and forest. As a result, the air qual-

ity of the Atlanta region has decreased significantly, with increased amounts of ozone and volatile organic compounds polluting the air. Another impact has been the well-defined urban heat island that has evolved (Figure 5.13). As you examine this image, note that May temperatures in the central business district reached 45°C (113°F) in some areas, whereas they were as low as 22°C (71°F) in the surrounding countryside. No wonder the city is called *Hotlanta* by many residents.

KEY CONCEPTS TO REMEMBER ABOUT URBAN HEAT ISLAND EFFECT

1. The urban heat island effect has a measureable impact on local temperature patterns within cities. Cities tend to be warmer than surrounding rural landscapes for a variety of reasons related to land cover and energy consumption.

www.wiley.com/college/arbogast

WileyPLUS

Temperature and Location

To explore temperature patterns in another way, go to the *Geo Media Library* and select *Temperature and Location*. In this animation you can "visit" five cities that experience the kinds of seasonal temperature patterns discussed in this chapter. These cities are Cordoba, Argentina; Yakutsk, Russia; Manaus, Brazil; St. Louis, Missouri; and San Francisco, California. Visit each of

these cities during each season and access insolation and temperature data from each location. As you work through this simulation, note how the environmental factors we have discussed influence the seasonal range of temperature at these places. Once you complete the simulation, be sure to answer the questions at the end to test your understanding of this concept.

SUMMARY OF KEY CONCEPTS

1. The atmosphere contains four major layers: the troposphere, stratosphere, mesosphere, and thermosphere. Measurable temperature trends appear in each of these layers. One important change in temperature occurs in

the troposphere, which cools with increasing altitude at a rate of 6.4°C per kilometer or 3.5°F per 1000 ft. This rate is called the environmental lapse rate.

2. On a large scale, a very close relationship occurs among insolation, net radiation, and air temperature. Nevertheless, a temporal lag exists on an annual and daily basis between when the receipt of the highest amount of radiation occurs and when the warmest temperature takes place. This lag exists because insolation is first absorbed by the surface of Earth before it is re-radiated as longwave radiation.

3. The maritime vs. continental effect explains the temperature difference between coastal and land-locked locations that are on or about the same latitude. Cities such as San Francisco, California, have a relatively narrow annual temperature range due to the thermal characteristics of large bodies of water like the Pacific Ocean. In contrast, continental locations such as Topeka, Kansas, have a much broader annual temperature range because landmasses do not absorb or release heat as consistently as water bodies.

4. Measurable differences occur with respect to temperature between urban and rural localities. Cities are typically warmer than the countryside due to the urban heat island effect: Paved surfaces and buildings have relatively low albedo and thus absorb more solar radiation. In addition, the moderating effect of water is reduced in cities because urban surfaces are sealed. Rural locations, in contrast, are covered with soil that absorbs water, which cools the landscape relative to an urban environment. Wind patterns also vary markedly between the two regions and are another cause for the relative difference in temperature.

5. Distinct seasonal temperature patterns occur on Earth. This geography is dependent on the interaction of several factors, including net radiation, season, latitude, the maritime vs. continental effect, and altitude.

CHECK YOUR UNDERSTANDING

1. Describe the four layers of the atmosphere and the temperature patterns that occur within them.

2. Why is latitude an important consideration in determining air temperature?

3. Where does the greatest range in annual temperature occur, at high or low latitudes? Why does this pattern exist?

4. Which location—Los Angeles, California, or Phoenix, Arizona—would experience the warmest peak temperature? Why does this pattern occur?

5. Central Canada and the southern tip of South America are both located at high latitudes. Nevertheless central Canada has a much greater range in temperature than the southern tip of South America. Why?

6. Which place would have a greater range in temperature, central Australia or Hawaii? Why?

7. Why is the tropopause at a higher altitude at the Equator than in northern Canada?

8. List three reasons why cities are typically warmer than rural landscapes.

9. Which place would be colder, the top or bottom of a high mountain? Why?

10. Why are temperatures in the lower stratosphere warmer than in the upper part of the troposphere?

ANSWERS TO VISUAL CONCEPT CHECKS

VISUAL CONCEPT CHECK 5.1

1. The answer is *b*; temperature decreases as you rise through the troposphere. This decrease occurs because a negative environmental lapse rate exists from the surface of Earth to the tropopause.

2. The correct choice is *a*. This altitude occurs in the upper part of the troposphere.

3. The answer is *a*. At this altitude, the temperature will begin to increase because the airplane is climbing closer to the ozone layer, which is found in the lower part of the stratosphere.

VISUAL CONCEPT CHECK 5.2

The answer is *b*. Vancouver lies next to a large body of water (the Pacific Ocean) that has a consistent temperature over the course of the year. As a result, the air temperature at Vancouver remains cool in the summer and warm in the winter relative to continental locations.

VISUAL CONCEPT CHECK 5.3

The answer is *a*. This image represents summer conditions in the Northern Hemisphere, because the high latitudes in Canada, Europe, and Asia are warm. Given that these are continental locations, which have a large annual temperature range, it must be summer and the Northern Hemisphere is tilted toward the Sun.

CHAPTER SIX

ATMOSPHERIC PRESSURE, WIND, AND GLOBAL CIRCULATION

This chapter investigates the way that air flows through the process of atmospheric circulation. The circulatory processes described in this chapter have dramatic effects that will be seen in later chapters, including daily weather (Chapter 8) and global climate patterns (Chapter 9). Flowing air also has significant impact on the circulation of ocean currents, coastal erosion by waves (Chapter 19), and the formation of sand dunes (Chapter 18). Many factors influence the direction and speed of airflow, including pressure and temperature differences, the Earth's rotation, and surface friction. After these variables are discussed, the focus will turn to the geographic patterns of airflow around the globe. Toward the end of the chapter, we turn to how winds affect oceanic circulation.

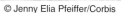
© Jenny Elia Pfeiffer/Corbis

Flowing air is usually invisible to the naked eye. However, in some cases, such as this spectacular sandstorm in Eritrea, Africa, strong winds blow dust in a way that allows us to visualize air movement. Flowing air is the focus of this chapter.

CHAPTER PREVIEW

Atmospheric Pressure

Atmospheric Pressure Systems

The Direction of Airflow

 Fluctuations in the Pressure Gradient

 The Coriolis Force

Global Pressure and Atmospheric Circulation

 Global Atmospheric Circulation

 The Asian Monsoon

 Global Atmospheric Circulation and Water Vapor Movement

Human Interactions: Harnessing Wind Energy

Local Wind Systems

Oceanic Circulation

 El Niño

LEARNING OBJECTIVES

1. Understand the concept of air pressure and how it is measured.

2. Compare and contrast high- and low-pressure systems and how air flows within and between them.

3. Describe the factors that influence large-scale wind patterns on Earth.

4. Discuss the various components of the global circulation model.

5. Discuss the seasonal migration of atmospheric pressure systems on Earth.

6. Understand why people are developing wind as a supplementary energy system and where in the United States the potential is high for this development, and identify the challenges facing the effort.

7. Describe the various local circulatory systems and why they occur.

8. Discuss the nature of oceanic circulation and how it compares with atmospheric circulatory patterns.

Atmospheric Pressure

Atmospheric circulation is an incredibly important process on Earth for a variety of reasons. The primary impact of airflow is to move heat energy around the globe in a way that moderates temperature on Earth. Airflow also affects global and local air quality and pollution levels. For example, winds can carry ash and gases from volcanic eruptions many kilometers from the volcano (Figure 6.1), or they can clear smog from large cities. Similarly, dust from eroded agricultural fields can be transported great distances before it settles back to Earth. Before we can begin to describe the nature of airflow, however, the concept of air pressure must first be introduced because it directly influences the character of large- and small-scale wind patterns.

As discussed in Chapter 4, the atmosphere is made up of a variety of gases that we collectively call "air." Like anything else, air is held to Earth by gravity and thus has weight. The weight of the air exerts pressure on the Earth's surface, which is measured as **air pressure** (also called *atmospheric pressure* or *barometric pressure*).

Factors That Influence Air Pressure

Atmospheric pressure is most closely associated with the temperature and density of air. The concept of air density is easy to understand if you consider that air expands or contracts depending on the environmental setting. Imagine, for example, you have a specific number of air molecules that are contained within a shoebox. If you can somehow move all the molecules in the shoebox to a larger box, the gas will expand to fit comfortably within the new container. In this simple example, the density of air molecules is greater in the shoebox than it is in the larger container.

You can see the effect of density on air pressure by noting how pressure changes with altitude (Figure 6.2). In general,

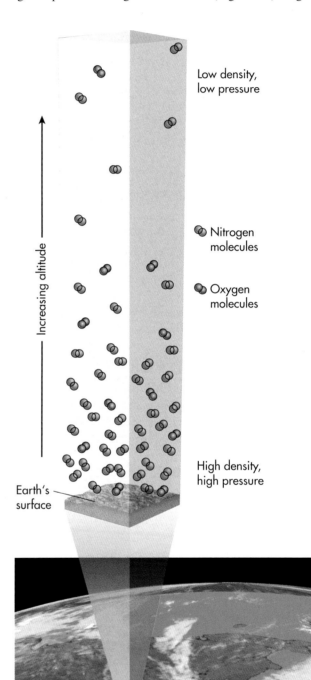

Figure 6.2 Air density, altitude, and atmospheric pressure. At low altitudes air molecules are held close to Earth by gravity and thus are denser, resulting in high atmospheric pressure. In contrast, the density of air molecules is low at high altitudes, and air pressure is thus relatively low.

Figure 6.1 Ash plume from Mount Etna in Sicily. Northwesterly winds carry volcanic ash from the erupting volcano toward the southeast, in an image taken on July 22, 2001, from space station *Alpha*.

Air pressure *The force that air molecules exert on a surface due to their weight.*

air pressure decreases with increasing altitude because most air molecules are held close to the surface by gravity. As a result, the density of air molecules is greater closer to the Earth's surface, which means that air pressure is relatively high in that part of the atmosphere. With increased altitude, the density of air molecules becomes progressively less, resulting in progressively lower air pressure.

In addition to the impact of gravity on the molecular density of air, atmospheric pressure is also strongly influenced by air temperature. The most obvious way that air temperature influences atmospheric pressure occurs when air close to the Earth's surface is warmed a great deal. Such warming causes air molecules to scatter and density thus decreases, resulting in relatively low atmospheric pressure. Consider the analogy of a hot air balloon. When heat is added to the air within the balloon, this causes the air to expand and lift within the relatively cooler (more dense) air that surrounds it. Low atmospheric pressure also results when air is forced to rise vigorously. Very cold surface air is usually associated with high atmospheric pressure because cold air is dense and thus sinks. In some instances, air from the upper atmosphere descends vigorously toward the Earth's surface, which also results in high pressure.

Measuring and Mapping Air Pressure

Air pressure is often measured in units called millibars (mb) with an instrument called a barometer (Figure 6.3). A common type of barometer consists of a long glass tube, closed at one end, which is filled with a liquid (usually mercury) and inverted into a dish containing the same liquid. The liquid in the tube drops down slightly, leaving a vacuum at the closed end of the tube. When the liquid in the tube comes to rest, the force due to atmospheric pressure, pressing down on the liquid in the dish, exactly balances the weight of the column of liquid. The tube can then be calibrated to measure atmospheric pressure in terms

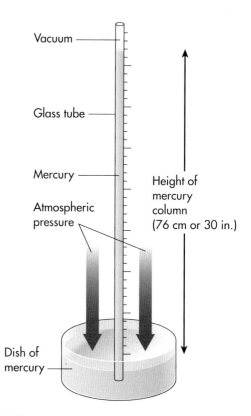

Figure 6.3 Measurement of atmospheric pressure. The pressure of the atmosphere is measured by the height of a column of mercury that can be supported by that pressure.

of inches or millimeters of mercury, which can be converted to millibars or any other pressure unit.

You can observe how air pressure changes with altitude by looking at three separate locations at different elevations (Figure 6.4). At sea level, for example, the average pressure

VISUAL CONCEPT CHECK 6.1

Mount McKinley is the tallest mountain in North America, rising 6194 m (20,320 ft) in Alaska's Denali National Park. How do you suppose air pressure changes from the base of the mountain to the top—does it increase or decrease? Why does this change occur?

Age Fotostock America, Inc.

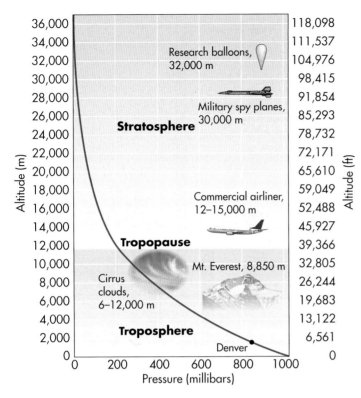

Figure 6.4 Atmospheric pressure and altitude. Average atmospheric pressure decreases with increasing elevation and altitude above the Earth's surface.

of the air is 1013.25 mb. In contrast, at Denver, Colorado (5280 ft—"the Mile High City"), the air pressure is 840 mb. At a still higher elevation, such as at the top of Mount Everest, the Earth's tallest mountain at 8850 m (29,035 ft), the average air pressure is only 320 mb.

Atmospheric Pressure Systems

In addition to the pressure changes that occur with respect to altitude, air pressure also varies horizontally across the Earth's surface, with distinct zone of high air pressure and low air pressure. A **high-pressure system** is a circulating body of air that exerts relatively high pressure on the surface of Earth because air descends (toward the surface) in the center of the system. In contrast, a **low-pressure system** is a circulating body of air where relatively less pressure exists on the Earth's surface because the air is rising (away from the surface) in the system's core. This portion of the chapter focuses on the nature of these large-scale pressure systems and describes how air flows within and between them.

Low-Pressure Systems

Low-pressure systems are often referred to as **cyclones**. If you observe a low-pressure system from the side (Figure 6.5), notice that the vertical flow of air consists of rising air, with the most vigorous upward flow in the center of the system. As a result, the central part of the system has the lowest pressure and is designated with the letter "L" on a weather map. In the Northern Hemisphere, the horizontal flow of air around the center of a low is counterclockwise (looking at it from above) as the air flows toward the core of the system. In the Southern Hemisphere, the horizontal flow of air is clockwise. This inward flow occurs because the rising air at the center of the system creates a void into which air must flow. Low-pressure centers are generally associated with cloudy or stormy weather because rising air cools with altitude and thus squeezes water out of the atmosphere. This relationship will be discussed in much greater detail in Chapter 7.

High-Pressure Systems

In contrast to low-pressure systems, large areas of relatively high air pressure are referred to as **anticyclones**. In these systems, the vertical flow of air consists of descending air, which diverges (spreads apart) at the surface (Figure 6.5). Because air sinks most vigorously in the center of a high, the highest pressure is in the central part of the system, which is designated with the letter "H." The horizontal airflow around the center of a high is clockwise in the Northern Hemisphere and counterclockwise in the Southern Hemisphere, just opposite to the direction of flow around a low-pressure system. High-pressure centers are generally associated with fair weather because descending air warms as it approaches the surface. As mentioned earlier, warm air can hold more moisture than cold air, and this makes precipitation less likely. Again, more will be said about this relationship in Chapter 7.

In general, high- and low-pressure systems (anticyclones and cyclones, respectively) create large-scale circulatory systems that are interconnected by airflow. Look again at Figure 6.5 and see how the air flows horizontally, in a process called **advection**, from the high-pressure system to the low-pressure system at the surface. You can see clearly the geographic pattern of low- and high-pressure systems on a barometric pressure map. Figure 6.6 illustrates pressure variations in the upper part of the atmosphere across the North Atlantic. The (gray) lines on the map are the isobars. The red arrows illustrate the way that air is moving relative to the pressure systems.

Notice in Figure 6.6 that the isobars in the eastern Atlantic form a rough oval or egg-shaped area. This region is the center of a low-pressure system (or the area of lowest air pressure) at this particular point in time. From the wind direction arrows, observe

---

High-pressure system *A rotating column of air that descends toward the surface of Earth where it diverges.*

Low-pressure system *A rotating column of air where air converges at the surface and subsequently lifts.*

Cyclones *Low-pressure systems.*

Anticyclones *High-pressure systems.*

Advection *The horizontal transfer of air.*

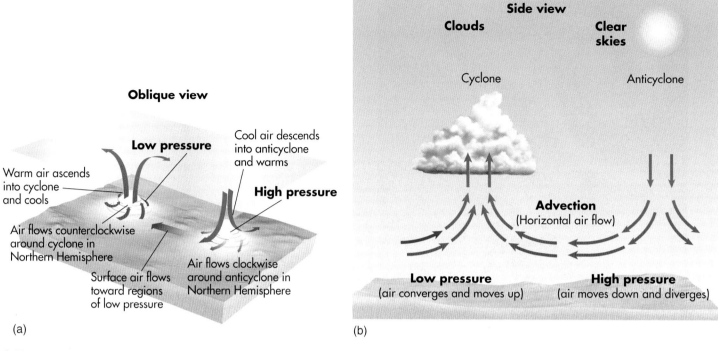

Oblique view

Low pressure

Warm air ascends into cyclone and cools

Cool air descends into anticyclone and warms

High pressure

Air flows counterclockwise around cyclone in Northern Hemisphere

Surface air flows toward regions of low pressure

Air flows clockwise around anticyclone in Northern Hemisphere

(a)

Side view

Clouds

Clear skies

Cyclone

Anticyclone

Advection (Horizontal air flow)

Low pressure (air converges and moves up)

High pressure (air moves down and diverges)

(b)

Figure 6.5 Atmospheric pressure systems. Oblique view (a) and side view (b) of typical low- and high-pressure systems. In a low-pressure system, air converging at the surface rises and forms clouds. In a high-pressure system, air descends and diverges at the surface; these systems are usually associated with clear skies.

that the air is moving counterclockwise around this low. Another low exists in the western part of the North Atlantic, over eastern Canada, where the winds are also circulating counterclockwise. In the central Atlantic, between the two lows, a broad area of high pressure exists that is rotating clockwise. On the western side of this system the air is flowing southwest to northeast, whereas on the eastern side of the system the air is moving northwest to southeast. Another way to describe these patterns is to state that the winds on the western side of the high are southwesterly and that they are northwesterly on the system's eastern side. Many people

mistakenly believe that the named direction of wind reflects the direction in which the air is moving. Instead, the name of the wind direction, such as "westerlies," reflects the direction from which the air is coming (Figure 6.7). Westerly winds, for example, originate in the west and flow toward the east. Similarly, a north wind in the Northern Hemisphere brings in air from the north toward the south.

Although atmospheric pressure maps such as Figure 6.6 are useful to illustrate detailed information, you can also observe basic barometric patterns by viewing satellite images such as the view of western Europe in Figure 6.8. Notice that France, Spain, and Germany are cloud-free, whereas Great Britain and Ireland are shrouded in a swirling band of clouds. Zones of clear sky,

Figure 6.6 Atmospheric pressure map of the North Atlantic. The red arrows represent the direction of the winds. Notice the pressure variability across the ocean and the way air flows.

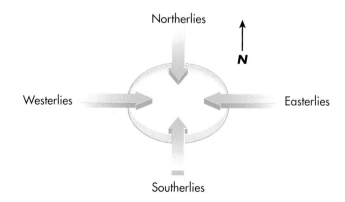

Northerlies

N

Westerlies

Easterlies

Southerlies

Figure 6.7 Compass headings and wind directions. Winds are named for the direction from which they originate. Westerly winds, for example, originate in the west and flow toward the east.

Figure 6.8 Atmospheric pressure systems in Europe. The low is indicated by the clouds that cover Great Britain and Ireland, whereas the high is the clear sky to the east over France, Spain, and Germany.

such as the large region in eastern Europe, are places dominated by high pressure. On the other hand, the cloudy skies over the British Isles reflect the influence of a low-pressure system. Remember that air pressure is low in the center of a cyclone and high in the middle of an anticyclone. Because air rises in the center of a low, the resulting lower pressure pulls air into the area from the adjacent high-pressure system. We feel this exchange of air as wind. Look at Figure 6.5 again to visualize how this process works and be sure you understand it thoroughly.

KEY CONCEPTS TO REMEMBER ABOUT ATMOSPHERIC PRESSURE SYSTEMS

1. Air pressure refers to the weight of air distributed on the surface of Earth. It generally decreases with increasing altitude.

2. Low-pressure systems are called cyclones and consist of rotating air masses that lift air from the surface. In the Northern Hemisphere, these systems rotate counterclockwise, whereas they rotate clockwise in the Southern Hemisphere.

3. Low-pressure systems are usually associated with clouds and precipitation.

4. High-pressure systems are called anticyclones and consist of rotating air masses that descend toward the surface. In the Northern Hemisphere, these systems rotate clockwise, whereas they rotate counterclockwise in the Southern Hemisphere.

5. High-pressure systems are usually associated with clear skies.

6. On weather maps, isobars indicate areas of equal air pressure.

7. Winds flow from high to low pressure.

The Direction of Airflow

Now that we have discussed the fundamentals of air pressure, let's look more closely at the concept of airflow and the factors that govern its movement. We will examine several factors, including unequal heating of land surfaces, the pressure gradient force, the Coriolis force, and various frictional forces.

Unequal Heating of Land Surfaces

The ultimate cause for all wind patterns on Earth is the unequal heating of surfaces that results from variations in the amount of solar radiation received between latitudes. This spatial variation in surface air temperature means that air density (and thus pressure) differs from place to place. At a fundamental level, surface air flows from areas of high pressure to low pressure because the atmosphere works to balance the difference between the two areas.

The best example of unequal heating on Earth is the difference that exists between the tropics and the poles. Recall from Chapters 4 and 5 that the tropics are much warmer than the poles because the equatorial regions receive the most direct insolation throughout the year. If no mechanism existed to balance this difference, then the tropical and polar regions would become excessively hot and cold, respectively.

Instead, this process of unequal heating causes motion of the atmosphere through the process of convection. Recall from Chapter 4 that convection is the vertical mixing of fluid material (in this case, air) due to differences in temperature. In contrast, remember that the term *advection* refers to the horizontal movement of air or water. Figure 6.9 shows how the processes of convection and advection relate to air movement. On a global scale, air heated near the Equator results in low air pressure as air rises within the upward-moving part of the convection process. Subsequently, it travels to higher latitudes in both hemispheres by advection, where the air cools and descends as a high-pressure system at some point in the downward part of the convection process. The combined processes of convection and advection represent the first stage of atmospheric circulation on Earth and

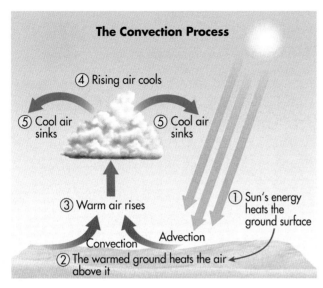

The Convection Process

④ Rising air cools

⑤ Cool air sinks

⑤ Cool air sinks

③ Warm air rises

① Sun's energy heats the ground surface

Convection

Advection

② The warmed ground heats the air above it

Figure 6.9 Atmospheric convection. Convection occurs when one portion of the Earth's surface is heated relative to another. When this happens, a large "bubble" of air lifts from the surface.

are the overall method through which heat is distributed around the planet. Once the air is set in motion in this manner, the other forces—pressure gradient, Coriolis, frictional—then directly influence the movement of the air.

Pressure Gradient Force

As we discussed earlier, air generally flows from areas of high pressure to regions of low pressure. The variable that drives the movement of air between two areas at different pressures is referred to as

the **pressure gradient force**. It is useful to think of a "gradient" as being analogous to the slope between two places on the surface of Earth. If the elevation of one place is significantly higher than the other, then the slope is said to be steeper than if the locations were located at or near each other. The pressure gradient force operates on the same principle: the greater the difference in surface pressures between two regions, the steeper the pressure gradient. The process of airflow is significantly affected by the pressure gradient because the greater the pressure difference between two places, the faster the air flows from high to low to equalize the pressure.

Figure 6.10 is a barometric map that illustrates the pressure gradient force. The map focuses on a hypothetical difference in surface air pressure between Oklahoma City, Oklahoma, and Nashville, Tennessee. In this image, the center of the high pressure (indicated by the "H") is located in Oklahoma City, whereas the center of the low (indicated by the "L") is located in Nashville. Air pressure in this figure ranges from 1028 mb at the center of the high to 992 mb at the center of the low. It should make sense to you that as a result of this variation, the air is generally flowing from west to east (left to right on the map)—that is, from the Oklahoma City area toward Nashville.

Close to the center of the high—say, from Oklahoma City to central Arkansas—the isobars are widely spaced. This means that very limited pressure change occurs across this portion of the Earth's surface at this particular point in time. Thus, the pressure gradient in this part of the circulatory system is shallow. If you were in this area on this particular day, you would find the

Pressure gradient force *The difference in barometric pressure that exists between adjacent zones of low and high pressure that results in airflow.*

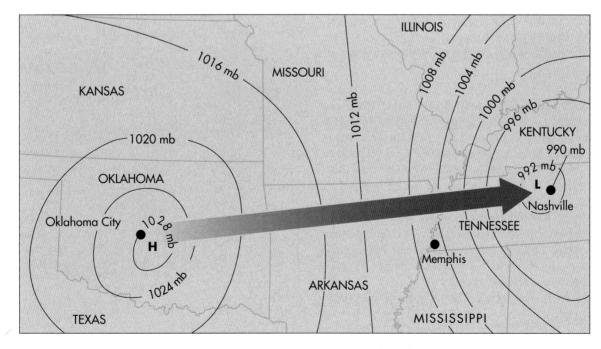

Figure 6.10 Hypothetical pressure gradient between Oklahoma City, Oklahoma, and Nashville, Tennessee. Notice the direction of airflow and the spacing of isobars across the region. A steep pressure gradient exists in regions where isobars are closely spaced together, whereas the gradient is shallow in places where the isobars are far apart.

Atmospheric pressure systems are often easy to see on satellite images. This particular image focuses on Europe in October 2001, with Spain in the lower left part of the image and clouds showing up in bright white. Given your understanding of pressure systems, where is the high-pressure system in this image? Where is the approximate center of the high? In what direction is this system spinning and where are the southerly and northerly winds in the system?

Provided by the SeaWiFS Project, NASA/Goddard Space Flight Center, and ORBIMAGE

winds to be light, because air flowing into that region would be moving slowly. As you approach the center of the low, however, the isobars become closer together. Given that isobars represent locations of equal atmospheric pressure, this close spacing can only mean that a rapid change in surface pressure occurs over a relatively small geographical space—for example, from Memphis to Nashville. In other words, this part of the circulatory system has a steep pressure gradient and the air flows faster as it moves toward the center of the low, to fill the relative void created by the less dense air at the surface. If you happened to be in this area on this particular day, you would notice strong winds.

Coriolis Force

In addition to the pressure gradient force, another factor that strongly influences the process of airflow in the atmosphere is the **Coriolis force**. In contrast to the pressure gradient force, which develops because of differences in atmospheric pressure between regions, the Coriolis force is related to the rotation of the Earth on its axis. Given this rotation, objects in the atmosphere, including air, appear to be deflected or pulled sideways as Earth rotates under them. In the Northern Hemisphere, the direction of deflection for an object moving toward the Equator is to the right when viewed from above the North Pole. In the Southern Hemisphere, the deflection of an object moving toward the Equator is to the left when viewed from below the South Pole.

A good way to visualize the Coriolis force is to examine the apparent path a hypothetical rocket would take if it was

Coriolis force *The force created by the Earth's rotation that causes winds to be deflected to the right in the Northern Hemisphere and to the left in the Southern Hemisphere.*

launched at the North Pole (90° N) toward New York City (about 40° N). Look at Figure 6.11 as you work through this discussion. Imagine that you are standing at the pole and are able to view the rocket's path as it flies south. During the initial period of

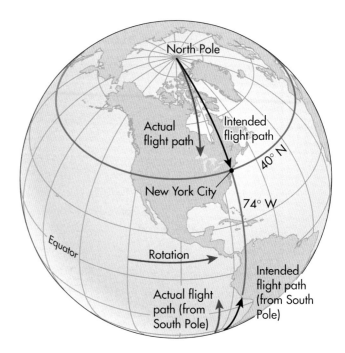

Figure 6.11 The Coriolis force. The Coriolis force influences the path of a rocket traveling from the North Pole to New York City, deflecting it to the west. Notice the direction of the Earth's rotation and how the rocket path is diverted more as it approaches the Equator. Also note that a rocket traveling from the South Pole to the Equator is deflected west as well.

Fluctuations in the Pressure Gradient

In order to better understand how the pressure gradient force influences airflow, go to the *Geo Media Library* and select *Fluctuations in the Pressure Gradient*. This simulation is based on Figure 6.10, which shows a hypothetical pressure gradient between Oklahoma City, Oklahoma, and Nashville, Tennessee. You can adjust the atmospheric pressure between the two places to see how these changes impact the flow of air between them. As you make these adjustments, notice the impact that they have on wind speed and the overall direction of the winds. After you complete the simulation, answer the questions at the end to test your understanding of this concept.

flight, the rocket follows the 74° W meridian as it flies toward the target. As it continues southward, however, it would appear to begin curving to the right (or west) relative to the target from your vantage point. Why? It is not because the rocket's direction actually varies from its original destination. Instead, it appears to curve because the Earth rotates *under* (or eastward of) the rocket while it is in the air. In this fashion, the rocket would land somewhere to the *west* of New York City unless the initial course was corrected in some way to account for the Earth's rotation. Figure 6.11 also shows the apparent path of a rocket fired from the South Pole (90° S) toward the Equator. Note that this rocket is deflected to the west as well, but this time from right to left relative to the path of initial motion as viewed from the South Pole.

Although we can observe how a rocket's path might be altered by the Earth's rotation, how does the Coriolis force influence the wind? Remember that at ground level the direction of airflow is influenced by the pressure gradient force and results in airflow that is perpendicular to the isobars (as in Figure 6.10). Once the air rises into the upper troposphere, however, its speed increases because it flows freely without obstruction. Once this altitude is reached, the air is influenced most directly by the Coriolis force, which causes winds to spiral to the right in the Northern Hemisphere and to the left in the Southern Hemisphere.

In this context, remember that convection loops consist of spiraling masses of descending or rising air that are linked horizontally by advection (Figure 6.12). This process is most pronounced

What you see on a weather map

What is happening in the atmosphere

① Converging winds   ⑥ Diverging winds

Air moves parallel to isobars at high altitudes

② Descending cool air that warms   **High**

⑤ Rising warm air that cools

Anticyclone at surface

Low

Cyclone at surface

Air moves perpendicular to isobars at Earth's surface

③ Diverging winds   ④ Converging winds

Figure 6.12 A dynamic convection loop. Cyclones and anticyclones are linked together in a convection loop consisting of air masses that spiral due to the Coriolis force. Note how the air masses move vertically within the high- and low-pressure systems and horizontally between them.

The Coriolis Force

To see how the Coriolis force looks in animation, go to the *Geo Media Library* and select *The Coriolis Force*. This animation examines how the Coriolis force influences atmospheric circulation. Be sure to watch how the rotation of

Earth on its axis impacts the way air flows on the planet. After you finish watching the animation, be sure to answer the questions at the end to test your understanding of this concept.

---

at higher latitudes and generally results in westerly airflow from the subtropics to the poles. Once again, you might be inclined to think that the air is flowing toward the west; instead, westerly winds are those that flow from west to east. At this point the Coriolis force and the pressure gradient force effectively balance each other, resulting in upper airflow that is *parallel* to the isobars rather than perpendicular, as seen at the surface. The net result of this flow pattern is that air moves *around* pressure systems in the upper atmosphere. Such winds are called **geostrophic winds**.

Frictional Forces

In addition to the pressure gradient and Coriolis forces, yet a third force influences the process of airflow in the atmosphere, one that occurs at ground level and that operates in direct opposition to the winds. This variable is the

force of friction and occurs because of the drag and impediments created by features on the surface of Earth, such as mountains, trees, and even buildings. As you can imagine, these features cause winds to slow down and move in irregular ways (Figure 6.13). The force of friction results in airflow that is somewhere between the flow driven by the pressure gradient (i.e., perpendicular to isobars) and the Coriolis force (which is parallel to isobars). As a general rule, the effect of friction is strongest at the surface and diminishes progressively to an altitude of about 1500 m (approximately 5000 ft). In response to this variability, the wind flows at an

---

Geostrophic winds *Airflow that moves parallel to isobars because of the combined effect of the pressure gradient force and Coriolis force.*

Figure 6.13 The effect of friction on wind flow near the Earth's surface. Compared to winds at higher altitudes, the flow of surface air is significantly modified by features on the Earth's surface.

114   Chapter 6   •   Atmospheric Pressure, Wind, and Global Circulation

Northern Hemisphere

Southern Hemisphere

(a) Pressure gradient force:

Out In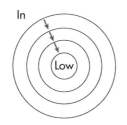

(a) Pressure gradient force:

Out In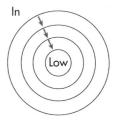

(b) Coriolis force:

Clockwise   Counterclockwise

(b) Coriolis force:

Counterclockwise Clockwise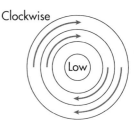

(c) Combined with frictional force:

Spiral out Spiral in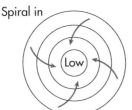

(c) Combined with frictional force:

Spiral out Spiral in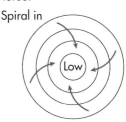

Figure 6.14 The factors that influence large-scale atmospheric circulation. (a) The pressure gradient force causes air to flow perpendicular to isobars. (b) The Coriolis force causes movement of air that is parallel to isobars. (c) In combination with the pressure gradient and Coriolis forces, frictional forces result in winds that flow somewhere intermediate between 0° and 90° of isobars.

angle relative to isobars at ground level. At higher altitudes, however, winds follow a geostrophic course that is parallel to isobars.

In an effort to integrate all the major factors that influence atmospheric circulatory processes, let's now turn to Figure 6.14 to review. Recall that the pressure gradient force causes winds to flow at right angles to isobars in the direction of the lower pressure (Figure 6.14a). If the influence of the Coriolis force is also included (Figure 6.14b), winds then flow parallel to the isobars as geostrophic winds. This process occurs because of the balancing effect that the Coriolis force and pressure gradient force have on one another. In other words, the Coriolis force keeps wind from flowing across isobars, whereas the pressure gradient force stops winds from curving up the pressure slope. Finally, when the force of friction is taken into account (Figure 6.14c), the end result is winds that follow an intermediate course relative to the isobars, somewhere between perpendicular (due to the pressure gradient force) and parallel (due to the Coriolis force) to those lines of equal atmospheric pressure.

KEY CONCEPTS TO REMEMBER ABOUT THE VARIABLES THAT INFLUENCE LARGE-SCALE WINDS

1. Air generally flows from high pressure to low pressure.

2. Large-scale atmospheric circulation is caused by the unequal heating of the tropics and poles. The process of airflow begins at the Equator because of convection.

3. Isobars are isolines that connect points of equal atmospheric pressure.

4. The speed of airflow is determined by the pressure gradient force. The steeper the gradient, the stronger the winds.

5. The Coriolis force causes air moving toward the Equator to be deflected to the right in the Northern Hemisphere and to the left in the Southern Hemisphere.

6. Features on the Earth's surface, such as mountains, forests, and buildings, create a frictional force that acts opposite to the wind's direction.

7. The combined effect of the pressure gradient force, Coriolis force, and frictional force causes a spiral motion of air in both low- and high-pressure systems.

Global Pressure and Atmospheric Circulation

In the preceding sections, we looked at the fundamental characteristics of air pressure systems and the variables that influence the process of airflow in the atmosphere. With these concepts in mind, let's now examine the general circulation of air around the globe. This discussion will refer to the flow of air both in the upper part of the atmosphere and at the surface. The difference between the flow in these different parts of the atmosphere is often noticeable on the ground. If you want to see this difference yourself sometime, look for a partly cloudy day where two distinct layers of clouds occur, one low and another high. When these conditions exist, the clouds in the upper part of the atmosphere are often moving in a slightly different direction or speed from those closer to the surface.

As discussed previously, the primary driver of global circulation is the unequal heating of the tropics and the poles. Because of this energy imbalance, the atmosphere works to balance the system through the process of airflow. If the Earth's surface had a uniform character (i.e., no distinction between continents and oceans), did not rotate, and its axis were not tilted, the circulatory system would be very easy to understand. In this simplistic scenario, which is illustrated in Figure 6.15, low pressure would occur at the Equator because the air is very warm and high pressure would occur at the poles because the air is very cold. Very simply, air would rise away from the surface at the Equator, within a low-pressure system, and would then flow toward the poles in the upper atmosphere by advection. Once it reached the poles, the air would descend toward the surface in a high-pressure system where it would then diverge and flow back toward the Equator by advection.

As you know, however, Earth does not have a uniform surface, it does rotate, and it is tilted with respect to the plane of the ecliptic. Given these factors, the global circulatory system is more complex than the simplified model presented in Figure 6.15. Figure 6.16 illustrates the general global circulation model as it truly functions, showing the major wind systems on Earth. Let's look at the model in more detail in order to explain how these wind patterns occur. We will begin at the Equator and proceed systematically toward higher latitudes, focusing on the directions of winds at both the surface and the upper part of the troposphere.

Tropical Circulation

The logical place to begin discussion of global atmospheric circulation is the tropics. As noted previously in this chapter,

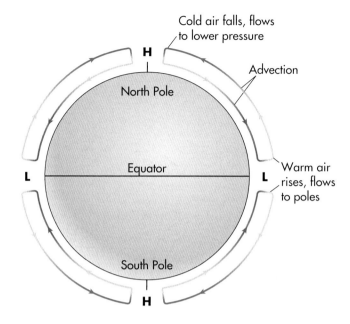

Figure 6.15 Global circulation on a nonrotating, untilted Earth with a consistent surface composition. In this model, a few simple convection loops would dominate the system, with rising air (low pressure) at the Equator, descending air (high pressure) at the poles, and horizontal flow of air by advection in between.

atmospheric circulatory processes are set in motion in this part of Earth because of the high Sun angles and extremely warm temperatures there. Although tropical air eventually finds its way into the midlatitudes, it is useful to think of the tropical circulatory system as being a convection loop, with air flowing between the Equator and about 30° N and S latitude. Each of these loops is referred to as a Hadley cell. In that context, the discussion of tropical circulation focuses on the Intertropical Convergence Zone and the Subtropical High Pressure System.

Intertropical Convergence Zone Tropical circulatory processes begin at the Equator, where air is warmed due to year-round receipt of direct sunlight. This warming helps create a zone of low pressure, called the **equatorial trough**, because warm air is less dense and more buoyant. As a result of the rising air mass, air from higher tropical latitudes in both hemispheres flows toward the equatorial trough along the surface by advection, converging in a narrow band known as the **Intertropical Convergence Zone** or **ITCZ** (Figure 6.16).

Equatorial trough   *Core of low-pressure zone associated with the Intertropical Convergence Zone.*

Intertropical Convergence Zone (ITCZ)   *Band of low pressure, calm winds, and clouds in tropical latitudes where air converges from the Southern and Northern Hemispheres.*

(a)

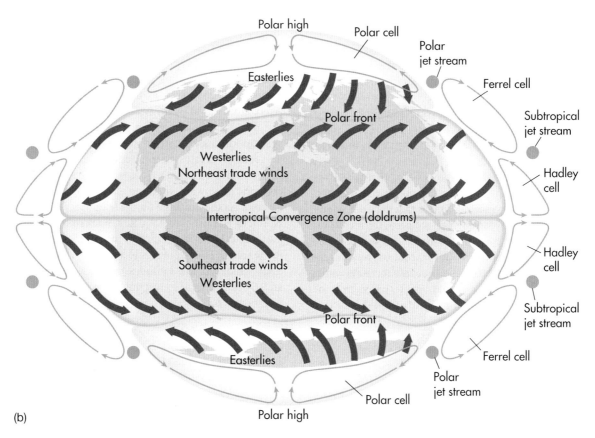

(b)

Figure 6.16 The global circulation model. (a) This three-dimensional diagram illustrates the basic flow of air in cross section as well as top views. Notice the various convection loops that exist around the surface of Earth. (b) A two-dimensional map of the world shows the directions of winds across various bodies of land and water.

Image Courtesy GOES Project Science Office

Figure 6.17 The Intertropical Convergence Zone. The ITCZ is a zone of low pressure, as indicated by the band of dense clouds that extends from Central America west into the Pacific Ocean in this satellite image.

The converging winds are known as **trade winds** because they were systematically used to power sailing vessels during the Age of Exploration between the 1400s and 1800s. Notice in Figure 6.16b that the trade winds north of the ITCZ are northeasterly, whereas trade winds south of the region are southeasterly. Together, they form the **tropical easterlies**. At the point where the northeasterly and southeasterly trade winds converge at the ITCZ, the winds can become relatively calm and highly variable because the pressure gradient is very weak. Sailors referred to this zone of calm air the doldrums because their wind-powered vessels could be stuck in this area for days or even weeks.

An important characteristic of the ITCZ is that it is a region of cloudiness, frequent rains, and even volatile weather. This occurs for two primary reasons. One is that the trade winds flow over warm oceans across much of their length; therefore, a great deal of evaporation from these surfaces takes place, and the air contains abundant moisture. In this context, high levels of atmospheric moisture can be reached because the air is warm and expanding. A second reason is that, due to the warmth of the air, the ITCZ is a place where the air lifts from the Earth's surface. As the air rises, it cools and promotes clouds and rainfall because water condenses at lower temperatures. Due to these combined processes, it is usually easy to find the ITCZ on a satellite image because it consists of a band of clouds and thunderstorms (Figure 6.17) that can be violent. Airline pilots flying through the ITCZ are thus wary of the weather because it can be quite turbulent.

Subtropical High-Pressure System   As air in the equatorial trough rises from the Earth's surface, it spirals upward to the upper part of the troposphere. During this process, the temperature of the air drops as it rises. Although this cooling process will be discussed more thoroughly in Chapter 7, for now it is sufficient to know the process occurs because the air is expanding as it rises.

High-altitude air on the northern side of the ITCZ flows northward by advection, while air on the southern side of the ITCZ flows to the south. Given that the air has cooled, it must sink, and it does so at approximately 25° to 30° N or S latitude. These zones of sinking air are referred to as the **Subtropical High (STH) Pressure System** (see Figure 6.16a). Like the doldrums, airflow in the core of the STH is often weak. Sailors referred to this area as the *horse latitudes*. According to legend, sailing ships there would often remain still for such a long time that sailors pitched their horses and cattle overboard to save provisions.

In contrast to the ITCZ, descending air within the STH is dry because much of its moisture was lost as precipitation over the Equator. The air is also compressed as it descends, making the air denser and warmer as it approaches the surface of the Earth. This compression creates a high-pressure zone of hot, dry air. On the Earth's surface, this zone is characterized by extensive deserts, such as the Sahara Desert in Africa and the Arabian Desert in Saudi Arabia (Figure 6.18).

The STH functions in the same way as the high-pressure systems discussed earlier in this chapter. That is, as the air descends, it rotates in a clockwise direction in the Northern Hemisphere and counterclockwise in the Southern Hemisphere. Once the downward-spiraling airstreams reach the surface, they diverge. As they diverge on the southern side of the STH in the Northern Hemisphere, they flow back to the ITCZ, forming the northeasterly trade winds. On the northern side of the STH in the Southern Hemisphere, the airstreams form the southeasterly trade winds. At low latitudes, this flow to and from the ITCZ and STH forms a convection loop known as a **Hadley cell** (see Figure 6.16 again).

Trade winds   *The primary wind system in the tropics that flows toward the Intertropical Convergence Zone on the equatorial side of the Subtropical High-Pressure System. These winds flow to the southwest in the Northern Hemisphere and to the northwest in the Southern Hemisphere.*

Tropical easterlies   *Band of easterly winds that exists where northern and southern trade winds converge.*

Subtropical High (STH) Pressure System   *Band of high air pressure, calm winds, and clear skies that exists at about 25° to 30° N and S latitude.*

Hadley cell   *Large-scale convection loop in the tropical latitudes that connects the Intertropical Convergence Zone (ITCZ) and the Subtropical High (STH).*

Which one of the following answers correctly describes the location of the ITCZ in this June 2004 GOES east infrared satellite image? White areas represent clouds.

a) The ITCZ is the zone of high pressure seen in the center of the Atlantic Ocean east of Florida.

b) The ITCZ is the band of clouds in the United States that extends from Texas into the Great Lakes region.

c) The ITCZ is the band of clouds that extends east and west through the northern part of South America.

d) The ITCZ is the large zone of clear skies in the southern part of South America.

Image Courtesy GOES Project Science Office

Midlatitude Circulation

Now let's examine the way the atmosphere circulates in the mid-latitudes. In general, the primary purpose of midlatitude circulation is to mix the cool polar air that originates at high latitudes and the warm tropical air that develops at lower latitudes. With this mixing in mind, the midlatitudes are the regions where these contrasting air masses converge. The following discussion describes the circulatory processes in the midlatitudes and how they function in the context of balancing temperature differences.

The focal point of midlatitude circulation is the **polar front**, which on average occurs at about 60° N and S latitudes (Figure 6.19). This atmospheric feature is the contact between cold air orginating at very high latitudes and the relatively warm air that streams northward from the tropics. You can see how tropical air is pumped into the midlatitudes by looking again at Figure 6.16. For the purposes of this discussion, focus your attention on the Northern Hemisphere and notice that the midlatitudes are north of the STH. Remember that this pressure system is centered at about 30° N and consists of warm, dry air that spirals downward in a clockwise fashion. As this descending air reaches the surface, it diverges, with air on the southern side of the system flowing toward the ITCZ in the form of northeasterly winds. In contrast, air on the north side of the STH flows northward in association with Ferrel Cells that are located between the STH and Polar front. These winds tend to be southwesterly. It is this southwesterly air that flows toward the polar front, where it converges with air flowing southward from the highest latitudes. This southwesterly flow of air contributes to the westerly winds, or **westerlies**, that prevail in the midlatitudes.

The midlatitude westerlies are especially important because they often flow at very high speeds, reaching velocities of between 350 km/h and 450 km/h (about 200 mi/h and 250 mi/h) in the upper troposphere. The winds reach these extreme speeds due to a distinct temperature gradient along the polar front, which

(a)

(b)

Figure 6.18 Global vegetation and the Subtropical High (STH) Pressure System. (a) Brown shades represent sparse vegetation in deserts, whereas progressively darker shades of green represent progressively denser vegetation such as the Amazon rainforest. The extensive deserts in northern Africa and the Arabian Peninsula exist because these regions are dominated by the STH. (b) The Rub al-Khali, or Empty Quarter, of the Arabian Peninsula is a large sea of sand, as large as all of France, with no roads and hardly any rainfall.

Image courtesy MODIS Land Group/Vegetation Indices, Alfredo Huete, Principal Investigator, and Kamel Didan, University of Arizona

© Filippo Picinni Leopardi/National Geographic Society/Corbis

Polar front *The contact in the midlatitudes between warm tropical air and colder polar air.*

Westerlies *Midlatitude winds that generally flow from west to east.*

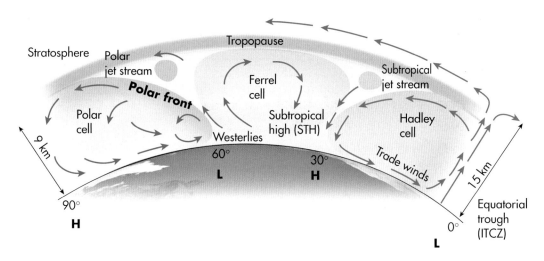

Figure 6.19 The polar front in the Northern Hemisphere. The polar front is the boundary between warm air (to the south) and cold polar air (to the north). Note how air from the STH flows toward the polar front.

results in a very steep pressure gradient as well. Given the seasonal effect created by the Earth–Sun geometric relationship, this temperature/pressure gradient is very weak during the summer months, but strengthens with the progression of winter. The resulting river of rapidly moving air is known as the **polar front jet stream** and occurs at altitudes of 10 km to 12 km (about 30,000 ft to 40,000 ft; see Figures 6.19 and 6.20). Like the ITCZ, the polar jet stream is often easy to see on satellite imagery because a band of clouds forms along this line (as in Figure 6.20b).

Although the term *westerly winds* implies that midlatitude winds are flowing straight from west to east, this is not always the case. In fact, the smooth westward flow in the upper-air circulatory system frequently develops distinct undulations, called **Rossby waves**. Rossby waves along the polar front are the mechanism through which significant temperature differences on either side of the front are moderated, especially during the winter months. Figure 6.21 shows how the process develops.

(a)

(b)

Figure 6.20 Jet streams. (a) Jet streams consist of rivers of air that flow at very high speeds. (b) Several jet streams flow in the upper atmosphere and can frequently be identified by a band of clouds, as in this image of the Nile/Red Sea area in the Middle East.

Polar front jet stream *River of high-speed air in the upper atmosphere that flows along the polar front.*

Rossby waves *Undulations that develop in the polar front jet stream when significant temperature differences exist between tropical and polar air masses.*

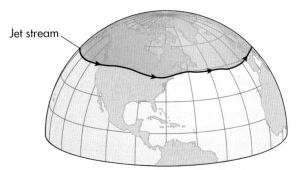

(a) Polar jet stream with small undulations

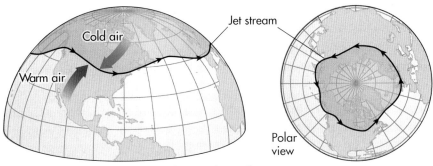

(b) Rossby waves develop, causing north–south motion of large masses of warm and cold air

Polar view

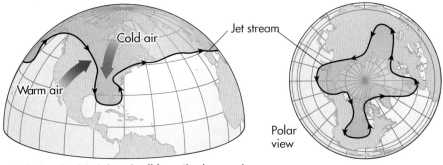

(c) Cells of cold air break off from the larger air mass, forming isolated cyclones of cold air

Polar view

Figure 6.21 Development of Rossby waves in the midlatitudes of the Northern Hemisphere. The progressive development of jet stream undulations along the polar front ultimately results in the pinching off of cold air pools and the reestablishment of zonal flow.

At any given location in the midlatitudes, the flow of the polar jet stream essentially follows a smooth west-to-east path for several weeks. This circulatory pattern is typically called **zonal flow** because cold arctic air is confined to a small zone at very high latitudes. As the temperature contrast on either side of the polar front increases, the midlatitude atmosphere responds by forming an undulation in the jet stream, beginning the Rossby wave. The core of this undulation becomes

a midlatitude cyclone rotating counterclockwise. With the continued development of this Rossby wave, the westerly winds no longer flow directly west to east, but have significant northerly and southerly components as well due to the counterclockwise circulation. In contrast to zonal flow, the variable flow associated with a developing Rossby wave is called **meridional flow**, because the air frequently flows parallel to the meridians.

Zonal flow *Jet stream pattern that is tightly confined to the high latitudes and is thus circular to semicircular in polar view.*

Meridional flow *Jet stream pattern that develops when strong Rossby waves exist and the polar front jet stream flows parallel to the meridians in many places.*

FRIGID TEMPERATURES ON MOUNT WASHINGTON

Where do you think you would find the worst weather on Earth, the North or South Pole? You might be surprised to learn that many meteorologists think the harshest conditions are found in New Hampshire on Mount Washington, which rises 1917 m (6,288 ft) in the White Mountains. Here the polar jet stream routinely slams directly into the highest point of land along the eastern United States, with results that rival the brutal conditions in Antarctica.

The highest wind speed ever recorded anywhere on Earth was measured at 372 km/h (231 mi/h) at the weather station on Mount Washington's summit in 1934.

Annual average temperature at the summit is only –3° C (26.5° F), with annual average snowfall of about 6.5 m (21 ft). One year the snowfall on Mount Washington was over 14 m (47 ft)! Visitors often marvel at the pictures showing the summit buildings covered in rime ice, which forms when fog droplets freeze as they come into contact with structures. The rime ice grows out into the wind, which is the direction from which the particles are coming, giving it a feathery appearance. Since the summit is covered with dense fog about 60% of the time, the ice builds up into thick layers.

(a)

(b)

(a) Tuckerman's Ravine is famous for spring skiing. (b) Rime ice on the observatory on Mount Washington. Note the view of clouds below the summit.

With the onset of meridional flow, warm air on the east side of the wave pushes poleward because the winds are southerly. At the same time that this northward push of warm air occurs, cold air from the north plunges southward on the west side of the system, where the winds are northerly. Such an influx of cold air into a region is called an *Arctic outbreak* and can result in extremely cold air reaching latitudes far south of its origin. If you have spent any winters in the northern part of the United States, you probably know that an Arctic outbreak can bring frigid temperatures into the region for a long time. An extreme Arctic outbreak can even bring freezing temperatures as far south as Florida, which can threaten its agriculture. As time progresses, the cold pool of air can literally be pinched off from the main body of arctic air, resulting in the reestablishment of zonal flow conditions. Such a pool of cold air can persist in

a region for several weeks, gradually warming because of the higher Sun angles at these somewhat lower latitudes.

Polar Circulation

In contrast to the complex patterns associated with tropical and midlatitude circulation, atmospheric circulation in the polar regions is associated with a simple circulatory loop known as a *polar cell*. Air that flows northward at the polar front cools considerably and subsequently descends at very high latitudes, producing a weak high-pressure system (see Figure 6.16 again). This system is called the **Polar High** and

---

Polar High *Zone of high atmospheric pressure at high latitudes.*

Global Atmospheric Circulation

Now that you have investigated atmospheric circulation and the seasonal migration of pressure systems, a good review is to observe the patterns in animated form. Go to the *Geo Media Library* and select *Global Atmospheric Circulation*. This animation demonstrates the specific components of the global circulatory system and how they move. Once you view this animation, be sure to answer the questions at the end to test your understanding of this process.

consists of a mass of descending air that rotates in a clockwise fashion in the Northern Hemisphere and counterclockwise in the Southern Hemisphere. These rotating systems direct cold, dry air toward the polar front in the form of **polar easterlies** and are strongest in the Northern Hemisphere, where large landmasses exist. If you want to review why this geographic pattern exists, return to Figure 5.11 to view the annual range of temperature on Earth and note the variation that exists between the hemispheres. As a result of this variability, the Polar High located over northern continental regions of the Earth makes northernmost Canada and Siberia bitterly cold and dry during the winter.

Seasonal Migration of Pressure Systems

So far in this discussion, the focus has been on the geography of air pressure systems and how they transfer heat energy in the atmosphere. Given your understanding of the Earth–Sun geometric relationship, however, you might also suspect that a distinct seasonal component occurs in association with global atmospheric circulation. In fact, a powerful seasonal component does come into play, one that is illustrated in Figure 6.22.

A good reference point for studying the seasonal pressure migration is the subsolar point and its relationship with the ITCZ. Remember that the ITCZ migrates with the subsolar point. For example, during the Spring and Fall Equinoxes, the Sun is directly overhead at the Equator. This geographic position results in the Equator receiving the most intense radiation. Because the Equator is generally the warmest zone on Earth during these months, the ITCZ is generally located there as well (Figure 6.22a). During winter in the Northern Hemisphere, the subsolar point is located at the Tropic of Capricorn because the Sun is directly overhead at that latitude. Because this zone receives the most intense radiation, this is where the ITCZ is generally located (Figure 6.22b).

Similarly, the pattern is reversed with the approach of the Northern Hemisphere spring and summer. With the coming of the Northern Hemisphere Spring Equinox, the ITCZ follows the subsolar point back to the Equator. Subsequently, the ITCZ migrates into the Northern Hemisphere during April and May, reaching the Tropic of Cancer in the latter part of June. As the days shorten in July and August, the ITCZ migrates back to the south.

Polar easterlies *Band of easterly winds at high latitudes.*

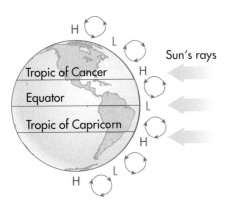

(a) Equinoxes, equatorial
low, ITCZ around Equator

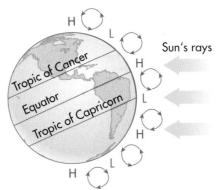

(b) December 21, equatorial high,
ITCZ around Tropic of Capricorn

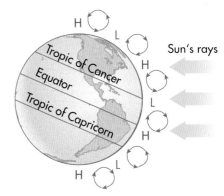

(c) June 21, equatorial high,
ITCZ around Tropic of Cancer

Figure 6.22 Seasonal migration of atmospheric pressure. The positions of the major high- and low-pressure systems during (a) Equinoxes, (b) Northern Hemisphere Winter Solstice, and (c) Northern Hemisphere Summer Solstice.

Like many geographic processes, the Asian monsoon is easy to visualize in animation. Go to the *Geo Media Library* and select *The Asian Monsoon*. This animation clearly illustrates the movement of pressure systems over the course of the year and shows how they contribute to precipitation patterns. After you view this animation, be sure to answer the questions at the end to test your understanding of this concept.

Remember, however, that the ITCZ is not the only large pressure system that moves seasonally. *All of them do.* This migration occurs because the pressure systems maintain a more or less consistent distance from one another across the surface of Earth. Thus, when the ITCZ migrates, all the systems move (see Figure 6.22). You will discover in Chapter 9 that this seasonal migration of pressure systems influences the distribution of global climates in a major way.

Monsoonal Winds One of the most important outcomes of the seasonal migration of air pressure systems is the **monsoon**. The monsoon is a cyclical shift of the prevailing wind direction that occurs at a subcontinental scale over the course of a year. This process is a good example of how closely geographic variables are integrated because it is related both to the seasonal migration of the ITCZ and the maritime vs. continental effect described in Chapter 5. Although a form of monsoon occurs in many subtropical places around Earth, including the southwestern United States, it is most pronounced in Southeast Asia

because this is where the greatest seasonal fluctuation of the ITCZ occurs.

In order to understand the monsoon, let's compare the geographic variability that exists with respect to the seasonal migration of the ITCZ (Figure 6.23). Beginning first with the Americas, notice that the ITCZ is located in South America throughout the year and migrates from the central part of the continent in January to the northern part of the continent in July. In other words, the ITCZ shifts across approximately 20° of latitude in South America over the course of the year. How does this migration range compare with Asia? Figure 6.23 shows that the position of the ITCZ varies dramatically in Asia, ranging from northern Australia in January to northern

---

Monsoon *The seasonal change in wind direction that occurs in subtropical locations due to the migration of the Intertropical Convergence Zone (ITCZ) and Subtropical High (STH) Pressure System.*

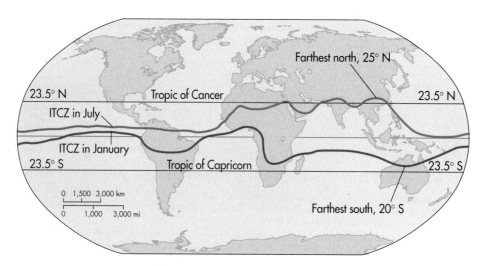

Figure 6.23 Position of ITCZ during the solstices. The red line marks the position of the ITCZ in January (bottom line) and the blue line indicates its position in July (top line). Note the extent of seasonal migration in Asia as compared to South America.

An excellent way to visualize global atmospheric circulation is to observe the movement of water vapor. In this context, view the simulation *Global Atmospheric Circulation and Water Vapor Movement*. This excellent simulation models the movement of water vapor on Earth over the course of the year as it relates to the wind patterns discussed so far. After you watch this simulation, be sure to answer the questions at the end to test your understanding of this concept.

India in July. This is a migration of about 45° latitude, or twice that which occurs in South America. The primary reason for this tremendous differential is the huge size or continentality of the Asian continent (as compared to the Americas), which results in the extreme range of temperature you saw in Figure 5.11.

How does this large temperature range and seasonal ITCZ migration create the Asian monsoon? A good place to examine the monsoon is India (Figure 6.24). In the winter a strong outflow of very dry, continental air moves from central Asia south across India. This *winter monsoon* occurs because the pressure gradient slopes steeply from the very strong Siberian high-pressure system (in central Asia) toward the ITCZ, which is far to the south over the Indian Ocean (Figure 6.24a).

During the summer, the direction of surface winds reverses, causing the *summer monsoon*. This reversal occurs because temperatures over the Asian landmass increase significantly due to the high Sun angle in summer. As a result, the ITCZ shifts northward over Asia (see Figure 6.23 again), bringing

low pressure into the region. Given the seasonal shift in the large pressure systems, a subtropical high is now located over the Indian Ocean and the pressure gradient slopes steeply toward the continent (Figure 6.24b). The warm winds from this high-pressure anticyclone pick up moisture as they pass over the Indian Ocean. As the air moves north along the pressure gradient, it is lifted initially by the warmth of the Indian landmass and then intensively by the high Himalaya Mountains. As the air rises, it begins the process of cooling and precipitation occurs. The high precipitation of the monsoonal rainy season lasts from June to September.

Human Interactions: Harnessing Wind Energy

People have used flowing air as an energy source for thousands of years. Ancient Egyptians, for example, used wind to sail ships on the Nile River. In the late 12th century,

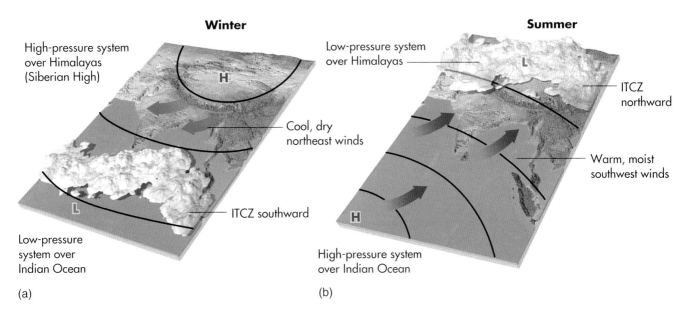

Figure 6.24 Monsoon flow in India during (a) winter and (b) summer. The flow of air is distinctly related to the seasonal position of the ITCZ and the relative temperature of the Asian landmass to the north.

Figure 6.25 A Dutch windmill. Windmills like this one were first built in the Netherlands in the 12th century to grind flour and pump water from swampy areas.

FRILET Patrick/hemis.fr/Getty Images

windmills were developed in northwestern Europe to grind flour and pump water out of low-lying areas. Many such windmills still dot the landscape in Holland (Figure 6.25). Wind energy was the power source behind the great sailing voyages, such as those by Magellan and Cook, during the Age of Exploration that set the stage for the modern world.

Given increasing demands for electricity and for renewable, clean sources of energy, wind power technology has advanced dramatically in the past 30 years. This shift began with the development of the modern wind turbine in the early 1980s following the oil shortages of the 1970s. Like the windmills of old, modern wind machines use large blades that slow wind speed and collect kinetic energy. As the wind flows over the top of the curved blade, it causes lift and a turning motion that spins a drive shaft connected to a turbine. This turbine generates electricity. To give you an idea of how much energy a turbine can produce, consider that a 5 megawatt turbine can produce more than 15 million kilowatt hours (kWh) of electricity in a year, which is enough to power more than 1400 households.

What are the pros and cons associated with wind energy production? On the positive side, wind energy is a clean and an efficient way to produce electricity. Unlike coal power plants, it produces no greenhouse gas emissions and thus does not contribute to climate change. Some problems do exist, however. The blades of wind turbines can strike migratory birds, and turbines can be damaged by strong storms and by lightning. To some people, wind turbines are aesthetically unattractive and are too loud, although quieter turbines are currently being developed. The biggest problem

associated with wind turbines is that they are useless if the wind does not blow. In order to reach maximum efficiency, wind speeds ≥10 mph are required for long periods of time. As a result, many parts of the world are not suitable for developing a sustainable wind-energy program. In the United States, for example, the southeastern part of the country is poorly suited for wind energy, as are significant parts of the American West. Much of the Midwest is also less than ideal (Figure 6.26).

On the other hand, many parts of the country, such as the Great Plains and along many mountain ranges, are excellent places to develop wind energy. Efforts are currently underway to dramatically increase wind energy production in these favorable areas. This upturn in wind energy production is part of a global pattern that accelerated in the first decade of the 21st century. Between 2005 and 2010, average annual growth of the industry was about 27%. As of 2011, over 200,000 wind turbines have been installed around the world, with Europe being a focus of rapid growth. In Denmark, for example, 28% of the nation's electricity is produced by wind. Within the United States, the number of units recently surpassed 50,000. The high potential for growth in the industry is nicely exemplified within the state of Kansas, where about 1.2 GW (gigawatts) of energy production has been installed. Estimates indicate, however, that about 952 GW of power generation is possible. Continued development of this energy source could bring thousands of jobs to the area and billions of dollars in economic development. Farmers and ranchers also benefit because they lease the ground on which turbines are installed, resulting in a steady stream of supplemental income for them.

To maximize the production of wind energy in these places, extensive wind farms are being developed throughout much of the central United States. A variety of incentives are in place to build these complexes, including tax credits and government subsidies. The world's largest wind farm is the Horse Hollow Wind Energy Center in north central Texas (Figure 6.27). This facility contains over 400 turbines spread over 190 km^2 (73 mi^2) and generates enough power for about 220,000 homes per year. Similar large wind farms occur elsewhere in the Great Plains, as well as California, which now ranks second behind Texas as the leading producer of wind energy in the United States. An interesting development to continue watching is the Cape Wind project off the coast of Cape Cod, Massachusetts. In November 2008, voters along the south shore of the state approved a nonbinding ballot to allow the construction of 130 large turbines far offshore. A local historic preservation group attempted to block construction, however, because the complex would be built in Nantucket Sound, which they claimed to be environmentally sensitive. In 2011, the Massachusetts Supreme Court ruled that development of the complex could proceed, and a purchase agreement between the group and the Massachusetts Department of Public Utilities was subsequently signed in November, 2012. As a result, installation of wind turbines could begin sometime relatively soon.

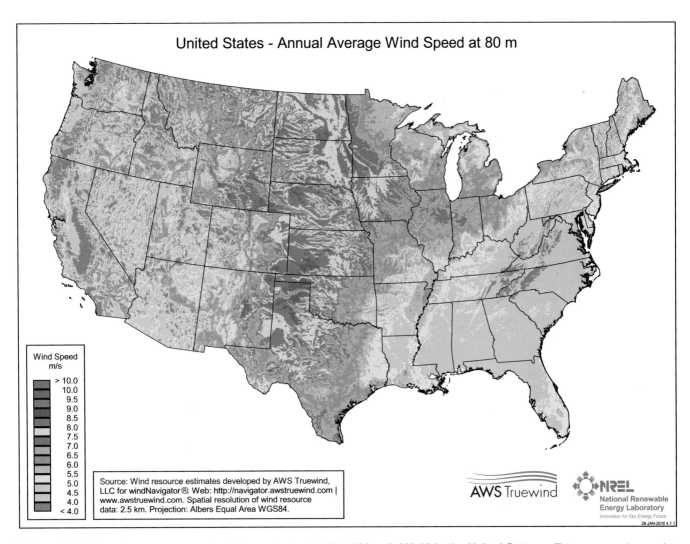

United States - Annual Average Wind Speed at 80 m

Wind Speed
m/s

> 10.0
10.0
9.5
9.0
8.5
8.0
7.5
7.0
6.5
6.0
5.5
5.0
4.5
4.0
< 4.0

Source: Wind resource estimates developed by AWS Truewind, LLC for windNavigator®. Web: http://navigator.awstruewind.com | www.awstruewind.com. Spatial resolution of wind resource data: 2.5 km. Projection: Albers Equal Area WGS84.

AWS Truewind

NREL
National Renewable
Energy Laboratory
Innovation for Our Energy Future

28-JAN-2010 4.1.1

Figure 6.26 Map of average annual wind speed at a height of 80 m (~260 ft) in the United States. This map can be used to predict the suitability of regions for wind-energy potential. Note that the central United States is particularly well suited for wind energy production, whereas much of the southeastern and western parts of the country have low potential.

Alamy

Figure 6.27 The Horse Hollow Wind Energy Center in Texas. This facility has over 400 turbines, which collectively produce enough energy for 220,000 homes per year.

1. The fundamental consequence of global air movement is to equalize the differences in temperature and pressure between the tropics and the poles.

2. The global circulatory system consists of a series of separate, but connected, pressure systems that distribute air around Earth.

3. Tropical circulation occurs in association with convection processes in Hadley cells, forming the Intertropical Convergence Zone (ITCZ) and the Subtropical High (STH) Pressure System. The ITCZ is a zone of low pressure where the winds are easterly. In contrast, the STH is a high-pressure system that rotates clockwise in the Northern Hemisphere and counterclockwise in the Southern Hemisphere.

4. Midlatitude circulation is driven by contrasting air temperatures on either side of the polar front. A major feature of midlatitude circulation is the jet stream, which consists of high-speed westerly winds that flow along the polar front. During the summer, the jet stream exhibits zonal flow that is confined to high latitude. As fall approaches, the jet stream begins to develop distinct undulations that are associated with Rossby waves.

5. The monsoon is a distinct circulatory feature that exists in tropical areas where large land bodies border oceans. Landmasses are warm relative to water in summer, resulting in low pressure over land and onshore airflow. During winter, water bodies are warm relative to land and the winds reverse their flow direction.

6. The production of wind energy is rapidly growing in the United States and around the world. Excellent places for wind-energy production within the United States occur within the central part of the country and along mountain ranges in the west.

Local Wind Systems

So far, we have examined circulatory systems and associated wind patterns that occur over regional and global scales. On a much smaller scale, local wind systems result from temperature differences created by topography, proximity to large bodies of water, or other geographic features.

Land–Sea Breezes

Land–sea wind systems are localized along the shores of major water bodies, such as oceans or large lakes. These winds form due to the different heating and cooling characteristics of continents and water, as we discussed in Chapter 5 (the maritime vs. continental effect). During the day, air over land heats up more rapidly than air over water, resulting in a small zone of low pressure that forms over the land near the shore. Over water, in contrast, the air pressure is relatively high because the air is cooler and denser as it sinks. This difference sets in motion a **sea breeze**: The air flows from the sea toward the land, that is, from high to low pressure (Figure 6.28a).

This local circulatory pattern reverses at night because air over the land cools more rapidly than air over the sea. As a result, the pressure systems switch locations, with

Sea breeze *Daytime circulatory system along coasts where winds flow from a zone of high pressure over water to a zone of relatively low pressure over land.*

(a) Daytime sea breeze

(b) Nighttime land breeze

Figure 6.28 Coastal wind systems and the maritime vs. continental effect. (a) A sea breeze develops when air flows inland from the water during the daytime. (b) A land breeze occurs at night when air flows toward the water.

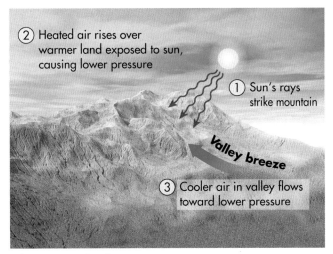

(2) Heated air rises over warmer land exposed to sun, causing lower pressure

(1) Sun's rays strike mountain

Valley breeze

(3) Cooler air in valley flows toward lower pressure

(a) Daytime valley breeze

(1) Heat lost quickly from mountain slope

Mountain breeze

(2) Cooler air sinks downward into valley

(b) Nighttime mountain breeze

Figure 6.29 Local wind flow caused by differences in elevation. (a) During the daytime, air flows upslope in a valley breeze. (b) At night, the flow reverses in a mountain breeze.

relatively higher pressure over land and low pressure over the water. When this adjustment occurs, the air reverses direction and moves from land toward sea in a **land breeze** (Figure 6.28b).

Topographic Winds

In addition to the winds that develop along coastlines, several kinds of winds form due to variations in topography (Figure 6.29). These winds generally form because large temperature differences can occur with elevation in hilly or mountainous landscapes. For example, a **valley breeze** is a wind system that develops when mountain slopes heat up due to re-radiation and conduction over the course of the day. When this occurs, a zone of relatively low pressure develops on the mountain slopes, whereas high pressure is found in the lowlands below. As a result, air can flow upslope to fill the pressure void created by the upward movement of air over the heated slopes. Like the coastline breezes, the situation in mountainous regions reverses at night, resulting in a downslope **mountain breeze**.

A more extreme example of mountain breezes is called **katabatic winds**, which form in particularly cold places such as

Greenland and Antarctica. From the Greek *katabatikos*, which means "descending," this process is most common during the winter months, when extremely cold air accumulates over higher-altitude regions covered by ice sheets. This extremely cold, dense air then flows downhill under the force of gravity. These winds sometimes flow at great speeds that can be quite destructive, especially where the air is funneled down a narrow valley. Although the air associated with katabatic winds typically warms when descending, it usually remains colder than the air that it replaced.

The last kind of topographically related wind process is the **Chinook winds** (also called *foehn winds*). These winds also form in mountainous regions, but occur only when a steep pressure gradient develops along the range (Figure 6.30). For this gradient to develop, a high-pressure system must be on the side of the range that faces the direction of oncoming winds; this side is called the **windward side**. On the other side of the range, or **leeward side**, a low-pressure system must be in place. When this combination of systems occurs, air flows along the pressure gradient—that is, from the windward to leeward side of the mountain range. The resulting airflow moves downslope on the leeward side, resulting in dry air that warms as it descends.

Land breeze *Nighttime circulatory system along coasts where winds from a zone of high pressure over land flow to a zone of relatively low pressure over water.*

Valley breeze *Upslope airflow that develops when mountain slopes heat up due to re-radiation and conduction over the course of the day.*

Mountain breeze *Downslope airflow that develops when mountain slopes cool off at night and relatively low pressure exists in valleys.*

Katabatic winds *Downslope airflow that evolves when pools of cold air develop over ice caps and subsequently descend into valleys.*

Chinook winds *Downslope airflow that results when a zone of high air pressure exists on one side of a mountain range and a zone of low pressure exists on the other.*

Windward side *The side of a mountain range that faces oncoming winds.*

Leeward side *The side of a mountain range that faces away from prevailing winds.*

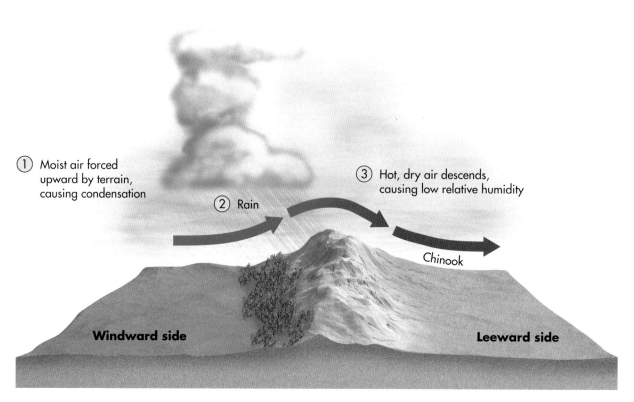

① Moist air forced upward by terrain, causing condensation

② Rain

③ Hot, dry air descends, causing low relative humidity

Chinook

Windward side

Leeward side

Figure 6.30 The development of Chinook (foehn) winds. For these winds to develop, the pressure gradient must slope from the windward to leeward side of a mountain range.

An excellent example of a valley wind is the *mistral wind*, which occurs along the eastern Mediterranean coast of France during the winter months. This wind develops when cold air sweeps across France from the English Channel, hits the Alps, and spills down the Rhône Valley into the coastal region. On days when this process is especially active, winds can gust to over 120 km/h (75 mph). Another great example of downslope winds is the *Santa Ana winds* in Southern California. These winds occur in winter when high-pressure systems dominate over the Great Basin to the east. On the southern side of the high, air flows west toward California, rising up and over the Sierra Nevada and other mountain ranges (Figure 6.31a). When the air descends on the western side of these ranges, it picks up speed as it rushes into the Los Angeles area.

Santa Ana winds are often beneficial because they clear smog from the Los Angeles metro area. On the other hand, they can fan wildfires during extremely dry conditions, sometimes resulting in catastrophic blazes that send people fleeing from the area. A recent very bad year for such fires was 2007 when a series of wildfires spread across Southern California in late October. Smoke plumes from these fires were visible from space and clearly showed the influence of Santa Ana winds (Figure 6.31b). At least 1500 homes were destroyed, and about 2000 km² (770 mi²) of land was burned. Unfortunately, nine people died as a result of these fires and over 80 were injured. Bad fires also occurred in May of 2009 and 2013, which is un-

usual because fire season usually occurs at the end of summer when the landscape is driest. The 2009 Jesusita Fire was noteworthy because it threatened Santa Barbara, a wealthy coastal enclave northwest of Los Angeles. Over 80 homes were destroyed by the fire, which lit up the hills above the city at night (Figure 6.31c).

Oceanic Circulation

Gyres and Thermohaline Circulation

So far in this chapter, we have seen how atmospheric circulatory systems operate and how they distribute heat energy around Earth. Another way that heat energy is moved is through oceanic circulation (Figure 6.32), which is strongly related to atmospheric circulation in many ways. Surface currents in the ocean are driven by winds through the transfer of energy from the air to the water by friction. As a result, the direction of surface ocean currents is related to the same pressure factors that influence atmospheric circulation. For example, the tropical easterly winds produce easterly oceanic currents in the low latitudes.

The Coriolis force also plays a major role in the movement of oceanic currents. Notice in Figure 6.32, for example, that ocean currents in the Northern Hemisphere generally move in a clockwise direction, moving warm tropical water into higher latitudes. A great example of such a current is the Gulf Stream, which transports warm water north from the tropical latitudes

(a)

(b)

NASA

(c)

© Lisa Vander Meulen/Corbis

Figure 6.31 Santa Ana winds. (a) Santa Ana winds develop when high pressure over the Great Basin forces air up and over the Sierra Nevada and other mountain ranges to the west. As the air descends on the west side of the range, it picks up speed and warms. (b) Satellite image of Southern California on October 23, 2007. Note the smoke plumes driven to the west (over the Pacific Ocean) by the Santa Ana winds. (c) Nighttime view of the Jesusita Fire in the hills near Santa Barbara in 2009. This fire burned 35.34 km² (13.5 m²) and destroyed 80 homes.

Figure 6.32 Global oceanic circulation. Ocean currents fundamentally move in the same direction as the winds. Note, for example, the easterly current along the Equator and how it relates to the easterly winds. The primary difference between ocean and air currents is that the continents interrupt the ocean currents, forming distinct gyres.

along the east coast of the United States (Figure 6.33). Currents in the Southern Hemisphere perform the same function but circulate counterclockwise. A great example of such a current is the Humboldt Current, which flows northward along the west coast of South America. This current delivers cold water from the Antarctic latitudes to the equatorial region where it warms and then flows west as the South Equatorial Current.

The primary difference between oceanic and atmospheric circulatory processes is that the continents block the movement of water. This blockage results in distinct circulatory systems, called **gyres**, such as the one in the Atlantic basin. In addition to the oceanic circulation across the surface, a slow mix of water also occurs vertically between layers of the ocean. These currents, which together make up the **thermohaline circulation**, also called the *oceanic conveyor belt*, which links all the world's oceans, are generated because of regional differences in water density that evolve through variations in temperature and salinity. In general, water at the surface of the ocean is typically warmer and less salty than deeper water. A good place to witness the beginning of this system is in the western part of the tropical Atlantic basin, shown in Figure 6.34. Here, as part of the North Atlantic gyre, warm water flows northward in the Gulf Stream (as illustrated in Figure 6.33). As this warm water travels to higher latitudes, extensive evaporation occurs, which increases the relative salinity of the current. Along with

NASA, Courtesy of Otis B. Brown, Robert Evans, and M. Carl, University of Miami, Rosentiel School of Marine and Atmospheric Science

Figure 6.33 Sea-surface temperatures in the western Atlantic. NOAA-7 satellite image showing sea-surface temperatures for a week in April. Red represents the warmest water, which is flowing northeastward along the east coast of North America in the Gulf Stream. Greens, blues, and purples are progressively cooler sea-surface temperatures.

Figure 6.34 Surface and deep water circulation loop in the world oceans. Note the locations where water upwells and downwells. These combined processes drive the circulation of ocean currents around the world.

Gyres *Large oceanic circulatory systems that form because currents are deflected by landmasses.*

Thermohaline circulation *The global oceanic circulatory system that is driven by differences in salinity.*

Downwelling current *A current that sinks to great depths within the ocean because water temperature drops and salinity increases.*

Upwelling current *A current that ascends to the surface of the ocean because water temperature increases and salinity decreases.*

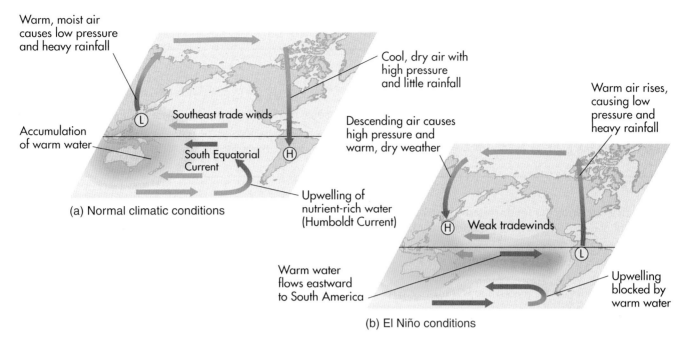

Warm, moist air causes low pressure and heavy rainfall

Cool, dry air with high pressure and little rainfall

Southeast trade winds

Accumulation of warm water

South Equatorial Current

Descending air causes high pressure and warm, dry weather

Warm air rises, causing low pressure and heavy rainfall

(a) Normal climatic conditions

Upwelling of nutrient-rich water (Humboldt Current)

Weak tradewinds

Warm water flows eastward to South America

Upwelling blocked by warm water

(b) El Niño conditions

Figure 6.35 Atmospheric circulation and surface water flow in the Equatorial Pacific. (a) Normal conditions occur when strong easterly flow pushes warm water into the western Pacific. (b) An El Niño occurs when easterly flow weakens, allowing warm water to collect along the South American coast. Note the relationship between precipitation and the location of pressure systems.

this increased salinity, the surface waters gradually cool as they continue northward.

Due to decreasing temperature and high salinity, the water density increases and the water sinks in the northern Atlantic in Figure 6.34. In other words, the sinking water becomes a **downwelling current**. This downward motion is enhanced by the development of sea ice at high latitudes, which pushes most salt into the water below the ice and thus further increases the salinity of the water and its density compared to the water around it. The resulting downwelling current subsequently flows at great depths to the southern part of the Atlantic basin. From there, it flows to the east between Antarctica and Australia to the Pacific Ocean, where it then flows northward toward Alaska. As the water moves into the tropical regions of the Pacific, it warms and becomes an **upwelling current** that moves back to the surface. This pattern is repeated in several places around the world's oceans. To give you an idea of the time involved for ocean currents like this to move, research indicates that it takes about 2000 years for water to flow from Great Britain in the northern Atlantic Ocean to the southern Pacific Ocean!

El Niño

Like many other Earth processes that can depart from the normal pattern, the oceanic circulatory system sometimes functions in unusual ways. One of the most important anomolies in oceanic circulation is the *El Niño* phenomenon, which is a reversal of the "normal" flow in the tropical Pacific Ocean. During most years, the dominant current in the tropical Pacific is easterly along the South Equatorial Current (see Figure 6.32). As you might suspect, this easterly flow is caused by the east-

erly trade winds that flow from strong subtropical high-pressure systems to the ITCZ (Figure 6.35).

When the easterly pattern is in place, the flow of water away from the South American coast allows cold water from the deep-flowing Humboldt Current (which flows northward from Antarctica along the coast of Chile) to upwell (rise to the surface) in the eastern Pacific near the Equator. Due to this upwelling, the overlying air is cool and descending, which, in turn, results in little evaporation of moisture from the ocean and, consequently, little precipitation. You can see the effect of this low precipitation in the lack of vegetation along the west coast of South America in Figure 6.18. As the water flows westward along the Equator, however, it warms because of the high Sun angle. By the time the current reaches the western Pacific, the water is quite warm, which results in overlying (atmospheric) low pressure and heavy rains in that part of the basin.

For some as yet unexplained reason, the "normal" circulatory pattern in the Pacific Ocean reverses every three to eight years. The apparent cause of this reversal is that the normally strong tropical easterlies weaken, allowing a distinct westerly flow to develop in surface ocean waters. This opposing pattern is called El Niño, derived from the expression *Corriente del Niño*, or the "Current of the Christ Child," used by Peruvian fishermen to describe the influx of warm waters into the eastern Pacific Ocean. This reversal of surface flow usually occurs around Christmastime (hence the name) and is particularly noticeable to fishermen because the warmer waters are relatively sterile when compared to the nutrient-rich upwelling current (Humboldt Current) that normally supports superb fishing. In addition to the depleted fish industry, an El Niño brings intense storms to the normally dry parts of

To investigate El Niño more thoroughly, go to the *Geo Media Library* and select *El Niño*. This animation has four parts that demonstrate how the patterns evolve, recover, and impact the coast of Peru. Once you complete the animation, be sure to answer the questions at the end of the module to test your understanding of this concept.

South America (and even the southwestern United States) and drought to the western Pacific. This weather reversal occurs because the atmospheric pressure in the eastern Pacific is relatively low, whereas it is relatively high in the western Pacific. Remember that air flows from areas of high pressure to low pressure, which, in this case, pushes the warm surface waters to the east side of the Pacific.

KEY CONCEPTS TO REMEMBER ABOUT OCEANIC CIRCULATION

1. Oceanic circulation is driven by winds and generally flows in the same direction as atmospheric circulation. Due to the presence of the continents, however, distinct circulatory cells, or gyres, exist in the oceans.

2. Like that of the atmosphere, the primary consequence of oceanic circulation is to redistribute heat energy around Earth.

3. Warm, generally low-salinity currents flow along the top of the oceans. In contrast, cooler, saltier water flows at great depths.

4. Normal equatorial circulation in the Pacific is easterly, resulting in extremely warm waters in the western Pacific and upwelling cold waters in the eastern part of the basin. This normal pattern causes precipitation in the west and dry conditions in the east.

5. An El Niño occurs when the normal (equatorial) circulatory system in the Pacific is reversed. This reversal results in drought in the western Pacific and strong storms in the eastern Pacific.

SUMMARY OF KEY CONCEPTS

1. Air pressure refers to the weight of air distributed on the surface of Earth. The two basic kinds of pressure systems are high (anticyclones) and low (cyclones). High-pressure systems in the Northern Hemisphere are columns of air that rotate clockwise and descend toward the surface of Earth within their core. In contrast, low-pressure systems in the Northern Hemisphere are columns of air that lift from the Earth's surface within their core as they rotate counterclockwise. The rotation of these respective systems is opposite in the Southern Hemisphere.

2. The speed and direction of airflow depend on many variables. The pressure gradient force refers to the difference in air pressure that exists between adjacent areas. The Coriolis force refers to the deflection that occurs in winds due to the rotation of Earth on its axis. Features on the Earth's surface, such as mountains, forests, and buildings, create a frictional force that acts opposite to the wind's direction. The combined effect of the pressure gradient force, Coriolis force, and frictional force causes a spiral motion of air in both low- and high-pressure systems.

3. The global circulatory system consists of a series of separate, but connected, pressure systems that distribute air around Earth. Humans are now trying to harness this energy source in a systematic way to supplement current energy supplies.

4. In addition to large-scale wind systems, several regional and local-scale circulation systems occur, including monsoon winds, land–sea breezes, katabatic winds, and Chinook/foehn winds.

5. Oceanic circulation is driven by winds and generally flows in the same direction as atmospheric circulation. Due to the presence of the continents, however, distinct circulatory cells, or gyres, exist in the oceans.

6. An El Niño occurs when the normal (equatorial) circulatory system in the Pacific is reversed. This reversal results in drought in the western Pacific and strong storms in the eastern Pacific.

CHECK YOUR UNDERSTANDING

1. Define the concept of air pressure and discuss how it is measured.

2. Why is air pressure greater at low altitudes than high altitudes?

3. Compare and contrast airflow in a high-pressure system with that of a low-pressure system in the Northern Hemisphere.

4. Discuss the three factors that influence large-scale atmospheric circulation.

5. Compare and contrast the various circulatory systems within the global circulation model.

6. Which place is more likely to be a zone of convection, the ITCZ or STH? Why does this pattern occur?

7. Why is the polar front an important feature in the midlatitudes?

8. What are Rossby waves, and why are they more likely to form during the winter months than the summer months?

9. How does Earth–Sun geometry influence the position of atmospheric pressure systems?

10. What are the challenges we face in our attempt to systematically develop wind energy?

11. How is a land–sea circulatory system like a tropical Hadley cell?

12. How are land–sea breezes and mountain breezes similar? What conditions are required for these circulatory systems to develop?

13. Within the Atlantic gyre, what is the primary purpose of the Gulf Stream?

14. Discuss the El Niño pattern, how it differs from the "normal" pattern, and how it impacts weather.

ANSWERS TO VISUAL CONCEPT CHECKS

VISUAL CONCEPT CHECK 6.1

Air pressure decreases from the base of Mount McKinley to the top of the mountain. This occurs because air molecules tend to be held closely to Earth due to the force of gravity. Given that more air molecules occur at the base of the mountain than the top, the air pressure (or weight of the air) is greater at the base of the mountain. Conversely, air pressure decreases with altitude because progressively fewer air molecules exist.

VISUAL CONCEPT CHECK 6.2

The high-pressure system in this image is centered over central Europe. This can be determined because that is the approximate center of the zone of clear (or sunny) skies. Given that this anticyclone is in the Northern Hemisphere, it spins clockwise. Thus, northerly winds exist on the eastern side of the system, whereas southerly breezes are on the west. Spain is under the influence of a low-pressure system, as indicated by the dense cloud cover.

VISUAL CONCEPT CHECK 6.3

The answer to this question is *c*. The ITCZ consists of the band of clouds that extends east and west through the northern part of South America.

CHAPTER SEVEN | ATMOSPHERIC MOISTURE AND PRECIPITATION

Chapter 6 focused on air pressure and the processes associated with atmospheric circulation. As air flows, it carries many things, including dust, volcanic ash, and even small organisms. One of the most important of these atmospheric constituents is water, which moves within the air in vapor, liquid, and solid forms.

Water is a critical part of the global ecosystem because it is necessary for all living things. In addition, water serves an important regulatory function in climate patterns and processes because it moderates temperature. Water vapor in the atmosphere is especially vital because it absorbs heat from the Sun. As water vapor moves around the planet in atmospheric currents, the effect is to moderate surface temperature.

The first part of this chapter focuses on the chemical and physical composition of water and how it changes from one phase to another. From there, the discussion moves on to the way that clouds form and how precipitation occurs. You will find this information useful the next time it snows or an intense storm occurs where you live.

© Ocean/Corbis

Have you ever wondered why intense thunderstorms develop? This beautiful image shows a beautiful thunderstorm moving over the eastern shore of Lake Albert in Uganda. This chapter focuses on the various ways that water changes physical state and how it can occur.

CHAPTER PREVIEW

Physical Properties of Water

 Latent Heat

The Hydrosphere and the Hydrologic Cycle

Humidity

 Atmospheric Humidity

Evaporation

Adiabatic Processes

 Adiabatic Processes

Cloud Formation and Classification

Precipitation

 Orographic Processes

 Convectional Precipitation

Human Interactions: Drought in the U.S. Farm Belt

LEARNING OBJECTIVES

1. Discuss the physical properties of water and its three phases.
2. Understand the concept of the hydrosphere and hydrologic cycle.
3. Compare and contrast the various kinds of humidity.
4. Explain the concept of dew-point temperature and why it is an important variable in the hydrologic cycle.
5. Describe the nature of adiabatic processes and why the dry and wet adiabatic rates are used in moving bodies of air.
6. Explain the various types of clouds and why they occur.
7. Compare and contrast orographic and convectional uplift.

Physical Properties of Water

Have you ever really thought about the character of water, the way it feels, flows, and freezes? Water is fascinating stuff if you really stop to think about it. It is found almost everywhere on Earth and is crucial to life as we know it. Water covers 71% of our planet's surface and even comprises about 70% of the human body by weight. It is a very important component of the atmosphere because it stores energy that contributes to global wind patterns. In addition, it plays a major role in the regulation of climate. A good place to start investigating how water behaves in the physical environment is by first reviewing its basic properties.

Hydrogen Bonding

The logical place to begin a discussion of water is the way that water molecules are bound to each other. A drop of water is made up of billions of molecules, with each consisting of two hydrogen atoms combined with one oxygen atom (Figure 7.1a). This combination is the basis for water's well-known chemical formula: H_2O. Within a water molecule, oxygen attracts the bonding electrons

(a) Water Molecules

Hydrogen bonds between unlike charges

(b) Ice

Figure 7.1 Chemical composition of water. (a) Water molecules are composed of two hydrogen atoms and one oxygen atom, which form partially positive ends of molecules (hydrogen) and partially negative ends (oxygen). Given the attraction between positive and negative charges, relatively weak hydrogen bonds are produced between molecules. (b) Hydrogen bonds are the reason liquid water molecules form solid ice at a comparatively high temperature; it takes more energy to break these bonds than is the case for most other common liquids.

more strongly than do the hydrogen atoms. Because of this unequal attraction, the oxygen end of the molecule has a partial negative charge, whereas the hydrogen side has a partial positive charge.

Water molecules are attracted to each other because of these contrasting positive and negative charges. The negative end of one molecule is attracted toward the positive end of an adjacent molecule. This is why water molecules stick together in a drop of water and why water occurs in liquid form at normal surface temperatures. (Most chemical compounds are solids or gases at normal surface temperatures.) The attraction between molecules of water is known as *hydrogen bonding* because hydrogen atoms form particularly good bonds between molecules. These bonds are strongest, of course, when water is frozen as ice. In these circumstances, the molecules firmly bond to each other in distinct hexagonal forms (Figure 7.1b).

Another interesting attribute of water is its high surface tension. Surface tension results when molecules at the surface of a liquid have a strong attachment to each other but not to the molecules of air above them. This attraction is particularly strong in water because of hydrogen bonding. Thus, the water molecules pull harder to the sides to create the smallest possible amount of surface area, forming spherical bubbles and droplets. The high surface tension also creates an elastic "skin" on the surface of water, strong enough to allow some kinds of insects (such as water striders) and even small lizards to walk on the surface.

Yet another important feature of water is its ability to move upward in thin openings (or capillaries) against the force of gravity within the soil and plants in a process called **capillary action**. This motion occurs because water molecules pull other water molecules along through hydrogen bonding. This process is very important because it enables plants to transport nutrients from their roots up into their stems and leaves.

Thermal Properties of Water and Its Physical States

One of the most important characteristics of water is that it absorbs and releases latent heat, which, if you recall from Chapter 4, is hidden energy stored in molecular bonds. This form of hidden energy, which cannot be detected, contrasts with sensible heat that is (or "sensed"). The capability of water to store and release latent heat is important because it contributes significantly to atmospheric circulation and helps regulate climate. Understanding this property is also important because it explains how water can exist in the three physical states, or phases, familiar to you: solid (as *ice*), liquid (as *water*), and gas (as *water vapor*). The movement of water from one phase to another occurs when hydrogen bonds are formed, loosened, broken entirely, or tightened.

The transformation of these bonds is directly related to the ways that heat energy interacts with water molecules. A simple rule of thumb is that heat energy must be applied to molecules if hydrogen bonds are to be loosened or broken, and extracted from

Capillary action *The process through which water is able to move upward against the force of gravity.*

molecules if they are to be formed or tightened. When bonds are loosened, energy is transformed from sensible heat to latent heat, which has a net cooling effect because less sensible heat is sensed. In contrast, when bonds are strengthened, energy is converted from latent heat to sensible heat, which has a net warming effect.

The thermal properties of water are important because they strongly influence weather patterns on Earth, providing over 30% of the energy required to drive atmospheric processes. For example, much of the energy contained within a thunderstorm is produced when latent heat is released because water changes rapidly from gas to liquid form (this will be discussed in more detail later). This section of the chapter focuses on the way that water changes physical states and the amount of latent heat energy released or absorbed during these transformations. As you follow this discussion, refer frequently to Figure 7.2, which shows the various pathways that water follows and the amount of latent heat that is absorbed or released.

Let's begin the examination of water phases and latent heat transfers by considering what happens when liquid water turns to ice in the process of **freezing**. Think of this process as one in which progressively more hydrogen bonds develop and tighten. It actually starts when water cools to a temperature below 4°C (39°F) because the motion of water molecules begins to slow considerably, which allows hydrogen bonds to strengthen. When liquid water cools down to 0°C (32°F), the motion of the water molecules slows even further, still more hydrogen bonds develop and tighten, and ice forms. The exact temperature at which water crystallizes to form ice depends on the air pressure. Water has to cool a bit more to freeze at higher air pressure, for example, because as density increases,

---

Freezing   *The process through which water changes from the liquid to solid phase.*

Photo credits: (a) Corbis Images; (b) ImageState/Age Fotostock America, Inc.; (c) Digital Vision/Getty Images

Figure 7.2 Physical states of water.   Arrows indicate the pathways that water follows in changing between various states, when latent heat is either absorbed or released. Arrow colors reflect the net cooling or warming effect due to the transfer of sensible heat to latent heat or latent heat to sensible heat, respectively. (a) Water exists as ice at temperatures below 0°C (32°F) under normal air pressure because the molecules slow down enough to lock into hexagonal crystals. (b) At temperatures between 0°C (32°F) and 100°C (212°F), water molecules move freely, but slowly enough to remain attached to one another in the liquid phase. (c) When temperatures exceed 100°C (212°F), all water molecules move fast enough to become completely detached from one another in the vapor phase. (Note that clouds actually contain liquid water as well as water vapor; we see them because the liquid water reflects light.) The number of calories shown as absorbed or released in any given phase change assumes 1 g of water.

Let's review the concept of latent heat and how it relates to the phases of water. To do so, view the animation *Latent Heat*. This feature nicely illustrates the various phases of water and the way that latent heat is either absorbed or released. After you view the animation, be sure to answer the questions at the end to test your understanding of this concept.

increasing amounts of nitrogen and oxygen molecules go into solution. During the process of freezing, 80 calories (cal) of heat energy are released for every gram of water as the *latent heat of freezing*.

Have you ever wondered why ice floats on liquid water, even though it is solid? The reason is that ice is actually less dense than liquid water. This decreased density occurs because as more hydrogen bonds form through freezing, the medium expands and volume increases. Water can increase by 9% in volume, which is why water pipes sometimes break in homes during the winter. As you will see in Chapters 14 (Weathering and Mass Movement) and 17 (Glacial Geomorphology: Processes and Landforms), the ability of water to expand and contract has had a profound impact on shaping Earth's surface.

When heat energy is applied to and absorbed by ice at 0°C (32°F), the motion of the water molecules increases and some of the hydrogen bonds begin to break. As a result, the ice melts and becomes liquid water, which is a noncompressible fluid that conforms to the shape of its container (such as a glass). During this transformation, 80 cal of latent heat are absorbed as the *latent heat of melting* to change 1 gram (g) of ice to 1 g of water. At this point the water molecules are still bound together, but the bonds are sufficiently weak that water flows.

Now let's examine what happens when liquid water changes to water vapor. This process begins at room temperature when added energy causes hydrogen bonds to further loosen, resulting in the liberation of some water molecules to the vapor phase in the process of **evaporation**. The added heat is referred to as the *latent heat of vaporization* and amounts to 600 cal for every 1 g of water. If the temperature of water increases to 100°C (212°F), the tremendous amount of added energy causes all molecular bonds to break, and all water molecules thus move into the atmosphere as vapor. Conversely, when water vapor changes back to liquid form through the process of **condensation**, 600 cal of energy are released as the *latent heat of condensation*.

In somewhat rare circumstances, it is possible for water to change directly from the ice phase to vapor in a process called **sublimation**. This transformation occurs if the temperature of ice rapidly increases from 0°C (32°F) to 100°C (212°F). Under these conditions, 680 cal of energy known as the *latent heat of sublimation* are added to the medium. This amount of energy is derived from the addition of the 600 cal associated with vaporization *plus* the 80 cal absorbed when ice changes to liquid water. In other words, the energy associated with both the melting and vaporization processes must be accounted for. Conversely, water can change directly from the vapor phase to ice through the process of **deposition**. When this transformation occurs, 680 cal of latent heat are released, which reflects the addition of the energy released by *both* condensation and freezing.

The Hydrosphere and the Hydrologic Cycle

The previous discussion implies that water is stored in various places on Earth such as the atmosphere, lakes, and oceans. Overall, the total water realm of Earth is known as the **hydrosphere**. Within the hydrosphere, approximately 97.2% of water is stored in the oceans as salt water (Figure 7.3). The remaining 2.8% is largely freshwater, most of which is stored in massive polar ice sheets and smaller mountain glaciers. This frozen water accounts for about 2.15% of the total global water volume. Approximately 0.63% is stored within Earth as groundwater.

What about the remaining 0.02% of hydrosphere water? Although this amount is analogous to *only one drop* of liquid in a

Evaporation *The process through which water changes from the liquid to vapor phase.*

Condensation *The process through which water changes from the vapor to liquid phase.*

Sublimation *The process through which water changes directly from ice to the vapor phase.*

Deposition *The process by which water vapor changes directly to ice.*

Hydrosphere *The water realm on Earth.*

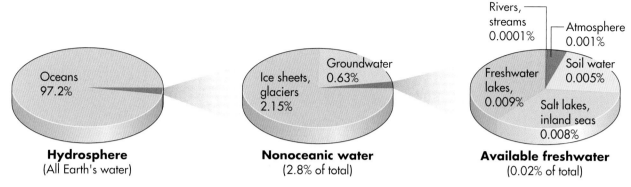

Figure 7.3 Geographic distribution of water within the hydrosphere. Most water is contained within the oceans, with proportionately smaller amounts stored in polar ice sheets and as fresh surface and soil water.

liter of water, it is the most important to us because it includes the water available for plants, animals, and human use. The right-hand pie chart in Figure 7.3 shows the geographic distribution of this part of Earth's water realm. Of this small amount, most (0.009% of the total volume) is stored in freshwater lakes. Another 0.008% is contained within salty lakes and inland seas. Soil water, which is held at shallow depths within reach of plant roots, comprises 0.005%. Of the remaining 0.002% of the total water volume, the atmosphere and streams contain 0.001% and 0.0001%, respectively.

As you probably know on some level, water moves from one place to another within the hydrosphere. You have no doubt seen water move from the atmosphere to the ground as rain to form street puddles. And then, over the course of several warm days, the water evaporates from the street as it moves back to the atmosphere. Water not only moves back and forth from the air to the ground, but also to and from places on and within the ground. Taken together, this movement of water between the various storage locations (or reservoirs) constitutes the **hydrologic cycle**.

Hydrologic cycle *A model that illustrates the way that water is stored and moves on Earth from one reservoir to another.*

Overall, the hydrologic cycle is balanced in the sense that Earth has a finite amount of water, and the amount evaporated equals the amount precipitated on a global scale. However, a variety of local and regional imbalances occur. To see how the hydrologic cycle fits into the context of the global water balance, examine Figure 7.4, which shows the flow of water on Earth in thousands of cubic kilometers (km^3) per year. With the notion of a global water budget in mind, consider that water flows to the land and ocean surfaces are positive inputs, whereas those leaving are negative. As you examine these geographic patterns, note the positive balance ($+36$) on landmasses. This surplus reflects the fact that land gains 36,000 km^3 more water through precipitation than it loses through evaporation. In contrast, oceans have a negative balance (-36). This negative balance exists because oceans lose 36,000 km^3 more water by evaporation than they gain through precipitation. Obviously, if there were no direct connection between the land and the oceans, the landmasses would gradually become submerged by water and the oceans would empty. This reversal does not occur because 36,000 km^3 of water run off the land per year and return to the oceans.

VISUAL CONCEPT CHECK 7.1

The hydrologic cycle has many different components. This image shows a typical pond in New England. Explain how this pond is part of the hydrologic cycle. What are two ways in which water could fill the pond? What happens to the water once it becomes stored within the pond? How does water return to the atmosphere from the pond?

Dennis Hallinan/Alamy

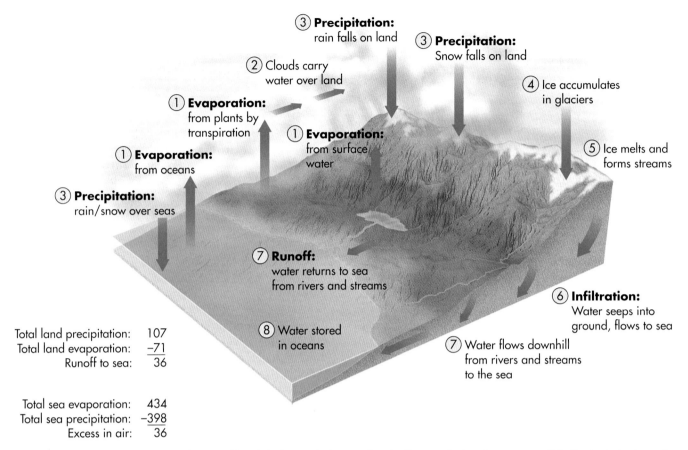

③ **Precipitation:**
rain falls on land

③ **Precipitation:**
Snow falls on land

② Clouds carry
water over land

① **Evaporation:**
from plants by
transpiration

④ Ice accumulates
in glaciers

① **Evaporation:**
from oceans

① **Evaporation:**
from surface
water

⑤ Ice melts and
forms streams

③ **Precipitation:**
rain/snow over seas

⑦ **Runoff:**
water returns to sea
from rivers and streams

⑥ **Infiltration:**
Water seeps into
ground, flows to sea

⑧ Water stored
in oceans

⑦ Water flows downhill
from rivers and streams
to the sea

Total land precipitation:	107
Total land evaporation:	−71
Runoff to sea:	36

Total sea evaporation:	434
Total sea precipitation:	−398
Excess in air:	36

Figure 7.4 The global water balance. The global water balance refers to the way that water moves within the hydrologic cycle and the relationship among evaporation, precipitation, and runoff. Oceans are a net source of evaporation, whereas precipitation processes dominate on landmasses. The system stays balanced because water runs off the land and returns to the ocean. Values are in cubic kilometers (km^3) and indicate average annual water flows to and from the Earth's land and ocean areas.

Humidity

Let's now look more closely at how water actually changes physical state from vapor to liquid and the factors that govern this process. In the atmosphere, this change of state ultimately results in precipitation.

> **KEY CONCEPTS TO REMEMBER ABOUT THE PHYSICAL COMPOSITION OF WATER AND THE HYDROLOGIC CYCLE**

1. Water molecules in the liquid or solid state are attracted to one another through hydrogen bonding.

2. When water changes its physical state, latent heat energy is either lost or gained.

3. The vast majority of water on Earth (the hydrosphere) is stored in the oceans.

4. Humans collectively depend on less than 1% of all water on Earth.

5. The hydrologic cycle refers to the movement of water from one reservoir to another by various processes such as precipitation, evaporation, and overland flow in streams.

Maximum, Specific, and Relative Humidity

This discussion begins with the concept of **humidity**, which is the concentration of water vapor within the air. Vapor concentration is expressed in three ways: *maximum humidity, specific humidity*, and *relative humidity*. Although at first it may seem confusing that three different humidity types are recognized, their definitions and interrelationships are really quite easy to follow. Here is a simple analogy to keep in mind as you work your way through this section. Imagine that you are about to fill a drinking glass to some level with water. Given that the glass can hold only so much water, you probably fill it to some level below the maximum amount of water that the glass can hold. If you try to put more

---

Humidity *A measure of how much water vapor is in the air. The ability of air to hold water vapor is dependent on temperature.*

water in the glass than it can hold, the water will spill over the top. This analogy is useful because the atmosphere works in somewhat the same way as water moves from vapor to liquid. Sometimes the air can hold a lot of vapor, similar to a large glass, and other times the amount of vapor the air can contain is less.

Maximum humidity refers to the maximum amount of water vapor that a definable body of air can hold. Although the amount of liquid a glass can hold depends on the size of the glass, the maximum humidity of an air parcel is dependent on the air temperature. An important rule to remember is that *warmer air can hold much more water vapor than colder air*. This relationship exists because air expands in the process of warming, which leaves room for more water vapor, and contracts when cooling, leaving less room. This is a very simple but important rule in physical geography that explains the processes of evaporation and precipitation. Figure 7.5 presents a simplified example of this concept.

The relationship between air temperature and maximum humidity is described by a *saturation curve*, such as the one presented in Figure 7.6. In the case of the atmosphere, the term *saturation* refers to the point where the air cannot hold any more water vapor based on the temperature at that time. To help make sense of this concept, simply think of what happens when a sponge absorbs water. The sponge absorbs water until it is saturated and unable to hold any more. The atmosphere works on much the same principle. The line on the saturation curve merely illustrates the respective intersections of maximum humidity and air temperature. Notice that maximum humidity is measured as grams of water per kilogram of air, or g/kg in abbreviated form.

You can see, for example, that at an air temperature of $-10°C$ ($14°F$) the maximum humidity is approximately 2 g/kg.

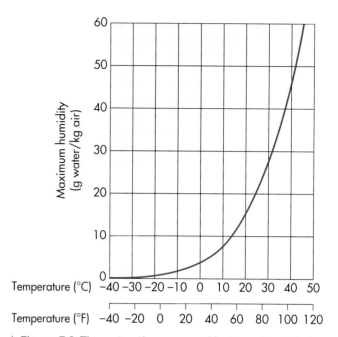

Figure 7.6 The saturation curve. Maximum humidity increases as temperature increases. In other words, warmer air can hold more water vapor than cooler air.

Notice that the maximum humidity increases exponentially with increasing temperature. At 20°C (68°F) the maximum humidity is about 15 g/kg, and at 30°C (86°F) the maximum humidity nearly doubles again to about 26 g/kg. If you are having difficulty relating this concept to the water glass, try to imagine what would happen if the glass could expand and contract with warmer and colder temperatures, respectively. In this scenario, the glass can hold a larger amount of *total* water when it is warm, and less as it contracts with progressively decreasing temperature. The atmosphere operates on somewhat the same principle in that it expands when warm and contracts when cold.

Whereas the term *maximum humidity* refers to the maximum amount of water vapor that a mass of air can hold, the term **specific humidity** refers to how much water vapor is *actually* in the air. Think back to the glass analogy. Recall that the glass can hold a maximum amount of water, which, in fact, rarely occurs in the course of everyday use. Instead, you usually fill the glass to some point that is short of the container lip. The volume of this *specific* amount of liquid can be measured; for example, the glass may contain 1 cup of water, even though the glass can actually hold 2 cups.

Finally, we have the concept of **relative humidity**, which is the ratio of specific to maximum humidity—in other words, how close the air is to being saturated. In the case of the glass, you could say that the glass is half full if it contains 1 cup of water,

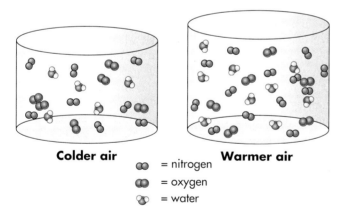

Colder air   **Warmer air**

   = nitrogen

   = oxygen

   = water

Figure 7.5 Maximum humidity and air temperature. A given parcel of warmer air can hold more water vapor than a similar parcel of colder air. The reason is that molecules at higher temperature are farther apart, so more water molecules can fit in a given amount of air.

Maximum humidity *The maximum amount of water vapor that a definable body of air can hold at a given temperature.*

Specific humidity *The measurable amount of water vapor that is in a definable body of air.*

Relative humidity *The ratio between the specific and maximum humidity of a definable body of air.*

even though it can hold 2 cups. With respect to the atmosphere, this ratio is usually expressed as a percentage and is calculated as

$$\text{Relative humidity } (\%) = \left(\frac{\text{Specific humidity}}{\text{Maximum humidity}}\right) \times 100$$

$$\text{RH } (\%) = (\text{SH} / \text{MH}) \times 100$$

The following example should make these humidity concepts clearer. Let's say that a given mass of air at a certain temperature can potentially hold 2 units of water vapor, as seen in the simplified diagram in Figure 7.7a. In other words, its maximum humidity is 2 units.

But how much water vapor is actually there—in other words, what is the specific amount of vapor present? Well, imagine that the mass of air contains a 1/2 unit of water vapor, as in Figure 7.7b. Under these circumstances, you could say that the specific humidity is a 1/2 unit. In this simplified example, you can conclude that the volume of air actually has a 1/2 unit but has a potential of 2 units, so the air contains the water vapor that it can *potentially* hold (see Figure 7.7c). In other words, *the relative humidity is 25%*.

Humidity is usually measured with an instrument known as a *sling psychrometer* (Figure 7.8). This measurement system uses two thermometers—a wet bulb and a dry bulb—mounted side by side on a small platform. The wet bulb is covered with a cotton sleeve and moistened with water of room temperature, whereas the dry bulb remains dry. At this point, the temperature of the two bulbs is the same. The two thermometers are then whirled in the air. If the surrounding air is not saturated, water from the moist

Joseph M. Moran, American Meteorological Society

Figure 7.8 Sling psychrometer. This system uses a pair of thermometers that detect the amount of cooling due to evaporation. The dry-bulb thermometer shows the actual temperature, whereas the wet-bulb thermometer indicates the amount of cooling.

cotton sleeve on the wet bulb will evaporate. As a result, heat is absorbed from the bulb, and thus the temperature of the bulb drops. In these conditions, the wet-bulb and dry-bulb temperatures are set on a scale, and the relative humidity is then calculated. The amount of evaporation, and subsequent cooling of the wet bulb, increase when the air is progressively drier. On the other hand, if the air is saturated, then no cooling of the wet bulb takes place because no water is evaporated from the cotton sleeve, and the thermometers show the same temperature. Other devices can measure relative humidity directly. One such system absorbs an amount of water vapor that depends on the relative humidity. No matter how relative humidity is measured, however, when the relative humidity is low the air is relatively dry. In contrast, when relative humidity is high, the air is close to being saturated. And always remember that warm air can hold more water vapor than cold air.

You can see how specific humidity changes with air temperature by examining Figure 7.9a, which shows the geographic relationship of specific humidity to latitude in the middle atmosphere during January 2004. Specific humidity is greater in the lower latitudes where the temperatures are generally warm. In contrast, the specific humidity is relatively low at the higher latitudes because air temperature is relatively cold.

A common mistake many students make is to assume that the relationship illustrated in Figure 7.9a means that relative humidity varies in the same geographic pattern, with relative humidity being higher at lower latitudes than at high latitudes. Although this pattern may sometimes follow, it is not necessarily so. In fact, the relative humidity at higher latitudes is typically higher than in warmer zones. Figure 7.9b shows the nature of this geographic distribution. Why do you suppose this inverse relationship exists? The reason is that cold air can hold less water vapor than warm air. So even though the high latitudes typically contain less water vapor, the inherently colder air is proportionately closer to saturation than the warmer air at lower latitudes. Thus, the relative humidity is usually greater at high latitudes compared to low latitudes.

(a) Maximum humidity = 2 units

(b) Specific humidity = $\frac{1}{2}$ unit

(c) Relative humidity = 25%

Figure 7.7 Hypothetical humidity values. (a) In this example, the maximum humidity is 2 units of water vapor. (b) In this example, the specific humidity is ½ unit. (c) The mass of air can hold 2 units of water vapor, but only ½ of a unit is present. Thus, the relative humidity is 25%.

Jan 2004 mean sp-hum 500mb NCEP

(a)

Jan 2004 mean RH 500mb NCEP

(b)

Figure 7.9 Humidity maps. (a) A direct correlation appears between specific humidity and latitude in the lower atmosphere (as reflected at the 500-mb pressure surface). Reds and yellows indicate high values of average specific humidity, whereas greens, blues, and purples are progressively lower levels of specific humidity. (b) Higher latitudes usually have higher relative humidity, as indicated by the orange and yellow shades. Greens, blues, and purples represent areas with lower amounts of relative humidity.

The same kinds of changes in relative humidity associated with latitude can also be seen over a typical day at any location that experiences strong diurnal changes in temperature—in other words, changes that occur over the course of a day and night. Figure 7.10 shows air temperature warming over the course of the day from 4:00 a.m. to 3:00 p.m. As the warming process continues, the maximum humidity increases because the capacity to hold vapor becomes greater. In contrast, the specific humidity, represented by the blue shading, does not change between the morning and afternoon hours. What happens to relative humidity as the day progresses? It gradually decreases as the ratio of

Figure 7.10. Relationship among air temperature, time of day, and relative humidity. Although specific humidity does not change in this example, maximum humidity increases while relative humidity decreases.

specific humidity to maximum humidity becomes less. If you lived in a place where such a daily temperature/humidity change occurred, you might notice that the air became slightly less sticky as it warmed over the course of the day.

Dew-Point Temperature

As shown in Figure 7.10, the relative humidity of a hypothetical air mass decreased over the course of an imaginary day because the temperature warmed from 5°C (41°F) to 30°C (86°F), creating a larger capacity to hold water vapor in the afternoon. This decrease in relative humidity occurred despite the fact that the specific humidity *did not change*. Assuming that the air mass would cool over the course of the evening and by early morning would be back to 5°C (41°F), without any change in specific humidity, what would the relative humidity be? To calculate this value, simply work backward in the process and watch how the capacity of the air mass to hold moisture decreases until it is saturated. In this particular example, if the air cooled to 15°C (59°F), the relative humidity would increase to 50%. If the air continued to cool to 5°C (41°F), then the relative humidity would rise to 100%.

Another way to look at this example is to say that the **dew-point temperature** of this air mass is 5°C (41°F). The

---

Dew-point temperature *The temperature at which condensation occurs in a definable body of air.*

Figure 7.11 Formation of dew. Water condenses on grass when nighttime temperatures cool to the dew-point temperature.

dew-point temperature, therefore, is an important designation and is defined as the temperature at which a mass of air becomes saturated. The term *dew point* originates from the formation of dew on grass (Figure 7.11). This occurs when air at ground level cools to the dew-point temperature at night and water condenses on blades of grass. As the year progresses, dew typically forms during the early fall and later part of the spring, when nighttime temperatures cool sufficiently for condensation to occur at ground level, but not so much that water crystallizes as it does in winter. Dew is less likely in the summer months because nighttime temperatures remain relatively high and the dew-point temperature is not reached at ground level.

Higher in the atmosphere, the dew-point temperature is important because it marks the temperature at which water begins to condense into liquid form as clouds, fog, and ultimately various forms of precipitation. This condensation occurs because the air mass can no longer hold any more water vapor, either because the air cools to the dew-point temperature or because more water vapor is added to the air mass through evaporation of surface water. Whereas the maximum humidity of an air mass is a function of its temperature, the dew-point temperature depends on the specific humidity, which is another way of referring to how much water vapor is actually in the air.

You can observe the trend in dew-point temperature by reexamining the saturation curve in Figure 7.6. This time, first consider the humidity values on the vertical (y) axis and then temperature data on the horizontal (x) axis. In this fashion, you are simply reversing the association you made earlier when you determined the maximum humidity at certain temperatures. For example, if the specific humidity of the air is 2 g/kg, then the dew-point temperature is $-10°C$ ($14°F$). If the specific humidity increases to 15 g/kg or 26 g/kg, then the dew-point temperature also rises, in this case, to $20°C$ ($68°F$) and $30°C$ ($86°F$), respectively.

In an effort to integrate all of these new concepts into a coherent mathematical model that explains how precipitation occurs, let's look at some examples in Figure 7.12. First consider an air mass with a temperature of $20°C$ ($68°F$), which you can see in Figure 7.12a. At this temperature, the maximum humidity is 17.3 g/kg. That is, the air mass has the *potential* to contain 17.3 g/kg of water vapor. If the specific humidity is also 17.3 g/kg, that means the parcel is saturated, the relative humidity is 100%, and the dew-point temperature is $20°C$ ($68°F$). On the other hand, if the specific humidity is only 9.4 g/kg (Figure 7.12b), then the relative humidity is about 54% and the dew-point temperature is approximately $10°C$ ($50°F$). If the air temperature were to increase without a change in specific humidity, the relative humidity would decrease further.

How could we get the relative humidity of the drier air mass to increase? This kind of increase can happen in one of two ways. One way would be to somehow increase the amount of water vapor in the air mass, perhaps through the process of evaporation at a nearby ocean source. For example, if another 4 g/kg of vapor was added to the air mass through this process (Figure 7.12c), then the specific humidity would be 13.4 g/kg and the relative humidity would increase to 78%. At the same time, the dew-point temperature would increase to about $16°C$ ($61°F$). The second way would be to lower the temperature (Figure 7.12d), let's say to $15°C$ ($59°F$). If this cooling process occurred at the same time that specific humidity remained 9.4 g/kg, then the relative humidity would increase to about 74% because the air could hold less moisture; in other words, the maximum humidity would be less.

An important reason why understanding atmospheric humidity is relevant to your life is that it allows meteorologists to predict when and how precipitation will occur. You may have encountered such a discussion on your local weather station, when the weatherperson talks about how much moisture is in the air at a given time, especially when thunderstorms are possible. When these scenarios occur, the meteorologist may say that the dew-point temperatures in your region are "currently high." This is simply another way of saying that the air is laden with water vapor and it would take much cooling to set off a storm.

At another level, understanding atmospheric humidity is critical for agriculture because it is closely associated with crop production. Farmers are keenly aware of atmospheric moisture conditions because their crops depend on the appropriate amount of water, which, in turn, is relevant to you because the price of food is directly related to supply. If an extended period of drought occurs in a major agricultural region, production yields decrease dramatically and the cost of food increases.

Atmospheric humidity conditions also have ramifications for human comfort and health, especially during the warm, humid days of summer when dew-point temperatures are elevated. When the heat index is high, most people are very uncomfortable working outside for any length of time. The reason for this discomfort is that our bodies do not regulate internal temperature effectively when the air is warm and muggy. The primary way that we regulate internal temperature is through the production of sweat. When the air is dry, sweat evaporates and body heat is removed in the same manner that the wet bulb cools on a sling psychrometer. On warm/humid days, however, sweat evaporates less efficiently and our core temperature increases. If the sweat-producing activity continues, body temperatures can rise to the point that heat stroke occurs and the temperature regulation system shuts down completely.

(a) Air mass temperature = 20°C (68°F); MH = 17.3 g/kg

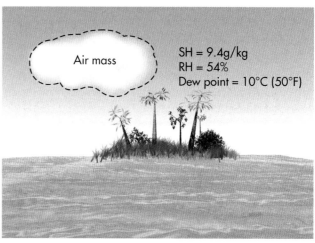

(b) Air mass temperature = 20°C (68°F); MH = 17.3 g/kg

(c) Air mass temperature = 20°C (68°F); MH = 17.3 g/kg

(d) Air mass temperature = 15°C (59°F); MH = 12.83 g/kg

Figure 7.12 Humidity examples. (a) Hypothetical air mass with air temperature of 20°C (68°F) and relative humidity of 100%. (b) Same air mass, now with a relative humidity of 54%. (c) Same air mass, but now with increased relative humidity due to higher specific humidity. (d) Same air mass with a cooled air temperature of 15°C (59°F) and relative humidity of 74%.

www.wiley.com/college/arbogast

WileyPLUS

Atmospheric Humidity

Although the concept of humidity is fundamentally a simple one, students often have difficulty with it because the terms can seem to be interchangeable. Another problem is that it is sometimes hard to understand these processes when we are simply reading about them. Like many concepts in physical geography, humidity is particularly well suited for simulation to enhance understanding. In this context, go to the *Geo Media Library* and select *Atmospheric Humidity*. This simulation is designed so that you can interact with the concept of humidity and assess how

it varies depending on environmental circumstances. You will be presented with a variety of scenarios in which you can manipulate the amount of maximum humidity and specific humidity in a hypothetical air mass so that changes in the relative humidity and dew-point temperature occur. As you work your way through these scenarios, remember that they are representative of an almost infinite number of combinations. When you complete the simulation, be sure to answer the questions at the end to test your understanding of this important concept.

1. Three kinds of humidity exist: maximum, specific, and relative. Maximum humidity refers to how much vapor a parcel of air can hold. This variable depends on temperature, with warm air having a higher maximum humidity than colder air. Specific humidity measures how much water vapor is in the air. Relative humidity is the ratio of specific humidity to maximum humidity.

2. When the relative humidity is 100%, the air is saturated and can hold no more water vapor.

3. Although the specific humidity may not change on any given day, the relative humidity can change dramatically due to rising and falling temperature. In this case, the warmer the temperature, the lower the relative humidity.

4. The dew-point temperature is the temperature at which a mass of air becomes saturated. This is the temperature at which condensation occurs in that air mass. As specific humidity increases, so does dew-point temperature.

5. Once the dew-point temperature is reached, the specific humidity begins to decrease as water vapor condenses to liquid form. This decrease in specific humidity continues as long as the process of cooling does.

When geographers consider the humidity characteristics of a particular region, they typically combine the processes of evaporation and transpiration into the singular concept of **evapotranspiration**. Evapotranspiration rates are dependent on many factors, including:

- **Net Radiation:** Heating of plants or ground surface. Evaporation rates are higher when a lot of energy (sunlight) exists.

- **Air Temperature:** Warm air can hold more moisture than cold air.

- **Relative Humidity:** How much moisture the air is already holding relative to its capacity to contain water vapor. Evaporation rates are higher when the air is dry because plenty of room exists for more water vapor molecules in the air.

You have previously seen that air flows along a pressure gradient from high pressure to low pressure. Similarly, water vapor flows along a vapor pressure gradient from areas of high vapor pressure to areas of low vapor pressure (Figure 7.13). Close to the water surface, the air contains more molecules of water vapor. As you move higher above the water, however, the air contains fewer water vapor molecules. Water vapor flows upward from the areas with more water vapor to the areas with less water vapor.

A final factor that influences evapotranspiration rates is wind speed. Wind moves moist air away from the source of the water vapor. Evaporation rates are higher in windy conditions when the air containing water vapor molecules can be moved

Evaporation

Where does water vapor in the atmosphere come from? Liquid water is transformed into water vapor through two processes, evaporation and transpiration. As described earlier in the chapter, the term *evaporation* is used when water is lost from a surface such as soil, water, or pavement. Liquid water is also transformed into vapor when it flows through leaf pores in plants to the atmosphere. This process of water loss is called **transpiration** and can result in the transformation of tremendous amounts of water. In tropical areas, for example, some trees can transpire over 300 liters (L) (~80 gal) of water a day! You can indirectly see the process of transpiration by noting the rigid upright posture of plants. This posture is maintained by water flowing upward through the plant by capillary action toward the leaf surface. A plant wilts when the amount of water in the soil declines and therefore less flows upward through the plant.

Figure 7.13 Water vapor gradient. Water vapor flows from areas of high-density toward areas of low-density vapor content. This flow creates room for additional water to be evaporated from the source area.

Transpiration *The passage of water from leaf pores to the atmosphere.*

Evapotranspiration *The combined processes of evaporation and transpiration.*

No wind

High relative humidity over water

Lower relative humidity over land

(a)

With wind

Water molecules blown over land, away from water

(b)

Figure 7.14 Wind and evaporation. (a) When the wind is calm, little evaporation occurs because the relative humidity of the air mass above the water is high. (b) More evaporation takes place during windy days because water vapor continually flows away from the moisture source.

away from the source, making room for more water molecules to move into the air. Figure 7.14 shows an example of this concept. This is also why fans are useful cooling devices; they create a wind to increase evaporation.

Adiabatic Processes

When water evaporates into vapor, it moves into the atmosphere where it exists in the gas phase. It can, however, recondense into liquid water to form clouds where it is temporarily stored until it either vaporizes again or falls to Earth as precipitation. This section deals with the processes through which condensation occurs, with the ultimate goal of providing the foundation for understanding the precipitation process. Although many of the terms in this section

may seem like jargon to you at first, they are everyday terms to many geographers because they are used when we monitor atmospheric conditions and predict the weather on any given day.

From the discussion about relative humidity, you might surmise that condensation must be somehow related to the cooling of air to the dew-point temperature. This is a correct assumption, but, as you might also imagine, the process is more complex than it first appears. After all, nighttime cooling can cause condensation of near-surface water vapor if the saturation point of the air is reached immediately above the ground, resulting in dew or even frost if vapor changes directly to ice crystals. Neither of these processes, however, is associated with precipitation.

Precipitation occurs only when a substantial parcel of air experiences a steady drop in temperature below the dew-point temperature, which can only occur if the air parcel is somehow lifted enough for it to cool adiabatically. For purposes of discussion, think of a parcel of air as having a diameter of about 300 m (1000 ft) with uniform humidity and temperature characteristics.

Recall from Chapter 6 that the dominant way in which a parcel of air rises or descends is through changes in atmospheric pressure. When this kind of movement occurs, the air either expands as it rises or compresses as it descends (Figure 7.15 on next page). Of particular relevance here is that air cools as it rises and expands and warms when it descends and compresses. It is important to note that these temperature changes are *internal* to the rising or descending air parcel and occur without any exchange of heat with the surrounding atmosphere. These kinds of temperature changes are called **adiabatic processes** because they result solely from expansion and compression of air due to pressure fluctuations.

The Dry Adiabatic Lapse Rate

In most cases that involve a parcel of surface air, the air is not saturated, and the relative humidity is thus less than 100%. Imagine that this parcel begins to rise and expand because it is warmer than the surrounding air. When these conditions occur, the air temperature decreases at what is called the **dry adiabatic lapse rate (DAR)** as the air starts to rise. This rate is not to be confused with the *environmental lapse rate* that was discussed in Chapter 5. The environmental lapse rate refers to the average change in temperature of *still* air with altitude, which

---

Adiabatic processes *Changes in temperature that occur due to variations in air pressure.*

Dry adiabatic lapse rate (DAR) *The rate at which an unsaturated body of air cools while lifting or warms while descending. This rate is 10°C/1000 m (5.5°F/1000 ft).*

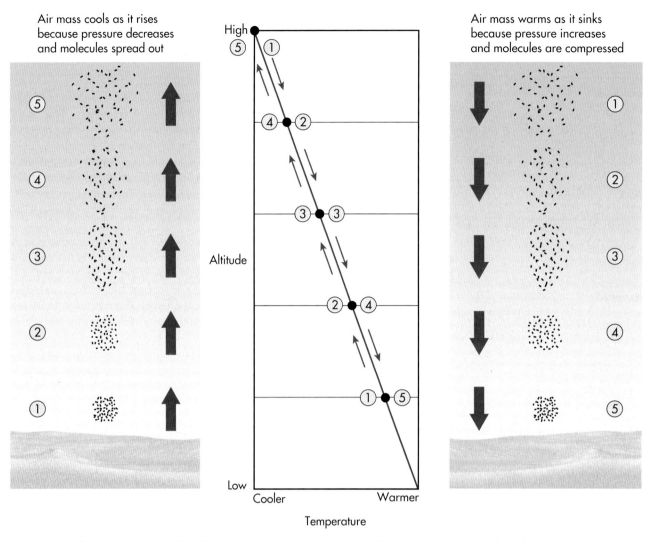

Air mass cools as it rises because pressure decreases and molecules spread out

Air mass warms as it sinks because pressure increases and molecules are compressed

Figure 7.15 Adiabatic heating and cooling. Air cools as it rises because the pressure drops and air molecules disperse. When air descends, it warms adiabatically because the air molecules are compressed by higher pressure.

can change with time or place. In contrast, the DAR is always constant and is applied to a parcel of rising air according to physical laws. The DAR has a value of about 10°C/1000 m (5.5°F/1000 ft) of vertical lift and is applied *only* when the air mass is not saturated. *This is a very important rule to remember.* You can see this rate of change if you examine Figure 7.16 and focus on the area at the bottom of the graph, which indicates altitudes of less than 1000 m.

The DAR is relevant not only to lifting air masses that are not saturated but also to descending ones. In the case of a descending air mass, the air *warms* at the DAR. That is, for every 1000 m that the air descends, it warms 10°C (5.5°F/1000 ft) due to compression of the air under pressure.

Figure 7.16 Adiabatic cooling in a hypothetical air parcel. The air cools at the dry adiabatic lapse rate (DAR) as it rises until it reaches the level of condensation. At that point, the rate of temperature change switches to the wet adiabatic lapse rate (WAR) due to the release of latent heat.

In an effort to integrate these concepts into an animated format, go to the *Geo Media Library* and select *Adiabatic Processes.* This animation reviews all the concepts that have been discussed in this section of the text. As you watch the animation, pay particularly close attention to when the DAR and WAR are

used. Also, watch what happens to the air parcel as it descends and warms at the DAR. When this process occurs, the effect is to produce a warmer air mass at the surface than the original parcel of air. After you complete the animation, be sure to answer the questions at the end to test your understanding of this concept.

The Wet Adiabatic Lapse Rate

The DAR applies when air is not saturated and the relative humidity is less than 100%. How does temperature change once the dew-point temperature is reached, saturation occurs, and the relative humidity is 100%? The altitude at which the saturation point is reached is known as the **level of condensation**, which varies depending on the particular temperature and humidity characteristics in any given air parcel. Air parcels that have relatively high humidity when they begin to lift reach the level of condensation at a lower elevation than air parcels that are comparatively dry.

Once the level of condensation is reached, water droplets begin to form if the air continues to cool. In this fashion, the air is analogous to a saturated sponge that releases water as you squeeze it at the same time that you lift it. Give it a try. All you have to do is take a sponge and let it absorb a small amount of water in the sink. Starting at the surface of the sink, slowly raise the sponge and squeeze it slowly as you lift it. At some point, the sponge will begin to release water. After you reach this point, start over but let the sponge absorb more water before you begin. Raise the sponge as before and try to exert the same amount of force as you did previously. You should notice that the sponge begins to release water at a lower elevation than before. This occurs because the sponge was closer to being saturated the second time you ran this simple experiment. The atmosphere operates on much the same principle as this sponge test.

If the air parcel continues to rise and cool, increasing amounts of water vapor condense into liquid water. During this process, another principle comes into effect—the release of latent heat of condensation that was described earlier in this chapter. This heat energy has the effect of warming the air at the same time that it is cooling due to the reduction in atmospheric pressure and expansion. The cooling effect is stronger overall, however, so the overall temperature continues to decrease if the

air continues to rise. Nevertheless, the rate of cooling in this saturated state is less than in the unsaturated state because of the release of latent heat.

 KEY CONCEPTS TO REMEMBER ABOUT ADIABATIC PROCESSES

1. Adiabatic processes refer to the temperature changes that occur in an air parcel due solely to changes in air pressure. When air pressure increases, a parcel of air is compressed and warms adiabatically. In contrast, when air pressure decreases, a parcel of air expands and thus cools internally because air molecules are spaced farther apart.

2. The level of condensation is the elevation at which condensation occurs, that is, when the relative humidity is 100%.

3. Air cools at the dry adiabatic lapse rate (DAR) until condensation occurs. This rate is 10°C/1000 m (5.5°F/1000 ft). When air warms adiabatically, it always does so at the DAR because condensation is not occurring.

4. Once the level of condensation is reached, the air temperature begins to decrease at the wet adiabatic lapse rate (WAR), which is 5°C/1000 m (2.7°F/1000 ft). This lesser rate of cooling occurs because latent heat of condensation is released when water changes from vapor to liquid.

Look at Figure 7.16 again and this time focus on the area above the condensation level to see how this change appears graphically. The cooling rate of saturated air is called the **wet adiabatic lapse rate (WAR)**. It is also known as the moist or saturated lapse rate. Although the WAR varies with moisture content and temperature, the average rate is about 5°C/1000 m (2.7°F/1000 ft); this value will be used hereafter in this text.

Level of condensation *The altitude at which water changes from the vapor to liquid phase.*

Wet adiabatic lapse rate (WAR) *The rate at which a saturated body of air cools as it lifts. The average rate is about 5°C/1000 m (2.7°F/1000 ft).*

Cloud Formation and Classification

As noted in the preceding discussion on adiabatic processes, once air reaches the dew-point temperature, through cooling at the DAR, water vapor begins to condense into liquid droplets. We see the results of this process through the development of clouds, with the base of any particular cloud representing the level of condensation in the atmosphere.

Clouds are visible masses of suspended, minute water droplets or ice crystals. Recall from Chapter 4 that clouds have a very important role in regulating the way solar radiation interacts with Earth. Some insolation is reflected immediately back into space from bright cloud tops. At the same time, clouds cause insolation to be scattered and absorbed. They also absorb longwave radiation from the Earth's surface. In addition, the character of clouds is a clue to atmospheric dynamics at any given point in time.

Figure 7.17 shows the close relationship between atmospheric water vapor and the geographic distribution of clouds in the United States on April 24, 2005. Notice in part (a) that high concentrations of vapor (shown in white) are present in the western part of the country, as well as over the Great Lakes region. These concentrations of vapor are closely associated with the band of clouds in part (b) that cover the same geographic areas.

Two conditions are necessary in order for clouds to form:

1. The air must be saturated. Air can become saturated two different ways: when air cools below the dew-point temperature or when more water vapor is added to the air.

2. There must be a substantial quantity of small particles, such as dust or pollutants, about which water vapor can collect or condense. These particles are called **condensation nuclei**.

Condensation nuclei *Microscopic dust particles around which atmospheric water coalesces to form raindrops.*

(a)

(b)

Figure 7.17 Satellite images showing atmospheric moisture on April 24, 2005. (a) GOES water vapor image. Notice the particularly high concentrations of water vapor (in white) in the west and northeast. (b) GOES visible image. Note how the concentrations of clouds are closely associated with the vapor patterns in (a).

VISUAL CONCEPT CHECK 7.2

Water vapor is produced in the exhaust of jet engines, such as those on large airliners. As the vapor cools in the air behind the plane, it often forms condensation trails, or contrails. What conditions must be present for the contrail to form?

a) Air must be close to saturation.
b) Condensation nuclei (dust) must be present.
c) The plane must be above the condensation level of the air.
d) All of the above.

Cloud Classification

Like many other things in nature, it is possible to classify clouds based on particular characteristics, specifically with respect to altitude and form (Figure 7.18). The international cloud classification scheme recognizes three categories based on their form. **Cirrus clouds** are thin and wispy clouds composed of ice crystals rather than water droplets. **Cumulus clouds** consist of individual, puffy clouds with a flat, horizontal base. **Stratus clouds** consist of layer-like, grayish sheets that cover most or all of the sky. As you can see in Figure 7.18, these clouds occur in various combinations. For example, stratocumulus clouds have characteristics of both stratus and cumulus clouds because they are individual clouds that spread out more than regular cumulus clouds.

Clouds can be further classified based on their altitude, with three categories present that you can see in Figure 7.18. *High clouds* are usually found at altitudes greater than 6 km (20,000 ft) and generally consist of cirrus clouds because very little water vapor is present and the temperatures are cold. *Middle clouds* range between approximately 2 km and 6 km (6500 ft and 20,000 ft) and include cumulus and stratus clouds that form during intervals of stable and changing weather, respectively. *Low clouds* typically occur below 2 km (6500 ft), usually in association with stratus and cumulus clouds. These low clouds are the source of most precipitation, with *nimbostratus clouds* affiliated with long-term rain or snow events and *cumulonimbus clouds* growing to great heights during short-term, severe storms.

With a fundamental understanding of clouds and their formation, it is possible to determine atmospheric conditions and even to predict weather a few days in advance. For example, if the sky is filled with individual puffy, cumulous clouds, it means that the atmosphere is fairly stable and fair weather will probably last a little while. If you happen to notice that high-level cirrus clouds are becoming more numerous and dense, it probably means that a storm system is approaching. These clouds often give way to altocumulus clouds, which indicate that the storm system is closing in and that moisture is flowing into the region from hundreds of kilometers away. If the clouds begin to thicken and expand into nimbostratus clouds, it means that some form of precipitation will most likely occur soon.

Fog

Although condensation of water vapor usually occurs at fairly high levels of the atmosphere, sometimes conditions are right for it to happen very close to the ground. When this relationship occurs, the hazy, moist air is known as *fog* and is one of the more distinctive atmospheric phenomena people encounter. As you probably know, foggy conditions can be rather eerie because visibility is poor and the Sun and street lights have an unusual glow. For these same reasons, fog is a travel hazard because drivers have difficulty seeing oncoming cars and sharp bends in the road.

Fog can form in several different ways. Perhaps the most common type of fog on autumn or spring days is **radiation fog**, which develops at night when a temperature inversion exists in the lower troposphere. Recall from Chapter 5 that temperature generally cools at a consistent rate (the environmental lapse rate) from the surface up into the higher levels of the troposphere. Occasionally, a temperature inversion develops whereby a body of cooler air lies beneath warmer air. If this cooler air reaches the dew-point temperature, radiation fog develops. This kind of scenario sometimes develops in deep valleys when cool air collects at the bottom of the valley (Figure 7.19a). You can see this type of fog when you are driving and popping in and out of fog while passing through hilly country.

In addition to radiation fog, another type of fog is **advection fog**. This type of fog develops when warm air flows over a cooler surface, such as snow or a body of water. When this process occurs, the warmer air cools to the dew-point temperature. Yet a third type of fog is **sea fog**, which develops when cool marine air comes into direct contact with the colder ocean water. In the United States, sea fog frequently develops on the coast of California, along which the cool California current flows. If you happen to live near the Bay area, or have visited in the summer, you know that San Francisco can be socked in by sea fog (Figure 7.19b).

Regardless of the type of fog that develops, you have probably noticed that it typically dissipates or *burns off* during the day. This process occurs because the cool air in which the fog forms warms above the dew-point temperature as the day progresses. Next time you experience morning fog, notice that by the late morning or middle of the day it will probably be gone.

Cirrus clouds *Thin, wispy clouds that develop high in the troposphere.*

Cumulus clouds *Individual puffy clouds that develop due to convection.*

Stratus clouds *Layered sheets of clouds that have a thick and dark appearance.*

Radiation fog *Fog that develops at night when a temperature inversion exists.*

Advection fog *Fog that develops when warm air flows over cooler air.*

Sea fog *Fog that develops when cool, marine air comes into direct contact with colder ocean water.*

Cirrocumulus: appears as white patches; usually formed from cirrus or cirrostratus torn by winds; indicates approaching surface winds

Cirrus: thin and whispy high-level clouds; known as mares' tails

Altocumulus: puffy with dark undersides; sometimes known as a mackerel sky (resembling fish scales); usually a sign of fair weather

Altostratus: thick sheets of pale gray; enough to obscure Sun or Moon; usually a sign of rain or snow

Cumulus: puffy, fleecy, fair-weather clouds, most common in summer, with blue sky visible between clouds; often evaporate at night

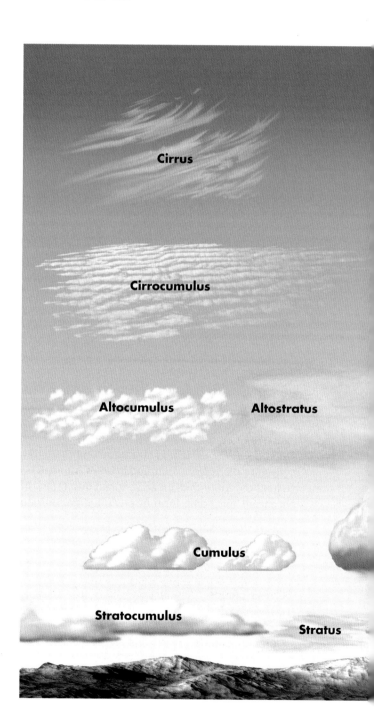

Cirrus

Cirrocumulus

Altocumulus   Altostratus

Cumulus

Stratocumulus

Stratus

Cirrostratus: thin, gauze-like sheets; usually gray and featureless but produce spectacular colors at sunsets; usually a sign of approaching rain

Cumulonimbus: thunderheads, extending to the tropopause or beyond, with typical anvil-shaped top; most storm clouds are cumulonimbus

Tropopause

Cirrostratus

High level
above 6 km
(above 20,000 feet)

Cumulonimbus

Middle level
2–6 km
(6,500 to 20,000 ft)

Nimbostratus

Low level
below 2 km
(0 to 6,500 ft)

Stratus: thick, dull gray, low-lying layers; occurs as fog at ground level, often produces mist or drizzle

Nimbostratus: dark rain clouds; form near the ground but usually extend upward

Stratocumulus: forms in patches, sheets, or layers of white or gray; can produce overcast winter skies but clears quickly in summer

Figure 7.18 Cloud classification. Clouds are classified according to their height and form. Most of these categories are variations on the three basic forms of clouds: cirrus, stratus, and cumulus.

(a)

(b)

Figure 7.19 Types of fog. (a) An excellent example of radiation fog in the Blue Mountains of Australia. This fog developed when nighttime cold air collected in the valley bottom. It was entirely gone by late morning because the air in the valley had warmed above the dew-point temperature. (b) Sea fog at San Francisco, California. This type of fog develops when moisture-laden air flows over cold water, causing the air temperature to drop.

Precipitation

As we just discussed, condensation of water vapor leads to the formation of clouds. Cloud formation does not always lead to precipitation, however, as clouds usually appear in the sky in one form or another, and it certainly does not rain or snow every time they do. This logically leads to the question, *why does precipitation occur?*

You probably already know that precipitation must involve some kind of a phase change from vapor to liquid or from liquid or vapor to solid. On the other hand, the nature of the process might be a mystery to you, as it is to most people. The bottom line is that precipitation does not occur until droplets of water are heavy enough for them to fall under the influence of gravity. Until then they remain suspended in the atmosphere.

Types of Precipitation

Droplets grow in size and become heavier by two processes: ice-crystal formation and coalescence of water droplets. Ice crystallization is the dominant process in most places outside the tropical regions and occurs because most clouds or portions of clouds extend to altitudes where air temperatures are below the freezing point of liquid water. In these places, ice crystals occur in a matrix of water vapor and supercooled water droplets that form around condensation nuclei (Figure 7.20a). The ice crystals "feed" off of the water droplets in that they cause droplets to rapidly evaporate when they are in close proximity to one another. Subsequently, the ice crystals absorb the water vapor that was released through the evaporation of the droplets, causing them to grow. Assuming the crystals grow sufficiently large in order to fall, they precipitate either as snowflakes or melt on their way to the surface and become raindrops that can grow further through the coalescence process.

In contrast to ice crystallization, which is a process of vapor attraction, raindrop coalescence is a process that merges small water droplets into large ones (Figure 7.20b). This process occurs mostly in tropical regions where the cloud tops do not reach sufficiently high for liquid water to freeze. Raindrop coalescence is the simple growth of water droplets through the persistent collision of small droplets in the upper atmosphere. One way in which this occurs is that large raindrops fall faster than smaller ones, allowing the larger raindrops to absorb the smaller droplets as they overtake them on their fall to Earth.

Precipitation can occur in several forms:

- Rain—large, unfrozen water droplets
- Snow—ice crystals that do not melt before reaching the surface
- Sleet—rain that freezes before hitting the ground
- Freezing rain—rain that freezes on impact with ground that is below the freezing point of water
- Hail—ice crystals that are continually pulled back up and grow in size during violent thunderstorms

This process is discussed in more detail in Chapter 8.

Precipitation Processes

As just described, precipitation occurs when water drops or ice crystals become sufficiently large to fall under the force of gravity. The only way that this can happen is for an air mass to rise sufficiently high to condense large quantities of water. After all, the simple formation of clouds does not necessarily mean that precipitation will occur. How, then, does an air parcel rise to the point where precipitation does occur?

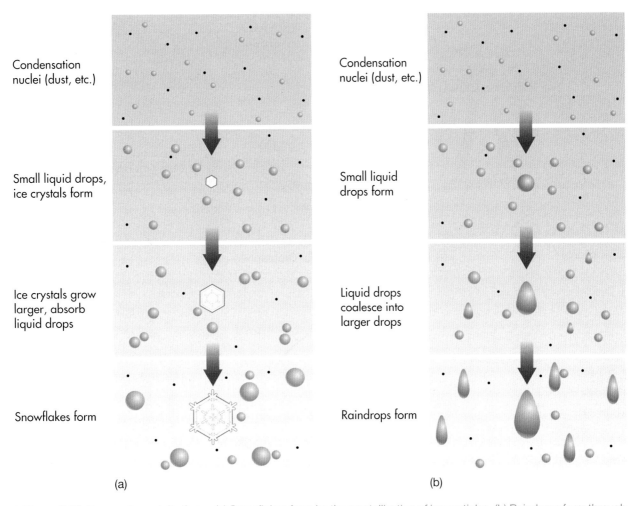

Condensation nuclei (dust, etc.)

Small liquid drops, ice crystals form

Ice crystals grow larger, absorb liquid drops

Snowflakes form

(a)

Condensation nuclei (dust, etc.)

Small liquid drops form

Liquid drops coalesce into larger drops

Raindrops form

(b)

Figure 7.20 Types of precipitation.   (a) Snowflakes form by the crystallization of ice particles. (b) Raindrops form through coalescence of small water droplets.

This kind of air lifting takes place in one of four ways (Figure 7.21). One way that air rises is through the process of **convectional uplift**, which results in distinct bubbles of air rising through denser surrounding air. Another way is called **orographic uplift**, which occurs when air is forced to flow up and over mountains. A third way is through the collision of large air masses along frontal boundaries. This process is called **frontal uplift** and typically occurs when contrasting bodies of air collide. The fourth way in which large parcels of air can uplift is through the process of **convergent uplift**. Convergent uplift occurs whenever bodies of air meet at a central location, forcing air upward at that point. Although this process is sometimes associated with low-pressure systems, it is most common in the low latitudes along the Intertropical Convergence Zone (ITCZ), where air from the Northern and Southern Hemispheres converges.

The remainder of this chapter focuses on convectional and orographic processes because they are relatively limited in a geographic extent and beautifully exemplify adiabatic processes in a way that is easy to understand. Frontal and cyclonic precipitation processes are more complex because they are associated with major weather systems that influence entire regions of the country in many different ways. These larger systems are discussed in Chapter 8.

Orographic Uplift   A good place to begin a holistic discussion of rising air, adiabatic processes, and precipitation is with the concept of orographic uplift because it makes so

Convectional uplift   *Uplift of air that occurs when bubbles of warm air rise within an unstable body of air.*

Orographic uplift   *Uplift that occurs when a flowing body of air encounters a mountain range.*

Frontal uplift   *Uplift of air that occurs along the boundary of contrasting bodies of air.*

Convergent uplift   *Uplift of air that occurs when large bodies of air meet in a central location.*

(a) Convectional

(b) Orographic

(c) Frontal

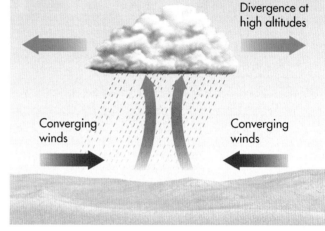

(d) Convergent

I **Figure 7.21 Uplift processes that can cause precipitation.** (a) Convectional, (b) Orographic, (c) Frontal, and (d) Convergent.

VISUAL CONCEPT CHECK 7.3

This satellite image shows the clouds over the Florida peninsula in January 2002. East of Florida is the Atlantic Ocean; west is the Gulf of Mexico. Notice that a distinct pattern of clouds is specifically associated with the Florida peninsula. Given the pattern that you see, which one of the following statements makes the most sense?

a) No convection is occurring over the Florida landmass.
b) Lots of condensation is occurring over the Gulf of Mexico.
c) Strong convection is occurring over the Atlantic Ocean.
d) The temperature of the Florida landmass is warmer than that of the surrounding oceans.

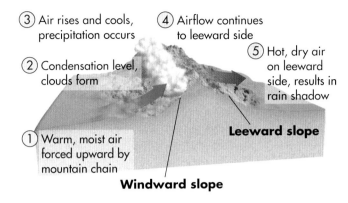

③ Air rises and cools, precipitation occurs

④ Airflow continues to leeward side

② Condensation level, clouds form

⑤ Hot, dry air on leeward side, results in rain shadow

Leeward slope

① Warm, moist air forced upward by mountain chain

Windward slope

(a)

(b)

Figure 7.22 Orographic uplift. (a) When flowing air encounters a mountain range, it must flow over the top. As it does, the air adiabatically cools on the windward side and warms on the leeward side of the range. (b) Graph of the associated changes in temperature with altitude.

much visual sense. Simply put, when airflow is interrupted by a mountain range, the air must flow up and over the barrier. For point of reference, the side of the mountain range that faces the direction of the oncoming wind is called the *windward side*, whereas the opposite, or downwind side of the range, is referred to as the *leeward side*. If the range is sufficiently high for the air parcel to reach its dew-point temperature through adiabatic cooling on the windward side of the range, clouds form and precipitation can potentially occur.

Figure 7.22 shows a hypothetical flow of air over a mountain and a graph of the associated temperature changes, which occur with altitude. The numbers in both images represent specific points for comparisons in the process. In this example, wind is flowing off the sea, let's say the Pacific Ocean, on the left side of the diagram (point 1). As the air encounters the mountain range, it begins to rise up the windward slope, cooling at the DAR until it reaches the level of condensation at point 2.

From point 2, the air begins to cool at the WAR because condensation, cloud formation, and precipitation are occurring, resulting in a landscape that may look very much like that pictured in Figure 7.23. The air continues to cool at this lesser rate, due to the release of latent heat of condensation, until the mountain crest is reached at point 3 (Figure 7.22). Here, the specific humidity of the air mass is less than it was at the level of condensation because moisture was lost through precipitation. This moisture loss is somewhat analogous to the water that is squeezed out of a sponge. Subsequently, the parcel of air continues over the mountain and descends on the leeward side (point 4).

As the air descends down the leeward side of the mountain range, a very interesting thing happens— the air dries rapidly. Recall that when air descends, it warms adiabatically at the DAR, which is 10°C/1000 m (5.5°F/1000 ft). This rate of warming takes place because as soon as the air begins its descent

(a)

(b)

Figure 7.23 Orographic clouds. (a) These clouds formed on the windward side of a small mountain range along the northeastern coast of Australia. (b) Orographic precipitation along the Inyo Mountains in eastern California.

from the mountain crest, its maximum humidity increases while the specific humidity remains constant. In other words, the air is no longer saturated and thus warms at the DAR. The air warms at this rate from the crest of the mountain range at point 3, to an elevation of 0 m (0 ft) in the leeward valley at point 5. Given the high temperature of the air mass at this point, coupled with the very low relative humidity, the leeward side of the mountain range is dry and therefore called the **rain shadow**.

The orographic pattern just described can result in a fascinating range of landscape variability in a relatively small geographical area, from windward slopes that receive abundant precipitation to leeward slopes that are very dry. A good place to see the results of these processes over a short distance is in California. Figure 7.24a shows the assemblage of landform regions of the state, including (from west to east) the low Coast Ranges, broad Great Valley, and high Sierra Nevada (mountains). The prevailing winds are westerly in the region, bringing in moisture-laden air with high dew points off the Pacific Ocean to the west. Notice that air flows over the Coast Ranges, down into the Great Valley, and then up and over the Sierra Nevada.

Figure 7.24b shows mean annual precipitation in the region. What patterns do you see? For one, observe that annual precipita-

tion is high along the Coast Ranges and Sierra Nevada, with values reaching 180 cm (about 70 in.) in the central Sierras. Much of this precipitation falls as snow, which can reach depths of over 20 m (68 ft) at the highest elevations during a snowy winter! In contrast, the Great Valley is a distinct rain shadow, with only 40 cm (16 in.) compared to the 200 cm (80 in.) that falls in some parts of the mountains to the west. This valley will be discussed in more detail in Chapter 20 when agriculture in California is examined. Another rain shadow exists east of the Sierras, with only 25 cm (10 in.) of precipitation in northeastern California.

Probably the best known example of a snowy winter in the Sierras occurred in 1846–47 when the ill-fated Donner Party tried to cross the mountains in October on their way to California in a wagon train. Just before they reached the summit at what is now Donner Pass, an unseasonally intense snow fell and they became stranded short of the pass. A series of powerful winter storms subsequently swept across the range that winter, forcing them to stay put and resulting in a snow pack about 4.5 m to 6 m (15 ft to 20 ft) thick. Of the 82 pioneers who were stranded, 34 died either at their camps or trying to escape the mountains in the harsh conditions. Many of the survivors, who were rescued in April, had lived only because they resorted to cannibalism.

Rain shadow *The body of land on the leeward side of a mountain range that is relatively dry and hot (compared to the windward side) due to adiabatic warming and drying.*

Convectional Uplift As just described, one way in which the uplift of air can result in precipitation is through the process of convection. Recall from the discussion of atmospheric

(a)　　　　(b)　　　　(c)

Figure 7.24 Airflow and precipitation patterns in California. (a) Map showing the major landform regions in California. (b) Geography of mean annual precipitation. The pattern of precipitation follows the geography of the land, with greater rainfall on the windward sides of the mountain ranges and less rainfall on the leeward sides. (c) You can see these patterns on the satellite image, with green areas representing relatively dense vegetation where high rainfall occurs, and brown zones reflecting less vegetation in areas of relatively low precipitation.

DEATH VALLEY

Do you feel physically comfortable right now? Regardless of how you feel right now, it is safe to say that it would feel decidedly uncomfortable if you were in Death Valley, California, in the summer. Located in the Mojave Desert, Death Valley lies in the rain shadow of the Sierra Nevada and has the lowest elevation in the United States at 86 m (282 ft) below sea level. It is located only 136 km (84.6 mi) east/ southeast of Mount Whitney, which has the highest elevation in the contiguous 48 states at 4421 m (14,505 ft) above sea level. Given this large elevation difference over such short geographical space, the air temperature warms dramatically as it descends from the Sierra peaks to the floor of the valley. As a result, average temperature exceeds 27°C (80°F) eight months of the year and 38°C (100°F) for five. In fact, Death Valley holds the official record for highest reported temperature on Earth at 56.7° C (134°F) in 1913. As you can imagine, the relative humidity is low, often at or around 35%. This combination of high air temperature and low humidity frequently makes conditions extreme. In July 1966, for example, the air temperature was 119°F with 31% relative

humidity. With the heat index in mind from Chapter 5, this combination of high air temperature and very low humidity meant that it felt like 64°C (148°F) that day! No wonder the place is called *Death Valley*.

Art Wolfe/The Image Bank/Getty Images

www.wiley.com/college/arbogast

WileyPLUS

Orographic Processes

Now that you have a fundamental understanding of orographic processes, we can integrate all the concepts that have been discussed to this point in the text in a simulated format. At this time, go to the *Geo Media Library* and select *Orographic Processes*. This animation will give you an opportunity to interact with the flow of air across a mountain range and the fluctuating humidity levels that occur on its ascent on the windward side and descent on the leeward side. The first part of the animation provides a review of the orographic process that shows the basic patterns. The second part is a simulation whereby you can interact with the orographic process along a fictional mountain range by adjusting variables such

as temperature and specific humidity before an air mass begins its ascent on the windward slope. You can also change the elevation of the mountains to see how this factor influences adiabatic processes. In this context, you will be presented with several scenarios that represent a minute fraction of the nearly infinite number of possibilities in nature. As you interact with this simulation, notice how the level of condensation changes as you adjust the variables. Also, pay attention to the changes that occur with respect to the various forms of humidity on both the windward and leeward mountain slopes. After you complete the animation and simulation, be sure to answer the questions at the end to test your understanding of this process.

circulation in Chapter 6 that convection is the process through which heated air rises. Convection is the initial driver of global atmospheric circulation due to the unequal heating that exists between low and high latitudes. In addition to being a primary cause of atmospheric airflow, convection is a process that contributes greatly to cloud formation and precipitation.

Have you ever noticed that, on a hot summer day, the surface of an asphalt parking lot is much hotter than the surrounding grass-covered ground? This unequal heating creates a bubble of warm air over the parking lot. Warm air rises within the larger body of air, which is relatively still (Figure 7.25a). If the air cools sufficiently, the dew-point temperature will be reached, clouds will form, and precipitation may occur. A good indicator that convection is occurring on any given day is the presence of cumulus clouds, which are the individual puffy clouds pictured in Figure 7.25b. When you see these kinds of clouds, think of each as representing a bubble of air that lifted from the surface into the atmosphere, with the base of the clouds being the level of condensation. In most circumstances, a cumulus cloud will float along for some distance downwind and then evaporate.

(a)

(b)

Figure 7.25 Convection. (a) When a warm bubble of air forms due to unequal heating at the surface, it rises within a body of relatively still air, causing the formation of cumulus clouds. (b) Cumulus clouds formed by convection on the Great Plains.

Although most cumulus clouds eventually dissipate before rain falls, some grow into cumulonimbus clouds that produce rain because convection continues until the water drops become sufficiently large to fall. If convection is very strong, a thunderstorm may develop. Whether or not this evolution occurs depends on the overall stability of the air in which the convection bubble rises. At a fundamental level, air stability refers to the potential for convection within a body of air, with **stable air** being air in which strong convection cannot occur. In contrast, **unstable air** is air in which strong convection of air bubbles can occur. As we work through this part of the discussion, refer to Figure 7.26, which illustrates air stability in graphical form.

Let's first consider an unstable air mass in which convection can occur to the point that clouds develop (Figure 7.26a). In other words, a bubble of warm air rises within a main body of relatively cool air that is stationary. This larger body of air cools with increasing altitude at the environmental lapse rate, which, on average, is 6.4°C/1000 m (3.5°F/1000 ft). Imagine, however, that the environmental lapse rate is 12°C/1000 m (6.6°F/1000 ft) in this particular body of air. If we further imagine that the temperature of this body of air is 32°C (89.6°F) at ground level, this lapse rate means that the temperature is 20°C (68°F) at 1000 m and 8°C (46.4°F) at 2000 m. In Figure 7.26a, you can see this temperature change with elevation in the green graph line on the left as well as in the vertical list of blue temperature readings on the right.

Now, imagine that a bubble of relatively warm air develops over a large dark surface. This bubble has a ground temperature of 33°C (91.4°F), as you can see in the black temperature readings. Because it is warmer at ground level than the surrounding air, the bubble begins to convect, which means that it cools at the dry adiabatic lapse rate of 10°C/1000 m (5.5°F/1000 ft) due to its decreasing air pressure as it lifts. Given that this rate of cooling is less than the rate of cooling in the primary body of air that surrounds the bubble, the bubble continues to lift. Note, for example, that the convecting bubble represented by the arrows in Figure 7.26a has a temperature of 23°C (73.4°F) at 1000 m. If we assume that this bubble of air had a specific humidity of 20.6 g/kg at ground level, it would reach the level of condensation at 1000 m and a cloud would begin to form at that elevation.

Once condensation begins in our convection bubble, the air will then start to cool at the wet adiabatic lapse rate of 5°C/1000 m (2.7°F/1000 ft), which is still a lower rate than that observed in the surrounding body of air (that is cooling at a high environmental lapse rate). As a result, the discrepancy between the temperature of the two air bodies increases further

---

Stable air *A body of air that has a relatively low environmental lapse rate compared to potential uplifting air; thus, strong convection cannot occur.*

Unstable air *A body of air that has a relatively high environmental lapse rate compared to uplifting air within it; thus, strong convection can occur.*

(a) Unstable air

(b) Stable air

Figure 7.26 Graphical representations of unstable and stable bodies of air. (a) In this hypothetical body of unstable air, strong convection occurs when the environmental lapse rate is high and a bubble of relatively warm air forms at the surface. The graphical representation of temperature on the left shows changes with respect to altitude in the bubble and surrounding air mass; these changes can also be seen in numerical form to the right within the convecting air bubble and main air body. (b) In this hypothetical body of stable air, the environmental lapse rate is relatively low and the convecting air cannot lift because it cools more quickly than the surrounding air.

and convection continues. This is why the air temperature of the convective bubble is 18°C (64.4°F) at 2000 m, whereas it is only 8°C (46.4°F) in the surrounding air.

Now, let's look at what happens in a stable body of air by examining Figure 7.26b. Imagine in this scenario that the environmental lapse rate of the main body of air is 5°C/1000 m (3.3°F/1000 ft) and that the temperature at ground level is 32°C (89.6°F), which is the same as in the previous example. In this case, the temperature of the main body of air would be 27°C (80.6°F) at 1000 m and 22°C (71.6°F) at 2000 m.

Consider what happens over the warm parking lot described in the example of unstable air. Let's imagine that the air immediately above this parking lot warms to the same temperature it did previously, that is, 33°C (91.4°F). Given that this bubble of air is warmer than the surrounding air at the surface, it begins to convect. As it does, it cools at the dry adiabatic lapse rate of 10°C/1000 m (5.5°F/1000 ft) because the air pressure decreases as the air lifts. In this example, however, this rate of cooling is *greater* than that observed in the surrounding body of air. Thus, at 500 m the temperature of the convecting air bubble

Now that you have completed the discussion of convectional uplift, let's see what the effects of this process are in the real world. Go to the *Geo Media Library* and select *Convectional Precipitation*. This animation consists of a video that shows how convection results in the formation of clouds and precipitation. After you complete the video, be sure to answer the questions at the end to test your comprehension of this concept.

---

is 28°C (82.4°F), whereas it is 29.5°C (89.1°F) in the surrounding body of air. Under these conditions, convection cannot continue because the air bubble is cooler than the main body of air. As a result, the air sinks toward the surface and cloud formation cannot occur.

Human Interactions: Drought in the U.S. Farm Belt

As you can imagine, discussion of humidity and precipitation is highly relevant to people because food production is dependent upon ample rainfall and available moisture for plants and livestock. In normal circumstances, sufficient amounts of precipitation fall within the agricultural regions of the United States to produce healthy crop and livestock yields. At other times, however, conditions develop in these zones where atmospheric humidity levels remain low for months and sometimes even years. As a result, below-average precipitation falls for a lengthy period of time and drought conditions develop.

An excellent example of an intensive drought in the agricultural heartland of the United States can be seen since 2011, especially in the Great Plains states of Texas, Oklahoma, Kansas, and Nebraska. As you will see in the discussion of climate in Chapter 9, this part of the country has a sub-humid to semi-arid climate that results from being in the rain shadow of the Rocky Mountains, as well as far from the Gulf of Mexico, which serves as the primary moisture source for the eastern United States. Although the Great Plains are less humid than the moist eastern United States, enough precipitation nevertheless falls in most years to produce healthy wheat, corn, soybean (Figure 7.27), and other crop yields. Thick stands of range grass fatten cattle and other livestock for market. Since 2011, however, a searing drought has gripped this part of the country, one that is reminiscent of the infamous dust bowl that decimated the region in the 1930s. It is certainly the most intense drought in the past 50 years, with over 60% of U.S. farmland experiencing moderate to extensive drought.

Atmospheric scientists have learned that the development of drought conditions in the central United States is related to a variety of interrelated global environmental factors that have been discussed in the past two chapters. A particularly important variable, which was in place in 2011 and 2012, is La Niña conditions. If you recall from Chapter 6, La Niña occurs when the western Pacific Ocean is warmer than normal. This pattern produces ample rainfall in the western Pacific due to high evaporation rates, low air pressure, and unstable air. On the other side of the Pacific, however, La Niña brings lower than average rainfall due to cooler ocean temperatures, increased air stability, and less evaporation. La Niña even has an impact in the central United States because shifting atmospheric circulatory patterns cause the development of a strong high pressure system in the region. This ridge of high

(a)

(b)

Figure 7.27 Healthy corn and soybean fields. Scenes like this are common throughout the U.S. farm belt in most years due to plentiful rainfall and good growing conditions.

(a)

U.S. Drought Monitor — July 24, 2012
Valid 7 a.m. EDT

(b)

Intensity:
D0 Abnormally Dry
D1 Drought - Moderate
D2 Drought - Severe
D3 Drought - Extreme
D4 Drought - Exceptional

Drought Impact Types:
↜ Delineates dominant impacts
S = Short-Term, typically <6 months (e.g. agriculture, grasslands)
L = Long-Term, typically >6 months (e.g. hydrology, ecology)

The Drought Monitor focuses on broad-scale conditions. Local conditions may vary. See accompanying text summary for forecast statements.

http://droughtmonitor.unl.edu/

Released Thursday, July 26, 2012
Author: Richard Heim, NOAA/NESDIS/NCDC

© Marvin Dembinsky Photo Associates/Alamy

(c)

© Seth Perlman/AP/Corbis

(d)

Figure 7.28 Drought in the central United States. (a) Drought conditions in the region develop when strong high-pressure systems develop and persist, causing air to compress, temperatures to rise adiabatically, and relative humidity to fall. (b) U.S. drought monitor on July 24, 2012. Severe to exceptional drought conditions existed over much of the central United States at this time (*Source*: The U.S. Drought Monitor is produced in partnership between the National Drought Mitigation Center at the University of Nebraska-Lincoln, the United States Department of Agriculture, and the National Oceanic and Atmospheric Administration. Map courtesy of NDMC-UNL.) (c) Dead corn plants in the farm belt. Scenes like this were common in the drought areas due to high temperatures and reduced available moisture. (d) Dried farm ponds like this one were common in the central United States during the drought due to lack of water and high evaporation rates.

pressure pushes the midlatitude jet stream and associated storm track to the north (Figure 7.28a).

Given the descending air associated with the high, the air over the central United States becomes remarkably stable and temperatures rise due to compression of the air molecules. Dallas, Texas, had over 100 days with temperatures >100°F between 2011 and 2012. Temperatures in central Kansas exceeded 105 for several days in July 2012. Overall, July 2012 was the hottest month ever recorded in the contiguous United States, with over 1000 daily temperature records broken the first week of the month. With such persistent searing temperatures widespread, rainfall was scarce because specific humidity was low and maximum humidity was high. Relative humidity was often only 20% to 30%, which decreased the dew-point temperature to the point where the chance of measurable rain was effectively zero. In many places accumulated rainfall was less than 50% of normal. The problem of water availability was further magnified because the high temperatures increased evaporation rates considerably, causing soils to crack. Overall, the U.S. Department of Agriculture recognized severe to exceptional drought conditions throughout much of the heartland (Figure 7.28b).

The recent drought has had enormous impacts on agriculture in the heartland. Corn and sorghum production was down in 2012 by about 27% and 7%, respectively, from average. This decrease was no surprise considering that many fields were baked dry (Figure 7.28c). By the time of the fall harvest, only 35% of the soybean crop was rated excellent. Water levels in farm ponds were very low (Figure 7.28d) due to evaporation. Given reduced water supplies and degradation of grass cover on ranchlands, ranchers were forced to sell cattle at a much greater rate than normal to reduce the size of their herds. Given the impact of the drought on crop and livestock production, the price of many farm commodities rose dramatically for consumers in the latter part of 2012. As of May, 2013, much of the region remained in severe to exceptional drought (Figure 7.29). Although the drought had eased considerably in the Mississippi and lower Ohio River valleys, it had intensified in the central and northern Great Plains. Although the drought may break in the coming months, concerns abound that dry conditions may be the "new normal" in much of the U.S. farm belt due to ongoing climate change. This relationship will be discussed in more detail in Chapter 9.

U.S. Drought Monitor

May 7, 2013
Valid 7 a.m. EDT

Intensity:
- ☐ D0 Abnormally Dry
- ☐ D1 Drought - Moderate
- ☐ D2 Drought - Severe
- ☐ D3 Drought - Extreme
- ☐ D4 Drought - Exceptional

Drought Impact Types:
~ Delineates dominant impacts

S = Short-Term, typically <6 months
(e.g. agriculture, grasslands)

L = Long-Term, typically >6 months
(e.g. hydrology, ecology)

The U.S. Drought Monitor is produced in partnership between the National Drought Mitigation Center at the University of Nebraska-Lincoln, the United States Department of Agriculture, and the National Oceanic and Atmospheric Administration. Map courtesy of NDMC-UNL.

Figure 7.29 U.S. Drought Monitor on May 7, 2013. When compared to the map in Figure 7.28b, note that the drought had eased considerably in the eastern part of the affected area. On the other hand, this map shows intensification of the drought in the central and southwestern Great Plains.

KEY CONCEPTS TO REMEMBER ABOUT PRECIPITATION AND PRECIPITATION PROCESSES

1. In order for precipitation to occur, condensation nuclei must be present in the form of dust particles or other small solids.

2. Water droplets remain in suspension as clouds until they grow sufficiently large to fall under the force of gravity.

3. For precipitation to occur, some mechanism must be present to cause sufficient uplift of an air parcel such that high amounts of condensation take place.

4. Convectional uplift occurs when a bubble of air rises due to unequal heating of the Earth's surface. The bubble of air continues to rise as long as the air is unstable, that is, as long as the temperature of the bubble is warmer than the surrounding air. When these temperatures become equal, the air is stable.

5. Orographic uplift refers to the rising air that occurs due to a blocking mountain range. The windward side of the range faces the oncoming winds, whereas the leeward side is the downwind side. Condensation occurs on the windward side; the leeward side is typically in the rain shadow and is dry.

SUMMARY OF KEY CONCEPTS

1. Water molecules are attracted to one another through hydrogen bonding. Given the unique nature of this bond, water can exist in three physical states: liquid (water), solid (ice), and gas (water vapor).

2. Latent heat is stored in water molecules. When water changes phase, latent heat is either absorbed or released, depending on the transformation.

3. The hydrologic cycle refers to the ways in which water moves on Earth through the combined processes of evaporation, precipitation, and runoff. Most water is stored in the oceans, which are a net source of evaporation in the context of the global water balance. On landmasses, more precipitation occurs than evaporation.

4. Three kinds of humidity occur: maximum, specific, and relative. Maximum humidity refers to how much water vapor a parcel of air can hold. This variable depends on temperature, with warm air having a higher maximum humidity than colder air. Specific humidity measures how much water vapor is in the air. Relative humidity is the ratio of specific humidity to maximum humidity. The dew-point temperature is the temperature at which a mass of air becomes saturated. This is the temperature at which condensation occurs in that air mass. As specific humidity increases, so does dew-point temperature.

5. Adiabatic processes refer to the temperature changes that occur in an air parcel due solely to changes in air pressure. When air pressure increases, a parcel of air is compressed and warms adiabatically. In contrast, when air pressure decreases, a parcel of air expands and thus cools internally because air molecules are spaced farther apart. Air cools at the dry adiabatic lapse rate (DAR)

until condensation occurs. This rate is 10°C/1000 m (5.5°F/1000 ft). When air warms adiabatically, it always does so at the DAR because condensation is not occurring. Once the level of condensation is reached, the air temperature begins to decrease at the wet adiabatic lapse rate (WAR), which is 5°C/1000 m (2.7°F/1000 ft). This lesser rate of cooling occurs because latent heat of condensation is released when water changes from vapor to liquid.

6. Air is uplifted such that cloud formation and precipitation occur in any of four different ways. The extent of uplift and condensation depends greatly on the stability of the main body of air. Orographic lifting occurs when air flows over a mountain range. Convection occurs when bubbles of warm air develop in a body of relatively cool air that has a high environmental lapse rate. Frontal uplift occurs when two contrasting bodies of air collide. Convergence occurs when air flows into the center of a low-pressure system.

CHECK YOUR UNDERSTANDING

1. Describe the physical properties of water and its three phases.

2. Compare and contrast the various ways that water changes phase.

3. Discuss the hydrologic cycle and the concept of a *reservoir*.

4. What are the various kinds of atmospheric humidity?

5. Which body of air would have the higher maximum humidity: an air parcel that has a temperature of 25°C (77°F) or one that has an air temperature of 15°C (59°F)?

6. Of the two air masses described in the previous question, which of the two will likely have the higher specific humidity? Which one will probably have the higher relative humidity?

7. Imagine that the dew-point temperature of an air mass is 30°C (86°F). What would the specific humidity be? (Hint: Look at the saturation curve in Figure 7.6.) Would this be a humid air mass or dry? Would the actual temperature of the air mass likely be more or less than the dew point?

8. Imagine that the temperature of an air parcel is 20°C (68°F) and the specific humidity is 10 g/kg. What would the relative humidity be? To what temperature would the air have to cool to reach a point of saturation?

9. Why are condensation nuclei important for precipitation to occur?

10. Compare and contrast the wet adiabatic lapse rate (WAR) with the dry adiabatic lapse rate (DAR). When do you use the DAR as opposed to the WAR?

11. Discuss the three fundamental cloud types—cirrus, cumulus, and stratus—and the kinds of atmospheric conditions in which they form.

12. Why is the specific humidity of an air parcel less on the leeward side of a mountain range than on the windward side?

ANSWERS TO VISUAL CONCEPT CHECKS

VISUAL CONCEPT CHECK 7.1

The pond is part of the hydrologic cycle because it is a place where water is temporarily stored. Water can enter the pond either through precipitation or from within the ground itself. Once the water is stored in the pond, it can either flow back into the ground or evaporate into the atmosphere.

VISUAL CONCEPT CHECK 7.2

The answer is *d*, "all of the above." For contrails to form, the air must be close to saturation. Condensation nuclei must be present for water droplets to form, and the plane must be above the level of condensation.

VISUAL CONCEPT CHECK 7.3

The answer is *d*. Given that the cloud pattern is distinctly associated with the peninsula, strong convection must be occurring over the land. For this to happen, the Florida landmass must be warmer than the surrounding water.

CHAPTER EIGHT

AIR MASSES AND CYCLONIC WEATHER SYSTEMS

Chapter 7 focused on the concept of atmospheric humidity and included a discussion of orographic and convectional processes, which are two ways that uplift of air can result in precipitation. These processes involve air parcels contained within a spatially defined area, such as along a mountain range or over a relatively warm surface. The discussion in this chapter broadens geographically to large air masses and complex atmospheric systems that influence weather on regional and subcontinental scales. It does so by focusing on the large tropical and midlatitude cyclonic weather systems that frequently spawn the newsmaking thunderstorms, tornadoes, and hurricanes you hear about, or even experience yourself. As a result, most students find this to be an especially fun chapter to read.

Hurricanes are the largest storms on Earth. This satellite image shows Hurricane Katrina shortly before it struck the Gulf Coast in August 2005. Katrina caused billions of dollars in damage and killed thousands of people, many of whom died when New Orleans flooded. This chapter focuses on midlatitude and tropical circulatory processes, including the evolution of hurricanes and tornadoes.

CHAPTER PREVIEW

Air Masses and Fronts

Evolution and Character of Midlatitude Cyclones

 Formation of a Midlatitude Cyclone

 Migration of a Midlatitude Cyclone

Thunderstorms

 Formation of Thunderstorms

 Tornadoes

Tropical Cyclones

 Tracks of Hurricane Katrina and Hurricane Sandy

▌LEARNING OBJECTIVES

1. Define the concept of an air mass and describe the various air masses that influence North American climate.

2. Compare and contrast warm and cold fronts, including their character and movement.

3. Explain the evolution of midlatitude cyclones, their component parts, and how they typically migrate.

4. Describe the evolution of thunderstorms and how they can become severe.

5. Discuss how tornadoes form, how they are classified, and how people monitor them.

6. Define a tropical cyclone and explain how it forms and migrates.

7. Explain the evolution of a hurricane in the Atlantic Ocean, including its various stages, migration, and how it is classified and monitored.

Air Masses and Fronts

Have you ever experienced hot and humid weather for a few days and then, *boom*, a strong storm blasted through the area, and the temperature subsequently dropped sharply? You might have even talked with your friends about how the weather changed so fast. What you probably experienced was the passage of a cold front through your region and the air masses associated with it. The first air mass was warm with high water vapor content that could have made it uncomfortable to be outside. After the storm occurred, you noticed that a new air mass dominated the area, one that was relatively cool (maybe even cold) and dry, with sunny skies. Although you may not have known why these weather fluctuations occurred, you certainly noticed the change.

Air Masses

Before we can properly investigate midlatitude circulatory processes and how they influence weather, an air mass must first be defined. This definition is essential because the formation of midlatitude cyclones depends on the interaction of very large bodies of air, covering huge geographical regions thousands of square kilometers in size, that have distinctly contrasting physical properties.

An **air mass** is any large body of the lower atmosphere that has fairly uniform conditions of temperature and moisture. Although the terminology and definition sound like scientific jargon, they actually represent conditions that you have experienced and probably have noticed. The source region of an air mass is any large body of land or water from which the air derives these characteristics. Imagine, for example, that a maritime air mass moves from the northern Pacific Ocean to northern Canada, where it stagnates for a couple of weeks in January. What will happen to that air mass? How will it change character? From previous discussions, you should realize that the temperature of the air mass will drop while it is over Canada because the landmass is cold during winter. In addition, the air mass would lose a lot of its moisture because cold air holds less water vapor than warm air. If the air mass then moved south into the United States, perhaps due to changes in the pressure gradient or the overall circulatory system, it would be cold and dry by the time it reached where you live.

Air masses are categorized by a combination of letter designations. The first letter is lowercase and designates the moisture source of the air mass:

c = continental (dry)

m = maritime (moist)

The second letter is uppercase and designates the latitude position of the source region:

A = Arctic

P = Polar—from between 50° and 60° N or S latitude

T = Tropical—from between 20° and 35° N or S latitude

Using these designations, five principal types of air masses can be identified that most often affect North America (Figure 8.1). The most common cold air mass is the *continental Polar* (cP), which forms over northern Canada and consists of stable, dry air. You have probably experienced this type of an air mass when a cold snap occurred in the middle of January or February. Such an air mass can be further cooled when it is infused with *continental Arctic* (cA) air, which is extremely cold. This kind of air mass usually originates over Siberia and drifts over the North Pole on its way south into the United States. An air mass of this kind makes you run from one building to another because it is so cold. In northern Minnesota, for example, cA air has produced a record low temperature of −43°C (−46°F). Although cP air masses are most intense in winter, they sometimes affect the weather in the summer. Such an influence is particularly noticeable when it has been hot and muggy for a few days and then the temperature cools to perhaps 24°C (75°F) with a very comfortable northwest breeze.

Another dry, stable air mass that influences North America is the *continental Tropical* (cT). This air mass forms over places like Mexico and Arizona and thus is hot and arid. Temperatures in these air masses can reach 45°C (~115°F), with low relative humidity of ~40%. If you have ever been to the southwestern deserts of the United States in the summer, at places such as Phoenix, Arizona, or Las Vegas, Nevada, you probably experienced such an air mass that just . . . takes your breath away because it is so hot.

In contrast to the continental air masses, two maritime air masses periodically influence parts of the continent. One of these air masses is the *maritime Polar* (mP), which consists of cool and moist bodies of air that form over the northern Pacific Ocean and northwest Atlantic Ocean. Given the influence of westerly winds in the midlatitudes, this kind of air has a particularly strong influence in the Pacific Northwest. If you happen to live there, or have visited places such as Seattle, Washington, or Portland, Oregon in the winter, you certainly experienced the cool, damp air associated with an mP air mass. This air often has temperatures of about 12°C (~53°F) and relative humidity of 70% to 80%. The other maritime air mass is the *maritime Tropical* (mT), which forms over places like the Gulf of Mexico and is distinctive because it is warm and moist. You have most likely experienced this air in the summer when it was hot and sticky outside, with temperatures of about 33°C (~95°F) and ~60% relative humidity. This was an mT air mass and was the kind of air that likely kept you in an air-conditioned building much of the day.

Fronts

We have just discussed the character of air masses, the geographical regions where they originate, and how they move into North America. Now let's examine how they interact

---

Air mass *A large body of air in the lower atmosphere that has distinct temperature and humidity characteristics.*

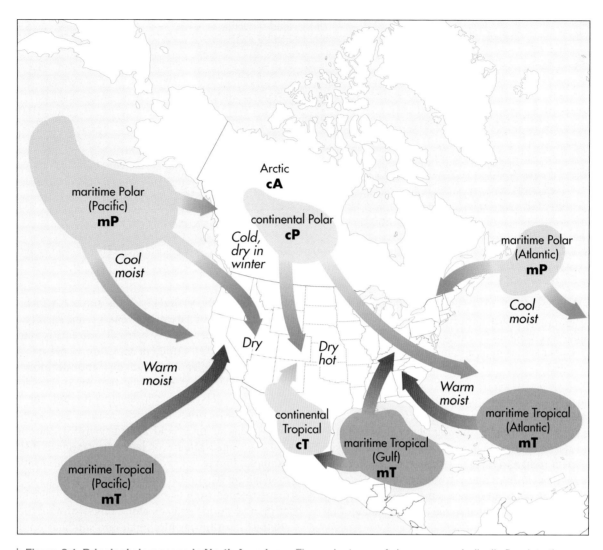

Figure 8.1 Principal air masses in North America. Five major types of air masses periodically flow into the continent. Note the character of these air masses and the geographical regions in which they originate.

with each other. The logical place to begin is by investigating the boundaries of these air masses on the Earth's surface. These boundaries are called *fronts*. The dominant front in the Northern Hemisphere midlatitudes is the polar front, which, if you recall, is the boundary between cold, dry (cP) air to the north and warmer, moist (mT) air to the south. These contrasting air masses flow parallel to one another along a **stationary front** most of the time. Sometimes, however, atmospheric conditions arise that cause one particular air mass to advance into another along distinct frontal boundaries. This part of the chapter focuses on these atmospheric boundaries and the types of precipitation that occur along them if air is uplifted sufficiently. Recall from Chapter 7 that this type of uplift is called *frontal uplift*.

Warm Fronts **Warm fronts** occur in places where warm air advances into relatively cool air. Because the warm air has the lesser density of the two bodies, it slowly slides over the top of the underlying cooler air (Figure 8.2). This process begins in the upper troposphere when warm air at this altitude gradually overrides the cooler air ahead of the front and causes the lifting mT air to cool adiabatically. As the air cools, clouds form, beginning with high-level cirrus clouds at the top of the upwardly moving air mass. As the surface warm front approaches, these clouds change to progressively lower stratus clouds, culminating in rain-producing nimbostratus clouds at the surface front. When these conditions evolve, the sky may be overcast with a slow but steady rate of precipitation that may last a day or two. This precipitation can be in the form of drizzle, light rain,

---

Stationary front *A boundary where contrasting air masses are flowing parallel to one another.*

Warm front *A frontal boundary where warm air is advancing into relatively cool air. This front is typically associated with slow, steady precipitation.*

Figure 8.2 Uplift of air at a warm front. At a warm front, the advancing warm air gradually slides over the top of underlying, passive cooler air, causing the formation of stratiform clouds as the warm air slowly cools.

and even snow. These kinds of conditions are common in the Midwest during spring when slow-moving mT air from the Gulf of Mexico interacts with the cooler air to the north. When this kind of weather settles in, it is sometimes hard to imagine that it will ever end.

Cold Fronts **Cold fronts** occur when cool air moves into a region that was previously dominated by warmer air. Because the cold air is denser and heavier than the warm air ahead of it, the warm air is forced to rise. In this fashion, a cold front is significantly different from a warm front because the cold air hugs the surface along a cold front and vigorously drives the warm air ahead of it aloft. Notice in Figure 8.3 that the edge of a cold front is very steep when compared to the angular nature of the warm front in Figure 8.2.

Once the warm air begins to lift along a cold front, cooling starts and vapor condenses, forming clouds. Because the

air rises rapidly, it quickly cools further, which means that large amounts of latent heat energy are rapidly released as the vapor condenses. As a result, rainfall is intense and of short duration. If sufficient moisture is present and enough latent heat is released, an intense **thunderstorm** can form, with the latent heat being the fuel for its development. Although thunderstorms can develop through the simple process of convection, they are prone to be more severe and widespread along a cold front because of the rapid cooling and associated release of latent heat release that occur. For example, Figure 8.4 shows how widespread frontal precipitation can be. This radar image shows a portion of a cold front that stretched from Canada southwest into Oklahoma. A line of thunderstorms extends along this section of the front from southwestern Lower Michigan into southeastern Kansas. We will describe the evolution of thunderstorms in more detail later in the chapter.

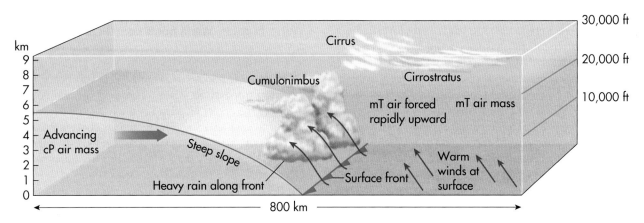

Figure 8.3 Uplift of air along a cold front. In contrast to a warm front, warm air ahead of a cold front is abruptly lifted and quickly cooled, causing cumulonimbus clouds to form.

Cold front *A frontal boundary where cold air is advancing into relatively warm air. This front is typically associated with intense rain of short duration.*

Thunderstorm *A brief but intense storm that contains strong winds, lightning, thunder, and perhaps hail.*

Figure 8.4 Frontal precipitation. This radar image shows a line of thunderstorms extending from southwestern Lower Michigan to southeastern Kansas. This line of storms developed along a strong cold front that passed through the Midwest. Orange and yellow colors represent areas of heavy rain, whereas the areas of lightest precipitation are in blue.

KEY CONCEPTS TO REMEMBER ABOUT AIR MASSES AND FRONTS

1. An air mass is a large body of air that has distinctive characteristics and forms in specific geographic regions.

2. Five principal types of air masses affect North America. Continental air masses include continental Polar (cP), continental Arctic (cA), and continental Tropical (cT). Maritime air masses are maritime Tropical (mT) and maritime Polar (mP).

3. Air masses have distinct boundaries called fronts. A stationary front is a place where contrasting air masses are flowing parallel to one another without one advancing into another.

4. A warm front is an atmospheric feature that forms where warm air advances into relatively cool air. Warm air slowly slides over the top of the cooler air along a warm front, causing slow and steady rainfall.

5. A cold front is an atmospheric feature that forms where cold air advances into relatively warm air. Given the higher density of colder air, the warm air ahead of the front is rapidly forced aloft, where it cools quickly. As a result, rainfall is intense and of short duration along the front.

Evolution and Character of Midlatitude Cyclones

With the concept of weather fronts in mind, let's now examine the atmospheric circulatory systems in which they develop. Such a system is called a **midlatitude cyclone** and is a well-organized low-pressure system that spins across a region. These systems are important because they are the mechanisms through which contrasting warm and cold air masses interact and mix. In order to understand how these systems evolve and influence weather, it is useful to review briefly how the polar front and associated jet stream move from a seasonal perspective. This

Midlatitude cyclone *A well-organized low-pressure system in the midlatitudes that contains warm and cold fronts.*

geography was examined in Chapter 6, so you might want to review Figure 6.21.

Let's begin this review in the summer when high latitudes receive high amounts of solar radiation. During this time of year, the polar front and jet stream are confined to the poles and exhibit a zonal-flow pattern much like that illustrated in Figure 6.21a. This confinement of the polar front to these high latitudes occurs, of course, because temperatures are warm well into the higher middle latitudes. Although midlatitude cyclones may develop along the polar front during this time of year, they are generally weak systems because no strong temperature contrast occurs north and south of the polar front.

As fall approaches and daily insolation at high latitudes decreases, large masses of cP air begin to develop north of the polar front. This increase of cold, dry air causes the polar front and associated jet stream to move south where it begins to encroach upon the warmer air. As this migration proceeds, the path of the jet stream gradually changes from a zonal pattern to an undulating meridional pattern with well-developed Rossby waves in the upper atmosphere (see Figure 6.21 again). This development occurs because the atmosphere is reacting to the growing temperature difference on either side of the polar front. These waves allow tongues of warm air (south of the front) to penetrate northward at the same time that wedges of cold air (north of the front) expand to the south.

On a global scale, the development of Rossby waves is how the atmosphere begins the process of mixing cold and warm air. The most vigorous mixing of these air masses occurs on a more regional scale, within midlatitude cyclones that form along distinct sections of the polar front in any given Rossby wave. The term *cyclone* is sometimes confusing because tropical storms near southern Asia are referred to by the same name. In addition, tornadoes are sometimes called *cyclones*. Although these specific kinds of storms are investigated later in this chapter, for now the term *cyclone* is used only in the context of midlatitude circulatory processes.

Formation of midlatitude cyclones is called **cyclogenesis** and encompasses a complex set of processes most often associated with an undulating polar jet stream pattern in the upper atmosphere and the interaction of these winds with airflow at ground level. To better understand this interaction, consider how the barometric pressure within the atmosphere changes from the surface to the top. In this context, it can be viewed as containing many different pressure surfaces, with the highest air pressure exerted at ground level and progressively lower air pressure at progressively higher altitudes. Just as a particular isobar (a line of equal air pressure) can be followed on the ground, it can also be traced horizontally within different levels of the atmosphere.

Cyclogenesis *The sequence of atmospheric events along the polar jet stream that produces midlatitude cyclones.*

Upper Airflow and the 500-mb Pressure Surface

An important example of a pressure surface within the atmosphere is the 500-mb surface, which occurs at a specific altitude at any given place on Earth. This altitude typically varies from one place to another. Recall that the average surface air pressure at sea level is 1013.2 mb and that air pressure decreases with altitude. Thus, the 500-mb pressure surface is at a relatively high altitude compared to the surface air pressure. From a meteorologic standpoint, the 500-mb pressure surface is important because it vertically divides the atmosphere in two, if we assume that the average surface air pressure is 1013 mb and 0 mb of air pressure occurs at the top of the atmosphere. In addition, the wind patterns at the 500-mb surface exert a strong steering influence on the circulatory patterns at the surface.

To understand the significance of the 500-mb pressure surface and how it is important to airflow, consider Figure 8.5. Figure 8.5a shows two columns of air (Air columns 1 and 2) that contain the same mass of atmosphere. Given that each column contains the same number of molecules, the surface pressure is the same. In addition, the altitude of the 500-mb surface in both columns is equal, in this case 5460 m (17,900 ft).

What happens if we warm one of the columns of air, say Air column 2, relative to the other (Figure 8.5b)? When this warming occurs, the column of air stretches vertically—because warm air rises—and the height of the 500-mb surface in that column moves to a higher altitude; let's say 5760 m (18,900 ft). At that same altitude in Air column 1, the air pressure is 480 mb. Air column 1 has this relatively lower pressure at 5760 m (18,900 ft) because the elevation of the 500-mb surface in that column did not change from the previous example (Figure 8.5a). This pressure/altitude difference between the two air columns reflects the fact that the warmer column of air now has a lower overall density than the first (cooler) one. No change occurs in the mass of Air column 2, which is why the surface air pressure is still 1000 mb. Another way of comparing the two air masses in the second example is to say that Air column 2 has a shallower vertical pressure gradient than Air column 1.

The same kind of altitude change in the 500-mb surface also occurs when air pressure at the surface varies between two regions. Using Figure 8.5a again as a hypothetical baseline scenario, let's say that the surface air pressure in Air column 1 decreases to 990 mb at the same time it increases to 1005 mb in Air column 2 (Figure 8.5c). What happens? Notice that the altitude of the 500-mb surface drops to, let's say, 5350 m (17,500 ft) in Air column 1, whereas it rises to 5600 m (18,400 ft) in Air column 2. This change occurs because the mass of air beneath the 500-mb surface decreases in Air column 1 at the same time it increases in Air column 2.

This example shows how pressure can vary within two distinct columns of air. Now examine how the altitude of the 500-mb surface might vary horizontally in the continuous medium of the actual atmosphere. Figure 8.6a presents a

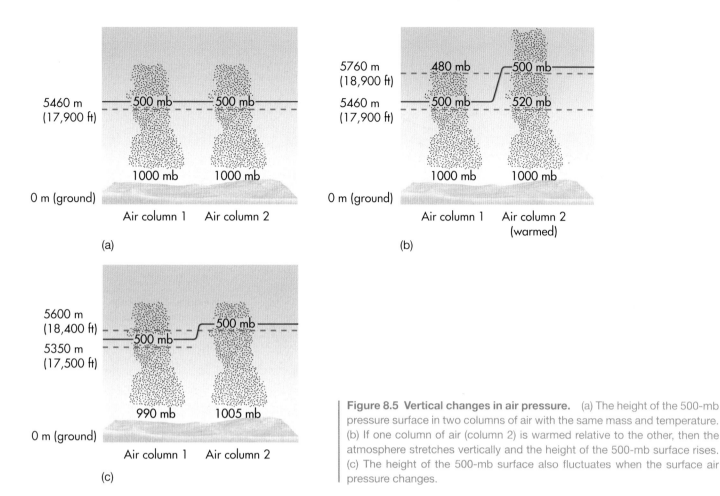

Figure 8.5 Vertical changes in air pressure. (a) The height of the 500-mb pressure surface in two columns of air with the same mass and temperature. (b) If one column of air (column 2) is warmed relative to the other, then the atmosphere stretches vertically and the height of the 500-mb surface rises. (c) The height of the 500-mb surface also fluctuates when the surface air pressure changes.

typical isoline map that shows the variation in altitude of the 500-mb surface across the United States on a particular day. This reflects the fact that surface air pressure varies across the nation in a manner similar to Figure 8.5c. The map shows a strong high-pressure system (anticyclone) in the western United States and two areas of low pressure—one over the northeastern part of the country and another off the Pacific Northwest coast. You can see how this information translates into an actual topographic map of the 500-mb surface in Figure 8.6b. Notice that the height of the pressure surface is higher and forms a ridge-like feature where the anticyclone occurs. Meteorologists call such an atmospheric feature a

Figure 8.6 Mapping the 500-mb pressure surface. (a) Isolines of constant pressure across the United States on a particular day. (b) A topographic map showing the variation in the 500-mb surface.

(a)

(b)

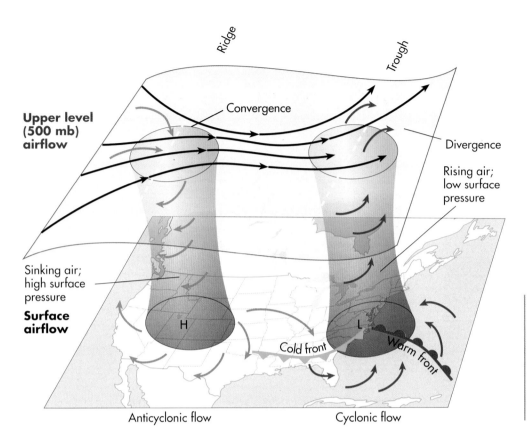

Figure 8.7 Linkage of upper airflow and surface airflow in a midlatitude cyclone. These systems form through a complex interaction of upper airflow and surface airflow. Under the high-pressure system, the weather is stable with clear sky. Where the cyclone forms, however, the atmosphere is unstable.

high-pressure ridge. In contrast, the height of the pressure surface is lower in the areas associated with the low-pressure systems along each coast. Meteorologists call these valley-like features in the 500-mb surface low-pressure troughs. Regardless of whether the upper atmospheric pressure system is a ridge or a trough, each shares the common reference of an axis, which is the imaginary line that extends along their length, that is, from north to south.

Interaction of Upper Airflow and Surface Airflow

In the context of understanding how midlatitude cyclones form, it is useful to examine the interaction of airflow on the ground with airflow at the 500-mb pressure surface. As we begin this part of the discussion, follow along by studying Figure 8.7, which shows airflow both on the ground and at the 500-mb altitude. This figure is very similar to Figure 6.12 because it shows a dynamic convection loop in the atmosphere, but differs because it is viewed within the context of the 500-mb surface.

Let's imagine a scenario in which a prominent Rossby wave exists in the upper atmosphere and a low-pressure trough has developed poleward of the jet stream at the 500-mb level.

When this condition develops, the upper airflow becomes disturbed, with a region of converging airflow developing behind (to the left of) the trough axis in Figure 8.7. At the same time, a zone of diverging air forms east (to the right) of the trough axis. As these airflow patterns continue to develop, the air pressure beneath them (at ground level) changes. At places where the air converges aloft, more air flows into a given unit of space at the 500-mb level. This convergence forces air toward the surface as a zone of high pressure. In contrast, where the air diverges aloft on the downwind side of the trough, less air is occupying a given unit of space, which has the effect of pulling air up from beneath to fill the void. When this situation occurs, the surface and upper airflows become closely linked as columns of spinning air—in other words, a dynamic convection loop—that are set in motion by the Coriolis force, as described in Chapter 6.

To understand better how midlatitude cyclones form and how they influence the weather on the ground, let's investigate the system evolution at the surface by studying the section of the polar front outlined in Figure 8.8. Notice that prior to the development of a midlatitude cyclone, the polar front is linear, with parallel-flowing and opposing winds—cold, easterly winds to the north of the front and warm westerly winds to the south.

High-pressure ridge *An elongated area of elevated air pressure in the upper atmosphere that is typically associated with sunny skies and calm winds.*

Low-pressure trough *An elongated area of depressed air pressure in the upper atmosphere that is typically associated with cloudy skies and rain.*

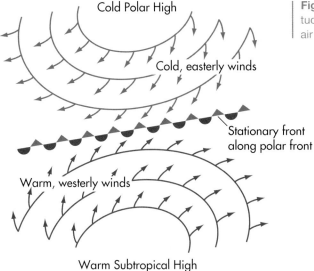

Cold Polar High

Cold, easterly winds

Stationary front along polar front

Warm, westerly winds

Warm Subtropical High

Figure 8.8 Section of the polar front. Prior to the development of a midlatitude cyclone, a stationary front exists along the polar front. Winds in contrasting air masses flow parallel and opposite to each other.

Cyclogenesis

As previously described in the discussion about vertical airflow within Rossby waves, cyclone development begins when the upper airflow and surface airflow along a stationary front become linked through the rise of warm air and descent of cold air (Figure 8.9a). At the surface, this developing linkage causes an undulation to form in the polar front (Figure 8.9b). As this undulation becomes better defined, the convergence of surface winds intensifies in a counterclockwise pattern, with cold air flowing south on the backside of the system, while warm air flows northward to the east of the circulatory core. In this situation, the leading edge of the inflowing cold air mass is the cold front, whereas the leading edge of the warm air mass is the warm front. Notice in Figure 8.7 how these fronts relate to the dynamic convection loop.

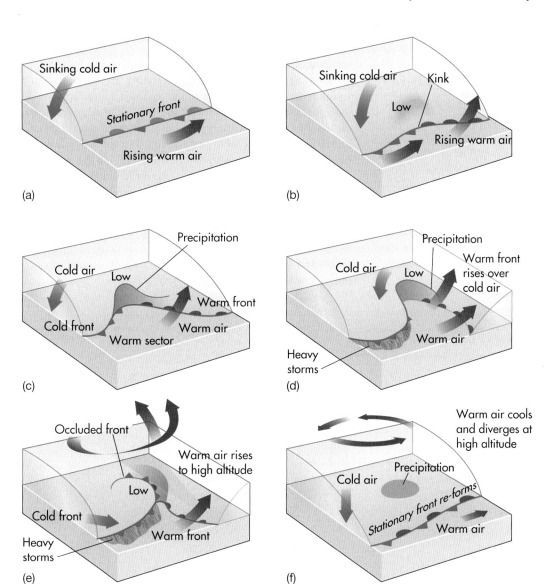

Figure 8.9 Evolution of a midlatitude cyclone. (a) Stationary front stage; (b) early stage when undulation develops along stationary front; (c) open-wave stage with prominent cold and warm fronts; (d) closing of open wave due to rapid speed of advancing cold front; (e) occluded stage when cold front begins to overtake the warm front; (f) dissolving stage when the warm sector is completely pinched off by overtaking cold front.

Figure 8.10 A classic midlatitude cyclone. A cyclone centered over the midwestern United States. Notice the prominent line of clouds, which developed along a strong cold front, extending from Illinois to eastern Texas. To the east of the cold front the air was warm and muggy, whereas west of the cold front it was relatively cold and dry, as indicated by the clear skies in western Kansas and eastern Colorado.

The developing cyclone is said to be in the *open stage* when the warm and cold fronts are fully separated; that is, when warm air exists at the surface all the way to the system's core (Figure 8.9c). This geographical area of warm air is referred to as the *warm sector*. Imagine that such a system develops over the northern Great Plains in the United States. In this scenario, the air behind the cold front (i.e., to the north) would consist of cold, dry (cP) air that originated in Canada, while the air in the warm sector that lies between the cold and warm fronts is relatively warm and moist (mT) air that may have come from the Gulf of Mexico. At this stage the cyclone may very well look like the one pictured in Figure 8.10.

When the cyclone is in the open stage, it is fully mature. This stage does not last very long, however, because the cold front moves faster than the warm front ahead of it. These contrasting frontal speeds occur because the air behind the cold front is denser than the warm air ahead of it and thus flows more quickly. In this fashion, the cold front drives into the warm, moist air ahead of it, whereas the warm front gradually slides over the top of the stationary cold air ahead of it (Figure 8.9d). As a result of this difference in speeds, the cold front begins to overtake the warm front in the *occluded stage* of development (Figure 8.9e). Although a distinct warm sector still exists in the southerly parts of the system, the overtaking cold air lifts warm air aloft near the center of the low. An **occluded front** develops when the cold front overtakes the warm front and begins to drive the warm air at the surface to a higher altitude. This process of occlusion continues as the system enters the *dissolving stage* (Figure 8.9f), at which time all the warm air is pinched aloft and a stationary front is reestablished.

Remember that the evolution of midlatitude cyclones occurs within the context of the overall westerly wind pattern in the jet stream. You can think of the jet stream as the medium that steers cyclones across the continent. Figure 8.11 shows an example of this migration, beginning with an open-stage cyclone centered in the central United States (Figure 8.11a). This view of the system shows the position of the warm and cold fronts and their relationship to the air mass source regions. In the open stage of development, mT air in the warm sector is clearly flowing north from the Gulf of Mexico into Louisiana and Arkansas. If you were there on that day, the weather would be warm and muggy, with southerly winds. To the north and west of the cold front, in contrast, the air is flowing in a southerly direction from Canada into places like the panhandle of Texas and Oklahoma. In these places, the temperature would probably be cold, with clear sky and northwest winds.

Occluded front   *The area where a cold front begins to overtake a warm front and thus lift warm surface air aloft.*

This image shows temperature in the United States. Where is the approximate location of the cold front?

a) Western Texas.
b) Florida.
c) It cuts across the states of Washington and Oregon.
d) It extends from western Michigan to central Texas.

Formation of a Midlatitude Cyclone

We can now examine how the development of midlatitude cyclones occurs in an animated way by viewing the animation *Formation of a Midlatitude Cyclone*. This module allows you to see the process of cyclogenesis in motion. Figure 8.9 is the foundation of this animation. As you watch the animation, follow how an initial kink in the atmosphere evolves into a mature cyclone that spins counterclockwise in the Northern Hemisphere. This animation will help you better understand how these systems and associated processes cause highly variable weather when they migrate through a region. Once you have completed the animation, be sure to answer the questions at the end of the module to test your understanding of this concept.

Now let's move on in our hypothetical sequence of cyclonic evolution to Day 2 (Figure 8.11b). Two important things are seen in this diagram. First, notice that the cold front is beginning to overtake the warm front, as you might have predicted. As a result, the temperature in Louisiana and Arkansas, which had been warm the previous day due to southerly airflow, is now colder because the winds are now northwesterly. Second, the center of the system has migrated slightly to the northeast along with the wave in the jet stream. Eastward migration continues into Day 3 (Figure 8.11c), with the onset of the occluded stage as the system moves into the northeastern part of the country.

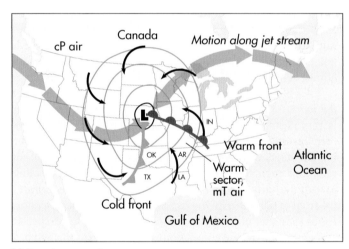

(a) Day 1, open stage

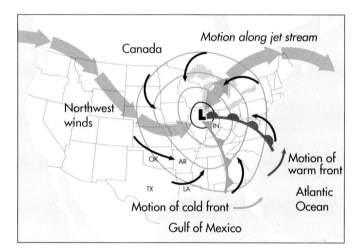

(b) Day 2, cold front overtaking warm front

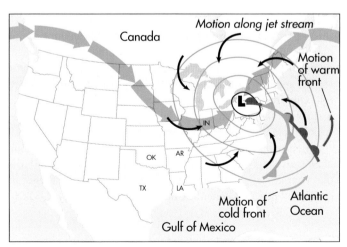

(c) Day 3, start of occluded stage

Figure 8.11 Evolution and migration of a cyclone. Midlatitude cyclones migrate from west to east as they spin counterclockwise and evolve from initial formation, through the mature stage, and into the occluded stage.

It is now time to visually integrate the various moving components of a midlatitude cyclone. View the animation *Migration of a Midlatitude Cyclone*. This module contains a video that shows the migration of a typical midlatitude storm system in North America over the course of several days. As you watch this system evolve, pay attention to *both* its counterclockwise circulation and its movement from west to east. See if you can identify the center of the system and associated cold and warm fronts. After you complete the video, be sure to answer the questions at the end of the module to test your understanding of this concept.

KEY CONCEPTS TO REMEMBER ABOUT THE FORMATION OF MIDLATITUDE CYCLONES

1. Midlatitude cyclones spin in a counterclockwise fashion in the Northern Hemisphere as seen from above.

2. Midlatitude cyclones occur when undulations form in the polar front jet stream. These undulations are related to the altitude of the 500-mb pressure surface.

3. As a cyclone spins in the Northern Hemisphere, it pulls in mT air from the south on its eastern side. This warm, moist air encounters cold air as it moves to the north.

4. As a cyclone spins in the Northern Hemisphere, it pulls cP air down from the north on its western side. This cold, dry air encounters warm air as it moves to the south.

5. As the system spins, it also migrates from west to east.

Thunderstorms

You have probably experienced a severe thunderstorm at some point in your life. Maybe you had a sense that a storm was brewing and then suddenly it hit with heavy rain, strong winds, booming thunder, lightning, perhaps even some hail. The storm lasted for a while and then gradually tapered off as it moved away. If you have ever been impressed by such a storm, talked about it with your friends, or wondered why it happened, then this next section will be interesting because it focuses on the evolution and characteristics of thunderstorms. It immediately follows the discussion about midlatitude cyclones because the most severe storms in the midlatitudes are most often associated with low-pressure systems. As you work your way through, just keep in mind that storm development is related to the way air rises and cools, and its interactions with latent heat.

Evolution of Thunderstorms

Like midlatitude cyclones, thunderstorms have a distinct life cycle, illustrated in Figure 8.12. The first stage in the development of a thunderstorm is the *cumulus stage* (Figure 8.12a), which begins with convection or the advancement of a cold front into mT air. This kind of disturbance causes air to rise, which, in turn, results in the formation of ice crystals and raindrops in growing cumulus clouds. Given their small size in this early stage of thunderstorm development, the ice crystals and raindrops continue to be uplifted. Condensation adds more latent heat energy to the air during the *developing stage,* and the atmosphere becomes very unstable with strong updrafts.

As the evolving thunderstorm develops further, it enters the *mature stage* (Figure 8.12b). In this stage, the atmosphere is very unstable, and continued convection or frontal advancement results in the massive uplift of air. Winds are blustery during this stage, with perhaps some lightning and thunder. As cold raindrops begin to fall to the ground, however, they pull cold air toward the surface with them at high speeds in a **downdraft** at the front of the storm. If you happen to be in this place as a storm forms, you can often feel a distinct increase in wind speed at this *gust front*. Where convection is most intense, an anvil head cloud forms at very high altitude (Figure 8.12c), which may be 6 km to 12 km (20,000 ft to 40,000 ft). This cloud completely develops when the top of the convection bubble is pulled (or sheared) downwind by upper airflow. You can see an example of a forming anvil cloud in Figure 8.12d. If thunderstorm development occurs along an advancing cold front, a group of mature storms and associated anvils will form a feature called a *squall line* (Figure 8.12e). These features show up clearly on radar images and are one of the many atmospheric phenomena about which meteorologists become excited.

Following the mature stage of storm development, thunderstorms begin to weaken. This weakening occurs during the *dissipation stage* and really begins as strong downdrafts develop when the storm has its maximum intensity. Look for

Downdraft *A rapidly moving current of cool air that flows downward in a thunderstorm.*

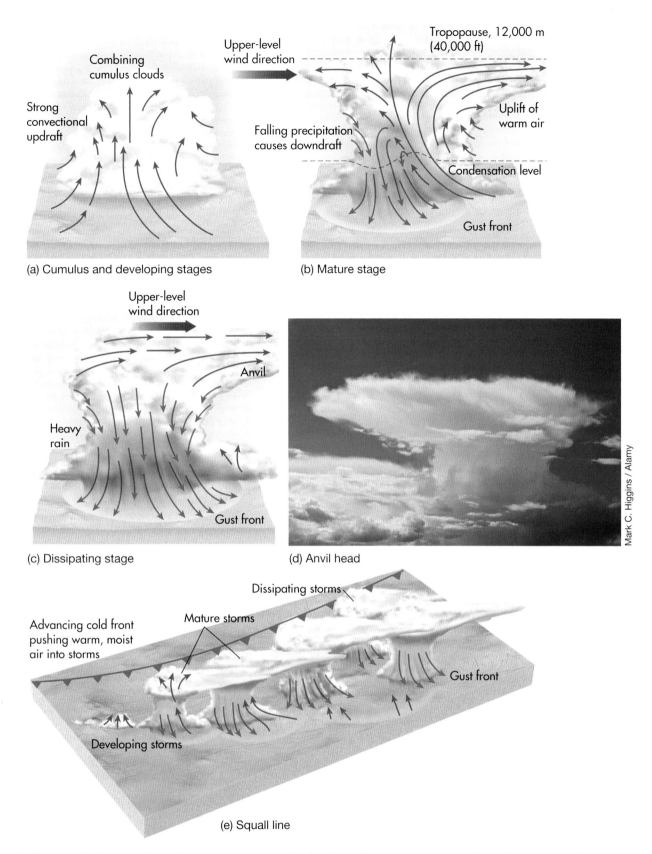

Figure 8.12 Stages of thunderstorm development. (a) Convection or advancement of a cold front forces air to rise during the cumulus stage. As the storm evolves, convection strengthens and the system moves into the developing stage. (b) The storm is most intense during the mature stage when strong updrafts occur next to strong downdrafts of cold air. (c) Formation of an anvil head. This feature develops in strong storms at the altitude where convection can no longer continue, causing the upper clouds to be sheared downwind. (d) Anvil head of a thunderstorm. Note how the cloud appears to be spreading outward at the top. (e) Squall lines consist of several individual thunderstorms that form along an advancing cold front.

Thunderstorms are fascinating atmospheric phenomena that involve complex patterns of airflow. These patterns, including updrafts and downdrafts, are particularly well suited for animation to visualize how they really move. To do so, view the animation *Formation of Thunderstorms* to observe these processes in motion and to watch the life cycle of a thunderstorm. Be sure to answer the questions at the end of the module to test your understanding of this concept.

these downdrafts in Figure 8.12c. These downdrafts cause the lower part of the atmosphere to cool, which serves to stabilize the air such that further convection cannot occur on a large scale. When these circumstances develop, rain can continue to fall for a short period of time, but will consist of a shower that gradually tapers off.

Severe Thunderstorms

In many cases, thunderstorms are short-lived events that are not particularly intense, consisting of brief downpours of rain, short gusts of moderately strong winds, some isolated lightning, and a few low rumbles of thunder. At other times, however, thunderstorms develop into very severe storms that include extremely heavy rain along with strong winds, intense lightning, numerous loud cracks of thunder, and hail. The most intense thunderstorms contain tornadoes with incredibly strong winds that can cause great damage. Although these kinds of events are relatively rare, it is nonetheless important to understand their development because you may encounter such a storm in the future. At the very least, you will appreciate them more the next time one occurs where you live.

Severe midlatitude storms form during the mature stage of thunderstorm development. Along with very strong winds, one of the characteristics of a severe thunderstorm is abundant lightning. The production of lightning begins when collisions among ice crystals and rain droplets cause a separation to develop in the electrical charge within clouds (Figure 8.13a). The result is that the tops and bottoms of clouds have positive and negative charges, respectively, while the ground is positively charged. In stage 2 the negative charge on the bottom of the cloud increases to the point where it overcomes the air's resistance to electrical flow. At this point, negatively charged electrons then begin flowing toward Earth along a zigzag, forked path, called a *leader,* at about 97 km/sec (60 mi/sec). As these electrons approach the ground, a positive charge collects in the ground. These positive charges are attracted to the negative charges of the downward-flowing leader and thus move upward through a channel (called a *streamer*) that begins at any conducting object, such as a tree, a house, and even people. Both leaders and streamers are invisible to the naked eye.

The electrical circuit becomes complete in stage 3 (Figure 8.13c), when the upward-flowing charges meet those moving downward. This meeting typically occurs at an altitude of about 30 m (~100 ft) and is the place where the actual lightning bolt you see begins. Less than a millisecond later, millions of volts of electricity reach the ground. The visible bolt, however, is the return stroke from this initial ground impact (Figures 8.13d and 8.13e). This stroke moves upward at a

VISUAL CONCEPT CHECK 8.2

This Doppler radar image shows a line of thunderstorms along a cold front extending across Kentucky and Tennessee. Which one of the following choices best explains the geography of air along the front?

a) The warmest air temperatures were northwest of the front.

b) The cold front caused mT air to rapidly uplift along the front.

c) The air southeast of the front consisted of cP air.

d) The most humid air was located northwest of the front.

Courtesy NOAA/National Climatic Data Center

Positive charges at tops of clouds

Charge separation

Negative charges (electrons) at bottoms of clouds

Positive charges in ground attracted to negative charges in clouds (a)

Negatively charged electrons flow to ground (leader)

Positive charges rise into objects on ground (streamer)

(b)

Leader

Streamer

Positive and negative charges meet in air, completing a circuit

(c)

Negative charges drained from cloud

(d)

Lyle Leduc /Getty Images, Inc.

(e)

Figure 8.13 Evolution of a lightning strike.   (a) Lightning begins when a separation develops in the electric charge between the bottom and top of a cloud. (b) Negatively charged electrons begin to move down toward the ground from the base of the cloud. At the same time, positively charged electrons begin to move upward through conducting material on the ground. (c) The circuit completes when the upward- and downward-moving electrons meet in the air. The process begins the visible lightning strike. (d) The lightning strike drains the cloud of excess negative electrons. (e) A massive lightning storm.

Figure 8.14 Hail damage. Hail can cause extensive damage to homes, crops, and automobiles, as this photo clearly demonstrates.

velocity of about 57,910 m/sec (190,000 ft/sec) and can reach altitudes of about 5 km to 6.4 km (~3 mi to 4 mi). Once the return stroke reaches the cloud, it drains the cloud of its excess negative charge and the cycle begins again.

Thunder occurs in association with lightning because lightning heats the air in the path of the stroke to above 45,000°F. This incredible heat creates a shock wave, which disintegrates within a few meters of the stroke but generates sound waves

heard as thunder. Have you ever noticed that thunder seems to rumble? This occurs because sound waves travel from different distances within individual lightning bolts, from lightning strikes, and from sound bouncing off of clouds. If you want to approximate your distance to a lightning strike, count the number of seconds between when you see the lightning flash and hear the thunder. A good rule of thumb is that a lightning strike is about 1.6 km (1 mi) away for every 5 sec it takes between when you see the flash and subsequently hear the thunder.

Another indicator that a thunderstorm is strong is hail, which consists of fragments of ice that plummet from the sky to Earth. Hail indicates that strong **updrafts** are present at the rear of the storm, which repeatedly pull ice crystals back into the upper part of the storm, where they grow larger by adding layers of ice through condensation. Once these hailstones grow sufficiently large that they cannot be pulled up anymore, they fall to Earth by the force of gravity, with large hailstones reaching speeds of 160 km (100 mi) an hour. Usually, hail ranges in size from *pea-sized* to *golf-ball-sized*. Occasionally, updrafts are sufficiently strong to produce *softball-sized* hail. The largest hailstone ever recovered in the United States fell in south-central Nebraska during a severe thunderstorm in June 2003. This hailstone was 17.8 cm (7 in.) wide, which is almost as large as a soccer ball! A severe hailstorm can be very destructive, causing millions of dollars in damage to agricultural crops, buildings, and automobiles (Figure 8.14).

Tornadoes

When certain atmospheric conditions are present, very strong **supercell thunderstorms** can develop. These extremely violent storms are unique because they contain large rotating up-

(a)

(b)

Figure 8.15 Supercell thunderstorms. (a) Typical development of a supercell thunderstorm. (b) Seen from the outside, a supercell thunderstorm is an ominous sight. The lower (circular) part of the cloud complex is rotating cyclonically.

Updrafts *An area of rapidly flowing air that is moving upward within a thunderstorm.*

Supercell thunderstorms *Large thunderstorms that contain winds moving in opposing directions and that are associated with strong winds, lightning, thunder, and sometimes hail and tornadoes.*

FASCINATING CLOUDS ASSOCIATED WITH THUNDERSTORMS

In addition to strong wind, lightning, and thunder, powerful thunderstorms sometimes produce fascinating cloud patterns diagnostic of distinct meteorological processes. One such cloud formation is a *shelf cloud* like the one pictured here. A shelf cloud develops when a cold downdraft within a storm surges outward along the ground after it reaches the surface. As it does so, it lifts the relatively warm and moist air ahead of the storm up to the level of condensation, forming this shelf-like feature. Such a cloud is often accompanied by a cold gust front.

Another interesting and diagnostic cloud formation associated with thunderstorms is *mammatus clouds*. Also known as *mammatocumulus*, which means "mammary cloud," mammatus clouds look like pouches or lobes hanging eerily upside down in the air like those shown here. These clouds are usually found on the base of a spreading anvil downwind of a large thunderstorm. Their formation is poorly understood but may be somehow related to overturning pockets of colder air that descend into warmer air at high altitudes. Although they are usually composed of ice, they also contain liquid water and can seem translucent. Mammatus clouds are more common during warm months and are more often seen in the Midwest and the eastern parts of the country.

Mike Hollingshead / Science Source

Tom Fleming/Photo Researchers/Getty Images

drafts known as **mesocyclones** (Figure 8.15) that range from 3.2 km to 9.7 km (2 mi to 6 mi) in diameter. When these atmospheric conditions occur, they can easily evolve to include tornadoes, which are small, but deep, low-pressure cells surrounded by a violently spinning mass of air. Although generally less than 400 m (0.25 mi) in diameter, tornadoes are the most destructive of all atmospheric phenomena, with wind speeds ranging from 160 to 480 km/h (100 to 300 mph). These high speeds occur because tornadoes have extremely tight pressure gradients that may differ by 100 mb between the inside and outside of the funnel. Between 1971 and early 2007, tornadoes were classified according to the Fujita scale based on

---

Mesocyclones *Strong updrafts that rotate within a supercell thunderstorm.*

the amount of damage caused. Acknowledging that the quality of building construction can vary and thus give misleading conclusions about wind speed, the *Enhanced Fujita Scale* was implemented in February 2007. This classification system is based on the speed of winds at 3-sec gusts at the point of damage (Table 8.1), ranging from the EF0 tornado to the rare EF5 tornado.

Figure 8.16 shows the evolution of a tornado. Tornadoes usually form at the rear of a storm when strong updrafts occur in conjunction with wind shear at higher altitudes. Wind shear occurs when winds are moving in different directions at variable altitudes. This wind shear causes a horizontal vortex of air to form (Figure 8.16a) that is then pulled vertically in the updraft (Figure 8.16b). Assuming this process is not interrupted, a funnel develops that becomes a full-fledged tornado if it reaches

TABLE 8.1

TABLE 8.1   The Enhanced Fujita Scale for Tornado Classification

EF Number	3-sec Gust Speed (mph)	Potential Damage
0	65–85 mph	*Minor damage*: Some shingles peeled from roofs and some damage to gutters and siding; branches broken off trees; shallow-rooted trees blown over.
1	86–110 mph	*Moderate damage*: Roofs stripped significantly; mobile homes turned over; broken windows.
2	111–135 mph	*Considerable damage*: Roofs ripped off homes; frame homes moved on foundation; mobile homes destroyed; large trees uprooted.
3	136–165 mph	*Severe damage*: Entire portions of well-constructed homes destroyed; severe damage to shopping centers; trains overturned and cars lifted from ground.
4	166–200 mph	*Devastating damage*: Well-constructed homes entirely destroyed; cars thrown and small missiles created.
5	Over 200 mph	*Extreme damage*: Strong frame houses leveled and blown away; cars thrown farther than 100 m (300 ft); steel-reinforced concrete damaged.

Figure 8.16 Evolution of a tornado.   (a) Air begins to rotate due to wind shear; (b) updrafts pull a rotating cylinder of air upward; (c) a tornado develops when the funnel reaches the ground. (d) Supercell tornado near Spearman, Texas, in 1990.

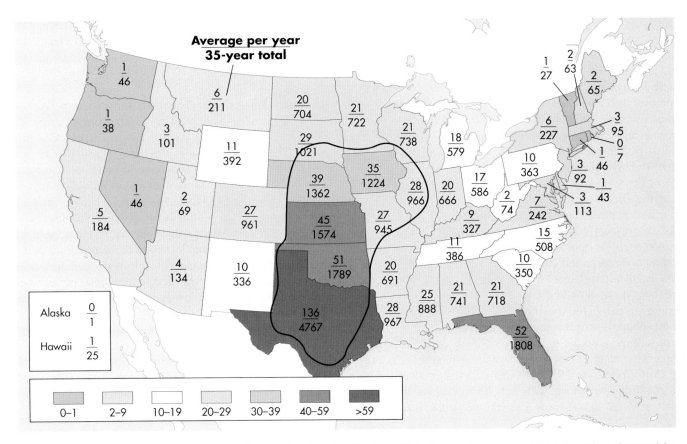

Average per year
35-year total

Alaska $\frac{0}{1}$

Hawaii $\frac{1}{25}$

0–1	2–9	10–19	20–29	30–39	40–59	>59

Figure 8.17 Distribution of tornadoes in the mainland United States. Note the yearly average and total number observed for over 35 years per state. The approximate boundary of Tornado Alley is outlined in black.

the ground (Figure 8.16c). Figure 8.16d shows a supercell tornado (arrow) that formed near Spearman, Texas, in May 1990. Notice how similar this tornado and spawning supercell system are to the diagram in Figure 8.16c. The airflow in a fully developed tornado can generate fierce winds of over 480 km/h (300 mph). Fortunately, tornadoes seldom last very long, only 15 min on average. The air sucked into the funnel slows down the winds—the tornado becomes clogged with the air it takes in. Eventually, the funnel stretches out into a thin shape and disappears.

On rare occasions, however, a tornado can be on the ground for up to an hour, or longer. The best-known example of such a storm was the famous Tri-State Tornado in 1925 that devastated Missouri, Illinois, and Indiana. This tornado was on the ground for over 3 h straight and cut a swath over 370 km (~230 mi) long! The funnel was about 1.6 km (1 mi) wide and killed nearly 700 people.

Although tornadoes occur every month of the year in the continental United States, they are most frequent between April and July. These months are peak tornado season because the strongest contrast in air masses exists at this time in the midlatitudes. Although tornadoes statistically occur at least once per year in every state of the United States (except Alaska), they are usually associated with the central part of the country. This region, which is known as *Tornado Alley* (Figure 8.17), is prime tornado country because the relative-

ly flat terrain provides an ideal place for cool, dry Canadian air to mix violently with warm, moist air from the Gulf of Mexico. These conditions are perfect for the development of the massive thunderstorms that contain tornadoes. Some of the most intense tornadoes in history have struck this region, such as the F-5 storm that struck my hometown of Topeka, Kansas, in June 1966 (Figure 8.18) when I was young. This

Figure 8.18 The Topeka Tornado. This storm was an F-5 tornado that traveled through much of the city of Topeka, Kansas, in June 1966. This tornado remains one of the most costly tornadoes in U.S. history.

© Bettmann/Corbis

storm killed 16 people, caused $100 million in damage (1966 dollars), and led to my lifelong interest in geography. Sometimes atmospheric conditions are so ripe that a *tornado outbreak* occurs, with numerous funnels over several hours or a couple of days. Such an outbreak occurred in early May 2007 when over 100 tornadoes were reported in the central Great Plains. The most severe of these storms hit Greensburg, Kansas, and caused damage of over $153 million. This storm was the first EF5 tornado ever reported, with estimated winds of 330 km/h (205 mph) in the core of the system. Approximately 95% of the town was destroyed in this storm (Figure 8.19). Since the Greensburg tornado, catastrophic EF5 tornadoes also struck Joplin, Missouri, and Moore, Oklahoma, in 2011 and 2013, respectively.

Although the central part of the country is most known for tornadoes, large and dangerous twisters are also common in the southeastern United States in states such as Arkansas, Alabama, Mississippi, and Georgia. Tornadic storms usually occur in this part of the country during the early spring because the jet stream is still flowing over the southern states. Some of the strongest storms in the country have occurred in this region, with the most recent intensive tornado outbreak occurring from April 25 to 28, 2011. This outbreak produced over 350 confirmed tornadoes, with four EF5s. Perhaps the most widely reported tornado was a storm that struck Tuscaloosa,

Alabama, which is home to the University of Alabama. This tornado was on the ground for 130 km (81 mi) and killed 43 people. Overall, more than 325 people died within the region due to the outbreak.

Human Interactions: Tornado Monitoring Although studying tornadoes is interesting purely from a meteorologic perspective, the primary goal of tornado research is to improve public safety. In this context, the National Weather Service constantly monitors atmospheric conditions and makes public announcements as weather events unfold on any given day. A *tornado watch* is issued for an area, such as the eastern part of Oklahoma or central Texas, when atmospheric conditions are favorable for the development of tornadoes. When an actual funnel or tornado has been identified, a *tornado warning* is issued for a specific place or, perhaps, county. At this time sirens are sounded, and people are advised to seek shelter, either in a low-lying ditch or, preferably, in a basement.

With respect to tornado monitoring, the most significant technological development in the past few decades has been Doppler radar, such as the Next Generation Weather Radar (NEXRAD) used by the National Weather Service (Figure 8.20). Prior to this development, tornado warnings were based largely on direct observations, which were often too late to save lives.

(a)

(b)

(c)

Figure 8.19 The Greensburg Tornado. (a) In May 2007, an EF5 tornado struck Greensburg, Kansas, killing 11 people and destroying 95% of the town. (b) Air photo of Greensburg before the tornado. (c) Air photo of Greensburg after the tornado. Note that most of the town was essentially destroyed.

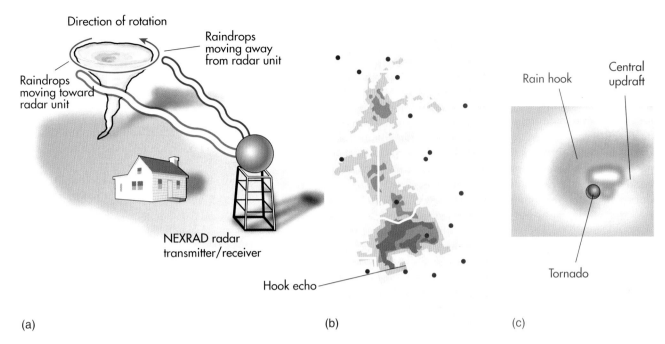

(a) (b) (c)

Figure 8.20 Doppler radar. (a) Doppler radar "sees" the rotation of a tornado because of the Doppler Effect, which, in this case, reflects the variation in returning wave frequency from either side of the system. (b) The rotation of a tornado is displayed as a hook echo on the computer screen. (c) A hook echo is indicative of aggressive cloud rotation, which usually means that a tornado is present.

In contrast to conventional radar, which shows the intensity of precipitation in any given storm, Doppler radar can detect *both* rainfall patterns and the actual rotation of a tornado. It does so by making use of the Doppler Effect, described first by the 19th-century physicist Christian Doppler, which states that the frequency of energy waves generated by a moving source changes relative to an observer.

A classic example of this effect is the sound a train makes as it approaches you. Have you ever noticed what happens? The pitch of the train's whistle rises with the approach of the train and then lowers after it passes. Using this effect, Doppler radar can determine that raindrops on one side of a tornado are moving toward the radar detector, while those on the other side are rotating away because of the shift in energy frequency that occurs relative to each side of the twister (Figure 8.20a). This rotation is seen by the meteorologist in the weather laboratory as a distinct **hook echo** on the computer screen (Figure 8.20b, c). If such a feature is identified, a tornado warning is quickly issued for the area in the path of the storm and people usually have sufficient time to seek shelter before the storm strikes.

Hook echo *The diagnostic feature in Doppler radar indicating that strong rotation is occurring within a thunderstorm and that tornado development is thus possible.*

Tornadoes are fascinating natural phenomena that are often caught on film by storm chasers and regular citizens alike. To learn more about tornadoes, view the animation *Tornadoes*. The first part of this module demonstrates how tornadoes form and shows real storm video. The last part of the module is an interactive exercise, involving the Enhanced Fujita scale, which allows you to determine the strength of a tornado based on the damage it caused. After you interact with this module, be sure to answer the questions at the end to test your understanding of this concept.

Tropical Cyclones

Now it is time to turn our attention to cyclonic storms that form in the tropical regions. Although tropical systems are similar to midlatitude systems because they are rotating low-pressure systems, they differ significantly because they form in geographical regions where there are no fronts or contrasting air masses. Instead, a tropical cyclone develops entirely within a homogeneous air mass at very low latitudes. As a result, abundant water vapor and latent heat are present to fuel these systems. About 80 tropical cyclones develop around the world every year, with the strongest becoming extremely powerful storms that, depending on the part of the world, are called hurricanes, typhoons, or cyclones.

Most Atlantic hurricanes form in association with an atmospheric feature called an **easterly wave**, which is a slow-moving trough of low pressure that migrates along the belt of tropical easterlies. Recall that this belt is associated with the easterly trade winds and occurs between 5° and 25° N and S latitude. This low-pressure system causes air to converge on the windward (east) side of the trough axis, while air diverges in the lee (west) side of the system. The convergence of upper air on the eastern side of the wave causes further convectional uplift within the system, and a zone of showers develops parallel to the trough axis. Figure 8.21 shows the basic structure of an easterly wave, including the axis, wind flow patterns, and precipitation zone.

If the storm strengthens into a **tropical depression**, it begins to rotate cyclonically around a definable center of low pressure at the same time that a complex of strong thunderstorms develops. At this time, sustained winds in the center of the system are 20 knots to 34 knots (23 mph to 39 mph). If the storm continues to strengthen and maximum sustained winds are between 35 and 53 knots (35 and 73 mph), it becomes a **tropical storm**. The storm at this point begins to become more circular in shape and is assigned a formal name (such as *Charlie*) for monitoring purposes as it proceeds on its easterly migratory path.

Hurricanes

If a tropical storm in the Atlantic Ocean or eastern Pacific Ocean strengthens to the point that the maximum sustained winds are greater than 63 knots (73 mph), then it officially becomes a **hurricane**. A storm of similar strength in the western Pacific Ocean is called a *typhoon* or a *cyclone* if it forms in the Indian Ocean. Australians call such a storm a *willy nilly* because they twist about and wreak havoc. For purposes of discussion here, we use the term *hurricane* and focus on Atlantic storms.

Although it is common to hear the word *hurricane*, storms of this magnitude are really quite rare because they require a

Figure 8.21 Characteristics of a tropical wave. Tropical waves consist of a center of surface low pressure that results in the convergence and divergence of air aloft. Showers develop on the eastern side of the wave axis because converging air causes convectional uplift.

Easterly wave *A slow-moving trough of low pressure that develops within the tropical latitudes.*

Tropical depression *A tropical low-pressure system with central sustained winds ranging between 20 knots and 34 knots (23 mph and 39 mph).*

Tropical storm *A tropical low-pressure system with maximum sustained winds between 35 knots and 63 knots (39 mph and 73 mph).*

Hurricane *A tropical circulatory system with maximum sustained winds greater than 63 knots (73 mph).*

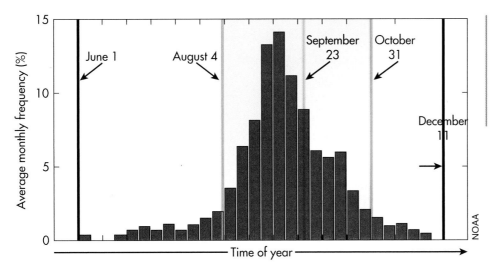

NOAA

Figure 8.22 Average frequency of monthly hurricanes in the Atlantic Ocean from 1896 through 1996. Hurricanes are most likely to develop from the beginning of August to the end of October because sea-surface temperatures are warmest during this time of year in the Northern Hemisphere.

particular combination of environmental variables. Most importantly, they require an abundance of latent heat energy to fuel their development. Such an abundance of energy can occur only when a great deal of evaporation from the underlying ocean takes place, which means that the surface waters must be warmer than about 27°C (80°F). This required temperature is the primary reason why hurricane season in the Atlantic Ocean largely occurs from August to October (Figure 8.22), in other words, in the period following the highest Sun and insolation.

In addition to the warm ocean waters, another requirement for hurricane development is that the upper air wind pattern must be favorable, with high air pressure aloft. This relationship may seem strange because hurricanes are zones of low air pressure, which might lead you to wonder how both low pressure at the surface and high pressure in the upper atmosphere could exist at the same time. Remember that the atmosphere has many different layers and that air can flow differently in each of them. In the case of a hurricane, high pressure aloft is required

because it serves to "cap" the top of the storm so that it can continue to strengthen at the surface. Otherwise, if the surface air can freely convect into the upper atmosphere, or if shearing winds aloft blow the top off the storm, then the system does not fully organize.

If all the necessary ingredients are in place for hurricane development, the system can grow to immense size and strength. Depending on the specific atmospheric conditions, hurricanes can range in diameter from 150 km to 500 km (100 mi to 300 mi). Hurricanes acquire their strength because of the steep pressure gradient that develops from the outside of the storm to a center of extremely low pressure, which can depress to 900 mb or even lower. As a result of this steep pressure gradient, winds spiral into the core of the system at very high speed. Figure 8.23 shows an example of the pressure gradient and storm size.

Hurricanes are classified based on their wind-speed intensity on the Saffir–Simpson scale. Table 8.2 shows this scale and how it relates to the central pressure of a storm and its wind speed. Notice the inverse relationship between these two variables. The fourth column of the table shows storm surge, which is the rise in sea level that occurs because strong winds cause water to pile up ahead of the storm. Storm surge is typically highest from a geographical perspective on the right front quadrant of a hurricane, where onshore winds flow as the storm approaches. These winds tend to push the water ahead of the storm and onto the coast. In addition to wind speed, the height of a storm surge depends on the timing of the tides and configuration of the coastline, with shallow coastlines typically more susceptible because water has a tendency to pile up at those locations.

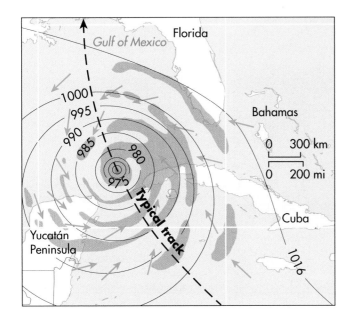

Figure 8.23 Hurricane weather map. This simplified weather map of a typical Atlantic hurricane illustrates the cyclonic rotation and pressure gradient associated with the system. Shaded areas are regions of heavy rainfall.

Tropical Cyclones • 191

TABLE 8.2	Saffir–Simpson Scale of Hurricane Intensity		
Category	Central Pressure, mb	Mean Wind, m/sec (mph)	Storm Surge, m (ft)
1 Weak	>980	33–42 (74–95)	1.2–1.7 (4–5)
2 Moderate	965–979	43–49 (96–110)	1.8–2.6 (6–8)
3 Strong	945–964	50–58 (111–130)	2.7–3.8 (9–12)
4 Very strong	920–944	59–69 (131–155)	3.9–5.6 (13–18)
5 Devastating	<920	>69 (>155)	>5.6 (>18)

Figure 8.24 shows the typical anatomy of a hurricane. Note the distinct bands of cumulonimbus clouds, which can drop prodigious amounts of rainfall, surrounding the core of the system. Also observe that air flows inward at the surface to a point where it rapidly flows upward at the eye wall. Remember that the entire time the air is flowing inward and upward, it is spinning cyclonically at intense speed. A unique characteristic of a well-developed hurricane is the eye, which occurs in the center of the storm and is associated with high-pressure air that rapidly flows toward the surface and warms adiabatically as it descends. As a result, this region of the storm is cloud-free with no winds and can be seen in satellite images of well-developed hurricanes.

Because hurricanes evolve in the easterly trade belt, they migrate in a fairly predictable way. Figure 8.25 shows typical tracks that hurricanes take in August. Notice that many of the storms originate in the eastern Atlantic Ocean. These storms actually begin to organize when they exit western Africa as easterly waves in the trade-wind belt. When they reach the open warm waters of the eastern Atlantic, they begin to intensify, first as tropical depressions, then as tropical storms, and finally, if conditions are favorable, into hurricanes. The systems continue to migrate westerly, sometimes strengthening, at other times weakening. If the storm remains organized for several days, it ultimately becomes influenced by the midlatitude westerlies and is deflected to the northeast.

Human Interactions: Coping with Destructive Hurricanes

As you can see in Figure 8.25, many storms remain over open water throughout their entire history. A few, however, strike land at some place in the Caribbean Sea, Gulf of Mexico, or the Atlantic seaboard in the United States. In most cases these landfall hurricanes are category 1 or 2 storms that, while packing a strong punch, usually blow through fairly quickly without causing too much chaos. Occasionally, however, stronger storms hit that result in catastrophic damage over a broad area and numerous human fatalities. Some of these storms are infamous, such as the 1900 hurricane that killed over 6000

High-level winds spiral outward

Descending dry air

Eye wall

Spiral bands of rain

High-speed winds spiral upward around eye

Concentric rings of cumulonimbus clouds

Figure 8.24 Components of a typical hurricane. Hurricanes contain numerous spiral rain bands of cumulonimbus clouds. Air spirals inward until it reaches the eye wall, where it circles upward. The eye of the storm is clear and calm, which is the result of air descending and adiabatically warming in the core of the system.

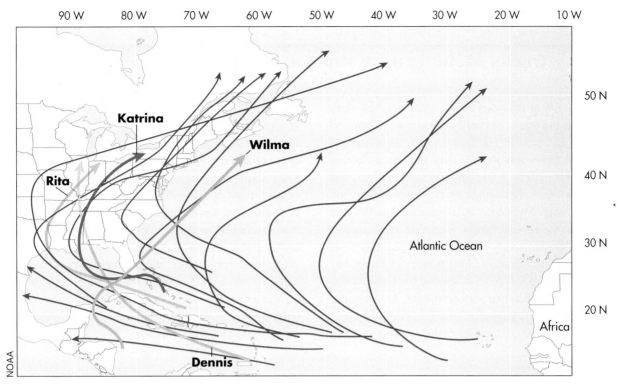

Figure 8.25 Typical track of August hurricanes in the Atlantic Ocean. During this time of year, hurricanes begin to develop over the warm waters of the eastern Atlantic. Subsequently, they migrate with the easterly trade winds until they become influenced by the midlatitude westerlies. This map also includes the tracks of the four major hurricanes that struck the United States in 2005.

people in Galveston, Texas. Similar legendary storms include the Okeechobee (Florida) hurricane in 1928, Donna (1960), Camille (1969), Hugo, (1989), and Andrew (1992). These storms, and others like them, cause many billions of dollars in damage and stretch the ability of people and government at all levels to cope. Two recent hurricanes, Katrina in 2005 and Sandy in 2012, are excellent examples illustrating how people

are catastrophically impacted by such intense storms and subsequently struggle to deal with their aftermath.

Hurricane Katrina struck the Gulf Coast near New Orleans on August 29, 2005. Katrina formed over the Bahamas on August 23 and first made landfall on the east coast of Florida as a Category 1 storm. It subsequently moved west over the Gulf of Mexico, where surface water temperatures were a very

VISUAL CONCEPT CHECK 8.3

This figure is a satellite image of Hurricane Floyd in September 1999, which at one point developed into a Category 5 storm. Note the position of Floyd in the central Atlantic Ocean in this image. Imagine that you are a hurricane forecaster who has become aware of the hurricane's presence for the first time. Describe how you would expect Floyd to migrate in the future, beginning at this position and ending when the storm would most likely dissipate.

Data from NOAA GOES satellite. Image produced by Hal Pierce, Laboratory for Atmospheres, NASA Goddard Space Flight Center

An excellent example of the need for monitoring can be seen in videos showing the tracks of Hurricane Katrina in 2005 and Hurricane Sandy In 2012. To see these tracks, view the animation *Tracks of Hurricane Katrina and Hurricane Sandy*. The first part of this module is an animated cartoon illustrating Hurricane Katrina's path. Subsequently, two videos follow the path of this catastrophic hurricane as it moved across the Gulf of Mexico. Watch for several things. First, look for the overall rotation of the system and how its path changed over time. Also note the rapid intensification of the system as it migrated across the Gulf. This intensification is obvious because the eye of the storm suddenly becomes very prominent. At this point, the storm was at its most powerful, specifically a Category 5 hurricane. Next, observe how the size of the eye fluctuated as it approached the coast; this occurred because the storm's strength varied somewhat before it made landfall. After you watch the Katrina animation, turn your attention to the track of Hurricane Sandy. Watch how this storm moved on a northerly track through the Caribbean and then followed the southeastern coast of the United States. Notice the approach of the midlatitude jet stream as the storm moved farther north and the pronounced westward turn the storm made prior to landfall as the storm was incorporated into a strong extratropical low-pressure system. As you watch these videos, think about why satellite images such as these are essential for good monitoring and how they save lives. Once you finish the video, be sure to answer the questions at the end of the module to test your understanding of this concept.

---

warm 32°C (90°F). Katrina thus rapidly intensified to Category 5 status with sustained winds of 280 km/h (175 mph). At this time, the storm had an extremely well-defined eye, as you can see in Figure 8.26a. The storm subsequently weakened slightly to a strong Category 3 shortly before it struck the Louisiana coast on a Monday morning. The coastline was devastated as far east as Florida, with thousands of homes and businesses literally flattened (Figure 8.26b) from the combined power of the strong winds and a storm surge that spread up to 24 km (15 mi) inland. The storm surge was highest in this region, up to 9 m (30 ft) in some places, because it was in the right front quadrant of the storm as it approached.

Although much of the Gulf Coast was blasted by the storm, the most tragic consequences were experienced in New Orleans. Established by the French in 1718 as a port on the lower Mississippi River, this famous American city is actually quite poorly located because it lies below sea level and therefore floods very easily. This problem has been unevenly managed since the town's birth by a continuously evolving complex of canals, pumping stations, and levees (high walls) along the Mississippi River (to the south) and Lake Pontchartrain (to the north). Given the poor geography of the area, experts in many fields had long known that New Orleans was a disaster waiting to happen. Disaster struck when Hurricane Katrina created a storm surge in Lake Pontchartrain that caused several levees to fail, resulting in severe flooding (Figure 8.26c) and the displacement of tens of thousands of people. Of the 1800 people who died in the storm along the Gulf, over 1400 fatalities occurred in New Orleans alone. The economic cost of the storm was about $100 billion, easily making it the most damaging storm in U.S. history.

In the aftermath of the storm, an intense national controversy erupted over who was to blame for the catastrophe in New Orleans. Fingers pointed everywhere. Some believed that the residents of the city who remained as the storm approached were at fault, even though about 100,000 people were poor and did not have personal vehicles to evacuate. Others blamed politicians from the mayor (a Democrat) to the president (a Republican) for perceived inadequate planning and reactions. Still others felt that the primary responsibility lay with the slow response of the Federal Emergency Management Agency (FEMA), which was established in the late 1970s to provide disaster relief in such an event. Regardless, the city was a mess for months afterward, and some of it remains uninhabited to this day. Although much of New Orleans has been rebuilt, a thick scar remains from Hurricane Katrina and nervous eyes turn to the Gulf every time a storm approaches.

Following Hurricane Katrina, the next disastrous storm to strike the United States was Hurricane Sandy in late October 2012. Sandy was an unusual storm for several reasons related to its path, time of year, size, landfall location, and interaction with the midlatitude storm track. The storm originated in the southeastern Caribbean as an easterly wave and proceeded on a northerly track across Jamaica and eastern Cuba (Figure 8.27a) as a Category 1 storm. After weakening slightly to a tropical storm, it hit the Bahamas again as a Category 1 hurricane. Once Sandy emerged into the Atlantic, it shadowed the eastern seaboard, causing extensive coastal damage in the southeastern United States, while remaining a Category 1 storm. Sandy then combined with a strong midlatitude surface low over the eastern United States to form a superstorm that extended over 1000 km (600 mi) in width (Figure 8.27b). Given the size of the storm and the broad wind field associated with it, Sandy is thought to have contained the highest kinetic energy of any Atlantic storm *ever* recorded.

New Orleans•

(a)

(c)

(b)

Figure 8.26 Hurricane Katrina. (a) Satellite image of Hurricane Katrina shortly before it struck Louisiana. Note the incredibly well-defined eye, which developed when the storm reached maximum strength. (b) Extensive damage along the Mississippi coast was caused by a combination of strong winds and storm surge, which in some places was over 9 m (30 ft) high. (c) Failure of protective levees in New Orleans caused widespread flooding.

Although most storms are deflected northeasterly when they reach the midlatitudes, Sandy was drawn to the west by the midlatitude low, causing a landfall along the coast of southern New Jersey. Northern New Jersey and New York City (particularly Long Island and Staten Island) were blasted (Figure 8.27c) by strong winds and a 4-m high (13-ft) storm surge that resulted, in part, because the storm struck at precisely the highest tide of the month. This surge flooded the subway system in the area (Figure 8.27d) and devastated coastal communities on the barrier islands that front the mainland. About 5 million people lost electrical power. Over 18 cm (7 in.) of rain fell in many places, causing rivers to swell dangerously. Because Sandy had combined with an early winter storm, up to 0.6 cm (2 ft) of snow fell in the Appalachian Mountains to the west. This heavy snowfall was driven in large part by orographic processes. Strong winds were felt as far away as Michigan. Overall, Sandy caused about $50 billion in damage, which makes it the second costliest storm ever in the United States, and killed 185 people along its path.

In contrast to the federal government's response to Hurricane Katrina, its reaction to Sandy was swift. Ironically, the storm hit just a few days before the 2012 presidential election, which greatly intensified the spotlight. President Obama quickly declared much of the eastern seaboard a Federal Disaster area, which meant that FEMA descended rapidly onto the scene to deliver clothing, food, and shelter. As of late April, 2013, however, thousands remained homeless and many coastal communities were struggling to recover from the effects of Sandy. The impact of this hurricane raised consciousness about the vulnerability of the greater New York area to catastrophic storms and fueled a national debate about the relationship of hurricanes to global climate change. As we will discuss in Chapter 19, one of the projected impacts of a warming world is rising sea levels. Assuming such changes occur in the future, the impact of storms such as Sandy will be even greater.

The intensity of Hurricane Katrina and Hurricane Sandy, as well as other large tropical storms, underscores the need for

(a)

May 21, 2009

November 5, 2012

(b)

(c)

(d)

Figure 8.27 Hurricane Sandy. (a) Track of Hurricane Sandy. Note the irregular track the storm took from the Caribbean to its landfall on the U.S. east coast. (b) Visible infrared satellite image of Sandy after it made landfall. The storm was huge, affecting most of the northeast United States. (c) Repeat photography of the coastline at Mantoloking, New Jersey. The top photo shows the area prior to the storm in May 2009, whereas the lower image was taken shortly afterward. Storm waves and surge cut across the island, eroding the beach, destroying houses and roads, and depositing sand. The yellow arrow in each image points to the same feature for reference. (d) Flooded subway system in New York City. High storm surge and waves flooded the southern part of Manhattan, including the subway system.

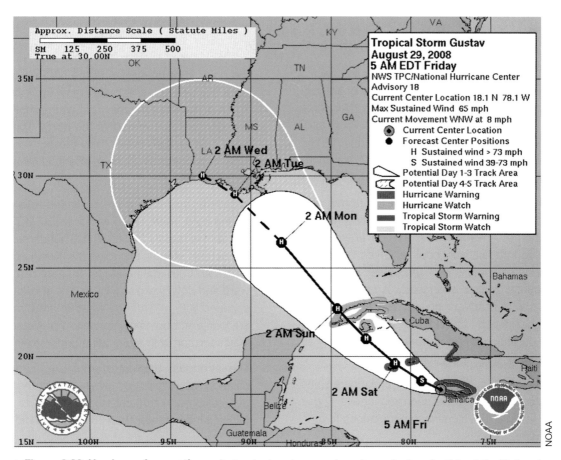

Figure 8.28 Hurricane forecasting. As tropical systems evolve, atmospheric scientists at the National Hurricane Center run statistical models based on known conditions to project the strength and path of a storm forward through time. This forecast cone illustrates the projected path of Tropical Storm Gustav several days forward from August 29, 2008. Note that the estimated location of the storms center widens with time, reflecting increased statistical uncertainty. Gustav ultimately made landfall as a Category 2 storm in western Louisiana. Given this forecast, approximately 3 million people successfully evacuated before the storm arrived.

accurate hurricane forecasting and monitoring. The death toll was so high in the 1900 Galveston hurricane in large part because no one knew it was coming. People living on the Atlantic and Gulf coasts today depend on hurricane forecasts to gauge the strength of hurricanes, where landfall will likely occur, and the potential height of storm surge. Fortunately, in this day of real-time satellite imagery, it is possible to monitor the development of any individual storm as it migrates across the Atlantic or Caribbean. As a result, forecasters at the National Hurricane Center can predict the general path a storm will take several days in advance and provide warning of the storm's approach to residents in its path (Figure 8.28).

KEY CONCEPTS TO REMEMBER ABOUT SEVERE MIDLATITUDE AND TROPICAL STORMS

1. In general, the most severe midlatitude storms form along strong cold fronts when warm, moist (mT) air ahead of the front is rapidly forced aloft.

2. Thunderstorms evolve in predictable stages, including the cumulus stage, mature stage, and dissipating stage. These stages are related to the upward and downward flow of air.

3. Strong winds, lightning, and thunder typically accompany severe storms. Lightning occurs when opposing electronic charges develop between the base of clouds and the ground. Thunder is created by the shockwave produced when lightning superheats the atmosphere locally.

4. The strongest storms associated with midlatitude weather are tornadoes, which are localized centers of intense low pressure that develop in association with supercell thunderstorms.

5. The strongest tropical storms are hurricanes, which develop when easterly waves strengthen beyond the depression and tropical storm phases to produce sustained winds greater than 63 knots (73 mph).

SUMMARY OF KEY CONCEPTS

1. An air mass is a large body of air that forms in specific geographic regions and thus has distinctive characteristics. Five principal air masses affect North America. Continental air masses include continental Polar (cP), continental Arctic (cA), and continental Tropical (cT). The maritime air masses are maritime Tropical (mT) and maritime Polar (mP).

2. Air masses have distinct boundaries called fronts. At a stationary front, contrasting air masses are flowing parallel to one another. A warm front is a place where warm air is advancing into relatively cool air. Given that warm air slowly slides over the top of the cooler air along a warm front, rainfall is slow and steady. A cold front is a place where cold air is advancing into relatively warm air. Given the higher density of colder air, rainfall is intense and of short duration along the front because warm air cools quickly when it is rapidly forced aloft.

3. A midlatitude cyclone in the Northern Hemisphere spins in a counterclockwise fashion as seen from above. These atmospheric features develop when undulations form in the polar front jet stream at the 500-mb pressure level. As a cyclone spins, it pulls warm (mT) air up from the south on its eastern side. This warm, moist air encounters cold air as it moves to the north. The cyclone also pulls cold air (cP) down from the north on its western side. This cold, dry air encounters warm air as it moves to the south.

4. In general, the most severe midlatitude storms form along strong cold fronts when warm, moist (mT) air ahead of the front is rapidly forced aloft. Thunderstorms evolve in predictable stages, including the cumulus stage, mature stage, and dissipating stage, that are related to the upward and downward flow of air. The strongest storms associated with midlatitude weather are tornadoes, which are localized bodies of intense low pressure that develop in association with supercell thunderstorms.

5. The strongest tropical storms are hurricanes, which develop when easterly waves strengthen beyond the depression and tropical storm phases to produce sustained winds greater than 63 knots (74 mph).

CHECK YOUR UNDERSTANDING

1. Define an air mass.

2. What are the specific characteristics of an mT air mass, and how do they differ from the characteristics of a cP air mass?

3. Which air mass is most likely to be associated with precipitation—an mT air mass or a cP air mass? Why?

4. Discuss the evolution and migration of a midlatitude cyclone.

5. Why are midlatitude cyclones a mechanism through which contrasting air masses are mixed?

6. How does the formation of an upper air trough at the 500-mb level result in the development of a midlatitude cyclone?

7. What is the basic difference between a warm front and a cold front? Why is the term *front* used in association with these concepts?

8. Precipitation along a warm front is gradual and long-lasting, whereas it is short-lived and often violent along a cold front. Why does this difference exist?

9. Describe the evolution of a thunderstorm and the various stages it goes through during its life cycle.

10. What is a downdraft, and why is it the first step in the dissipation of a thunderstorm?

11. Discuss the evolution of a hurricane in the Northern Hemisphere, including its various stages, movement, and relationship with ocean temperature.

ANSWERS TO VISUAL CONCEPT CHECKS

VISUAL CONCEPT CHECK 8.1

The answer is *d*. The cold front extends from western Michigan to central Texas. You can tell the location of the cold front because of a sharp temperature difference between the east and west sides of the frontal boundary.

VISUAL CONCEPT CHECK 8.2

The answer is *b*. The cold front caused mT air to uplift rapidly. Remember that a cold front plows aggressively into relatively warm, moist air. This interaction causes the mT air to lift rapidly.

VISUAL CONCEPT CHECK 8.3

One would expect Hurricane Floyd to move to the west because the prevailing winds are easterly at low latitudes. At some point, the hurricane would begin to interact with the midlatitude westerlies, causing the hurricane to veer northeastward. In fact, Hurricane Floyd migrated in just this manner, moving westward toward Florida. Shortly before it reached Florida, however, Floyd veered to the northeast and just grazed the mainland of the United States.

CHAPTER NINE | GLOBAL CLIMATES AND GLOBAL CLIMATE CHANGE

The previous chapters examined features associated with Earth's atmosphere, including the flow of radiation within it, air circulation, and weather systems. Although these concepts can be viewed individually, they can also be considered holistically because they interact to form predictable patterns. The development of midlatitude cyclones, for example, is related to atmospheric circulation, air pressure, precipitation, and the seasonal effects associated with Earth–Sun geometry. This chapter integrates all of the processes discussed so far in this text by focusing on the broad global patterns associated with world climate regions, past climate fluctuations, and future climate change.

James Randklev/Getty Images

Climate refers to the general temperature and precipitation characteristics of an area. This image of the Maroon Bells near Aspen, Colorado, shows two climate regions—one in the high mountains where cold, snowy conditions prevail, and another at a lower altitude that is sufficiently warm for trees to grow.

CHAPTER PREVIEW

Climate and the Factors That Affect It

Köppen Climate Classification

Geography of Köppen Climates

 Tropical Savanna Climate (Aw)

 Humid Subtropical Hot-Summer Climate (Cfa, Cwa)

 Marine West-Coast Climates (Cfb, Cfc)

 Global Climates

 Remote Sensing and Climate

Reconstructing Past Climates

 Reconstructing Past Climates Using Oxygen Isotopes

Causes of Past Climate Change

 The Milankovitch Theory

Human Interactions and Future Climate Change

 The Greenhouse Effect and Global Climate Change

LEARNING OBJECTIVES

1. Define the concept of climate and describe the large-scale factors that influence it on Earth.

2. Explain the Köppen climate system and how it is used to differentiate climate regions on Earth.

3. Discuss the characteristics of the various Köppen climate regions.

4. Describe the various methods used to reconstruct past climate change, the patterns of climate change in the past two million years, and why it is important to understand those fluctuations.

5. Understand the factors that have caused climate change in the past.

6. Discuss the issue of global warming and why most climate scientists feel it is caused by human industrial activity.

7. Describe the projected rates of warming in the 21st century and the impact it is expected to have in the future.

Climate and the Factors That Affect It • 201

Climate and the Factors That Affect It

When most people think of what "climate" means, they frequently think it is another word for "weather." This is a common misconception. The term **weather** refers to the state of the atmosphere at a specific place and time on the Earth's surface. For example, the weather is described when you say something like "it's cold and rainy today." Think about it for a moment—what *is* the weather as you are reading this chapter? The concept of **climate** is different from that of daily weather because it refers to the long-term average values of weather elements, such as temperature and precipitation. These averages are based on a 30-year period of time.

The concept of climate can be viewed geographically because Earth exhibits distinct climate patterns. The study of this geographic distribution and the character of Earth's climates is a subfield of geography called *climatology*. This subfield is not to be confused with *meteorology*, which is the study of short-term atmospheric phenomena such as thunderstorms that constitute day-to-day weather in any particular place. In contrast to this relatively short-term view, climatologists study long-term temperature and precipitation patterns that extend from years to centuries. Their research is important to many disciplines outside the field of geography, including agriculture, architecture, ecology, forestry, and economics, because climate is a factor that influences human behavior and natural processes in a variety of ways.

Before investigating the nature and distribution of world climates, we should review all the factors that play a role in the long-term weather patterns for a region (Figure 9.1). Given what has been covered so far in this book, these variables should come as no surprise to you.

1. *Latitude*—The two latitudinal variables critical to climate are the intensity of radiation and the length of the day, which are intimately related to the Earth–Sun geometric relationship. On average, lower latitudes receive more insolation over the course of the year than high latitudes because the Sun shines more directly on these locations. This insolation variability has a direct impact on ambient (overall) air temperature.

Earth–Sun Relationships
Position of Earth in orbit influences amount of radiation received

Latitude
Influences Sun angle and length of day

Air circulation
Influences flow of air and position of high- and low-pressure systems

Marine/Continental
Large water bodies moderate temperature, influence airflow, supply water vapor

Topography
Mountains affect precipitation on windward and leeward sides, influence airflow

Figure 9.1 Variables that influence global climates. The climate of any given place is determined by the interaction of many factors, such as Earth–Sun relationships, latitude, marine/continental relationships, air circulation, and topography.

Weather *Day-to-day changes that occur in temperature and precipitation.*

Climate *Average precipitation and temperature characteristics for a region that are based on long-term records.*

2. *Seasonality*—Seasonality is a critical variable in the character and distribution of global climates for two reasons related to Sun angle: (a) the number of hours of sunlight in a day changes in many places over the course of the year, and (b) insolation and temperature can be highly variable between the winter and summer seasons. In places like the tropical regions, however, very little annual change occurs with respect to these variables, which is reflected in the relatively consistent temperature and precipitation characteristics of tropical climates.

3. *Air mass circulation*—As discussed in Chapter 6, the atmosphere flows in predictable ways, with distinct pressure systems associated with specific zones on Earth. The unique distribution of these pressure systems results in regions with heavy or persistent precipitation, such as those associated with the Intertropical Convergence Zone (ITCZ), whereas others (such as those near the Subtropical High [STH] Pressure System) are relatively dry. Still other regions have distinct weather-producing systems, such as midlatitude cyclones that result in highly variable weather from day to day.

4. *Maritime vs. continental relationships*—Proximity to a large body of water can have a big impact on temperature and the direction of airflow, as we saw in Chapter 7. In the interior of continents the annual temperature range can fluctuate a great deal, whereas it is moderated along ocean coasts. Large bodies of water can also be great sources of atmospheric water vapor through the process of evaporation, and airflow can transport water vapor inland where it falls as precipitation. Evaporation of large amounts of water vapor is usually associated with warm oceans, whereas cold oceans can have a drying effect on adjacent continental locations.

5. *Topographic effects*—Topography can dramatically influence atmospheric processes in a region, as shown in Chapters 6 and 7. For example, the windward side of mountain ranges is often a place of heavy precipitation because air cools adiabatically there. In contrast, rain shadows commonly form in the leeward side of mountain ranges due to the descent of air and associated adiabatic warming. Similarly, the topography of a region can also influence the flow of air, resulting in Chinook winds and cool-air drainages that otherwise would not occur.

All these variables interact with one another in a holistic way to influence climate. As you study the next section on the geography of global climates, remember to look for the ways in which these factors interact to form a regional climate pattern. To assist you with this process, you will frequently be prompted with questions that are intended to make you think about the interrelationship of variables. If you have trouble seeing how any particular variable fits into the picture, review parts of the preceding chapters.

Köppen Climate Classification

The purpose of climate classification is to identify certain characteristics, such as temperature and precipitation, which have observable regional patterns. Given the discussions in the previous chapters, you may already be thinking of some distinctive regional climate patterns. For example, tropical regions tend to be wet due to the presence of the ITCZ, whereas latitudes around 30° N and S are drier due to the influence of the STH Pressure System. Although this pair of climate zones is easy to visualize, it is often difficult to classify many other climate zones because their geography depends on a number of variables. It is also difficult to classify climate zones because they form a spatial continuum; in other words, sharp breaks from one climate region to the next rarely occur.

Despite these difficulties, a variety of helpful climate classification systems have been devised. One such system is the Thornthwaite system, which was developed by the American climatologist C. Warren Thornthwaite. This system focuses on the local scale and is used frequently in the United States to assess soil moisture characteristics as they relate to vegetation and agriculture. It is based on the concept of **potential evapotranspiration (potential ET)**, which is an estimate of the amount of water used by plants with an unlimited water supply. Potential ET increases with increasing temperature, daylight length, and wind strength and decreases when humidity increases. Can you determine why these changes take place? Think about factors such as insolation and the vapor gradient. In contrast, **actual evapotranspiration (actual ET)** reflects the actual amount of water used by plants.

The Thornthwaite system identifies three basic climate zones on Earth. Low-latitude climates have potential ET >130 cm (51 in.), whereas middle-latitude climates are those with potential ET that ranges between 130 cm and 52.5 cm (20.5 in.). High-latitude climates are those with potential ET that is <52.5 cm. The Thornthwaite system can also be subdivided on the basis of the length of time, and by what amount, actual ET falls below potential ET. In this context, moist climates are those with a surplus or small deficit (15 cm; 6 in.) of water, whereas dry climates have a yearly deficit >15 cm.

Although the Thornthwaite climate system is used frequently in the United States, it is not commonly used elsewhere in the world. Instead, the most widely used classification system is the Köppen (pronounced *Kepun*) Climate Classification System, which was developed by the German botanist and climatologist Wladimir Köppen, who recognized the relationship

---

Potential evapotranspiration (potential ET) *A measure of the maximum possible water loss from a given land area assuming sufficient water is available.*

Actual evapotranspiration (actual ET) *The quantity of water actually removed from a given land area by evaporation and transpiration.*

TABLE 9.1   Köppen Climate Designations

First Letters	Derivation	Distinguishing Characteristic
A	Alphabetical	Mean temperature each month >18°C (64°F).
B	Alphabetical	Mean annual precipitation <76 cm (30 in.).
C	Alphabetical	Mean monthly warmest temperature >10°C (50°F) in warmest month; mean monthly coldest temperature between 18°C and –3°C (64°F–27°F) in coldest month.
D	Alphabetical	Mean temperature >10°C (50°F) in 4–8 months.
E	Alphabetical	Mean temperature <10°C (50°F) in all months.
H	Alphabetical	Significant climate changes due to altitude variations.
Second Letters		
F	German feucht, "moist"	6 cm (2.5 in.) of mean rainfall in each month.
M	Monsoon	Only 1–3 months with mean rainfall <6 cm (2.5 in.).
S	Summer dry	Summer dry season, with driest month having $<\frac{1}{3}$ the mean precipitation of wettest winter month.
S	Steppe (semi-arid)	Mean annual precipitation in low latitudes is 38 cm–76 cm (15 in.–30 in.), whereas it is between 25 cm and 64 cm in midlatitudes. No distinct seasonal trend in either latitude range.
T	Tundra	At least 1 month with mean temperature between 0°C and 10°C (32°F–50°F).
W	Winter dry	Winter dry season, with 3–6 months of <6 cm (2.5 in.) of mean rainfall in A climates. In C and D climates, the driest month has $<\frac{1}{10}$ the mean precipitation of the wettest summer month.
W	German wüste, "desert"	Mean annual precipitation <38 cm (15 in.) in low latitudes and <25 cm (10 in.) in midlatitudes.
Third Letters		
A	Alphabetical	Warmest month has mean temperature <22°C (71.6°F).
B	Alphabetical	Warmest month has mean temperature <22°C (71.6°F), but has 4 months with mean temperature >10°C (50.0°F).
C	Alphabetical	Warmest month has mean temperature <22°C (71.6°F); fewer than 4 months with mean temperature >10°C (50.0°F).
D	Alphabetical	Same as C, but coldest mean monthly temperature <–38°C (–36.4°F).
H	German heiss, "hot"	Mean annual temperature >18°C (64.4°F).
K	German kalt, "cold"	Mean annual temperature <18°C (64.4°F).

between major vegetation regions on Earth and regional climate characteristics. The Köppen system describes world climates based on average monthly temperature, average monthly precipitation, and total annual precipitation. It is such a widely used system in part because these variables are measured more frequently in many more places around Earth than any other climate variables. Although the system is great for characterizing regional climate conditions, it ignores more local factors such as wind speed and cloud cover, which often influence conditions dramatically in specific areas.

The best way to view the Köppen system is as a hierarchical classification scheme of categories and subcategories.

At its most general level, the system recognizes six major climate groups (*A, B, C, D, E, H*). Climate groups *A, B, C, D,* and *E* range from tropical latitudes near the Equator (*A*) to high latitudes at the poles (*E*) and cover huge geographical areas. Climate group *H* is the designation for high-altitude regions, regardless of latitude, and may be found in isolated places. Each of these major climate groups is distinguished on the basis of temperature, except for category *B*, which is based on moisture characteristics. The Köppen *A* climate zones, for example, have a mean temperature each month >18°C (64°F), whereas the *D* climate regions have average temperature >10°C (50°F) in four months to eight months. *B* climate regions are recognized as those with mean annual precipitation <76 cm (30 in.).

Although it is often sufficient to note the fundamental climate characteristics over broad areas, geographers are usually interested in the more specific attributes of smaller regions. To accomplish this task, the major Köppen groups are further subdivided on the basis of temperature and moisture by use of a second letter, and sometimes even a third. Climate regions with a second letter *s*, for example, have a distinct summer dry season. In contrast, areas with a second letter *w* have an observable winter dry season. Those areas further classified with a third letter *a* have a warmest month that has an average temperature >22°C (71.6°F). Table 9.1 outlines the various Köppen letter designations and their meanings as far as specific environmental characteristics are concerned. The full range of Köppen climate types and subtypes is presented in Table 9.2, with the geographic distribution of these classes shown in Figure 9.2.

As you might imagine, the Köppen classification system can be very cumbersome to use because it contains a lot of jargon. At this introductory level, your goal should be to understand the fundamental characteristics of the 23 major climate regions and avoid becoming overwhelmed by all the specific reasons why the many subcategories occur. With this strategy

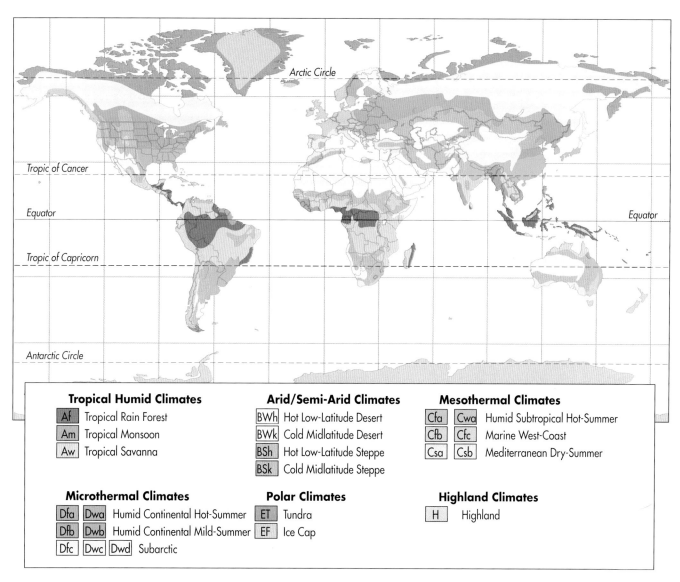

Figure 9.2 Map of Köppen climates on Earth. The Köppen system contains six major climate groups that are divided into as many as 23 subcategories.

TABLE 9.2   Climate Types and Subtypes in the Köppen Climate System

First Letter	Subcategory	Köppen Designation	Köppen Climate Name	Defining Characteristic
A	Tropical humid	Af	Tropical Rainforest	No dry season
		Am	Tropical Monsoon	Short dry season; heavy monsoonal rains in other months
		Aw	Tropical Savanna	Winter dry season
B	Arid/semi-arid	BWh	Hot Low-Latitude Desert	Low-latitude desert
		BSh	Hot Low-Latitude Steppe	Low-latitude dry
		BSk	Cold Midlatitude Steppe	Midlatitude dry
		BWk	Cold Midlatitude Desert	Midlatitude desert
C	Mesothermal	Cfa	Humid Subtropical Hot-Summer	Mild with no dry season, hot summer
		Csa	Mediterranean Dry-Summer	Mild with dry, hot summer
		Csb	Mediterranean Dry-Summer	Mild with dry, warm summer
		Cwa	Humid Subtropical Hot-Summer	Mild with dry winter, hot summer
		Cfb	Marine West-Coast	Mild with no dry season, warm summer
		Cfc	Marine West-Coast	Mild with no dry season, cool summer
D	Microthermal	Dfa	Humid Continental Hot-Summer	Humid with severe winter, no dry season, hot summer
		Dfb	Humid Continental Mild-Summer	Humid with severe winter, no dry season, warm summer
		Dwa	Humid Continental Hot-Summer	Humid with severe, dry winter, hot summer
		Dwb	Humid Continental Mild-Summer	Humid with severe, dry winter, warm summer
		Dfc	Subarctic	Severe winter, no dry season, cool summer
		Dwc	Subarctic	Severe, dry winter, cool summer
		Dwd	Subarctic	Severe, very cold and dry winter, cool summer
E	Polar	ET	Tundra	Polar tundra, no true summer
		EF	Ice Cap	Perennial ice
H	Highland	Created after Köppen system was devised		High elevation with cool to cold temperature

in mind, the following discussion is a simplified presentation of the Köppen system that focuses on the defining characteristics of each climate region, such as the annual range of temperature and precipitation patterns, along with the environmental variables that cause them. The description of each Köppen region will include a photograph of the landscape to represent how the climate affects the natural vegetation of the area.

Geography of Köppen Climates

This section of the chapter focuses on the global geography of Köppen climates. The discussion begins with the climates found in tropical latitudes and progresses to those occurring in very cold regions of the world. As you work your way through this discussion, refer frequently to the global map of Köppen regions in Figure 9.2 to improve your understanding of climate geography. You should also go to the Geo Media Library at the end of this discussion to view the climograph for each region. This site also includes a selection of photographs of each area to give you a better idea of the landscape and, in some cases, how people interact with it.

Tropical (A) Climates

Tropical *A* climates are the climates found at low latitudes that straddle the Equator, extending to approximately 25° latitude in both the Northern and Southern Hemispheres (Figure 9.2). *A* climates are warm in that the average monthly temperature exceeds 18°C (64°F). This climate region is further divided into three subcategories on the basis of precipitation.

The tropical rainforest climate (*Af*) occurs at very low latitudes and is most closely associated with consistently high amounts of monthly solar radiation and the strong influence of the ITCZ. As a result, the relative humidity is always very high, abundant rainfall occurs every month due to afternoon convection, and vegetation consists of dense rainforest (Figure 9.3).

The tropical monsoon climate (*Am*) is very closely related to the tropical rainforest climate (*Af*) because it also receives abundant precipitation. This climate region differs, however, because it has a more prominent seasonal pattern that is associated with a migrating Sun position and the flow of air. Thus, average monthly precipitation drops below 6 cm (2.5 in.) for a month or two. Nevertheless, this slightly reduced season precipitation is not sufficient to change the nature of the vegetation, which is also rainforest in these regions.

The tropical monsoon climate (*Am*) is directly related to the onshore flow of mT air that occurs in places like Southeast Asia and southwest India (Figure 9.2), which was discussed in detail in Chapter 6. This onshore flow occurs in summer when the ITCZ is overhead and reverses in winter when the landmass cools. Because the onshore monsoon winds are southwesterly, the climate is largely associated with the west coasts of landmasses. The tropical monsoon climate (*Am*) also occurs on the eastern side of landmasses in some places, such as the northeast coast of South America (Figure 9.2), where northeast trade winds bring humid air into the region on a seasonal basis.

The third subdivision of the tropical *A* climates is the tropical savanna climate (*Aw*). This climate region is poleward of the tropical rainforest climate (*Af*) in both hemispheres and is found over large regions of the world (Figure 9.2). The primary difference between this climate zone and the tropical rainforest (*Af*) and tropical monsoon (*Am*) climates is that the tropical savanna climate (*Aw*) has very distinct wet and dry seasons. As you might imagine, this cycle is distinctly related to the impact of Earth–Sun geometry on the migration of large-scale atmospheric pressure systems. The wet season occurs during the summer months because the ITCZ is overhead due to the high Sun angle that time of year. In contrast, the dry season occurs during the winter months because the ITCZ is in the opposite hemisphere during that time of year and is replaced by the STH. As a result of this distinct wet/dry cycle, the vegetation is a mix of grass and trees known as *savanna* (Figure 9.4).

Figure 9.3 The tropical rainforest climate (*Af*). Tropical rainforest in the African Congo. Due to the very warm temperatures, high humidity, and persistent rainfall in this region, the vegetation is very dense.

Figure 9.4 The tropical savanna climate (*Aw*). Tropical savanna in northern Australia consists of a mix of trees and grass due to the seasonal wet and dry cycle associated with the tropical savanna climate.

1. Three major tropical climates are recognized: (a) tropical rainforest climate (*Af*); (b) tropical monsoon climate (*Am*); and (c) tropical savanna climate (*Aw*).

2. The tropical rainforest climate (*Af*) has a very narrow temperature range due to persistent high Sun angle. Annual precipitation is consistently high throughout the year due to the influence of the ITCZ.

3. The tropical monsoon climate (*Am*) has a distinct wet and dry season that is related to seasonal landward and seaward flow of air that occurs at low latitudes. This climate region has a slightly more seasonal temperature range than the tropical rainforest climate (*Af*).

4. The tropical savanna climate (*Aw*) lies poleward of the tropical rainforest climate (*Af*). It has distinct wet and dry seasons that are related to the seasonal dominance of the ITCZ and STH, respectively.

Arid and Semi-Arid (*B*) Climates

Arid and semi-arid *B* climates are regions with both hot and cold temperatures that are relatively dry, with annual precipitation <76 cm (30 in.). These climate regions are poleward of the *A* climates and are the most widespread on Earth, with *B* climates occurring virtually on every continent (Figure 9.2). From a geographical perspective, the arid climates can be first subdivided on the basis of latitude into low-latitude arid climates and middle-latitude arid climates. The low-latitude arid climates tend to be the hottest and driest and are closely associated with the subsiding air of the subtropical high-pressure belt between 20° and 30° N and S latitude. To see this distribution, examine Figure 9.2 and note the extensive deserts in Australia, southern and northern Africa, and Asia. In contrast, middle-latitude arid climates are less dry, semi-arid climates that are related both to the STH and to rain shadow effects in continental interiors.

Low-Latitude Dry Climates The Köppen system recognizes two arid climate regions in the low latitudes of Earth. The most extreme of this pair of climate zones is the hot low-latitude desert climate (*BWh*). This climate zone is found in the center and eastern sides of the STH Pressure Systems, spanning 15° to 30° latitude in both hemispheres, and is characterized by mean annual precipitation <38 cm (15 in.) and average annual temperature >18°C (64.4°F). These conditions fundamentally occur due to subsiding air in the center of the STH that promotes extensive adiabatic warming and very little rainfall. The ground is further heated because the skies are usually clear and solar radiation strikes the surface directly due to the high Sun angle. On the ground, the hot low-latitude desert climate (*BWh*) is the classic tropical desert that you have seen pictured in many places (Figure 9.5a). The vegetation in these regions is extremely sparse and drought-tolerant, with waxy and spiny leaves designed to reduce water loss and trunks able to store water.

The second arid climate zone in the low latitudes is the hot low-latitude steppe climate (*BSh*) region. This climate zone is very closely related to the hot low-latitude desert climate (*BWh*) and results from similar air mass patterns. Because these regions are located slightly poleward of the hot low-latitude desert climate (see Figure 9.2 again), they have a greater annual temperature range and an annual precipitation that ranges between 38 cm and 76 cm (15 in. and 30 in.). The greater temperature range results from the occasional incursion of cP air masses, related to strong midlatitude cyclones, in the winter. As a result, the periodic influence of these storm systems causes some precipitation to fall during the winter months. Vegetation in this climate region looks very similar to what occurs in the Mojave Desert (Figure 9.5b) in the southwestern part of the United States.

Midlatitude Dry Climates In addition to the pair of arid climates that occur in the low latitudes, two dry climates are also recognized in Earth's middle latitudes. The most extensive of these middle-latitude dry zones is the cold midlatitude steppe climate (*BSk*). This climate region is most often found

www.wiley.com/college/arbogast

WileyPLUS

Tropical Savanna Climate (*Aw*)

To better visualize how the tropical humid climate functions, go to the *Geo Media Library* and select *Tropical Savanna Climate (Aw)*. This animation presents an oblique view (an elevated view from the side) of Timbo, New Guinea, from the Atlantic Ocean. In this fashion, you can see how the migration of pressure systems influences the climate of the region. You will also be able to manipulate the migration of the ITCZ, which will allow you to better see the importance of this pressure system on precipitation in any specific geographic region. Once you complete this simulation, be sure to answer the questions at the end to test your understanding of this concept.

(a)

(b)

Figure 9.5 Low-latitude desert climates. (a) Vegetation in the hot low-latitude desert climate (*BWh*) is very sparse due to the drying effects of subtropical high-pressure cells. (b) Vegetation in the hot low-latitude steppe climate (*BSh*) is very similar to that seen in the Mojave Desert in the southwestern Unted States. Although the vegetation here is sparse, some trees like this Joshua tree can grow.

deep within the continental interiors of North America and Eurasia (Figure 9.2) and is most often associated with the rain shadow effect that occurs in the lee of major mountain ranges, such as the Rocky and Himalayan Mountains. As discussed in Chapter 7, this effect occurs because maritime air masses are effectively blocked by the mountains, allowing relatively dry continental air masses to dominate the region. Most precipitation falls in summer and is related to convectional processes. When precipitation does fall in the winter, it is usually associated with the midlatitude storm track. Given the relatively high amount of precipitation in this dry climate, compared to the desert-like environments examined earlier, vegetation in this region consists of short-grass prairie (Figure 9.6a). Given your understanding of continental air masses, it should be no

surprise that the annual temperature range in the dry midlatitude climate is large, ranging from hot to quite cold. (*Do you know why this pattern occurs?*)

The second midlatitude dry climate on Earth is the cold midlatitude desert climate (*BWk*). This climate zone encompasses one of the smallest overall areas of any climate group, occurring only in central Asia, north-central China, Patagonia in Argentina, and at high elevations in the southwestern United States. The primary distinguishing characteristic of this climate is that it receives 15 cm (5.9 in.) or less of rainfall per year. The annual monthly maximum temperature range in this zone can be high, ranging from ~0°C in the winter months to ~30°C (85°F) in July. The dry character of this climate results in typical desert vegetation (Figure 9.6b).

(a)

(b)

Figure 9.6 Midlatitude dry climates. (a) Short-grass prairie in eastern Colorado. This low-growing vegetation is a response to the low rainfall in this climate region (*BSk*). (b) The Hongoryn Els sand dunes in Mongolia's Gobi Desert. This sparsely vegetated landscape occurs because of cold/dry conditions in the interior of Asia (*BWk*). This dune field is about 200 km (125 mi) long.

THE ATACAMA DESERT

The driest place on Earth is the Atacama Desert on the west coast of South America. This region is on the northern margin of the cold midlatitude desert climate (*BWk*) and extends about 1000 km (600 mi) from the southern border of Peru into northern Chile. To give you an idea of the Atacama's extreme aridity, consider that some areas within the desert have *never* received any measurable precipitation in recorded history.

Several interrelated factors explain the aridity of the Atacama Desert. The region lies on the west side of the Andes Mountains, which are a major orographic barrier to moisture-laden easterlies that circulate across tropical South America. Thus,

the Atacama Desert lies in the rain shadow of the Andes, where descending air under high pressure compresses and warms so much that precipitation is rare. In addition, the region borders the Humboldt Current (also called the Peru Current). As a result, the air above the current is very stable with little convection. Given these combined factors, the Atacama Desert looks otherworldly. It seems so alien, in fact, that it has been used to simulate the Martian landscape in science fiction movies.

Mark Chivers/Robert Harding World Imagery/Getty Images, Inc.

| A typical scene in the Atacama Desert. Note the very sparse vegetation in this extremely dry climate.

| This map shows the geographic relationship of the Atacama Desert, Humboldt Current, and Andes Mountains.

KEY CONCEPTS TO REMEMBER ABOUT THE DRY ARID AND SEMI-ARID (*B*) CLIMATES

1. Four major dry arid and semi-arid climates are recognized: (a) hot low-latitude desert climate (*BWh*); (b) hot low-latitude steppe climate (*BSh*); (c) cold midlatitude steppe climate (*BSk*); and (d) cold midlatitude desert climate (*BWk*).

2. The hot low-latitude desert climate (*BWh*) occurs in places like northern Africa and is dominated largely by the STH Pressure System. Vegetation is very sparse.

3. The hot low-latitude steppe climate (*BSh*) lies slightly poleward of the hot low-latitude desert climate (*BWh*) in places like Saudi Arabia and Iraq. It thus has a wider

temperature range than the *BWh* climate and slightly more precipitation. This greater precipitation occurs because periodic midlatitude circulatory systems move through the region in winter. Vegetation is sparse, consisting of some grasses, cactus, and isolated trees.

4. The cold midlatitude steppe climate (*BSk*) occurs in places like Denver, Colorado, which lie in the rain shadow of major mountain ranges. Large temperature ranges and seasonal precipitation variations occur in these areas. Vegetation consists largely of short to medium grasses.

5. The cold midlatitude desert climate (*BWk*) occurs in places like the Gobi Desert in north-central China. These regions are dry because they either lie adjacent to cold ocean currents (like northern Chile) or are deep within continents. Vegetation is limited.

This satellite image focuses on the Middle East, including northern Saudi Arabia, Iraq, Jordan, and Israel. The land bodies are tan-colored because they are deserts with very little vegetation. Which of the following statements best explains the presence of deserts in northern Saudi Arabia and southern Iraq? (*Hint:* The latitude of Baghdad is 33° N.)

    a) The region is predominantly influenced by the ITCZ.
    b) The region is influenced largely by low-pressure systems.
    c) The region is dominated by the STH Pressure System.
    d) The region lies on the windward side of a major mountain range.

Jeff Schmaltz, MODIS Rapid Response Team, NASA/GSFC

Mesothermal (*C*) Climates

Now that the tropical and desert climates of Earth have been discussed, let's turn our attention to the climate regions characterized by a somewhat more seasonal temperature pattern. These temperature zones are known as mesothermal (*C*) climates because they have both a warm and cool season, with fairly abundant precipitation. Mesothermal (*C*) climates are situated in the midlatitudes, typically between about 20° and 60° N and S latitude (Figure 9.2). Most people live in this climate zone.

If you look again at the global climate map in Figure 9.2, notice that the distribution of *C* climates varies noticeably between the Northern and Southern Hemispheres. (*How can this variability be related to the hemispheric land–water differential that exists?*) In the Southern Hemisphere, *C* climates extend across much of the South American landmass south of 40° S latitude; this occurs because the continent is narrow. North of the Equator, in contrast, the east–west belt of *C* climate is interrupted by a zone of arid climate in the lee of the Rocky Mountains in North America and by the Himalayas in Asia. These breaks occur in the Northern Hemisphere because the continents in that part of the world are very large, which limits the amount of oceanic-derived moisture that can reach the interior.

Three broad mesothermal climate zones occur on Earth. The most humid of these climate regions are the humid subtropical hot-summer climates (*Cfa, Cwa*), which are found on the eastern side of continents between about 20° to 35° N and S latitude (Figure 9.2). The distinction between *Cfa* and *Cwa* is based on the fact that *Cwa* climate regions have a distinct winter dry period—hence, the *w* designation. Otherwise, the two regions are climatically similar because they have a hot and humid summer season.

Cfa regions are humid, with average annual precipitation ranging from 100 cm to 200 cm (40 in.–80 in.). This humid climate arises because these regions are affected by mT air masses on the western side of the adjoining ocean's subtropical high-pressure zone. A good example of this geographical pattern oc-

curs in the southeastern United States, ranging from the Carolinas to east Texas (Figure 9.2). During a typical summer, a strong subtropical high-pressure system known as the "Bermuda High" develops off the southeast coast of the United States. Because of the anticyclonic circulation of the system, it pumps warm, moist air from the Atlantic into the *Cfa* region, where it falls as convectional rain due to land heating. Winter precipitation is caused by frontal systems associated with strong midlatitude cyclones. (*Why is it logical that in the winter midlatitude cyclones bring precipitation to the American Southeast, whereas they do not in summer?*) Total annual rainfall is about 162 cm (63 in.), with ample precipitation every month. Although winters are mild, a distinct annual high-temperature cycle is present, ranging from 16°C (62°F) in January to 33°C (91°F) in July.

Vegetation in *Cfa* regions is largely characterized by dense forests of broadleaf deciduous trees (like oak and elm) that drop their leaves during extended cold or dry periods. In particularly warm places, such as southern China and the Gulf Coast of the United States, the native forest consists of broadleaf evergreen trees that remain green throughout the year (see Figure 9.7 as an example).

The second climate within the *C* classification is the Mediterranean dry-summer climate (*Csa, Csb*). In contrast to the humid subtropical hot-summer climates (*Cfa, Cwa*), *Cs* climates are located along the west coast of continents, at an average latitude of about 35° N and S (Figure 9.2). The difference between the *Csa* and *Csb* climate zone is that the former has a hotter summer than the latter, with the warmest month having an average temperature >22°C (71.6°F). Otherwise, they are climatically very similar because the wet season occurs in the winter, as opposed to the summer when the air is dry. This pattern occurs because in the winter strong midlatitude cyclones bring moisture into the region from the nearby ocean to the west. During the summer months, however, the midlatitude storm track migrates north and is replaced by the STH Pressure System due to the high Sun angle. The amount of rainfall any particular *Cs* region receives is essentially related

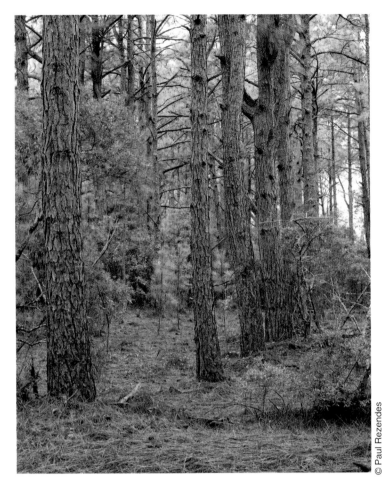

Figure 9.7 The humid subtropical hot-summer climate (*Cfa, Cwa*). Along the Gulf Coast of the United States, the vegetation in many places is the kind of broadleaf evergreen forest shown here.

© Paul Rezendes

to latitude, with lower-latitude locations usually being arid to semi-arid and those at higher latitudes more humid. (*Why do you think this pattern occurs?*) Regardless of the amount of rainfall received, the temperature in these regions is usually moderate with a very mild seasonal variation that may range from about 13°C (55°F) in January to 28°C (83°F) in August. The vegetation in this region is characterized by shrubs and trees that have thick, hard leaves (Figure 9.8) designed to retain moisture during long summer droughts.

The third type of mesothermal climates are the marine west-coast climates (*Cfb, Cfc*), which are found dominantly on the west coast of continents between 35° to 60° N and S latitude (see Figure 9.2). In North America this climate occurs from Oregon to northern British Columbia and also occurs on the east side of the continent in the Appalachian Mountains. Elsewhere, this climate is found in the British Isles, the west coast of Europe, New Zealand, the southern tip of Australia, and southern Chile. The distinction between the *Cfb* and *Cfc* subcategories is that the *Cfb* zones have four months with temperatures >10°C (50.0°F), whereas *Cfc* climate zones have fewer than four months that warm.

Given the higher latitude of these locations, they are more directly, and more frequently, in the path of the westerly storm track that brings cool, moist mP air masses from the large oceans to the west. Therefore, marine west-coast locations receive more precipitation than Mediterranean regions. This precipitation is magnified by orographic processes when mP air masses encounter large mountain ranges, such as the Cascades in the northwest United States or the Andes in Chile. In regions where this geographical interplay occurs, the amount of annual rainfall is prodigious, exceeding 254 cm/year (100 in./year). As you might expect, marine west-coast regions have a greater annual temperature range than their Mediterranean counterparts at lower latitudes, but the winters are still very mild compared to continental locations. Due to this heavy precipitation, the native vegetation in these areas consists of dense needle-leaf forests (Figure 9.9).

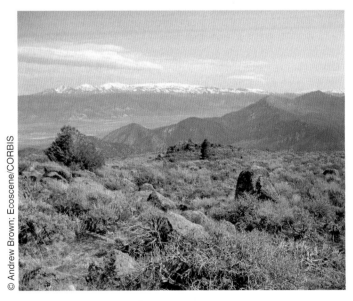

© Andrew Brown; Ecoscene/CORBIS

Figure 9.8 The Mediterranean dry-summer climates (*Csa, Csb*). The California chaparral is associated with the *Cs* climate and consists of wiry shrubs that retain moisture during summer dry periods.

Terry W. Eggers/Corbis Images

Figure 9.9 The marine west-coast climates (*Cfb, Cfc*). Needle-leaf forest in the Hoh Rainforest at Olympic National Park, Washington. Here, abundant precipitation associated with the marine west-coast climate (*Cfb, Cfc*) supports a dense forest of Sitka spruce, hemlock, and various ferns and mosses.

Humid Subtropical Hot-Summer Climate (*Cfa, Cwa*)

In order to get a better feel for the variables that influence the humid subtropical hot-summer climate, go to the *Geo Media Library* and select *Humid Subtropical Hot-Summer Climate (Cfa, Cwa)*. This module shows how the STH known as the Bermuda High pumps moisture into the southeastern United States during the summer. The animation will begin with a map that shows the rotating pressure system that is migrating on a seasonal basis due to the effects of Earth–Sun geometry. As spring progresses into summer, the Bermuda High forms off the southeastern coast. This system is rotating clockwise and, in so doing, pumps moisture into places like Louisiana, Georgia, and North Carolina. This can be shown by clouds forming over land that result in precipitation. Once you complete this animation, be sure to answer the questions at the end to test your understanding of this concept.

VISUAL CONCEPT CHECK 9.2

This map shows the average annual precipitation in the contiguous United States. Explain the patterns for these three regions: (1) the Southeast, including Mississippi and Tennessee; (2) the Intermountain West, including Nevada and Utah; and (3) the Pacific Northwest coast, including western Washington and Oregon.

Marine West-Coast Climates (*Cfb, Cfc*)

In order to visualize how the marine west-coast climate functions, go to the *Geo Media Library* and select *Marine West-Coast Climates (Cfb, Cfc)*. In this animation, you will be able to see how the migrating midlatitude storm track brings moisture off the eastern Pacific Ocean into the region. This animation is similar to the humid subtropical hot-summer climates (*Cwa, Cwb*) animation

you previously viewed in that it will show airflow and associated cloud development. As you watch the animation, notice the relationship between cyclones and moisture flow, as well as the effect that the Cascade Mountains have on precipitation intensity. Once you complete the animation, be sure to answer the questions at the end to test your understanding of this concept.

KEY CONCEPTS TO REMEMBER ABOUT MESOTHERMAL (C) CLIMATES

1. Three major mesothermal *(C)* climate subcategories are recognized: (a) humid subtropical hot-summer climates (*Cfa, Cwa*); (b) Mediterranean dry-summer climates (*Csa, Csb*); and (c) marine west-coast climates (*Cfb, Cfc*).

2. The humid subtropical hot-summer climates (*Cfa, Cwa*) occur on the southeastern side of continents in places like South Carolina and China. These regions have ample year-round precipitation, but have a distinct summer peak due to the influx of mT air by offshore STH Pressure Systems. Vegetation consists predominantly of pines and deciduous trees.

3. The Mediterranean dry-summer climates (*Csa, Csb*) occur on the southwest coast of North America and in the Mediterranean region. These climates have a warm but narrow temperature range due to their relatively low-latitude coastal location. The summer dry season is associated with dominance by the STH Pressure System. Vegetation consists of a mix of trees and grass.

4. The marine west-coast climates (*Cfb, Cfc*) occur in coastal zones poleward of their Mediterranean counterparts. Thus, they have a wider temperature range that is, on average, cooler. These regions receive ample annual precipitation, but have a distinct winter wet season due to passage of midlatitude cyclones. Vegetation consists largely of needle-leaf trees.

Microthermal (*D*) Climates

At this point, let's turn to the climate zones that are more continental in their character by focusing on *D* climates. These climate zones are typically located poleward of *C* climates, ranging from about 35° to 60° N and S latitude (Figure 9.2). *D* climates are known as "microthermal" because they have winters sufficiently cold to ensure that snow can remain on

the ground for extended periods of time. In contrast, summer temperatures are usually warm and can even be hot for brief periods of time. *D* climates are found almost exclusively in the Northern Hemisphere because no large landmasses are present at corresponding latitudes in the Southern Hemisphere. The only places you will see *D* climates in the Southern Hemisphere are in highland areas. (W*hy does this distribution of D climates make sense?*)

Humid Continental Climates Like the mesothermal *C* climates, microthermal *D* climates can be initially subdivided into three categories. The most southerly of these subcategories are the humid continental hot-summer climates (*Dfa, Dwa*), which are located in the central and eastern parts of North America and Eurasia (Figure 9.2). Two subclassifications are recognized in these climates: one is moist (≥ 6 cm [2.5 in.] of average rainfall each month) over the entire year (*Dfa*), whereas the other is dry (<6 cm [2.5 in.] of rainfall per month) in the winter (*Dwa*). The primary weather-producing system in humid continental hot-summer climates is the midlatitude jet stream, which steers cyclones through these regions. Given the continental locality of this climate, the annual range of temperature is high, ranging from warm, periodically hot summers to cold winters. Most precipitation falls during the summer, when mT air masses invade from the south. Winters are relatively dry because cP and cA air masses dominate the region. In these climate zones, the dominant vegetation is deciduous forest (Figure 9.10).

Poleward of the humid continental hot-summer climates (*Dfa, Dwa*) are the humid continental mild-summer climates (*Dfb, Dwb*). These climate regions range across central and eastern North America, western Europe, central Asia, and the far east of Asia (Figure 9.2). These regions are known for having a wide annual temperature range due to the effects of Earth–Sun geometry. In places like the upper Midwest in the United States, average high temperatures in July are >27°C (80°F), with about 10 cm (4 in.) of rain. In contrast, average high temperatures in January are about −5°C (23°F) and most precipitation is snow. In northern Michigan this climate is associated with northern hardwood and pine forests (Figure 9.11).

Figure 9.10 The humid continental hot-summer climates (*Dfa, Dwa*). Deciduous forest in the humid continental hot-summer climate, including various oak and maple species, covered much of this climate region in the United States prior to European settlement.

Subarctic Climates The third major subdivision of the *D* climates is the subarctic climates (*Dfc, Dwc, Dwd*). These climate zones occur poleward of the humid continental climates (*Dfa, Dwa, Dfb, Dwb*), stretching across all of northern North America and from northern Scandinavia east across all of Asia (Figure 9.2). It is a continental climate characterized by long, bitterly cold winters and short, cool summers. In northwestern Canada, where the *Dfc* subdivision occurs, the average high January temperature is about –25°C (–13°F), whereas it approaches 20°C (69°F) in July. The impact of this extreme temperature range can be seen in the vegetation, which is dominated by boreal forests consisting of needle-leaf trees such as pine, spruce, and fir (Figure 9.12). The *Dwc* variant is found only in far eastern Russia and is characterized by a distinct winter drought. The *Dwd* variant occurs only in extreme northeastern Siberia and is distinctive because, in addition to a winter drought, its coldest month is below –38°C (–36.4°F).

The subarctic climates (*Dfc, Dwc, Dwd*) are the source regions of cold, dry cP air masses and are often invaded by more frigid cA air masses that originate at still higher latitudes. Given that this region is often north of the midlatitude storm track, the annual amount of precipitation is relatively low. Most precipitation occurs during the summer months, when the belt of cyclonic storms migrates into the region from the south. Otherwise, the polar high-pressure system dominates the region, resulting in little rainfall. The annual range of temperature is greater within this climate than any other. (*Do you remember why this extreme range occurs? Think back to previous discussions about Earth–Sun geometry, seasonality, and insolation.*)

Polar (*E*) Climates

The most extreme climates in nonmountainous areas are the polar *E* climates, which occur at latitudes higher than 70° N and S latitude. The tundra climate (*ET*) is one of the two polar climate subcategories and is found along the arctic coastal fringes, including such places as the island region of northern Canada, the north slope of Alaska, the Hudson Bay region, coastal Greenland, and all of northern Eurasia (Figure 9.2). Given this geographical location, a moderating climate effect occurs from the nearby oceans, resulting in winter temperatures less severe than in the more continental regions to the south. Nevertheless, winters are very long and cold and summers are short and cool. In northern Alaska, for example, the average high in January is –11°C (13°F) and an average high in July is 15°C (59°F). Very little

Figure 9.11 The humid continental mild-summer climates (*Dfb, Dwb*). Pine forest in northern Michigan. This type of vegetation is closely associated with the wet-winter variant (*Dfb*) of this climate.

Figure 9.12 The subarctic climate (*Dfc*). Typical vegetation in the subarctic climate (*Dfc*) consists of needle-leaf trees and scattered openings of birch and aspen.

(a)

(b)

George F. Mobley/National Geographic/Getty Images

Doug Allan/Tartan Dragon Ltd./Getty Images, Inc.

Figure 9.13 The polar climates (E). (a) Vegetation in the tundra landscape consists of low-lying lichens and mosses, which are the main diet of caribou. (b) A scene from the Antarctic ice cap where temperatures are frigid year round.

precipitation falls in this region, largely because it is dominated by the polar high-pressure system the vast majority of the year. The vegetation in this climate region is characterized by short mosses and lichens that dominate the landscape (Figure 9.13a). These areas also contain well-developed bogs and zones of permafrost, which will be investigated in more detail in Chapter 17.

KEY CONCEPTS TO REMEMBER ABOUT MICROTHERMAL (D) CLIMATES

1. Three major subcategories of microthermal (D) climates are recognized: (a) humid continental hot-summer climates (*Dfa*, *Dwa*); (b) humid continental mild-summer climates (*Dfb*, *Dwb*); and (c) subarctic climates (*Dfc*, *Dwc*, *Dwd*).

2. The humid continental hot-summer climates (*Dfa*, *Dwa*) occur in places like the midwestern United States and eastern Asia. Vegetation consists largely of deciduous trees. These regions have wide annual temperature ranges and moderate yearly precipitation. Most precipitation is associated with midlatitude cyclones.

3. The humid continental mild-summer climates (*Dfb*, *Dwb*) occur in places like the north-central and northeastern United States and eastern Asia. Vegetation consists dominantly of needle-leaf trees. These regions also have a wide annual temperature range, but with colder winters and milder summers than the humid continental hot-summer climates (*Dfa*, *Dwa*). Average annual precipitation is moderate and most closely associated with the midlatitude storm track.

4. The subarctic climates (*Dfc*, *Dwc*, *Dwd*) occur in high-latitude, continental locations that span North America and Asia. Vegetation is boreal forest. These regions have a very wide annual temperature range and have a summer wet season associated with the midlatitude storm track.

In contrast to the tundra climate, the ice-cap climate (*EF*) occurs in the interior of Greenland and Antarctica where enormous glaciers cover the landscape. Sun angle is consistently very low in these areas, resulting in little annual insolation. In addition, the ice-covered surfaces have high albedo, which means that most radiation is reflected and thus provides no heat. As a result, temperatures are consistently brutally cold to frigid (Figure 9.13b). At McMurdo Station in Antarctica, for example, average annual high temperatures range from −5°C (25°F) in December to −27°C (−17°F) in July and August. (*Why do the coldest temperatures occur in July and August?*) Annual precipitation is very low, <8 mm (0.3 in.). The reason for this low precipitation is that the air over the Antarctic ice cap is continental antarctic (cAA) air, which is very cold and dry. In short, this place is a polar desert.

Highland (H) Climates

Thus far the discussion about Köppen climates has largely focused on climate zones influenced by changes in latitude. The system also recognizes highland (*H*) climates, which are those associated with the world's large mountain ranges such as the Andes in South America, the Himalayas in Asia, the Swiss Alps in Europe, the Southern Alps in New Zealand, and the Rockies in North America (Figure 9.2). *H* climates are largely determined on the basis of the temperature changes associated with the environmental lapse rate as it relates to altitude. Recall from Chapter 5 that this value is 6.4°C/km (3.5°F/1000 ft). Perhaps you have observed when hiking in mountains that temperatures usually become cool with increased elevation. Or perhaps you have noticed that mountain peaks may be covered with snow while it is hot in the surrounding lowlands.

Global Climates

Now that the geography of global climates has been discussed, it is a good time to review them in an interactive way. Go to the *Geo Media Library* and select *Global Climates*. This animation is a tour of Köppen climate regions. It provides an example of each climate zone along with a representative city from that region. The discussion includes a climograph from that locality, as well as photographs of the city and associated vegetation. As a result, you should be able to better visualize the characteristics of each climate zone. Once you complete the exercise, be sure to answer the questions at the end to test your understanding of Köppen climate regions.

Remote Sensing and Climate

Global climates and the factors associated with them can be analyzed in many different ways, including by satellite remote sensing. With such technology, we can see the pattern of global climates around Earth and some of the variables associated with them, such as atmospheric pressure systems and vegetation. To see some of these patterns, go to the *Geo Media Library* and select *Remote Sensing and Climate*. This animation allows you to see how remote sensing can be used to observe climate patterns on Earth. You will be asked, for example, to identify a variety of atmospheric features, such as the ITCZ, the STH, and midlatitude cyclonic systems. You can also test your ability to locate hurricanes and determine the effect that cold ocean currents have on climate. As you work your way through this module, try to integrate all the concepts discussed so far in this book and visualize how they fit together to form distinctive geographic patterns. Once you complete the module, be sure to answer the questions at the end to make sure you understand these concepts.

This climate zone also reflects the fact that the environmental characteristics in mountainous areas are extremely complex, with abrupt changes taking place over short horizontal distances and vertical slopes. These complexities are related to such factors as slope orientation (does the slope face the Sun?), time of day (does the Sun face the slope in the morning or afternoon?), and direction of airflow (windward or leeward?). As you recall from Chapter 7, orographic processes may be a very important part of a highland area, with abundant precipitation on the windward side of a mountain range and relatively dry conditions in the lee. Perhaps the best way to view highland (*H*) climates is to observe that they can vary dramatically over short distances and that local conditions can change very quickly over time. If you plan to spend time in the mountains, you should be aware of these characteristics.

Reconstructing Past Climates

The global climates discussed in the previous section on the Köppen system are essentially stable entities on human time scales, with fairly predictable conditions occurring from one year to the next. For example, if you live in a continental mid-latitude location, chances are that the summers will be very warm and the winters cold every year. However, climate conditions on Earth have fluctuated in the past for a variety of reasons. Short-term fluctuations on decadal scale have been caused by volcanic eruptions, in which resulting ash in the atmosphere blocks insolation, causing lower temperatures. Climate changes have also occurred over much longer periods of time. For example, the global climate during the Jurassic period (a part of the dinosaur age), approximately 175 million years ago, was generally warmer and more humid than it is today.

KEY CONCEPTS TO REMEMBER ABOUT POLAR (ET, EF) AND HIGHLAND (H) CLIMATES

1. Three major subcategories of very cold climates are recognized: (a) the tundra climate (*ET*); (b) the ice cap and sheets climate (*EF*); and (c) highland climates (*H*).

2. The tundra climate (*ET*) is a high-latitude maritime climate associated with tundra vegetation. It has a narrower average annual temperature range than

lower-latitude subarctic climates (*Dfd*, *Dwd*). Average annual precipitation is low.

3. The ice cap and sheets climate (*EF*) is associated with large glaciers at high latitudes. Thus, average annual temperatures are very cold, and very little or no vegetation is present.

4. The highland climate (*H*) occurs at high elevations in mountain ranges such as the Rockies, Himalayas, and Andes. Although temperature in these areas varies annually, the range is much narrower and colder than nearby localities at lower altitudes.

As you probably know, an important environmental issue currently receiving a lot of attention is *global warming*, which is the belief that human industrial activity is gradually causing the planet to warm. Although over 90% of publishing climatologists believe human-induced global warming and associated climate change is occurring, the issue has become greatly politicized because numerous special interest groups are now involved in the debate. This part of the chapter focuses on the widely accepted science associated with ongoing climate change, and as a result, will provide you with the scientific background required to consider the issue. Although the political dimensions of the problem are fascinating in their own right, they are beyond the scope of this chapter. A good place to begin the discussion of ongoing and future climate change is by investigating the magnitude and frequency of climate fluctuations in recent Earth history, particularly during the *Quaternary Period*. This period of time encompasses the past ~2 million years of Earth's history and is a time of frequent climate changes, ranging from intensive ice ages to intervals similar to current conditions. You might be thinking, why do we look at Quaternary climate change to understand the future? The reason is that by understanding the nature of past climate changes, we gain insight into the range of possibilities that might occur in the future.

The challenge facing scientists investigating Quaternary or other past climates is that no direct temperature and precipitation data exist to reconstruct conditions at that time. After all, humans have been keeping detailed climate records for only about 150 years. In some parts of the United States, such as the western states where European settlement occurred later, these records span an even shorter amount of time. Although 150 years sounds like a long time in the context of the average human life span, it is really quite short from a geological perspective. In order to reconstruct the climate conditions of any given place, say, 20,000 years ago, it is necessary to investigate scientific clues from that era about the character of the environment at that time. Such clues are called **proxy data** because

they are indirect evidence for past climate conditions. Scientists who are interested in this kind of prehistoric detective work are called *paleoclimatologists*. This part of the chapter focuses on the kind of evidence paleoclimatologists use to reconstruct past climates.

Pollen Records

One kind of clue that provides indirect evidence of past climate conditions is ancient plant pollen, which reflects the kind of vegetation that occurred in a region at any given time. Given the demonstrable link between climate and vegetation, as we have seen earlier in this chapter, it is logical to assume that changes in the prehistoric plant assemblage of an area over time would indirectly reflect climate change. Such change in a plant community can be reconstructed in regions (usually cool and moist) where pollen is well preserved in the deposits that slowly accumulate in lakes and marshes. Scientists who study such records are called *palynologists*.

The premise of palynology is that plants produce microscopic pollen that are unique in shape and size to specific species. You can see this kind of variation in Figure 9.14. Each year, plants within an area produce pollen as part of their reproductive cycle, which is then spread by wind as pollen rain. If you happen to have seasonal allergies, you are acutely aware of airborne pollen because of the discomfort it may cause. Some of this pollen accumulates at the bottom of lakes and marshes in sediments that progressively thicken with time. Given that the deposits at the base of the deposit are the oldest, and progressively decrease in age toward the top of the sequence, the change in pollen composition through time can be reconstructed. To do

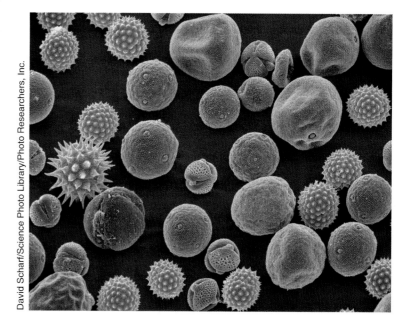

David Scharf/Science Photo Library/Photo Researchers, Inc.

Figure 9.14 Pollen grains. Different plant species produce distinctive forms of pollen that can be seen under a microscope. Note the different forms of pollen in this image, which is magnified 400 times.

Proxy data *Indirect evidence of an event. For example, fossil pollen is a proxy indicator of climate change because vegetation reflects climate.*

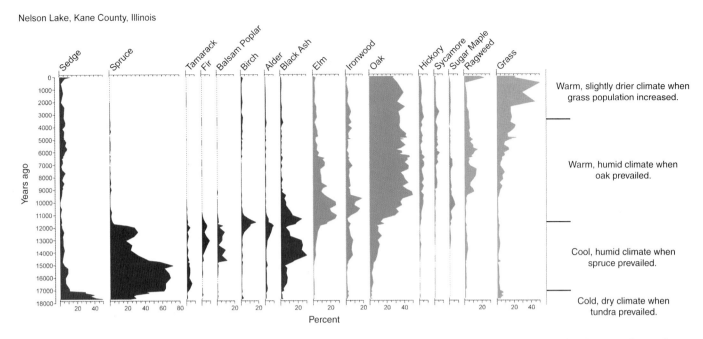

Figure 9.15 Pollen diagram from Nelson Lake, Illinois. Note the vegetation change through time and how it relates to climate fluctuations presented on the right-hand side of the graph. (*Source*: Eric Grimm, Illinois State Museum.)

so, a palynologist acquires a core (vertical sample) of the lake-bottom sediments and returns it to the laboratory where the pollen can be extracted from each layer of sediment and then analyzed.

Once the data are acquired from each layer of sediment, a diagram is created that shows the change in pollen in the area over time. This pattern of change is then used to reconstruct the climate history. The pollen diagram from Nelson Lake, Illinois (Figure 9.15) is a good example and should be interpreted in this way. Begin by looking at the vertical axis, which shows the age in years before the present, beginning with 17,000 years ago and ending today (0). From left to right across the top of the graph are the various plant species identified in the core samples, ranging from sedge on the left to grass on the right. The relative percentages of each of these plant species are shown at the bottom of each plant column. For example, around 15,000 years before the present, spruce trees accounted for about 60% of the overall sample, whereas the amount of oak pollen was very small. Approximately 10,000 years before the present, however, spruce pollen made up 0% of the sample, whereas oak was over 40%.

Using these percentage changes through time, a palynologist can reconstruct the climate for the Nelson Lake area for the past 17,000 years. The high amount of sedges between 17,000 years and 14,000 years before the present reflects a cold, dry climate when tundra prevailed in the area. As you will learn in Chapter 17, this climate/vegetation combination existed because an enormous glacier was nearby. Some 14,000 years ago, the climate became warmer and more humid, which caused tundra to be replaced by spruce, which is a needle-leaf tree. Spruce dominated the local vegetation until about 11,000 years before

the present, when the climate apparently warmed sufficiently to allow the influx of oak (a deciduous tree) at the expense of spruce. Oak has dominated for the past 10,000 years at Nelson Lake, although grass has increased in the past 2000 years. This expansion of grass is probably related to the climate becoming slightly warmer and drier.

Tree Ring Patterns

Another way that past climate change can be reconstructed is through the analysis of tree rings. This method is based on the fact that most trees produce annual growth rings, such as those shown in Figure 9.16a, over the course of their lives. In trees that have such ring patterns, each ring consists of a dark and light wood couplet, with the dark wood representing the winter period when growth is slow or nonexistent and the light wood reflecting rapid growth in summer. When local climate conditions are favorable, the rings are wide, whereas they are narrower (reflecting less growth) when it is colder or drier, depending on the type of tree.

The study of tree rings is called **dendrochronology** and can be used to reconstruct the history of regional climate change. A scientist begins such a study by obtaining a small core from a tree with a device called an increment borer. To obtain such a core, the implement is first screwed horizontally into the center

---

Dendrochronology *The dating of past events and variations in the environment and climate by studying the annual growth rates of trees.*

(a)

(b)

(c)

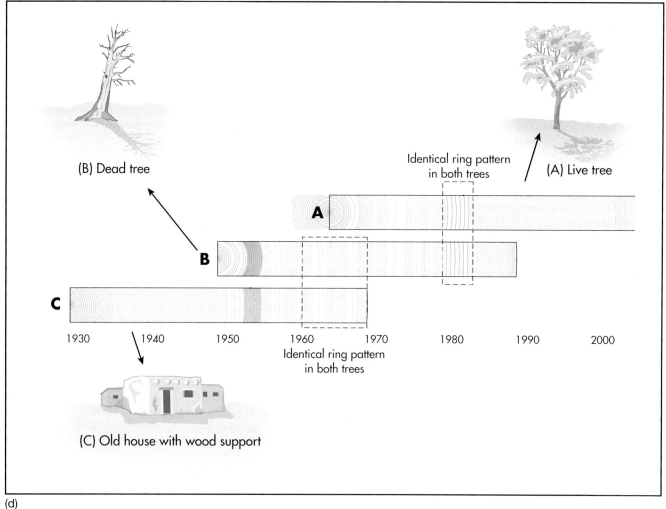

(d)

Figure 9.16 Using trees to reconstruct climate change. (a) The theoretical foundation of dendrochronology is that many trees produce annual growth rings, with dark and light wood representing winter and summer seasons, respectively. (b) Using an increment borer to core a tree. (c) A 4000-year-old bristlecone pine in the White Mountains of California. (d) The premise of cross dating is that the tree ring record can be extended back in time by overlapping ring patterns from a sequence of trees. In this simplified diagram, ring set A comes from a living tree, ring set B comes from a tree that recently died, and ring set C comes from wood that was used in an old house. This particular record would permit climate reconstruction between about 1930 and the early 2000s.

of the tree (Figure 9.16b) and then an approximately 6-mm (0.2-in.) wide core is extracted that cuts across the tree rings, thus showing all of them. The resulting hole is plugged with a small stick, and it then seals shut, resulting in no harm to the tree. Some trees live a very long time and thus produce long, continuous tree ring records. An excellent example of such a tree is the famous bristlecone pines in the White Mountains of California (Figure 9.16c), which are the oldest living things on Earth. Some of these trees are over 4000 years old and still growing! Thus, by looking at the rings of one tree alone, it is possible to reconstruct the growth cycles, and therefore the history of climate change, for the past 4000 years in the area where the pine trees have lived.

Although the living bristlecone pines have yielded an excellent record of climate change, paleoclimatologists are constantly looking for ways to extend the record farther and farther back in time. In the context of dendrochronology, this goal is accomplished through a method called *cross dating*. The premise of cross dating is that ring patterns from similar trees that overlapped in their life spans can be matched to extend the record of ring variability back in time.

To see how this method works, look at Figure 9.16d. Imagine that a core is extracted from a living tree (A) that produces a distinctive ring record. Near this living tree is another tree (B) that is dead but still standing. Upon coring (or, in the case of a dead tree, cutting) the second tree, it is discovered that the ring pattern in the latter half of its life was the same as that during the early years of the first tree's (A) life. It is thus logical to conclude that the trees overlapped in their life spans and that they responded to the same climate conditions as far as their growth was concerned. Imagine then that another

tree (C) is discovered that was used as a support structure in an old house structure. Analysis of its ring pattern indicates that in the latter part of its life it overlapped with the first part of the second (B) tree's life. We now have a record of climate change that extends back to a period before the life of the intermediate tree. Using this method, paleoclimatologists have extended the history of climate change in the White Mountain area to beyond 9000 years ago!

Ice Core Analysis

Yet another way that prehistoric climate records can be reconstructed is by investigating oxygen isotopes contained in ice cores recovered from glaciers in Greenland and Antarctica. These glaciers contain annual layers of ice that accumulated over thousands of years. Assuming again that the more deeply buried ice layers are the oldest, it is possible to count layers to determine when they formed. The premise of using oxygen isotopes is that oxygen contains two primary isotopes, O-16 and O-18, that differ in their atomic weight. Although both isotopes are found in liquid water, O-16 water is almost 500 times more common and more easily evaporated than its heavier counterpart, O-18 water. This variation depends somewhat on the water temperature at the time it evaporates. When water is warmer, for example, proportionately more O-18 water is evaporated than when the water temperatures are colder. Thus, it is possible to examine the ratio of O-16 to O-18 in layers of ice and yearly accumulations of marine sediments to indirectly reconstruct past climate cycles.

To see an example of how this method of climate reconstruction works, examine Figure 9.17, which illustrates the way

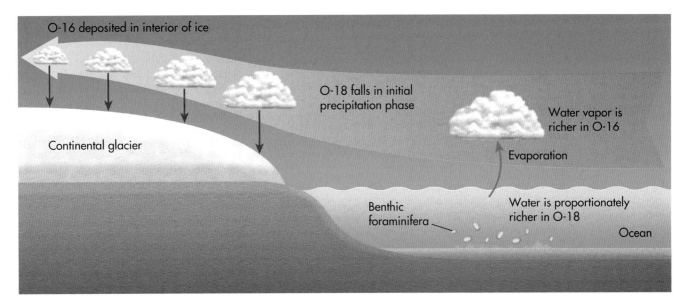

Figure 9.17 Using oxygen isotope ratios to reconstruct glacial and interglacial cycles. Seawater contains both O-16 and O-18 isotopes, which are preferentially evaporated (O-16) and precipitated (O-18) relative to each other during glacial cycles. Thus, microscopic marine organisms (benthic foraminifera) absorb relatively more of the heavier O-18 isotope during glacial cycles. At the same time, moisture-laden air flows toward growing ice sheets, where O-18 is precipitated first because it is heavier than O-16. By the time the air mass reaches the center of the ice sheet, it is enriched in O-16 relative to O-18.

(a)

(b)

19 cm long section of GISP 2 ice core from 1855 m showing annual layer structure illuminated from below by a fiber optic source. Section contains 11 annual layers with summer layers (arrowed) sandwiched between darker winter layers.

(c)

Figure 9.18 Ice cores. (a) Coring the Greenland ice cap. The scientists pictured here are drilling into the ice to obtain a sample for detailed analysis. (b) An example of an ice core extracted from the Greenland ice cap. When this portion of the core was extracted, scientists marked its depth so its position in the ice would remain known. For preservation, it is kept in a storeroom at a consistent temperature of –25°C (–13°F). (c) An example of ice layers, showing distinct bands within the ice. Each of these bands represents an annual layer of snow that transformed to glacial ice. Each layer contains a sample of the atmosphere at the time the snow fell, including the relative amount of O-16 and O-18.

in which O-18 and O-16 move relative to each other within the hydrologic cycle during a glacial period. The diagram shows a hypothetical air mass that flows away from its tropical source area toward a growing glacier at high latitudes. Note that relatively more O-16 is evaporated from the ocean than O-18 during this glacial cycle. This leaves proportionately more O-18 in the seawater, which is absorbed by microscopic marine organisms called *benthic foraminifera* that live on the ocean bottom. As the air mass flows toward the center of the growing glacier, precipitation occurs and O-18 is the first oxygen isotope to be lost because it is the heavier isotope. By the time the air mass reaches the center of the ice mass, the oxygen isotopes are mostly O-16. In this way, the marine sediments (which contain the benthic foraminifera) become relatively enriched in O-18 (because O-16 is preferentially evaporated) at the same time that thickening glacial ice contains proportionately more of the O-16 isotope (because most O-18 was left behind or precipitated earlier).

So how do scientists reconstruct this record of climate change? In a method somewhat similar to acquiring pollen data, they take cores of the marine sediments and of large ice masses to acquire samples of the various layers. Both the marine sediments and glacial ice masses gradually thicken with time, and the annual records of this growth are preserved. Although the coring process is essentially the same in both marine and ice settings, the method is much easier to visualize on a glacier (Figure 9.18). Members of a coring expedition pick a place where the ice is expected to contain the best annual sequence of isotopic data. Once they arrive there, they set up a station (Figure 9.18a) where a core tube can be drilled vertically into the glacier, through a large number of ice layers, to a great depth. With this tube, they can extract a cylinder of ice (Figure 9.18b) that contains a sample of what is, in effect, all the annual snowfalls (and thus the annual ratio of O-16/O-18) they can penetrate. Once the core is extracted, scientists merely have to count the layers (Figure 9.18c) to record the length of time represented in the cylinder of ice and then determine changes that occur in the ratio of O-16 and O-18 over time.

A composite record of cores demonstrates a sequence of distinct **oxygen isotope stages** (Figure 9.19) that reflect major climate changes that have occurred through the Quaternary Period. This record is particularly relevant to reconstructing glacial cycles, indicating that massive ice sheets advanced and

Oxygen isotope stages *Periods of time that have distinct O-18/O-16 ratios, which are used to reconstruct prehistoric climate change.*

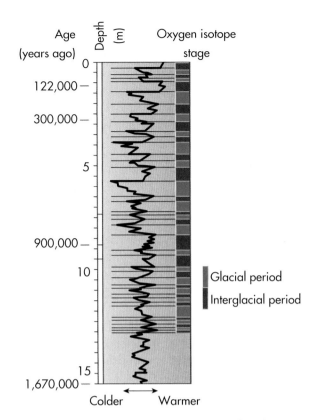

Figure 9.19 Oxygen isotope stages during the past approximately 1.7 million years. Each oxygen isotope stage reflects a glacial or interglacial period, as indicated through oxygen isotope ratios in marine sediments. This record indicates that glacial periods have been quite common during the past ~1.7 million years. (*Source*: Adapted from N. J. Shackelton and N. D. Opdyke, 1976. Oxygen-Isotope and Paleomagnetic Stratigraphy of Pacific Core v28–239, Late Pliocene to Latest Pleistocene. Geological Society of America Memoir 145.)

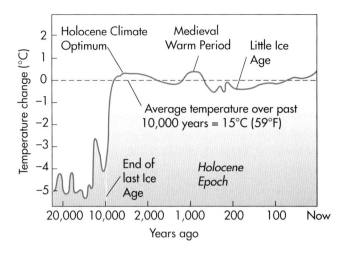

Figure 9.20 Holocene climate changes. This graph shows the changes in average surface temperature for the past 20,000 years relative to mean global temperature for the past 10,000 years. Note the dramatic temperature change at the end of the most recent ice age. Several notable climate fluctuations have occurred since that time, including the Holocene Climate Optimum, Medieval Warm Period, and Little Ice Age.

retreated across much of Earth's high latitudes more than 20 times during the Quaternary Period. The most recent of these glaciations reached its maximum extent about 18,000 years ago and ended about 10,000 years ago. We will discuss this history more thoroughly in Chapter 17.

The most recent 10,000 years of Earth history are generally considered to be the modern climate epoch known as the *Holocene*. Although Holocene climates have been remarkably stable compared to earlier glacial cycles, changes in average Earth temperature have nevertheless occurred (Figure 9.20). Although several fluctuations occurred during the Holocene, three are particularly notable. The *Holocene Climate Optimum* (a.k.a. Hypsithermal) occurred between 9000 and 6000 years ago and was a time when average summer temperatures in the Northern Hemisphere were 2°C to 4°C (~4°F–7°F) warmer than the present. Another Holocene warm interval was the *Medieval Warm Period*, which occurred between the 9th and 13th centuries. During this period, the average global temperature was about –0.5°C (~1°F)

warmer than the present and the Vikings were able to establish colonies in southern Greenland. Following this warm spell, Earth cooled again during the *Little Ice Age*, which occurred from the 13th to 19th centuries. Average global temperatures during this period were about 1°C to 1.5°C (2°F–3°F) cooler than now.

KEY CONCEPTS TO REMEMBER ABOUT RECONSTRUCTING CLIMATE CHANGES

1. Understanding past climate change is important because it provides a window into the potential nature of future climate changes.

2. The extent and timing of past climate changes can be reconstructed using indirect clues derived from ancient pollen, tree rings, and ice cores. Fossil pollen reflects the kind of vegetation that once lived in an area. Tree rings provide climate clues because they vary in width depending on environmental conditions. Ice cores can be used because they contain a record of oxygen isotopes (O-16 and O-18).

3. The oxygen isotope record indicates that many different glacial and interglacial periods have occurred during the Quaternary Period.

4. The past 10,000 years are referred to as the Holocene Epoch and have consisted of "modern climate." Nevertheless, several distinct climate episodes occurred, including the Holocene Climate Optimum, Little Ice Age, and Medieval Warm Period.

The sorting of oxygen isotopes across space by their size is a process that needs to be viewed in animated form to fully appreciate and understand. To do so, select the animation *Reconstructing*

Past Climates Using Oxygen Isotopes. Once you complete the animation, be sure to answer the questions at the end to ensure you understand this concept.

Causes of Past Climate Change

As you have seen, a variety of proxy climate data indicate that Earth's climate has fluctuated greatly during the Quaternary Period. Why did these fluctuations occur? One factor directly related to global cooling is extensive volcanic activity because increased levels of volcanic ash in the atmosphere block solar radiation. This effect was observed most recently in 1991 when Mount Pinatubo in the Philippines erupted catastrophically, resulting in a noticeably cool year globally in a sequence of otherwise warm years. Another factor that can cause climate change is asteroid impacts that dramatically increase levels of atmospheric dust, which also block sunlight. The best known example of such an impact occurred 65 million years ago when an asteroid about 10 km (6 mi) wide struck Earth near the modern Yucatan Peninsula in Mexico. Most geologists believe this impact, and subsequent extensive global cooling, caused the extinction of the dinosaurs. Still another factor that can cause global climate change is the movement of landmasses. A good example of how this variable influenced climate is the way North and South America became connected when the Isthmus of Panama formed. This joining of landmasses disrupted ocean circulation in the tropics, which, in turn, changed the circulation of the atmosphere.

Although the factors just discussed are linked with known periods of climate change on Earth, they are not considered to be responsible for the numerous glacial cycles that have occurred during the Quaternary Period. These cycles have been linked to rhythmic changes in the Earth–Sun geometric relationship. Milutin Milankovich, a civil engineer and geophysicist, first proposed this theory in the 1920s and argued that fluctuations in the Earth's tilt, rotation, and orbit best explained the glacial cycles. This theory was largely ignored until the mid-1970s when, as discussed earlier, isotopic analyses of ice cores revealed that many more glacial periods occurred than was previously believed. Most importantly, scientists discovered a close correlation between the orbital fluctuations proposed by Milankovitch and the timing of glacial and interglacial periods. Thus, the **Milankovitch theory** is now widely accepted

---

Milankovitch theory *The theory that best explains Quaternary glacial/interglacial cycles through long-term variations in Earth's orbital eccentricity, tilt, and axial precession.*

as best explaining Quaternary climate fluctuations. Let's take a closer look at this important theory by discussing its three components:

1. *Orbital Eccentricity*. The *eccentricity* of Earth's orbit refers to the variation in the shape of the orbit over time (Figure 9.21). Recall from Chapter 3 that Earth's orbit is not perfectly circular in its current configuration, but is instead slightly elliptical. The elliptical nature of the orbit results in Earth being about 5 million km (about 3.1 million mi) closer to the Sun at perihelion on or about January 3 than it is during aphelion on or about July 4. This difference amounts to an approximate 6% increase in incoming solar radiation from July to January. Over the course of cycles that last approximately 90,000 to 100,000 years, however, the orbit becomes even more elliptical. When the orbit is more elliptical, the amount of insolation received at perihelion is about 20% to 30% greater than at aphelion.

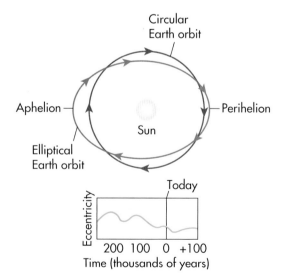

Figure 9.21 Orbital eccentricity. When Earth's orbit is highly elliptical, we receive more radiation at perihelion than at aphelion. This orbital variation occurs on cycles that last about 100,000 years and causes Earth to be approximately 18.27 million km (11.35 million mi) closer to the Sun during orbital perihelion than at aphelion.

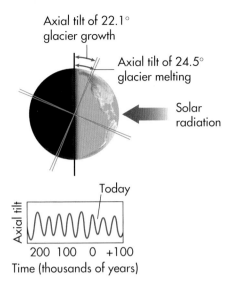

Axial tilt of 22.1°
glacier growth

Axial tilt of 24.5°
glacier melting

Solar
radiation

Today

Axial tilt

200 100 0 +100

Time (thousands of years)

Figure 9.22 Tilt obliquity. The amount of Earth's axial tilt varies over 40,000-year cycles, ranging from 22.1° to 24.5°. When tilt is less, glacial source areas experience cooler summers, allowing glaciers to grow. Seasonality is greatest when axial tilt is high.

This kind of variability would result in a substantially different climate from what we experience today, with cooler summers in the high latitudes of the Northern Hemisphere than currently occur.

2. *Tilt Obliquity.* The term *obliquity* refers to the variations that occur with respect to the axial tilt of Earth. As you know, the axis of Earth is tilted 23.5° from the plane of its orbit around the Sun (the ecliptic). This tilt fundamentally explains the seasonal differences that we experience. During a cycle that averages about 40,000 years, the tilt of Earth's axis varies between 22.1° and 24.5° (Figure 9.22). As tilt increases, the seasonal contrast becomes greater, so that winters are colder and summers are warmer. A tilt of 24°, for example, results in about 8% more solar radiation received at high latitudes in summer. This greater variation occurs because the subsolar point reaches higher latitude when Earth is tilted more. Conversely, when tilt decreases, the difference in seasons is less extreme because

the subsolar point is confined to a narrower latitude range. In the context of continental glaciation, more tilt means warmer summers, which makes it less likely that glaciers can become established. This variable has been linked to the warm Climatic Optimum in the early Holocene (Figure 9.20). On the other hand, conditions are more favorable for glacier growth when tilt is less because summer temperatures are cooler at high latitudes.

3. *Orbital Precession.* The *precession* of Earth's orbit refers to the way Earth slowly wobbles on its axis as it orbits the Sun. Over the course of an approximately 23,000-year cycle, the axis slowly migrates, making it appear as if the planet is wobbling, as when a spinning top slows down (Figure 9.23). This change alters the orientation of Earth with respect to perihelion and aphelion. If a particular hemisphere is pointed toward the Sun at perihelion, that hemisphere will be pointing away at aphelion, and the difference in seasons will be more extreme. This seasonal effect is reversed for the opposite hemisphere. Currently, northern summer occurs near aphelion.

As discussed previously, isotopic data from ice cores indicate a close correlation between orbital variations and periods of glaciation. Expansion of glaciers seems to occur on 23,000-, 46,000-, and 100,000-year cycles, with one, two, or all of the orbital variables playing a major role within any given glaciation. The 23,000- and 46,000-year cycles are minor fluctuations and are apparently most related to eccentricity of orbit and precession of the equinoxes—in other words, the time of year when a particular hemisphere is tilted toward or away from the Sun. Major glaciation seems to occur every 100,000 years, and although this 100,000-year cycle mirrors the length of the orbital eccentricity cycle, there appears to be no distinct relationship between the two. Instead, the 100,000-year rhythm seems to be most closely related to fluctuations in axial tilt and to the difference in the long time it takes a continental glacier to grow and the relatively short time it takes for the glacier to melt. When ice masses began to expand about 110,000 years ago, for example, axial tilt was 22°. Subsequently, it took 80,000 or 90,000 years for the ice sheet to reach its maximum size during the Wisconsin glaciation. About

www.wiley.com/college/arbogast

The Milankovitch Theory

WileyPLUS

To see how the Milankovitch theory looks when animated, go to the *Geo Media Library* and select *The Milankovitch Theory*. This animation shows the way that the Earth–Sun geometric relationship changes through time and how it contributes to glaciation. As you watch the animation, notice how

variations in orbital eccentricity, axial tilt, and orbital precession result in cooler summers at high latitudes where continental glaciers begin to grow. After you complete the animation, be sure to answer the questions at the end to test your understanding of this concept.

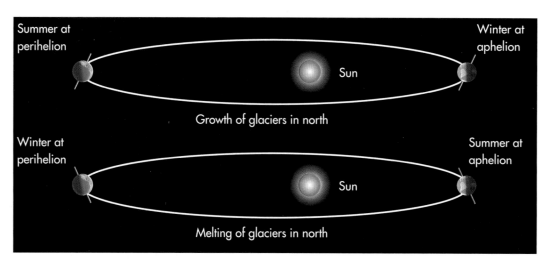

Figure 9.23 Orbital precession. Over the course of a 23,000-year cycle, the Earth's axis slowly wobbles. The impact of this wobble is that a gradual change occurs with respect to the hemisphere that is tilted toward or away from the Sun at aphelion and perihelion.

15,000 years ago, tilt had increased to 24°, and the high latitudes received much more summer-time radiation. In the ensuing 10,000 years, the great mass of ice melted almost entirely.

Human Interactions and Future Climate Change

As we have now documented, climate conditions on Earth have fluctuated dramatically in the recent past. We have also seen how scientists can confidently know that these changes occurred by investigating ancient pollen records, tree rings, and oxygen isotope data derived from ice cores and marine sediments. Understanding these relationships is especially important in the context of ongoing and predicted future climate change on Earth. Although you may have heard many points of view about this issue, the fact is that the *vast majority* of climate scientists around the world agree that some kind of global climate change is occurring and will continue for the foreseeable future. They further believe that the present warming trend is mainly due to **anthropogenic** (caused by humans) influences associated with greenhouse gas production. The remainder of this chapter is devoted to the scientific basis for this belief.

Anthropogenic *Caused by humans.*

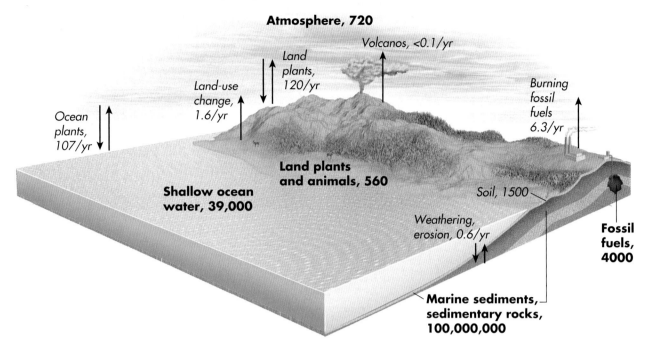

Figure 9.24 The carbon cycle. The carbon cycle represents the flow of carbon that occurs on Earth as it moves from one reservoir to another. Note the various places where carbon is stored and the amount of carbon contained within these reservoirs. (Reservoirs are noted in boldface type, transfers in italic type.)

The Carbon Cycle

In order to understand the science of the global warming issue, it is first necessary to briefly review the carbon cycle and the "greenhouse effect." Recall from Chapter 4 that the various components of the atmosphere, including the constant and variable gases, were investigated. One of the variable gases is carbon dioxide, CO_2. Although CO_2 comprises less than 1% of the atmosphere, it is an important natural regulator of temperature because it traps longwave radiation emitted by Earth, similar to how a greenhouse functions.

Carbon is one of the essential elements of life and is a part of all living things. It is stored in many different places on Earth and, over time, moves from one *reservoir* to another. Figure 9.24 is an idealized diagram of how the carbon cycle works, with storage and transfer amounts in billions of metric tons. A simple example of a transfer is the flow of carbon that occurs when a blade of grass dies and decomposes. During its life, carbon was stored within the grass fiber. After the blade of grass dies, however, it decomposes and the carbon within it (among other things, such as minerals) is gradually incorporated into the soil, where it is stored for a period of time.

The carbon cycle progresses fairly rapidly, by geologic standards of time, and can be measured in years to centuries. Nevertheless, it is thought that less than 1% of the total amount of Earth's carbon is actively within the cycle at any moment in time. If you examine Figure 9.24 again, you can see, for example, that about 120 billion metric tons of carbon transfer back and forth between plants and the atmosphere every year. (A metric ton is 1000 kg; multiply by 1.1 to get the equivalent

number of English tons, each of which is 2000 lb.) Although this sounds like a large number, it is small when you consider that 100 million billion metric tons of carbon are stored in marine sediments and rocks, mostly as calcium carbonate ($CaCO_3$). These carbonates are largely the remains of dead organisms that accumulated on ocean bottoms millions of years ago. Other large quantities of carbon are contained within coal and petroleum deposits that essentially consist of decomposed plant and animal matter.

Fossil fuels are carbon-containing energy sources such as coal, natural gas, and petroleum that are extracted from the ground in mines and wells. The term *fossil fuels* comes from the fact that these compounds were once living organisms that died and slowly fossilized due to intense pressure and heat, turning into other carbon compounds over thousands and millions of years. The vast majority of the energy required in this industrial and technological age is derived from the consumption of these fuel sources in one form or another. As you can imagine, a tremendous amount of energy is required to power the complex and interconnected web of factories, automobiles, shopping malls, home appliances, and every other conceivable modern convenience in the world today (Figure 9.25). To put this thought in perspective, consider that 98,000 kg (98 tons/216,000 lb) of prehistoric plant material is required to produce 3.8 L (1 gal) of gasoline. Viewed another way, you need

---

Fossil fuels *Carbon-based energy sources, such as gasoline and coal, that are derived from ancient organisms.*

(a)

(b)

Figure 9.25 Powering the global economy. (a) Times Square, New York City, at night. A tremendous amount of electricity is required to power the urban landscape. (b) Traffic in Saõ Paulo, Brazil. This type of scene, with thousands of cars burning gasoline, is played out in countless places around the world at any given point in time.

to load the prehistoric equivalent of 16.2 hectares (40 acres) of wheat—stalks, roots, and all—into the tank of your car just to drive 20 mi. In this context, the United States currently leads the world by consuming about 25% of the annual global energy supply. This lead exists despite the fact that the United States is home to only 4% of the world's approximately 6 billion people. Of the total amount of energy consumed in the United States in 2012, about 82% was derived from fossil fuels.

Through this global reliance on fossil fuels, humans have dramatically increased the amount of atmospheric carbon di-

oxide over time. In 1950, for example, approximately 5 billion metric tons of carbon was released into the atmosphere through industrial pathways. By 2010 this transformation of fossil fuel to atmospheric carbon dioxide had increased to over 30 billion metric tons. Approximately 50% of these total emissions has moved from the atmosphere and is stored in reservoirs such as the oceans and forests. The remaining 50%, however, has remained in the atmosphere.

To see the changes in atmospheric CO_2 concentrations in the industrial era, look at Figure 9.26. In 1850, the concen-

Figure 9.26 Changes in atmospheric carbon dioxide since 1850. Carbon dioxide levels have increased over 30% since 1850 and continue to rise. The black line represents atmospheric concentrations derived from ice core data. The red line is derived from actual atmospheric measurements taken by the Scripps Institute of Oceanography. (*Sources*: Pre-1958: A. Neftle et al., Historical CO_2 record from the Siple Station ice core, in Trends: A Compendium of Data on Global Change, 1994, U.S. Department of Energy. Post-1958: Earth Systems Research Laboratory (ESRL) National Oceanic and Atmospheric Administration (NOAA).)

tration of atmospheric CO_2 was 288 parts of carbon dioxide per every million part (ppm) of the overall atmosphere. as indicated by the ice core record. By 2013, this proportion had increased to almost 400 parts ppm. This may not seem like much of an increase, but when you consider the importance of atmospheric carbon dioxide in the natural regulation of global temperature, it is important indeed. Some good news is that U.S. carbon emissions in 2012 were at a 20-year low due to the ongoing transition from coal-fired power plants to those that burn natural gas. Nevertheless, global atmospheric CO_2 emissions will likely continue to rise in the foreseeable future as densely populated developing countries such as India and China, which between them contain about 2.3 billion people, continue to industrialize and use the same fossil fuel power sources that have worked so well for us.

Is Anthropogenic Climate Change Really Occurring?

The "million-dollar question," of course, is whether or not human-induced climate change is really occurring. You may first ask: "Is the climate really warming?" With respect to this initial question, the evidence strongly suggests that Earth is indeed in a warming phase. You can see the evidence of this warming by examining Figure 9.27, which shows the global surface temperature anomalies, relative to the 1951–1980 mean, since 1880. The graph represents the composite trend derived from meteorological stations around the world and clearly demonstrates that the average global surface air temperature increased over 1°C (1.6°F) since 1900. The first decade of this century was the warmest ever observed, with nine of the ten warmest years on record. In this decade, 2010 tied

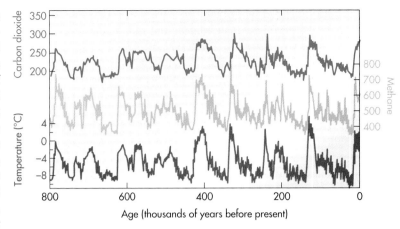

Reprinted by permission from Macmillan Publishers Ltd: E. Brook.
Paleoclimate: Windows on the greenhouse. Nature 2008; 453: 291–292.

Figure 9.28 Greenhouse gases and prehistoric temperature for the past 800,000 years from the Dome C ice core in Antarctica. This graph shows the close relationship between atmospheric temperature (red line), carbon dioxide (blue line), and another (lesser) greenhouse gas (methane), which is the green line. Temperature is the number of degrees above or below the average modern surface temperature of about 14°C (55°F).

2005 as the warmest year ever recorded globally. In the continental United States, 2012 was the hottest year the nation had ever seen, with much of the country locked in a ferocious drought during the summer and fall months. These increases in temperature on both global and national scales correlate very closely with the overall increase in atmospheric carbon dioxide that has occurred at the same time. Much of the increase in global temperature is attributed to an increase in average low temperatures because more longwave radiation is held in at night.

With recent temperature trends in mind, it is pretty clear that Earth's average temperature is warming. The more controversial issue for some is the cause of this warming trend. Skeptics argue that the apparent correlation between increased atmospheric CO_2 and temperature *could* be coincidental and that warming may be related to increased solar activity. In fact, the geologic record shows that at other times in prehistory temperature increased rapidly *and* naturally, and we might just happen to be living during another such period.

On the other hand, if we again examine the Quaternary, it is obvious that the relationship between atmospheric CO_2 and average global temperature is very close. This record is again derived from ice cores where measurements of past levels of atmospheric CO_2 can be measured from air bubbles trapped in layers of glacial ice. The graph in Figure 9.28 shows this comparison vividly for the past 800,000 years. These data were acquired from the Dome C core in Antarctica, which penetrated ice over 3200 m (~10,500 ft) thick. Look carefully and notice how well oscillations in temperature (the red line) compare with levels of atmospheric carbon dioxide (the blue line) and methane (another greenhouse gas, represented by the green line). As you can plainly see, there has been a very close relationship between global climate and greenhouse

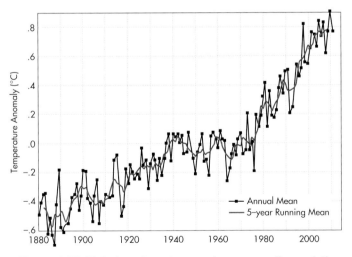

Figure 9.27 Global surface temperature anomalies, relative to the 1951–1980 mean, since 1880. The black line represents the annual mean, whereas the red line is the 5-year running average. Note the rapid warming in the past 40 years. (*Sources*: J. Hansen, R. Ruedy, M. Sato, K. Lo. "Global Surface Temperature Change." *Reviews of Geophysics* 2010; 48:29p; 2011 Update by NASA.)

gases for the latter half of the Quaternary Period as the global climate fluctuated between glacial and interglacial cycles. It is believed that warming at the end of glacial episodes was initially caused by increased insolation associated with Milankovitch cycles. Such warming was then amplified by rising atmospheric CO_2 levels as carbon was released from warming oceans.

After viewing these data, let's focus more closely on the past millennia, which encompass much of recent human history. Look at Figure 9.29, which shows surface temperature anomalies relative to the 1961–1990 average, in the Northern Hemisphere. This graph reflects two sets of data. The blue area from 1000 years ago to 1850 is the reconstructed record (e.g., from ice cores and pollen studies) that contains a range of uncertainty. The red line from 1850 to the present, in contrast, is based on actual measurements. If we assume that the average temperature during the reconstructed record was somewhere in the middle of the range, then it was about 0.2°C (0.4°F) cooler than the 30-year mean about 1000 years ago. Subsequently, between 1000 and 1900, the average temperature *further cooled* about another 0.2°C (0.4°F). This cooling was most intensive during the *Little Ice Age*, which occurred from the 14th to 19th centuries.

Now, look at the change that occurred around 1900. At this time the temperature began to warm quickly to its present state of being 0.7°C (1.3°F) warmer than the past 30-year average. In other words, the natural climate progression for the 850 years prior to the Industrial Revolution was one of *gradual cooling*, with a rapid warming in the past 100 years. Given the apparent link between temperature and greenhouse gases for the past 800,000 years (Figure 9.28), coupled with the fact that the climate had been *cooling* between 1000 and 1850 A.D., the rapid

warming of the past 100 years seems to be best explained by human-induced greenhouse gas emissions. At least, that is what the vast majority of climatologists believe. The effort to explain this relationship is led by the Intergovernmental Panel on Climate Change (IPCC), which is an international body of scientific experts established in 1988 by the World Meteorological Organization (WMO) and the United Nations Environmental Programme (UNEP).

Predicting Future Climate Change

If the one-million-dollar question is whether or not human-induced climate change is occurring, then the two-million-dollar question is "How much future warming will occur?" Remember that the current proportion of atmospheric carbon dioxide is about 400 per million parts of the atmosphere, and that this proportion exists with the developed world producing 73% of the excessive carbon dioxide. The big unknown is how climate will respond when the developing countries increase their carbon dioxide emissions by the predicted 25% in the next few decades.

In the context of understanding future change in an enhanced greenhouse world, the scientific challenge is somehow to predict how the climate system will respond. This complex task is best accomplished with general circulation models (GCMs), which are mathematical models that incorporate known climate variables, such as cloud cover, insolation, latitude, and proximity to oceans, to forecast future climate events. Using these variables, these models run simulations involving scenarios of low or high carbon dioxide levels. Models that run moderate-case scenarios use a doubling of atmospheric carbon dioxide from preindustrial levels (~300 ppm) to ~ 600 ppm by 2050 as a plausible scenario. Figure 9.30 illustrates the geography of warming by the 2050s using this scenario. Note that above-average levels of warming are predicted over much of Earth, with much greater than average warming expected at high latitudes of the Northern Hemisphere. This model further projects that warming will increase on the order of 2°–4° C (4°F–7°F) by 2100.

Although further warming is projected with high confidence for the rest of this century, the actual amount and geographical variability remain uncertain. Much depends on the nature of climate feedbacks, which are indirect or secondary changes that take place within the overall climate system in response to the initial forcing mechanism—in this case, increased atmospheric carbon dioxide. For example, it is possible that increased warming will cause more water to evaporate from oceans, resulting in more atmospheric vapor. This increase, in turn, can have a larger effect on the global energy balance, both as a *positive climate feedback* that further increases warming or potentially as a *negative climate feedback* that causes cooling. Increased atmospheric water vapor could cause a positive feedback because it traps solar energy, which would further warm Earth. On the other hand, increased atmospheric vapor will increase the number of clouds, which could increase the amount of sunlight that reflects back into space, thus resulting in a net cooling effect.

The ultimate positive feedback is a runaway greenhouse effect where significant warming causes the polar ice caps to

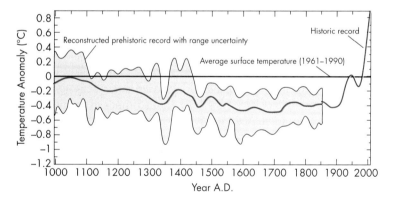

Figure 9.29 Global surface temperature anomalies in the Northern Hemisphere, relative to the 1961–1990 mean, in the past 1000 years. Temperature cooled slightly between 1000 years ago and approximately 1850. Since that time, temperature has been warming dramatically. (*Source:* M. E. Mann, Z. Zhang, M. K. Hughes, R. S. Bradley, S. K. Miller, S. Rutherford, and F. Ni. Proxy-based reconstructions of hemispheric and global surface temperature variations over the past two millennia. Proceedings of the National Academy of Sciences 2008; 105: 13252–13257. Copyright 2008 National Academy of Sciences, U.S.A.)

NOAA GFDL CM2.1 Climate Model

Surface Air Temperature Change [°F]
(2050s average minus 1971-2000 average) — SRES A1B scenario

-20 -16 -13 -11 -9 -7 -5 -3.6 -2.8 -2 -1.2 -0.4 0.4 1.2 2 2.8 3.6 5 7 9 11 13 16 20 °F

NOAA

Figure 9.30 Predicted climate change by the 2050s. This map shows the projected change in global average surface temperature from the late 20th century (1971-2000 mean) to the middle 21st century. The map is based on IPCC scenario A1B that assumes concentrations of atmospheric CO2 will rise to a little over 600 ppm by the 2050s.

melt and thus less radiation is reflected to space, which, in turn, results in even more warming. In contrast, the ultimate negative feedback is that melting ice caps cause warm-water currents such as the Gulf Stream to stop flowing as far north because of the rapid influx of cold water at high latitudes. Some models suggest that Earth might suddenly shift into another ice age if this feedback occurs. Fortunately, neither the ultimate positive nor negative feedback is considered likely.

Although the response of the Earth's future climate to feedbacks is uncertain, the fact is that the IPCC and the vast majority of publishing climatologists believe that warming will continue for the next few generations. Some argue that no major environmental consequence will happen—that, in fact, a warmer world might be a good thing. Those who live in the northern tier of states in the United States, for example, might find some warming a welcome respite from the usually harsh winters that occur there. A potential benefit may have already

occurred in the Great Lakes region, where the annual length of the growing season increased by about 1 week in the 1990s.

Although some "environmental benefits" may take place in a warmer world, the data suggest that many negative results will probably occur. One of these negative effects is a rising sea level, which would result from increased melting of polar ice caps and thermal expansion of water. According to current predictions, sea level will rise 15 cm to 90 cm (5.11 in. to 35.4 in.) by 2100, which will place 92 million people around the globe within the risk of coastal flooding. This issue will be discussed in more detail in Chapter 19.

Another potential negative factor is that climate change may promote the spread of insect-borne diseases such as malaria. Climate boundaries may also shift, making some regions drier while others become wetter. This boundary shift may cause significant displacement of human populations due to changes in agricultural patterns. Still another potential change is that climate variability may be enhanced, with a higher frequency of extreme events such as snowstorms, rainstorms, and intense spells of hot and cold weather. Data suggest that these kinds of extreme events have been occurring more frequently since 1980. Although it is difficult to link any individual event to human causes, a number of scientists believe that the searing temperatures and associated drought in the U.S. heartland during the summer of 2012 were related in part to ongoing climate change. Similar patterns have been seen elsewhere on Earth.

Because it appears that global warming really is occurring and that many negative consequences will likely result, world leaders have made efforts to control greenhouse emissions. The first truly global effort occurred at the 1992 Earth Summit in Brazil, where 150 nations signed a treaty that limited the emissions of greenhouse gases. A subsequent, more comprehensive, global treaty was proposed in 1997 at Kyoto, Japan. The resulting *Kyoto Protocol* called for the 38 leading industrial nations, including the United States, nations of the European Union, and Japan, to reduce greenhouse emissions to 5% to 8% below 1990 levels. These reductions could be accomplished either through

www.wiley.com/college/arbogast

WileyPLUS

The Greenhouse Effect and Global Climate Change

The issue of ongoing and future climate change is complex and can be seen more clearly in animation. To do so, open the animation *The Greenhouse Effect and Global Climate Change*. This module demonstrates how greenhouse gases such as carbon dioxide trap longwave radiation emitted by Earth. It subsequently walks you through the logic behind climate change science. The last part of the module is an excellent visualization of average global temperatures since the late 19th century and what they are projected to be in one scenario by 2100. It is impressive. After you complete the exercise, be sure to answer the questions at the end to ensure you understand this concept.

outright curtailment of emissions or through use of carbon credits as a currency between countries. In other words, each country would be allowed a certain number of carbon credits that could be bought, sold, or traded.

Although the United States signed the treaty, it was rejected by the U.S Senate (where international treaties are ratified) because it did not include guidelines for how developing countries should adjust their greenhouse gas emissions. The U.S. House of Representatives subsequently passed the *American Clean Energy and Security Act* in 2009, which sought to establish a cap-and-trade system that limits the amount of greenhouse emissions companies can produce. Like the Kyoto Treaty, this latter piece of legislation failed to pass in the U.S. Senate because opposing senators argued that the law would hurt the national economy by forcing businesses and corporations to develop new technologies that would reduce carbon emissions. Developing these technologies would add cost to products made in the United States, which would theoretically put us at a competitive disadvantage compared to developing countries that have no carbon standards. The opposing view is that climate change will add costs in the long run due to increased occurrence of drought, fire, and sea level rise (to name a few) that will far exceed the costs incurred now to mitigate the process. Regardless, you can expect to hear about the climate change issue for the rest of your life.

KEY CONCEPTS TO REMEMBER ABOUT GLOBAL CLIMATE CHANGE

1. A major issue facing the world today is global warming, which most climatologists agree is related to human enhancement of the natural greenhouse effect.

2. The carbon cycle refers to the way that carbon is stored in various reservoirs (such as the atmosphere, ocean, and rocks) and how it moves between them.

3. In order to understand the extent of potential future climate change, it is necessary to understand prehistoric climate variability. This can be reconstructed by indirect methods, such as pollen analysis and dendrochronology.

4. Since the onset of the Industrial Revolution in the mid-1800s, atmospheric carbon dioxide has increased about 30%. This increase is almost entirely due to human activities.

5. Between the years A.D. 1000 and 1850, global climate cooled slightly. Since 1850 it has been warming.

6. General circulation models predict that average global temperature will increase 2°C to 4°C (4°F–7°F) in this century.

SUMMARY OF KEY CONCEPTS

1. Climate refers to the average precipitation and temperature characteristics of a region. This depends on many interrelated variables, including latitude, insolation, seasonality, atmospheric circulation, and topographic effects.

2. The Köppen climate system categorizes global climates based on their average annual temperature and precipitation characteristics. Six major categories are recognized: tropical humid climates (*A*), arid/semi-arid climates (*B*), mesothermal climates (*C*), microthermal climates (*D*), polar climates (*E*), and highland climates (*H*).

3. The humid tropical climates (*A*), such as the tropical rainforest climate (*Af*), have a low annual temperature range and abundant precipitation caused by the Intertropical Convergence Zone (ITCZ). Precipitation becomes distinctly seasonal with increased latitude in these regions.

4. The arid/semi-arid climates (*B*) are regions of very low rainfall and sparse vegetation. This low precipitation is caused by the Subtropical High (STH) in places such as the Sahara Desert, by the rain shadow effect in places like Denver, Colorado, and by the continental effect in places like north-central China.

5. Mesothermal climates (*C*) are typically found in the midlatitudes. These climate regions have a distinct seasonal temperature and precipitation pattern. The humid sub-

tropical hot-summer climates (*Cfa*, *Cwa*), for example, have hot summers and mild winters. The wet season occurs in summer due to the influx of mT air pumped in by offshore STH pressure systems.

6. Microthermal (*D*) climates are largely continental climates that occur only in the Northern Hemisphere. These climate regions have hot to mild summers and cold winters. Average annual precipitation is moderate and is usually associated with midlatitude cyclones.

7. Polar (*E*) climates occur at very high latitudes and are associated with tundra and ice caps. Although these regions have a distinct annual temperature range, they are generally cold with low to moderate precipitation. Highland climates occur in mountainous areas and have the characteristics of subarctic and polar climates.

8. The issue of climate change is highly visible on the current national and international stage. In order to understand the nature of future climate changes, it is important to understand changes that have occurred in the recent geologic past.

9. The Quaternary Period includes the most recent ~2 million years of Earth's history. Several scientific methods, such as palynology, dendrochronology, and oxygen isotope analyses, can be used to reconstruct Quaternary climate changes.

10. Since the onset of the Industrial Revolution, atmospheric carbon dioxide has increased about 30% and is expected to increase further. In that same time, average global temperature has increased about 1°C (1.8°F). In the next century, average global temperature is predicted to increase about another 2°C to 4°C (4°F–7°F).

CHECK YOUR UNDERSTANDING

1. Compare and contrast the concepts of weather and climate.

2. How does latitude influence the distribution of global climates?

3. Discuss the variables that are included within the Köppen climate system.

4. Why does the wet-equatorial climate receive consistent rainfall throughout the year?

5. In contrast to the wet-equatorial climate, the tropical wet-dry climate has a distinct dry season. At what time of year does the dry season occur, and why does it happen?

6. The Mediterranean and marine west-coast climates are similar in that both have moderate annual temperatures. However, the marine west-coast climate receives more precipitation. Why?

7. Why is the tropical desert climate dry over the course of the entire year?

8. The Subtropical High (STH) Pressure System is usually associated with dry climates, but, in the case of the humid subtropical climate, it contributes to summer rainfall. Why does this pattern occur?

9. Both the moist-continental and boreal-forest climates have a large annual range of temperature. Why does this variability occur?

10. How can ancient plant pollen be used to reconstruct past climate changes?

11. How can tree rings be used to reconstruct past climate changes?

12. Describe the logic behind the use of oxygen isotopes to reconstruct Quaternary climate changes. What does this record indicate for the past ~1.9 million years?

13. Describe the three components of the Milankovitch theory. How does a change in each of these variables increase the possibility of glaciation?

14. Why do increased levels of atmospheric carbon dioxide contribute to global warming?

15. What is the evidence for a close relationship between atmospheric carbon dioxide and climate change in the prehistoric past?

16. What role has the United States played in the enrichment of atmospheric carbon dioxide?

17. What are the projected levels of future global warming and some of the potential impacts?

ANSWERS TO VISUAL CONCEPT CHECKS

VISUAL CONCEPT CHECK 9.1

The answer is *c*; the area is dominated by the Subtropical High (STH) Pressure System. This pressure system consists of descending air that warms adiabatically. Thus, it is dry in the Middle East.

VISUAL CONCEPT CHECK 9.2

The lower 48 states can be subdivided into three general zones of precipitation. The southeastern United States is a zone of high precipitation that originates from moisture from the Gulf of Mexico. This moisture is pumped into the southeastern states by the Subtropical High (STH) Pressure System. The western part of the United States is fundamentally dry, mostly because this area lies in the rain shadow of the Rocky, Cascade, and Sierra Nevada Mountains. The West Coast of the United States is an area of high precipitation because orographic processes release moisture that flows inland from the Pacific Ocean.

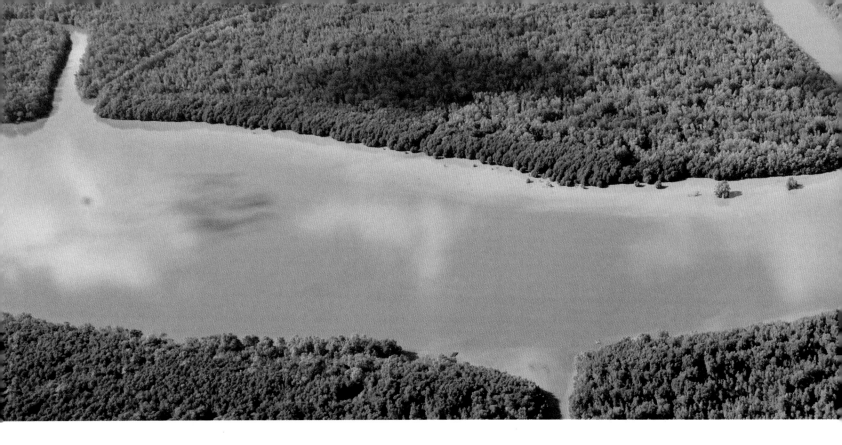

CHAPTER TEN   PLANT GEOGRAPHY

The preceding chapter focused on the geographical distribution of global climates. One way this distribution was illustrated was by showing the type of vegetation in a particular climate zone. Dense forests occur in humid regions, for example, because it is sufficiently wet in these zones to support extensive stands of trees. In contrast, vegetation is quite sparse in a place like the Sahara Desert because it is very dry.

With the foundation of climate in mind, this chapter focuses more specifically on the geographical distribution of plants on Earth and the variables that influence these patterns. As a result, this point marks an important transition within this text because the discussion turns entirely to the patterns of environmental variables *on the ground* rather than within the atmosphere or oceans. The remainder of this book focuses on these kinds of geographic distributions through the study of variables such as soils, rivers, and the landscapes shaped by tectonic forces, wind, and ice. For now, this chapter will illustrate why particular kinds of plant communities occur where they do on Earth.

Juan Carlos Muoz/age fotostock/Getty Images

Some physical geographers study the spatial distribution of plants, including why some plants grow in some places and not in others. This aerial photograph shows the dense rainforest along the Kinabatangan River on Borneo, Malaysia. What geographical variables result in this kind of forest at this location? Certainly, the role of climate is critical, as you saw in Chapter 9. Other factors are important as well, which you will investigate in this chapter.

CHAPTER PREVIEW

Ecosystems and Biogeography

The Process of Photosynthesis

 Photosynthesis and Respiration

The Relationship of Climate and Vegetation: The Character and Distribution of Global Biomes

Local and Regional Factors That Influence the Geographic Distribution of Vegetation

 Plant Succession

Human Interactions: Human Influence on Vegetation Patterns

 Deforestation

 Remote Sensing and the Biosphere

LEARNING OBJECTIVES

1. Define the concept of an ecosystem and its various components.

2. Understand the process of photosynthesis and how it relates to biomass construction.

3. Explain the concept of biomes.

4. Describe the geography and fundamental characteristics of the 12 global biomes.

5. Discuss how local and regional factors, including slope, aspect, vertical zonation, and riparian environments, influence vegetation.

6. Explain the concept of plant succession.

7. Describe the impacts humans are having in forests and grasslands on Earth.

Ecosystems and Biogeography

Recall from Chapter 1 that one of the subdisciplines of physical geography is *biogeography*. Biogeography examines the geographical distributions of organisms, their habitats, and the environmental or historical factors that produce them. As a result, biogeography is very closely related to ecology and thus focuses a great deal on **ecosystems**, which are communities of organisms that interact in interdependent ways (Figure 10.1). On a broad scale, the organisms within a particular land ecosystem can range from the smallest microbe living within the soil to the complex animals that roam the landscape. Ecosystems are not just confined to terrestrial settings, however, as they also occur within marine environments.

When considering the nature of any particular ecosystem, biogeographers often try to understand its various components. One important factor in an ecosystem is the abiotic or nonliving part of the system, which sets the environmental context for all the living organisms in the area. An excellent example of an abiotic factor is the climate of a particular region, or the types of rocks and minerals found in a specific place, to name but two.

As far as the living things are concerned within an ecosystem, biogeographers may first consider the role of **autotrophs**, which are organisms that make their own food from light energy or chemical energy without eating. Although phytoplankton and some bacteria are autotrophs, the most important self-feeding species are plants. Yet a third component of an ecosystem, **heterotrophs**, are animals that consume plants or other animals for survival. Heterotrophs, of course, are subdivided into three categories, including herbivores that feed on plants, carnivores that eat other animals, and omnivores, which consume both plants and animals.

The final component of any ecosystem is made up of **decomposers**, which feed on dead plant and animal material.

Figure 10.1 Components of a forest ecosystem in northern North America. Grizzly bears hunt for salmon along a stream in Alaska. What are the various components of this ecosystem, and how do they fit together within a holistic perspective?

Ecosystems *A community of plants, animals, and microorganisms linked by energy and nutrient flows.*

Autotrophs *Organisms that synthesize their own food using heat or light as the source of energy.*

Heterotrophs *Organisms that consume complex organic substances for food.*

Decomposers *Organisms that consume dead or decaying organic substances for nutrition.*

Food Chain

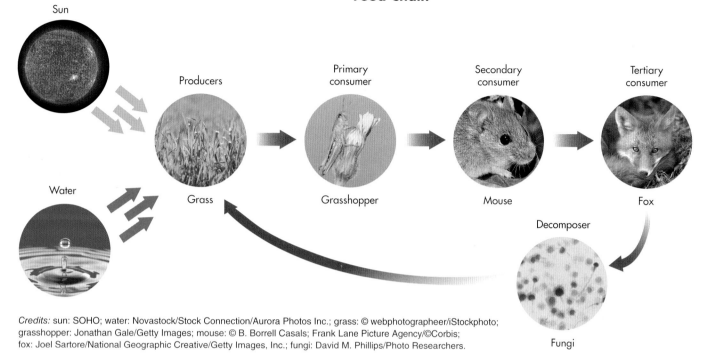

Credits: sun: SOHO; water: Novastock/Stock Connection/Aurora Photos Inc.; grass: © webphotographeer/iStockphoto; grasshopper: Jonathan Gale/Getty Images; mouse: © B. Borrell Casals; Frank Lane Picture Agency/©Corbis; fox: Joel Sartore/National Geographic Creative/Getty Images, Inc.; fungi: David M. Phillips/Photo Researchers.

Figure 10.2 An example of a food chain. Note the various levels of consumption in this chain and how it ultimately cycles back to the basic producers.

Decomposers are critical because they promote the return of minerals and organic material to soils and water that other plants and animals can subsequently consume. When ecosystems are viewed holistically, ecologists and biogeographers often consider the feeding pattern (or trophic structure) within the community. The central concept within this pattern is the food chain, which can include many levels. An example of a food chain with five levels would start with grass followed by grasshopper, mouse, fox, and fungi that consume the fox when it dies (Figure 10.2).

Although all components of ecosystems are important, plants are usually the most visible living part of any landscape. They are certainly the most important autotrophs and are critical to life on Earth because they convert light energy to chemical energy through the process of **photosynthesis**. This production is the foundation of the food chain. Given the important role that plants play on Earth and their highly visible nature, the remainder of the chapter focuses on the global distribution of plant communities. The discussion begins with a more in-depth examination of the way that solar energy is converted to organic molecules through the process of photosynthesis.

The Process of Photosynthesis

Photosynthesis requires the presence of water, sunlight, and carbon dioxide. Carbon dioxide is essential because it is con-

verted to carbohydrate compounds known as sugars. Light energy is necessary because it is the power that drives the conversion of carbon dioxide. This process requires a water medium, which is contained within the plant fibers, because water provides the electrons required for the conversion to take place. In this fashion, light energy from the Sun is locked up in chemical forms that are subsequently used elsewhere in the food chain. This conversion is represented in the following equation:

$$6H_2O + 6CO_2 + \text{light energy} \rightarrow C_6H_{12}O_6 + 6O_2$$

water + carbon dioxide + glucose (sugar) → oxygen

In other words, six molecules of water, plus six molecules of carbon dioxide, plus light energy produce one molecule of glucose (sugar), plus six molecules of oxygen. Although the result of this chemical reaction may not seem impressive at such a scale, consider that approximately 91 billion metric tons (100 billion tons) of carbon dioxide are removed from Earth's atmosphere each year and that each plant produces millions of these sugar molecules each second.

Photosynthesis *The conversion of solar radiation into chemical energy. Sugars and starches are produced from carbon dioxide and water through the interaction of light and chlorophyll in green plants. The process releases oxygen into the atmosphere.*

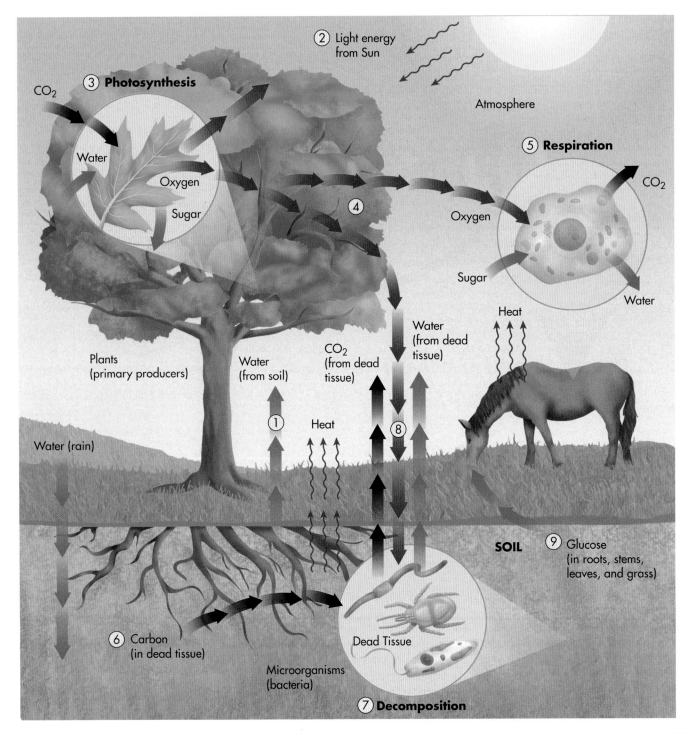

Figure 10.3 Photosynthesis, respiration, and decomposition cycles. Water, oxygen, and carbon dioxide take various pathways within the biosphere. Plants are part of these pathways through the processes of photosynthesis and decomposition, while animals (including people) are part of respiration and decomposition processes.

Figure 10.3 shows the photosynthetic and respiratory cycles, including the various pathways that oxygen, water, and carbon dioxide take within the system. As you work through this discussion, follow the numbered stages in the diagram. These numbers are not meant to imply that stages occur in a specific order, but merely to help organize the flow of the presentation.

For simplification, think of the cycle as broadly consisting of primary producers, such as plants, and decomposers, such as bacteria. Plant consumers are also present, but they are not really relevant to the photosynthetic cycle.

A logical place to begin to trace the various pathways is in the soil, which provides water to the living plant (1). At the same time the roots of the plant are absorbing water, the green leaves of the plant absorb solar radiation (2). Compounds in the plant's leaves called chlorophylls and carotenoids are important ingredients in the process because they absorb sunlight of specific wavelengths. Chlorophylls absorb wavelengths in the blue and red part of the electromagnetic spectrum, whereas carotenoids absorb blue-green light (carotenoids appear orange or yellow).

Figure 10.4 shows how these patterns of electromagnetic absorption take place. Notice that plants do not absorb energy in the green and orange parts of the spectrum. Instead, these wavelengths are either reflected or pass through the plant's leaves. The green wavelength is most actively reflected, which is why plants appear to be green to the human eye. This energy absorbed in the blue and red wavelengths are important to the photosynthetic process because it excites chlorophyll and carotenoid electrons within the plant.

Turn back to Figure 10.3. Stage 3 (upper left) shows the photosynthetic process where carbon dioxide is absorbed by the plant and converted to sugars by light energy in the water medium. A by-product of this process is oxygen, which is released into the atmosphere (4). Excess CO_2 is also released to the atmosphere or used in the photosynthetic process. Oxygen that is released is then absorbed by animals, which produce CO_2 as a by-product of their own respiration (5). When plants and animals die, their remains fall to the ground and are consumed by decomposers (6). As the decomposers absorb oxygen through respiration from the atmosphere or soil (7), they combine the oxygen with decomposing carbohydrates, releasing energy contained within CO_2 and water vapor that are emitted into the atmosphere (8).

Although most of the solar energy absorbed by plants is used for respiration, a great deal of it is used in photosynthesis to produce glucose, which is retained in the plant to build and sustain leaves, fruits, and seeds. In addition, plants convert glucose to cellulose, which is the structural material used to build cell walls. Although much of the produced glucose is used to sustain the plant, a great deal of it is stored as starches and other carbohydrates in roots, stems, leaves, and grass (9), where it can be used later. Some of these carbohydrates subsequently provide energy for animals after they are consumed.

Scientists who study plant production are called *plant ecologists*. Plant ecologists are often interested in the net amount of vegetation produced by photosynthesis. An important indicator of plant productivity is **biomass**, which is the dry weight of all living organisms in a given area. This measure is important because it reflects the amount of chemical energy that has been stored and is thus available for consumption. Biomass is typically measured in kilograms per square meter of ground or as metric tons per hectare. Given the distribution of global climates, it should be no surprise that biomass varies greatly across geographic regions. Forests have the highest biomass, whereas grasslands have less.

In the context of global biomass, examine Figure 10.5, which is a composite satellite image that shows the geography of global biomass as constructed from data collected from 1978 to 1986. The ocean portion of the figure is a composite of more

Biomass   *The amount of living matter in an area, including plants, animals, and insects.*

Figure 10.4 **Preferential absorption of solar wavelengths in plants.** Note that blue and red wavelengths are actively absorbed, whereas the green and orange parts of the spectrum are not.

Figure 10.5 The global biosphere. This composite image was acquired by the NIMBUS-7 and AVHRR satellites over a period of eight years. It shows the geographic distribution of potential global biomass.

GSFC/NASA

than 60,000 images collected by a sensor on the NIMBUS-7 satellite. This sensor measured the distribution and amount of phytoplankton, which are microscopic plants that grow in the upper part of the ocean where the Sun penetrates. Red and orange colors represent high concentrations of phytoplankton, whereas yellow to violet colors reflect progressively lower concentrations. The land vegetation image is a composite collected during 15,000 orbits by the Advanced Very High Resolution Radiometer (AVHRR). This sensor measured radiation from the land surface, which is then used to estimate vegetation production. The dark green areas are rainforests that have the highest growth potential, whereas the lighter greens reflect the potential vegetation growth in other tropical and subtropical forests, as well as midlatitude forests and farmland. The lightest shades of yellow represent deserts where little growth potential exists. Snow and ice-covered regions have no vegetation potential.

KEY CONCEPTS TO REMEMBER ABOUT ECOSYSTEMS AND PHOTOSYNTHESIS

1. Biogeography is a subdiscipline of physical geography that focuses on the geographical distribution of living organisms and the environmental factors that influence these patterns.

2. Ecosystems are communities of organisms that interact in interdependent ways.

3. The major components of ecosystems are abiotic factors, autotrophs, heterotrophs, and decomposers.

4. Photosynthesis is the process through which solar radiation is converted to chemical energy in plants.

www.wiley.com/college/arbogast

WileyPLUS

Photosynthesis and Respiration

Now that the process of photosynthesis has been examined in a static way, the concept can be reinforced by watching it in animated form. To do this, go to the Geo Media Library and select *Photosynthesis and Respiration*. This animation

nicely illustrates the various pathways involved in the photosynthetic process. After you view this animation, be sure to answer the questions at the end to test your understanding of this concept.

The Relationship of Climate and Vegetation: The Character and Distribution of Global Biomes

Many different factors influence the geographic distribution of plants and biomass. Remember that environmental variables affect one another in holistic ways to form distinct geographic patterns. The study of plant geography is no exception. In order to understand these interactions, however, it is first essential to fully appreciate the impact certain individual variables have on these distributions.

The most important factor influencing the geographic distribution of vegetation is climate. Secondary factors such as geology, soils, landscape position, and human behavior also influence the pattern of vegetation on Earth and are discussed later in the chapter. For now, however, the focus is on the relationship between climate and vegetation. You saw in Chapter 9 how vegetation reflects the geographic distribution of climates. These patterns, in turn, are related to many other primary factors, including latitude, proximity to oceans, and the global atmospheric circulatory system. In this context, to say that the geographic distribution of vegetation is related to climate is, in effect, the same as acknowledging the many different variables that influence the climate system. Remember these relationships in the following discussion.

A good way to study the relationship between climate and vegetation is through the concept of global biomes. A **biome** is a definable geographic region that is classified and mapped according to the predominant vegetation and organisms of that particular environment. Each biome is a distinct ecosystem and thus has a particular community of plants, animals, and microorganisms that are linked by energy and nutrient flows. Although ecosystems can be viewed at many different scales, this part of the chapter focuses on the plant communities in large areas that typically cover thousands of square kilometers. Although the biome concept also includes information about the character of soils, the major biomes are described here because vegetation is the most visible component of a biome. The relationship of soils to vegetation and climate will be covered in Chapter 11.

This section of the chapter focuses on geographic units as they relate to **potential natural vegetation**—that is, the vegetation you would find in a region if humans did not modify the landscape. Figure 10.6 is a map showing the distribution of the 12 biomes, and Table 10.1 outlines their fundamental characteristics. You should refer to this map and table fre-

---

Biome *The complex of living communities maintained by the climate of a region and characterized by a distinctive type of vegetation.*

Potential natural vegetation *The vegetation that would occur naturally within a specific area if no human influence occurred.*

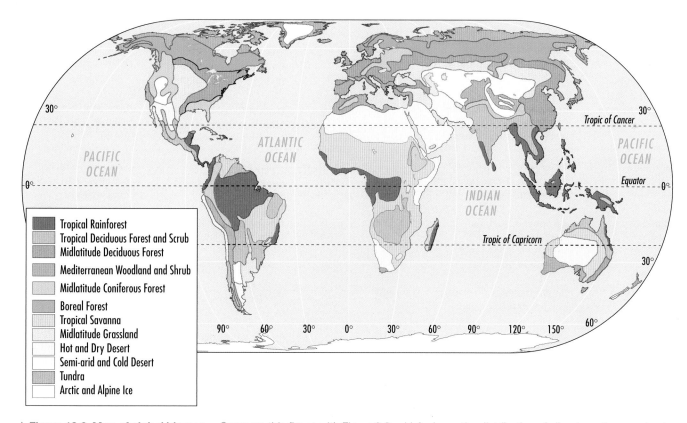

Figure 10.6 Map of global biomes. Compare this figure with Figure 9.2, which shows the distribution of climate regions, and note some of the similarities and differences.

TABLE 10.1 Characteristics of the Major Global Biomes

Biome	Dominant Vegetation	Köppen Climate	Average Annual Precipitation and Moisture Patterns	Average Annual Temperature Patterns
Tropical Rainforest	Broadleaf evergreen trees. Dense canopy with open forest floor. Lianas (vines) and epiphytes.	Af, Am (wet side)	180 cm–400 cm (71 in.–158 in.)	20°C–30°C (68°F–86°F).
Tropical Deciduous Forest and Scrub	Ecotone between rain-forest and grasslands; broadleaf trees with seasonal leaf fall.	Am, Aw	130 cm–200 cm (51 in.–79 in.), with distinct winter dry season.	Seasonal, but always warmer than 18°C (64°F).
Midlatitude Deciduous Forest	Broadleaf trees with seasonal leaf fall.	Cfa, Cwa, Dfa, Dwa, Dwb	75 cm–150 cm (30 in.–60 in.), with summer maximum; no seasonal water deficits.	Hot summers and cool to cold winters.
Mediterranean Woodland and Shrub	Individual short shrubs separated by grassy patches.	Csa, Csb	25 cm–65 cm (10 in.–26 in.); winter surplus and summer deficit.	Warm summers and cool winters.
Midlatitude Coniferous Forest	Needle-leaf evergreen trees, including spruce, pine, fir.	Cfa, Cfb, Cfc, Cwa, Dfb,	75 cm–150 cm (30 in.–60 in.) in southeastern pines; 30 cm–100 cm (12 in.–39 in.) in northern conifers and highlands; 150 cm–500 cm (60 in.–197 in.) in west coast rainforests.	Hot summers with cool winters in southeast; cold winters and cool summers in northern conifers and highlands. Cool winters and warm summers in west-coast rainforest.
Boreal Forest	Needle-leaf conifers, including spruce, pine, fir.	Dfc, Dwc, Dfd	30 cm–100 cm (12 in.–39 in.); no seasonal water deficit; ground frozen most of year.	Long, cold winters and short, cool summers.
Tropical Savanna	Ecotone between tropical forests and deserts; grass-lands with isolated trees.	Aw	9 cm–150 cm (4 in.–60 in.); short summer wet season and long dry season.	Seasonal, but always warmer than 18°C (64°F).
Midlatitude Grassland	Tall-grass prairie in humid areas to short-grass steppes in more arid regions.	BSk, Dfa, Dfb, Dwa, Dwb	25 cm–75 cm (10 in.–30 in.); summer water stress.	Seasonal, with hot summers and cool to cold winters.
Hot and Dry Desert	Xerophytes such as succulents and cacti trending to large patches of bare ground.	BWh, BWk	Less than 2 cm (0.7 in.); perpetual moisture deficits.	Perpetually hot.
Semi-Arid and Cold Desert	Short-grass and dry shrubs; patches of bare ground.	BSh, BSk	2 cm–25 cm (0.7 in.–10 in.).	Average annual temperature = 18°C (64°F).
Tundra	Stunted shrubs and sedges; mosses, lichens.	ET, Dwd	15 cm–180 cm (6 in.–71 in.); no seasonal moisture deficit; frozen ground virtually all year.	Above freezing only 2 or 3 months of year, with warmest month <10°C (50°F).
Alpine and Arctic Ice	Some algae perhaps.	EF, H	Less than 10 cm (4 in.).	Frigid.

quently as you read this section. These biomes are organized into four categories: (1) forest biomes, (2) grassland biomes, (3) desert biomes, and (4) tundra biome. The alpine and arctic ice biome is not discussed in detail because it simply consists of landscapes covered by bare rock and glaciers and therefore has no plants.

As you study the remaining 11 biomes, remember that the boundary between vegetation communities is rarely distinct. Instead, a transition usually occurs from one vegetation type to another that may take place over several hundred kilometers. Such a vegetation transition is called an **ecotone** because it contains elements of two distinct plant communities. These transitions occur because the boundaries between climate regions are usually difficult to define precisely.

Forest Biomes

Forest biomes are geographic regions where the vegetation is dominated by trees. These vegetation assemblages occur in areas where a net surplus of available moisture occurs, either because of high average annual precipitation or because average temperatures are sufficiently cool that little evaporation takes place. Forest biomes occur from the tropical regions to the high midlatitudes and can broadly be categorized as the tropical forest biomes, midlatitude forest biomes, and the boreal forest biome. The tropical and midlatitude forest biomes are further subdivided into distinct geographical zones.

Ecotone *The transition area where two or more ecosystems merge.*

Tropical Rainforest Biome The tropical rainforest biome straddles the equatorial region between 23.5° N and 23.5° S (Figure 10.6) and is very closely related to the tropical rainforest climate (*Af*) zones. It is also associated with small areas of tropical monsoon climate (*Am*) that have a very limited dry season, such as in central Mexico. Average annual precipitation in the tropical rainforest biome ranges from 180 cm to 400 cm (70.1 in. to 158 in.), with monthly rainfall consistently greater than 6 cm (2.4 in.). These areas are always warm, with an average temperature ranging from 20°C to 30°C (68°F to 86°F).

The tropical rainforest biome contains a staggering assortment of trees and plants that form a very dense biomass. The trees are broadleaf and experience no seasonal leaf fall because of the consistently high temperature and abundant moisture; thus, these trees, like the needle-leaf trees that you are probably familiar with, are *evergreen*. The rainforest also contains a variety of plants called *lianas* and *epiphytes*. Lianas are woody vines that use trees for support but are rooted in the ground. Epiphytes, in contrast, are plants that are rooted in a nook of a host plant and thus use the plant for support.

A mature tropical rainforest consists of four definable layers that can be seen vertically (Figure 10.7). The uppermost layer consists of randomly spaced emergent leaf crowns that exist at the tops of the tallest trees, some of which may be up to 60 m (200 ft) tall. Many of these trees are so large that they have *buttress roots* (Figure 10.8) that help support them. Directly below the emergent layer is the canopy, which is the densest layer of the tropical rainforest and consists of a near-continuous web of branches and leaves. This part of the tropical rainforest is from 15 m to 40 m (50 ft to 130 ft) above the forest floor.

(a) (b)

Figure 10.7 The tropical rainforest.   (a) The layered structure of the rainforest creates a continuous canopy that densely shades the forest floor. (b) Do you see the emergent trees in this tropical rainforest?

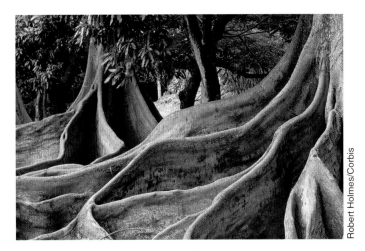

Robert Holmes/Corbis

Figure 10.8 Buttress roots at the base of an emergent tree in the rainforest. These roots give support to the tall trees that extend to heights of 40 m to 60 m (130 ft to 200 ft).

Beneath the rainforest canopy is a third layer, called the *understory,* consisting of another dense network of interwoven branches and leaves that is 5 m to 15 m (15 ft to 50 ft) high (Figure 10.7). In conjunction with the upper canopy, this dense layer blocks sunlight from reaching the forest floor where the ground layer occurs. Given the shady character of the lower forest, the vegetation in this fourth layer of the rainforest is unevenly distributed and ranges from 0 m to 5 m (0 ft to 15 ft) above ground level. Contrary to the general assumption most have about the character of the ground-level rainforest, many places are open—that is, you could walk through them more easily than you might think. Only where distinct openings in the rainforest canopy are present, such as along streams, does sufficient light reach the forest floor so that dense *jungle* vegetation can grow.

Tropical Deciduous Forest and Scrub Biome

The second major subdivision within the tropical forest biome is the tropical deciduous forest and scrub plant community. Extensive areas of tropical deciduous forest and scrub vegetation occur in Africa, South America, India, and Southeast Asia (see Figure 10.6). This plant subdivision is associated with the relatively dry tropical environments where annual precipitation is from 130 cm to 200 cm (50 in. to 80 in.), which occur due to a distinct seasonal dry cycle. These areas are found within both the tropical monsoon climates (*Am*) and the more humid parts of the tropical savanna climates (*Aw*). Although the temperature varies more in these areas than in the tropical rainforest regions, it is always warmer than 18°C (64°F).

The tropical deciduous forest and scrub plant assemblage reflects the progressive transition to a drier climate at higher tropical latitudes. In the wetter areas, where average annual precipitation is closer to 200 cm (80 in.), the vegetation is dominated by tropical deciduous forest (see Figure 10.6 again). Although these areas contain many of the same species seen in the tropical rainforest, they are not as tall and are not evergreen. In other words, they drop their leaves during the dry season

Alan Arbogast

Figure 10.9 Example of the tropical deciduous forest. Although this is a low-latitude region, note that the trees are relatively short and the canopy is more open than in the tropical rainforest biome.

(hence, the name tropical *deciduous* forest). This environment also has less overall species diversity and total biomass than the rainforest. Nevertheless, this biome has a greater array of shrublike species and smaller plants at ground level because more sunlight reaches the forest floor (Figure 10.9). This enhanced ground-level sunlight is due to the upper canopy of the forest being less dense than in the tropical rainforest.

The vegetation gradually shifts from tropical deciduous forest to tropical scrub in areas where average annual rainfall is closer to 130 cm (50 in.). Tropical scrub consists mostly of low-growing, deciduous trees that reach heights ranging from 3 m to 9 m (10 ft to 30 ft; Figure 10.10). These trees tend to be ragged-looking due to the relative lack of moisture, and they vary in density from being closely bunched to widely spaced. With increased distance between the trees, a lower level of tall

Geostock/Photo Disc Green/Getty Images

Figure 10.10 Tropical scrub vegetation in the Australian Outback. Note the ragged-looking trees and the scrubby understory.

bushes and grasses appears. In contrast to the tropical rainforest and tropical deciduous forest, the species diversity within the tropical scrub regions is relatively low, with a few species dominating many of the taller growth areas.

Midlatitude Deciduous Forest Biome The midlatitude deciduous forest biome is found across the midlatitudes of North America, northeastern Asia, and western and central Europe (Figure 10.6) where the climate has a distinct seasonal temperature cycle with cold winters. This plant community consists of a dense network of broadleaf trees that forms a nearly continuous canopy in the summer. In the eastern United States, the dominant species are oak, elm, and maple, to name a few trees you may be familiar with (see Figure 9.10). These trees have a distinct seasonal cycle in which the leaves change color and fall during the autumn months.

The midlatitude deciduous forest is associated with a variety of climate regions where average annual precipitation ranges between about 75 cm and 150 cm (30 in. and 60 in.), including the humid subtropical hot-summer climates (*Cfa, Cwa*), the humid continental hot-summer climates (*Dfa*), and the warmer parts of the humid continental mild-summer regions (*Dwa, Dwb*). Although these climates vary in some significant ways, each has sufficient moisture to support trees, either because average annual rainfall is abundant or evaporation is low because average annual temperatures are cool. In eastern North America, the midlatitude deciduous forest biome extends across a broad area from the Gulf of Mexico into southeastern Canada. Elsewhere, it occurs in an arc that extends from Europe to eastern Asia. This plant assemblage is also found in southeastern Australia and northern New Zealand (see Figure 10.6 again).

Mediterranean Woodland and Shrub Forest Biome A second subdivision in the midlatitude forest biome is the Mediterranean woodland and shrub forest. This plant assemblage is closely associated with the Mediterranean

Rich Reid/National Geographic RF/Getty Images

Figure 10.11 The Mediterranean woodland and shrub biome in Southern California. This biome is adapted to dry summers and wet winters and consists of shrubby trees with intervening grass.

dry-summer climate (*Csa, Csb*) and is found on the west coast of continents in the lower middle latitudes (see Figure 10.6 again). Average annual precipitation in these areas ranges from 25 cm to 65 cm (10 in. to 26 in.). Winters are cool and wet, with a moisture surplus, whereas summers are hot and dry, with a water deficit.

The vegetation in the Mediterranean woodland and shrub biome is dominated by a dense cover of woody shrubs, known as the *chapparal* in North America (Figure 10.11). In many other places, the trees are somewhat larger and more widely scattered, with some patches of grass present. Although the trees vary greatly among continents—oak is most common in California—the overall appearance is quite similar regardless of the specific region. Given the distinct dry season that occurs

VISUAL CONCEPT CHECK 10.1

This is an image of the tropical deciduous forest biome. Which one of the following statements is best associated with this biome?

a) The trees in this forest are evergreen.
b) This forest is located in the midlatitudes.
c) This biome receives less average annual rainfall than the tropical rainforest biome.
d) The canopy in this forest is closed.
e) The wet season occurs in this biome when it is influenced by the Subtropical High (STH) Pressure System.

Paul A. Souders/Corbis

in summer, many of the species in this biome are adapted to fire, which, as in the case of tropical savanna, recycles nutrients in the system. Some plant species in the Mediterranean woodland and shrub biome have seeds that germinate only after they are stimulated by the heat of a fire.

Midlatitude Coniferous Forest Biome The last major subdivision in the midlatitude forest biome is the midlatitude coniferous forest. This plant community consists largely of needle-leaf evergreen trees and is found in places like the west coast and mountains of North America, the southeastern United States, and eastern Europe (see Figure 10.6). Coniferous forests in North America's mountainous areas are associated with humid continental mild-summer climate (*Dfb*) and have short growing seasons. Some of these coniferous forests occur at relatively low latitudes because temperatures are cool at high elevations. In North America, mountain forests include species such as Douglas fir and ponderosa pine (Figure 10.12a). On the west coast of North America, these forests are associated with the marine west coast climates (*Cfb, Cfc*), where annual precipitation is between 150 cm and 500 cm (60 in. and 197 in.) and average annual temperatures are mild for a midlatitude location. Given the cool, moist environment in this region, the forest is often referred to as *temperate rainforest* (see Figure 9.9). This forest also contains significant stands of deciduous trees, such as the giant redwoods in northern California and southern Oregon (Figure 10.12b).

Another kind of midlatitude coniferous forest exists in the southeastern part of the United States within the humid subtropical hot-summer climate (*Cfa, Cwa*) zone. This forest is called the *southeastern coniferous forest* and stretches across southeastern Louisiana, southern Mississippi, southern Alabama, central and southern Georgia, and the Florida panhandle on sandy sites that do not hold much water. The biological diversity in this region is very high, with the dominant tree species being longleaf pine that has an understory of wiregrass (see Figure 9.7 again). Needles in these trees come in bunches of three and are remarkably long, ranging from 20 cm to 45 cm (8 in. to 18 in.) in length.

Boreal Forest Biome The last major forest biome is the boreal forest biome. The global distribution of this biome is closely associated with the subarctic climates (*Dfc, Dwc, Dwd*), and is the broad expanse of coniferous forest that occurs across Alaska, Canada, and Russia (see Figure 10.6). This vegetation assemblage is located entirely within the higher midlatitudes in the Northern Hemisphere because the large continents have very cold winter temperatures and relatively low rainfall. If you recall, the Southern Hemisphere contains no large landmasses, other than frigid Antarctica, at high latitudes. As a result, this hemisphere has no large expanse of subarctic climate and thus, no boreal forest biome.

The boreal forest biome, or *taiga* in the Russian language, has the least diverse assemblage of plants of any

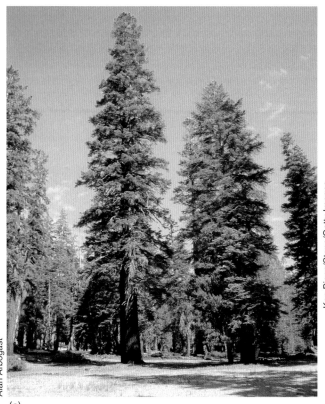

(a)

(b)

Figure 10.12 Examples of the midlatitude coniferous forest biome. (a) This stand of ponderosa pine in California's Sierra Nevada is a great example of an alpine midlatitude coniferous forest. (b) Giant redwoods in northern California. These trees are over 100 m (350 ft) tall and can be over 6 m (20 ft) in diameter!

Alan Arbogast

Ken Biggs/Stone/Getty Images

Figure 10.13 The boreal forest. This scene is from Alaska. Coniferous trees dot this landscape because it is near the northern margin of this biome.

Figure 10.14 The tropical savanna. This photograph shows savanna in Brazil. Here, the vegetation consists of patches of grass that are dotted by isolated trees and woody shrubs.

biome. Most of the trees in the boreal forest biome are coniferous needle-leaf evergreens such as spruce, pine, and fir, with large regions covered almost entirely by one or two of these species (Figure 10.13). These expanses of conifers are occasionally broken by small stands of broadleaf deciduous trees, specifically birch, poplar, and aspen. These trees are adapted to fire in that they are the first trees to move into a freshly burned area. Subsequently, they are succeeded by the coniferous trees, which also include the needle-dropping species tamarack and larch. The lower level of the boreal forest biome is sparsely vegetated, consisting of low shrubs, mosses, and lichens.

Although this region appears to be biologically productive, it is not due to the short growing season and the persistently wet soils in which the plants grow. The stress of this environment on plant growth is best seen in the northern parts of the boreal forest biome, where the growing season is very short and cold. Here, the trees are quite scraggly and stunted in comparison to their counterparts that live in the southern, warmer parts of the biome.

Grassland Biomes

Now let's turn our attention to the grassland biomes. These plant communities are distinct because they consist largely of grass with few trees. Two major grassland biomes are recognized: (1) tropical savanna biome and (2) midlatitude grassland biome.

The Tropical Savanna Biome The tropical savanna consists of grasslands at low latitudes that contain isolated trees. This biome is most closely associated with the tropical savanna climate (*Aw*) and is present in Africa, South America, India, and Australia (see Figure 10.6). In these regions, average annual precipitation is from 9 cm to 150 cm (4 in. to 60 in.), and average monthly temperature is greater than 18°C (64°F).

A good place to examine this biome is in Africa, where tropical savanna covers about 13,000,000 km² (5,000,000 mi²) of the continent. Recall from Chapter 9 that this region is most directly influenced by the seasonal migration of the tropical Hadley cell and associated ITCZ and STH pressure systems, with a summer wet season and winter dry season. In response to this distinct wet/dry cycle, the vegetation is a mix of forest and grass (Figure 10.14). Forest patches are denser where the conditions are more humid and thin considerably where it is drier. During the wet season, the savanna springs to life as grasses turn green and grow tall at the same time that leaves develop on trees. When the dry season comes, however, the grass withers and browns while the trees drop their leaves. A major factor that affects the savanna biome is the occurrence of wildfires, which are common in the dry season. These wildfires are critical to the ecosystem because they reduce the dead vegetation and recycle nutrients back into the system (see the Discover feature in this chapter).

Midlatitude Grassland Biome Midlatitude grasslands are broad areas where the dominant vegetation is grass. The climate of these areas is seasonal, with warm to hot summers and cool to cold winters. Average annual precipitation ranges from 25 cm to 75 cm (10 in. to 30 in.). Several climate regions fall within this general description, including the dry portions of the humid continental hot-summer (*Dfa*, *Dwa*) climates, the drier side of the humid continental mild-summer climates (*Dfb*, *Dwb*), and most of the cold midlatitude steppe climate (*BSk*). In many cases, the relative dryness of these climate zones is due to the continental effect in that the regions are located far from moisture sources. In other places, such as the interior of North America, distinct rain shadows exist in the lee of large mountain ranges such as the Rockies.

WILDFIRES

When most people see a wildfire, they assume it to be a bad thing that must be suppressed at all costs. Plant geographers and ecologists know, however, that fire is a natural part of many ecosystems and produces important ecological benefits. Regular fires reduce the amount of dead wood and thus lower the likelihood of potentially larger wildfires by removing their fuel. Fires often remove alien plants that compete with native species for nutrients and space. Fires also remove undergrowth in forests, which allows sunlight to reach the forest floor, thereby supporting the growth of native species. The organic debris produced by a fire adds nutrients to the soil to be used later by plants and animals. Fires can also provide a way of controlling insect pests by killing off the older or diseased trees and leaving the younger, healthier trees.

Overall, fire is a catalyst for promoting biological diversity and healthy ecosystems. Jack pine forests of the upper Midwest, for example, are highly dependent on fire for renewal because they have serotinous cones that require intense heat to open and release their seeds. If these forests did not burn periodically, the trees would gradually be replaced with less fire-adapted species and other plants and animals within the ecosystem would be threatened. Grasslands are also highly dependent on fire because it prevents brush and trees from invading the prairie. Prairie grasses and flowers can survive a fire because their roots extend deeply into the ground.

Under natural circumstances, wildfires are usually started by lightning and are particularly likely in drought conditions. In grasslands, ranchers often set controlled fires to reduce brush and rejuvenate the prairie. Most of the wildfires you hear about, unfortunately, are the catastrophic events such as those in California nearly every fall that result in the loss of homes and even human life. Many of these begin with lightning, but some have been linked to arson. These fires are particularly intense due to a combination of factors, including previous fire suppression and buildup of dead wood that fuels the fire. Another factor is that more people have moved into drier areas that are naturally fire-prone, especially in hilly country where downsloping winds such as the Santa Ana winds described in Chapter 6 fan a fire. Although these fires can be tragic, remember that the overall ecological benefits of fire far outweigh the negative aspects often portrayed.

National Office of Fire and Aviation Bureau of Land Management, National Interagency Fire Center

Stephen J. Krasemann/Photo Rsearchers

Wildfires are one of the most impressive natural phenomena on Earth, as the left-hand photograph from a fire in Montana attests. Although wildfires are most commonly associated with forests, they are a critical part of the prairie ecosystem because they recycle nutrients and remove woody vegetation. The right-hand image is of a prairie fire in the Great Plains.

Extensive midlatitude grasslands occur in South Africa, eastern Europe, central Asia, and central North America (see Figure 10.6). In South America, grasslands are known as the *Pampas* (Figure 10.15a), whereas in North America they are most closely associated with the *Great Plains* (Figure 10.15b). In Russia, grasslands are referred to as the *steppe*. Although it may not look like it, the biomass in these areas is quite high, with a nearly continuous cover (except in the drier places) of grass and flowering plants that have an extensive root network. Most grasses are perennials that lie dormant through the winter and sprout in spring. This landscape is also adapted to fire, which usually occurs during the summer dry season. The common recurrence of fire historically contributed to the lack of trees and woody shrubs in this region, as did grazing by large animals such as bison (also known as "buffalo") in the Great Plains. Given that fire is suppressed by humans in many grasslands today, forest vegetation is actually expanding into the fringe of these areas.

Desert Biomes

Desert biomes occur in areas where average annual precipitation is less than 25 cm (10 in.) and cover about one-fifth of Earth's land area. Although deserts most commonly occur in low latitudes, they also appear in colder regions such as in Nevada and Utah in the western United States and in portions of western Asia. Two major subdivisions within the desert biome are recognized: (1) the hot and dry desert biome, and (2) the semi-arid and cold desert biome.

Hot and Dry Desert Biome The hot and dry desert biome occurs in the subtropical regions and is centered on 30° latitude (see Figure 10.6). Major deserts of this kind include the Sahara in northern Africa, the Kalahari Desert in southern Africa, and the Mojave and Sonoran Deserts in the southwestern United States. Another large area of hot and dry desert occurs in central Australia. These areas are associated with the hot low-latitude desert climate (*BWh*), which is dominated by the STH Pressure System. Average annual precipitation can be less than 2 cm (0.8 in.), and average monthly temperature is greater than 18°C (64°F). As a result of the combined low rainfall and high temperatures, these areas often experience chronic moisture deficits.

The vegetation in the hot and dry desert consists of bare ground that grades into **xerophytic plants**, including cacti, low-growing shrubs, and scattered clusters of grass (Figure 10.16).

Xerophytic plants *Plants in very dry places that have a number of survival mechanisms in response to prolonged periods of drought.*

(a)

Figure 10.16 Example of the hot and dry desert biome. The Sonoran Desert in the southwestern United States features tall, columnar saguaro cactus, as well as a variety of smaller cacti and hard-leaved shrubs.

(b)

Figure 10.15 The midlatitude grassland biome. (a) The Pampas of Argentina is one of the great grasslands on Earth. (b) Mixed grass prairie in the central Great Plains. Note the combination of short and midlevel grasses on this landscape. Such a mix of grasses occurs because this region lies on the eastern part of the Rocky Mountain rain shadow, with annual precipitation of about 63 cm (25 in.). Grasses are taller to the east and shorter to the west of this area.

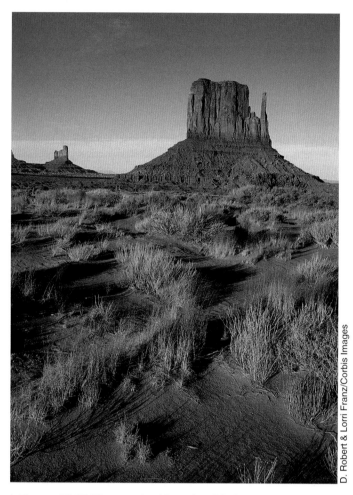

Figure 10.17 The semi-arid and cold desert biome. This sagebrush landscape in Monument Valley, Utah, is a nice example of vegetation in the semi-arid and cold desert biome.

These plants have evolved in ways that ensure their survival in very dry regions. One survival strategy used by many plants is to be *drought-resistant*, which means that they can conserve moisture either by developing very small waxy leaves that store moisture (these plants are called *succulents*), or by having roots that penetrate deeply into the ground. Other plants are *drought-evading* in that they rapidly flower and reproduce during the rare times of relatively high precipitation.

Semi-Arid and Cold Desert Biome The semiarid and cold desert biome occurs extensively in western North America and central Asia (see Figure 10.6). This biome is most closely associated with the cold midlatitude desert climate (*BWk*) but is also found in areas of the cold midlatitude steppe climate (*BSk*). In North America, this climate occurs in places like Utah and Nevada where extensive rain shadows form in the lee of the Sierra Nevada. This climate also occurs in places like the Gobi Desert in central Asia, which is far from a moisture source. Both places feature a distinct seasonal temperature pattern, with hot, dry summers and cold winters.

The vegetation of the semi-arid and cold desert consists of bare patches that grade to low-growing spiny and woolly plants that are randomly spaced. These uneven and sharp plant surfaces protect the plants from grazing animals and significantly reduce moisture loss. Some plants have a silver or glossy leaf, which give them a higher albedo and thus keep the plants cooler by reflecting more solar radiation. In the western United States, plants in this vegetation community include creosote bush, sagebrush, and mesquite (Figure 10.17).

Tundra Biome

The last of the major biomes on Earth is the tundra biome, which borders the Arctic Ocean at very high latitudes across North America and Russia (see Figure 10.6). The *arctic tundra* is found largely in association with the tundra climate (*ET*), which means that winters are long with an average temperature of −34°C (−29°F). It is also found in the colder and drier parts of the subarctic climate (*Dwd*). Although the summer growing season is only 60 to 80 days long, average temperature can reach 12°C (54°F) due to the very long daylength and higher Sun angle. Average annual precipitation is 15 cm to 25 cm (6 in. to 10 in.). Tundra also occurs to a limited extent at very high elevations within the highland climate (*H*) as *alpine tundra*. In this context, alpine tundra is found as far south as the equatorial Andes Mountains.

Regardless of whether the tundra is alpine or arctic, the plant assemblage and thus the appearance are essentially the same. The tundra biome consists of a diverse array of grasses, shrubs, sedges, mosses, and lichens (Figure 10.18) that tend to be very short because the growing season is limited and the landscapes are wind-swept. Tundra landscapes are very fragile and are easily disturbed by human impact. Although people have historically avoided tundra because of the harsh conditions, it is now a growing source of interest because of untapped reservoirs of fossil fuels and other minerals that may exist there.

KEY CONCEPTS TO REMEMBER ABOUT BIOMES

1. A biome is a definable assemblage of plants that occupy a large region of geographic space with generally recognizable boundaries.

2. The global distribution of biomes is closely related to the geography of climates.

3. Trees are typically found in humid regions, whereas dry to semi-arid regions are populated by shrubs and grasses.

4. The vegetation of deserts and that of the tundra biome are very similar in that both consist of low-growing plants that have adjusted to extreme environmental conditions.

Ron Sanford/Corbis

(a)

Alan Arbogast

(b)

Figure 10.18 The tundra biome. (a) The vegetation here consists of low-growing plants, primarily very short grasses, mosses, and lichens, which have adapted to the short, cold growing season. (b) Lichens are a mix of algae, bacteria, and fungi that favor rocky surfaces such as those pictured here. The central rock, which is about 15 cm (6 in.) wide at its thickest point, is surrounded by low-growing moss.

VISUAL CONCEPT CHECK 10.2

This image shows a typical scene in one of the major biomes on Earth. Which biome is pictured here, and how do migrating atmospheric pressure systems influence the seasonal climate and thus the vegetation in this place?

National Geographic Society/Getty Images

Local and Regional Factors That Influence the Geographic Distribution of Vegetation

In addition to the geographic distribution of biomes, other factors can influence the types of plants that live in a region or area. It is useful to think of these variables as being local or regional in scale because they may cause a particular pattern within the overall biome. The following discussion of local factors is not meant to be a complete review of all the variables that can influence the distribution of vegetation, but rather seeks to provide a good sense of the most important of these factors.

Slope and Aspect

The **slope** of an area is its degree of steepness, with high slopes being places that are steep and low slopes being relatively flat. As the slope of a landscape increases, the amount of water that runs off the surface, rather than soaking into the ground, usually increases. Steep slopes thus typically have less available soil moisture, and the associated vegetation can be different than on shallow slopes. Soils also tend to be thinner on steep slopes than on shallow slopes, which influence the amount of nutrients available for plants.

Another factor related to the slopes within a particular landscape is the concept of *aspect*. Slope aspect refers to the orientation of the slope—that is, whether the slope faces north, south, east, or west. Aspect is an important variable in plant geography

---

Slope *The degree of steepness of a portion of the landscape.*

Figure 10.19 Slope aspect and Sun angle. Slopes that directly face the Sun (adret slopes) have higher Sun angles and thus a warmer and drier microclimate than opposite slopes (ubac slopes).

because it is related to how intensely the Sun strikes a given slope (Figure 10.19). In this context, the **adret slope** is the slope that faces the Sun most directly; that is, the slope at which solar radiation arrives at a high angle. Where this geographic situation occurs, the **microclimate** (climate over a small region) is usually warmer, more evaporation subsequently occurs, and the slope can be relatively dry. The opposing, or low-Sun slope, is called the **ubac slope**. This aspect is characterized by a relatively cooler microenvironment with less evaporation because less direct solar radiation is received. If you live in a region that receives abundant snowfall, notice the effect of aspect on hills that slope down on the north and south sides of highways. One slope will be free of snow due to melting, whereas the other slope will have snow on it for a longer period of time.

Now think about how aspect affects plant growth. In the Northern Hemisphere, where would you expect warmer temperatures and higher rates of evaporation: north- or south-facing slopes? It is warmer, with more evaporation, on the south-facing slopes because they face the Sun more directly. As a result, trees can grow (or are more frequent) on the north-facing ubac slope while they do not occur (or are less frequent) on the south-facing slope. Figure 10.20 shows a great example of this pattern by focusing on the growth of bristlecone pine in California.

Vertical Zonation

Another way that vegetation can vary locally from what you might initially expect is through the concept of vertical zonation. **Vertical zonation** refers to the change in vegetation that occurs with respect to elevation rather than latitude. Recall that temperature usually decreases at a consistent rate (the en-

Figure 10.20 Slope aspect and growth of bristlecone pine in the White Mountains, California. (a) Growth on the south-facing (adret) slope of a mountain. Note the wider spaces between trees. (b) Growth on the north-facing (ubac) slope. Bristlecone pine favors this slope, rather than the adret slope, because less evaporation on this surface leads to more available water.

vironmental lapse rate) as you gain altitude in the troposphere. This overall trend shows up most clearly on the world climate map (see Figure 9.2) as highland climate (*H*) regions. Recall that these climate zones are known for tremendous variability over short horizontal and vertical distances. They are associated with the Earth's large mountain chains, including the tropical Andes where near-arctic conditions are found at high elevations.

In the context of vertical zonation, the key relationship is that environmental changes that occur with respect to elevation correlate to the same ones you see with latitude. Figure 10.21 illustrates how vertical zonation works in the European Alps. In the lower elevations of the mountains, the vegetation often

---

Adret slope *The slope that faces the Sun most directly.*

Microclimate *Average temperature and precipitation characteristics of a small area within a larger climate region.*

Ubac slope *The slope that faces away from the Sun.*

Vertical zonation *The change in environmental characteristics that occurs with respect to altitude.*

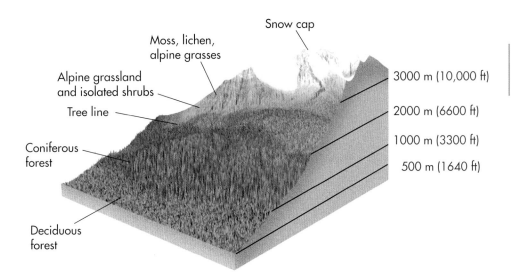

Snow cap

Moss, lichen, alpine grasses

Alpine grassland and isolated shrubs

Tree line

Coniferous forest

Deciduous forest

3000 m (10,000 ft)

2000 m (6600 ft)

1000 m (3300 ft)

500 m (1640 ft)

Figure 10.21 Vertical zonation of vegetation in the European Alps. Note the changes in vegetation that occur with elevation.

consists of deciduous forest. As you move to about 1000 m (3300 ft), the vegetation gradually shifts to coniferous trees that are adapted to cooler conditions.

At about 2000 m (6600 ft), the vegetation shifts from trees to alpine grassland and scrub vegetation. The elevation where trees can no longer grow because conditions are too harsh is called the **tree line**. At still higher elevations, the vegetation becomes tundra (Figure 10.22), and, ultimately, the landscape is covered with snow and ice.

When do environmental conditions become too harsh to support trees? The critical variable appears to be that seasonal mean temperature registers at least 6.5°C (43.7°F) at a depth of 10 cm below the ground where roots grow. You are probably most familiar with *alpine tree* line that occurs at high elevations in the mountains, above which are only scattered tundra, bare rocks, and even perpetual snow. Although alpine tree line exists

in all major mountain ranges on Earth, a distinct pattern exists with respect to latitude and altitude, with the elevation decreasing progressively with increased latitude. Another kind of tree line is called *circumpolar tree line* and is the highest latitude at which trees can grow. Regardless of whether the tree line is alpine or circumpolar, the appearance is the same, consisting of scattered trees that are stunted and deformed. Such an area is referred to as the *Krumholz* zone, which literally means "crippled wood." In these areas, trees are often bent in a particular direction that reflects the prevailing winds (Figure 10.23).

Tree line *The line that represents the upper limit in mountains and high latitudes where environmental conditions support the growth of trees.*

Figure 10.22 Alpine vegetation in the Snowy Range in Wyoming. The scattered trees, which are similar to high-latitude boreal forest, grade upward into tundra. This is an excellent example of vertical zonation.

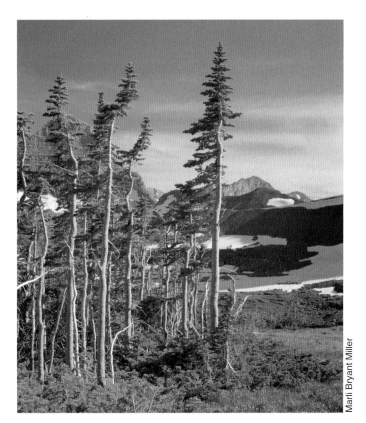

Figure 10.23 A tree line landscape. Trees at the circumpolar or alpine tree line are often dwarfed and scraggly because of the harsh environmental conditions. Can you tell the direction (left to right or right to left) of the prevailing winds in this photo?

Plant Succession

Plant succession refers to the natural changes that occur in a biome of a particular place over time. Generally speaking, these natural changes progress to the point where the most complex array of vegetation possible arises, given the regional climate, water, and soil variables. If the succession begins on freshly deposited sediment, then we use the term *primary succession*. If, on the other hand, the succession occurs in an area that was disturbed through fire or some other catastrophic event, then the term *secondary succession* applies.

Regardless of whether the succession is primary or secondary, the sequence of plant stages is usually predictable for a given area. The first plants that occupy the area are called *pioneer species*. These plants set the stage for future succession because they initiate the gradual change of the host soil through root penetration and the cycling of organic material and minerals as they spread across the area. In this fashion, bacteria and animals begin to inhabit the area, and the site vegetation gradually becomes more complex. In time, the local microenvironment becomes hospitable for larger plants, which change the shading and associated microclimate at ground level. These adjustments in humidity and soil temperature make it possible for additional plants to move into the area. When this progression reaches the ultimate complexity of an area, the vegetation is said to be **climax vegetation**.

A good example of primary succession can be seen on a sequence of coastal sand dunes, which are mounds of sand that grow through wind deposition. They are highly mobile landforms that move due to a complex interaction of wind and sand supply. These landforms are relatively infertile microenvironments because they are very well drained and nutrient-poor, due to high amounts of quartz and feldspar and low amounts of nitrogen, rcalcium, and phosphorus.

In the context of plant succession, the first plants to inhabit a fresh deposit of wind-blown sand are beach grasses. These grasses are adapted to the relatively infertile conditions of a new sand dune and continue to grow even if they are slightly buried by fresh deposits of sand. As they spread across the dune and develop more complex root systems, they cause it to stabilize, which means that the sand stops moving. At this time, new plants move in that are adapted to the relatively infertile conditions of the dune. These plants, however, cannot tolerate the stress of being buried as the initial plants could. When this more established successional stage occurs, the soil chemistry and ground microenvironment continue to evolve, culminating in the establishment of trees on the older dunes.

Figure 10.24 Primary succession on Great Lakes coastal dunes. Beach grass grows on the low dune in the foreground, which is the most recent deposit of windblown sand. The higher dunes in the background are covered by forest because these dunes are older and have been stable for a longer period of time.

Figure 10.24 is a scene from a coastal dune field along Lake Michigan and is an excellent example of this succession. Compare the vegetation on the low (young) dune in the foreground with that on the higher (older) dunes in the background. The younger dune is inhabited by beach grass, while those in back are covered with forest. In the Great Lakes region, this succession sequence is thought to take about 150 years.

Riparian Zones

Another important factor that can influence the geographic distribution of vegetation is proximity to water. This factor is especially important in semi-arid regions where water is often scarce. An example of this kind of relationship is the **riparian zone**, which is defined as the area immediately adjacent to a stream. Such zones are suitable for trees because more water is available for plant growth near the river than away from it, especially because elevation increases with distance away from the valley bottom. A good place to see how riparian vegetation may differ from what occurs in more upland areas can be seen in parts of the Great Plains of North America (Figure 10.25). Notice in this image that trees are growing only along the river, while the surrounding landscape consists of grass and sagebrush.

Plant succession *The predictable vegetation transition that occurs within a landscape over a period of time.*

Climax vegetation *The vegetation within an area that has reached its ultimate complexity.*

Riparian zone *The strip of land bordering a body of water that supports plants and animals adapted to water systems.*

Plant Succession

Let's take a closer look at the concept of plant succession by watching the sequence of plants that populate a more complex landscape than the dunes just described. Go to the Geo Media Library and select *Plant Succession*. This animation illustrates the process of secondary plant succession in a cool, humid landscape in the higher midlatitudes. The environmental stage is set with a small stream that flows through a forested area about 7000 years ago. Subsequently, a colony of beaver builds a dam across the stream, causing a pond to form. From this point, watch how the succession of plants proceeds in this environment, culminating in the complete disappearance of the pond and return of climax vegetation. Once you complete the animation, be sure to answer the questions at the end to test your understanding of plant succession.

Figure 10.25 An example of riparian vegetation. An overlook in Wyoming provides a great view of riparian vegetation along a small stream. In the surrounding hills the plant community consists of very short grass and sagebrush, while the relatively moist valley bottom supports this dense stand of trees.

Human Interactions: Human Influence on Vegetation Patterns

As noted earlier, the discussion about plant biomes and local vegetation factors assumed the potential natural vegetation of an area or region. In other words, it was based on the character of plant communities as they would naturally occur without human impacts. The reality is, however, that "natural vegetation" in its purest form is rather rare today in this era of intense human interactions with the environment. People have had a growing impact on local, regional, and global plant communities for the past two centuries, which has intensified since the onset of the industrial age. In the contiguous United States, for example, it is very difficult to find vegetated landscapes that have not been altered in some way by humans, either through clearing of forest or plowing.

To see where we are today, examine Figure 10.26, which is a map that expresses as a percentage the relative

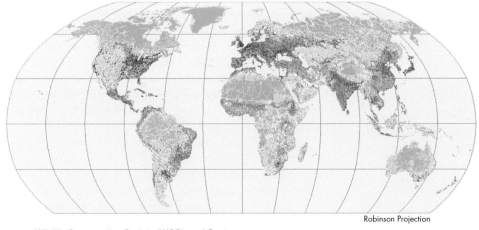

Wildlife Conservation Society (WCS), and Center for International Earth Science Information Network (CIESIN)/Columbia University. 2005. Last of the Wild Project, Version 2, 2005 (LWP-2): Global Human Footprint Dataset (Geographic). Palisades, NY: NASA Socioeconomic Data and Applications Center (SEDAC).

Robinson Projection

Human Footprint Index

1 10 20 30 40 60 80

Figure 10.26 Human footprint on global biomes. This map expresses as a percentage the relative human footprint on global biomes. Values range from 0 (least influenced) to 100 (most influenced). Note the major footprint in places like Europe and the eastern United States, whereas places like the Sahara Desert and northern Canada have experienced low human impacts.

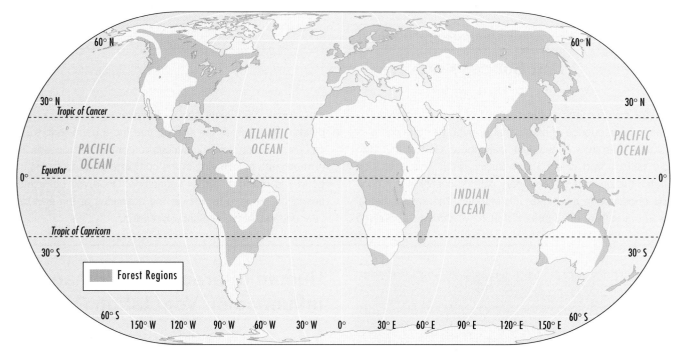

Figure 10.27 The extent of forest regions on Earth. Forests have adapted to every zone of latitude on Earth except for the polar regions.

human footprint on biomes around the world. Values range from 0 (least influenced) to 100 (most impacted). As you can clearly see, eastern China, India, Europe, and the eastern United States have extensive human footprints in biomes. The portions of Earth that remain relatively untouched by humans are usually remote, extreme places like the Sahara Desert in North Africa, northern Russia into Siberia, and northern Canada. This part of the discussion focuses on some of the ways that people have impacted vegetation on Earth. It is not meant to be comprehensive, but will provide a sense of how humans have modified the natural vegetation in biomes around the world.

Deforestation

Perhaps the most obvious way that people impact vegetation is through **deforestation**, which is the process where large tracts of forest are cleared by humans, either for commercial use or to make way for agriculture. A good place to begin the discussion of deforestation is to examine the current extent of forest cover around the world. According to the United Nations *Global Forest Resources Assessment*, approximately 3.9 billion hectares (9.6 billion acres) of Earth were covered by forest in 2010. Of this total, 47% was in the tropics, 33% was in the boreal forest, 11% in the middle latitudes and temperate areas like the Pacific Northwest, and 9% in the subtrop-

ics (Figure 10.27). Overall, this extent of coverage amounts to about one-third of Earth's total landmass, excluding Antarctica and Greenland.

Ongoing monitoring indicates that the extent of global forest cover has decreased extensively in the past 20 years due to human activities. This kind of land cover change can easily be detected with satellite remote sensing, as can be seen in Figure 10.28. High amounts of loss have occurred even though forest is actually expanding in some areas due to natural processes or through *afforestation*, which is the planting of trees by people in previously nonforested areas. As you can see in Table 10.2, Europe and Asia had net increases in forest cover between 1990 and 2010. The pattern in Asia is particularly interesting, as the net gain this century reverses a trend of overall forest loss in the 1990s. These gains are largely the result of aggressive planting of trees in China, which have offset high amounts of deforestation in places such as Borneo and Indonesia. In North America, the rate of forest loss appears to have stabilized in the past 20 years.

Although some modest increases in forest cover have occurred recently in some areas, the average annual rate of global deforestation between 2000 and 2010 was about 13 million hectares (32 million acres). At a continental scale, this high rate of deforestation is clearly most associated with Africa and South America. Over the course of the past 20 years, Africa has lost about 12% of its forest, whereas about 5% of the forest has been cut in South America (Table 10.2), with about 60% to 70% of this deforestation related to cattle ranches. Another reason that tropical rainforests are cut is by subsistence farmers who practice *slash* and *burn agriculture*. This technique first involves cutting

---

Deforestation *The removal of trees for economic or agricultural purposes.*

(a)

(b)

Figure 10.28. Monitoring deforestation rates. (a) Infrared satellite image acquired in August 1992 with the Landsat Thematic Mapper of a part of Mato Grosso state in Brazil. (b) Infrared image of the same area acquired in July 2006 with the ASTER instrument on NASA's Terra satellite. Note the extensive amount of deforested area in this image compared to that observed in 1992.

trees in a tract of land and then burning the leftover woody debris (Figure 10.29). Burning this debris not only clears the land so that it can be tilled and planted; it also increases the soil fertility by recycling organic nutrients. Although this method produces good crop yields for a few growing cycles, soil fertility decreases pretty quickly. As a result, new land must be cleared and burned to begin the cycle again.

Deforestation of tropical rainforests is a major concern because of the negative impact it has on biodiversity within the tropical forest biomes. Given the biocomplexity of these regions, it is unknown how the effect of cumulative clearing ripples through the system. Many scientists believe it is possible that deforestation will cause extinction of plant and animal species that have yet to be discovered. In addition to the extinction loss, which in turn has unknown consequences for

the overall ecosystem, such species are potentially important because they could yield the next wave of pharmaceutical drugs for human medical use. Deforestation may also be very important with respect to global warming. The tropical forest regions are sometimes called the *lungs of Earth* because they absorb carbon dioxide and release oxygen through photosynthesis. If forest loss continues, less atmospheric carbon dioxide will absorbed, which means that the net level of carbon dioxide in the air will increase.

Given these concerns, it is good news to know that the rates of deforestation in tropical regions appear to be slowing somewhat. This trend is clearly obvious in the Amazon where the rate of forest loss has fallen nearly 80% since 2004. Of particular significance, this decline has occurred during a period of robust economic growth in Brazil, suggesting that the

TABLE 10.2	Change in Forest Cover (in millions of hectares), 1990–2010			
Continent	Total Forest 1990	Total Forest 2000	Total Forest 2010	Percent Change 1990–2010
Africa	702	650	620	−11.7
Asia	551	548	580	+0.5
Oceania	201	198	195	−3.0
Europe	990	1000	1005	+1.0
North and Central America	555	549	550	−1.0
South America	923	886	875	−5.2
Total World	**3963**	**3869**	**3825**	**−3.5**

Source: U.N. Food and Agriculture Organization, Global Forest Resources Assessment 2010 (France, 2010).

(a)

(b)

Figure 10.29 Slash and burn agriculture in the Amazon rainforest. (a) Subsistence farmer in Brazil setting a fire in a cleared plot. (b) The forest in this image has been cut and burned to prepare the ground for cultivation. Organic matter from the burned forest temporarily increases soil fertility.

prior strong link between deforestation and the economy may be weakening. Although the reasons for this decoupling are complex, they appear to belated to such factors as improved satellite monitoring (see Figure 10.28), increased enforcement of environmental laws, and growing awareness of the Amazon ecosystems' value, to name a few. It is hoped that this trend continues.

Although deforestation is often considered a tropical issue, the history of forest management has also been mixed in the higher latitudes. In the Great Lakes region of North America, for example, virtually all the native pine forest was cut in the late 19th and early 20th centuries (Figure 10.30) to provide timber for buildings in growing cities such as Chicago, Milwaukee, and St. Louis. The forest in this region is still heavily managed, with large jack and red pine plantations in many parts of the region that are grown for pulpwood. A similar history has also transpired in the needle-leaf forests of the Pacific Northwest. Most of the old-growth forests are gone in this area, and the second- and third-growth forests continue to be extensively logged (Figure 10.31). The ongoing

Figure 10.30 Stump prairie in northern Michigan. These stumps are the remnants of old-growth pine trees on the Kingston Plains that were cut in the 1890s. Following this period of deforestation, many dried limb and branch fragments lay on the ground. This *slash* subsequently burned in a very hot fire that torched seedlings and essentially cooked the soil. As a result, forest has not regenerated over much of the area.

Figure 10.31 Clear cutting in the Pacific Northwest. Deforested areas on mountain hillslopes south of Olympic National Park in Washington.

Figure 10.32 Breaking prairie in the Great Plains. Scenes like this were common as settlement spread in the Great Plains region in the late 1800s and early 1900s. By the 1930s most of the native grassland had been plowed under and converted to farmland.

Figure 10.33 Wheat field near harvest in the Great Plains. Given that wheat is a type of grass, it is a perfect agricultural crop for the grassland biome.

demand for old-growth timber in this region has controversial overtones regarding habitat loss and species extinction, most notably with respect to the spotted owl. Environmentalists fervently believe that the remaining forests should be preserved to protect important owl habitat. Logging firms, in contrast, argue that further cutting is necessary to meet timber demand and to protect jobs.

Agriculture in the Midlatitude Grassland Biome

In addition to the tremendous impact that people have had on global forests, another biome that has been severely modified by humans is the midlatitude grassland biome. You can see the potential natural distribution of grassland on Earth in Figure 10.6. Notice how much of central North America, for example, was covered by grass. Prior to European settlement of North America, as much as 162,000 hectares (400,000 acres) of the continental interior were grassland.

As you will learn in the next chapter, the soils in the midlatitude grassland biome are some of the most fertile in the world. This fertility is ideal for the development of large-scale agriculture, which began in this part of the United States during the era of westward expansion in the middle and late 1800s. Most of the settlers during this time were farmers who "broke" the native grasslands with ox teams and plows. In the early part of the 20th century agriculture became mechanized with tractors that could turn soil over more quickly and effectively (Figure 10.32). Today, large-scale agriculture dominates the region today in the form of extensive corn, soybean, and wheat operations (Figure 10.33). Although the benefit of this form of intense agriculture to people is obvious, the impact on the

www.wiley.com/college/arbogast

Deforestation

The topic of deforestation is of particular concern to many scientists. One way to monitor the rates and patterns of deforestation is through satellite remote sensing. To see some examples of how this kind of monitoring occurs, select the animation *Deforestation*. This interactive module illustrates the kind of patterns that geographers observe when they study deforestation around the world. As you work your way through the animation,

you will be asked to identify deforested and uncut areas using both true-color and infrared satellite imagery. A short animation also illustrates the rate of rainforest deforestation in the Amazon and Borneo. After you complete this exercise, be sure to answer the questions at the end of it to test your understanding of this concept.

midlatitude grassland biome has been immense. In that context, current estimates are that only 1% to 4% of virgin (unplowed) prairie remains throughout the entire biome.

Overgrazing

Another way that humans have impacted the network of natural vegetation is through overgrazing of the landscape by livestock, such as cattle and sheep (Figure 10.34a). This impact is most extensive in marginal landscapes that are extremely sensitive to disturbance, such as the more arid parts of the American West. (Figure 10.34b). Prior to European settlement, this region contained very few large herbivores that would have grazed on the vegetation there. In the 20th century, however, this region became available for ranchers because the U.S. Forest Service, Bureau of Land Management, and National Park Service classified much of it as public lands. According to federal legislation, it is permissible to graze sheep and cattle in these areas at a very low cost. This practice continues to the present day and has dramatically affected this biome. Some of the potential impacts are weed invasion into disturbed areas, reduction in forest due to grazing of young saplings, and the increased likelihood of more intense and widespread fires.

(a)   Rick & Nora Bowers/Alamy   (b)   UIG via Getty Images

Figure 10.34 Impact of grazing on vegetation in the American West. (a) This image of a fence line in Arizona shows the difference between a plot heavily grazed (left) versus a healthy grassland (right). (b) Overgrazing reduces the cover of vegetation on hillslopes, which can result in severe erosion when rain falls.

This pair of infrared satellite images is of the same region in the Amazon rainforest at different times. Remember that red on this kind of image represents vegetation. The (a) image was acquired in 1975, whereas the (b) image dates to 2000. The difference that you see reflects extensive deforestation along road networks in the region. Which one of the following statements is accurate with respect to the impact that this kind of deforestation has on global climate change?

a) Fewer trees will result in the production of more oxygen.

b) The amount of atmospheric carbon dioxide will increase because fewer trees are present to absorb it.

c) Global temperature should decrease given that fewer trees are present in the rainforest.

d) There will be less atmospheric carbon dioxide because trees produce carbon dioxide as a by-product of respiration.

(a)

(b)

www.wiley.com/college/arbogast

WileyPLUS

Remote Sensing and the Biosphere

Plant ecologists and geographers are very interested in the patterns and rates of vegetation change on Earth. In this context, one of the ways that plant geography is studied and monitored is with remote sensing. To provide a sense of how this form of monitoring can be done, go to the Geo Media Library and select *Remote Sensing and the Biosphere*. Within this interactive module, you will be able to analyze (1) *Land Productivity*, (2) *Ocean Productivity*, and (3) *Fire Patterns*. In each of these submodules, you will be presented with a series of satellite images and prompts to identify specific features on imagery. Once you complete the interactivity, be sure to answer the questions at the end to test your understanding of the concepts contained within it.

1. Biogeography examines the geographical distributions of organisms, ecosystems, and the environmental contexts in which they occur.

2. The photosynthetic, respiratory, and decomposition cycles explain how biomass is produced and consumed through the interaction of solar energy, carbon dioxide, water, and oxygen in plants, animals, and soil.

3. A biome is a definable geographic region that is classified and mapped according to the predominant vegetation and the adaptations of organisms to that particular environment. Twelve major biomes are recognized on Earth.

4. Six major forest biomes in the world occur in places where average annual precipitation is high. They also occur in places that have moderate amounts of precipitation and average temperatures that are sufficiently cool to maintain a good water balance.

5. The two major grassland biomes in the world are (1) tropical savanna and (2) midlatitude grassland. The tropical savanna occurs in the tropical ecotone where the ITCZ dominates in the summer (bringing the wet season) and the STH dominates in winter (bringing the dry season). Midlatitude grasslands occur deep within continents or within rain shadows.

6. Two major desert biomes are recognized in the world: (1) hot and dry desert, and (2) semi-arid and cold desert biome. The hot and dry desert occurs in the higher tropical latitudes, dominated by the STH Pressure System. The semi-arid and cold desert biome occurs deep within continents and rain shadows within the midlatitudes.

7. The tundra biome occurs at very high latitudes and elevations where environmental conditions are too harsh for trees to grow. The plants here consist of low-growing woody shrubs, lichens, and mosses.

8. The human impact on the distribution and character of natural vegetation around the planet has been dramatic, including deforestation, extensive agriculture, and overgrazing.

CHECK YOUR UNDERSTANDING

1. Define the concept of ecosystems and the various components within them.

2. Explain the process of photosynthesis and why it is important to life on Earth.

3. Why is biomass higher in tropical regions than in desert regions?

4. Describe the concept of global biomes and the factors that influence them.

5. Why is the floor of the tropical rainforest biome considered to be "open"?

6. Many of the trees in the tropical deciduous forest biome can be found in the tropical rainforest biome. Nevertheless, the trees in the rainforest are evergreen, whereas they drop their leaves in the tropical deciduous forest biome. Why does this difference exist?

7. The ITCZ and STH Pressure System are very closely associated with the tropical savanna biome. What is this relationship?

8. Describe the change in the character of grasses in the Great Plains from the eastern part of the grassland to the west. Why does this difference exist?

9. Where would the warmest/driest slopes be in the Southern Hemisphere: on the north- or south-facing slopes?

10. Where is overgrazing likely to have the biggest impact: in the Great Plains or in the eastern United States? Why does this difference exist?

11. Why is there likely to be a denser forest on the windward side of a mountain range than on the leeward side?

12. Explain the concept of plant succession.

13. Tundra is found in the very low latitudes of South America, even though this region is commonly thought to be "tropical." Why?

14. Discuss the rates and geography of deforestation on Earth.

ANSWERS TO VISUAL CONCEPT CHECKS

VISUAL CONCEPT CHECK 10.1

The answer to this visual concept check is *c*, which states that this biome receives less annual precipitation than the tropical rainforest biome. This answer makes sense because the relative lack of rainfall causes a more open forest to develop than in the rainforest.

VISUAL CONCEPT CHECK 10.2

This biome is the African Serengeti, which exists in the part of Africa that has a distinct wet and dry season. The wet season occurs in summer when the insolation is high and the ITCZ dominates the weather. In contrast, the dry season occurs in winter when the STH Pressure System moves into the region.

VISUAL CONCEPT CHECK 10.3

The answer to this question is *b*; there will be more atmospheric carbon dioxide as a result of deforestation. This increase will occur because trees absorb carbon dioxide in the context of the global carbon cycle.

CHAPTER ELEVEN

THE GLOBAL DISTRIBUTION AND CHARACTER OF SOILS

If asked to name the most important natural resources on Earth, you might say they are water, forests, fossil fuels, and the air we breathe. Certainly, these resources are very important, but you might be surprised to learn that many people believe our most important natural resource is soil. Why? Because our food directly or indirectly comes from it, we build structures within it, and complex ecosystems are supported by it. This chapter will help you appreciate the character of soil and the important role it plays on Earth.

This chapter examines how soils form, what properties affect soil fertility and development, and where the basic types of soils are found on Earth. This chapter naturally follows the preceding chapters on climate and vegetation because a close relationship is usually found between these variables and soil characteristics. In this sense, this chapter integrates many components of the atmosphere, hydrosphere, biosphere, and lithosphere in a holistic way. It also marks the point where we begin to examine the geographic patterns associated with parts of Earth that are within Earth, or the *lithosphere*, rather than being on top of it, as in the case of the *biosphere*, or above it in the *atmosphere*.

Soil is one of the most important natural resources on Earth and comes in many different forms. This image nicely shows the relationship between Earth's surface and the underlying soil. Notice the color of the sediment and the plant roots, and consider how water would flow downward. This chapter focuses on the evolution of soils and how they can differ around Earth.

CHAPTER PREVIEW

What Is Soil?

Measurable Soil Characteristics

Soil Chemistry

 Soil Colloids and pH

Soil Profiles (Reading the Soil)

 Soil Horizon Development

Soil Science and Classification

 African Climate, Vegetation, and Soils

 Regional Pedogenic Processes

 North American Climate, Vegetation, and Soils

Human Interactions with Soils

■ LEARNING OBJECTIVES

1. Discuss the nature of soil and the factors that influence its development.

2. Describe the measurable soil characteristics and why they are important.

3. Explain the various aspects of soil chemistry and why they are important for soil fertility.

4. Define the concept of a soil horizon and explain how the classic sequence of horizons forms.

5. Describe the concept of soil classification and differentiate between the various soil orders in a holistic way.

6. Explain why soils are important natural resources and how they are managed in the United States.

What Is Soil?

Did you ever play in the dirt as a child? Did you ever stop to think about where the dirt had come from and how long it had been there? Although you generally referred to the material as "dirt," in all probability you were playing in soil. You may wonder what difference it makes whether it is called "dirt" or "soil": Are they not the same thing? In everyday language the two terms can be used interchangeably, but in the context of physical geography and soil science, the concepts are indeed very different. The fact is that dirt is not necessarily soil, but soil *forms* in dirt. Another way to look at it is that soil is what happens to dirt.

Soil is the uppermost layer of the Earth's surface and contains mineral and organic matter capable of supporting plants (Figure 11.1). Another way to think about soil is that it is the outermost rind of Earth, similar to how an orange peel is the outermost part of the orange. Still another way to consider soil is as the transition between the atmosphere and the rocky Earth.

Although all soils can be viewed in these various ways, they nevertheless vary dramatically in terms of their character and thickness, ranging from a few centimeters to several meters in depth.

Soils are indeed one of our most important natural resources. They are fundamental to life on Earth because of their interaction with climate and vegetation. Soils provide plants with physical support as well as nutrients and water for growth. Plants, in turn, support soils by anchoring them to Earth. Without a vegetative cover, soil is prone to erosion, which leads to reduced fertility for agriculture and can contribute to famines in the world's poorest nations.

Basic Soil Properties

The different variables that, taken together, comprise soil can be generally classified into the following four groups (Figure 11.2).

1. **Inorganic materials** Inorganic materials are naturally occurring chemical elements or compounds that possess a crystalline structure. Each specific mineral (element or compound) owes its structure to the particular rock from which it came. Some of the most common minerals found in soil contain the elements silicon, aluminum, iron, calcium, potassium, and magnesium. The element calcium, for example, is part of the rock limestone. As limestone disintegrates through a process called *weathering* (discussed in Chapter 14), the minerals become reduced in size enough for plants to use them as nutrients. The key point for our present purposes is that, until the plants absorb the minerals, they are stored in the soil.

2. **Organic matter** Organic matter forms from living and decayed organisms and accumulates in the upper part of soil. As plants and animals die and decay, bacteria and fungi decompose their remains and produce **humus**, which is partially decomposed organic matter. Humus has a sponge-like quality that allows soil to hold water. In addition, humus increases soil fertility because (a) it helps stabilize the soil

Figure 11.1 Analyzing soil in the field. Soil scientists investigating a soil profile in a backhoe trench. Understanding soils is an important part of geography. Many geography students spend at least some time doing "hands-on" work like this.

Alan Arbogast

Figure 11.2 General composition of soil Approximately 50% of soil is mineral and organic matter. The remaining 50% consists of pore spaces between grains. These spaces hold water and air.

Soil *The uppermost layer of the Earth's surface that forms by the influence of parent material, climate, relief, and chemical and biological agents.*

Humus *Decomposed organic matter, typically dark, that is contained within the soil.*

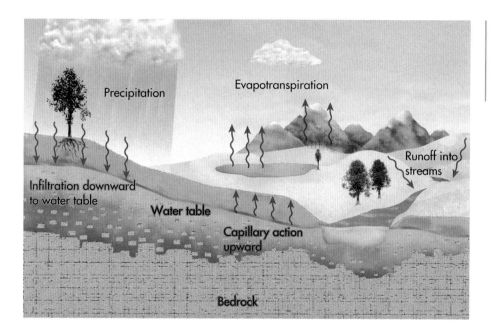

Figure 11.3 Delivery of water to soil. Soils receive most of their water from rain and melting snow, which move downward into the soil by infiltration. The soil also receives water stored below the ground by capillary action.

Precipitation

Evapotranspiration

Runoff into streams

Infiltration downward to water table

Water table

Capillary action upward

Bedrock

by helping to hold it together and (b) microorganisms within humus give off compounds of nitrogen, phosphorus, sulfur, and other mineral substances that are nutrients for plants.

3. **Water** Water is critical to soils for a variety of reasons. The most obvious reason is that plants require water to grow. In turn, high plant production contributes to increased biomass, which generally results in more humus production and stable, more fertile soils. Water is delivered to soils in the form of rain and melting snow (Figure 11.3). After precipitation occurs, water can take several pathways with respect to the soil. If the precipitation is heavy, or if it falls on soils that are already very wet, some or most of the water will flow across the surface into local rivers and streams as runoff and never be absorbed into the soil.

When snow melts slowly, rain falls relatively slowly, or the soil is dry, water has a tendency to more readily infiltrate into soils. Once infiltration begins, some water will either evaporate into the atmosphere directly from the ground or when plants absorb it through their roots and carry it to their leaves through the process of transpiration. In the context of soil-water loss, the combined processes of evaporation and transpiration are often lumped together in the single term *evapotranspiration*, which we described in Chapter 9.

Let's examine what happens to the soil water that is not lost immediately by evapotranspiration. A soil is *saturated* when water mostly fills all available pore spaces, except for a few small pockets of air here and there (Figure 11.4). Subsequently, within a few days, all excess water drains downward under the force of gravity toward the underlying groundwater zone, leaving larger pockets of air behind. Although most of this downward-moving water is lost as far as the soil is concerned, some of it will be pulled back up into the soil at a later time through the process of **capillary action** (see Figure 11.3). This occurs when the molecular attraction at the boundary between the underlying water and soil particles is stronger than

the attraction between water molecules. To see how capillary action works, fill a transparent glass about halfway with water. If you look at the side of the glass, where the water meets the inside surface, you can see that the edge of the water actually climbs a short way up the side of the glass. Water molecules have a slightly stronger attraction to the glass than they do to other water molecules. Capillary action in soils works the same way.

Although most water is flushed from the soil pores during the draining process, some remains in the soil-water belt because it is held to soil particles by the force of **surface tension**. Observe for yourself how surface tension works. Splash a few

Capillary action *The force that causes water to rise in the small tubular conduits within the soil.*

Surface tension *The contracting force that occurs when the water surface meets the air and acts like an elastic skin.*

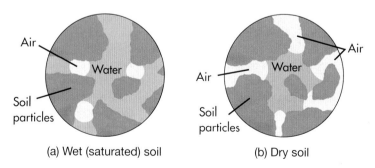

(a) Wet (saturated) soil

Air

Water

Soil particles

(b) Dry soil

Air

Water

Air

Soil particles

Figure 11.4 Wet vs. dry soils. (a) A soil is considered to be saturated when pore spaces are nearly full of water, which occurs after extended periods of precipitation or snowmelt. (b) Subsequently, water drains under the force of gravity, leaving water held to soil particles by surface tension. When most of the downward-moving water is lost, the soil is dry.

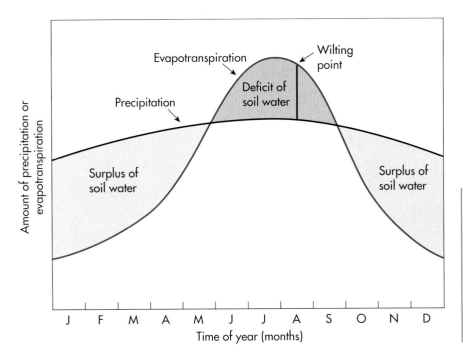

Figure 11.5 Hypothetical soil–water budget for a location in the midlatitudes of the Northern Hemisphere. Although precipitation is relatively low from October to May, it is greater than the amount of soil water lost through evapotranspiration, causing a soil–water surplus. During the summer months, however, the wilting point is sometimes reached because more soil water is lost by evapotranspiration, even though more precipitation actually occurs during that time of year.

drops of water onto the surface of an upright bottle. Notice that each drop is enclosed within a film that pulls water inward into a rounded shape that withstands the force of gravity. Within soils, surface tension works in the same way, holding water onto soil particles until they are lost through evapotranspiration.

After the excess water has drained, the soil is left at **field capacity**, which refers to the maximum amount of water that a soil can hold after free gravitational drainage ceases. If additional precipitation occurs shortly after field capacity is reached, then the soil will again become saturated and the cycle begins anew. On the other hand, if drought conditions begin and little rain falls, soil water will not be replenished. Even so, soil water will continue to be lost through evapotranspiration, resulting in progressively less water held by surface tension. If dry conditions persist, the soil reaches the **wilting point**, which represents the time when no soil water is available for plant use and plants literally begin to wilt.

The term **soil–water budget** reflects the balance between soil–water gains and losses. To understand this concept, Figure 11.5 presents a simple example that reflects changes over the course of the year in a typical continental region in the midlatitudes in the Northern Hemisphere. During most of the year, the region has a surplus of soil water, which occurs from October through April. Interestingly, this surplus happens to correlate with the time of year when the least precipitation falls. Why does it happen at this time of year? First, very little

evaporation occurs because the Sun angle is low, temperatures are cool to cold, and it is frequently cloudy. Second, plants are not active during this time of year, so little soil water is lost due to transpiration. During the summer months, in contrast, deficits of soil water frequently develop, even though precipitation is greater at that time of year, because the amount of evapotranspiration exceeds rainfall.

4. **Air** The air in soil is mostly carbon dioxide, which plants give off during respiration and then take in during photosynthesis. Soils contain so much carbon dioxide that the soil layer is sometimes considered to be part of the atmosphere and thus a transition to the solid Earth below. The amount of air in the soil fluctuates depending on how wet the soils are. As you can see in Figure 11.4, when soils are saturated, the pore spaces are full of water and thus contain little room for air. In contrast, when soils dry out, the pore spaces contain less water and more air.

Soil-Forming (Pedogenic) Processes

A key concept to understand at this point is that soils form and evolve through a complex sequence of interrelated **pedogenic processes**. It is useful to think of these processes as consisting of *additions*, *depletions*, *translocations*, and *transformations* that occur within the soil (Figure 11.6). The following discussion illustrates how these processes operate individually.

Field capacity *The amount of water remaining in the soil after the soil is completely drained of gravitational water.*

Wilting point *The threshold amount of soil water below which plants can no longer transpire water.*

Soil–water budget *The balance of soil water that involves the amount of precipitation, evapotranspiration, and water storage and loss.*

Pedogenic processes *The natural processes of soil formation that involve additions, translocations, transformations, and losses.*

1) Soil Additions
- Organic matter
- Water (as precipitation, condensation, runoff)
- Oxygen, carbon dioxide (from atmosphere)
- Minerals (from precipitation)
- Sediments (from wind, water)
- Energy (from Sun)

2) Soil Translocations
- Clay, organic matter (carried by water)
- Nutrients (circulated by plants)
- Dissolved minerals (carried by water)
- Sediments (carried by animals)

3a) Upper Soil Depletions
- Water (by evapotranspiration)
- Carbon dioxide (from oxidation of organic matter)
- Nitrogen (by biological and chemical means)
- Sediments (by erosion)
- Energy (by longwave radiation)

3b) Lower Soil Depletions
- Water and dissolved minerals

4) Soil Transformations
- Organic matter (decomposed into humus)
- Particles made smaller (by weathering)
- Particles made into structures (by accretion)
- Minerals transformed (by weathering)

Figure 11.6 Basic soil-forming processes. Soils form through the complex interaction of several processes. Although each of these processes usually operates to some extent in all soils, the rate at which any given one operates varies from place to place.

KEY CONCEPTS TO REMEMBER ABOUT SOIL PROPERTIES

1. Soils contain organic material and a variety of minerals. Soils also contain voids, called pore spaces, between grains of sediment.

2. Pore spaces contain both air and water. Following heavy rains or periods of snowmelt, soil pore spaces usually fill with water. This water drains within a few days, leaving the soil at field capacity with water held by surface tension.

3. The soil–water budget refers to the balance between water gained by precipitation and water lost through evapotranspiration.

1. **Soil additions** Soil additions consist of material and energy that are added to the soil, such as organic material, water from precipitation, various minerals, solar radiation, oxygen and carbon dioxide from the atmosphere, or sediments deposited by wind or water. A good example of a soil addition is the leaf litter that accumulates during the

fall under trees. In forest settings where the leaves are not raked, organic decomposers gradually break them down and their remains are incorporated into the soil by simple decay or with the aid of burrowing animals and worms. A similar process occurs when grasses die at the end of the growing season. Animal remains and animal excreta that accumulate on the ground also rapidly become part of the soil.

2. **Soil depletions** Soil depletions occur when some aspect of the soil is lost for one reason or another (see Figure 11.6). These losses can occur both in the upper and lower part of the soil. An example of a loss in the upper part of the soil is when erosion by wind, water, or glaciers removes sediment. Depletion in the upper part of the soil also occurs when plants take up nutrient minerals or water through evapotranspiration. Another form of depletion takes place when carbon is lost to the atmosphere as carbon dioxide when organic matter is oxidized. In a similar vein, nitrogen is lost to the atmosphere through biological and chemical means. The loss of longwave radiation to the atmosphere is another form of loss in the upper part of the soil. In the lower part of the soil, losses occur when water drains out of the soil and carries soluble minerals with it.

3. **Translocations** Translocations refer to the vertical movement of materials through the soil. Some movement occurs as nutrients are cycled through plants or when animals move soil around. A significant amount of additional movement occurs in the presence of water, which percolates downward through the soil during wet conditions. When water moves in this fashion, it dissolves minerals and carries them deeper within the soil, where they

Living plant shoots/grass

Additions of dead plant matter from previous year

Transformation of organic matter into humus by decay

Translocation of dissolved minerals by eluviation

Accumulation of eluviated minerals by illuviation

(a)

Randall Schaetzl

(b)

Figure 11.7 How additions and translocations affect soil development. (a) One way that additions occur in soil is through the accumulation of dead plant matter, which slowly transforms into humus by decay. The processes of eluviation and illuviation move materials vertically through soil. Eluviation moves material, such as dissolved minerals, out of the upper soil. These minerals recrystallize and accumulate in the lower part of the soil by illuviation. (b) In this photo of actual soil, the dense-looking whitish layer toward the base of the soil is calcium carbonate that was eluviated by water moving downward from the upper part of the soil.

recrystallize and collect. When water moves materials, usually in the form of minerals, clays, and organic matter, *out* of the uppermost part of the soil, the process is called **eluviation**, or leaching. When leached materials from the upper part of the soil recrystallize *within* the lower part of the soil, the process is called **illuviation**. Figure 11.7 shows the results of the combined processes of eluviation and illuviation.

4. **Transformations** Transformations refer to the decomposition of minerals and organic matter into other forms within the soil. An excellent example of a transformation is when organic matter, such as leaf and animal litter, decomposes into humus. Another example is when minerals are transformed into other minerals or are made smaller by weathering. In addition, soil particles begin to clump together in a recognizable structure.

VISUAL CONCEPT CHECK 11.1

Which one of the following choices describes how these fall leaves are related to soil development?

a) Soil losses
b) Soil additions
c) Soil transformations
d) Soil depletions

Brand X Pictures/Punchstock

Eluviation *The dissolution and downward mobilization of minerals by water in the soil.*

Illuviation *The recrystallization of minerals that occurs directly below the zone of eluviation.*

Soil-Forming Factors

Let's now examine the factors that generally influence soil development. Five of these variables, traditionally called the **soil-forming factors**, consistently affect how soils evolve. These factors are climate, organisms, relief, parent material, and time; these variables can be remembered with the simple acronym *CLORPT*. Although you need to consider these factors in every soil, their character and relative significance vary considerably from a geographic perspective. In order to understand the discussion of specific soil types presented later in this chapter, you must first comprehend how soil-forming factors influence soil development individually and collectively. As you examine these variables, keep in mind that all of them contribute to soil development in some way in each and every soil.

1. **Parent material** An important concept to consider in soil science is that soils always form *within* sediment of some kind. This sediment is referred to as **parent material** and is closely related to the geology of the area in which a particular soil forms. The character of the parent material at any given locality is important in the context of soils because it sets the stage for future development and

affects the way in which additions, translocations, depletions, and transformations occur.

In general, soils form in two very broad types of parent material, residual and transported. **Residual parent material** is sediment that develops when underlying rock weathers *in place*; in other words, it has not been moved from someplace else. Such rock may be millions, or even hundreds of millions, of years old. The concept of weathering will be discussed more thoroughly in Chapter 14, but for now, think of it as the process by which rock slowly breaks up into smaller fragments and particles. This transformation takes place through chemical and physical changes that are largely associated with water.

As this weathering occurs—in other words, as the rock fragments become smaller and change their chemical composition—a material called **regolith** begins to form. Figure 11.8 shows the relationship among bedrock, regolith, and soil. Notice how the rocks in the upper part of the bedrock break up into smaller pieces. Somewhere in this part of the vertical sequence is a transition in which the bedrock is sufficiently altered through weathering to be fundamentally different from the underlying (solid) rock from which it came. At the same time, it has not been modified by additions and translocations from above that occur in soil. In other words,

(a)

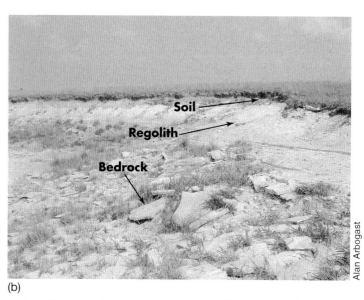

(b)

Figure 11.8 Relationship of regolith to soil. (a) Regolith consists of weathered bedrock that forms parent material in which soil forms. (b) Limestone bedrock and overlying regolith in the Great Plains. Note how the size of rock fragments becomes progressively smaller above the horizontal slab of solid rock. The soil is the dark zone at ground level.

Soil-forming factors *The variables of climate, organisms, relief, parent material, and time that collectively influence the development of soil.*

Parent material *The mineral or organic material in which soil forms.*

Residual parent material *Parent material that forms by the weathering of bedrock directly beneath it.*

Regolith *The fragmented and weathered rock material that overlies solid bedrock.*

regolith is not rock and is not soil. Rather, it lies somewhere between the two and is the material in which the soil forms. Often, regolith production at the bedrock interface occurs at the same time that soil development is occurring in the upper part of the weathered zone.

The second kind of parent material is **transported parent material**. As the name implies, transported parent material differs from residual parent material because it has been moved from one place to another. Remember that residual parent material is regolith that forms *in place* as bedrock weathers. Although the sediments that comprise this bedrock were deposited after being moved from someplace else, the deposition happened such a long time ago (millions of years ago) that, as far as regolith production is concerned, it is considered to have occurred in one place. On the other hand, transported parent material is sediment that was moved relatively recently by water or glaciers. Or perhaps wind blew sand to form dunes like those pictured in Figure 11.9. If this sediment is left undisturbed for many years, plants and animals will begin to live in the uppermost part of the deposit and a soil will form through the combined processes of additions, depletions, translocations, and transformations.

2. **Climate** Climate is a critical variable in the context of soil development. Temperature and precipitation are the

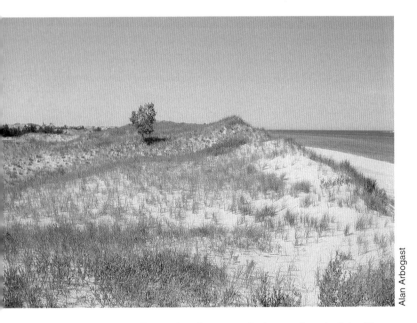

Alan Arbogast

Figure 11.9 Example of transported parent material. These sand dunes along the shore of Lake Michigan consist of sand deposits that were blown there by the wind. Note the grass growing on parts of the dunes. The establishment of grass begins the process of soil formation by providing organic material to be added to the sand parent material.

most important aspects of climate that affect soil formation because they influence the kind and rate of biological and chemical reactions that occur in the soil. In general, more reactions occur in warmer and wetter places than in colder and drier regions. As a result, soils tend to be better developed and thicker in warm, humid regions than in cold, dry locations.

Climate also strongly influences the process of translocation in a soil. Remember that translocation refers to the movement of organic and mineral matter from the upper part of the soil into a deeper zone. Most of this movement occurs when water percolates through the soil after it falls at the surface. To visualize this process, imagine, for example, the gradual melting of snow that occurs after a big winter storm passes through an area. As the snow melts, some of it evaporates into the atmosphere, but much of it is absorbed into the ground. When spring arrives, plants rapidly take up some of this soil water, but a certain percentage of the water continues to move slowly downward within the parent material under the force of gravity. As this water comes into contact with minerals in the soil or sediment, it dissolves them—much as water dissolves sugar—and carries the dissolved particulates deeper into the soil through the process of eluviation. At some point, the water stops flowing downward and the dissolved minerals recrystallize in the zone of illuviation. Simply put, more eluviation occurs in humid climates than in arid regions because more water percolates through the soil. Figure 11.10 shows in general terms the depth to which eluviation penetrates and the resulting depth of the illuvial zone in humid, semi-arid, and arid climate regimes.

3. **Organisms** Organisms collectively are made up of the plant and animal life that resides within soil. The importance of organisms in soil formation simply cannot be overestimated. At a fundamental level, organisms give life to soil and contribute greatly to the notion that soil is significantly more than *just dirt*. At a more complex level, organisms have a symbiotic relationship with soil in that they not only acquire their food from it, but also help regulate the soil environment in which they live.

Organisms contribute to soil health and formation in many different ways. One way is when plant roots penetrate deeply into the ground. This penetration creates passageways in the soil in which water and oxygen can flow. Many small animals, such as ants, ground squirrels, prairie dogs, and other rodents, spend the majority of their lives burrowing through the soil (Figure 11.11). Through this ongoing process, the soil is continually mixed in a process called **bioturbation**. The mixed soil is softened and aerated,

---

Transported parent material *Parent material such as glacial or stream sediments that has recently been deposited and in which soil forms.*

Bioturbation *The mixing of soil by plants or animals.*

(a) Humid climate   **(b) Semi-arid climate**   **(c) Arid climate**

Figure 11.10 **Eluviation and illuviation in three climate regimes.** More eluviation takes place in (a) humid climates than in (b) semi-arid and (c) arid climates. In addition, the zone of illuviation (where minerals precipitate from solution) is deeper and thicker in more humid climates. Typically, this zone is progressively thinner and less deep as the climate becomes increasingly dry.

which makes it easier for plants to grow. Earthworms are important within the system because they recycle soil elements through consumption and waste production. Another important biological component is the huge number of

microorganisms, such as algae, fungi, and bacteria, that live within the soil. The primary function of these microorganisms is to assist with the decomposition of organic matter into humus.

(a)   (b)

Figure 11.11 **Soil bioturbation.** (a) Prairie dogs in the Great Plains constantly burrow in the soil, which recycles nutrients and keeps soil loose and well aerated. (b) Termite mounds in central Brazil are composed of earth brought to the surface by the termites. This process mixes the soil.

4. **Relief** Recall that topography refers to the configuration of the Earth's surface—in other words, the position, shape, and orientation of hills, valleys, and mountains. In a related vein, the term **relief** refers to the differences between the highs and lows of a landscape (Figure 11.12a). The Rocky Mountains, for example, is a region of very high relief because tall peaks are separated by lower valleys. In contrast, much of the Midwest has low relief because the landscape is relatively flat.

The most important variable associated with relief and soil development is slope. The slope of the landscape is loosely defined as the ground surface that connects the higher areas with lower areas. Slope is significant in the context of erosion because soils are generally thinner and less well developed where slopes are steeper. Figure 11.12b shows a hypothetical landscape with a steep slope next to a relatively flat surface. Note that the soil on the steep slope is thinner; that is, the bedrock is closer to the surface. This pattern occurs because more erosion takes place on the steeper slope, predominantly due to water flowing downhill under the force of gravity. As a result, the rate of erosion on the steeper hillslope may be equal to, or even greater than, the rate of soil formation (Figure 11.12c). In contrast, on the more level surface, the soil can develop more fully because it lies essentially undisturbed from erosion. In addition, the soil is thicker because sediments eroded from the steep slope are added where the slope is more level.

5. **Time** Time is a critical variable because soils need it to develop. If it is assumed that all the other soil-forming factors (parent material, climate, organisms, relief) are equal, then the best-developed soils occur on surfaces that have been stable for the longest period of time. For example, where would you expect to find the better-developed soil: in a 3000-year-old sand dune or in a 100,000-year-old stream deposit? The answer is that the older stream deposit most likely has a better-developed soil because it had more time to develop. The significance of this greater amount of time is that collectively more additions, translocations, depletions, and transformations have taken place in the older soil than in the younger soil.

KEY CONCEPTS TO REMEMBER ABOUT PEDOGENIC PROCESSES AND SOIL-FORMING FACTORS

1. Four pedogenic processes—additions, translocations, deletions, and transformations—collectively promote soil development.

Relief *The difference between the high and low elevation of an area.*

(a)

(b)

(c)

Figure 11.12 Relief and soil formation. (a) The term *relief* refers to the differences between the high and low parts of a landscape. In this scene, the area in the foreground is a zone of low relief, whereas the mountain in the background is an area of high relief. (b) Soils are generally thinner on steeper slopes than on flatter surfaces because steep slopes are more prone to erode. (c) Relief affects soil thickness on slopes, such as the exposed bedrock (arrow) on the steep slope shown here. The soil is very thin on this slope because of erosion; soil is thicker on top of the ridge and in the valley.

2. Some pedogenic processes cause physical changes to occur within the soil, whereas others cause soil components to move within the soil.

3. Pedogenic processes occur within the context of five soil-forming factors: parent material, climate, organisms, relief, and time.

4. Each soil-forming factor contributes to all soil development in some way.

Measurable Soil Characteristics

To understand soil formation thoroughly, remember that the soil-forming factors just discussed are individual parts of a holistic system in the sense that each contributes something to the development of any particular soil. As a soil forms under a particular set of environmental conditions, it develops properties that are unique to that particular setting. By using people as an analogy, it becomes clear that each person whom you know has different characteristics that you can readily identify, such as brown hair, blue eyes, and height. With respect to soils, the properties are physical characteristics that can be measured to distinguish different soils from one another. By examining the physical properties of a soil, it is possible to learn something about how the soil formed. These properties are color, texture, structure, soil chemistry, and soil pH. Although these properties will now be discussed individually, remember that all work together to give any one soil a particular overall character.

1. **Color** Soils vary considerably with respect to color, both within and between soils. Color can be an indicator of the composition of the soil. For example, soils with dark colors usually have high humus or organic matter content (Figure 11.13a), which might indicate that the biomass of the region is high or that the environment is cool and wet or that organic matter decomposes slowly. Soils with a high content of iron-bearing minerals are often a reddish color, whereas soils (or parts of soils) that contain abundant calcium are whitish in color. These colors might reflect the color of the local parent material, or they may indicate the amount and type of eluviation that has occurred over time.

Soil color is typically measured with a Munsell Color Chart (Figure 11.13b). This color-system uses three elements of color—hue, value, and chroma—to designate a specific color. *Hue* is a measure of the chromatic composition of light that reaches the eye. The Munsell system is based on five principal hues: red (R), yellow (Y), green (G), blue (B), and purple (P). In addition, five intermediate hues represent midpoints between each pair of principal hues, resulting in ten major hue names. The intermediate hues are yellow-red (YR), green-yellow (GY), blue-green (BG), purple-blue (PB), and red-purple (RP). *Value* indicates the degree of lightness or darkness of a color in relation to a neutral gray scale under natural lighting, ranging

(a)

NRCS/USDA

(b)

Photo Courtesy of Munsell Color

Figure 11.13 Differences in soil color. (a) The lower part of this soil in Oklahoma is red because it formed in residual parent material (regolith) derived from iron-rich bedrock. In contrast, the upper part of the soil is relatively dark because of the addition of humus to the parent material. In other words, all the parent material was reddish at one time, but the upper part was transformed to a darker color through the addition process. (b) Soil color is measured with a Munsell Color Chart that specifies hue, value, and chroma.

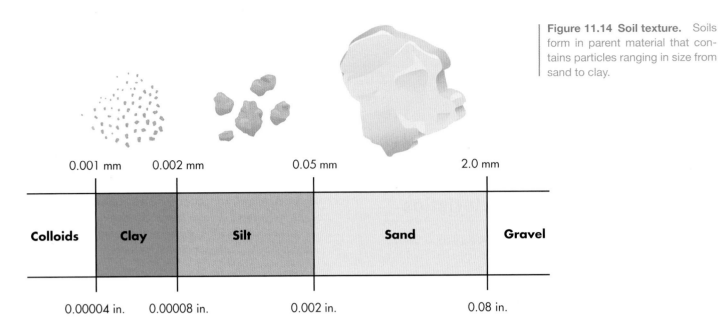

0.001 mm   0.002 mm   0.05 mm   2.0 mm

| Colloids | Clay | Silt | Sand | Gravel |

0.00004 in.   0.00008 in.   0.002 in.   0.08 in.

Figure 11.14 Soil texture. Soils form in parent material that contains particles ranging in size from sand to clay.

from pure black (0/) to pure white (10/). *Chroma* is the relative strength of the spectral color. The scales of chroma for soils extend from /0 for neutral colors to a chroma of /8 for the strongest soil color. To put it all together, a soil that is pale brown, for example, is designated 10YR 6/3. In contrast, a soil that is very dark brown has a Munsell designation of 10YR 2/2. Still another example is a soil that has a yellowish red (7.5YR) hue and is strong brown in color. This soil would have a 7.5YR 5/6 color designation.

2. **Texture** As discussed earlier, soil contains a multitude of inorganic particles that are typically related to the parent material. Most soils contain a continuum of three distinct size categories (Figure 11.14), which are (in order from largest to smallest) sand, silt, and clay. Some larger (gravel) and smaller (colloid) particles may also be present. The combined percentages of these soil particulates are referred to as the textural class of the soil, as shown in the soil textural triangle in Figure 11.15.

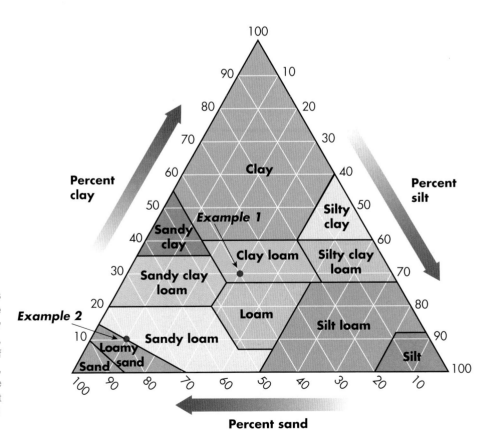

Figure 11.15 Soil textural triangle. Soils can be texturally classified based on the relative proportion of sand, silt, and clay in the parent material. For percent sand, use the lines slanting from the bottom of the triangle up to the left; for percent silt, use the lines slanting from the right side of the triangle down to the left; for percent clay, use the horizontal lines. Examples 1 and 2 described in the text are shown.

SOIL CONSERVATION ON STEEP SLOPES

Soils on steep slopes are prone to erosion because flowing water has a lot of energy and thus moves sediment. You can tell that hillslope erosion has been a problem in places where conservation terraces are found. Farmers build these features to reduce the effects of soil erosion by creating a series of stair-steps on steep hillslopes. These steps produce a series of essentially horizontal surfaces on which farmers plant their crops. Each of the terraces is separated by a built-up ridge of dirt, called a berm, which reduces the velocity of water and traps sediment. Although some soil erosion still occurs where terraces exist, the amount is significantly reduced. Perhaps the most impressive conservation terraces on Earth are the rice paddies shown here in the mountainous area of Yunnan province, China. Note that each of these surfaces can hold a great deal of water, which falls as precipitation during the summer monsoon.

Yann Layma/The Image Bank/Getty Images, Inc.

Water

Crops

Berms

Sediments collect at the base of berms instead of washing away

So how does the textural triangle work? Imagine, for example, that a soil contains 40% sand, 30% silt, and 30% clay. To determine the textural class, simply follow the relative percentages of each variable to where the lines intersect within the triangle. For percent sand, use the lines slanting from the bottom of the triangle up to the left; for percent silt, use the lines slanting from the right side down to the left; for percent clay, use the horizontal lines. In this particular case, the soil would be texturally classified as *clay loam*. Now imagine that a soil has the relative percentages of 80% sand, 10% clay, and 10% silt. This soil would be texturally classified as *loamy sand*.

You may wonder why it matters whether a soil is clay loam, loamy sand, or any other textural class? These kinds of classifications may seem mind-numbing rather than informative, but in the case of soils individual textural classifications are important because they provide a rough measure of how well water flows through a soil or how appropriate the soil is for agriculture, just to name two examples. Regarding water movement, soils that are sandy generally drain better—that is, water flows through them faster—than clay-rich soils. Thus, soil texture has significant implications for the soil–water budget in any given locality. Although clay-rich soils hold more water than sandy soils, they are frequently difficult to work with because clays stick together so well. In general, the best agricultural soils are the loamy soils because they are nicely balanced. On the one hand, they contain some sand, which allows water to drain through. On the other hand, they contain some clay, which holds the soil together and prevents water from draining through the soil too fast. These clays are also important with respect to overall soil fertility, as you will learn later. A high percentage of silt allows the soil to be easily mixed and aerated.

3. **Structure** **Soil structure** refers to the way in which soil particles naturally clump together in *peds*, which are natural soil aggregates. Another way of looking at structure is that it is related to how clumps of soil break apart and their resulting shape, size, and arrangement. Soil structure is a difficult concept to illustrate; nevertheless, four primary types of structure are recognized (Figure 11.16).

Soil structure *The way soil aggregates clump to form distinct physical characteristics.*

Granular structure

Blocky structure

Platy structure

Columnar structure

Figure 11.16 Basic soil structures. Particles within the soil clump together in various ways, resulting in a distinct classification system for soil structure.

- *Granular* structure occurs when the peds break up into very small, rounded clumps. You can recognize granular structure when the soil is slightly cohesive but still easily spills between your fingers.
- *Blocky* structure occurs when the clumps break up into distinct blocks that can be easily held.
- *Platy* structure occurs when the peds are stacked in distinct plate-like forms.
- *Columnar* structure occurs when the peds are organized vertically.

Again, you might be saying to yourself, "Why does it matter how soil particles group together?" The fact is that it is extremely important to how water percolates through the soil and how easily plant roots can penetrate it. Which soil structure presented in Figure 11.16 would be the most difficult for plants to penetrate? The answer is the platy structure because the roots cannot drive straight down through the soil. As far as water's infiltration is concerned, it flows best in soil with granular and blocky structure.

Soil Chemistry

Given that soils contain a variety of mineral, organic, water, and atmospheric components, each soil has a distinct chemical composition that is important regarding its formation and fertility. The chemical composition of any given soil is determined by a complex set of chemical reactions that occur within the soil solution. As noted before, when water comes into contact with minerals under the right conditions, it causes many minerals to dissolve. Similarly, when water mixes with carbon dioxide and various forms of organic matter, it creates carbonic and organic acids, respectively, that promote further changes within the soil. A primary product of these kinds of transformations is chemical ions that are free to be absorbed by plants.

Soil pH

One of the ways to characterize the chemical composition of soil is by measuring its **pH**, which is the relative concentration of hydrogen ions (H^+) present in solution. The pH scale ranges

pH *The measure of acidity or alkalinity of a solution, ranging from 0 to 14, based on the activity of hydrogen ions (H^+).*

VISUAL CONCEPT CHECK 11.2

What is the texture of a soil that is 50% sand, 40% clay, and 10% silt?

a) Loam
b) Silt
c) Sandy clay loam
d) Sandy clay

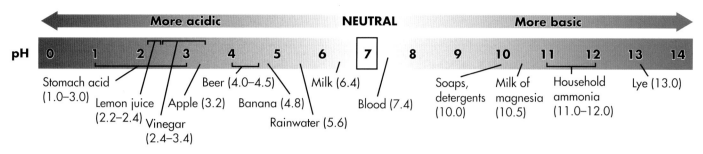

More acidic NEUTRAL More basic

pH 0 1 2 3 4 5 6 7 8 9 10 11 12 13 14

Stomach acid
(1.0–3.0)
Lemon juice
(2.2–2.4) Apple (3.2) Beer (4.0–4.5) Milk (6.4)
Vinegar
(2.4–3.4) Banana (4.8) Blood (7.4)
Rainwater (5.6)
Soaps,
detergents
(10.0) Milk of
magnesia
(10.5) Household
ammonia
(11.0–12.0) Lye (13.0)

Figure 11.17 The pH scale. pH refers to the concentration of hydrogen ions (H⁺) in solution. Note that most natural foods are slightly acidic and most common cleaning agents are basic.

from 0 to 14, with lower values indicating acidic conditions and higher values indicating basic (alkaline) conditions. Neutral pH is about 7, which is the pH of pure water. To get a feel for pH and how it relates to various common products, see Figure 11.17.

In the context of soils, pH is an important indicator of soil fertility. Highly alkaline soils, for example, are not efficient with respect to dissolving minerals and making them available as nutrients for plants. Acidic soils, on the other hand, result in extensive leaching of minerals—so much so that many nutrients are completely lost before they can ever be consumed by plants. In general, most plants, including agricultural crops, are adapted to soils that have a neutral or near-neutral pH. It is possible, however, to treat slightly acidic soils with alkaline fertilizers, such as slaked lime (calcium carbonate, $CaCO_3$), to raise the pH to the appropriate level.

Colloids and Cation Exchange

An important part of soil chemistry is the concept of **colloids**, which are extremely small (<0.001 millimeter) particles that exist in inorganic and organic form. Inorganic colloids consist of crystalline clays that are thin and platy. These microscopic bodies are created when larger particles are chemically altered through weathering. Organic colloids, in contrast, consist of humus derived from organic matter. Regardless of whether colloids are inorganic or organic, they are critical to soil chemistry because they are chemically active and have a high water-holding capacity.

Colloids are also important to soil chemistry because they hold and exchange cations. **Cations** are positively charged (atomic) ions that exist in soil solution due to the dissolution of soil minerals, such as calcium, magnesium, iron, and potassium. Because soil colloids are negatively charged, they attract the positively charged cations that are suspended in the soil solution. Although colloid particles are individually small, they collectively have a very large surface area on which cations can become attached. Some ions, such as metal and hydrogen ions, are bound tightly to colloids, whereas basic ions such as calcium move with relative freedom. In many instances, the loosely bound (basic) cations are replaced by the metallic and hydrogen

ions through the process of cation exchange. Soils that have a high **cation exchange capacity (CEC)** are typically the most fertile because they contain an abundance of colloids on which cations can be held and exchanged.

In the context of soil fertility, colloids are important because they keep minerals from being completely leached from the soil. Often, cations are held by the colloid until a plant root comes into contact with the colloid and absorbs the cation as a nutrient. Figure 11.18 illustrates how individual negatively charged colloid particles (in this case, clays) attract the positively charged cations surrounding them until a plant root "consumes" the stored nutrients.

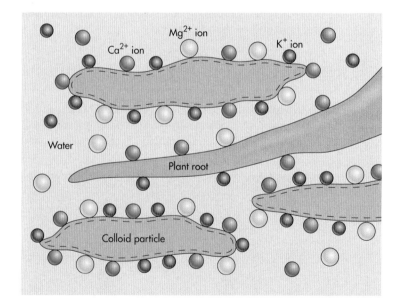

Figure 11.18 Microscopic view of colloids and soil fertility. Colloids are microscopic clays and organic particles that are negatively charged. Cations, which are positively charged atoms that exist in soil solution, are thus attracted to the colloid particles, which hold them until a plant root absorbs them.

Colloids *Very small (10 nanometers to 1 micrometer), evenly divided solids that do not settle in solution.*

Cations *Positively charged ions, such as sodium, potassium, calcium, and magnesium.*

Cation exchange capacity (CEC) *The total amount of exchangeable cations that a soil can absorb.*

The chemical composition of any given soil is determined by a complex set of reactions taking place within the soil solution. The ultimate outcome is often dependent on the nature of soil colloids, which are microscopic particles that occur in organic and inorganic form. These particles have an important role as far as cation exchange and soil pH are concerned. To better understand these relationships, it is useful to view them in animated form. Go to the *Geo Media Library* and select *Soil Colloids and pH*. Once you complete the exercise, be sure to answer the questions at the end to test your understanding of this concept.

KEY CONCEPTS TO REMEMBER ABOUT MEASURABLE SOIL CHARACTERISTICS AND SOIL CHEMISTRY

1. All soils have physical and chemical properties that can be measured, including color, texture, structure, soil chemistry, and soil pH.

2. Physical and chemical properties differ among and within soils.

3. Soil pH refers to the acidity or basicity of the soil.

4. Colloids are collections of microscopic soil particles that hold and exchange cations.

Soil Profiles (Reading the Soil)

With the various soil components, processes, and chemistry in mind, let's now investigate the progressive development of soil that results from additions, transformations, translocations, and depletions. The nice thing about these processes is that they cause the soil to become progressively organized through time in a way that can be easily seen by looking at a **soil profile**. Just as the profile of your face refers to a side view that ranges from the top of your head to your chin, a soil profile displays all the soil characteristics in a vertical "image." A soil profile can be obtained by cleaning a natural exposure such as a creek bank, or, more commonly, by excavating a soil pit with a shovel or backhoe. Although it may not seem like it, this kind of work is really a great deal of fun. Figure 11.19 shows examples of exposing a soil profile.

As soils form, they organize internally into distinct **soil horizons**. If the term "horizon" seems confusing, simply think of the term as being equivalent to "layer" or "band" for the time being. Look again at the soil profile in Figure 11.19a. Do you

---

Soil profile *A vertical exposure in which all soil components can be seen.*

Soil horizons *The distinct layers within a soil that result from pedogenesis.*

(a)

(b)

Figure 11.19 Soil profiles.   (a) After a soil pit is excavated you can then see the soil profile and its specific characteristics, such as the dark zone near the surface of this particular soil and the reddish zone beneath it. (b) Sometimes soils are exposed naturally, as along this stream cutbank. In these situations, they can easily be studied.

see the distinct layers that occur in the profile? These layers are referred to as soil horizons because they typically blend into one another; in other words, it is very rare to see abrupt boundaries between different parts of the soil.

The specific character of soil horizons varies dramatically among different parts of the world, but, in general, any well-developed soil contains, from top to bottom, O, A, E, B, C, and R horizons (Figure 11.20). These horizons form by different processes and have the following characteristics.

1. **O horizon**—The **O horizon** is the uppermost horizon of the soil, existing at ground level and consisting mainly of undecomposed organic matter. A good example of an O horizon is the leaf litter that accumulates on the forest floor during the fall season. Thus, an O horizon forms through the addition of organic matter.

2. **A horizon**—The **A horizon** is the uppermost mineral horizon of the soil. This horizon is distinctive because it contains abundant humus, which consists of decomposed organic matter added to the soil. Given their relatively high humus content, A horizons are typically a darkish gray color. In general, A horizons have granular structure due to the high concentration of humus. In Figure 11.13, for example, the dark horizon at the top of the soil profiles is an A horizon.

3. **E horizon**—The **E horizon** lies immediately below the A horizon (see Figure 11.20) and is the zone of eluviation; thus, this horizon forms largely through the process of translocation. In other words, this horizon evolves due to the downward movement of minerals leaching from it. As a result of this process, the E horizon is usually relatively light in color compared to the horizons above and below it.

4. **B horizon**—The **B horizon** lies below the E horizon and is the zone of illuviation and thus forms largely through the process of translocation. In other words, this part of the soil evolves because minerals recrystallize at this level. Recall that the zone of illuviation is deeper in more humid climates than in relatively dry ones, which is to say that the B horizon is deeper and thicker where precipitation is greater (see Figure 11.10). Given the accumulation of eluviated minerals in the B horizon, this horizon usually has stronger structure (such as "blocky") than the overlying horizons. The base of this horizon corresponds

O horizon
(leaf and plant litter)

A horizon
(enriched with humus)

E horizon
(zone of eluviation)

B horizon
(zone of illuviation)

C horizon
(unaltered parent material)

R horizon
(bedrock)

Solum

Regolith

Bedrock

Figure 11.20 Soil horizons in a fully developed soil on a bedrock-dominated landscape. With time, soils evolve to produce a predictable sequence of horizons. Keep in mind that the soil in this illustration formed within regolith.

to the base of the soil. The full thickness of the soil, from the top of the A horizon to the bottom of the B horizon, is referred to as the **solum**.

5. **C horizon**—The **C horizon** consists of unaltered parent material that has yet to be affected by soil-forming processes. In other words, this horizon is regolith and is not part of the solum.

6. **R horizon**—Where soils form in landscapes dominated by bedrock, such as those you find in the Appalachian Mountains or parts of the Great Plains, an R horizon occurs. Very simply, the **R horizon** is the bedrock that is the source of the regolith/parent material, which is the sediment in which the soil forms.

O horizon *The uppermost soil horizon that consists of undecomposed plant litter.*

A horizon *The soil horizon that is enriched with humus.*

E horizon *The soil horizon that is progressively depleted of minerals through dissolution and translocation.*

B horizon *The soil horizon that forms below the E horizon because translocated minerals recrystallize.*

Solum *The A, E, and B horizons of a soil, which form through pedogenic processes.*

C horizon *Unaltered soil parent material.*

R horizon *Unweathered bedrock that underlies soil.*

Time and Soil Evolution

Although the profile in Figure 11.20 is typical, it is far from being the profile observed in every soil. In fact, not all horizons are present in every soil. Horizons also vary in color and configuration among different soils. These differences are due not only to the five soil-forming factors discussed previously, but also to geographic variability in the way that soil processes function.

To see how soil horizons generally evolve, consider Figure 11.21. This figure presents the formation of a soil in a hypothetical profile observed at four separate times: today (0), 500, 1000, and 5000 years into the future. Imagine that "time 0" begins with the deposition of sediment deposited along a stream during a big flood. This sediment is thus transported parent material within which a soil will form given the additions, translocations, transformations, and depletions that will ultimately occur.

If you could visit this site at three "future" times, you would see that the soil evolves in a fairly predictable way. This is not to say that each soil would change in the same way with respect to horizon thicknesses or colors or that it would evolve at the same rate given climate and textural differences. However, the overall pattern very well could be consistent. Following deposition of the fresh parent material, the first horizon that would become obvious in the first 500 years or so would likely be the A horizon. This horizon would develop due to the additions of organic matter that resulted from the plant and animal life at or near the surface. (Given your understanding of soil horizons, what would be the diagnostic characteristic of this horizon?)

At the same time that the A horizon is forming, translocations and transformations would also be occurring. The cumulative result of these combined processes might begin to be seen about 1000 years in the future by the development of a Bw horizon immediately beneath the A horizon. The "w" means that the horizon is weakly developed, and will probably be identified because the structure has become slightly blocky due to the illuviation of clays and other minerals at this level of the soil. Another thing to notice is that the A horizon has become thicker because of the continual additions that have occurred over the millennia. It is probably darker as well.

If you returned to examine the soil pit after 5000 years of development, you might very well discover that a complex set of horizons had formed. Remember: This development has occurred because plants, microbes, and animals have been interacting within this soil for a lengthy period of time. In addition, water has been percolating through the soil for the same length of time. As a result of these processes operating day by day, month by month, and year by year for 5000 years, the soil has a distinctive E horizon and a thick Bt horizon. The E horizon is the zone of the soil that has been extensively leached of its mineral and clay components. These minerals and clays moved downward and recrystallized in the evolving B horizon. In this

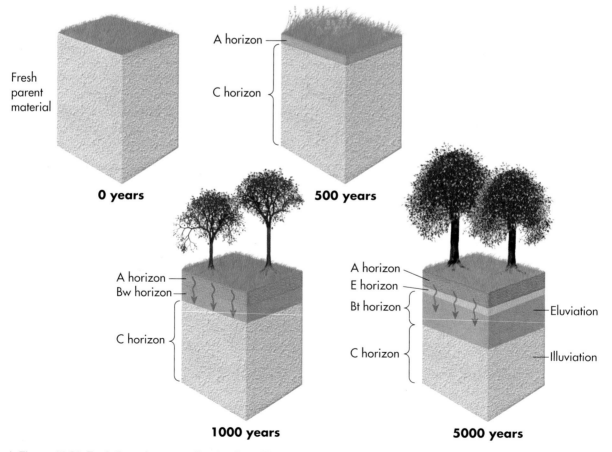

Figure 11.21 Evolution of a generalized soil profile through time. In an idealized situation, the development of A, B, and E horizons might take around 5000 years. Note the progression of horizon development at different periods of time.

The formation of soil horizons is complex because it involves the internal organization of sediment through a combination of additions, depletions, transformations, and translocations over a period of time. This organization is best understood when it is viewed in an animated way that demonstrates the progressive development of soil. To do so, go to the *Geo Media Library* and select *Soil Horizon Development*. Once you complete the animation, be sure to answer the questions at the end to test your understanding of this concept.

---

case, the inclusion of the lowercase "t" means that the horizon is texturally distinctive, through the illuviation of clay at that level, from the rest of the soil. Another way to make this distinction is to say that the B horizon is likely to be an **argillic horizon** with a strong blocky structure. This structure developed because abundant clay was added to this part of the soil through eluviation.

Soil Science and Classification

The previous section discussed the basic model of soil formation and the various horizons that evolve over time. Given that model, you can now examine some of the soil variability on Earth, which is known because of the research conducted by scientists working in the field of **soil science**. This scientific discipline specifically deals with soils as natural resources, including their physical, chemical, and biological properties.

Soil science is a discipline particularly relevant to humans because we use soils in countless ways. Of particular importance is that soils are the medium in which we grow our food. Thus, understanding the properties of soils is essential to industrialized agriculture and directly affects the quality and price of the food we buy. Soils are also the fabric in which we build our homes, buildings, highways, and myriad other structures. As a result, engineers must have a good understanding of soil characteristics in order to build structures that are properly supported. Soils are also important filters for surface and groundwaters and are thus highly relevant to hydrology, which focuses on the way that water moves on Earth. People who study soils are called *soil scientists* (Figure 11.1). The goal of soil scientists is to improve our understanding of the Earth's soils in order to preserve and efficiently utilize them.

One of the ways in which soil scientists study and manage soils is in association with the concept of soil classification. In general, classification is a common practice in many scientific disciplines because it makes it easier to remember attributes about the phenomena under study. Classification also provides a basis from which one category can be compared with another. Given the predictable pedogenic processes and factors that operate on Earth, soils are particularly well suited for classification.

Although many different classification systems are used around the world, the one used in the United States is called **soil taxonomy**. This system is based on the existing properties of a soil, such as color, texture, structure, and mineral content, which can be measured. Many other soil classification systems are genetic; that is, the soil scientist attempts to reconstruct how the soil evolved, even though the initial environmental conditions may not be known. Because soil taxonomy is a generic, logically based system, the classification scheme is hierarchical with several levels of general classifications, each with sublevels below it. With each successively lower level, fewer and fewer similarities appear between soil types.

The highest level within soil taxonomy is the **soil order**. Twelve soil orders occur around the world, each one distinguished on the basis of diagnostic horizons (Table 11.1) that meet certain criteria. These diagnostic horizons are the **epipedon** (from the Greek word *epi*, meaning "over" or "upon"), which is the horizon formed at the soil surface, and the subsurface horizons formed by removal or accumulation of material.

Soil orders are subdivided into several categories, which, from highest to lowest, are *suborders, great groups, families,* and *series.* To give you an idea of this hierarchy's

---

Argillic horizon *A B horizon that is enriched in eluviated clay.*

Soil science *The study of soil as a natural resource through understanding of its physical, chemical, and biological properties.*

Soil taxonomy *The method of soil classification that is based on the physical, chemical, and biological properties of the soil.*

Soil order *A group of 12 distinctive soils differentiated at the most general level.*

Epipedon *The uppermost horizon or horizons of a soil that is used for diagnostic purposes.*

TABLE 11.1 Well-Developed Mineral Soils

Order Name	Derivation	Characteristics
Alfisols	Nonsense syllable	Soils with high-base-content B horizon rich in eluviated clay. Found from low to subarctic latitudes.
Aridisols	Latin: *aridus* = dry	Soils in dry climates that are low in organic matter and frequently have subsurface horizons rich in calcium carbonate or soluble salts.
Mollisols	Latin: *mollis* = soft	Soils in semi-arid/subhumid grasslands in midlatitudes that have humus-rich A horizon and a B horizon that has a high base status.
Oxisols	French: *oxide* = oxide	Highly weathered soils in tropical environments that have low base status and subsurface horizon that has a high oxide concentration.
Ultisols	Latin: *ultimus* = last	Highly weathered soils in tropical and subtropical environments that have low base content and a subsurface horizon rich in eluviated clay.
Spodosols	Greek: *spodos* = wood ashes	Soils in cool, moist environments that have a B horizon rich in eluviated sesquioxides.
Vertisols	Latin: *verto* = turn	Tropical and subtropical soils with high base status that contain an abundant amount of expandable clay that swells when wet and shrinks when dry.
Organic Soils		
Gelisols	Latin: *gelare* = freeze	Largely organic soils that form in extremely cold environments where permafrost is thick. Due to repeated freezing and thawing, soil horizons and surface expression are chaotic.
Histosols	Greek: *histos* = tissue	Very dark soils consisting mostly of organic matter. Typically found in cool/moist environments where organic decomposition is slow.
Weakly Developed Soils		
Andisols	Spanish: Andes	Weakly developed soils formed within glassy volcanic sediments ejected by active volcanoes.
Entisols	Latin: *recens*-fresh	Horizonless soils usually formed within recently deposited sediments.
Inceptisols	Latin: *inceptum* = beginning	Soils with poorly developed horizons, but which may evolve further.

extent, consider that 50 suborders and 180,000 soil series are recognized in the United States. It is well beyond the scope of this text to describe each of the soil series that has been identified, but we will outline the basic characteristics of the 12 soil orders. This will provide a good overview of how soils differ and how pedogenic processes and factors affect soil development.

The Twelve Soil Orders

This portion of the soils discussion focuses on the 12 soil orders that occur within the soil taxonomy classification system. These soils reflect the maximum development that occurs in these orders in association with various combinations of the soil-forming factors: (1) climate, (2) organisms, (3) relief, (4) parent material, and (5) time. Note that two of these factors, climate and organisms, are distinctly environmental in nature. In other words, they reflect the specific climate and vegetation relationships in any given place or region. Given this close correlation, it should not surprise you that the global distribution of soil orders reflects these environmental factors and thus builds directly on the previous two chapters. Figure 11.22 shows the geographic distribution of soil orders on Earth. Notice the similarity of the patterns on this map and the global climate and vegetation maps seen earlier (see Figures 9.2 and 10.6, respectively). As you work your way through this discussion, try to keep these geographic patterns in mind.

Global Soil Regions

December 2000

Figure 11.22 The global distribution of major soil orders. This map is color-coded to reflect the geography of soil orders. Areas mapped as "Rocky Land," such as in central Asia, are places where large expanses of rock are exposed at the surface and soils have yet to form. Areas designated as "Shifting Sands," such as in the Sahara Desert, are places where the wind deposits sand in a way that precludes the development of soil. Areas mapped as "Ice Sheet/Glacier," such as in Antarctica and Greenland, are places where ice covers the ground and soils cannot form. (*Source:* U.S. Department of Agriculture.)

Soil Science and Classification • 285

As with the descriptions of climate and vegetation, the discussion of soil orders begins by starting at the tropics and proceeding to higher latitudes. A few soil orders can occur anywhere under the proper conditions and will be dealt with at the end of the discussion. While you are studying the various orders, refer back to Figure 11.22, which shows the geographic distribution of soil orders and is similar to the global climate and vegetation maps in Figures 9.2 and 10.6, respectively. You might want to familiarize yourself with these illustrations again before you begin your tour of soil orders. Remember that many different variations of these soil orders arise, given the unique interactions of the five soil-forming factors; this discussion provides only a basic overview of their character and distribution. The discussion of each order is supported by a figure that includes a map of its global distribution, an example of a typical landscape associated with the order, and an example of a soil profile of that order.

Oxisols **Oxisols** are soils that form in tropical environments and are found in about 8% of the ice-free land area on Earth (Figure 11.23a). These soils form through a specific soil-forming process associated with tropical environments called **laterization**, which means "to become brick-like." The premise of laterization is that the hot and wet environment causes a great deal of mineral weathering to occur within soils. Bases such as calcium, magnesium, and potassium ions are rapidly leached because they are easily mobilized and water consistently percolates through the soil. Percolating water flow is so steady that even otherwise-resistant minerals such as silica are leached. The only minerals not leached are iron and aluminum sesquioxides (minerals containing three

(a)

(c)

(b)

Soils of the Great Plains Slide Set, University of Nebraska.

NRCS/USDA

Figure 11.23 Oxisols. (a) Generalized map of Oxisols on Earth. These soils occur on about 8% of Earth's ice-free area. (*Source:* U.S. Department of Agriculture.) (b) Oxisol landscape in Hawaii. Note the reddish color of the soil, which occurs due to extensive weathering in this tropical environment. These soils are used to grow sugarcane. (c) A typical Oxisol. Notice the distinctive orange color of this soil, which occurs because soluble minerals such as calcium and magnesium are completely leached, leaving iron behind.

Oxisols *Mineral soils in tropical and subtropical environments that form through laterization and thus have an oxic horizon within 2 m of the surface.*

Laterization *A regional soil-forming process in tropical and subtropical environments that results in extensive eluviation of minerals except for iron and aluminum.*

atoms of oxygen for every two atoms of the metal), which results in highly weathered A and B horizons that together are called an **oxic horizon** because the sesquioxides are oxidized (i.e., atoms lose an electron). In addition, very little organic matter is added to the soil because decomposition of surface litter is rapid.

Oxisols are the most weathered of the soils formed by laterization. Given the oxidized character of these soils, they are usually red, yellow, or yellowish brown in color (Figures 11.23b and c). Well-developed Oxisols typically have a very thin A horizon and a thick Bo horizon, with the "o" reflecting the presence of oxidized sesquioxides. Although Oxisols are associated with the rainforest, they actually have very low fertility because most nutrients have been weathered out of the system. In addition, Oxisols have low cation exchange capacity, which further reduces fertility. Given this soil infertility, plants derive most of their nutrients from standing and decomposing vegetation.

As noted in Chapter 10, farmers in Oxisol regions can temporarily improve soil fertility by cutting and burning the rainforest. Once the plant debris is burned, the resulting organic residue collects in the upper part of the soil, providing nutrients for crops. Unfortunately, this method of soil fertilization is effective for only a few years in any given field, which requires that additional forest stands be cut to maintain crop production. This cause and effect is one of the principal driving forces leading to tropical deforestation.

Ultisols   **Ultisols** form in warm, moist, subtropical environments like those on the southeastern coasts of the United States and China. They also occur in the more humid parts within the tropical wet/dry climate regions. Overall they occur on about 9% of the ice-free land surfaces on Earth (Figure 11.24a). Recall from Chapter 10 that the vegetation in these tropical and subtropical areas consists of forest. As a result of this

(a)

(b)

(c)

NRCS/USDA

Alan Arbogast

Figure 11.24 Ultisols.   (a) Generalized map of Ultisols on Earth. These soils are found on 9% of Earth's ice-free land area. (*Source:* U.S. Department of Agriculture.) (b) An Ultisol landscape in central Brazil. Again, note the reddish color of this soil, which is due to extensive weathering in this subtropical environment. This region was once covered by forest and is now farmed. (c) A typical Ultisol. Note that most of the profile has an orange hue, somewhat similar to an Oxisol, because it also forms by laterization. Note the A horizon (dark zone) at the top of the soil and the increased structure in the underlying Bt horizon (reddish zone) due to eluviated clays.

Oxic horizon   *A diagnostic soil horizon in tropical and subtropical environments rich in iron oxides.*

Ultisols   *Mineral soils in subtropical environments that formed through laterization and thus are depleted of calcium and have an argillic horizon.*

environmental association, these soils form through a variant of the laterization system and are similar to Oxisols in their reddish color (Figure 11.24b) and low base concentration.

The primary difference between Oxisols and Ultisols is that Ultisols are a bit more fertile because they are not as heavily leached. This relative lack of weathering happens because Ultisols occur in regions that experience a dry season. Ultisols also contain a distinct argillic (Bt) horizon, resulting from the translocation of clays. Some Ultisols have Bt horizons that contain as much as 70% clay. They support productive forests but require the frequent addition of lime fertilizers if they are used for continuous agriculture. In this manner, many Ultisols in the southeastern United States have been farmed for over 200 years.

Vertisols　A third soil order that frequently occurs in the tropical and subtropical regions is **Vertisols.** These soils are found on about 2% of the ice-free land surface on Earth, including some areas outside of the tropics (Figures 11.25a, b). The primary soil-forming factor associated with Vertisols is parent material because they contain an abundance of expandable clays, which are clays that swell like a sponge when they are wet and shrink when they dry. Because these soils contain so much clay, fields that lie within Vertisol soils are typically dark with very strong structure.

Vertisols　*Soils that contain an abundance of expandable clay and thus swell and shrink during wet and dry cycles, respectively.*

(a)

(b)

(c)

(d)

Figure 11.25　Vertisols.　(a) Generalized map of Vertisols on Earth. These soils occur on about 2% of Earth's ice-free land area. (*Source:* U.S. Department of Agriculture.) (b) A Vertisol landscape in South Dakota. The small humps and ridges are known as "gilgai" and form due to frequent expansion and contraction of clays. (c) Ground cracks that resulted from contraction of expandable clays in a Vertisol. (d) A Vertisol in New Mexico. Note the strong structure (due to high clay content) in the upper part of the profile and the cracks in the ground. The horizontal grooves in the lower part of the profile were made by the backhoe that excavated the trench.

One of the distinctive characteristics of Vertisols is that prominent cracks can form in the ground (Figure 11.25c, d) during long dry periods because the clays within the soil shrink so much. Because of the constant shrinking and swelling that occurs in these soils, they are slowly but constantly mixed, with much of this *pedoturbation* occurring when fragments of soil fall into the cracks. The results of this shrinking and swelling can be seen in a profile through the presence of shiny clay skins on peds. These glossy surfaces, or *slickensides*, are created when the long-term expansion and contraction of the soil cause clays to be polished as they rub against one another. As a result, a diagnostic horizon of a Vertisol is a Bss horizon, with the "ss" reflecting the presence of slickensides.

If you have ever noticed that cracks form in the ground during long droughts around your home, then the soils in your neighborhood are likely Vertisols, or closely related to them. In most places, Vertisols occur in isolated pockets that are too small to show up on the global soils map. Significant regions of Vertisols form in the tropical and subtropical regions of the world, however, including the Deccan Plateau of western India,

along the Nile River in eastern Africa, and much of eastern Australia. Although these soils are often high in base content, they are difficult to farm because they are hard to cultivate.

Alfisols   **Alfisols** form in cooler parts of the moist continental climate zones that lie poleward of the subtropical regions. These soils occur on about 10% of Earth's ice-free land area over an extremely wide range of latitude, extending from as high as 60° N in North America to the equatorial zone in South America and Africa (Figure 11.26a). Large areas of these soils occur in places like the American Midwest (Figure 11.26b), northern Canada, Europe, Asia, and the east coast of China. Smaller pockets are found in Central America, Australia, and Africa. Given the relatively high moisture of these regions, the natural landscape is largely covered with deciduous forest, but extensive areas of boreal forest occur as well.

Alfisols   *Soils generally found in seasonal midlatitude regions that formed through podzolization and have an alkaline argillic horizon.*

(a)

(b)

Randall Schaetzl

(c)

Figure 11.26 Alfisols.   (a) Generalized map of Alfisols on Earth. These soils occur on about 10% of Earth's ice-free land area. (*Source:* U.S. Department of Agriculture.) (b) A typical Alfisol landscape in central Missouri. Prior to cultivation, the vegetation of this region was a mix of forest and tall-grass prairie. (c) An Alfisol from the northern Midwest. Note the light E horizon (white arrow), which is present in well-developed Alfisols. The reddish zone below the E is a Bt horizon rich in eluviated clays.

Now that you have examined the three tropical soils, you should review the relationship of climate, vegetation, and soil orders by going to the *Geo Media Library* and selecting *African Climate, Vegetation, and Soils*. This brief animation illustrates how these variables are interrelated to form a geographically distinct landscape. As you watch it, be sure to look at the spatial patterns that exist on the African continent. Once you complete the animation, be sure to answer the questions at the end to test your understanding of this concept.

Alfisols form by a specific soil-forming process called **podzolization**. Podzolization is similar to laterization because both processes are associated with abundant leaching in their respective humid environments. The primary difference between the two processes is that organic matter decomposes slowly where podzolization occurs because the environment is cooler. Thus, a distinct O horizon forms in the regions where podzolization is the dominant pedogenic process. This O horizon consists mainly of deciduous leaves and coniferous needles. This litter layer is slightly acidic, which causes infiltrating soil water to become even more acidic. As a result, organic acids are incorporated into the upper part of the soil to form a distinct A horizon. From there, they are leached to the B horizon.

The leaching of organic acids causes translocation of the iron and aluminum sesquioxides, which are not moved where laterization occurs. In this environment, podzolization results in a very well-defined E horizon (Figure 11.26c), which is ashy gray in color and is called an **albic horizon** because it is so light. Below this horizon, a reddish B horizon forms where eluviated iron and aluminum collect. This horizon is called a **spodic horizon** because of the high sesquioxide content.

Alfisols are the soils that form in the warmer regions where podzolization occurs, specifically in the areas where the litter layer typically consists of deciduous leaves. Alfisols are similar to Ultisols in that they contain an argillic Bt horizon that is rich in clay. These soils are not as heavily weathered as Ultisols, however, which results in Alfisols generally being more fertile because they have higher concentrations of base ions. Nevertheless, Alfisols contain a distinct E horizon that is lighter, through extensive eluviation of minerals and clays, than the overlying A and underlying B horizons. Because the rate of decomposition is much less in this cooler environment, the A horizon of Alfisols is generally better developed and darker than in Ultisols.

Spodosols **Spodosols** are acidic soils that are also associated with the podzolization process. These soils occur in about 3% of the ice-free land surfaces on Earth (Figure 11.27a). In contrast to Alfisols, Spodosols form in somewhat cooler, more humid regions. As a result, the vegetation in these regions is typically coniferous and hardwood forest (Figure 11.27b), and the litter layer is thus a mix of needles and leaves. As with Alfisols, this litter layer is acidic, which lowers the pH of percolating water sufficiently to mobilize organic acids and sesquioxides. In places where the parent material is sandy, soil formation occurs more rapidly because water percolates quickly through the solum.

Spodosols have a very distinct sequence of soil horizons that are readily visible (Figure 11.27c). The uppermost part of the soil usually contains a thin O horizon (needle-leaf litter) over a thin A horizon, which is very dark due to the slow decomposition of organic matter. In well-developed Spodosols, the E horizon is prominent because iron, aluminum, and base ions have all been eluviated, resulting in a distinct white or light-gray albic horizon. Directly beneath the E horizon is a particular type of spodic horizon, which is designated as a Bs horizon. This horizon is dark brown or reddish in color because it contains high amounts of

---

Podzolization *A regional soil-forming process in cool, humid environments that results in the eluviation of iron, aluminum, and organic acids to form well-developed E and Bs horizons.*

Albic horizon *A diagnostic horizon of podzolization from which clay and free iron oxides have been removed, resulting in a light-colored E horizon.*

Spodic horizon *A mineral soil horizon characterized by the illuvial accumulation of aluminum, iron, and organic carbon.*

Spodosols *Soils in cool, humid regions that form through podzolization and contain a spodic horizon enriched in eluviated iron, aluminum, and organic carbon.*

(a)

Randall Schaetzl

(c)

Alan Arbogast

(b)

Figure 11.27 Spodosols. (a) Generalized map of Spodosols on Earth. These soils occur on about 3% of Earth's ice-free land area. (*Source:* U.S. Department of Agriculture.) (b) A typical Spodosol landscape in the Great Lakes region. (c) A well-developed Spodosol in the Great Lakes region. Note the thick E horizon, which has distinct tongues (red arrows) that extend deeper into the soil. These features developed because water preferentially followed tree roots as it moved downward. The dark horizon (white arrows) directly below the E horizon is a Bhs horizon rich in eluviated organic acids and iron.

illuviated organic matter, aluminum, and iron. If sufficient time or eluviation has transpired, the Bs horizon will actually become cemented with eluviated organic material called *ortstein* (Figure 11.27c). In these instances, the B horizon will have a Bhs designation to reflect the presence of both humic acids and sesquioxides.

Histosols   Although the vast majority of soils possess a distinct mineral component, some soils are almost entirely organic. The organic soils usually fall into the soil order **Histosols**. Histosols are found on about 1% of Earth's ice-free land surfaces (Figure 11.28a), most of which are in cool, moist environments. These soils form in isolated shallow lakes, ponds, wetlands, and bogs (Figure 11.28b) where the decomposition of organic

matter is very slow as a result of the cool and moist environmental conditions.

Given the organic nature of Hisotols, the major horizons of Histosols are O horizons that are very dark in color (Figure 11.28c). Because these soils are extremely fertile, they are actively cultivated for crops, such as mint or cranberries, in places where they are easily accessible. They are also excellent soils for sod farms and certified organic farms. These soils are also used as mulch and a low-grade fuel source. In the American Midwest, Histosols are commonly referred to as *peat* or *muck*.

Gelisols   In 1998 the soil taxonomy system was revised to include a new soil order: **Gelisols**. Gelisols develop in very cold climates at high latitudes or high mountain elevations and

Histosols   *Organic soils that form in cool, wet environments where organic carbon decomposes very slowly.*

Gelisols   *Soils in subarctic and arctic environments that contain permafrost within 2 m of the surface.*

(a)

Randall Schaetzl

(c)

Alan Arbogast

(b)

Figure 11.28 Histosols. (a) Generalized map of Histosols on Earth. These soils occur on about 1% of Earth's ice-free land area. (*Source:* U.S. Department of Agriculture.) (b) A typical Histosol landscape in Michigan forms in bogs like this where organic matter decomposes slowly. (c) A Histosol from Upper Michigan. Note the very dark color of this soil, which reflects the high amount of organic matter. The lighter sediment visible at a depth of about 40 cm is a mineral deposit on which the soil formed.

are estimated to occupy about 9% of Earth's ice-free land area, including most of the state of Alaska (Figure 11.29a). Thus, these soils are frozen most of the year, resulting in permafrost within 2 m of the surface. Because these soils are very cold, the decomposition of organic matter proceeds even more slowly than where Histosols occur. As a result, Gelisols contain large amounts of organic carbon, some of which is in the form of muck or peat at the surface, and are thus dark in color (Figure 11.29b). Given that Gelisols regularly freeze and thaw, they are actively mixed and are thus frequently associated with deformed topography (Figure 11.29c).

Mollisols   Moving away from the humid zones, we next look at soils that form in more arid regions, beginning with **Mollisols**.

These soils occur over about 7% of Earth's soil network, including most of the American Midwest and Great Plains, a broad belt across Eurasia (including the Russian steppe), and the Pampas in South America (Figure 11.30a). Mollisols are closely associated with the more humid parts of the dry midlatitude climate and associated grassland vegetation (Figure 11.30b). They are the most extensive soil order in the United States and are found on 21.5% of the land area (Figure 11.30a). In the Great Plains, the presence of Mollisols is directly linked to the fact that the region lies within the Rocky Mountain rain shadow.

Mollisols form in association with a specific pedogenic process called **calcification**. As with podzolization, vegetation plays an important role in the calcification process. Where podzolization occurs, deciduous and coniferous trees produce a needle-leaf

---

Mollisols   *Soils that form through calcification and have a mollic epipedon that overlies mineral matter, which is more than 50% saturated with base ions.*

Calcification   *A regional soil-forming process in which calcium carbonate is cycled within the soil.*

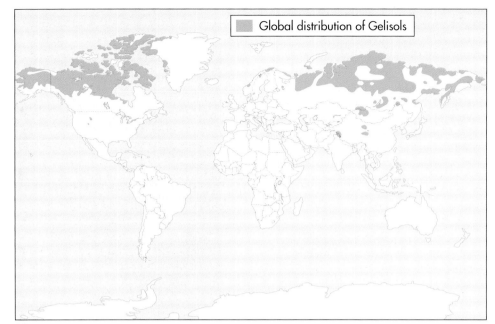

(a)

Figure 11.29 Gelisols. (a) Generalized map of Gelisols on Earth. These soils are found in about 9% of Earth's ice-free land area. (*Source:* U.S. Department of Agriculture.) (b) A Gelisol in Alaska. The lighter horizon at the base of the soil is permafrost. (c) A typical Gelisol landscape in Alaska. The distinct polygons on the ground, called *patterned ground*, form as a result of repeated freezing and thawing.

(b)

(c)

litter layer that depresses the pH sufficiently to mobilize iron and aluminum sesquioxides. In grassland environments, where calcification takes place, the grasses play an important role in soil formation because they are base cyclers. The first part of the cycle occurs when grasses absorb bases as nutrients during the growing season. When the grass dies, however, the bases contained within the grass are cycled back into the soil as it decomposes.

Here is where the role of climate becomes important. If this base cycling somehow occurred in a humid environment, the bases would be leached completely out of the soil because they are so easily mobilized. Given the general moisture deficit that occurs in places like the Great Plains or the Russian steppe, however, the bases are not completely leached. Instead, they are merely translocated to the zone of illuviation, or B horizon, just like iron and aluminum where podzolization occurs. Since this part of the soil is rich in calcium, it is called a **calcic horizon**. This process can ultimately result in the complete cementation (solidification) of the B horizon by calcium carbonate, resulting in material called *caliche*.

Mollisols are also distinctive because they have thick, dark A horizons with high humus content (Figure 11.30c). This horizon forms in part because grasses decompose slowly in the relatively dry environment, therefore slowly and consistently adding humus to the soil. Grass also has a dense underground root system that contributes to the high content of organic matter and calcium minerals in these soils. Soil organisms, like earthworms, ants, and moles, mix the soils, carrying organic matter deeper below the surface. Given the association with the calcification process, the B horizons of Mollisols are rich in calcium; the B horizon of a Mollisol is saturated at least to a level of 50% by bases such as calcium ions. It also often contains an abundance of translocated clays. As a result, the B horizon of a well-developed Mollisol is designated as a Btk ("k" meaning abundant calcium carbonate) horizon.

Calcic horizon *A diagnostic soil horizon of calcification enriched in illuviated calcium carbonate.*

(a)

(c)

(b)

Figure 11.30 Mollisols. (a) Generalized map of Mollisols on Earth. These soils are found on about 7% of Earth's ice-free land area. (*Source:* U.S. Department of Agriculture.) (b) A typical Mollisol landscape in the central Great Plains. Note the tall-grass prairie. (c) A Mollisol in North Dakota. Note the thick A horizon near the surface and the prominent Bk horizon (arrow) that is rich in eluviated calcium.

VISUAL CONCEPT CHECK 11.3

Which one of the regional pedogenic processes is contributing to the development of this soil? What soil horizons are present? How can you tell? How did each of them form? Here's a hint: This soil is not fully developed, but the evolving horizons can be seen.

The combination of high amounts of organic matter and minerals makes Mollisols among the most fertile and productive soils on Earth because they contain abundant nutrients that plants need. In the United States, Mollisols support a wide variety of agricultural crops, including corn, wheat, and soybeans. The next loaf of bread you buy, in fact, will most likely contain wheat grown in Mollisols.

Aridisols   As the name implies, **Aridisols** are soils that form in arid environments. These soils are found over about 12%

of Earth's ice-free land area, including the Sahara Desert and significant parts of Australia, South America, the Middle East, Asia, and the southwestern United States (Figure 11.31a). These dry environments have very sparse vegetation (Figure 11.31b), which means that Aridisols have poorly developed A horizons, if they have them at all (Figure 11.31c).

Aridisols can form through either the process of calcification or another process called **salinization**, which entails the movement of sodium (Na) in the soil. In the other pedogenic processes, such as laterization and podzolization,

(a)

Global distribution of Aridisols

(c) — H. Curtis Monger, New Mexico

(b) — Alan Arbogast

(d) — H. Curtis Monger, New Mexico

Figure 11.31 Aridisols.   (a) Generalized map of Aridisols on Earth. These soils occur on about 12% of Earth's ice-free land area. (*Source:* U.S. Department of Agriculture.) (b) A typical Aridisol landscape in the southwestern United States. (c) An Aridisol in New Mexico. Note the lack of a well-developed A horizon, which is a reflection of the desert environment. (d) An Aridisol with a calcic horizon (arrow) in New Mexico. This horizon developed through extensive eluviation of calcium over a long time.

Aridisols   *Mineral soils that form in arid environments and thus are poorly developed.*

Salinization   *A regional soil-forming process in which soluble salts are cycled within the soil.*

At this point we have completed our discussion of soils that generally form in association with regional environmental variables and pedogenic processes. These processes are particularly well suited for animation to more completely illustrate how they work. Go to the *Geo Media Library* and select *Regional Pedogenic*

Processes. This animation illustrates how the soil-forming processes laterization, podzolization, calcification, and salinization influence the development of soils at a regional level. When you complete the animation, be sure to answer the questions at the end to test your understanding of this concept.

---

sodium is easily leached from soils because it is one of the more soluble minerals, and water is sufficiently abundant to move it out of the solum. In semi-arid to arid environments, however, sodium is not completely leached from soils and instead is moved only part way down the soil profile when infrequent rains do fall. When the soil dries out during the extended dry periods that occur in this environment, the partially leached minerals rise back up through the profile in a process called *wicking*.

This wicking process is most active in enclosed basins where pools of water collect when it does rain. Sodium in these low spots can be leached many centimeters when the water percolates through the profile following a hard rain. When the site dries out, however, the salt rises back up and forms a salty crust at the surface called an Az horizon. If this crust reaches a thickness of 15 cm (6 in.) and contains at least 20 g/kg of salt, it is called a **salic horizon**. These salts are highly toxic to most plants. In regions where calcification is the dominant process associated with Aridisols, the soils have thick Bk, or calcic, horizons that can become completely cemented with time. If they do, the horizon is designated as a Bkm horizon (Figure 11.31d).

At this time, each of the three major soil orders that occurs within the midcontinental United States—Alfisols, Mollisols, and Aridisols—have now been discussed. This juncture thus marks a good point to examine holistically how soils vary along a study transect that extends from Ohio to central Colorado (Figure 11.32). Which side of this transect has the most moisture? Which side is drier? Do you know why? You should know that the eastern side of this transect is more humid due to the influx of mT air from the Gulf of Mexico. As you proceed westward, however, the climate generally becomes drier due to the influence of the Rocky Mountain rain shadow. The response of vegetation is that trees populate the eastern side of the transect. As you move westward, the vegetation shifts to tall grass and then to short grass.

How do soils differ along this transect? The primary agent of change is that more leaching takes place on the eastern side of this transect than on the west. This geographic pattern should make sense because the climate in Ohio is generally more humid, with more precipitation, than in eastern Colorado. As you can see in Figure 11.32, this distribution of soil water causes a distinct E horizon in Ohio, which fades away as you move farther west. The thickest A horizons are in the tall-grass region of the Mollisol belt because the biomass is so high, and thus decomposition and leaching are less than to the east. In this zone, the depth of illuviated carbonate (lime, $CaCO_3$) is the greatest and the B horizon is the thickest. As you move deeper into the rain shadow, in the western part of the Mollisol belt, the A horizon thins significantly and the depth of the B horizon becomes shallower. This pattern culminates within the zone of Aridisols with a very thin or nonexistent A horizon and a very shallow, carbonate-rich B horizon. This discussion is an excellent example of how many geographic variables come together to produce a distinctive pattern—in this case, regarding soils.

Entisols   Finally, it is time to examine the soils that are not related to any specific environmental variable or pedogenic process. In other words, these soils can be found in many different places on Earth and are a product of specific soil-forming factors. The most widespread of these more randomly distributed soils are **Entisols**, which means *young soils*. These soils occur on about 16% of Earth's ice-free land area (Figure 11.33a).

Although Entisols are soils in the sense that they can support plants, they are the least developed soils because they have no diagnostic horizons such as an E or a B horizon (Figure 11.33b). They may contain a thin A horizon. This overall lack of development can occur for several reasons: (1) the climate is extremely dry, making the processes of soil additions and translocations very weak; (2) the parent material is quartz sand, which is a resistant deposit that does not

---

Salic horizon *The diagnostic horizon of salinization that forms due to the recrystallization of secondary salts.*

Entisols *Soils that are very weakly developed and thus have no distinct horizonation.*

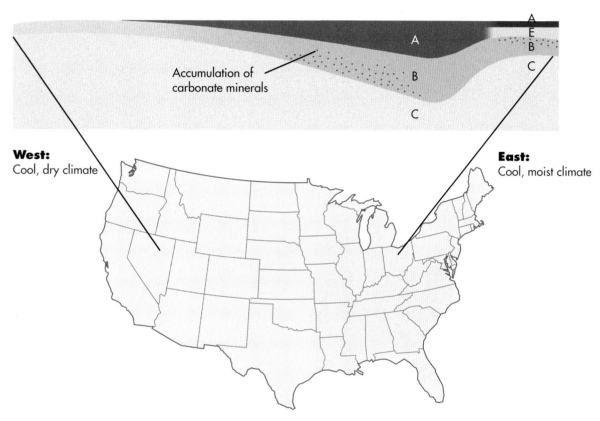

West:
Cool, dry climate

East:
Cool, moist climate

Figure 11.32 Midcontinent soil transect in the United States. The change in soil characteristics is reflected by soil order, from central Nevada in the west to Ohio in the east. (*Source:* Modified from C. E. Millar, L. M. Turk, and H. D. Foth, *Fundamentals of Soil Science*. New York: John Wiley & Sons, 1990.)

www.wiley.com/college/arbogast

WileyPLUS

North American Climate, Vegetation, and Soils

At this point, you should have a basic understanding of midlatitude soil orders and where they occur. In this context, it's a good time to review these relationships by going to the *Geo Media Library* and selecting *North American Climate, Vegetation, and Soils*. This contains a brief animation that illustrates how these variables are interrelated with one another to form a geographically distinct landscape. As you watch the animation, be sure to pay attention to the geographic patterns and their similarities. Once you complete the animation, be sure to answer the questions at the end to test your understanding of this concept.

(a)

NRCS

(b)

(c)

Figure 11.33 Entisols. (a) Generalized map of Entisols on Earth. These soils occur on about 16% of Earth's ice-free land area. (*Source:* U.S. Department of Agriculture.) (b) An Entisol in southern Idaho. Note the very thin, if existent, A horizon in this soil. (c) An Entisol landscape in Michigan. The vegetation is growing on a recent deposit of windblown sand along Lake Michigan.

Alan Arbogast

easily yield nutrients through weathering; or (3) the parent material has recently been transported, resulting in little time for a soil to develop (Figure 11.33c). Given that this range of soil-forming factors can occur just about anywhere, Entisols appear at all latitudes.

Inceptisols **Inceptisols** are soils with weakly developed horizons. These soils are found on about 10% of Earth's ice-free land area (Figure 11.34a), and can occur just about anywhere, even in places like the Appalachian Mountains (Figure 11.34b). It is useful to think of Inceptisols as being

---

Inceptisols *Soils that have one or more weakly developed horizons due to some alteration and removal of soluble minerals.*

a step in formation above Entisols, which have limited horizon development. The primary soil-forming factor associated with Inceptisols is time. On the one hand, these soils are better developed than Entisols because they have had more time to develop. Thus, Inceptisols may have a weak E horizon (Figure 11.34c). On the other hand, insufficient time has elapsed for the soil to develop horizons that meet the specific criteria of a better-developed soil order such as Mollisols, Spodosols, or Ultisols. Another important soil-forming factor associated with Inceptisols is parent material because most of these soils form in sediments that have been recently transported, but not so recently that the soil shows no sign of development. As in the case of Entisols, the soil-forming factors that contribute to the formation of Inceptisols can occur just about anywhere.

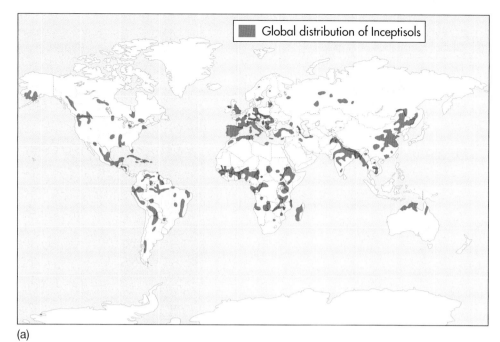

Global distribution of Inceptisols

(a)

Randall Schaetzl

(c)

Alan Arbogast

(b)

Figure 11.34 Inceptisols. (a) Generalized map of Inceptisols on Earth. These soils occur on about 10% of Earth's ice-free land area. (*Source:* U.S. Department of Agriculture.) (b) An Inceptisol landscape. Soils on these steep slopes within the Appalachian Mountains have weakly developed horizons because they are easily eroded. (c) An Inceptisol in Wisconsin. Note the weak E horizon (arrow) that indicates some eluviation of iron has occurred.

Andisols   **Andisols** are soils that form in parent material that is more than half volcanic ash. Like Entisols and Inceptisols, Andisols occur at scattered places on Earth, occupying about 1% of Earth's ice-free landmass (Figure 11.35a). Most of Hawaii, for example, lies within areas classified as Andisols (Figure 11.35b). Given their volcanic origin, it is accurate to say that the most important soil-forming factor associated with Andisols is parent material because fresh deposits of volcanic ash are deposited on the landscape following a volcanic eruption. These deposits consist mostly of glass-like shards that frequently contain a high proportion of carbon. As a result, these soils are usually dark in color (Figure 11.35c) and quite fertile. Given the high frequency of volcanic eruptions in some areas, Andisols may be poorly developed because the soils have had insufficient time to develop.

Human Interactions with Soils

Throughout much of recorded history, soil has been a key environmental factor around which human societies revolved. For most people living in the United States today, however, soil is a vague

Andisols   *Soils formed in parent material that is at least 50% volcanic ash.*

thing that lies beneath the ground on which we walk. This lack of awareness is due in large part to major demographic changes in the country over the past 200 years. In 1800 approximately 90% of the labor force was farmers intimately familiar with the soil because they worked with it every day. By 1900 the percentage of the labor force devoted to agriculture had dropped to about 38%. Today, farmers comprise less than 2% of the American workforce. In contrast, over 80% of Americans live in urban environments where soils are rarely considered by the average person.

In spite of the indifference most people have toward the ground beneath them, the fact is that soils are very important to everyone. In terms of agriculture, understanding the character of soil is essential for a variety of reasons directly related to crop yields and food production (Figure 11.36). For example, the texture of soil is important because it influences how rapidly water drains through the soil. In general, farmers desire loamy-textured soils like a silt-rich Mollisol because they hold draining water long enough (but not too long) for crops to be consistently well watered. Likewise, soils with good structure are usually preferred because they have higher porosity and better aeration, and crop roots can spread easily through the ground.

Another important factor that farmers consider relating to soils is their fertility. As we have seen in tropical regions, soil fertility is an

(a)

NRCS/USDA

(c)

Marbut Collection/Soil Science Society Collection

(b)

Figure 11.35 Andisols. (a) Generalized map of Andisols on Earth. These soils are associated with volcanic regions and occur on about 1% of Earth's ice-free land area. (*Source:* U.S. Department of Agriculture.) (b) Typical Andisol landscape in Hawaii. The Hawaiian Islands are covered in most places with volcanic sediments. (c) Andisol in Hawaii. Note the banding in this soil, which reflects periodic deposition of new volcanic parent material.

important issue there because Oxisols and Ultisols are heavily leached of their nutrients by the abundant rainfall. As a result, subsistence farmers clear their land and burn the forest residue (Figure10.29) to temporarily increase soil fertility. Soil fertility is also a major concern in the agricultural regions of the United States, such as the Midwest and California's Central Valley. These regions are very important in the United States because they provide much of the food Americans consume. Substantial amounts of food are also exported from these regions to countries around the world. As you will see in Chapter 20, the Central Valley of California is known as the nation's "salad bowl" because most of our vegetables are grown there. In a similar vein, the Midwest is the location of the important corn and wheat belts (Figure 10.33) that provide much of our food. A large part of this zone overlaps with the fertile Mollisols (as well as some Alfisols) shown in Figure 11.22. If you consider that the average American consumes over 55 kg (~125 lb) of wheat per year, as well as over 23 kg (50 lb) of corn, it is easy to see why soils in this region are heavily farmed.

Given the intense pressure on the nation's agricultural soils, it is essential to maintain high levels of fertility to insure sustainable levels of inexpensive food and to meet increased demand as population grows. During the commercial agricultural era, which intensified in the United States after World War II, the

primary way to increase fertility and crop yields has been to apply commercial fertilizers loaded with calcium, magnesium, potassium, and nitrogen (Figure 11.37a). Another technique used to raise fertility is the application of animal manure. Such

Figure 11.36 Soils and agriculture. (a) Workers harvesting lettuce near Yuma, Arizona. In order to produce food efficiently, farmers must have a keen understanding of the character and variability of soils.

UIG via Getty Images

(a)

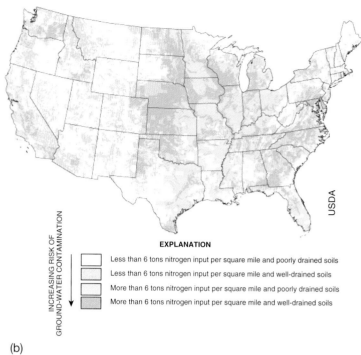

(b)

Figure 11.37 Fertilizing soils. (a) Typical scene of a farmer fertilizing the soil In a farm field. In an effort to boost production, fertilizers such as potassium, nitrogen and magnesium are added to a field to maintain and enhance soil fertility. (b) Map of risk for groundwater contamination by nitrates from the application of fertilizer in the contiguous United States. Note that the agricultural heartland of the country is particularly at risk for groundwater contamination.

EXPLANATION

INCREASING RISK OF GROUND-WATER CONTAMINATION

- Less than 6 tons nitrogen input per square mile and poorly drained soils
- Less than 6 tons nitrogen input per square mile and well-drained soils
- More than 6 tons nitrogen input per square mile and poorly drained soils
- More than 6 tons nitrogen input per square mile and well-drained soils

applications have proven very effective in enhancing soil fertility, increasing yields by as much as 50% in some places. In conjunction with these commercial fertilizers, chemicals are also routinely applied to control weeds and pests. Unfortunately, there is a down side to such heavy application of chemicals to farms. One major problem, for example, is nitrate contamination of groundwater supplies. Nitrate comes from nitrogen, which is a plant nutrient supplied by inorganic fertilizer and manure. As we will discuss in Chapter 16, groundwater is water stored below the Earth's surface, within the ground. These supplies are often important sources of drinking water in many communities.

As you can see in Figure 11.37b, the extent of nitrate groundwater contamination is very high in California's Central Valley and throughout the agricultural Midwest. Also note that it is highest in well-drained soils, which makes sense given the way that soil texture influences water infiltration. Given concerns about nitrate groundwater contamination, many farmers are looking for other ways to maintain and increase soil fertility without applying chemical fertilizers. One method employed is the practice of crop rotation, which involves alternating the kinds of crops grown on fields over the course of time. Wheat has a tendency to rigorously pull nutrients from soils and thus deplete them, but if such a field is occasionally converted to legumes such as soybeans, alfalfa, or peas, nitrogen is naturally added to the soil and fertility is enhanced. Another successful method used to reduce reliance on chemical fertilizers is thorough recycling of crop residues following harvest. These stalks and leaves are then plowed into the soil the next growing season, enhancing soil fertility.

Aside from the obvious significance of soils as the medium in which our food grows, soils are also important from an engineering perspective because the foundations of buildings and other structures are often built within them. As we have seen, a critical variable is the amount and character of the clays that may be contained within the soil. Vertisols, in particular, contain a high percentage of expandable clays that swell and shrink during wet and dry periods,

respectively. Structures built on Vertisols often have weak foundations and walls because the soil can be expanding or collapsing around them. This process can be so strong that it actually causes a basement wall to collapse or a wall to crack like the one shown in Figure 11.38, which costs thousands of dollars to repair.

Given the obvious significance of soils to people, it is important to understand soils well and care for them. The agency responsible for mapping and managing soils in the United States is the Natural Resources Conservation Service (NRCS), which is part of the United States Department of Agriculture (USDA). The origins of the NRCS can be traced to the infamous *Dust Bowl* era of the 1930s when soils in the central Great Plains were devastated by the combined impacts of overplowing, drought, and wind erosion. The impacts of the Dust Bowl will be discussed in Chapter 18, but for now it is relevant to consider that perhaps 800 million tons of topsoil were blown from the region because the native (and protective) grassland had been eliminated by unsustainable farming practices.

Figure 11.38 Cracked outside wall in a home built on clay-rich soils. This house was severely damaged because the surrounding soil contains expandable clays, which can either expand like a sponge when wet or lose cohesion and collapse when they dry.

Soil Survey of Pulaski and Wilcox Counties, Georgia

USDA NRCS

(a)

(b)

U.S. Department of Agriculture

Figure 11.39 County soil surveys. (a) The cover of the soil survey for Pulaski and Wilcox counties, Georgia. Survey guidebooks such as this one are available for most counties in the United States. (b) A typical soil map in a soil survey. Note the labels in distinct map areas, or polygons. Each label represents a specific soil type (or series) in the area. Each series has distinctive characteristics, such as the degree of development, color, texture, and structure, to name a few.

As a result of this environmental catastrophe, the federal government established the Soil Erosion Service in 1933, which later became the Soil Conservation Service in 1935. In 1994 the agency acquired its current name, the Natural Resources Conservation Service.

The NRCS has an office in each county within the country and employs field agents who consult with farmers, engineers, and other residents who have questions about how they are utilizing soil. A key reference published by the NRCS is the county *Soil Survey*, which discusses the character of soils in the county and presents its geographic distribution in a series of accessible maps (Figure 11.39). Soil surveys are often updated using refined mapping techniques that allow farmers to practice precision agriculture, which maximizes yields by targeting specific crops to particular soil variations within fields. These kinds of applications improve agricultural efficiency, which, in turn, enables farmers to produce increasingly greater amounts of food in a way that keeps costs low for you.

KEY CONCEPTS TO REMEMBER ABOUT SOIL SCIENCE AND CLASSIFICATION

1. Soil science is a distinct subdiscipline of geography that focuses on the chemical, physical, and biological properties of soils.

2. One of the primary ways that soil scientists study and compare soils is through soil classification. Soils are classified (in the United States) through a hierarchical system called *soil taxonomy*. The most general classification in soil taxonomy is the soil order, of which 12 orders are recognized. They are distinguished on the basis of measured physical properties and horizons.

3. Oxisols and Ultisols form as a result of laterization, which involves the intense weathering of soils in tropical and subtropical environments. These soils are typically reddish in color because minerals besides oxidized iron and aluminum are leached.

4. Alfisols and Spodosols form due to podzolization, which involves extensive translocation of organic acids and sesquioxides in cool, humid environments. Alfisols and Spodosols have distinctive E horizons that are light gray in color. Spodosols also have a distinctive Bs horizon that forms due to the illuviation of iron and aluminum.

5. Histosols are soils that are entirely organic because they form in cool, humid places where the decomposition of organic matter is very slow.

6. Mollisols and Aridisols form in drier environments. Both can form because of calcification, which occurs where base ions are partially translocated in the soil. Aridisols can also form due to salinization, which involves the movement and surface recrystallization of sodium.

7. Andisols, Entisols, Gelisols, Inceptisols, and Vertisols are directly related to local environmental factors. Andisols form in volcanic environments, whereas Gelisols occur in subarctic regions where the soil is frozen most of the year. Entisols and Inceptisols can form anywhere and represent soils that are very weakly to slightly better developed, respectively. Vertisols are soils that contain abundant expandable clays that swell when wet and shrink when dry.

1. Soil consists of the outermost layer of Earth and forms through the complex interaction of additions, transformations, translocations, and losses. These processes occur in various combinations that depend on the five soil-forming factors: (a) climate, (b) organisms, (c) relief, (d) parent material, and (e) time.

2. Soils have a variety of distinctive characteristics, including color, texture, structure, pH, and cation exchange capacity, that can be measured and compared from one soil to another.

3. Soils are organized into horizons that form by distinctive processes. The O horizon is the uppermost horizon and consists of freshly added organic matter, such as leaves.

The A horizon contains decomposed organic matter called humus. The E horizon forms through the eluviation of minerals, which recrystallize and collect through the process of illuviation in the underlying B horizon. The C horizon is unaltered parent material, which, in bedrock landscapes, is known as regolith.

4. Soils in the United States are classified on the basis of their genetic properties in a scheme called *soil taxonomy*. The highest level of this classification system is the soil order. Twelve soil orders are recognized: Oxisols, Ultisols, Alfisols, Spodosols, Histosols, Vertisols, Gelisols, Mollisols, Aridisols, Entisols, Inceptisols, and Andisols.

CHECK YOUR UNDERSTANDING

1. How is regolith related to bedrock and soil?

2. Describe the soil-forming factors and provide an example of how they interact to produce soil.

3. Compare and contrast the concepts of soil texture and soil structure.

4. Which texture will allow water to drain more rapidly: sand or clay? Why?

5. What are the various aspects of soil chemistry that are used to characterize soil?

6. Which parent material will have a higher cation exchange capacity—one that is sandy, or one that is rich in clay? Explain your answer.

7. Explain how the various soil horizons form.

8. Which pair of horizons is created through the process of translocation? In what way does this evolution occur?

9. It is conceivable that a soil could be classified as an Entisol even though it is very "old." In what environment—tropical, subtropical, or arid—would this kind of weak development most likely occur? Why?

10. What is a soil profile, and why is it useful for determining the characteristics of a soil?

11. How are Spodosols and Oxisols similar? How are they different?

12. Why is grass an important part of the calcification pedogenic model?

13. In which parent material would you most likely find an Inceptisol: a sand dune that is 500 years old or one that is 1000 years old? Why?

14. Provide an example of how soils are important to agriculture and how they are managed.

ANSWERS TO VISUAL CONCEPT CHECKS

VISUAL CONCEPT CHECK 11.1

The answer is *b*. The photograph shows an example of soil additions. When leaves fall to the ground, they slowly decompose and become incorporated into the soil.

VISUAL CONCEPT CHECK 11.2

The answer is *d*; this soil is classified as a sandy clay.

VISUAL CONCEPT CHECK 11.3

This soil is forming by the podzolization pedogenic process, which results in the eluviation of iron, aluminum, and organic acids. The soil contains developing A, E, and Bs horizons. You can identify the A horizon because it is dark, which occurs through the addition of humus. The E horizon is whitish and forms through the eluviation of iron and aluminum. The Bs horizon is visible beneath the E horizon as the light orange zone. This color reflects the illuviation of iron.

CHAPTER TWELVE

EARTH'S INTERNAL STRUCTURE, ROCK CYCLE, AND GEOLOGIC TIME

The previous chapter discussed the formation and character of soils, which can be viewed as the Earth's skin. Now it is time to look more intently at the solid Earth by focusing on its internal structure and composition. The inner Earth is important to understand because processes far below the surface drive the movements of our planet's outermost layer, the crust. Geographers are interested in the composition of the crust because it is the part of the solid Earth that most directly affects people, such as when an earthquake devastates a community. This chapter will also give you an appreciation for geologic time and why it is relevant, especially when you consider how quickly Earth is now changing due to human impacts.

The Grand Canyon is one of the best known landscapes on Earth. In addition to being a beautiful place, the canyon is interesting because it contains rocks that span much of the Earth's history. This chapter focuses on the geologic history of Earth, the evolution and composition of rocks, and the character of continents and ocean basins.

CHAPTER PREVIEW

Earth's Inner Structure

Rocks and Minerals in the Earth's Crust

 Identification of Igneous Environments

 Clastic Rocks

 The Rock Cycle

Geologic Time

 Geologic Time

Human Interactions with Geologic Time: The Anthropocene

▌LEARNING OBJECTIVES

1. Compare and contrast the various structural layers of Earth.

2. Explain the differences among igneous, sedimentary, and metamorphic rocks and discuss the various kinds of rocks that occur in each of these categories.

3. Discuss the concept of radiometric dating and how it is used to estimate the age of rocks.

4. Describe the geological timescale and how it is subdivided.

5. Holistically explain how the evolution of landforms is tied to the rock cycle and geologic time.

6. Describe the general history of petroleum exploration in the industrial period and how it may change in the future.

7. Discuss the Anthropocene and why it may soon become a formal epoch in the geological timescale.

Earth's Inner Structure

Have you ever wondered what the inside of Earth is like? Science fiction writers have often pondered this question, with the most famous example being Jules Verne's famous book *Journey to the Center of the Earth*. In this story, a professor leads a party of explorers on a wild adventure to the center of Earth through a volcano in Iceland. Although such an adventure is pure fiction, most geologists would certainly like to make the trip. Unfortunately, the only way to actually see a portion of the Earth's interior is to core into it with a large drill rig and collect samples. Although such studies are informative, the deepest geologists have ever been able to drill into Earth is only about 12 km (7.5 mi). The distance to the center of Earth is thousands of kilometers, however, which means that geologists must rely on indirect evidence to construct models about the Earth's interior.

Much of what is known about the Earth's structure is based on how **seismic waves** travel through Earth after earthquakes occur. Cooler areas within Earth transmit these waves at higher speeds than hotter zones because cooler regions are more rigid. Areas that are denser absorb seismic waves, whereas other variations in density cause seismic waves to be reflected or bent. These differences have allowed scientists to determine the nature of Earth's interior and map its basic structure, somewhat like the explorers in the Jules Verne novel. The following discussion outlines that character of the layers associated with the Earth's structure.

The Major Layers

With a cross-sectional view of Earth, you can see that the inner Earth contains several major layers (Figure 12.1). Because it is difficult to imagine the depths and distances associated with these layers, it is useful to compare them to the distances between places on a map of North America (Figure 12.2). For example, the distance to the center of the Earth's core—in other words, to the center of Earth—is about 6370 km (3963 mi), which happens to be the approximate distance from Anchorage, Alaska, to Miami, Florida.

The Earth's innermost layer is the **inner core**. This part of Earth has a radius of about 1220 km (760 mi),

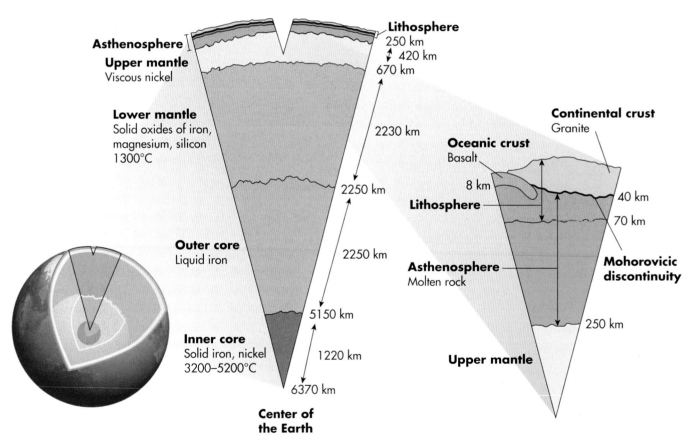

Figure 12.1 Earth's interior structure. The Earth's interior structure consists of several major layers, each with a distinct mineral composition and density. Note that the middle image is a blown-up diagram of the pie-shaped cross section in the left-hand diagram. The right-hand image, in turn, is a blown-up diagram of the pie-shaped cross section in the upper part of the middle image.

Seismic waves *Vibrations that travel through the Earth when stress is released in an earthquake.*

Inner core *The inner part of the Earth's core. This area is about 1220 km (760 mi) thick and consists of solid iron and nickel.*

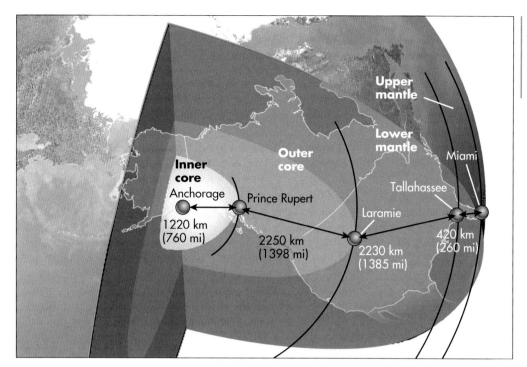

Figure 12.2 Distance to the center of Earth. The distance to the center of Earth is about the width of the North American continent between Anchorage, Alaska, and Miami, Florida.

which is the approximate distance from Anchorage to Prince Rupert, British Columbia (Figure 12.2). The inner core is composed mostly of solid iron, along with a little bit of nickel. The pressure in this part of Earth is enormous, about 3 million times greater than the air pressure you feel at the surface, due to the weight of the overlying material comprising the rest of Earth. As a result, the inner core is solid even though the temperature there is well above the melting temperature of iron. Rocks simply cannot melt under such high pressure. Geologists estimate that the temperature of the inner core ranges from about 3200°C to 5200°C (5800°F to 9400°F).

Surrounding the solid inner core is the **liquid outer core** (Figure 12.1). The top of this layer lies about 2900 km (1800 mi) below sea level. The liquid outer core is approximately 2250 km (1398 mi) thick, which is about the same distance as from Prince Rupert, British Columbia, to Laramie, Wyoming (Figure 12.2). In other words, it would take two or three days to drive that distance at a reasonable pace. The outer core is composed of the same material as the inner core, except the iron is molten; that is, it is mostly liquid. This variation occurs because the temperatures of the outer core approach those of the inner core, hot enough to melt solid rock. The difference between the two layers is that the outer core has less internal pressure, which allows rock to melt. The outer core is important because it generates at least 90% of the Earth's magnetic field and the resulting magnetosphere that protects Earth from the solar wind. The

magnetic field may exist because the Earth's inner core rotates 19 km/year (12 mi/year), which is faster than the rest of the planet. This rotational difference is thought to cause circulation patterns in the outer core that generate electrical currents. These currents, in turn, may generate the magnetic field. Scientists are not certain of the exact processes that lead to the Earth's magnetic field, but circulation of the outer core is most likely involved.

Working out from the inner core, the next layer encountered is the **mantle**, which surrounds the core (Figure 12.1). This part of Earth is composed largely of solid iron, magnesium, and silicon oxides and is divided into two parts: the *lower mantle* and *upper mantle*. The lower mantle is about 2230 km (1385 mi) thick—in other words, about the same distance as from Laramie to Tallahassee, Florida (Figure 12.2). Again, this distance would take two or three days to drive. Although temperatures in the mantle are still quite hot, over 1300°C (2370°F), they are much cooler than in the core. These cooler temperatures, coupled with intense pressure, cause the lower mantle to be solid. This pressure lessens gradually, along with temperature, into the upper mantle.

The upper mantle is approximately 420 km (260 mi) thick, which is about the same distance as from Tallahassee to Miami, Florida. In contrast to the lower mantle, which is solid, the upper mantle consists mostly of viscous nickel; that is, it is a material like very thick syrup or a slowly flowing plastic. The upper part of the upper mantle is called the

Liquid outer core *The outer part of the Earth's core. This area is about 2250 km (1400 mi) thick and consists of molten iron and nickel.*

Mantle *The layer of the Earth's interior that lies between the liquid outer core and the crust. This area is about 2900 km (1800 mi) thick and consists largely of silicate rock.*

asthenosphere, which occurs between about 40 km and 250 km (25 mi and 105 mi) below Earth's surface. This part of the inner Earth is the least rigid portion of the mantle because it contains scattered zones of high temperature due to radioactive decay. These zones comprise about 10% of the asthenosphere and consist of molten rock that moves very slowly by convection. This process is highly influential on the Earth's surface, causing earthquakes, volcanoes, and deformation of rocks to form mountain chains.

The uppermost layer of Earth is the **lithosphere** (Figure 12.1). This portion of Earth extends from the surface into the uppermost part of the asthenosphere at a depth of 70 km (44 mi). It contains the *crust*, which is the cool, stiff, and brittle exterior of Earth. The boundary between the crust and the asthenosphere is very well defined and is called the **Mohorovicic discontinuity** (pronounced Mo-ho-RO-vi-chich; Moho for short), named for the Yugoslavian scientist who first suggested its presence in 1909. This boundary is known because earthquake waves change speeds dramatically at the Moho level as a result of the great difference in density between the mantle and crust.

The crust is the thinnest of the major Earth layers, ranging in thickness from 8 km to 40 km (about 5 mi to 25 mi). This thickness represents about 1% of the Earth's overall structure and is equivalent to driving from downtown Miami to a distant suburb in the metropolitan area. It floats on top of the mantle because its overall density is less. The crust contains two parts: oceanic crust and continental crust. **Oceanic crust** is about 8 km (about 5 mi) thick and consists mostly of a rock called *basalt*, which is very fine-grained and high in silica, magnesium, and iron; thus, oceanic crust is often called *sima*, which is short for silica and magnesium. In contrast, **continental crust** is composed largely of granite, which is a coarse-grained rock high in silica, aluminum, potassium, calcium, and sodium. As a result, continental crust is often called *sial*, short for silica and aluminum. Continental crust is fundamentally thicker and less dense than oceanic crust, averaging about 40 km (25 mi). It is thicker beneath mountain areas, where it can reach depths of 50 km to 60 km (31 mi to 37 mi). In nonmountainous areas, however, continental crust is only about 30 km (19 mi) thick. As you will see in the next chapter, the difference in composition between continental and oceanic crust has played a critical role in the overall history of Earth.

With the layers of Earth in mind, take a closer look at the relationship among the asthenosphere, lithosphere, and crust.

Remember that these layers blend into one another, with the lithosphere being the transition between the asthenosphere and crust. Note in Figure 12.1 that the depth of the Moho level is greater beneath continents than under ocean basins. The reason for this pattern is fairly straightforward, as shown in the sequence of diagrams in Figure 12.3. Figure 12.3a represents a period of time during which intrusions of magma rise up from the mantle into the overlying continental crust. In this way, the overall elevation and mass of the continental crust increase, perhaps due to mountain building. This increased mass increases the pressure on the underlying asthenosphere, which is plastic, relative to the ocean basin. As a result, the depth of the Moho level increases beneath the continent.

Over a very long period of time, the elevation of the continent gradually decreases due to erosion (Figure 12.3b). This process transports sediments to the continental margin, where they are deposited. As a result, the mass of the continent slowly decreases and the underlying asthenosphere begins to bounce back (or rebound). At the same time, the asthenosphere beneath the continental margin begins to subside because of the weight of the new sediments. This process continues until the continent is nearly leveled (Figure 12.3c), which results in the almost complete rebound of the underlying asthenosphere. At the continental margin, however, the asthenosphere subsides even more due to the increased weight of sediments transported from the landmass. This overall process of subsidence and rebound of the asthenosphere is called *isostatic adjustment*, or *isostacy* for short.

KEY CONCEPTS TO REMEMBER ABOUT THE EARTH'S INTERNAL STRUCTURE

1. Earth's interior is divided into seven major layers. These layers are the inner core, outer core, lower mantle, upper mantle, asthenosphere, lithosphere, and crust.

2. The inner core and lower mantle are solid rock bodies composed mostly of iron. The outer core is liquid iron and the upper mantle viscous nickel.

3. The Earth's upper three layers are of most interest to geographers because these layers contribute directly to the development of surface landforms and thus affect

---

Asthenosphere *The layer of very soft rock that occurs in the upper part of the upper mantle. This region is about 40 km to 250 km (25 mi to 105 mi) below the surface of Earth. The soft character of this rock allows isostatic adjustments to occur.*

Lithosphere *The outer, solid part of Earth that is about 70 km (44 mi) thick and includes the uppermost part of the asthenosphere and the crust.*

Mohorovicic discontinuity *The boundary between the Earth's crust and the upper part of the asthenosphere; seismic waves change speed at this boundary.*

Oceanic crust *Basaltic part of the Earth's crust that makes up the ocean basins. Oceanic crust is about 8 km (5 mi) thick and is also called sima because it consists largely of silica and magnesium.*

Continental crust *Granitic part of the Earth's crust that makes up the continents. Continental crust averages about 40 km (25 mi) in thickness and is also called sial because it consists largely of silica and aluminum.*

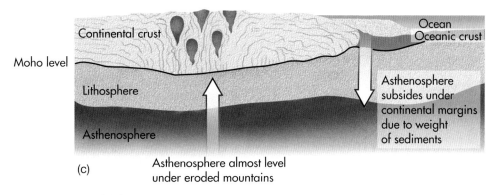

Figure 12.3 Isostatic adjustment. (a) Continental crust is thickest where mountain ranges occur. Given the extreme weight of these locations, the underlying asthenosphere subsides relative to the area beneath the oceanic crust. (b) As mountains weather, the weight of the landmass decreases, allowing isostatic rebound of the asthenosphere. At the same time, the asthenosphere beneath the continental margin begins to depress due to the increased weight of the freshly deposited sediments eroded from the continental interior. (c) As the continent weathers further, the isostatic rebound of the underlying asthenosphere continues. At the same time, the asthenosphere is further depressed due to the increased weight of margin sediments.

peoples' lives. The asthenosphere is the lowermost of these layers and extends to a depth of 250 km (155 mi). The upper part of the asthenosphere forms the lithosphere, which contains the Earth's crust.

4. The Earth's crust is divided into oceanic and continental crusts. The oceanic crust is composed mostly of dense, fine-grained basalt, whereas continental crust consists mostly of coarse-grained granite.

What two types of crust occur in landscapes that look like this? Which of the two is denser and why?

Digital Vision/Getty Images

Rocks and Minerals in the Earth's Crust

Have you ever picked up an interesting-looking rock when walking along the beach or hiking in the mountains? Have you ever noticed a fantastic exposure of rock along a highway? If so, perhaps you wondered how the rocks formed, where they came from, or how old they were. In most cases the rocks were very old and contained elements that had been recycled many times throughout Earth's history. Some of these minerals may have even been in the center of Earth at one time. This part of the chapter examines the basic kinds of rocks on Earth and how they form.

Rocks are composed of a variety of Earth elements. Figure 12.4 shows the approximate percentage by weight of the various elements, including oxygen and silicon, with lesser amounts of aluminum, iron, calcium, sodium, potassium, and magnesium. These elements combine in various ways to form **minerals**, which are naturally occurring substances with distinctive chemi-

cal and atomic configurations that usually take some kind of crystalline form. A great example of a crystalline mineral that you may be familiar with is quartz, which typically occurs as a clear, six-sided prism (Figure 12.5).

When minerals become bonded in a solid state, they form **rock**. Most rocks are very old by human standards, with the oldest being over 3.5 billion years old! Nevertheless, new rock is probably being created someplace at the very instant that you are reading this chapter. Is a major volcanic eruption occurring on

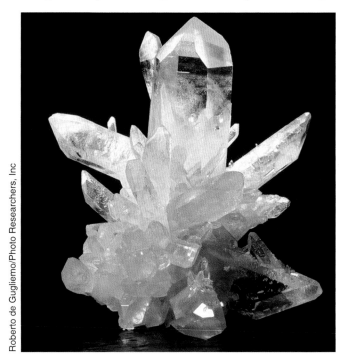

Roberto de Gugliemo/Photo Researchers, Inc

Figure 12.5 Quartz crystals. Quartz is an excellent example of a crystalline mineral. In small grains, it forms sediments in deserts, river deposits, and coastal areas. It is also a common part of most rocks formed in the continental crust.

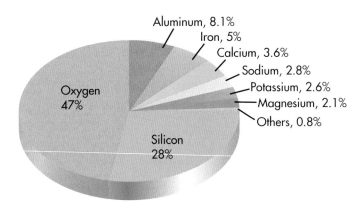

Figure 12.4 Chemical elements in the Earth's crust. Percentages refer to the percentage by weight of each element in the Earth's crust.

Oxygen 47%
Silicon 28%
Aluminum, 8.1%
Iron, 5%
Calcium, 3.6%
Sodium, 2.8%
Potassium, 2.6%
Magnesium, 2.1%
Others, 0.8%

Minerals *Naturally occurring substances with distinctive chemical configurations that usually manifest themselves in some kind of crystalline form.*

Rock *An amorphous mass of consolidated mineral matter.*

Earth right now that is spreading lava across the surface, or did a large flood recently happen that deposited mud in river valleys? If such an event is taking place, then the foundation is being set for the development of solid rock at those locations in the future.

Because of the many different chemical elements that occur on Earth, an extremely wide variety of rock types exist. It goes far beyond the scope of this book to describe each one of the many different kinds of rocks. Instead, we will focus on the three major rock classes—igneous, sedimentary, and metamorphic—and some fundamental kinds of rock that occur within these groups. After completing this section of the chapter, you should be able to identify these basic rocks, determining whether they lie randomly on the ground or are exposed at the surface in a **rock outcrop**. The presence of a particular rock type yields important clues about the geologic history of the area.

Igneous Rocks

One of the distinctive categories of rocks on Earth is that associated with liquid rock, known as **magma**, that originates in the mantle and moves upward into the Earth's crust or onto the surface (Figure 12.6a). When this magma cools, it solidifies to form **igneous rocks**. You are probably most familiar with magma that

flows across the surface of Earth as red-hot *lava* (Figure 12.6b) that erupts from volcanoes. This liquid rock ultimately cools to form **extrusive igneous rock**, which is so named because it results from magma that flows out of—or extrudes from—Earth.

In contrast to the magma that flows onto the Earth's surface and cools, the majority of magma cools within Earth (Figure 12.6a) to form **intrusive igneous rock**, so named because the magma intrudes into older rocks. This cooling process can occur in a variety of places beneath the Earth's surface, with any body of intrusive igneous rock referred to as a **pluton**. The largest plutons are called *batholiths*, which are features of no known depth that are up to several hundred kilometers long and 100 km (62 mi) wide. A smaller, somewhat rounded pluton that is attached to a batholith is called a *stock*.

In some instances, magma will move upward from a magma chamber into relatively small fissures within the overlying rock. One such feature is a *dike*, which is a pluton that consists of a wall-like feature that develops when magma intrudes into a vertical rock fracture. Sometimes magma moves upward through such a fracture and then spreads horizontally within the crust along zones of rock weakness, forming a pluton called a *sill*. In other cases, upwardly moving magma spreads horizontally along a zone of rock weakness, while at the same time pushing upward in such a way

(a)

R.W. Decker, courtesy USGS Hawaiian Volcano Observatory

(b)

Figure 12.6 Various kinds of igneous environments. (a) This diagram illustrates the different ways that intrusive and extrusive igneous rocks form. (b) Lava flow at Mauna Loa Volcano in Hawaii. This material will ultimately cool and darken, consistent with the dark material surrounding it, forming extrusive igneous rock.

Rock outcrop   *A place where rocks are exposed at the surface of Earth.*

Magma   *Molten rock beneath the surface of Earth.*

Igneous rocks   *Rocks that form when magma rises from the mantle and cools, either within the Earth's crust or on the surface.*

Extrusive igneous rocks   *Rocks that form when magma cools on the Earth's surface.*

Intrusive igneous rocks   *Rocks that form when magma cools within the Earth's crust.*

Pluton   *An extremely large mass of intrusive igneous rock that forms within the Earth's crust.*

This section on igneous rocks contained a number of terms that relate to specific processes and places within the Earth's crust. Now is a good time to review these terms in an interactive way. Go to the *Geo Media Library* and select *Identification of*

Igneous Environments. This animation is a drag-and-drop exercise that will test your ability to associate terms with respective igneous features. Once you complete the exercise, answer the questions at the end to test your understanding.

---

that the surface rocks are warped upward in a dome-like feature. When this warping magma cools, it forms a *laccolith*.

Regardless of whether the igneous rock is intrusive or extrusive, it consists largely of silicate minerals that contain

crystalline chemical compounds dominated by silicon and oxygen atoms. Although many different kinds of silicate minerals occur, seven are most common (Figure 12.7). Three of these minerals are *felsic*, which means (among

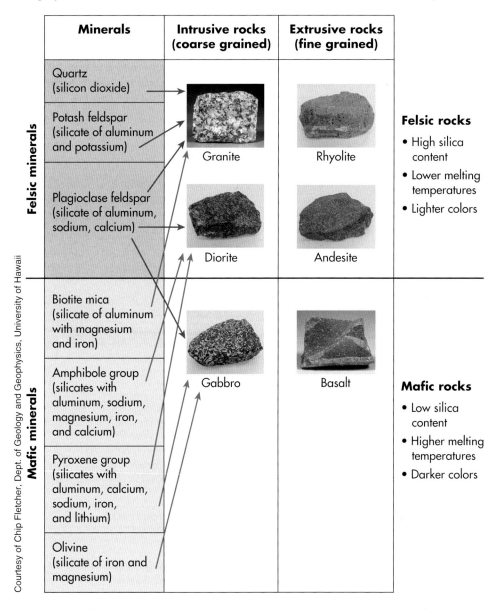

Figure 12.7 Silicate minerals and igneous rocks. This diagram shows the most important groups of silicate minerals, as well as some common igneous rocks. Notice that various combinations of minerals produce different rocks, both in extrusive and intrusive forms.

other things) that they have (1) high silica content; (2) relatively low melting temperatures; and (3) lighter colors. The remaining four of the seven silicate minerals are *mafic*, which essentially have opposite characteristics to felsic minerals.

Figure 12.7 indicates how silicate minerals can combine in different ways to form various kinds of igneous rocks. A good example on the mafic side of the diagram is the intrusive rock *gabbro*, which forms when plagioclase feldspar mixes with minerals from the pyroxene group and olivine. If the same minerals combine in an extrusive environment, the mafic rock is called *basalt*. On the felsic side of the diagram, a good example is the intrusive rock *granite*, which forms when potash feldspar combines with plagioclase feldspar, biotite mica, and quartz. In an extrusive setting, these minerals combine to form *rhyolite*. Note the other kinds of combinations that can occur.

You may wonder how you can tell the difference between extrusive and intrusive igneous rocks if they are made from the same materials? The answer is that extrusive and intrusive rocks differ fundamentally in their appearance. Intrusive igneous rocks have coarse crystalline grains that you can see with your unaided eye. These coarse textures evolve because magma cools relatively slowly, perhaps over several thousand years, when it is trapped within the Earth's crust, such as in a sill, dike, or laccolith (see Figure 12.6a). A good example of a coarse-grained igneous rock is *granite* (Figure 12.7). Granite is a felsic rock that has distinctive dark grains consisting of the minerals *amphibole* and *biotite*, pinkish-tan grains of *feldspar*, and clear grains of *quartz*.

In contrast to intrusive igneous rocks, extrusive igneous rocks reach the surface and cool relatively quickly, sometimes in a matter of days. A good example of an extrusive igneous setting can be seen in Figure 12.6b, which shows a lava flow in Hawaii. Given that the lava cools relatively quickly in this setting, silicate minerals do not have time to congregate with one another, as they do in intrusive igneous rocks. The result is a fine-grained rock. A common kind of extrusive igneous rock is *basalt*, which is a dark mafic rock (Figure 12.7). Perhaps the most distinctive extrusive igneous rock is *obsidian* (Figure 12.8). Obsidian is also called *volcanic glass* and forms when nongaseous lava cools so fast that crystals cannot form.

Sedimentary Rocks

Now let's turn to a second major rock class, which is **sedimentary rocks**. Although some forms of sedimentary rock have organic origins, most consist of vast quantities of formerly loose minerals that collected in some kind of depositional

Doug Martin/Photo Researchers, Inc.

Figure 12.8 Obsidian. Obsidian is an extrusive igneous rock that develops when nongaseous lava cools very quickly, before crystals can develop. Prehistoric Native Americans prized this rock for its rich, dark color and smooth glassy surface.

setting, such as a river floodplain, shallow seafloor, interior valley, lake, or marsh (Figure 12.9). These minerals can arise from any of the three major rock groups—igneous, sedimentary, or metamorphic (to be discussed later)—but their ultimate source is usually some form of igneous rock that was weathered in a way that liberated countless individual particles such as clay (from feldspar) or sand (from quartz). These particles could then be transported as **sediment** by wind, water, or glaciers to a new place, where they subsequently accumulate as sedimentary deposits. Most sedimentary rocks are the remnants of sediments that accumulated in an oceanic environment. Many of these sediments were eroded from the continents in various ways and then carried to the ocean where they settled to the bottom.

After the sediment accumulates, all the grains gradually become cemented to one another in a gradual process called **lithification** to form rock. This process occurs because water is squeezed out of the lower layers of sediments when they are compacted by the cumulative weight of the deposits above them. As a result of compaction and associated water loss, minerals such as calcium and silica recrystallize in the sediments and essentially glue them together to form rock.

Within the overall category of sedimentary rocks, three major classes of sediments occur: clastic, chemical, and organic. Table 12.1 lists the basic characteristics of these rocks.

Sedimentary rocks *Rocks that form through the deposition and lithification of small fragments or dissolved substances from other rocks or, in some cases, marine animals.*

Sediment *Solid fragments of rocks that are transported to some location and deposited by wind, water, or ice.*

Lithification *The process whereby sediments are cemented through compaction to form rock.*

1) Sediments erode from continent

2) Transport of sediment by streams

3) Sediments settle on sea floor

4) Burial and lithification

5) Eventual uplift of rock layers

Different kinds of sediment accumulate in distinct layers

Figure 12.9 Marine sedimentary environments and rock formation. This generalized diagram illustrates how clastic marine sediments ultimately become rock: (1) sediments erode from continents; (2) sediments are transported by streams into the ocean; (3) sediments settle to the ocean bottom; (4) the deposited sediments are buried and slowly lithify to become rock; and (5) the lithified sediments eventually uplift as new rock layers.

TABLE 12.1	Some Common Types of Sedimentary Rocks	
Class	**Rock Type**	**Composition**
Clastic: Formed from rock or mineral fragments	Conglomerate	Coarse-grained sandstone that contains rocks and pebbles of different sizes
	Sandstone	Sand grains cemented by minerals
	Siltstone	Silt particles cemented by minerals
	Claystone	Clay particles cemented by minerals
	Shale	Fine-grained particles that often contain fossils; most abundant sedimentary rocks; forms layers along which rock can split
	Limestone	Calcium carbonate, including that from animals and microorganisms; forms rock layers
Chemical: Formed from mineral precipitates from seawater or salty lakes	Dolomite	Magnesium carbonates, mostly formed by chemical replacement in limestone
	Evaporites	Minerals (gypsum, rock salt, calcite) left in deposits after evaporation of water
	Chert or flint	Extremely fine grains of silica (quartz) formed in layers or nodules within limestone
Organic: Formed from carbon-based organic matter	Coal	Carbon-based materials lithified by heat and pressure
	Oil (petroleum)	Liquid hydrocarbon trapped in sedimentary deposits; important fossil fuel
	Natural gas	Gaseous hydrocarbon trapped above oil in sedimentary deposits; important fossil fuel

Clastic Sedimentary Rocks   Perhaps the easiest type of sedimentary rock to understand is the category of clastic sedimentary rocks. These rocks contain mineral fragments derived from any of the three major rock groups that accumulate and lithify. The most common minerals in clastic sedimentary rocks are silicates, such as quartz and feldspar. Quartz is by far the most important clastic sediment, primarily because it is hard and very resistant to alteration. On the finer end of the textural scale, clays can be weathered from feldspars and carried by wind and water to some point where they are deposited.

Several types of clastic sedimentary rocks occur on Earth. **Sandstone** is created when individual sand grains are deposited in thick layers by water along a beach or river, or by the wind within dunes. This collection of sand grains then lithifies, forming rock. When the sandstone also contains abundant rocks and pebbles of various sizes, it is called **conglomerate** (Figure 12.10).

An excellent place to see beautiful sandstone outcrops is the southwestern United States, with one of the best known rock bodies being the Navajo Sandstone (Figure 12.11). The sediments comprising this rock accumulated in sand dunes approximately 175 million years ago when the region was a desert and sand was blown and deposited by the wind. Subsequently, the climate became more humid, resulting in rivers that buried the dunes with stream sediments. Still later, the region was covered by an ocean that deposited tremendous quantities of marine sediments. As a result of the compression and cementation of sands in the old dunes, they gradually became rock.

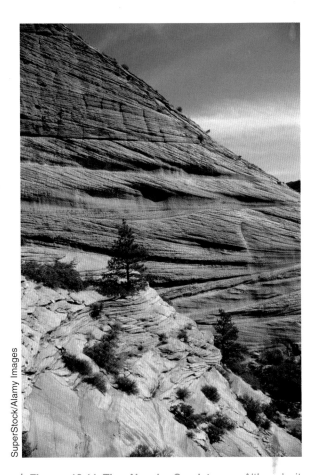

SuperStock/Alamy Images

Figure 12.11 The Navajo Sandstone. Although it may be hard to believe, this landscape was once an active dune field in what is now Zion National Park, Utah. You can tell that the sands accumulated in dunes because of the distinct bedding lines exposed in the outcrop. Over a long period of time, the sand grains in the dunes lithified to form this rock body.

Andrew J. Martinez/Photo Researchers, Inc

Figure 12.10 Conglomerate rocks. Conglomerate is a type of rock consisting of large clasts contained within a matrix of fine-grained particles. Note in this photo the mix of coarse grains, cobbles, and fine sediments. Such a mix of particles likely accumulated in a stream environment.

In addition to the sedimentary rocks created from individual sand grains, rocks can also be composed of finer grained silts and clays. When fine grains dominate a sedimentary deposit, it is referred to as *mud*. Silts and clays are easily suspended in and transported by moving water. Thus, mud typically accumulates in bodies of relatively still water, such as a coastal bay or lagoon (Figure 12.12a). After they are deposited in these settings, they lose a great deal of their original volume through compaction. The origin of these rocks and their associated names are logically based: *Siltstone* is dominated by silt-sized particles, and *claystone* consists largely of clays. The most common fine-grained sedimentary rock is **shale**, which is a claystone that easily breaks into small flakes and has a platy appearance (Figure 12.12b).

Sandstone   *A sedimentary rock created when individual sand grains are deposited in thick layers by wind or water and lithify.*

Conglomerate   *Sandstone that contains a wide variety of particle sizes.*

Shale   *Sedimentary rock consisting of lithified clay-sized sediments.*

Ron Erwin/All Canada Photos/Getty Images

(a)

(b)

Alan Arbogast

Figure 12.12 Formation and character of shale. (a) Mud collectively consists of individual clay particles, along with some silts, which settle out of suspension in places where water is very calm. If these sediments remain in place, they will gradually lithify to form shale. (b) Close-up of shale. Note the flaky appearance of this rock type.

www.wiley.com/college/arbogast

WileyPLUS

Clastic Rocks

Clastic rocks can be easily identified based on the size of sediment grains contained within them. In order to get a better feel for these rock types, go to the *Geo Media Library* and access *Clastic*

Rocks. This animation is a simple interactivity that allows you to better visualize the character of clastic rocks. To test your understanding, be sure to answer the questions at the end of the exercise.

Chemically Precipitated Sedimentary Rocks When environmental conditions are favorable, thick deposits of mineral compounds can accumulate at the bottom of the ocean or inland lakes. In ocean environments, this process is associated with the precipitation and recrystallization of calcium carbonate from seawater. This deposition can occur for a variety of reasons, including increase in water temperature, intense evaporation, upwelling of water, and bacterial decay, to name a few.

An excellent place to see the formation of chemically precipitated sedimentary rocks is the Bahama Islands, which lie off the eastern coast of Florida. Figure 12.13 is a satellite image that clearly shows the precipitation of calcium carbonate in this island chain. In this image, carbonate appears as the whitish plumes within the matrix of turquoise and bluish waters that lie between the islands. The turquoise and bluish colors appear because the water is shallow in the island complex and the satellite sensor can see the ocean bottom. Away from the islands, however, the water rapidly deepens, which is reflected by the dark ocean colors. Once the carbonates

accumulate on the ocean floor, they subsequently lithify to form rock. The Bahama Islands are made of calcium carbonate deposited in this manner.

In other situations, organisms rich in calcium carbonate such as clams, mussels, and oysters accumulate on the ocean bottom after they die. Subsequently, they become cemented when calcium carbonate precipitates and settles within the matrix of animal remains. The resulting rocks can contain abundant fossils (Figure 12.14), which enable geologists to reconstruct environmental conditions at the time the sediments were initially deposited. Regardless of the particular depositional process, whether it is entirely through carbonate precipitation or due to the accumulation of marine organisms, the resulting rocks are typically classified as **limestone**. If the deposits are rich in magnesium, the resulting carbonate rock is called

---

Limestone *Sedimentary rock consisting of over 50% calcium carbonate (CaCO₃).*

Figure 12.13 Carbonate depositional environments in the Bahama Islands. In this region, carbonate minerals are precipitating in the shallow waters of the island chain. They form the whitish haze that surrounds the islands.

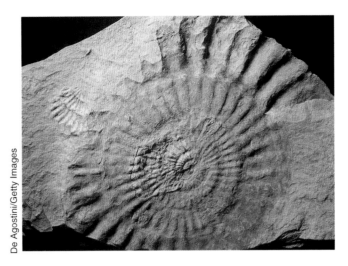

Figure 12.14 Fossils in limestone. This rock formed tens of millions of years ago when dead ammonites accumulated on the ocean bottom and calcium carbonate subsequently precipitated within the matrix of shells. The combined materials then lithified to form limestone.

dolomite. Under more unusual conditions, silica can precipitate from solution at the bottom of the ocean to form *flint* or *chert*.

In addition to forming in marine environments, chemically precipitated rocks also develop within continental locations when minerals evaporate from concentrated solutions (Figure 12.15a). These minerals collect at the surface and are called **evaporites**. As you can imagine, evaporates are most likely to evolve in semi-arid to arid regions such as the southwestern United States (Figure 12.15b).

(a)

(b)

Figure 12.15 Formation of evaporites. (a) Evaporites form when minerals recrystallize on the exposed surface after waters evaporate from shallow oceans or inland lakes. (b) The whitish deposits in this image are salts and other highly soluble minerals that recrystallize as a crust when evaporation occurs in this extremely arid environment of Death Valley.

Dolomite *Sedimentary rock consisting of over 50% calcium-magnesium carbonate (CaMg[CO₃]₂).*

Evaporites *Surface salt residue that collects through the evaporation of water and the crystallization of sodium.*

Organic Sedimentary Rocks As the name implies, organic sedimentary rocks consist of carbon-based materials that accumulate in thick deposits at the surface of Earth. An example of this process was shown in Chapter 11, specifically the accumulation of plant and organic matter as peat in cool and moist environments. Such thick accumulations of solid organic remains have occurred on a large scale over time, only to be buried by other sediments. When these organic deposits are buried progressively deeper, they slowly compact under the weight of the overlying sediments. In addition, they begin to be heated because temperature slowly rises with increased depth of burial. Over the course of millions of years, the amount of water and oxygen in the carbon progressively lowers and the deposit gradually solidifies, resulting in **coal**. Sometimes these deposits are cooked so much that they liquefy to form **petroleum** (crude oil). **Natural gas** comes from the remains of microscopic plants that live in the surface waters of the ocean. When they die, they settle to the bottom of the ocean where they decompose and form gas. This form of gas is typically about 85% methane and 10% ethane, with smaller amounts of propane and butane. In places where organic sedimentary rocks occur, they are usually found in distinct beds that are separated by layers of limestone, sandstone, or shale. These deposits frequently accumulate in domes that form when rocks are bent due to compression for one reason or another; this process will be discussed more thoroughly in Chapter 13.

In addition to coal and natural gas, petroleum is especially vital in modern civilization. Although humans have used petroleum in some way since at least the 4th century, when the upper classes in ancient Babylon burned it to provide light, the modern history of oil began in the mid-1800s when the process through which it could be distilled to kerosene was discovered in Canada. The first oil well in the United States was drilled in western Pennsylvania in 1859 and produced about 20 barrels of crude per day, with each barrel containing about 160 L (42 gal) of fluid. Demand for oil grew rapidly throughout the late 1800s, primarily for kerosene use in oil lamps. With the advent of internal combustion engines in the early 20th century, demand skyrocketed. Today it is by far the most dominant energy source for automobiles in the form of refined gasoline. In addition, many consumer products, such as computers, televisions, cell phones, and plastic bottles, to name a very few, are made with it. To meet this demand, over 80 million barrels of oil are required every day worldwide. The United States alone consumes over 18 million barrels of oil per day. In short, the global economy in its present form is utterly dependent on petroleum.

Metamorphic Rocks Igneous and sedimentary rocks can be altered after they develop. This alteration results in **metamorphic rock**, which occurs when a former igneous or sedimentary rock is subjected to intense heat or pressure within Earth over millions of years. Thus, the name of the rock comes from a Greek word meaning "to change form." (A good example of metamorphosis is when a caterpillar changes into a butterfly.) Figure 12.16 shows the various ways in which metamorphic rocks form. One way occurs when rocks are slowly compressed

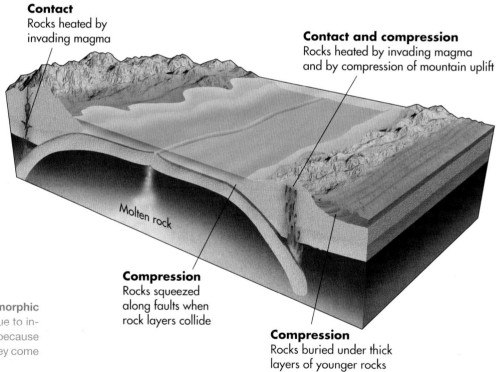

Contact
Rocks heated by invading magma

Contact and compression
Rocks heated by invading magma and by compression of mountain uplift

Molten rock

Compression
Rocks squeezed along faults when rock layers collide

Compression
Rocks buried under thick layers of younger rocks

Figure 12.16 Formation of metamorphic rocks. Metamorphic rocks evolve due to intense pressure of overlying rocks or because rocks are heated dramatically when they come into contact with igneous bodies.

Coal *A solid fossil fuel that consists of carbonized plants and animals.*

Petroleum *Naturally occurring oily liquid that consists of ancient hydrocarbons.*

Natural gas *Naturally occurring mixture of hydrocarbons that often occurs in association with petroleum and is found in porous geologic formations.*

Metamorphic rocks *Rocks that form when igneous, sedimentary, or other metamorphic rocks are subjected to intense heat and pressure.*

FOSSIL FUELS

From a scientific perspective, geology is a fascinating discipline because it provides important insights into the evolution of Earth. It also has many practical applications that are important to our industrial way of life. An excellent example of geology's relevance is our dependency on organic sedimentary rocks. These rocks are particularly important to humans because they are the source of fossil fuels—coal, natural gas, and petroleum—which are the overwhelming source of power for industrial activities.

World Energy Consumption by Type

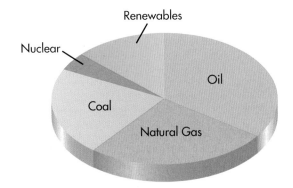

Renewables

Nuclear

Oil

Coal

Natural Gas

A typical coal-fueled power plant. Look at the amount of coal required to keep this plant humming.

Time & Life Pictures/Getty Images

If you use electrical devices of any kind, the "juice" used to power them was most likely derived from burning coal at your local power plant. The amount of coal consumed to power the electrical grid is staggering. For example, a typical 500 MW power plant requires over 3000 tons of coal per day to sustain operations. Another way to look at is that if you were to burn one 100 watt light bulb 24 hours a day for a year, the nearby power plant would require over 300 kg (~700 lbs) of coal to power that light! Natural gas is another important fuel source because it is also used to generate electrical power. In fact, most power plants in the United States built since the 1990s are powered by natural gas. You are probably most familiar with this fuel source because you use it to cook and heat your home. On the other hand, did you know that the typical American family probably consumes over 2800 m^3 (100,000 ft^3) of natural gas per year to keep their home comfortable?

How many times have you purchased refined petroleum in your life so far?

Bloomberg via Getty Images

due to the thick accumulation of overlying sediments or because they are deeply buried by crustal processes. Another way that alteration occurs is through *contact metamorphism*, which occurs when rocks are heated dramatically when they come into contact with invading magma in batholiths, dikes, or sills such as those illustrated in Figure 12.6. Whatever kind of alteration

(a)

(b)

Figure 12.17 Some metamorphic rocks. (a) Schist forms when the metamorphic rock slate is further altered. (b) Gneiss evolves when igneous or sedimentary rocks are dramatically heated. Note the banding in this rock where minerals aligned under the intense heat.

takes place, the result is that mineral components within the rock are rearranged to form different mineral varieties, or the minerals recrystallize into new mineral forms.

Although many different kinds of metamorphic rocks occur, just a few will be described here. *Slate* forms when shale is heated and compressed, forming a grayish, smooth rock that is quite hard. You may be familiar with slate because it is used as roofing shingles, patio flagstones, chalkboards, and the surface of the best pool tables. When slate is further heated and compressed, so that distinct planes and coarse texture develop, it becomes *schist* (Figure 12.17a). Limestone that has undergone metamorphism becomes *marble*, which has a distinctive appearance because mineral impurities show up as swirling bands. Marble is popular because it is frequently used in expensive tabletops and flooring. Finally, *gneiss* is a hard, banded metamorphic rock (Figure 12.17b) that forms when igneous or sedimentary rocks have been in close contact with intrusive magmas. The banding occurs because minerals become aligned during the metamorphic process. These kinds of rocks are described as being *foliated*, whereas those that lack banding are *nonfoliated* (Figure 12.17a).

The Rock Cycle

Recall from earlier chapters that water and carbon move from one reservoir on Earth to others as part of the hydrologic and carbon cycles, respectively. In a similar way, minerals contained within

VISUAL CONCEPT CHECK 12.2

This image shows rock outcrop along a highway. The lower arrow points to a shale bed, whereas the upper arrow focuses on a limestone rock unit. Which one of the following choices best explains this combination of rocks?

a) They are igneous rocks.
b) They were modified by intense heat and pressure.
c) These units reflect fluctuating sea levels in the past.
d) They formed when sandy sediments accumulated along a beach.

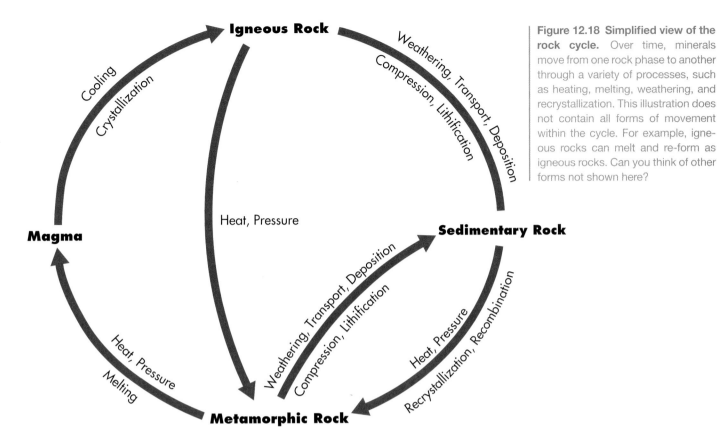

Figure 12.18 Simplified view of the rock cycle. Over time, minerals move from one rock phase to another through a variety of processes, such as heating, melting, weathering, and recrystallization. This illustration does not contain all forms of movement within the cycle. For example, igneous rocks can melt and re-form as igneous rocks. Can you think of other forms not shown here?

rocks move in a *rock cycle* (Figure 12.18). Although movement within this cycle is much slower than other cycles, often over millions of years, it nevertheless comprises a distinct series of steps where rock gradually moves from one physical state or location to another.

Let's examine the rock cycle from a holistic perspective by examining Figure 12.18 and starting with the formation of an extrusive igneous rock such as basalt. This rock originates from magma that flows out of the ground, usually from a volcano, but perhaps from just a fissure in the ground. This magma cools and crystallizes, forming basalt that lies at or close to ground

level. With time, the basalt gradually weathers due to various erosion processes, causing individual grains of the former rock mass to be transported and deposited someplace else. As the deposit of sediments thickens, those at the bottom of the stack become compressed, causing lithification and formation of sedimentary rocks. If these sedimentary rocks become deeply buried over time, or come into contact with the intense heat of an intrusive magma body, they will be altered to metamorphic rock because the original sediments will be recrystallized, recombined, or replaced. These metamorphic rocks, in turn, may be further heated so that they melt back to the magma state.

WileyPLUS

The Rock Cycle

Although the rock cycle moves very slowly, it is possible to animate the concept to capture its essence. Go to the *Geo Media Library* and select *The Rock Cycle*. This animation describes the ways in which rocks are created on Earth and how their minerals move from one reservoir to another. It begins with the

formation of igneous rock and then describes how these rocks can be altered to form new kinds of rock, whether it be a new kind of igneous rock, sedimentary rock, or metamorphic rock. After you complete the animation, be sure to answer the questions at the end to test your understanding of this concept.

If the sedimentary rocks remain near the surface, they too will be weathered, causing sediments to be loosened from the host rock mass. These sediments will then accumulate someplace else to form rock of a potentially different character. Remember that these are possible directions that the rock cycle might take. It is possible that sedimentary rocks may be re-melted directly to magma or that metamorphic rocks may be weathered to form new sedimentary rock.

A good example of how the rock cycle works is the carbonate environment of the Bahama Islands (see Figure 12.13). Recall that in this area recrystallized calcium carbonate accumulates on the shallow ocean floor. In the context of the rock cycle, this carbonate likely originated from continental limestones that were eroded, and the resulting calcium carbonate was carried in solution by rivers to the ocean. Some of this carbonate is currently recrystallizing in the Bahama Islands to form new rock. Some of that rock has been eroded and deposited on the shore to create the beautiful white, powdery beaches of the islands. Naturally, it has taken a long time for these combined processes to occur.

Geologic Time

In the context of the rock cycle, what does a long time mean? To answer this question requires considering the concept of **geologic time** (or *deep time*), which is radically different from the human timescale. Given our relatively short lifetimes, human beings tend to think that anything 100 years old, or older, is ancient. For example, the American Civil War seems like a very long time ago, does it not? After all, it occurred in the 1860s, long before your grandparents, or even great, great grandparents, lived. As far as the history of Earth is concerned, however, 100 years is a blink of an eye. Scientists accept that Earth is about 4.6 billion years old and that it lies within a much larger and older universe that is approximately 14 billion years old. This kind of timescale is naturally difficult to comprehend. It is important to understand the concept of deep time, however, because it provides a context within which we can study Earth. It also helps us realize the incredibly rapid nature of environmental change associated with ongoing human impacts on the planet.

In an effort to make the concept of geologic time more understandable, geologists have devised a geologic timescale that is subdivided on the basis of major events in Earth history. Figure 12.19 shows this timescale and its various subdivisions. The timescale is categorized into (from largest time expanse to shortest) eons, eras, periods, and epochs. All time earlier than 540 million years (my) ago is lumped into the division of *Precambrian* time, which includes both the *Archean* and *Proterozoic* Eons. This interval of time

---

Geologic time *The period of time that encompasses all of Earth history, from its formation to the present.*

represents 88% of Earth history, during which very little life of any kind existed. Notable events during early Precambrian time include the initial stages of atmospheric development, the emergence of the first cyanobacteria, and the onset of photosynthesis. Toward the end of Precambrian time the modern atmosphere evolved.

Beyond Precambrian time is the *Phanerozoic* Eon, which is subdivided into three major *eras*, the *Paleozoic*, *Mesozoic*, and *Cenozoic*. These eras are further subdivided into *periods*. The beginning of the Paleozoic Era (the *Cambrian* Period) is marked by the emergence of the first fish and shellfish about 540 million years ago, whereas the end of the Paleozoic Era is related to a major extinction during the *Permian* Period around 250 million years ago. Other significant events in the Paleozoic Era include the development of the first trees and amphibians at the beginning of the *Devonian* Period (~417 million years ago) and the first winged insects at the onset of the *Mississippian* Period (~354 million years ago).

The Mississippian Period is particularly noteworthy because organic deposits began to accumulate in sufficient quantity to ultimately form the extensive coal and other fossil fuel resources we use today. This period of massive organic deposition began because the average global temperature during the early Mississippian Period was quite warm, about 22°C (72°F). Thus, many parts of Earth were covered by steamy swamps that contained copious amounts of organic carbon. As plants died, they slowly sank to the bottom of the swamps and began to decay. These deposits were later buried by thousands of meters of sediment; the resulting compaction and heating caused thick beds of coal to form and many petroleum and natural gas reserves to be created. This period of massive organic deposition terminated at the end of the Pennsylvanian Period about 290 million years ago.

Following the Paleozoic Era is the Mesozoic Era. This era is a definable unit of deep time because it is the interval in which dinosaurs dominated life on Earth, beginning around 250 million years ago and ending about 65 million years ago. *Isn't it amazing that dinosaurs existed for over 180 million years?* In addition, the first birds and mammals evolved during this era of time, specifically at the onset of the *Jurassic* Period 206 million years ago. The Cretaceous Period is significant because dinosaurs reached their maximum dominance during this time, most notably in the form of the *Tyrannosaurus Rex*, and the first flowering plants appeared. In addition, two major episodes of mountain building occurred in North America.

The end of the Cretaceous Period at 65 million years ago is marked by the extinction of the dinosaurs. Most geologists believe that this extinction was caused by the catastrophic impact of a large asteroid with Earth. This impact is thought to have caused widespread fires and high levels of atmospheric dust that may have blocked sunlight for months. As a result, the vast majority of animals that depended directly on plants, either by eating them or by eating the animals that ate plants, became extinct. In contrast, it is thought that animals such as small rodents and insects survived because they fed on the

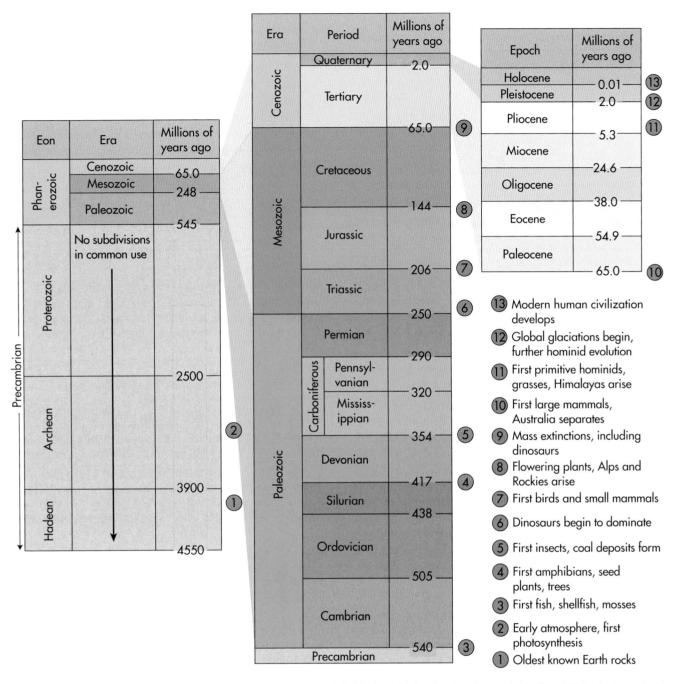

Eon	Era	Millions of years ago
Phanerozoic	Cenozoic	65.0
	Mesozoic	248
	Paleozoic	545
Precambrian — Proterozoic	No subdivisions in common use	
		2500
	Archean	
		3900
	Hadean	4550

Era	Period	Millions of years ago	
Cenozoic	Quaternary	2.0	
	Tertiary		
Mesozoic	Cretaceous	65.0	⑨
	Jurassic	144	⑧
	Triassic	206	⑦
Paleozoic	Permian	250	⑥
	Pennsylvanian	290	
Carboniferous	Mississippian	320	
	Devonian	354	⑤
	Silurian	417	④
	Ordovician	438	
	Cambrian	505	
	Precambrian	540	③

Epoch	Millions of years ago	
Holocene	0.01	⑬
Pleistocene	2.0	⑫
Pliocene		⑪
Miocene	5.3	
Oligocene	24.6	
Eocene	38.0	
Paleocene	54.9	
	65.0	⑩

⑬ Modern human civilization develops
⑫ Global glaciations begin, further hominid evolution
⑪ First primitive hominids, grasses, Himalayas arise
⑩ First large mammals, Australia separates
⑨ Mass extinctions, including dinosaurs
⑧ Flowering plants, Alps and Rockies arise
⑦ First birds and small mammals
⑥ Dinosaurs begin to dominate
⑤ First insects, coal deposits form
④ First amphibians, seed plants, trees
③ First fish, shellfish, mosses
② Early atmosphere, first photosynthesis
① Oldest known Earth rocks

Figure 12.19 The geologic timescale. This timescale is subdivided on the basis of major geological and biological events in Earth history, such as the appearance of flowers or major extinctions.

corpses and organic waste of the larger animals that were decimated.

The Cenozoic Era is significant because it represents all of post-dinosaur time, beginning 65 million years ago and continuing until the present. During this period most current landscape features such as mountain chains and river valleys developed, as well as life as we know it. The Cenozoic Era is subdivided into two periods, the *Tertiary and Quaternary*. Given the relative youth of these two periods, plus the fact that most current landforms developed within this time interval,

scientists know a great deal more about them than about earlier periods. Thus, these periods are further subdivided into *epochs*. The *Paleocene* Epoch, for example, is marked by the emergence of the first large mammals following the extinction of the dinosaurs 65 million years ago. Subsequently, the first primitive hominids developed at the beginning of the *Pliocene* Epoch 5.3 million years ago. The subsequent *Pleistocene* and *Holocene* Epochs are important because they are the intervals of time during which most human evolution has occurred. The Pleistocene Epoch is considered noteworthy because it

Understanding geologic time is difficult because it is hard to relate a diagram such as Figure 12.19 to the significant geologic events used to mark major intervals of time. Such events include major periods of mountain building, catastrophic extinctions, and the emergence of certain kinds of plants or animals in the fossil record. In an effort to improve this understanding, go to the *Geo Media Library* and select *Geologic Time*. This animation illustrates, with images and text, some of the major geologic events associated with past time intervals. The simulation is interactive because you can explore various time periods on your own and view their unique qualities. After you visit all the time intervals within this simulation, be sure to answer the questions at the end to test your understanding of the geologic timescale.

---

is generally considered to represent the time of the great ice ages, which will be discussed in Chapter 17. The subsequent Holocene spans the past 10,000 years and, as discussed in Chapter 9, is generally considered to be the time of so-called modern climate.

"Telling" Geologic Time

You might wonder, How do geologists really know the age of Earth and when major events occurred? The answer lies in our understanding of the way in which rock elements radioactively decay through time. Rocks are composed of countless atoms, all of which contain protons and neutrons in their nuclei. Although many of these atoms remain stable indefinitely, some do not. These unstable atoms are called **radioactive isotopes**. In these atoms, particles within the nucleus break apart and the atom decays into a different element. For the purpose of analogy, imagine a pet dog that suddenly woke up on Monday morning and had become a cat, then the following Monday changed again into a mouse.

In the context of calculating rock age, isotopic decay is important because radiation is emitted when this process (known as *radioactivity*) occurs. Each isotope decays at a different constant rate that is known by geologists. The reference timeframe for the decay rate for any isotope is called its *half-life*; this is the amount of time required for one-half of the isotopes in any given sample to decay. For example, thorium-232 requires 14.1 billion years to change through its decay series (including radium-228 and radon-220, among others) into the stable isotope lead-208. In contrast, the half-life of the radioactive carbon isotope (carbon-14) is 5730 years, during which time it converts to the stable isotope nitrogen-14. In this fashion, radioactive isotopes provide a running time clock for the history of Earth, and geologists use *radiometric dating* to calculate age.

If a geologist is interested in the age of a rock sample, he or she simply compares the amount of the original isotope with the amount of the decayed end product. Using radiocarbon dating, for example, a geologist can compare the amount of carbon-14 to the amount of nitrogen-14. If the ratio between the two is 1:1—that is, equal parts—then it means that one-half of the radioactive isotope has decayed. In other words, about 5730 years have passed since the carbon was deposited.

Putting Geologic Time in Perspective

Although understanding radiometric dating provides confidence in the ages reported in Earth history, it does not really aid with the comprehension of deep geologic time. If you want to put geologic time in some kind of perspective, consider that virtually all scientists believe that anatomically modern humans—that is, people looking basically like you and me—have lived on Earth for about 150,000 years. Sounds like a long time, yes?

Here is another analogy that helps explain deep time. Imagine that all geologic time is contained within a single calendar year, with time beginning on January 1 and extending until December 31, as shown in Table 12.2. In this analogy each day is equivalent to 12.6 million years, each hour is 525,000 years, each minute is 8750 years, and each second is 146 years. The year begins with the formation of Earth. From that point until the middle of April, no life-forms inhabit Earth. In other words, Earth is a lifeless planet for the first 4.5 months of the "year." *Nothing* except single-cell organisms such as amoebas and simple bacteria exist from early May until late November, when the first vertebrate animals (fish) develop. Subsequently, the first land plants emerge on December 3. Reptiles develop on about December 13, followed by the first mammals on December 18. Dinosaurs dominate Earth until late December, when they become extinct and mammals such as shrews and other small rodents emerge and more fully evolve. Early hominids do not evolve until 6 p.m on December 31, which means that anatomically modern people have existed for an extremely short period of time within the year. Your life occurs within the last 0.05 sec of the year. No wonder people have difficulty understanding the concept of geologic time!

Radioactive isotopes *Unstable isotopes that emit radioactivity as they decay from one element to another.*

TABLE 12.2   The Geologic Timescale Viewed Analogously Within a Calendar Year

Event	Age*	Analogous Time in Calendar Year
Earth formed	4.6 by	Jan. 1
First single-celled organism	3.2 by	mid-April
Oxygen in atmosphere	2.0 by	mid-July
First cell with nucleus	1.0 by	Oct. 12
First vertebrates	625 my Nov. 10	Nov. 10
First land plants	340 my Dec. 3	Dec. 3
First reptiles	220 my Dec. 13	Dec. 13
First mammals	155 my Dec. 18	Dec. 18
Grand Canyon downcutting	10 my	Dec. 31, 5 A.M.
Early hominids	3 my	Dec. 31, 6 P.M.
Your birth	~18	Dec. 31 (0.05 sec before midnight)

*by = billion years ago; my = million years ago.

A Holistic View of Geologic Time and the Rock Cycle: The Grand Canyon and the Spanish Peaks of Colorado

Although time analogies are fun and useful to consider, it is possible to sense the essence of deep time more directly when you travel. An excellent place to feel the incredible depth of time is the Grand Canyon in Arizona (Figure 12.20) because it contains rocks that span much of Earth history (Figure 12.21). As a result, the rocks tell us a great deal about Earth's evolution and the environmental changes that have occurred over time in what is now northwestern Arizona.

The oldest rocks are Precambrian in age and are found in the inner gorge of the canyon. The best known of these oldest rocks is the Vishnu Schist, which is a metamorphic rock about 2.0 to 1.7 billion years old. This rock layer was the foundation of an ancient mountain chain that was subsequently eroded between about 1.7 and 1.25 billion years ago. Additional rocks of Precambrian age formed between approximately 1.25 billion and 825 million years ago.

(a)

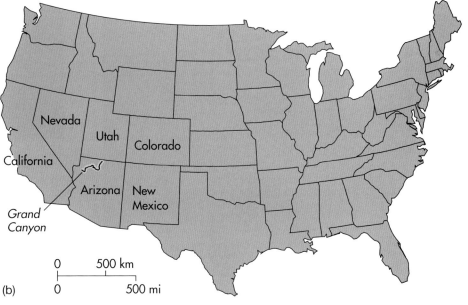

(b)

Figure 12.20 The Grand Canyon.   One of the reasons that the Grand Canyon is such a fascinating place is because it exposes rocks that span the past 2 billion years of Earth history.

0   500 km

0   500 mi

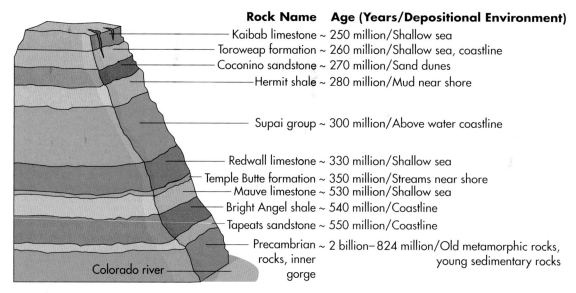

Rock Name	Age (Years/Depositional Environment)
Kaibab limestone	~ 250 million/Shallow sea
Toroweap formation	~ 260 million/Shallow sea, coastline
Coconino sandstone	~ 270 million/Sand dunes
Hermit shale	~ 280 million/Mud near shore
Supai group	~ 300 million/Above water coastline
Redwall limestone	~ 330 million/Shallow sea
Temple Butte formation	~ 350 million/Streams near shore
Mauve limestone	~ 530 million/Shallow sea
Bright Angel shale	~ 540 million/Coastline
Tapeats sandstone	~ 550 million/Coastline
Precambrian rocks, inner gorge	~ 2 billion–824 million/Old metamorphic rocks, young sedimentary rocks

Colorado river

Figure 12.21 Rock layers in the Grand Canyon. Many different kinds of rocks are exposed in the Grand Canyon, Arizona, reflecting several kinds of depositional environments that have occurred in the region over time. Note the relationship between the rock type and the kind of place in which the sediments originally collected.

After the last Precambrian sediments accumulated, another lengthy period of erosion occurred. This period lasted about 300 million years and is called the *Great Unconformity* because no rocks from this period exist. Geologists believe that rocks were almost certainly there, but they were eroded by an advancing sea sometime late in that interval of time. The rocks above the Great Unconformity consist of alternating layers of sandstone, shale, and limestone that collectively accumulated between 550 and 250 million years ago. These rocks are testament to fluctuating sea levels that resulted in the deposition of

Jacques Descloitres, MODIS Land Rapid Response Team, NASA/GSFC

(a)

Jim Wark/AirPhoto

(b)

Figure 12.22 The Spanish Peaks in southern Colorado. (a) This true color satellite image of Colorado and adjacent states shows the location of the Spanish Peaks. This image was acquired in September 2002 and shows the first snow of that year, which can be seen as white on the mountain crests. (b) Looking south at the Spanish Peaks. The crest of East Spanish Peak (left) is 3866 m (12,683 ft) high, whereas the top of West Spanish Peak (right) is 4153 m (13,626 ft) in elevation. Note the prominent radial dike that extends to West Spanish Peak. Several dikes like this one occur in the area.

sandstone when the area was above sea level, shale when the area was a coastal bay or lagoon, and limestone when the region was beneath a shallow sea. These rocks have been progressively exposed in the past 10 million years because the Colorado River has been downcutting. To put this period of downcutting in the context of the time analogy just discussed, consider that it began on December 31 at about 5 A.M. (Table 12.2).

Another place to appreciate deep Earth history is a place called the Spanish Peaks in southern Colorado because this area has a complex geologic history that spans a relatively recent interval of time. In addition, you can see how the rock cycle fits within the context of the geologic timescale here. The Spanish Peaks lie on the eastern flank of the Sangre de Cristo mountain range (Figure 12.22a) and rise about 2134 m (7000 ft) above the

Great Plains (Figure 12.22b), which lie to the east. The evolution of this landscape began sometime before 75 million years ago when thick sedimentary deposits accumulated in a marine environment that fluctuated in depth. These deposits, consisting of alternating layers of sand, clay, and calcium carbonate, may have originated from other rocks and slowly compressed and lithified over millions of years to form sandstone, shale, and limestone, respectively, after the ocean receded.

Approximately 75 million years ago, during the late Cretaceous Period, the nearby Sangre de Cristo Mountains began to form (Figure 12.23). Over the course of the next 50 million years, until about 25 million years ago, this mountain range slowly rose above the surrounding plains. The force of this uplift caused large fractures to appear in the preexisting

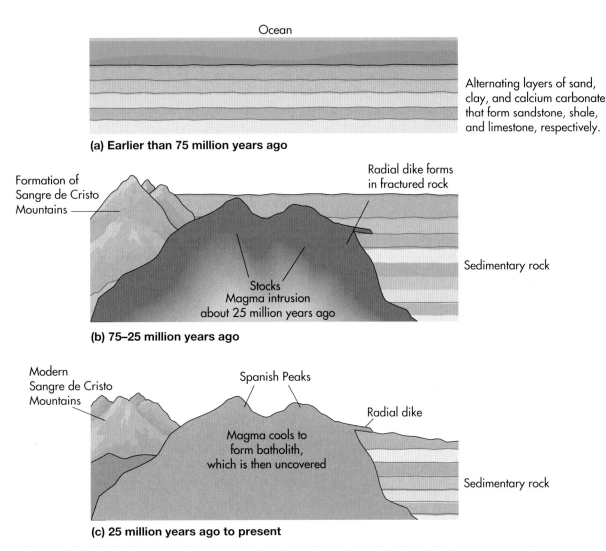

(a) Earlier than 75 million years ago

(b) 75–25 million years ago

(c) 25 million years ago to present

Figure 12.23 Evolution of the landscape at the Spanish Peaks. (a) Before 75 million years ago, the region was covered by a fluctuating ocean, which deposited alternating layers of sediment that became sandstones, shales, and limestones. (b) Formation of the Sangre de Cristo Mountains occurred between about 75 and 25 million years ago. The force of this uplift bent and fractured the sedimentary rocks to the east. Between about 25 and 14 million years ago magma intruded into these fractures. The magma subsequently cooled, forming a batholith, two stocks, and a number of radial dikes. During this same interval of time, the Sangre de Cristo range began to erode intensively and wear down. (c) In the past 5 million years the sedimentary rocks overlying the igneous features have been eroded, which has exposed the stocks and radial dikes seen in Figure 12.22b.

layers of sedimentary rock. Between about 25 million and 14 million years ago, tremendous quantities of magma began to intrude into this portion of the Earth's crust, creating a giant chamber within the sedimentary rock body and filling smaller cracks in that same mass of rocks.

Sometime after 14 million years ago the intruded magma cooled to form a granite batholith with an associated pair of stocks at the top. Extending outward several kilometers from the stocks were **radial dikes** that also consisted of granite. In the past 5 million years, the sedimentary rocks that were once present were completely eroded away from this part of Colorado, exposing the granite stocks and dikes on the surface of Earth. The igneous bodies remained intact because these kinds of rocks are much harder and more resistant to erosion than sedimentary rocks. The end result is that the Spanish Peaks now tower over the landscape.

With the formation of the Spanish Peaks in mind, think about the broader questions related to geologic time and the rock cycle. In the context of the rock cycle, note that the sediments once contained within the layers of sandstone, limestone, and shale, which in turn covered the stocks and radial dikes, have been transported to other locations where new rock formations developed or are now developing. What is particularly impressive is that these layers of sedimentary rock must have collectively been greater than 2134 m (7000 ft) thick! Otherwise, the magma that formed the batholith, stocks, and radial dikes could not have been intruded within them. It is even more impressive that the process of erosion subsequently removed *all* the sedimentary rock above the stocks. In the time since the stocks and dikes were exposed, they were being slowly worn down by erosion. The resulting liberated sediments have also been transported elsewhere to form new rock.

In the context of deep time, the entire sequence of events that led to the development of what we now call the Spanish Peaks took over 75 million years, beginning with the deposition of more than 2134 m (7000 ft) of sand, calcium carbonate, and clay in a marine environment. These sediments slowly lithified to form rock. Next, the formation of the Sangre de Cristo Mountains lasted 50 million years and caused large fractures to develop in the sedimentary rocks. These fractures, in turn, filled with magma some time after 25 million years ago that intruded from a pair of developing stocks associated with an underlying batholith. This magma then cooled, forming granite. Then *all* the sedimentary rocks were eroded away, exposing the granite stocks and dikes, and leaving a pair of mountains. Imagine how much time *that* process took. And perhaps the most impressive thing of all in this history is that the past 75 million years is encompassed within the last 2 weeks in the calendar analogy used to represent geologic time (see Table 12.2). In other words, the Spanish Peaks have existed for only a very short period of time within the geologic timescale. *Think about that!*

Human Interactions with Geologic Time: The Anthropocene

As you can see, the subdivisions of the geologic timescale are based on significant events that occurred at various times in Earth history. The onset of the Paleozoic Era, for example, is marked by the emergence of the first fish about 540 million years ago. Similarly, the boundary between the Paleozoic and Cenozoic Eras at 65 million years ago was set because the dinosaurs became extinct at that time. More recently, the beginning of the Holocene Epoch is set at 10,000 years ago because that is considered to mark the end of the last great ice age and the onset of the general climate conditions we know today.

In the context of significant time markers in Earth history, the geologic community is currently considering the addition of a new epoch to the timescale, one that reflects the dramatic impact that humans are now having on the environment. This new epoch is informally called the **Anthropocene** and is based on the recognition that humans have become the foremost agent of change on the Earth's surface. The term was first used in print in 2000, and momentum has since grown within the scientific community to make the epoch a formal part of the geologic timescale. A source of debate centers on when the beginning of the Anthropocene should be marked. Some believe that the onset of this proposed epoch should be set as far back as 8000 years ago when the origins of human agriculture occurred. Others argue that the most logical time to begin the Anthropocene is the beginning of the Industrial Revolution in the mid-18th century. This latter time marker may make the most sense because it shows up clearly in the ice core record as elevated levels of pollutants.

Regardless of when the beginning of the Anthropocene is defined, we are clearly living at a time when the human footprint on the planet is pervasive and growing. As we saw in Chapter 9, the evidence strongly suggests that increased concentrations of atmospheric CO_2 related to human industrial activity (Figure 9.26) are causing global climate change. In Chapter 10, we saw that the human footprint on vegetation patterns around the globe is extensive (Figure 10.26), with deforestation, commercial forestry, and agriculture dominating over much of Earth's land surface (see Figures 10.29 and 10.33). Only the most remote places remain in what can be considered their natural state. In future chapters of this text, you will see the extensive impact that humans are having on groundwater supplies (Chapter 15), stream behavior (Chapter 16), melting ice caps and glaciers (Chapter 17), amount of windblown dust (Chapter 18), and the world's coastlines (Chapter 18).

One way to view this issue is simply by assessing the amount of dirt, soil, and rock moved by people as they adjust Earth's surface to meet their needs (Figure 12.24). At any given point in the present time, countless trenches are being

---

Radial dikes *A long, wall-like feature of intrusive igneous rock that forms when magma is injected into thin cracks within older rock and cools.*

Anthropocene *A proposed new epoch in the geological timescale based on the recognition that humans are now the foremost agent of change on the Earth's surface.*

(a)

(b)

Figure 12.24 Humans moving rocks and dirt. (a) An obvious place where people move a lot of Earth is a mine. Although many mines are located far below ground, a large number of them are strip mines such as this one that took off the top of a hill. (b) This scene of a bulldozer moving Earth is replayed in countless places at any given point in time.

excavated, mines enlarged, roads built, hillslopes bulldozed, and agricultural fields shaped over much of Earth's surface. So much of this activity occurs today in our complex industrialized world that humans are now the primary shapers of Earth's landscapes, with over 100 gigatons of sediment and rock estimated to be moved every year (Figure 12.25). This amount is believed to be greater than the combined amount of sediment moved by wind, glaciers, streams, and ocean currents. The total amount of Earth moved so far by humans is thought to be sufficient to build a 4000-m-high (~13,000-ft-high) mountain range, 40 km (25 mi) wide and 100 km (~60 mi) long!

In this context, imagine for a moment if geologists from another solar system came to study Earth millions of years in the future, after humans became extinct. They would discover that dinosaurs disappeared after an asteroid impact at a point in deep geological time and that other significant geological events occurred at distinct moments in Earth history. And, from their view in the distant future, one of those key moments would certainly be when humans began to dramatically shape the environment at the onset of our Industrial Revolution. No wonder that contemporary geologists are strongly considering making the Anthropocene a formal epoch in the geological timescale.

Figure 12.25 Assessing the amount of Earth moved by humans in the past 5000 years. The red line represents the amount of Earth intentionally moved by people in activities such as mining, road constructing, and bulldozing, whereas the green line reflects the amount of sediment that has moved indirectly as a result of agricultural activities. A good example of such an indirect impact is sediment eroded from a plowed field because plants are no longer present to protect the soil. Notice the dramatic increase in the amount of Earth moved in gigatons in the very recent past. (*Source*: R. LeB. Hooke, On the history of humans as geomorphic agents. Geology 2000; 28:843–846.)

KEY CONCEPTS TO REMEMBER ABOUT ROCKS, GEOLOGIC TIME, AND OIL

1. Three kinds of basic rocks occur: igneous, sedimentary, and metamorphic.

2. Igneous rocks form when magma cools within Earth (forming intrusive igneous rocks) or on Earth's surface (forming extrusive igneous rocks).

3. Sedimentary rocks consist of large layers of formerly loose sediments that slowly compacted and lithified.

4. Metamorphic rocks form when igneous, sedimentary, or other metamorphic rocks are altered by intense heat and pressure.

5. The age of rocks can be confidently dated because radioactive isotopes decay at a known rate, called the half-life.

6. The geologic timescale involves a concept of deep time that transcends the normal confines of the human timescale.

7. Our current economy depends a great deal on fossil fuels for energy. Many scientists are concerned that we are reaching the peak in global oil production.

8. As a result of human activities, Earth scientists are considering formal adoption of the Anthropocene as a new epoch on the geological timescale.

1. Four major layers occur in the internal structure of Earth, including (from lowest to highest): (a) inner core, (b) outer core, (c) lower mantle, (d) upper mantle (including the asthenosphere, lithosphere, and crust). These layers are identified on the basis of their chemical composition and solid or molten character. Of particular interest is the oceanic and continental crust. Oceanic crust consists largely of silica and magnesium (*sima*) and is denser than continental crust, which is composed of silica and aluminum (*sial*).

2. Three basic kinds of rocks occur on Earth: igneous, sedimentary, and metamorphic. Igneous rocks form when magma cools within Earth (forming intrusive igneous rocks) or on the Earth's surface (forming extrusive igneous rocks). Sedimentary rocks consist of large layers of formerly loose sediments that slowly compacted and lithified. Metamorphic rocks form when igneous, sedimentary, or other metamorphic rocks are altered by intense heat and pressure.

3. The rock cycle describes the ways in which sediments move from one place to another on Earth over time to become new kinds of rock, or how they are transformed to become new kinds of rock. A key concept in this cycle is that rocks gradually break apart as time goes by. The liberated rock fragments are then transported to another place, where they are deposited and gradually lithify to form new rock.

4. Earth is about 4.6 billion years old. Geologic time refers to the length of time from the origin of Earth to the present and is hierarchically subdivided into (from longest to shortest time) Eons, Eras, Periods, and Epochs.

5. Fossil fuels are derived from organic remains that have been buried within Earth. We use these remains to power our economy.

CHECK YOUR UNDERSTANDING

1. Compare and contrast the four major layers of Earth.

2. Describe the concept of isostatic depression and rebound. Why do each of these processes occur, and why does this concept reflect the relationship between the crust and asthenosphere?

3. Describe the nature of the Earth's inner core and why it has the characteristics that it does.

4. Describe the two types of crust found on Earth. What is the basic composition of each type of crust?

5. What is the difference between intrusive and extrusive igneous rock? What are some examples of each kind of rock, and how do you tell the difference between them?

6. What are the various kinds of sedimentary rock, and why do they occur? How does the process of lithification fit into the context of sedimentary rock?

7. What are the two ways that rocks become metamorphic rocks?

8. Describe the flows that occur within the rock cycle. Where does the cycle begin? How do igneous rocks become sedimentary rocks? How do sedimentary rocks become metamorphic rocks? How do the processes of erosion, deposition, and lithification fit within the rock cycle?

9. How is the geologic timescale configured? Using the terms Epoch, Era, Period, and Eon, arrange them into the hierarchy of geologic time.

10. How does the concept of radioactivity fit within the context of calculating the age of ancient rocks? Why are the terms *radioactive isotope* and *half-life* relevant?

11. Describe how the formation of the Grand Canyon is related to the geologic timescale and rock cycle.

12. How is our economy dependent on organic sedimentary rocks, and why are we vulnerable as far as petroleum resources are concerned?

ANSWERS TO VISUAL CONCEPT CHECKS

VISUAL CONCEPT CHECK 12.1

Two types of crust are commonly present in an image like this—oceanic and continental crust. Oceanic crust, which is also called *sima*, is denser than continental crust because it consists largely of silica and magnesium. Continental crust, which is also called *sial*, is less dense because it consists largely of silica and aluminum.

VISUAL CONCEPT CHECK 12.2

The answer to this question is *c*. The rocks collectively indicate fluctuating sea levels in the past. Limestone reflects the accumulation of carbonates in a shallow sea, whereas shale reflects the deposition of mud in a very shallow coastal bay or lagoon. So, at this site, the environment was once a shallow coastal bay or lagoon and then changed to a shallow sea when sea level rose. The many layers in this outcrop indicate that these fluctuations occurred several times.

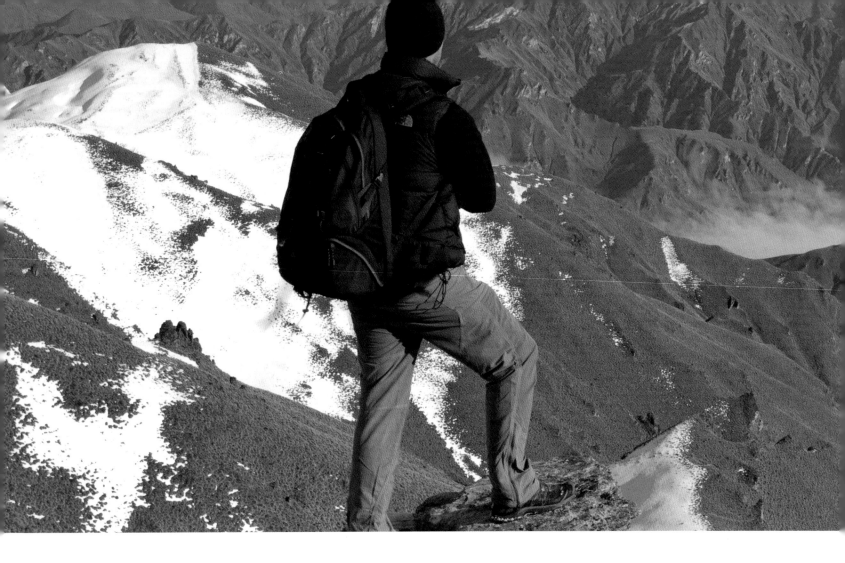

CHAPTER THIRTEEN

TECTONIC PROCESSES AND LANDFORMS

The previous chapter discussed the internal structure of Earth, as well as the way rocks form and evolve in the lithosphere. Now our attention turns to the various ways that interactions among the crust, mantle, and asthenosphere shape the landscape to form mountains and volcanoes, as well as cause destructive earthquakes. These processes fall under the general framework of plate tectonics and are associated with the kinds of destructive geologic events frequently covered in the news. The next time you hear about a major earthquake or volcanic eruption, remember this chapter to understand what happened.

© Doug Pearson/JAI/Corbis

Mountainous landscapes are some of the most spectacular on Earth, as this image from the Southern Alps of New Zealand clearly shows. Mountains such as these pictured here form by tectonic processes, which consist of movements within the Earth's crust that deform rocks, build volcanoes, and cause earthquakes. This chapter focuses on those processes and the landforms that result.

CHAPTER PREVIEW

Plate Tectonics

 Continental Drift

Types of Plate Movements

 Folding

 Plate Tectonics

Earthquakes

Human Interactions: Earthquakes as Natural Hazards

 Types of Faults

Volcanoes

 Walk the Pacific Ring of Fire

 Volcanoes

LEARNING OBJECTIVES

1. Describe the theory of plate tectonics and the evidence supporting the concept of continental drift.
2. Explain the various types of plate movements that occur as a result of continental drift.
3. Discuss how plate convergence promotes folding of rock, resulting in the formation of anticlines, synclines, and other structural features.
4. Explain the general development of the Appalachian Mountains and why segments of the Ridge and Valley Province may be structurally more complex than they appear.
5. Compare and contrast the various kinds of faults that occur on Earth.
6. Describe why earthquakes occur, the various features associated with them, and how their location can be precisely determined.
7. Compare and contrast the various types of volcanoes and why they occur.
8. Discuss the Pacific Ring of Fire and explain the geography of the Cascade Volcanic Arc.

Plate Tectonics

If you frequently travel in the United States, or live in the eastern or western parts of the country, you are probably somewhat familiar with the Appalachian and Rocky Mountains. These regions contain prominent mountain chains that have formed over the course of geologic time. Although many of these mountains are partially composed of igneous rocks that represent some kind of volcanic origin, many contain marine sedimentary rocks that are now thousands of meters above the modern seafloor. How is that possible? You are also probably aware that earthquakes tend to occur in distinct places such as California, and that volcanoes are common in the Pacific Northwest of the United States, as well as other regions on Earth. Why is this so?

This chapter focuses on these questions by investigating the concept of **plate tectonics**, which is the now accepted theory that Earth's crust consists of individual pieces (plates) that move individually and collectively. Understanding plate tectonics is important because it explains the geographic distribution of mountain chains and why earthquakes and volcanoes occur along plate boundaries. The first part of the chapter closely examines this theory and how it explains the formation and geography of mountain ranges. The second part of the chapter focuses on the specific processes and landforms associated with volcanism and earthquakes.

The Lithospheric Plates

A good place to begin a discussion of plate tectonics is the early 20th century, when a German geophysicist named Alfred Wegener noticed that many of the continents look as if they could have fit together at one time as a single landmass (Figure 13.1). Wegener noticed that the east coast of South America, for example, mirrors the southwestern edge of Africa. Wegener looked more deeply into the problem and discovered that portions of some continents share the same kind of rocks and fossils with other continents (Figure 13.1), which is precisely what one would expect if the continents were indeed connected at some point in the past. As a result of these studies, Wegener proposed in 1915 that a *supercontinent* existed about 300 million years ago that he named **Pangaea** (meaning *whole*

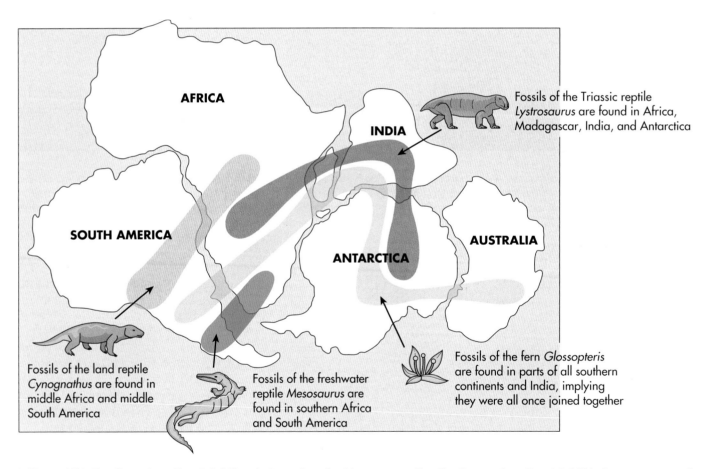

Fossils of the Triassic reptile *Lystrosaurus* are found in Africa, Madagascar, India, and Antarctica

Fossils of the land reptile *Cynognathus* are found in middle Africa and middle South America

Fossils of the freshwater reptile *Mesosaurus* are found in southern Africa and South America

Fossils of the fern *Glossopteris* are found in parts of all southern continents and India, implying they were all once joined together

AFRICA

INDIA

SOUTH AMERICA

ANTARCTICA

AUSTRALIA

Figure 13.1 Fossils and continental drift. A strong line of evidence supporting the theory of continental drift is the occurrence of specific fossil plant and animal species on continents that are now far apart. (*Source*: U.S. Geological Survey.)

Plate tectonics *The theory that the Earth's crust is divided into a number of plates that move because they float on the asthenosphere.*

Pangaea *The hypothetical supercontinent, composed of all the present continents, that existed between 300 and 200 million years ago.*

Earth). He further argued that the continents had drifted apart since that time in a process he called **continental drift.**

The reaction to Wegener's idea in the geologic community was openly hostile, as it directly challenged the long-standing notion that the continents had been connected by a variety of land bridges at one time. In addition, his explanation for how the continents moved, specifically that they "plowed" through the denser oceanic crust, was physically impossible. As a result, Wegener's ideas were largely ignored until the 1950s, when a systematic study began of the ocean basins around the world. At the same time, geologists began to investigate the cause of earthquakes more closely. These combined studies demonstrated that the Earth's crust consists of a series of interconnected plates (Figure 13.2) that indeed move relative to one another. This new understanding was the foundation for the theory of plate tectonics that emerged in the 1960s. This theory was revolutionary because it accurately explained a great deal of volcanic and earthquake activity on Earth, as you will see later in this chapter. It also validated Wegener's basic premise of continental drift because it provided the mechanism for such migration. As for Wegener himself, he never experienced personal vindication for his ideas because he froze to death in 1930 on an expedition to Greenland, some 30 to 40 years before his hypothesis was accepted by the geological community.

Continental drift   *The theory that the continents move relative to one another in association with plate tectonics.*

Plate Movement

The next obvious question is *why* does continental drift occur; how is it possible that large landmasses move around Earth over time? At a fundamental level, it happens because tectonic plates float on the asthenosphere and are moved by convection loops driven by *geothermal* activity. Recall from Chapter 6 that convection occurs in the atmosphere when solar energy heats the ground surface and the warmed ground warms the overlying air. As the warm air rises, the cool air sinks. The same loop of rising and sinking currents also happens inside Earth. In contrast to atmospheric convection, which is driven by an outside source (the Sun), convection within Earth occurs because the radioactive decay of elements such as uranium, deep within Earth, creates very high temperatures between 1000°C and 2000°C (1800°F to 3600°F).

These high temperatures cause plumes of magma to rise slowly by convection within the mantle and into the asthenosphere (Figure 13.3). As these magma plumes reach the base of the crust, they spread horizontally and cool, moving segments of the crust in the process. To visualize this movement, consider a mechanized people mover in the floor of an airport terminal or mall. This device works somewhat like a treadmill that continuously loops around. If you stand on it where the tread rises out of the floor, it carries you forward. The process of continental drift works in somewhat the same way, with plates carried forward by the moving magma beneath. Ultimately, the magma slowly cools, solidifies, and sinks back into the mantle, where it melts again. In this sense, a magma convection current is somewhat similar to a tropical Hadley cell in the atmosphere.

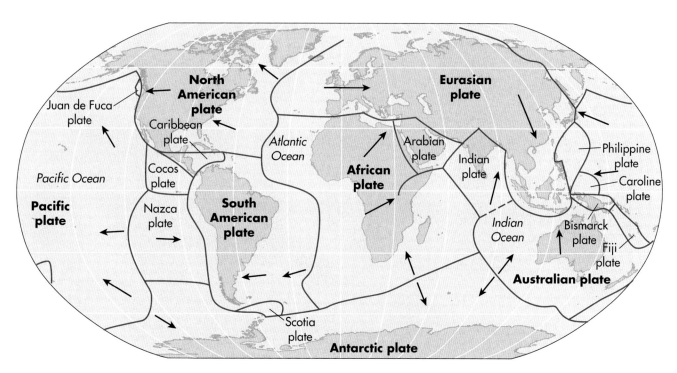

Figure 13.2 Geographic distribution of lithospheric plates.   The seven major plates (boldface type) cover 94% of Earth; about a dozen smaller plates cover the remaining 6% of Earth. Notice the "jigsaw" appearance of the map and the way that the plates move (arrows) relative to one another. (The boundary between the Indian and Australian plates is uncertain; many sources refer instead to the "Indo-Australian" plate.)

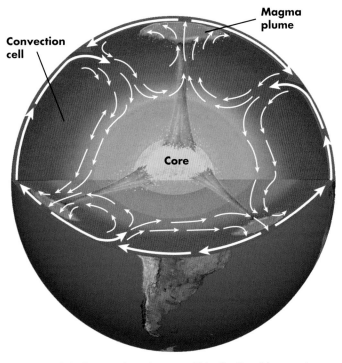

Figure 13.3 Convection plumes within Earth. Magma plumes rise from deep within Earth and spread apart at the crust. Rock created as the magma cools subsequently sinks back into the mantle, where it melts again.

How do we know that the plates are moving? For one thing, scientists can now measure the motion through remote sensing and Global Positioning Systems discussed in Chapter 2. In California, for example, sensors placed on either side of major faults demonstrably move relative to one another over the course of time, indicating that the underlying plates are shifting. Evidence of plate movement also appears in the age of the seafloor. Convection in the mantle brings magma up through fractures in the crust, where it extrudes onto the seafloor, cools, and forms new oceanic crust. Figure 13.4 shows isochrons (lines of equal age) for the ocean floors around the world and clearly indicates areas where oceanic plates have been spreading apart. Notice that the oldest part of the oceanic crust in the Atlantic Ocean basin is along the east coast of North America, where it is about 180 million years old. In contrast, the seafloor in the center of the Atlantic Ocean is less than 10 million years old. This age geography supports the theory that new rocks are being constructed in the center of the ocean basin and are gradually spreading away from there. An additional piece of evidence supporting the theory of continental drift is that magnetic stripes run parallel to the Mid-Atlantic Ridge across the seafloor in a pattern similar to the isochrones shown in Figure 13.4. These stripes reflect the fact that the Earth's magnetic field has periodically reversed through time, with magnetic north becoming south, and vice versa. About 800,000 years have passed since the last such

Figure 13.4 Seafloor ages. In this map colors are associated with seafloors, whereas the continents are gray. Notice how the youngest seafloor rocks (in red) occur in the zones of seafloor spreading, such as in the Atlantic Ocean, with progressively older rocks (yellows, greens, and blues, respectively) away from these regions.

To see how the plates have drifted over time, go to the *Geo Media Library* and access *Continental Drift*. This animation shows the position of continental landmasses during different intervals of geologic time. Use the controls in the animation to play it, to step forward or backward in time, or to rewind to the beginning of the animation. You can also put the individual continental positions into the context of the geologic timescale by accessing specific time periods such as the Triassic or Cretaceous. After you complete the animation, be sure to answer the questions at the end to test your understanding of continental drift.

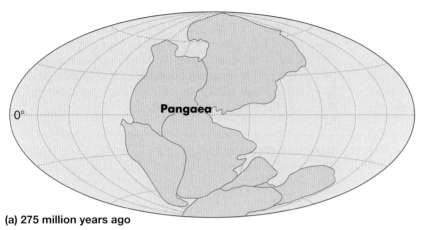

(a) 275 million years ago

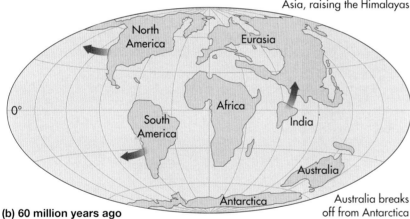

India slams into southeast Asia, raising the Himalayas

Australia breaks off from Antarctica

(b) 60 million years ago

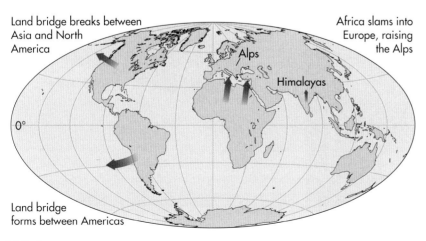

Land bridge breaks between Asia and North America

Africa slams into Europe, raising the Alps

Land bridge forms between Americas

(c) Today

reversal. The ocean floor contains a record of magnetic reversals because, as new seafloor is progressively created, its magnetic orientation is aligned with the direction at that time. Overall, the isochrones and magnetic strips indicate that the rate of crust movement appears to range from 1.0 cm to 15.0 cm (0.4 in. to 5.9 in.) annually.

Given the overwhelming evidence for continental drift, it is possible to reconstruct the patterns of movement through time. Geologists generally agree that the continents began to coalesce to form Pangaea as early as 300 million years ago, during the Pennsylvanian Period. If you recall from Chapter 12, this is the period of time when Earth was much warmer and massive amounts of carbon began to accumulate in steamy swamps. Figure 13.5a shows how Pangaea probably looked around 275 million years ago. Notice that South America and Africa were indeed joined at this time, as were the eastern coast of North America and Africa. By about 200 million years ago, most of the continents had begun to slowly drift apart to reach their present positions. An important exception is the Indian landmass, which moved relatively quickly (as far as deep time is concerned) across what is now the Indian Ocean to collide

Figure 13.5 Continental drift in the past 275 million years. (a) Geography of the Pangaea supercontinent during the Permian Period about 275 million years ago. At this time, all the continents were connected to form one giant landmass. (b) Geography of continents at the beginning of the Cenozoic Era about 60 million years ago. Note the relative directions of continental movement. (c) Configuration of continents today. The continents continue to move around the surface of Earth and will eventually change the world's appearance again.

with the southern side of the Asian continent. This collision created the Himalaya Mountains, which will be discussed in more detail later in the chapter.

Types of Plate Movements

With the concept of plate movement in mind, let's now look more closely at the various types of plate movements and the impact they have on the Earth's surface. For the sake of simplicity, consider that the range of potential plate movements includes (1) moving away from each other, (2) sliding past each other, and (3) colliding head on. This section discusses these types of plate movements and how they shape Earth's surface.

Passive Margins

The simplest kind of tectonic interaction occurs at **passive margins**. As the name implies, passive margins are relatively stable localities from a tectonic perspective. These regions frequently occur in places where the continental crust and the bordering oceanic crust are actually on the same tectonic plate. One such place is the eastern seaboard of North America, which has a passive margin with the oceanic crust in the western part of the Atlantic Ocean basin. Even though the continental crust that forms the North American landmass is a separate rock body from the adjoining oceanic crust, they are both a part of the North American plate (Figure 13.6). As a result, they have a passive margin because it is not geologically active.

Transform Plate Margins

Another type of plate boundary is the **transform plate margin**. In contrast to passive plate margins, transform boundaries are places where plates slide horizontally past each other (Figure 13.7a). At transform margins, the plane of motion is along a nearly vertical break (or fault) that extends through much of the lithosphere. The San Andreas Fault in California and the Dead Sea Fault along the border of Israel and Jordan (Figure 13.7b) are excellent examples of transform faults. The San Andreas Fault will be discussed in more detail later in this chapter in the context of earthquakes.

Passive margin   *A place where the continental crust and the oceanic crust are on the same tectonic plate and thus do not move relative to each other.*

Transform plate margin   *A plate boundary where opposing plates move horizontally relative to each other.*

Figure 13.6 Passive margin along the eastern edge of North America.   The continental crust of North America meets the oceanic crust of the Atlantic Ocean. Nevertheless, they are part of the same tectonic plate and are slowly moving as one body to the northwest.

Transform fault

(a)

(b)

Figure 13.7 Transform plate margins. (a) Plates that meet at transform margins move horizontally past each other along transform faults. (b) Satellite image of the Dead Sea Fault. This transform fault occurs along the boundary of the Arabian and African plates and extends from the Gulf of Aqaba on the southern tip of Israel through the Dead Sea and north into Lebanon. The Arabian plate is generally moving to the north, whereas the African plate is moving in a southerly direction in this region.

Plate Divergence

In some parts of the world, lithospheric plates move away from one another in a process called *plate divergence*. This type of movement occurs in regions where rising magma plumes within Earth move upward and outward between plate fractures, spreading Earth's plates apart in a process called **rifting** (or divergence). A great place to see the process of rifting is on the seafloor where magma extrudes through fractures in the oceanic crust (Figure 13.8). This extrusion of magma produces a ridge-like feature, appropriately called a **mid-oceanic ridge**, which lies parallel to the rift zone. The Mid-Atlantic Ridge is perhaps the best-known oceanic ridge and is located, as its name suggests, at the middle of the Atlantic Ocean along the Atlantic Rift Zone. If these plates are diverging, what do you suspect is happening to the width of the Atlantic Ocean? The answer is that it is slowly increasing every year. This process began when Pangaea began to break apart about 200 million years ago, and it continues to this day.

Plates can also diverge within continents, causing a gradual split in the landmass to occur. One of the best-known locations

---

Rifting *The spreading apart of the Earth's crust by magma rising between fractures in the Earth's plates.*

Mid-oceanic ridge *A ridge-like feature that develops along a rift zone in the ocean due to magma upwelling.*

Figure 13.8 Oceanic rifting, seafloor spreading, and plate divergence. In ocean basins, the seafloor spreads where plumes of magma interact with the crust, creating a rift. One of the best examples of seafloor spreading exists in the middle of the Atlantic Ocean.

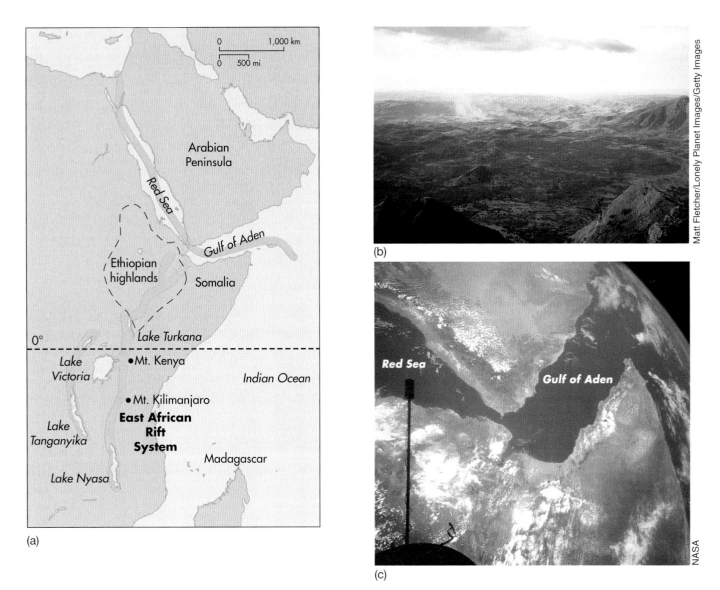

(a)

(b)

(c)

Figure 13.9 The Great Rift system in eastern Africa and the Middle East.   (a) Map showing the various components of the East African Rift system. (b) The Suguta Valley in Kenya is part of the East African Rift; note the steep bluffs bordering the valley. (c) Satellite image of the triple junction with view to east/southeast. The Red Sea is to the upper left, the Gulf of Aden is to the upper right, and the East African Rift is in the foreground.

of continental plate divergence is in eastern Africa where the East African Rift is located (Figure 13.9a). Continental rifting in this area has produced a distinct valley landscape bordered by steep canyon walls (Figure 13.9b), as well as several large lakes such as Lake Victoria (Figure 13.9a). In the northern part of the rift zone, plate divergence is occurring in three places that merge at a single place known as a *triple junction* (Figures 13.9a and c). This junction is in close geographic association with the Arabian, African, and Indian plates. Two of the rifts are oceanic, with one opening the Red Sea and the other causing enlargement of the Gulf of Aden on the southern side of Saudi Arabia. On land, the third rift, the East African Rift, extends in a southerly direction from the juncture of the Red Sea and the Gulf of Aden. Along this rift, the eastern part of the African continent is slowly splitting in two, resulting in several large lakes, such as Lake Victoria.

Plate Convergence

Having just seen what happens when plates diverge from one another at rift zones, let's turn now to the places where two plates collide directly into each other in a process called *plate convergence*. Plate convergence occurs in three general settings, including (1) oceanic crust to continental crust, (2) oceanic crust to ocean crust, and (3) continental crust to continent crust. The type of convergence associated with each of these settings results in specific processes that produce distinctive landforms on the Earth's surface.

Collision of Oceanic Crust with Continental Crust   The first type of plate collision discussed occurs when two plates collide, one consisting of oceanic crust and the other of continental crust. Recall from Chapter 12 that oceanic crust is generally denser than continental crust because oceanic crust consists

340   Chapter 13   •   Tectonic Processes and Landforms

of basaltic rock, whereas continental crust tends to be granitic. In convergence zones where this difference occurs, the oceanic crust can be viewed like a heavy slab of lead that encounters the more spongy continental plate and thus sinks beneath it in a process called **subduction** (Figure 13.10). Subduction is actually initiated when oceanic plates diverge due to seafloor spreading. As noted previously, this spreading forces oceanic crust to move horizontally away from the spreading zone and toward continental margins. When the oceanic crust converges with the continental crust, the more dense oceanic crust is slowly forced beneath the less dense continental crust and down into the upper mantle. As the slab is forced more deeply, the internal temperature gradually increases in a distinct *geothermal gradient*. When the oceanic crust ultimately encounters the upper mantle, it melts and is recycled into magma as part of the rock cycle.

The process of subduction is not continuous, but is instead episodic. In other words, the oceanic plate does not flow smoothly beneath the continental crust. Instead, the oceanic crust episodically stops and starts in its downward movement below the continental crust. It moves only when the buildup of stress behind it—which is ultimately derived from the zone of seafloor spreading—exceeds the force of friction holding it in place beneath the continental crust. This stress can cause earthquakes and significant deformation of the rocks above the subduction zone, which also results in the construction of mountain ranges and even volcanoes in the places where magma works its way to the surface. This relationship will be discussed in more detail later in the chapter.

Collision of Oceanic Crust with Oceanic Crust

A second type of plate convergence occurs when two bodies of oceanic crust collide with one another. Although both plates have similar densities in these settings, one plate will typically be subducted beneath the other, forming a deep *oceanic trench* (Figure 13.11a). As previously discussed, this subduction will produce stress in the overriding seafloor that may cause it to crumple upward to form mountains below the ocean surface. More often, subduction of one plate of oceanic crust beneath the other will result in the formation of volcanoes (due to melting of the crust in the upper mantle) that grow upward in the water. If these volcanoes enlarge sufficiently, they will breach the surface of the ocean to form distinctive island chains, such as Northern Marianas Islands of the Pacific Ocean (Figure 13.11b). Another great example of such an island chain formed by the collision of two plates of oceanic crust is the islands associated with Japan. Again, the formation of volcanoes will be discussed in more detail later in the chapter, including Japan's Mount Fuji.

Continent-to-Continent Collision

The third type of plate collision is one that occurs when continents on separate plates collide with one another. This type of convergence is distinctive because both plates consist of continental crust that has similar densities. In these situations, one plate does not ride over or below the other, which means they smash together like two cars in a head-on collision. The crust, like the cars, crumples, causing folding of the formerly horizontal bedrock through compression (Figures 13.12a and b).

You can see evidence for past compression and folding of rocks by looking at the orientation of rock layers or **rock structure**. Figure 13.12 illustrates how rock layers are folded and the various features that result. In general, a close correlation exists between the amount of compression and the nature of the fold. A **monocline** is a one-sided slope where beds of horizontal rock are inclined in a single direction over a large

---

Subduction *The process by which one lithospheric plate is forced beneath another. This usually happens when oceanic crust descends beneath continental crust, but it can occur where two plates of oceanic crust meet.*

Rock structure *The internal arrangement of rock layers.*

Monocline *A geologic landform in which rock beds are inclined in a single direction over a large distance.*

5. Volcano erupts

4. Magma rises through cracks in continental crust

1. Oceanic crust moves toward continental crust

2. Subduction of dense oceanic crust below lighter continental crust

3. Oceanic crust melts into magma

Figure 13.10 The subduction process. Subduction occurs where dense oceanic crust is forced beneath less dense continental crust. This process causes the oceanic crust to move down into the upper mantle and melt. When the resulting magma rises to the surface through cracks in the overlying crust, it frequently causes volcanoes.

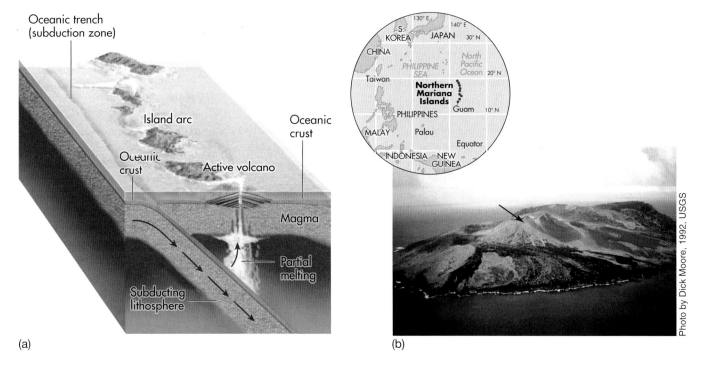

(a)

(b)

Figure 13.11 Collision of two plates of oceanic crust. (a) When two plates of oceanic crust collide, one of the plates will subduct beneath the other despite the fact they both have similar densities. An oceanic trench forms along the subduction zone. An island arc also develops due to volcanic activity along the plate boundary. (b) Guguan Island in the northern Marianas Islands in the Pacific Ocean. This island is a volcano (note the cone in the middle of the island) that formed due to subduction along the Marianas Trench. This trench is the deepest location on the surface of Earth's crust, with a maximum depth of 9636 m (31,614 ft) below sea level.

area. This kind of fold occurs when the amount of compression is relatively low. An excellent example of a monocline occurs in southeastern Utah, near the town of Mexican Hat (Figure 13.12c). More extensive compression causes anticlines and synclines to form (Figure 13.12a). An **anticline** is a portion of the fold where the rock layers arc upward to form a concave arch along the fold axis. One-half of such a fold is called a *limb*. In contrast, a **syncline** is the portion of the fold where rock layers dip downward to form a convex trough. If the collision is especially intense, the rocks can be folded so much that an **overturned fold** (or recumbent fold) results. Still more compression results in an **overthrust fold**, which occurs when one part of the rock mass is shoved up and over the other.

Geomorphology and the Evolution of the Appalachian Mountains Large-scale compression of rocks is frequently associated with the growth of mountains. The study of

mountain formation, as well as the shape of the Earth's surface in general, falls within the subdiscipline of physical geography called **geomorphology**. Geomorphology is the study of the formation, shape, spatial distribution, and evolution of landforms on Earth—thus, the name *geo* (Earth) *morphology* (shape). In contrast to a *landscape*, which is the overall appearance of a place in terms of its vegetation, topography, or human modifications, a **landform** is a distinct geographic feature, such as a mountain, river valley, coastline, or sand dune, to name but a very few. Each of these landforms evolved due to distinct geomorphic processes, which will be discussed in later chapters. As you might imagine, the study of geomorphology is rooted deeply in geology because it requires a thorough understanding of how sediments are eroded, transported, and deposited. The discipline is also geographically based because geomorphologists are interested in the way landforms change across space and time.

Anticline *A convex fold in rock in which rock layers are bent upward into an arch.*

Syncline *A concave fold in rock in which rock layers are bent downward to form a trough.*

Overturned fold *A structural feature in which the fold limb is tilted beyond vertical, which results in both limbs inclined in the same direction, but not at the same angle.*

Overthrust fold *A structural feature where one part of the rock mass is shoved up and over the other.*

Geomorphology *The branch of physical geography that investigates the form and evolution of the Earth's surface.*

Landform *A natural feature, such as a hill or valley, on the surface of Earth.*

Layers of
sedimentary
rock

Compression   Compression

Limb

Compression   Compression

Monocline   Syncline   Anticline   Overturned   Overthrust
fault

(a)

(b)

Martin Bond/Photo Researchers, Inc.

(c)

Courtesy Prof. Allen F. Glazner, Dept. of Geological
Sciences, University of North Carolina

Figure 13.12 The folding process. (a) The vast majority of rocks lithify within horizontal beds. If these rocks are subsequently compressed by plate collision, they are deformed in some way. In many places, folded rocks are associated with mountains and valleys. (b) Folds in a rock outcrop in the United Kingdom. These rocks were once horizontal but have been bent in several different directions by compressional forces. (c) The Comb Ridge Monocline near Mexican Hat, Utah. Compressional forces produced a single fold here that connects two beds of relatively horizontal sedimentary rock. One of these beds of horizontal rock is on the top of the mountain, whereas the other underlies the landscape in the foreground.

As far as the growth of mountains is concerned, they typically form during a distinct interval of time called an **orogeny**. Many excellent examples occur on Earth of how continental collision and compression cause orogenies in which rocks are intensively folded. One such example is the Appalachian Mountains, which extend approximately 2500 km (~1600 mi) along the eastern United States. Several phases of mountain building have occurred in this region, including the *Taconic Orogeny* (~450 million years ago), *Acadian Orogeny* (~375 million years ago), and *Allegheny Orogeny* (~290 to 248 million years ago).

Given that the Allegheny Orogeny occurred most recently, it is the best understood because the modern landforms are remnants from that time. It also appears to have impacted the most extensive area, ranging from what is now Pennsylvania to Alabama. This area was intensively folded when the northwest corner of Africa collided with the east coast of North America at the time Pangaea formed. Geologists believe that rugged mountains formed during this interval of time that may have been between 5000 m and 8000 m (~16,400 ft and 26,200 ft) high. In other words, if you could go back to that time, you would see mountains in what is now Virginia that looked much like the modern Rocky Mountains in Colorado, or maybe even the modern Himalayas in Tibet. Over the past 200 million years these once towering mountains have been eroded to well-rounded landforms, such as those seen in the Great Smoky Mountains in North Carolina, that are less than 2000 m (6550 ft) high.

The Appalachian Mountains are subdivided into several different subregions. Perhaps the best known of these regions is the Ridge and Valley Province, which extends along much of the length of the Appalachian range. As the name implies, the landscape here consists of a series of prominent ridges and intervening valleys. Figure 13.13a shows a satellite view of this pattern. A characteristic ground-level view is seen in Figure 13.13b.

When you look at the Appalachian landscape, it is tempting to assume that the ridges are anticlines because they arch up and the valleys are synclines because they bend downward. However, the relationship of surface landforms and the underlying geologic structure is often much more complex than you might think. It is well beyond the scope of this text to examine every such complexity, but to see some simple examples, look at Figure 13.14. This diagram shows a general time sequence illustrating how portions of the landscape evolved. Part (a) reflects what the topography might have looked like after it was initially folded during the Allegheny Orogeny. As you would expect, the landscape at that time consisted of well-developed ridges underlain by structural anticlines and prominent valleys associated with the synclines.

Orogeny *A period of mountain building, such as the Allegheny Orogeny.*

(a)

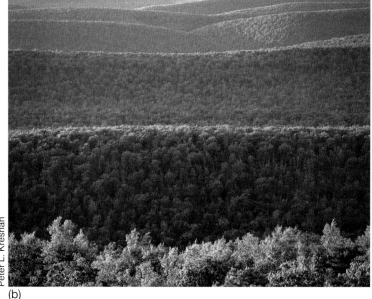

(b)

Figure 13.13 Appalachian Ridge and Valley. (a) Infrared satellite image of a portion of the Ridge and Valley Province in central Pennsylvania. The ridges (arrows) are covered by forest, which makes them stand out particularly well. (b) Oblique air photograph of the Ridge and Valley Province. Note the well-developed ridges and intervening valleys in this panoramic image.

The expected relationship between the shape of the Earth's surface and the underlying structure began to change in some cases due to the behavior of streams. Notice in Figure 13.14a how a network of parallel streams developed on the landscape after it was compressed. Do you see how the path of streams follows the structure? These streams began to vigorously erode the uplifted ridges, but they did so at variable rates that depended on the resistance (or hardness) of the rocks on which they flowed. The behavior of streams will be discussed in more detail in Chapter 16, but, for now, understand that streams erode some rocks (such as shale) more easily than

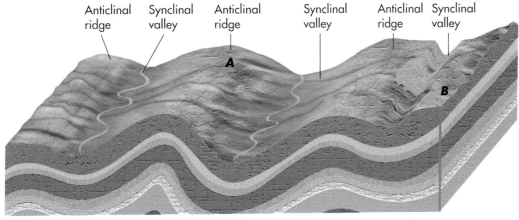

Anticlinal ridge Synclinal valley Anticlinal ridge Synclinal valley Anticlinal ridge Synclinal valley

A

B

(a) Initial folding during Allegheny Orogeny

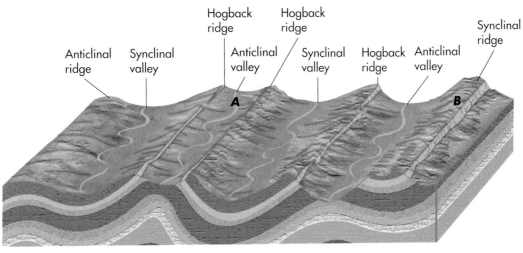

Hogback ridge Hogback ridge Synclinal ridge

Anticlinal ridge Synclinal valley Anticlinal valley Synclinal valley Hogback ridge Anticlinal valley

A

B

(b) After erosion of softer rock layers

Figure 13.14 General evolution of the Ridge and Valley Province. (a) The landscape shortly after folding. Anticlines form the high ground, whereas valleys occur in synclines. (b) As time progresses, erosion modifies the landscape. Anticlines at **A** and **B** are eroded by streams, forming valleys in upward-arching structures. Hogback ridges exist on both sides of the anticlinal valley at **A**, where limbs of the former anticline form the high ground. To the right at point **B**, extensive erosion has inverted the topography so that the ridge is underlain by a syncline.

others (such as limestone). As a result of this preferential erosion, some parts of the Appalachian landscape were eroded extensively, whereas others were not. In addition, stream erosion focused on the highest points, which were the crests of the original anticlines.

Because of this variability, the relationship between surface topography and underlying structure in the Appalachians became less clear over time. Look at the ridge marked A in Figure 13.14b, for example. This ridge was initially underlain by an anticline (Figure 13.14a). Erosion removed the crest of the ridge, however, forming a valley parallel to the underlying structure. The ridges on each side of the valley are remnants of the limbs of the anticline. Such a ridge is called a **hogback ridge** because one side of the ridge is steeper than the other. Also note that the valley formed between the two hogbacks at point A. In Figure 13.14b the ridge is underlain by the original anticline; that is, the rocks arch up there as you would expect. In other words, the stream eroded into the core of the anticline, forming a valley within it along its length. Such a landform is called an **anticlinal valley**.

Hogback ridge *A ridge underlain by gently tipped rock strata with a long, gradual slope on one side and a relatively steep scarp or cliff on the other.*

Anticlinal valley *A topographic valley that occurs along the axis of a structural anticline.*

Figure 13.15 Location of major alpine chains on Earth. Many of the Earth's major mountain chains have resulted from compression along crustal plates. Others have developed due to other processes, such as volcanism. The Andes is the longest mountain chain, at 7200 km (4500 mi); the Rockies are not far behind, at 6000 km (3700 mi). However, the tallest mountains are in the Himalayas.

A **synclinal valley** lies immediately to the right of the hogback ridge-and-valley network at point *A* and is what you would expect since it lies below the adjacent ridge. To the right of that valley is another hogback ridge, which borders yet another anticlinal valley. See how the valley is cut into rocks that arch up?

Now look at point *B* in Figure 13.14b. A ridge exists at the surface here, but it is actually underlain by a syncline. See the U-shape of the rock layer that caps the ridge? This is the same rock body that forms the hogback ridge to the left and now is on top of the ridge here. Such a pattern is interesting because it means that the topography at point *B* has been *inverted*. In other words, what was once the valley floor (point *B*, upper image) is now the crest of a ridge (point *B*, lower image). Such complex relationships between the underlying geologic structure and the way the landscape now appears demonstrate that landscapes are sometimes more complex than they appear. They also reflect how much *time* must be involved for this kind of landscape to evolve. First, there had to be uplift and then *substantial* amounts of erosion. Overall, it took about 250 million years!

Synclinal valley *A topographic valley that occurs along the axis of a structural syncline.*

Although the Appalachian Mountains are a great example of how collision produced a landscape, it is certainly not the only such example on Earth. Crustal collision continues to uplift mountains in several prominent places around the world, creating many of Earth's distinct mountain chains shown in Figure 13.15. The famous Swiss Alps, for example, formed by compression. This mountain range forms the border between Austria and Italy and began to develop during the Cretaceous Period (144 my to 65 my ago) when the continental plate of Africa collided with the southern side of the European plate. A second orogeny occurred during the Tertiary Period (65 my to 2 my ago). An interesting aspect of the Alps geology is that part of the mountain range consists of rock material from the African plate. In other words, part of the European landmass is actually rock from Africa. Another mountain range that has formed largely through continental collision is the Himalaya Mountains (Figure 13.16), which border India and China. Most people probably know that Mount Everest is the highest mountain in the world. What you may not have known is that this peak, as well as the entire Himalayan chain, has been uplifted because the northeastern part of the Indian plate has been colliding into the southern side of the Eurasian plate for about the past 30 million years. As a result of this collision, the mountains have shot out of the ground, at least as far as geologic time is concerned. In so doing, metamorphic and sedimentary rocks (including marine fossils) have been uplifted thousands of meters above sea level.

Folding

A pair of animations on the *Geo Media Library* nicely illustrates the process of crustal folding. The first one is *Folding*, which animates the fundamental process of rock deformation by compression. As you watch this animation, be sure to think about geological structure and note how anticlines, synclines, and other folds develop.

After you watch *Folding*, turn your attention to *The Southern Appalachians*. This animation is a movie focusing on the geomorphic evolution of places like the Smoky Mountains. As you watch this movie, think about the impact that continental drift and compression has had on this landscape. Also pay particular attention to changes that have occurred on this landscape since the last orogeny in the region. After you complete each of these animations, be sure to answer the questions at the end to test your understanding of these concepts.

Plate Tectonics

To review the process of plate tectonics and the movements that occur along plate boundaries, go to a pair of animations in the *Geo Media Library*. The first one is *Active and Passive Margins*, which compares the movements that occur on unstable and stable tectonic boundaries. The second is *Tectonic Plate Boundary Relationships*, which focuses on subduction, transform, and collision boundaries. As you watch these animations, be sure to associate the boundary process with the type of landscapes discussed in the preceding text. After you complete each animation, be sure to answer the questions at the end to test your understanding of plate tectonics and plate margins.

Getty Images, Inc.

Figure 13.16 The Himalaya Mountains. Mount Everest is the tallest mountain in the world at 8850 m (29,035 ft). The 25 tallest mountains in the world are all in the Himalayas, including all 14 peaks that are over 8000 m (~26,250 ft). This mountain chain extends in an arc about 2400 km (1500 mi) long and forms the border between India and Tibet. The Himalayas formed due to collision of the Indian plate with the Asian landmass.

Which one of the following statements is accurate with respect to the geologic structure in this roadcut?

a) The rocks are horizontal.
b) This is a nice example of a rift system.
c) The structure shows the limb of an anticline.
d) The structure shows the base of a syncline.

Scott Drzyzga

KEY CONCEPTS TO REMEMBER ABOUT PLATE TECTONICS

1. The Earth's crust is not a single sheet, but rather a jigsaw puzzle of interconnected plates.

2. A variety of evidence (location of fossil remains and actual measurement of motion) strongly supports the theory of continental drift, indicating that plates have shifted through geologic time.

3. Plate margins can be passive, transform, converging, or diverging.

4. At diverging plate margins, magma plumes from the asthenosphere cause rifting, either on continents or on the seafloor.

5. At converging plate margins, plates collide or subduct in a variety of settings, causing geomorphic features such as alpine chains and volcanoes.

Earthquakes

The previous discussion described the various ways in which plate margins move relative to one another and how mountains can form in these areas. Now it is time to discuss more specific processes associated with plate boundaries—volcanism and earthquakes. The logical place to begin this discussion is with the concept of earthquakes because they occur both as independent events and frequently in association with volcanoes.

Earthquake *Shaking of the Earth's surface due to the instantaneous release of accumulated stress along a fault plane or from underground movements within a volcano.*

Seismic Processes

You have no doubt heard about a major earthquake that happened someplace in your lifetime, perhaps one that devastated a city and unfortunately killed hundreds or even thousands of people. Examples abound of recent earthquakes that have made the news in such a fashion. For example, an earthquake occurred in Java, Indonesia, in May 2006 and killed over 5000 people. In May 2008 a massive earthquake occurred in central China and killed over 65,000 people. Yet another horrific earthquake struck Haiti in January 2010, killing over 300,000 people and leaving 1,000,000 homeless. It is quite possible another one will have occurred by the time you read this text. Clearly, earthquakes are important natural events that humans should understand.

An **earthquake** occurs when the sudden release of accumulated tectonic stress results in an instantaneous movement of the Earth's crust. Although this type of movement is usually associated with plate boundaries, it can also occur due to the rapid movement of magma within a volcano. Such movements produce shock waves that radiate through the lithosphere, causing the ground to shake in sometimes dramatic ways. The strongest earthquakes are usually most associated with plate boundaries because stress builds at these locations as plates grind against each other. This stress creates a fracture between adjoining plates called a **fault**, along which the adjacent rock bodies are displaced relative to each other. At some locations, faults extend deeply into the Earth's crust, whereas at other places they are very shallow. Faults and earthquakes can also occur in the middle of plates, but this association is less common. Although most large earthquakes occur along major plate boundaries, smaller earthquakes can take place whenever a small fault occurs within a rock body.

Fault *A crack in the Earth's crust that results in the displacement of one lithospheric plate or rock body relative to another.*

To see the relationship of earthquakes and plate boundaries, look at Figure 13.7a, which shows the plate movements along a transform margin. The important thing to remember is that these plates do not slide by one another gradually. Instead, these opposing masses of rock are locked by friction for long periods of time, causing elastic stress to build on both sides of the fault. This stress builds to a critical point when a rupture occurs along the fault plane. The stress released is somewhat analogous to the stress instantly released in a stretched rubber band when it is cut, causing the rock layers on either side of the fault immediately to adjust. The place within the lithosphere where the fault breaks is called the *focus* and usually occurs a few kilometers deep within the ground (Figure 13.17). In California, for example, the average focal depth is between 6.4 km and 9.5 km (4 mi to 6 mi). The point on the surface directly above the focus is known as the **epicenter**. When the rocks finally break free of each other, they do so in an abrupt movement and release powerful seismic waves into the surrounding crust.

Locating the Epicenter When an earthquake occurs, the first goal of geologists is to determine the location of the epicenter. This search is accomplished through the process of triangulation, in which the distance to the epicenter is compared among seismographs stationed at three separate locations. This methodology is based on the fact that earthquakes produce two kinds of seismic waves, or *body waves*, that radiate through the Earth's interior. These body waves consist of *P waves* (or *primary waves*) and *S waves* (or *secondary waves*) that move in different ways and travel at different speeds. P waves are compressional waves that cause the

---

Epicenter *The point on the Earth's surface that lies directly over the focus of an earthquake.*

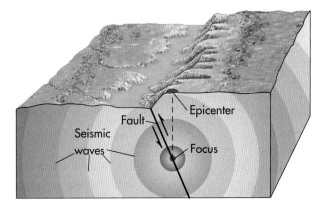

Figure 13.17 Schematic diagram of earthquake processes. Earthquakes occur when tension is abruptly released along a fault, causing opposing rock bodies or plates to move in opposite directions relative to one another. This movement generates seismic waves that radiate through the crust. Uplift along a fault can result in a distinct cliff or bluff along a fault, known as a fault scarp. The motion of plates that causes an earthquake can come from stress buildup associated with a transform fault, subduction, and collision.

Earth's crust to expand and contract rapidly in a horizontal manner as the waves radiate out from the epicenter. P waves typically move at about 1.5 km to 8 km (~1 mi to 5 mi) per second through the crust. S waves, however, move about 60% to 70% slower than P waves and shake the ground in a vertical fashion that is similar to sending a wave pulse along a rope or string when you whip it up and down. In an earthquake, S waves cause the ground to vibrate in a rolling fashion that can be quite noticeable and frightening. The magnitude of both these waves is measured at observation stations with a device called a seismograph (Figure 13.18).

Figure 13.18 Measuring seismic waves. Seismic waves associated with an earthquake are measured with a seismograph. When the Earth shakes, a platform moves back and forth beneath a stationary pen that records the movement on a continuous roll of paper. (a) Primary waves (P waves) consist of compressional waves that cause vibrations parallel to their travel direction. (b) In contrast, secondary waves (S waves) cause vertical vibrations to develop. P waves move more quickly than S waves and are thus the first to be felt.

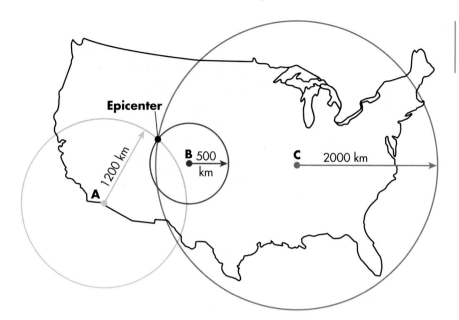

Figure 13.19 Seismic triangulation. The epicenter of an earthquake is the point where the distance measurements from three seismic observation stations overlap.

Given the known speed difference between P waves and S waves, it is possible to calculate the distance to an earthquake epicenter at any observation station by noting the time lag between the passing of each wave. When the passing of seismic waves is viewed in a regional context, it is logical to assume that a wave will require more time to reach an observation station that is farther away from the epicenter than one that is closer. With these time patterns in mind, geologists can determine the earthquake epicenter by triangulating the time lag between three separate observation stations. The epicenter is located at the point where the respective distance measurements overlap (Figure 13.19).

Yet a third kind of wave is a surface wave, which moves only through the upper crust. Although *surface waves* move much more slowly than the P and S waves within Earth, and actually arrive after the two body waves, they are responsible for most of the damage associated with an earthquake.

Measuring Earthquake Magnitude The strength of earthquakes is measured on the **Richter scale**, which is related to the amplitude of seismic waves moving through the Earth's crust as determined by a seismograph. The Richter scale is logarithmic and is represented by whole numbers and decimal fractions. An increase of one magnitude unit (such as 4 to 5) corresponds to 10 times greater ground motion; an increase of two magnitude units corresponds to 100 times greater ground motion; and so on (Figure 13.20). The magnitudes of earthquakes range between 0 (weakest) and 9 (strongest) on the Richter scale. The actual ground motion for, say, a magnitude 5 earthquake is about 0.04 mm (0.001 in) at a distance of 100 km (62 mi) from the epicenter; it is 1.1 mm (0.04 in) at a distance of 10 km (6.2 mi) from the epicenter.

Types of Faults In addition to the production of seismic waves, earthquakes cause deformation of the rocks both within the crust and on the Earth's surface. This deformation is usually associated with the various kinds of faults that occur when opposing rock bodies move relative to each other in association with plate tectonics (Figure 13.21).

One type of fault is a **normal fault**, which is a vertical fault in which one slab of the rock is displaced up and the other slab down (Figures 13.21a and 13.22). This kind of fault is created by tension forces acting in opposite directions. Where such faults occur, the opposing blocks are pulled away from one another by gravity, which causes one of the fault blocks to

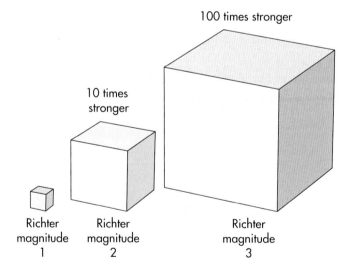

Figure 13.20 The Richter scale. These cubes show the logarithmic nature of the Richter scale, with the size of each cube representing relative power. For example, a magnitude 2 earthquake is 10 times stronger than a magnitude 1 earthquake. Similarly, a magnitude 3 earthquake is 100 times stronger than a magnitude 1 earthquake.

Richter scale *The logarithmic scale used to measure the strength of an earthquake.*

Normal fault *A steeply inclined fault in which the hanging rock block moves relatively downward.*

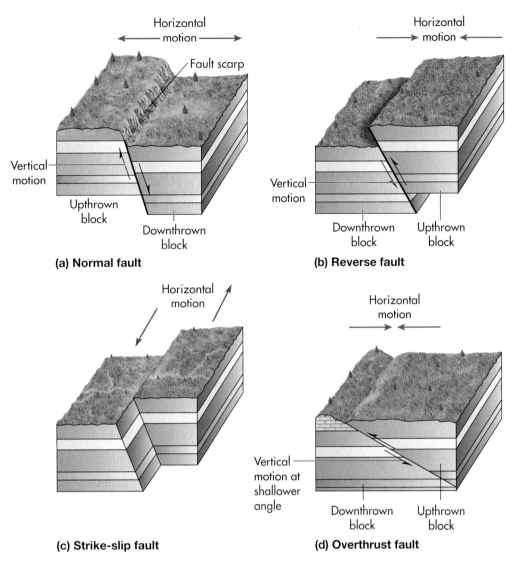

(a) Normal fault

(b) Reverse fault

(c) Strike-slip fault

(d) Overthrust fault

Figure 13.21 Types of faults. (a) Normal faults and (b) reverse faults result when one block moves up while the other moves down. (c) Strike-slip faults result when blocks move horizontally relative to one another. (d) Overthrust faults occur when the upthrown block also slides over the downthrown block.

Figure 13.22 A normal fault. This roadcut exposure shows a small normal fault in surface rocks near Death Valley, California. The fault plane is clearly visible as the diagonal contact between the two blocks of rock.

Alan Arbogast

Figure 13.23 The Sierra Nevada. The Sierra Nevada forms the western boundary of the Basin and Range Province. Uplift and westward tilting of the horst has formed an approximately 2700-m (~9000-ft) escarpment into the Owens Valley graben (foreground).

slip up relative to the fault plane as an upthrown block (called a **horst**), while the other slips down as a downthrown block (called a **graben**). The exposed side of the upthrown block forms a cliff-like feature known as a **fault escarpment** (or *scarp* for short). A fantastic example of a landscape created by normal faulting is the awesome Sierra Nevada in eastern California. As you approach this mountain range from the east, the horst forms an abrupt escarpment that rises over 2700 m (~9000 ft) above Owens Valley, the graben to the east (Figure 13.23). Given that the Sierra horst is tilted to the west, the slope toward the west is much more gradual.

Another kind of fault is a **reverse fault** (Figure 13.21b). Although a reverse fault looks very similar to the normal fault, the cause and nature of the movement differ. Whereas a normal fault entails movement of blocks away from each other, a reverse fault results when rock blocks move toward each other, causing one block to ride up steeply over the other.

As noted previously, overthrust faults result when one rock body is thrust up and over another, usually in association with folding. These faults differ from normal and reverse faults because their fault planes are usually at a shallower angle in comparison.

Of the four types of faults illustrated in Figure 13.21, three are directly associated with some kind of uplift—specifically the normal, reverse, and overthrust faults. In contrast to these types of faults, the **strike-slip fault** entails purely horizontal movement of the two plates past each other. Strike-slip faults and transform faults are closely related because they share the same kinds of horizontal movement. The primary difference between transform

and strike-slip faults is that transform faults are associated with large, tectonic plate boundaries, whereas strike-slip faults occur where small rock blocks move horizontally relative to one another.

As indicated previously, one of the best-known transform faults is the San Andreas Fault in California, which occurs along the boundary of the Pacific and North American plates. The Pacific plate is moving in a northwesterly direction along this boundary relative to the North American plate at an average rate of about 5 cm/year (2 in./year). Figure 13.24 shows

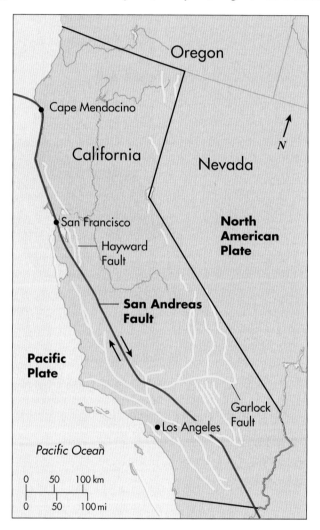

Figure 13.24 California fault systems. The San Andreas Fault is a transform fault that forms the boundary between the North American (to the east) and Pacific (to the west) plates. Arrows on either side of the fault reflect the relative movement of plates. Note the various subsidiary faults in the system.

Horst *An upthrown block of rock that lies between two steeply inclined fault blocks.*

Graben *A downthrown block of rock that lies between two steeply inclined fault blocks.*

Fault escarpment *A step-like feature on the Earth's surface created by fault slippage.*

Reverse fault *A steeply inclined fault in which the hanging rock block moves relatively upward.*

Strike-slip fault *A structural fault along which two lithospheric plates or rock blocks move horizontally in opposite directions and parallel to the fault line.*

R.E. Wallace/USGS

Figure 13.25 The San Andreas Fault. In Southern California, the Pacific plate (foreground) is moving toward the northwest (left) relative to the North American plate. Note the change in stream direction that occurs at the fault boundary. This distinct pattern evolved because the stream flows onto the Pacific plate from the North American plate. Because the Pacific plate has been gradually moving in a northwest direction relative to the North American plate, the stream has slowly changed direction as well.

the fault geography, including the major subsidiary faults in the San Francisco region. Many of these faults, such as the Hayward and Garlock faults, are strike-slip faults. In many areas, the San Andreas Fault is clearly visible as a large trough on the Earth's surface, as pictured in Figure 13.25. In this photograph, the Pacific plate is in the foreground, whereas the North American plate is the hilly ground in the background. You can also see how the movement of the fault has occurred through time by looking at the stream that flows from the North American plate onto the Pacific plate. Notice that the stream bends sharply at the fault and then flows directly on it for a short distance before turning onto the Pacific plate. In other words, the Pacific plate has moved at a sufficiently slow speed to allow the stream to bend with it. Geologists say that the total accumulated displacement along this fault is at least 560 km (~350 mi) since it came into being about 15 to 20 million years ago.

Given the immense size of the San Andreas Fault, the earthquakes associated with it are sometimes quite violent and destructive. The largest earthquake on the San Andreas in recorded history is the 1906 San Francisco earthquake, which had a magnitude of 7.9 on the Richter scale and caused a rupture along 470 km (290 mi) of the fault. This extensive earthquake caused as much as 8.5 m (28 ft) of near instantaneous slip in some places and resulted in extensive damage in San Francisco and the surrounding area. Between the effects of the earthquake and the catastrophic fire that occurred afterward, 700 to 800 people were killed and about 28,000 buildings were destroyed.

Human Interactions: Earthquakes as Natural Hazards

As discussed at the beginning of this chapter, examples abound of recent earthquakes that have severely impacted people in many places around the world. The 2010 earthquake in Haiti is, unfortunately, an excellent example of such a natural disaster. This earthquake measured 7.0 on the Richter scale and occurred along the Enriquillo-Plantain Garden fault system. It was particularly horrific because the epicenter was near Port-au-Prince, which is a densely populated and impoverished city that is ill equipped to handle such an event. As the 1906 San Francisco earthquake attests, such earthquakes have even occurred in the United States. Another big earthquake (magnitude 7.1), called the *Loma Prieta* earthquake, occurred along a branch of the San Andreas Fault south of San Francisco in 1989.

You may be surprised to learn that one of the largest events recorded in the continental United States occurred in 1811 and 1812 in the central part of the country, far from the active continental margin. This event took place within the New Madrid Seismic Zone, which extends through southeastern Missouri, western Tennessee, and northeastern Arkansas. At least four earthquakes occurred from December 1811 to February 1812 that had reconstructed magnitudes of about 8.0 on the Richter scale. These earthquakes were felt over a huge area, with even reports of church bells ringing 1800 km (~1100 mi) away in *Boston* due to the Earth shaking! Fortunately, the central United States was lightly populated at that time, and so the damage was minimal. If such an event occurred today, however, it would be absolutely devastating. According to a 2008 report by the Federal Emergency Management Agency, a serious earthquake in this area could result in the highest economic losses by a natural disaster in the United States, with potentially catastrophic damage in states ranging from Kansas to Kentucky, Illinois, Indiana, and Alabama. Missouri and Tennessee would be particularly hard it, with as many as 4000 fatalities projected in Memphis alone.

In this context, an earthquake is viewed as a **natural hazard** that has the potential of having a negative effect on people. Although many different kinds of natural hazards occur on Earth, such as hurricanes, tornadoes, landslides, and floods, to name a few, earthquakes are probably the most intense because they occur without warning and can kill many thousands of people in a very short period of time. Given the devastating potential impact of earthquakes on people, it is no wonder that geologists are working hard to better predict when and where earthquakes will occur. Unfortunately, with current knowledge, the best predictions are nothing more than statements of probability. In northern California, for example, geologists now believe a 62%

Natural hazard *An actual or potentially occurring natural event, such as an earthquake or volcanic eruption, that has a negative effect on people or the environment.*

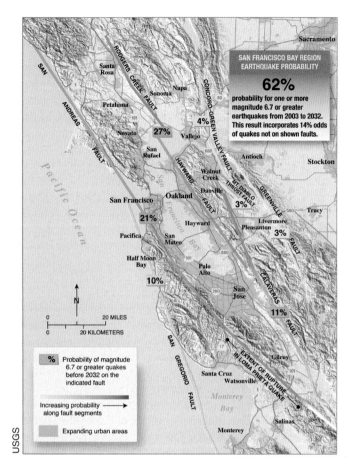

Figure 13.26 Earthquake probability map in the San Francisco Bay region. This map shows the probability of an earthquake with a magnitude 6.7 or greater on the San Andreas Fault and associated faults (such as the Hayward Fault) in the Bay region. Also shown are the extent of urban areas and the extent of the rupture that occurred along the San Andreas Fault during the Loma Prieta earthquake in 1989.

chance exists that an earthquake with a magnitude greater than 6.7 or greater will occur in the San Francisco region by 2032 (Figure 13.26).

In addition to the recent earthquakes already mentioned, a pair of additional examples demonstrates the extreme hazards that can be associated with earthquakes. In these cases, however, the primary hazard was not the earthquake itself, but rather the indirect effects that occurred minutes to hours after the shaking ended. Both of these events occurred on offshore faults near island chains that resulted in massive waves that blasted coastal communities.

The first of these events occurred in December 2004 when the second strongest earthquake ever recorded occurred near Sumatra, an Indonesian island in the eastern part of the Indian Ocean (Figure 13.27). This earthquake, known as the *Sumatra-Andaman* earthquake, occurred along the boundary between the Burma and Indian plates on the Sumatra Fault. The quake measured a staggering 9.1 on the Richter scale and continued for 10 minutes, which is a very long period of time considering that most earthquakes last no more than a few seconds. The quake was so strong that the *entire* Earth shook, with ground motions of at least 1.3 cm (0.5 in.) occurring *everywhere* on the globe as seismic waves spread from the epicenter. According to the United States Geological Survey, the quake caused the water level in a well in *Virginia* to suddenly increase 9 m (3 ft)! The length of the rupture along the Sumatra Fault associated with this quake was 1200 km to 1300 km (720 mi to 780 mi), making it the longest such rupture ever observed.

In addition to the substantial lateral movement of the adjoining plates, the seafloor on the Burma plate was suddenly thrust up and over the India plate about 4 m to 5 m (13 ft to 16 ft; Figures 13.28a and b). This thrust instantly displaced 30 km³ of water that resulted in a

VISUAL CONCEPT CHECK 13.2

This view of the San Andreas Fault is in central California. Which one of the following choices explains why there are hills along each side of the fault?

a) There is subduction occurring at the fault.
b) The San Andreas Fault is a normal fault.
c) There is rifting occurring along the fault.
d) There is compression of the rocks on either side of the fault.

Figure 13.27 The epicenter of the Sumatra-Andaman earthquake in December 2004. The initial earthquake (at red dot) measured 9.1 on the Richter scale and was the second most powerful quake on record. As a result of this earthquake, a massive tsunami was created that killed over 200,000 people along the coasts of the Indian Ocean. The yellow dots show the location of aftershocks.

tsunami, which is a kind of seismic wave that radiates outward from the disturbance zone. Tsunamis form and radiate outward in much the same way that water is displaced when you flick your hand upward while holding it beneath surface. Although tsunamis can develop due to a volcanic eruption or landslide, they are most commonly associated with earthquakes.

Once the water was displaced due to the Sumatra-Andaman earthquake, a tsunami rapidly radiated outward in the ocean (Figure 13.28c). The tsunami subsequently raced across the Indian Ocean at speeds ranging from 500 km/h to 1000 km/h (310 mi/h to 620 mi/h). It reached the northwest coast of Sumatra to the east in about 15 minutes and the southern

---

Tsunami *A destructive sea wave caused by a disturbance within the ocean, such as an earthquake, volcanic eruption, or landslide.*

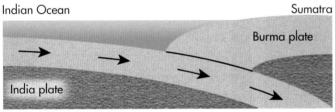

Stress gradually builds at subduction zone along India and Burma plate boundary.

(a) Before the earthquake

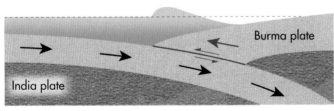

Earthquake causes Burma plate to thrust up and over the India plate several meters, causing massive displacement of overlying water.

(b) Earthquake strikes

Figure 13.28 Evolution of the 2004 tsunami in the Indian Ocean. (a) Tectonic stress builds up where the India plate subducts beneath the Burma plate along the Sumatra Fault. (*Source*: Geoscience Australia.) (b) Pent-up stress is suddenly released in a massive earthquake. In addition to significant horizontal movement, the Burma plate is thrust up and over the India plate. The vertical movement displaces water in the overlying Indian Ocean. (c) After the water is initially displaced, seismic waves rapidly radiate outward in the water.

Tsunami spreads rapidly, reaching the nearby coast of Sumatra to the east in 15 minutes and the more distant shore of Somalia to the west in 7 hours.

(c) Tsunami radiates outward along the length of fault rupture

EARTHQUAKE PREDICTION AND THE LEGAL SYSTEM IN ITALY

One of the primary goals of seismologists is one day to be able to predict when an earthquake is about to happen. So far, however, the ability to make that kind of warning about forthcoming seismic activity is beyond the reach of geologists. Instead, they think about the probability that an earthquake of a certain magnitude will occur in a certain place within some future interval of time (see Figure 13.26). In spite of this limitation, it is nevertheless important that scientific understandings about earthquakes as natural hazards be effectively communicated so that people can plan accordingly.

The tension between predicting when an earthquake might occur and the communication of scientific information was illustrated in 2009 in Italy. In April of that year, a magnitude 5.8 earthquake struck L'Aquila, which is in the central part of the country. The earthquake damaged over 5000 buildings, killed about 300 people, and left about 40,000 people homeless. In addition to this tragic loss of human life and resulting chaos, the aftermath of the earthquake was particularly noteworthy because seven members of the *Italian National Commission for the Forecast and Prevention of Major Risks* were criminally accused of not providing adequate warning before the event occurred.

The case against the commission members revolved around the seismic activity that occurred before the major event and the way that the potential risk was described to the community. A month before the earthquake a laboratory technician had measured elevated levels of radon in the area and had thus publicly predicted that a major seismic event would soon occur. Although some scientists believe a link may exist between radon levels and earthquakes, no correlation has been scientifically verified. As a result, the director of the commission accused the technician of being alarmist and inciting fear. A few days before the earthquake, a series of tremors occurred, including a pair of strong shocks the day before. After these events, commission members essentially said that there was no cause for concern, although some acknowledged a major earthquake could occur at some undetermined time in the future.

Given the consensus that earthquakes cannot be currently predicted, it was believed that the commission members would be acquitted. Instead, each of the defendants

Shake map of the L'Aquila earthquake in central Italy. The earthquake occurred at a depth of 9.46 km (5.9 mi) and was about 90 km (60 mi) northeast of Rome.

Earthquake damage to the local government office in L'Aquila. Over 5000 buildings were similarly damaged, and about 300 people were killed, making it the worst earthquake in Italy since 1980.

was convicted of involuntary manslaughter for "an assessment of the risks that was incomplete, inept, unsuitable, and criminally mistaken." As a result, they were sentenced to six years in prison and prohibited from ever holding public office again. The conviction caused a furor in the global scientific community because it appeared that the defendants were convicted of failing to predict an event that is unpredictable. Some saw much broader implications for science as a whole, arguing that penalizing scientists for making what turns out to be incorrect predictions or for not predicting something that does occur will restrict research to safe certainties. Others argued, however, that the defendants indeed did a poor job communicating the overall seismic risk in the area and had failed in that regard.

tip of South Africa, which is about 8500 km (5300 mi) to the southwest, in about 12 hours (Figure 13.29). Coincidentally, a French radar satellite happened to pass over the Indian Ocean 2 hours after the earthquake and measured the tsunami wave heights as they moved across deep water. These observations indicated that the maximum wave height in the open ocean was only 60 cm (2 ft), which would have scarcely been noticed by ships in the area. As the waves approached the shore, however, they grew rapidly due to the shallow water, reaching an estimated height of 24 m (80 ft) at Banda Aceh in the northwest coast of Sumatra. Some of the tsunami's energy spread into the Pacific Ocean, even creating a tsunami along the west coasts of North and South America that measured from 20 cm to 40 cm (7.9 in. to 15.7 in.) in height.

Following the Indonesian tsunami in 2004, yet another tsunami was triggered in March 2011 along the coast of Japan by a magnitude 9.0 earthquake. This quake, which is known as the *2011 Tōhoku earthquake*, occurred where the Pacific plate is subducting under the plate beneath northern Honshu. It was the fourth largest earthquake measured on Earth since 1900 and was preceded by a number of strong (up to magnitude 7.0) foreshocks in the previous couple of days. The epicenter of the main event was located about 70 km (40 mi) east of the Oshika Peninsula (Figure 13.30a) and occurred at a fairly shallow depth of about 29 km (19 mi). It lasted approximately six minutes and moved parts of northeastern Japan up to 2.4 m (7.9 ft) *closer* to the coast of North America. The amount of slippage along the fault was perhaps as much as 40 m (130 ft) in some places, and the seafloor was instantaneously raised up to 7 m (23 ft) along the fault.

As a result of the raised seafloor, water was quickly displaced above the fault in much the same manner as the uplift that produced the Indonesian tsunami. The resulting tsunami reached the shore of Japan in less than an hour, was up to 6 m (20 ft) high in places (Figure 13.30b), and inundated 560 km² (217 mi²) of the Japanese coast. The tsunami rippled across the Pacific Ocean and reached the coast of North America in about 9 hours. Surges up to 2.4 m (8 ft) occurred along the coasts of Oregon and California, damaging docks and marinas to the tune of about $10 million in losses. Tsunami waves reached the coast of Chile about 22 hours after the earthquake but caused relatively minor damage.

The impact of both the Indonesian and Japanese tsunamis on coastal communities was devastating. Video from numerous sites during both events illustrated the incredible power of moving water. In each case these clips showed several waves—best described as enormous surges of water—that completely engulfed coastal communities (Figure 13.30b) while blasting everything in the way (Figure 13.31). The energy released by the Indonesian tsunami is estimated to have been equivalent to about 5 megatons of TNT, which is *more than all* of the explosive energy used during World War II (including

Figure 13.29 Travel time in hours for the 2004 tsunami in the Indian Ocean. The tsunami reached the shores of Indonesia and Thailand within minutes of the earthquake. In contrast, it reached the southern tip of South Africa in about 12 hours. (*Source*: NOAA.)

USGS Community Internet Intensity Map
NEAR THE EAST COAST OF HONSHU, JAPAN
Mar 11 2011 02:46:23 PM local 38.3N 142.37E M9.0 Depth: 29 km ID:usc0001xgp

1415 responses in 149 ZIP codes and 329 cities (Max CDI = IX)   100 km

INTENSITY	I	II-III	IV	V	VI	VII	VIII	IX	X+
SHAKING	Not felt	Weak	Light	Moderate	Strong	Very strong	Severe	Violent	Extreme
DAMAGE	none	none	none	Very light	Light	Moderate	Moderate/Heavy	Heavy	V. Heavy

Processed: Tue Jul 23 01:17:46 2013

(a)

(b)

AFP/Getty Images

Figure 13.30 The 2011 Tōhoku earthquake and associated tsunami. (a) The earthquake occurred off the northeast coast of Honshu (star) and measured 9.0 on the Richter scale. (b) A tsunami was produced by the earthquake that surged into coastal communities.

the two atomic bombs). If the power of the Japanese earthquake and tsunami could be harnessed, it is estimated that it would have been sufficient to power the metropolitan Los Angeles region for *a year*. The devastation in both events was so extensive and widespread that it can be clearly seen in satellite imagery showing the landscape before and after the tsunamis (Figure 13.32).

The Indonesian and Japanese tsunamis offer an excellent opportunity to compare the state of earthquake and tsunami preparedness in each region when the respective events occurred and how people were impacted. At the time of the In-

donesian tsunami in 2004, no warning system existed in the Indian Ocean basin. As a result, people living in coastal communities were unaware of impending disaster as the tsunami approached. Ironically, the first hint of trouble came in the brief moments before tragedy when the ocean was drawn rapidly away from the coast, as much as 2.5 km (1.6 mi) in some places, as water was pulled from the shore into the rapidly approaching wave. Video shows that many people were excited by this sudden exposure of the ocean bottom and even ran onto it to collect stranded fish. Unfortunately, they were unaware that the sudden retreat of water and beach expansion meant that a horrific wave was about to engulf them. Because no warning was given, more than 200,000 people perished, with over 160,000 fatalities in Indonesia alone. Another 35,000 people were killed in Sri Lanka.

Michael L. Bak/Department of Defense/Getty Images

(a)

YOMIURI SHIMBUN/AFP/Getty Images

(b)

Figure 13.31 Coastal impacts associated with the 2004 and 2011 tsunamis. (a) Damage along the shore of the Indian Ocean due to the 2004 tsunami. (b) Damage on the northeast coast of Japan resulting from the 2011 tsunami.

(a)

(b)

(c)

Figure 13.32 Satellite imagery of coastal devastation associated with the 2004 and 2011 tsunamis. (a) Satellite image of Banda Aceh, Indonesia, before the tsunami. (b) Satellite image of Banda Aceh, Indonesia, after the tsunami. Note the change in the configuration of the shoreline and the reduction in island size. (c) Satellite image of a portion of Soma, Japan, before the tsunami (left) and after (right). Note the dramatic change in the shape of the coastline in the center part of the image.

In contrast to the lack of warning for people along the coast of the Indian Ocean, the Japanese had a very effective warning system in place when the 2011 event occurred. The Japanese are well acquainted with tsunamis (the name is Japanese: *tsu* meaning "strong" and *nami* meaning "wave") because about 200 major events have occurred in their country alone in the past 1300 years. Given this high frequency, as well the common occurrence of major earthquakes along the various countries of the Pacific Rim, a tsunami warning system was established in the Pacific Ocean in 1949. In association with this system, an extensive network of sensors was deployed throughout the Pacific basin in the 1950s and 1960s that monitors seismic events and associated fluctuations in sea level. Enhancements to this system were motivated, in part, by the great Alaska earthquake in 1964, which produced a wave that reached a height up to 6.3 m (21 ft) in northern California and killed 11 people.

Although this system is of little use during a sudden tsunami, it can provide warning within 15 minutes for events that originate far away. The effectiveness of the Japanese warning system was apparent during the 2011 event. In comparison to the incredibly high death toll associated with the Indonesian tsunami, "only" about 16,000 people died during the Japanese event. This relatively low death toll occurred because most people quickly moved to higher ground when the warning was sounded. Of the people who unfortunately perished, about 65% were over 60 years old, suggesting they could not move fast enough to reach safety. Clearly, the Japanese tsunami system saved lives. In this context, it is very good news that a tsunami warning system was activated in the Indian Ocean in 2006. The effectiveness of this system was tested when yet another large (magnitude 8.4) earthquake struck Banda Aceh, Indonesia, in 2012. Nearby islands in the Indian Ocean were warned within 8 minutes that a tsunami might be approaching. Fortunately, that particular earthquake did not produce such a wave.

A side story of the tragic Indonesian tsunami demonstrates why it is important to understand natural Earth processes. As previously discussed, most people had no idea that a tsunami was approaching. As mentioned earlier, the only hint of something unusual was the rapid recession of water shortly before the surge struck. To most people this withdrawal of water was a source of curiosity, or even an opportunity to catch fish easily. For a 10-year old British girl on a family holiday in Thailand, however, it was a warning that a tsunami was approaching because she had just studied tsunamis in her primary school. She keenly recognized that the sudden recession of water from the beach meant a tsunami was approaching and thus pleaded with her parents to run for safety. Her parents were understandably skeptical at first, but she was insistent. As they left the beach, the girl warned others nearby of the approaching disaster and convinced them to leave. In the end, the awareness and persistence of this little girl saved her life, as well as the lives of over 100 others.

To review the various kinds of faults, go to the *Geo Media Library* and access the animation *Types of Faults*. This animation contains a video presentation that illustrates the fault types discussed in this chapter and will demonstrate the various ways that faults move and the stresses that cause this movement. As you watch the media, try to relate the various fault movements to the character of the landscape. After you have completed the exercise, be sure to answer the questions at the end to test your understanding of fault types.

KEY CONCEPTS TO REMEMBER ABOUT FAULTS AND EARTHQUAKES

1. There are four major kinds of faults: normal, reverse, overthrust, and strike-slip faults.

2. Earthquakes occur when moving plates are stuck along fault planes, causing stress to increase until a rupture occurs along the fault.

3. The place within Earth where a fault breaks is called the focus. The location on the surface directly above the focus is called the epicenter.

4. Earthquakes are measured with a seismograph, which measures the intensity of P and S waves that radiate outward from the earthquake focus.

5. The strength of earthquakes is measured on the logarithmic Richter scale, which indicates the magnitude of seismic waves passing through the Earth's crust.

6. Earthquakes are perhaps the most intense natural hazard on Earth, often resulting in catastrophic damage and loss of human life. As a result, geologists are working to improve earthquake prediction.

7. Devastating tsunamis can result when earthquakes occur in the world's oceans. The Indonesian tsunami in 2004 and the Japanese tsunami in 2011 reflect the tremendous damage that can result from these events. Evidence indicates that tsunami warning systems save lives.

Volcanoes

After investigating the processes and geography of earthquakes, it is time to discuss volcanoes, which are the other major feature associated with plate tectonics. As briefly discussed in Chapter 12, a **volcano** is a mountain or large hill containing a conduit that extends down into the upper mantle, through which magma, ash, and gases are periodically ejected onto the surface of Earth or into the atmosphere (Figure 13.33). In most cases,

Volcano *A mountain or large hill containing a conduit that extends down into the upper mantle, through which magma, ash, and gases are periodically ejected onto the surface of Earth or into the atmosphere.*

(a)

(b)

Figure 13.33 Volcano eruptions. (a) Ash and steam cloud from Augustine Volcano in Alaska. (b) Magma eruption at the Pu'u 'O'o Volcano in Hawaii.

volcanoes are inactive for some time and erupt only when the pressure of material rising from the mantle becomes excessive. The length of inactivity varies dramatically between volcanoes. Some volcanoes lie dormant for hundreds or thousands of years before they erupt, whereas others are in a near-constant state of eruption.

Although no hard and fast classification scheme exists for volcanoes, most geologists distinguish three basic kinds: cinder cones, composite volcanoes, and shield volcanoes. This classification depends in large part on whether the eruption is explosive or fluid. The first part of the discussion will focus on the volcanoes associated with explosive eruptions.

Explosive Volcanoes

As the name implies, an explosive volcano is one that erupts very quickly and with great force. Of the two kinds of explosive volcanoes that occur on Earth, the easiest kind to understand is a **cinder-cone volcano**, which usually forms very quickly after a single eruption. These volcanoes are small relative to other types of volcanoes, have steep sides (~30°), and consist of solidified magma fragments, rock debris, and ash that are ejected from a central vent. A great example of a cinder-cone volcano can be seen in northeastern New Mexico at Mount Capulin (Figure 13.34). This volcano has erupted only once, about 62,000 years ago, and is approximately 300 m (1000 ft) high.

In contrast to cinder-cone volcanoes, **composite volcanoes** are volcanoes that build up and grow over the course of several eruptions. Composite volcanoes are typically inactive for long periods of time between eruptions, but when they do erupt, they tend to do so quite violently. Such an eruption occurs because the magma within composite volcanoes is rich in silicas, which are minerals containing silicon (Si), and therefore highly viscous (meaning sticky and slow flowing). As a result, gases are gradually trapped in the magma during the inactive phase and build up pressure within the volcano until it explodes. Such an eruption sends volcanic ash high into the atmosphere, and thick layers of volcanic debris will accumulate on the slopes of the volcano, causing it to enlarge. This debris may consist of alternating layers of lava, which is magma flowing on the surface, and fragmented rock debris called **pyroclastic material** (or *tephra*), such as volcanic ash, cinders, and boulders (Figure 13.35a). Also called *stratovolcanoes* because they contain strata (or layers) of volcanic debris, these volcanoes typically have moderately steep cones with a semi-horizontal

Figure 13.34 Cinder-cone volcanoes. Cinder-cone volcanoes are relatively small volcanoes that build up by the accumulation of solidified magma fragments, rock, and ash over a short period of time. Mount Capulin is a typical cinder-cone volcano and is about 305 m (1000 ft) high.

top containing the crater. Composite volcanoes are much larger than cinder cones, perhaps over 3000 m (10,000 ft) high. A beautiful example of a composite volcano is the famous Mount Fuji in Japan (Figure 13.35b).

Sometimes the eruption of a composite volcano is so explosive that it literally blows the top off of a mountain, creating a large crater. After this massive kind of eruption occurs, the crater may partially fill with a **lava dome**, which forms slowly in the ensuing decades. Such a volcanic feature consists of a steep-sided mound built by highly viscous magma that gradually oozes from the central vent. Although these features sometimes occur alone and are thus classified as a particular type of volcano by many geologists, the vast majority of the recently active lava domes occur in association with composite volcanoes.

Volcanic Arcs at Plate Boundaries Composite volcanoes are most commonly associated with plate boundaries in places where subduction is occurring. This geographical relationship occurs because the subducting plate melts when it encounters the upper mantle and the molten material gradually moves upward through the crust. Given that subduction tends to occur along the length of certain plate boundaries, these zones are places where a chain of volcanoes is typically located at the surface. Such a chain is called a **volcanic arc** and is genetically similar to the island arc

---

Cinder-cone volcano *A small, steep-sided volcano that consists of solidified magma fragments and rock debris that may form in only one eruption.*

Composite volcano *A large, steep-sided volcano that grows through progressive volcanic eruptions, which are usually explosive, and consists of layers of volcanic debris.*

Pyroclastic material *Fragmented rock materials resulting from a volcanic explosion or ejection from a volcanic vent.*

Lava dome *A steep-sided volcanic landform consisting of highly viscous lava that does not flow far from its point of origin before it solidifies.*

Volcanic arc *A chain of volcanoes created by rising magma derived from a subducting tectonic plate.*

(a)

(b)

Figure 13.35 Composite volcanoes. (a) Composite volcanoes consist of layers of lava and pyroclastic material that pile up around one or more conduits to subterranean magma chambers. (b) Mount Fuji in Japan, reverently called *Fuji-san* by the Japanese people, is a composite volcano.

shown in Figure 13.11. A well-known tectonic feature associated with a number of volcanic arcs is the **Pacific Ring of Fire**, which follows most of the outline of the Pacific plate (Figure 13.36). Using New Zealand as an arbitrary starting point, the Pacific Ring of Fire extends northwestward to the eastern coast of Asia, and north across the Aleutian Islands of Alaska. From there, it

continues south along the western coast of North and South America. This region contains about 75% of the world's volcanoes, including Mount Ruapehu in New Zealand, Mount Pinatubo in the Philippines, Mount Fuji in Japan (Figure 13.35b), Kliuchevskoi Volcano on Russia's Kamchatka Peninsula, Augustine Volcano in Alaska (Figure 13.33a), Mount Garabaldi in Canada, and Cotapaxi in Chile, just to name a few. Most of these volcanoes are composite structures and have formed because subduction is occurring at many places along the Pacific plate. Given the tectonic stresses that occur in these areas, the plate boundary is known for earthquakes.

Pacific Ring of Fire *The chain of volcanoes that occurs along the edge of the Pacific lithospheric plate.*

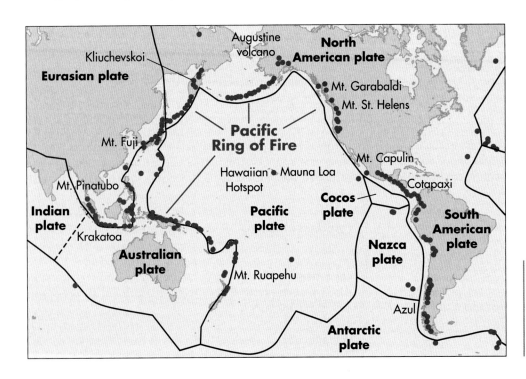

Figure 13.36 The Pacific Ring of Fire. Subduction along the Pacific plate and subsidiary plates (such as the Nazca and Cocos plates) has produced many prominent composite volcanoes. The locations of a few of them are shown here.

Cascade eruptions during the past 4000 years

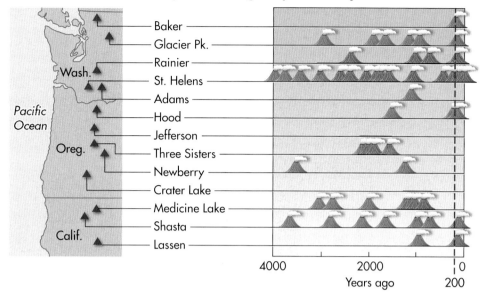

Figure 13.37 The Cascade Volcanic Arc. This diagram shows the geography and eruption history of volcanoes within the Cascade Volcanic Arc in the United States. The arc also extends for a short distance into British Columbia, Canada, where five volcanoes occur. Mount St. Helens in Washington has clearly been the most active volcano in the arc, with 15 eruptions in the past 4000 years. Mount Jefferson, in contrast, has not erupted once during this timeframe. The Canadian volcanoes have been largely inactive as well. (*Source*: U.S. Geological Survey.)

In the Pacific Northwest of North America, the Pacific Ring of Fire is associated with a chain of volcanoes known as the *Cascade Volcanic Arc* (Figure 13.37). This volcanic arc is present because the Juan de Fuca plate, which is located off the west coast of North America (see Figure 13.2), is subducting beneath the North American plate to the east. The best known of the Cascade volcanoes is Mount St. Helens, which lies in the southern part of the state of Washington. This volcano is so well known because the most recent major volcanic eruption in the mainland United States occurred here in May 1980. It has also been the most active volcano in the arc within the past 4000 years (Figure 13.37).

Prior to the eruption, Mount St. Helens had a classic conical shape (Figure 13.38a), and, as a result, it was frequently called the Mount Fuji of North America because it was such a picturesque volcano. Mount St. Helens had been inactive for 123 years when it slowly reawakened in the spring of 1980. Several small eruptions occurred throughout April and early May of that year that opened a new crater and sent periodic clouds of volcanic debris onto the surrounding landscape. On May 18 the volcano erupted catastrophically (Figure 13.38b). Much of the volcanic cone of Mount St. Helens was blown off in this eruption, and the

surrounding landscape was devastated (Figure 13.38c). Subsequently, a nice example of a lava dome formed in the center of the crater. Evidence for this eruption, such as the new pattern of streams and the partial filling of nearby Spirit Lake with volcanic debris, can still be seen on satellite imagery.

Mount St. Helens remains the most active volcano in the Cascade Volcanic Arc apparently because it continues to be more consistently fed with magma from the subducting Juan de Fuca plate than the other volcanoes in the region. Following a short period of increased seismic activity, Mount St. Helens erupted again in October 2004, creating a moderate ash plume. Volcanic activity continued sporadically into May 2005, including low levels of seismic activity, emissions of steam and volcanic gases, and minor productions of ash. None of this activity was catastrophic in nature and was merely associated with enlargement of the lava dome.

Given the active history of Mount St. Helens in the past 4000 years (Figure 13.37), this volcano continues to be closely monitored by the U.S. Geological Survey, as are other large volcanoes such as Mount Rainier and Mount Adams. It is possible that one of these volcanoes, or another one in the Cascade Volcanic Arc, will violently erupt in your lifetime. If such an eruption were to

---

www.wiley.com/college/arbogast

Walk the Pacific Ring of Fire

WileyPLUS

A good way to appreciate the Pacific Ring of Fire is to experience it on a virtual tour. To do so, go to the *Geo Media Library* and access the animation *Walk the Pacific Ring of Fire*. This animation allows you to work your way around the margin of the Pacific plate and observe the nature of its boundary

with adjacent plates. It will also provide a sampling of notable tectonic events that have occurred at individual places along the plate boundary. Once you complete this tour, be sure to answer the questions at the end to test your understanding of this concept.

(a)

(b)

(c)

Figure 13.38 Eruption of Mount St. Helens. (a) The near-perfect conical shape of Mount St. Helens before the May 1980 eruption. (b) The catastrophic eruption on May 18, 1980. (c) Mount St. Helens after the eruption. The blast was so explosive that much of the peak was blown off, reducing the height of the mountain from 2950 m (9677 ft) to about 2550 m (8364 ft) and leaving a wide crater. Note the lava dome (arrow) in the center of the crater.

occur on Mount Rainier, for example, it would truly be a catastrophic event. In addition to the destruction caused by the blast, the eruption would cause ancient glaciers on the mountain summit suddenly to melt, resulting in a massive flood of volcanic debris and water (called a *lahar*) that would sweep toward the populated areas near Tacoma, Washington. This flood would likely cause many deaths and massive property loss.

Fluid Volcanoes

In addition to the volcanoes that erupt explosively, a number of prominent volcanoes have very fluid eruptions with flowing rivers of lava. In contrast to the highly viscous magma associated with explosive volcanoes, the magma that results in fluid eruptions contains far less silica and is thus less viscous. As a result, the lava associated with such an eruption flows across the surface until it cools to form basalt. Such eruptions can occur for weeks and months at a time and generally take place more frequently than explosive eruptions.

Although fluid eruptions are occasionally associated with composite volcanoes, such as Mount Nyaragongo in the Demo-

cratic Republic of the Congo in eastern Africa, they usually result in the formation of a **shield volcano** that has shallow, sloping sides (Figure 13.39). This type of volcano develops because successive eruptions of fluid magma cause the volcano to build upward gradually over a much broader area than composite volcanoes. A number of prominent shield volcanoes occur on Earth in association with subduction zones and rift zones. One of the best known is Skjaldbreiður, which straddles the Mid-Atlantic Ridge in Iceland. The name of the volcano is derived from the Icelandic term for *broad shield* and is the source of the more general term that is applied to similar volcanoes around the world. In Alaska, a prominent shield volcano is Mount Sanford, which is part of the Wrangell Volcanic Field in the eastern part of the state. This volcano began to form about 900,000 years ago and was most recently active approximately 320,000 years ago. This shield volcano is unusual because one of the slopes is very steep, rising 2400 m (8000 ft) over the course of 1.6 km (1 mi).

Shield volcano *A very broad volcano with shallow slopes that forms in association with nonviscous lava flows.*

Lava fountain — Magma chamber — Central vent — Widespread lava flow — Layered basalt deposits — Mantle plume

Figure 13.39 Shield volcanoes. Shield volcanoes are broad, low features that surround a central magma chamber. Magma associated with shield volcanoes typically has a low viscosity, which allows it to flow freely in broad streams of molten lava.

Hotspots

The previous discussion focused on volcanoes that form along subduction zones and rift zones. In some cases, however, a volcano is associated with a stationary zone in the asthenosphere where upwelling magma from a mantle plume is released at the surface. Such a geologic feature is called a **hotspot**. Although hotspots occur more randomly than sub-

Hotspot *A stationary zone of magma upwelling that is associated with volcanism within the interior of a crustal plate.*

duction zones and rift zones, which can be traced for great distances, they are associated with some of the most intensive volcanic activity on the planet. These eruptions can be of either a fluid or explosive nature.

A great example of fluid eruptions associated with a hotspot can be seen in the state of Hawaii, which lies in the middle of the Pacific tectonic plate where the *Hawaiian hotspot* occurs. The hotspot currently underlies the island of Hawaii, which is the easternmost island in the chain (Figure 13.40a). This hotspot is releasing highly fluid magma that has built a variety of shield

(a) Pacific plate — Kauai (oldest) — Oahu — Maui — Hawaii (youngest) — Direction of plate motion — Fixed hotspot — Upwelling magma (mantle plume)

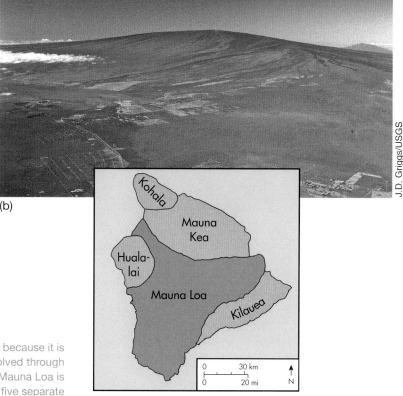

(b)

J.D. Griggs/USGS

(c) Kohala — Mauna Kea — Huala-lai — Mauna Loa — Kilauea

0 30 km
0 20 mi
N

Figure 13.40 The Hawaiian Hotspot. (a) This hotspot is fixed because it is at the top of a mantle plume. The Hawaiian island chain has evolved through progressive movement of the Pacific plate over the hotspot. (b) Mauna Loa is the largest volcano on Earth. (c) The island of Hawaii consists of five separate shield volcanoes.

volcanoes on the island, such as Mauna Loa (Figure 13.40b). Mauna Loa is the largest volcano on Earth and is one of five that collectively comprise the island (Figure 13.40c).

Evidence indicates that the Hawaiian hotspot has been active for a long time and has a fascinating history related to continental drift. Figure 13.41 shows a trail of islands and submerged highlands called *seamounts* that extend to the northwest from the island of Hawaii. This trail exists because the oceanic crust in the Pacific basin has rotated over the Hawaiian hotspot for the past 70 million years. Do you see the sharp turn in the island/seamount chain? The features north of that point, beginning with the *Emperor Seamount Chain*, must have formed when the plate was moving generally to the north as it flowed over the hotspot. Shortly after the Emperor Seamount Chain formed, approximately 43 million years ago, the plate began moving more toward the northwest. This movement has continued to the present day and explains why the state of Hawaii includes a string of islands northwest of the island of Hawaii. Each of these islands was formed when that particular part of the Pacific plate was over the hotspot. With an eye to the future, where would you expect the next island of the Hawaiian Islands to develop—to the northwest of Hawaii or to the southeast? Why?

Whereas the Hawaiian hotspot provides an excellent example of a fluid system that lies within the middle of the ocean, hotspots are also associated with explosive volcanic eruptions that occur within the interior of continents. Such a continental example is the Yellowstone hotspot, which lies beneath Yellowstone National Park in northwestern Wyoming. Similar to the Hawaiian hotspot, the Yellowstone hotspot is a fixed zone of upwelling magma. In this case, however, the hotspot lies beneath the North American plate and the magma is highly viscous. Geological studies indicate that the apparent location of the Yellowstone hotspot has moved over time (Figure 13.42a), reflecting the west/southwestern migration of the North American plate. About 16 million years ago the hotspot was located in what

is now the southeastern corner of Oregon. Between 12 million and about 2 million years ago, it migrated across what is now southern Idaho.

The Yellowstone hotspot has been in its present location for approximately the past 2 million years, with three major eruptions occurring during this time at 2.2 million years ago, 1.3 million years ago, and about 630,000 years ago. Each of these eruptions was cataclysmic, resulting in the formation of a giant caldera (Figure 13.42b). The Yellowstone caldera—that is, the crater from the most recent eruption—is over 50 km (~32 mi) wide. As you can imagine, eruptions of this magnitude must have ejected enormous amounts of volcanic debris. The first eruption in the area, which is the largest of the three, ejected 2500 km^3 (~1550 mi^3) of material. By comparison, only 1 km^3 to 2 km^3 (~0.6 mi^3 to 1.2 mi^3) of material was ejected by Mount St. Helens in 1980. Some more recent eruptions have occurred at Yellowstone, such as a lava flow around 70,000 years ago and a steam explosion about 14,000 years ago, but these events were relatively small and isolated.

Imagine what the large Yellowstone eruptions must have been like! If a similar eruption happened in the present time, it would impact the United States in a truly catastrophic way. It is conceivable that the eastern half of the country would *shut down* as long as the eruption occurred and volcanic ash fell across the landscape. The last major eruption of Yellowstone about 600,000 years ago, for example, resulted in ash that accumulated to a depth of about as much ~30 cm (~1 ft) in what is now Iowa, over 1200 km (~750 mi) away. A glimpse into the nature of the potential impact was seen in April 2010 when an erupting volcano in Iceland grounded air travel in Europe for about a week due to high concentrations of ash in the atmosphere. Given the truly gargantuan size of a potential Yellowstone eruption and the fact that it would likely last for weeks, it would have devastating consequences for the nation's economy and human health downwind. Global climate would also be impacted because the extensive ash plume would reflect solar

World Ocean Floor Panorama, Bruce C. Heezen and Marie Tharp, 1977 © by Marie Tharp 1977/2003

Figure 13.41 Migration of the Hawaiian hotspot over time. In addition to creating the modern Hawaiian Islands, the apparent migration of the hotspot formed the Emperor Seamount Chain. Note that the chain contains a sharp turn (arrow) that reflects a shift in the drift of the Pacific plate.

(a)

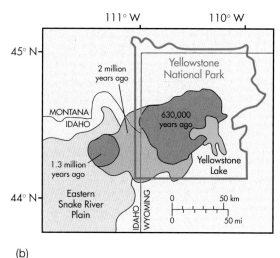

(b)

Figure 13.42 The Yellowstone Hotspot. (a) Migration of the Yellowstone hotspot between 16 million and 600,000 years ago. Although it appears that the hotspot has moved, it is the North American plate that has migrated toward the southwest. (b) Calderas in the vicinity of Yellowstone National Park in northwest Wyoming. Each of these calderas formed during individual eruptions.

radiation, potentially causing a series of very cool years. Fortunately for us, geologists believe that the next major eruption of Yellowstone is thousands of years away.

Nevertheless, if you have been to Yellowstone, you know that the underlying hotspot indeed remains active today. Geothermal features such as geysers, mudpots, and fumeroles are very common within the park. A **geyser** is a superheated fountain of water that suddenly sprays into the air on a periodic basis (Figure 13.43).

This process occurs because boiling water beneath Earth is constricted as it rises through a subterranean passageway. When the pressure builds sufficiently, the water bursts into the sky. When the geyser has used all the water in its reservoir, it dies away until the water is replenished from the surface or underground springs.

Geyser *A superheated fountain of water that suddenly sprays into the air on a periodic basis.*

(a)

(b)

Figure 13.43 Geothermal features at Yellowstone National Park. (a) Schematic illustration of geyser processes. Water below the Earth's surface is superheated and rises in small cracks in the overlying rock. If a constriction exists in this rock, the heated water backs up until the growing pressure causes it to blast toward the surface where it erupts. (b) Old Faithful Geyser is probably the most famous system of its kind in the world. It is so named because it erupts on average every 91 minutes.

Given your understanding of volcanoes, describe the volcano pictured here. Is it a cinder-cone, composite, or shield volcano? How can you tell? What kind of eruption will most likely occur when it does erupt?

Lyn Topinka/USGS

A **mudpot** consists of a bubbling mixture of gaseous mud and water. These systems form where hot water is limited and hydrogen sulfide gas is present, creating sulfuric acid. This acid dissolves the surrounding rock into fine particles of silica and clay that mix with what little water there is to form the mudpot. A **fumerole** is a steam vent that results because underlying groundwater is boiled away before reaching the surface.

KEY CONCEPTS TO REMEMBER ABOUT VOLCANOES

1. Volcanoes occur most frequently on plate boundaries. In particular, most of the world's volcanoes are found along the Pacific Ring of Fire.

2. Volcanoes are most frequently associated with the process of subduction because oceanic crust melts and then rises through cracks in overlying continental crust. The Cascade volcanic arc is an example of this relationship.

3. Volcanic eruptions can be broadly classified as explosive or fluid. Explosive eruptions occur when the magma is viscous, forming composite volcanoes (also called stratovolcanoes) such as those in the Cascades. When the magma is not viscous, it flows freely, resulting in broad rivers of lava that collectively form shield volcanoes such as those in the Hawaiian Islands. Cinder-cone volcanoes evolve through the accumulation of solidified magma fragments, rock debris, and ash that are ejected from a central vent.

4. Some volcanoes form over hotspots, which are places where upwelling magma reaches the surface. Examples of hotspot activity are the Hawaiian Islands and Yellowstone National Park.

Mudpot *A bubbling mixture of gaseous mud and water at the Earth's surface that is associated with geothermal activity.*

Fumerole *A steam vent that results because underlying groundwater is boiled away before reaching the surface.*

www.wiley.com/college/arbogast

Volcanoes

WileyPLUS

Now that we have discussed volcanoes thoroughly, you can see their associated processes in animated form. Go to the *Geo Media Library* and access *Volcanoes*. This animation contains a video that compares shield and composite volcanoes and shows various examples of each kind. As you watch the video, note how the concept of plate tectonics fits into the evolution of volcanoes, where they occur, and how they erupt. After you complete the animation, be sure to answer the questions at the end to test your understanding about volcanoes.

SUMMARY OF KEY CONCEPTS

1. The Earth's crust is not a single sheet, but rather a jigsaw puzzle of interconnected plates. A variety of evidence (location of fossil remains, apparent magnetic polar wander, and actual measurement of motion) strongly supports the theory of continental drift, indicating that plates have shifted through geologic time.

2. Plate margins can be passive, transform, converging, or diverging. At diverging plate margins, magma plumes from the asthenosphere cause rifting, either on continents or on the seafloor. At converging plate margins, plates collide or subduct, causing geomorphic features such as alpine chains and volcanoes.

3. Folding is the process by which rocks are compressed and deformed. Several kinds of geologic structure are associated with folded topography, including (a) monoclines, (b) synclines, (c) anticlines, (d) overturned folds, and (e) overthrust faults.

4. Four major kinds of faults occur on Earth: (a) normal, (b) reverse, (c) overthrust, and (d) strike-slip faults. Earthquakes occur when moving plates are stuck along a fault, which increases stress until a rupture occurs along the fault plane. The strength of earthquakes is measured on the logarithmic Richter scale, which indicates the magnitude of seismic waves passing through the Earth's crust.

5. A volcano is a mountain that erupts periodically when the buildup of pressure beneath it passes a critical threshold. Most volcanoes occur along subduction zones. Three primary kinds of volcanoes occur on Earth: (a) cinder-cone volcanoes, (b) composite volcanoes, and (c) shield volcanoes. The type of volcano depends on the character of the magma associated with the system.

CHECK YOUR UNDERSTANDING

1. How does fossil evidence support the theory of continental drift?

2. How does the age of the Atlantic seafloor support the theory that the mid-Atlantic plate boundary is a rift?

3. What is the difference between a converging and diverging plate boundary? Why do diverging plate boundaries occur?

4. How does continental collision cause mountains to grow? What are the two processes through which alpine chains evolve?

5. Describe the geologic structure of the Ridge and Valley Province in the Appalachian Mountains. How can it be that some mountains are underlain by structural synclines, whereas others are underlain by anticlines?

6. Compare and contrast the various kinds of faults on Earth.

7. Discuss earthquake processes, including why they happen, the nature of seismic waves, and how they are measured.

8. Why are earthquakes typically associated with plate boundaries?

9. How much stronger is a magnitude 5 earthquake than a magnitude 2 earthquake?

10. Describe the formation and history of a tsunami that develops due to an earthquake.

11. Why are volcanoes associated with subduction zones?

12. What are the differences between composite volcanoes and shield volcanoes?

13. What is the Pacific Ring of Fire, and why does it have that name?

14. What tectonic feature is associated with Yellowstone National Park, and what unique features does it contain? In the context of your answer, explain how the migration of the Yellowstone hotspot fits into the picture.

ANSWERS TO VISUAL CONCEPT CHECKS

VISUAL CONCEPT CHECK 13.1

The answer is c. The structure shows the limb of an anticline. You can recognize this as an anticline because the rock structure rises rather steeply.

VISUAL CONCEPT CHECK 13.2

The answer is d. There is compression of the rocks on either side of the fault. Movement along the fault produces incredible strain that is causing deformation of the rocks along the fault boundary.

VISUAL CONCEPT CHECK 13.3

This volcano is a composite volcano, specifically Mount Rainier in Washington. You can tell the nature of this volcano because it is much too large to be a cinder-cone volcano. It is not a shield volcano because it is tall with moderately steep sides. Mount Rainier will probably erupt violently because the magma is highly viscous; thus, a great deal of pressure will build before the eruption occurs.

CHAPTER FOURTEEN

WEATHERING AND MASS MOVEMENT

It is useful to think of the various geomorphic processes described in the rest of the text as part of a continuum of landscape evolution that begins with the tectonic processes covered in Chapter 13. Those large-scale processes elevate large blocks of land above sea level and, in so doing, provide high topographic relief. Once this relief is attained, other forces begin to gradually wear down the landscape over a long period of time. These smaller-scale processes produce the more subtle rolling hills, valleys, and floodplains that are common throughout the world. Before these landscapes can be investigated, however, we first need to examine the mechanisms that break apart rock and loosen sediment so that they can be subsequently eroded, transported elsewhere, and deposited. This chapter investigates the major components of this process: weathering and mass movement.

Guy Vanderelst/Getty Images

Look at this amazing rock outcrop. It is located in Arches National Park in Utah and is one of the many beautiful places in the southwestern United States. This landform is a remnant of a solid rock mass that once covered the area. Year after year, piece by piece, rock fragments broke off from this unit and were carried away by erosion to create this stunning landscape. This chapter focuses on the way that rocks break apart and wear down through the process of weathering.

CHAPTER PREVIEW

Weathering

Human Interactions: Acid Rain
 Weathering

Mass Wasting
 Weathering and Mass Movements

LEARNING OBJECTIVES

1. Describe the process of weathering and the role of rock resistance.

2. Compare and contrast mechanical and chemical weathering.

3. Discuss the various weathering processes.

4. Explain the concept of acid rain in the United States.

5. Define the concept of mass wasting.

6. Compare and contrast the various kinds of mass wasting.

Weathering

The logical place to begin a discussion of landscape evolution is with the concept of weathering because it is often the initial force that begins to wear rocks down. Recall from Chapter 11 that **weathering** is the process by which rocks break down or decay into smaller fragments. Two primary kinds of weathering processes occur on Earth: *mechanical* (also called *physical*) and *chemical*. Although each of these weathering processes can operate alone, they often work together to break rocks apart. Figure 14.1 shows the relationship between these weathering processes and climate. Notice that physical weathering is more pronounced in colder/drier environments, whereas chemical weathering is related more to warmer/wetter climates. The reasons for these environmental relationships will become clear as you work through the chapter.

An important factor that influences the amount or pattern of weathering is the relative resistance of different kinds of rock at an outcrop. Some rocks are simply harder than others and are thus more resistant to weathering and erosion.

Limestone, for example, is generally more resistant to physical weathering than shale. And within limestone rock masses, the dolomite units are typically more resistant than chalky rocks, which sometimes dissolve easily due to the chemical interactions discussed in Chapter 15. It is also possible that certain parts of an individual rock unit are more resistant to weathering and erosion than others within the same unit due to variations in the strength of the cementing materials. This kind of differential cementation is particularly common in sandstone. In places where this kind of variability exists within a rock outcrop or within an individual rock unit, long-term weathering can have a noticeable impact and can even result in very unusual landforms (Figure 14.2). Such a weathering pattern is called *differential weathering* because rock units weather at different rates. If a more resistant rock layer overlies a less resistant layer, it protects the underlying rock somewhat from erosion and is thus known as **caprock**. Keep these variations in mind as you work your way through this chapter.

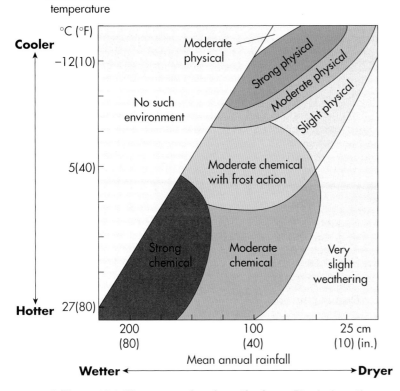

Figure 14.1 **The geography of weathering.** Physical weathering dominates in colder, drier environments, whereas chemical weathering tends to prevail in warmer, wetter regions (although some overlap occurs).

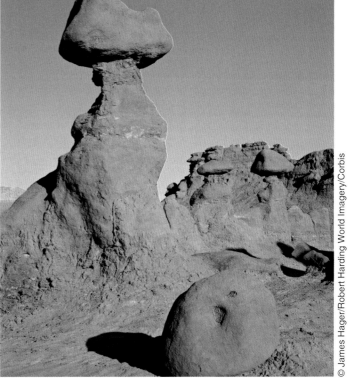

Figure 14.2 **The effects of differential weathering.** Differential weathering occurs when rock resistance varies between or within individual rock units. This form of weathering often leaves unusual rock forms, such as this *hoodoo* in Arches National Park in Utah. In this particular example, the caprock is a sandstone that is more resistant to weathering than the underlying rock layer.

Weathering *Physical or chemical modification of rock or sediment that occurs over time.*

Caprock *A unit of relatively resistant rock that caps the top of a landform and thus protects underlying rocks from erosion.*

Figure 14.3 Joints in rocks. Joints are fissures that develop along horizontal or vertical planes within rock. These columnar joints formed in basalts now exposed at Devil's Postpile National Monument in California.

Mechanical Weathering

Mechanical weathering, also called *physical weathering*, involves the destruction of rocks through physical stresses. When this kind of weathering occurs, rocks do not change their chemical composition. Instead, they simply break into progressively smaller pieces of the same kind of rock. As they do so, the overall surface area of the rock progressively increases, which facilitates even more weathering because more area within the rock mass is available for additional weathering to occur. This portion of the chapter describes the various kinds of mechanical weathering.

Frost Wedging A common type of mechanical weathering is **frost wedging**. It begins when water works its way into fractures in the rock, such as a **joint** (Figure 14.3). Joints can form when rock repeatedly expands and contracts with heating and cooling, respectively, causing fractures to develop along horizontal and vertical planes that essentially break the rock into large blocks. Joints can also form due to regional tectonic forces that produce stress within the rock mass. Although joint features may look like faults, they are not because no relative vertical or horizontal movement is occurring on either side of the fracture.

Once joints develop, they allow water to flow deeper into the rock mass at different points. If the rock body lies in a

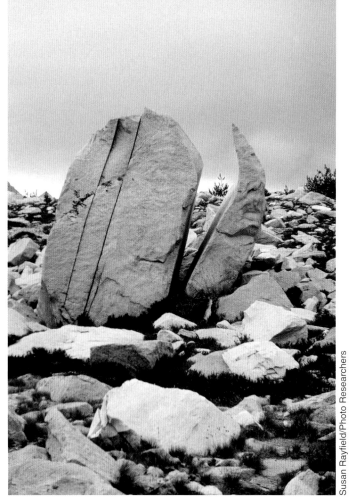

Figure 14.4 The effects of frost wedging. Frost wedging occurs when water enters cracks and joints within rocks and subsequently expands and contracts upon freezing and thawing, respectively. This process causes these fissures to widen over time. Frost wedging can cause even large boulders to fracture in two, such as this specimen in California's Sierra Nevada.

region that has distinct winter and summer seasons, the water will tend to go through numerous freeze and thaw cycles during the transitional months. When water freezes, it expands its volume as much as 9%. You can see this in the way that water in an ice cube tray freezes into cubes that are slightly larger than the liquid water level. If water subsequently thaws back to liquid form, it then contracts. (By the way, water is one of the very few substances that act this way; most materials are larger in volume as a liquid than as a solid.) Over the course of time, the stress produced by this freeze-thaw cycle causes the joints progressively to enlarge so that boulders slowly break away from each other (Figure 14.4). In places where frost action is most active, such as high mountain peaks or arctic

---

Mechanical weathering *The breakup of a body of rock into smaller rocks of the same type.*

Frost wedging *Expansion and contraction of water in rock cracks due to freezing and thawing.*

Joint *A crack or fissure along horizontal or vertical planes in a rock mass that divides the rock into large blocks.*

Figure 14.5 A rock sea. This rock sea of boulders (or *felsenmeer*) in the Colorado Rockies was created through the long-term process of frost action at high altitudes.

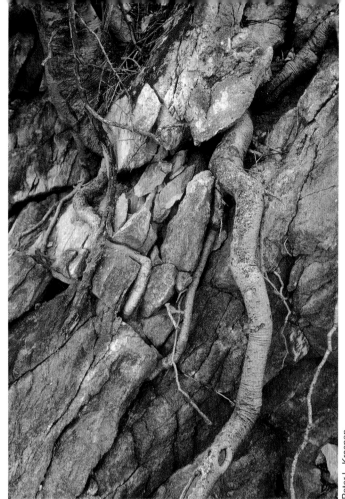

Figure 14.6 Plant roots and mechanical weathering. Plants can slowly break apart rocks when they grow in cracks such as this one and enlarge over time.

environments, large fields of fractured boulders form (Figure 14.5); they are called *felsenmeer* (German for "rock sea").

In addition to its effects on rock, the freezing and thawing of water can also cause modifications in soil. These modifications are most pronounced where soils are fine-textured with horizontal planes between silt and clay particles. Under these conditions, water collects in the planes and then expands upward upon freezing, causing the soil to rise upward in a process called **frost heaving**. If you happen to live in the northern part of the United States or in Canada, you can often see evidence of frost heaving in farm fields. In these areas, individual stones and cobbles (large, rounded rock fragments) can be slowly brought to the surface over time when water collects beneath them and freezes. After each winter, farmers pull any freshly surfaced rocks out of their fields to make plowing easier and stack them in a pile at the edge of the lot. You can also see the effects of frost heaving on northern paved roads when cracks develop in them because they are uplifted somewhat during the winter. Building contractors are acutely aware of the frost-heaving process. To avoid problems with structures such as decks settling in uneven ways, builders are required to install support footings below the depth of the frost line. The depth of this line is progressively greater with higher latitude or altitude.

Two other weathering processes have the same effect as frost wedging in that they break down rocks into progressively smaller fragments. One is related to the impact of plants when their roots grow into cracks and joints in rock. As the plants and associated roots grow, they can literally cause rock to split farther apart (Figure 14.6) in a process called *root wedging*. The second process takes place in association with temperature changes that occur over the course of seasons, between day and night, or due to changes in the direction in which the Sun faces a rock over the course of the day. These temperature fluctuations can cause rock to expand and contract when heated and

cooled, respectively. Over the course of countless such cycles, the rock will gradually break into progressively smaller pieces. This kind of mechanical weathering is very common in arid regions (Figure 14.7) or at high elevations where intense solar radiation occurs during the day and radiational cooling prevails at night.

Figure 14.7 Temperature changes and mechanical weathering. This rock in the Mojave Desert has split into four pieces due to thermal expansion and contraction.

Frost heaving *Upthrust of sediment or soil due to the freezing of wet soil beneath.*

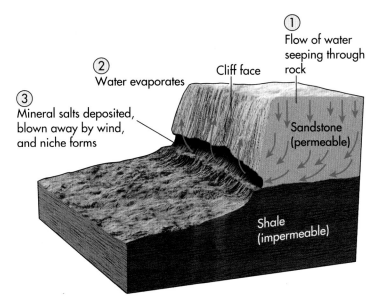

① Flow of water seeping through rock

② Water evaporates

Cliff face

③ Mineral salts deposited, blown away by wind, and niche forms

Sandstone (permeable)

Shale (impermeable)

Figure 14.8 Salt-crystal growth and niche formation. Salt-crystal growth is most prominent where permeable sandstones overlie relatively impermeable shales. As a result, water seeps out at the contact between the two rock types, and salt-crystal growth in the sandstone creates a niche.

Salt-Crystal Growth Another kind of physical weathering process is **salt-crystal growth**, which involves the buildup of salts on rock surfaces. Salt is a dispersing agent, and its concentration subsequently causes the minerals that cement the sediment grains within the rock to weaken their bonds; this, in turn, loosens sediment so that it can then be eroded. This process is similar to frost action in that it requires the presence of water moving within rock.

An excellent place to see the effects of salt-crystal growth is in the steep cliffs of the southwestern United States, where permeable sandstones often overlie relatively impermeable shales (Figure 14.8). Water stored within the sandstones tends to move downward fairly easily because these rocks have large pore spaces. This moving water easily dissolves salts such as halite (sodium chloride), calcite (calcium carbonate), or gypsum (calcium sulfate). When this downward-moving water meets the shale, it can no longer move vertically and thus flows horizontally toward the cliff wall, where it seeps out and rapidly evaporates.

As a result of this evaporation, salts build up on the rock surface at the contact between the sandstone and shale. This high concentration of salts dissolves the cementing mineral bonds, which, in turn, causes the sand grains within the rock to separate slowly from one another. These loosened grains are then washed or blown away by the water or wind, forming a niche at the contact between the two rock types. Although salt-crystal growth is a small-scale process, it has contributed greatly to the evolution of landscapes in the canyons of the desert Southwest by creating large niches and cave-like recesses in cliff walls.

Exfoliation Another way in which rock physically weathers is when it expands through the process of *unloading*. Unloading occurs when deep rocks are slowly uncovered because overlying rock gradually erodes off of them. As the overlying rock mass is gradually removed, the amount of compression on the deeper rocks progressively lessens. If unloading progresses to the point where the formerly deep rocks become surface rocks, they may expand so much that curved joints develop parallel to the surface. These joints are called *sheeting structure* and result in the flaking of loosened rock from the main rock mass in a process called **exfoliation** (Figure 14.9). Although unloading may be the dominant cause of exfoliation, the process is probably enhanced by daily and seasonal temperature cycles that cause the rock to expand and contract.

Exfoliation is most active in igneous and metamorphic rocks, but also occurs in some sedimentary rocks. A great place to see exfoliation on a large scale is Yosemite National Park in California, where several large exfoliation domes occur in granite monoliths. The most famous of these landforms is *Half Dome*, which towers over the adjacent Yosemite Valley (Figure 14.9b). A close look at this peak reveals the curved sheeting structure associated with exfoliation.

Chemical Weathering

Rocks are also altered by **chemical weathering**. In contrast to physical weathering, which causes rock simply to break apart into smaller fragments, chemical weathering decays the rock by changing its chemical composition. This decomposition occurs through the interaction of air, water, biological processes, and rock minerals in various ways. In particular, water is very important as a weathering agent because of its ability to dissolve minerals. As a result, chemical weathering is typically more active in warm, humid environments (see Figure 14.1) where abundant water is present and various chemical reactions freely occur. Several chemical weathering processes are common, each having a unique combination of environmental variables that cause rock to decompose.

Salt-crystal growth *A weathering process involving the buildup of salts on rock surfaces through evaporation. These salts weaken the cement that bonds rock particles, allowing them later to be washed or blown away.*

Exfoliation *Form of physical weathering in which sheets of rock flake away due to seasonal temperature changes or by the expansion of the rock due to unloading.*

Chemical weathering *The decomposition and alteration of rocks due to chemical actions of natural physical and biological processes.*

(a)

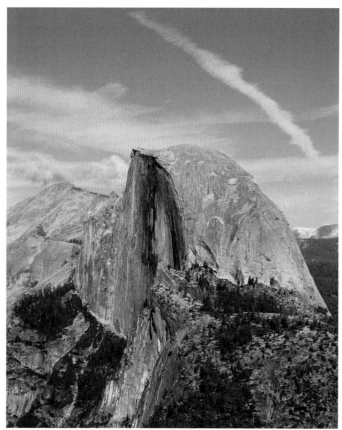

Gregory G. Dimijian/Photo Researchers

a. v. ley/Flickr RF/Getty Images

Figure 14.9 The impact of exfoliation. Exfoliation occurs when overlying rocks are removed through unloading and when temperature changes cause expansion and contraction of surface rocks. When these fluctuations occur, sheeting structure develops and rock flakes off the surface. (a) Exfoliation on Enchanted Rock, an exposed batholith in central Texas. Note how flaky the surface appears. (b) The peak of Half Dome Mountain in Yosemite National Park is also composed of granite. Its unique shape gives rise to its name. The peak shows the sheeting structure and jointing associated with the process of exfoliation. These sheets gradually flake off, revealing underlying rock that will then itself expand.

(b)

DISCOVER...

PREHISTORIC CLIFF DWELLINGS IN THE SOUTHWESTERN UNITED STATES

From an archaeological perspective, niches within cliffs of the southwestern United States are interesting because many contain the remains of ancient villages. These settlements were constructed by Native Americans, the ancestors of the modern-day Pueblo people who still inhabit the region. Although the history of the Pueblo extends back several thousand years, perhaps the most distinctive period occurred between about A.D. 1100 and 1300 when they built elaborate villages in canyon walls throughout much of the region. The niches in which the dwellings were constructed occur within the thick sandstones of the Mesa Verde Group, which overlies the relatively impermeable Mancos Shale. This rock sequence is very much like the one shown in Figure 14.8 and is the perfect setup for salt-crystal growth and niche formation.

Photodisc/Getty

(a)

Cliff dwellings such as these were occupied by the ancestral Pueblo people about 700 to 800 years ago.

In the time prior to the cliff-dwelling period, ancestral Pueblo people tended to live in small villages that were scattered around the upland areas. By about 1000 years ago, people had begun concentrating in larger villages along the canyon rims. Archaeologists believe that the move down into the canyons by the ancestral Pueblo around A.D. 1100 and the associated construction of elaborate cliff dwellings reflect the increased concentration of people into villages that were highly defensible. The buildings were constructed of bricks made of hard sandstone that were held together with a mortar mixed of mud, sand, and water called *adobe*.

Fantastic cliff dwellings can be seen in a variety of places in the southwestern United States, such as Canyon de Chelly in Arizona, Bandalier National Monument in New Mexico, and Mesa Verde in Colorado. Although these sites are 700 to 800 years old, they clearly were built by people living in a complex culture, with villages occupied by hundreds to thousands of people. The structures ranged in size from one-room storage units to large, multipurposed buildings with more than 150 rooms. Ceremonial rooms (called *kivas*), which were used for religious ceremonies, are present at many sites, as are open plazas where cultural and civic events took place.

(b)

This structure is a room for religious ceremonies called a *kiva*. People entered the room by way of the ladder poking through the opening at the top of the structure.

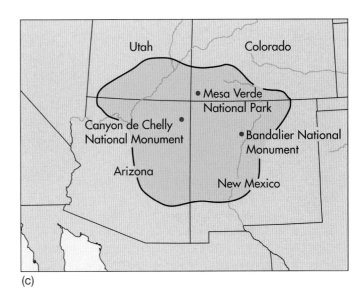

(c)

Location of famous sites in the southwestern United States where cliff dwellings are found.

If you visit such places, think about geologic time and how it relates to these amazing archaeological ruins. For the sandstone and shale to be present, sediments must have accumulated in a sea or ocean that varied in depth over a long period of time. This sea then disappeared, and sediments slowly lithified to become rock. Somewhere along the line, over millions of years, streams slowly cut into the rock to form the canyons in the area. At the same time, the niches gradually formed due to salt-crystal growth. Very late in this sequence prehistoric people moved into the area and built these structures. Although no one has lived in them for at least 700 years, the niches remain essentially unchanged. We know this because the ruins that lie within the niches are in phenomenal condition.

What prompted the ancestral Pueblo to leave their elaborate cliff dwellings? No one knows for sure. The best scientific guess is that the Pueblo were influenced by a variety of factors, including climate change, drought, deforestation, and competition from other cultural groups. Regardless, their descendants live on in the modern Tanoan and Hopi peoples who live in the region today.

Hydrolysis **Hydrolysis** is a weathering process that decomposes silicate minerals within rocks. This form of weathering occurs when silicate molecules are split after the addition of hydrogen and hydroxyl ions derived from water. In granitic

Hydrolysis *A chemical weathering process that results in the decomposition of silicate molecules within rock through the reaction of hydrogen and hydroxyl ions in water.*

Edges
Water can act on two surfaces

Hydrolysis

Corners
Water can act on three surfaces

Rounded rocks
Water can act on only one surface

(a)

© Alan Majchrowicz/Age Fotostock

(b)

Figure 14.10 Spheroidal weathering. (a) Corners and edges of rocks offer more surfaces on which water can act, so they are worn away faster than the rest of the rock. (b) Rounded rocks such as these in Joshua Tree National Park in California usually result from spheroidal weathering. Note the small trees for scale.

rocks, for example, hydrolysis causes feldspars to weather into clays, such as kaolinite and silica, which may be subsequently washed or blown away and deposited somewhere in a soil or stream system. If a sufficient amount of clay is removed in this fashion, more resistant quartz grains are exposed and, with further weathering, liberated from the host rock, to be transported by a stream or perhaps blown into sand dunes.

You can see the effects of hydrolysis in several different ways. The most prominent impact of hydrolysis on the landscape is that the sharp edges and corners of rocks become rounded in a process called *spheroidal weathering* (Figure 14.10). This process has the greatest impact where numerous joints occur in rocks, providing more corners and edges on which decomposition can occur.

Oxidation Another form of chemical weathering is called **oxidation**. In contrast to hydrolysis, which requires water, oxidation occurs when oxygen is added to chemical compounds and causes electrons within the compounds to be lost. In everyday life, we are most familiar with oxidation when tools and household equipment rust when left outside for a few days. Rusting occurs because oxygen reacts with iron to produce iron oxides, which are reddish-colored and flake off with time. The iron underneath is then exposed to the air and forms rust in its turn. The same process happens in rocks that contain abundant iron, such as those in the southwestern United States. These rocks have been heavily oxidized, resulting in their orange appearance (Figure 14.11).

Carbonation A third kind of chemical weathering occurs when water containing carbon dioxide causes minerals to

dissolve and wash away in solution. This form of weathering is called **carbonation** (because carbon is reacting with minerals) and occurs because carbon dioxide is easily dissolved by atmospheric water vapor, forming carbonic acid. As a result,

© David L. Brown /Design Pics/©Corbis

Figure 14.11 The effects of oxidation. Oxidized sandstones in the southwestern United States. Scenes like this at the Fiery Furnace in Arches National Park are very common in this part of the country because iron-bearing rocks crop out in many places due to the arid climate and lack of vegetation.

Oxidation *A form of chemical weathering in which oxygen chemically combines with metallic iron to form iron oxides, resulting in the loss of electrons.*

Carbonation *A type of chemical weathering caused by rainwater that has absorbed atmospheric carbon dioxide and formed a weak carbonic acid that slowly dissolves rock.*

Figure 14.12 Effects of carbonation in limestone. When carbon dioxide in the air dissolves in rainwater, it forms a weak carbonic acid that can erode limestone. This limestone bed in Switzerland was once a solid rock layer that has been extensively worn by acidic rainwater flowing over it.

precipitation contains this carbonic acid, which is sufficiently acidic to dissolve many minerals, especially limestone (calcium carbonate). When rainfall collects on the rock surface, it dissolves the limestone and forms pits and rounded surfaces (Figure 14.12). If you are interested in your local weathering rates, visit a cemetery and compare the decomposition in limestone versus marble or slate tombstones. Since marble and slate are much more resistant than limestone to weathering, the difference can be dramatic.

Human Interactions: Acid Rain

As just described, carbonation is the natural process through which rocks are chemically weathered when they come into contact with acidic water. In a similar way, by-products of human industrial activity cause chemical weathering of rocks through the process of acid rain. **Acid rain** is the broad term applied to the way in which industrial acids fall out of the atmosphere in association with precipitation.

In North America, acid rain is a particular problem in the northeastern part of the United States and eastern Canada. This area lies downwind of large industrial cities, such as Chicago, Detroit, Cleveland, and Pittsburgh, where large amounts of chemical pollutants are emitted into the atmosphere. In particular, sulfur dioxide and nitrogen oxides are the primary acids linked to acid rain. According to the United States Environmental Protection Agency, about two-thirds of all sulfur dioxide and

Acid rain *The precipitation by rain, fog, or snow of strong mineral acids, primarily sulfur dioxide and nitrogen oxides, that originated in factories.*

one-fourth of all nitrogen oxides in the atmosphere come from electric power generation plants that rely on burning fossil fuels such as coal.

After these acids are emitted from factories in the industrial Midwest, they are carried aloft and eastward by the prevailing westerly winds. Subsequently, they are deposited on the surface of Earth through precipitation or within fog. In these downwind areas the pH of rainfall is about 4, whereas it is generally above 5 in the western half of the country. Similar patterns have been identified in other parts of the world, particularly eastern Europe.

The impact of acid rain on the landscape and as a weathering agent has been significant. With respect to chemical weathering, acid rain dramatically accelerates the weathering process. Although some of this weathering would have naturally occurred by carbonation, it has been enhanced by acid rain. In addition to the acceleration of chemical weathering, acid rain is also associated with several negative environmental impacts especially on forests and lakes. Forests that lie within the belt of acid precipitation are often stressed (Figure 14.13). This stress occurs in part because supporting soils are acidified, which accelerates the leaching of nutrient cations before the trees absorb them. Evidence also exists that acid deposition on the needle leaves of conifers reduces their tolerance to cold. Similarly, the pH of lakes and ponds can be lowered due to acid rain, causing stress to the plants and animals that live within those water bodies. Many lakes in New England, for example, do not support fish because they contain toxic levels of inorganic acids.

Although acid rain is a serious environmental problem, recent studies indicate that environmental regulations imposed in association with the *Clean Air Act* are having a positive effect,

Figure 14.13 Forest stress due to acid rain. Acidic rainfall is concentrated in the eastern part of the United States. Forests in these areas, such as this stand of Fraser firs in the Appalachian Mountains, are affected because soils are acidified, which accelerates leaching of nutrient cations before the trees can absorb them. In addition, trees become less tolerant to cold weather.

This image is of a rock in the southwestern United States. Given what you see, and the location of the rock, which one of the following weathering processes produced the rock pattern visible here?

a) Hydrolysis
b) Carbonation
c) Freeze-thaw activity
d) Thermal expansion
e) Oxidation

Alan Arbogast

at least as far as the United States and Canada are concerned. Before the first federal regulatory controls on air quality were passed in 1970, the amount of pollutants emitted into the air had risen steadily since the onset of the Industrial Revolution in the 19th century. These emissions peaked at 28.8 million tons in 1973 in the United States alone. Given growing concerns about acid rain in the 1970s and 1980s, the Clean Air Act was amended in 1990 to give industries more pollution control options such as switching to low-sulfur coal or adding scrubbers to control acid emissions. In more extreme cases, plants were ordered closed because they could not comply. The 1990 amendments clearly had a positive effect on acid emissions, with a 32% decrease to about 18 million tons in 1995. By 2010 the amount had dropped to about 11 million tons. Although nitrogen dioxide emissions remain unchanged, we appear to have at least turned the corner with respect to acid rain in North America. Major problems remain in the former Soviet bloc of eastern Europe, however, where virtually no environmental regulations were in effect during the Communist era. As a result, the relatively new democratic governments are struggling with how to cope with the serious environmental degradation in these areas.

KEY CONCEPTS TO REMEMBER ABOUT WEATHERING

1. Two general kinds of weathering processes take place: mechanical (physical) and chemical. Mechanical weathering causes rock bodies to break into smaller rock fragments, whereas chemical weathering causes the rock to change chemical composition.

2. Although mechanical weathering and chemical weathering can occur anywhere, most mechanical weathering occurs in cold environments, whereas chemical weathering is more pervasive in warm/humid climates.

3. The three major kinds of mechanical weathering are frost wedging, salt-crystal growth, and exfoliation. Frost wedging involves water that expands and contracts upon freezing and thawing in joints and cracks in rock. Salt-crystal growth occurs when water seeps out of rock in arid environments. Exfoliation happens when rock expands because of unloading and temperature changes.

www.wiley.com/college/arbogast

Weathering

WileyPLUS

Now is a good time to review the various weathering processes discussed so far in this chapter. Given that they involve water movement and the gradual wearing down of the landscape, it is useful to observe them in an animated format. Go to the *Geo Media Library* and select *Weathering*. This animation shows, for example, how frost wedging gradually causes rock to split apart and how the surface area of a rock mass increases with progressive mechanical weathering.

The animation also reviews the various ways that rock decomposition occurs as a result of chemical weathering processes such as hydrolysis and carbonation. As you watch this animation, think about how weathering contributes to the wearing down of rock masses and the time involved for this erosion to occur on a noticeable scale. After you complete the animation, be sure to answer the questions at the end to test your understanding of weathering.

4. The three major kinds of chemical weathering are hydrolysis, oxidation, and carbonation. Hydrolysis occurs when water interacts with rocks, causing chemical reactions to occur. Oxidation occurs when oxygen combines with metals to form oxides. Carbonation takes place when carbonic acid in precipitation accumulates on rocks and causes them slowly to dissolve.

5. Humans have increased the amount of chemical weathering downwind of industrial cities through the process of acid rain. Most of these acids come in the form of sulfur dioxide and nitrogen oxides derived from coal-burning power plants.

Mass Wasting

Once sediment is liberated from rock in some fashion as a result of the weathering process, it can be moved and deposited by gravity, wind, waves, or flowing water and glaciers. In this way, the Earth's landscape is modified and shaped to form features such as hills, valleys, beaches, and sand dunes. The remainder of this chapter will focus on the effect of gravity and the role it plays in landscape evolution by moving large volumes of sediment through the process of **mass wasting**.

From a geographical perspective, the place to consider mass wasting is on hillslopes, which are land surfaces inclined at various angles from the horizontal. This process takes place when the downward force of gravity overcomes the resisting force of the slope material's shearing strength—in other words, its cohesiveness and internal friction. The point at which sediment becomes unstable is called the *threshold point* and depends a great deal on the material's *angle of repose*. This angle is the natural maximum slope that any deposit of a particular kind of sediment—for example, sand—can achieve without moving under the force of gravity. The angle of repose ranges between about 25° for fine sands and approximately 40° for angular pebbles and cobbles. You can see these relationships when you see piles of sediment deposited at construction sites or sand and gravel quarries. After the sediment

is dumped, its slope naturally moves to the angle of repose (Figure 14.14).

Generally speaking, most landscapes contain a variety of hillslopes that grade downward to a stream of some size (Figure 14.15). The *local relief* of the area—that is, the vertical distance between the top of the hills and the stream downslope—may have been created either through uplift of the landscape or through downcutting by the stream. In some cases, both processes occur.

Regardless of how the relief was created, the resulting hillslopes contain many distinctive components. In all probability, the core of the slope is some form of bedrock that dates to either an igneous or sedimentary process that occurred tens of millions or hundreds of millions of years ago. In relatively humid areas, the solid core is mantled by regolith, which, if you recall, is formed when rock is weathered. Also recall that regolith is further modified through various pedogenic processes at the surface to form soil. Depending on the steepness of the hillslope and the regional environmental conditions, the soil will be either well developed and thick, or poorly developed and thin. Most mass-wasting processes occur within the mantling regolith and soil, although where relief is particularly steep, they may also impact outcropping bedrock. Sediment that moves down the hillslope due to some mass-wasting process is called **colluvium**. Collovium tends to be deposited at the base of hillslopes because the gradient becomes less at those locations and the energy to transport sediment is lost.

Several different kinds of mass wasting occur on Earth, as shown in Figure 14.16. Although these variations tend to grade into one another, an association exists between specific

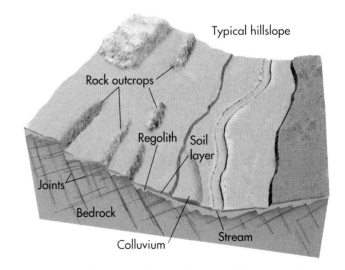

Figure 14.15 Generalized view of a typical hillslope. Hillslopes typically have distinctive characteristics that are easily visible. Note the relationships of bedrock, regolith, soils, and valley stream in the image, which are common in many hillslopes.

Fine sand   Coarse sand   Irregular pebbles

Figure 14.14 The angle of repose. The angle of repose is the natural maximum angle at which sediments of a given size can rest before they begin to slide under the force of gravity. In general, coarse sediments have a steeper angle of repose than finer sediments.

Mass wasting *Movement of rock, sediment, and soil downslope due to the force of gravity.*

Colluvium *Unconsolidated sediment that accumulates at the base of a slope.*

Wet conditions

1. Solifluction

2. Earthflow

3. Mudflow

4. Slump

5. Debris flow

6. Debris slide

7. Soil creep

8. Rockfall

Dry conditions

Slow speed ◄──────────────────► Fast speed

Figure 14.16 Major processes and forms associated with mass wasting. Movements to the right of this figure are relatively fast, whereas those to the left are relatively slow. Movements near the top of this figure are triggered by wet conditions, such as heavy rain; those near the bottom of the figure can occur under dry conditions.

environmental settings and individual mass-wasting processes. These processes are subdivided on the basis of:

1. The kinds of Earth materials being moved, such as rock or regolith, and their association with the presence of water.
2. Their physical properties, including whether they are hard, plastic, or fluid.
3. The kind of motion; for example, whether the material is falling or flowing.
4. How quickly the process occurs.

The following discussion focuses on the processes outlined in Figure 14.16.

Rockfall

The easiest kind of mass-wasting process to envision is **rockfall.** As the name implies, rockfall occurs when rocks fall quickly down a hillslope under the force of gravity. Rockfall is usually most common in places where extensive rock outcrops occur on steep hillslopes or canyon walls (Figure 14.17a). Frost wedging and large temperature changes contribute significantly to rockfall because these processes gradually split outcropping rocks apart from one another. Eventually, the gravity threshold is passed and stones and boulders break free to fall onto the lower part of the hillslope. Rockfall also occurs in association with differential weathering in places where relatively resistant rocks such as limestone and sandstone overlie softer rocks like shale. In these situations, the shale surfaces gradually retreat into the cliff through a combination of weathering and erosion, leaving the resistant rock as a ledge that juts out over the shale. Sooner or later the force of gravity exceeds the strength of the rock and it collapses. This process is very common in places like the Grand Canyon.

Extensive deposits of fallen rock, called **talus** (or *scree*), often accumulate through rockfall. These deposits can collect in relatively small amounts along the base of individual, small hillslopes or in larger quantities along high cliffs. Typically,

---

Rockfall *A mass-wasting process in which rocks break free from cliff faces and rapidly tumble into the valley below.*

Talus *A pile of rock fragments and boulders that accumulates below a cliff due to rockfall.*

talus slopes have a relatively high angle of repose, ranging between 35° and 40°, because boulders become lodged together and thus support one another. In most cases, talus accumulates in cone-shaped bodies called *talus cones* (Figure 14.17b). This pattern occurs because most canyon walls are interspersed with ravines and gullies that funnel boulders down where they accumulate. In places where numerous talus cones lie adjacent to each other, the resulting landform is called a *talus apron*.

Soil Creep

The slowest mass-wasting process is appropriately called **soil creep.** Soil creep occurs on virtually every soil-covered slope, as the persistent force of gravity slowly pulls surface particles downhill. In this fashion, the soil behaves as a plastic substance that is gradually molded over time. This process is enhanced by cycles of wetting, drying, freezing, and thawing in the soil, as well as by disturbances caused by burrowing animals and even people. The net result of soil creep is that the angles of hillslopes and the elevation of hilltops slowly decrease over time. Evidence for this process can be seen in artificial features such as fence posts and telephone poles that tend to shift slightly downhill (Figure 14.18).

A distinctive kind of creep is called **solifluction** (meaning "soil flowage"), which occurs above the tree line in tundra landscapes. This process is very straightforward and involves the

---

Soil creep *The gradual downhill movement of soil, trees, and rocks due to the force of gravity.*

Solifluction *A form of soil creep that occurs in arctic environments where freeze-thaw processes result in lobes of soil moving gradually downslope.*

Dave Bunnell

(a)

Ralph Lee Hopkins/Photo Researchers

Figure 14.18 Evidence of soil creep. Soil creep occurs when hillslopes gradually move downhill under the force of gravity. Notice here the tilted wooden columns, which are a good sign that soil creep is occurring.

Marli Bryant Miller

(b)

Figure 14.17 Rockfall. (a) A typical rockfall in a steep-walled canyon. Rockfall occurs when rocks in canyon walls are gradually loosened from the host rock mass by frost wedging and/or large temperature changes and thus quickly fall under the force of gravity. (b) Talus cones, such as these in Banff National Park in Canada, form along canyon walls and mountain escarpments where ravines and gullies funnel boulders into the adjacent valley.

Courtesy of Gabriel Guttierez-Alonso

Figure 14.19 The effects of solifluction. The distinctive lobes on this hillslope in Kyrgyzstan have evolved because unfrozen soil becomes saturated in the summer and sags downhill over underlying permafrost.

impact of seasonal temperature changes on the landscape. Most of the year, the soil is frozen from the surface to some depth. During the short summer months, the upper part of the soil thaws. Water within this part of the soil (the active layer) cannot percolate very deeply, however, because the sediment beneath it is continuously frozen as permafrost. As a result, the overlying (unfrozen) soil becomes saturated and slowly sags downhill in a very uneven way, producing distinct lobes that overlap one another (Figure 14.19).

Landslides

Another form of mass-wasting process is called a **landslide**. As the name implies, a landslide is a mass of rock, regolith, and soil that suddenly moves downhill. This movement is initiated when an event such as an earthquake or period of heavy rain causes the downward force of gravity to overcome the sediment's cohesiveness and internal friction. Slope shear strength depends on the frictional resistance and overall cohesion of the underlying sediment. When slope shear strength is reduced for some reason, a portion of a hillslope can move downhill at a rapid rate, either as a debris slide or a slump.

A **debris slide** occurs when slope failure occurs along a plane that is roughly parallel to the surface. A famous example of a debris slide is the *Madison slide*, which occurred in Montana near Yellowstone National Park on August 17, 1959 when a magnitude 7.1 earthquake rocked the region late in the evening for about 30 sec. This earthquake initiated a massive

landslide (Figure 14.20) on a mountain that borders the Madison River, which flows through the area. Approximately 28 million m³ (37 million yd³) of rock was instantly detached from the mountain in an area that measured about 600 m (about 2000 ft) in length and 300 m (about 980 ft) in thickness. During the ensuing landslide, the rock mass slid down the mountain slope, crossed the Madison River, and, in the process, killed 28 people who had been camping in the vicinity. The landslide was so large that it dammed the Madison River, creating a new lake, now known as Earthquake Lake, which is nearly 100 m (330 ft) deep.

In contrast to debris slides, which flow parallel to the surface, a **slump** occurs when rock and sediment rotate and move

USGS

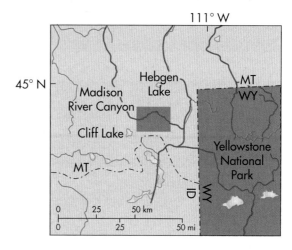

Figure 14.20 The Madison slide. Slopes can instantly fail when the threshold for cohesiveness is passed, causing tons of rock and sediment to move rapidly down the hill. The massive Madison slide, for example, was prompted by a large earthquake in the region. This slide blocked the Madison River, forming Earthquake Lake (arrow).

Landslide *An instantaneous movement of soil and bedrock down a steep slope in response to gravity, triggered by an event such as an earthquake.*

Debris slide *A mass-wasting process in which slope failure occurs along a plane that is roughly parallel to the surface.*

Slump *A mass-wasting process in which rock and sediment rotate and move down the slope along a concave plane relative to the surface.*

down the slope along a concave plane relative to the surface. Slumps typically occur when impermeable clay-rich deposits such as shale underlie more porous rock or sediment. Under such circumstances, water will flow through the upper deposits and then parallel to the clay surface upon encountering it. This water lubricates the contact between the rock layers and thus reduces frictional force, causing the rock and sediment mass to rotate as it slides down the hill (Figure 14.21a).

At the top of the slump, a wall-like escarpment is produced where the slump block breaks away, leaving a steep face. The sediment in the middle of the slump usually moves more quickly than near the scarp, resulting in a series of step-like features called *slump terraces*. At the base of the slump, the sediment flows more slowly and bulges out from the middle section of the flow as a distinctive *toe*. Although slumps happen virtually everywhere, they are by far more common in humid regions because the sediments are more frequently saturated there. They usually occur in isolated places, such as in Figure 14.21b, with very little impact on humans. Occasionally, however, a large slump occurs in a residential zone and causes severe damage to personal property.

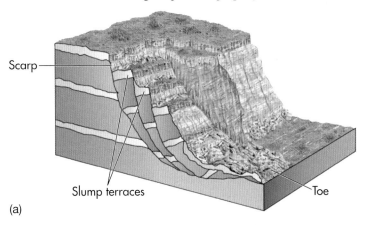

Scarp

Slump terraces

Toe

(a)

Alan Arbogast

(b)

Figure 14.21 Slump processes. (a) A typical slump includes a scarp, slump terraces, and a toe. (b) Slumps like this frequently occur in north-central Kansas because a thin shale (and associated regolith) at the base of hills becomes lubricated during heavy spring rains. Repeated slumps like this one have gradually caused the hillslopes to retreat throughout the area, forming these small hills.

Flows

As the name implies, flows are mass-wasting events involving sediments that are very wet. Such conditions occur when rainfall is high for a period of time, or when humans have stripped hillslopes of protective vegetation. These processes are catastrophic forms of mass wasting and are a serious environmental hazard in many parts of the world. Flows are common in Southern California, for example, and are especially dangerous in suburban areas where homes are built on steep hillslopes. Every two or three years some form of major flow occurs in this area, causing extensive property loss and even the occasional human death. Three basic kinds of flow occur: (1) earthflows, (2) mudflows, and (3) debris flows.

An **earthflow** is a slow-to-rapid type of mass movement that involves soil and other loose sediments, some of which may be coarse. This kind of flow may move downhill across a fairly broad surface. Although a **mudflow** is very similar to an earthflow, it differs because it consists of fine-textured sediments and is extremely fluid; thus, it moves more quickly. Mudflows are most frequently associated with steep canyons in mountainous regions and are initiated in humid zones when heavy rains fall on already saturated slopes. In arid zones, they occur when heavy rains fall on hillslopes that are unprotected by vegetation. Regardless of the environment, mudflows start when large volumes of water begin to run directly off the surface down the canyon. As the water flows, it picks up increasing amounts of mud from the underlying soil, resulting in a thick, viscous flow of water-laden sediment that has tremendous power. In large mudflows, the power is so great that large boulders are frequently transported great distances and human structures such as homes and bridges can be destroyed. Once the mudflow reaches the valley, it often spreads out across the landscape (Figures 14.16 and 14.22).

Earthflow   *A slow-to-rapid type of mass movement that involves soil and other loose sediments, some of which may be coarse.*

Mudflow   *A well-saturated and highly fluid mass of fine-textured sediment.*

Bettman/Corbis

Figure 14.22 Mudflow deposits. Mudflows occur when a significant amount of saturated fine sediments flows quickly. This mudflow deposit is located in the Never-Summer Range in Colorado. Note the lack of coarse fragments such as rocks and boulders.

Another kind of flow associated with heavy rainfall is a **debris flow**. Debris flows differ from mudflows because, in addition to mud, they also contain a mix of boulders, trees, and even buildings as they flow rapidly downhill. These rapidly moving and powerful flows can be triggered by extremely heavy rains in mountainous regions or by a volcanic eruption that instantly melts summit glaciers.

Debris flows are very serious natural hazards that can impact people catastrophically. An unfortunate example was the massive debris flow that swept down a steep hillslope into the coastal community of La Conchita, California, in January 2005 (Figure 14.23). This debris flow was triggered by a three-day storm that produced nearly 25 cm (10 in.) of rain in the region.

The slopes above La Conchita consist of ancient debris flow and landslide deposits that had failed to a smaller extent in 1995 when a large slump destroyed nine homes and blocked a county road. When these deposits became saturated again in early 2005, they failed on a much larger scale, releasing approximately 14,000 m³ (18,311 yd³) of material that flowed more than 100 m (330 ft) into the community. This debris flow destroyed 13 homes and killed ten people. Given that slope failures frequently occur in California during winter, this disaster demonstrates why it is important to understand how mass-wasting processes operate and how they can be triggered by passing weather systems.

Avalanches

Another mass-wasting process that you have probably heard of is an **avalanche**, which is a large mass of snow or rock that suddenly slides down a mountainside (Figure 14.24a). Avalanches are most commonly associated with snow and typically occur when thick deposits of snow accumulate at steep angles on mountain slopes. These steep-angled snow deposits can become unstable when a subsequent large storm produces another

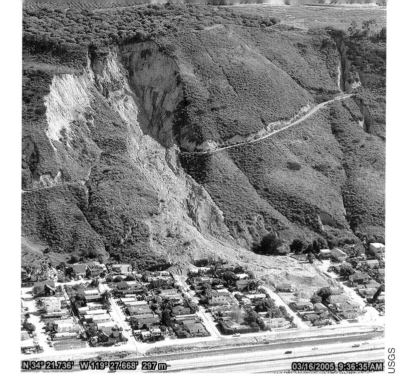

Figure 14.23 The La Conchita debris flow. This massive debris flow along the coast of Southern California was triggered by heavy rains over the course of three days. Thick deposits of mud, trees, and other debris swept into the community of La Conchita and unfortunately killed several people.

layer of snow of lighter density than the snow it buries. Under these conditions, an avalanche will occur because the two layers of snow are not firmly bound together to form a uniform mass. When the uppermost layer is somehow destabilized, a section will break free as a *slab avalanche*. Such an avalanche may release over 225 million m³ (~300 million yd³) of snow, which is equivalent to about 20 football fields covered with about 3 m (10 ft) of snow.

VISUAL CONCEPT CHECK 14.2

What type of mass-wasting process is pictured here?

a) Debris flow
b) Mudslide
c) Rockfall
d) Slumping
e) Avalanche

© Zoonar/Tim Heusinger/Age Fotostock

Debris flow *A rapidly flowing and extremely powerful mass of water, rocks, sediment, boulders, and trees.*

Avalanche *A large mass of snow or rock that suddenly slides down a mountainside.*

(a)

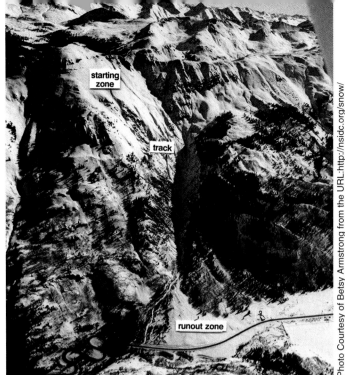

starting zone

track

runout zone

Photo Courtesy of Betsy Armstrong from the URL:http://nsidc.org/snow/avalance/#ANATOMY

Dominique Favre/Reuters/Landov

(b)

Figure 14.24 Avalanches. (a) Avalanches are usually associated with the sudden failure of enormous quantities of snow that cascade down steep mountain slopes. (b) Avalanches begin at the starting zone, travel down the track, and end at the runout.

Avalanches have three major components (Figure 14.24b). They begin in a place appropriately called the *starting zone*, which is usually located on the steepest slopes of a mountain where deep snows lie at precarious angles. After the avalanche begins, it typically follows a chute or ravine system on the mountain slope called a *track*. The track is where most destruction takes place, either through trees being knocked down or through the movement of boulders and other debris. As you travel through the mountains in summer, you can often see the track of a recent avalanche by looking for avalanche scars, which are treeless zones extending down the slope of a mountain. Ultimately, an avalanche loses its energy when it reaches the base of the mountain and the slope lessens. This area is called the *runout* and is the place where snow and debris are deposited.

Although most avalanches happen in remote places that have no impact on people, they can be a worrisome natural hazard at ski resorts and even along mountain highways. In 2009, for example, more than 25 skiers died (and many more were injured) around the world after being swept away in avalanches. Avalanche hazards were a particular problem in the Rocky Mountains and Pacific Northwest, but also caused fatalities in the European Alps and Southern Alps in New Zealand. People who live and work in mountainous areas are keenly aware that some areas are particularly prone to avalanches. In an effort to control the hazard, avalanches are often purposely started when conditions appear ripe. To minimize the risk, explosives are dropped from helicopters onto the unstable snow pack. The resulting explosion causes an avalanche to develop under somewhat controlled conditions.

www.wiley.com/college/arbogast

WileyPLUS

Weathering and Mass Movements

To better understand the process of mass movements, go to the *Geo Media Library* and access *Weathering and Mass Movements*. This animation allows you to visualize the various mass-wasting processes in action. As you work through this exercise, observe how these processes operate, how they differ from each other, and how they shape the landscape. In addition, the animation will give you a better comprehension of the impact these hazards have on people and the structures in which they work and live. Once you complete the exercise, be sure to answer the questions at the end to test your understanding of this concept.

1. Mass wasting refers to the downslope movement of sediment, soil, and rock under the force of gravity.

2. Mass-wasting processes are differentiated on the basis of how quickly they move and the relative saturation of sediments.

3. Rockfall, creep, solifluction, and landslides are mass-wasting processes that occur under the force of gravity. Rockfall and landslides occur when rocks and sediments break free for some reason, such as lubrication of underlying clays or earthquakes.

4. Flows occur when sediments become saturated and then move under the force of gravity. Flow processes include earthflows, mudflows, and debris flows.

5. Avalanches occur in mountainous areas when steep snow packs break free.

SUMMARY OF KEY CONCEPTS

1. Weathering refers to the decay and breakup of rocks on the surface of Earth. Two general kinds of weathering processes take place: mechanical (physical) and chemical. Mechanical weathering causes rock bodies to break into smaller rock fragments, whereas chemical weathering causes the rock to change chemical composition. These processes are closely associated with environmental conditions and may vary locally due to the relative resistance of different rock units.

2. The three major kinds of mechanical weathering are frost wedging, salt-crystal growth, and exfoliation. The three major kinds of chemical weathering are hydrolysis, oxidation, and carbonation.

3. Mass wasting refers to the downslope movement of sediment, soil, and rock under the force of gravity. Mass-wasting processes are differentiated on the basis of how quickly they move and the relative saturation of sediments.

4. Rockfall and soil creep are mass-wasting processes that are associated with unsaturated sediments.

5. Flows occur when sediments become saturated and then move under the force of gravity. Flow processes include solifluction, earthflows, mudflows, and debris flows.

CHECK YOUR UNDERSTANDING

1. Why is the term *weathering* appropriate for the process it describes?

2. What does the term *differential weathering* mean, and why does it occur?

3. Why is chemical weathering more likely to occur in warm/wet environments, whereas frost action mostly occurs in colder regions?

4. Why does frost wedging cause physical or mechanical weathering? How is it similar to the effect that plant roots and large temperature changes have on rock? Why do these processes increase the likelihood of further weathering of a rock mass?

5. How does the process of unloading lead to exfoliation?

6. How are the processes of hydrolysis and carbonation similar? How are they different?

7. What is acid rain, and why is it most closely associated with the northeastern part of the United States and eastern Canada?

8. Define the concept of mass wasting.

9. Why is the process of soil creep considered to be a mass-wasting process?

10. Describe the process of solifluction and its environmental context.

11. Why is slumping more likely to occur in humid regions than arid regions? What particular rock type—sandstone or shale—is most commonly associated with slumping? Why?

12. What is the difference between a mudflow and a debris flow? How would the process of human-induced deforestation make a mudflow or debris flow more likely?

ANSWERS TO VISUAL CONCEPT CHECKS

VISUAL CONCEPT CHECK 14.1

This weathering pattern was most likely produced by thermal expansion. It is certainly the result of mechanical weathering because the rock is merely breaking up into smaller fragments. Freeze-thaw is a mechanical process but is unlikely to have much of an effect in southern Texas.

VISUAL CONCEPT CHECK 14.2

The mass-wasting process pictured here is rockfall. The rocks at the base of the cliff broke free from the exposure above and fell to their current location.

CHAPTER FIFTEEN | GROUNDWATER AND KARST LANDSCAPES

Have you ever considered the significance of freshwater in your daily life? It is used in many ways, including for drinking, cooking, and washing, as well as for agriculture and recreation. Water is critical to life on Earth in many different ways, yet people often take it for granted. How often do you think about where freshwater comes from and the various places where it is stored? Do you ever even consider what it would be like to run out of water? If you happen to live in the humid eastern half of the United States, you may never think about these questions. In many other parts of the world, however, freshwater is a precious resource that people think about every day.

Recall from Chapter 6 that water covers about 71% of the Earth's surface. In subsequent chapters, we examined how water moves in the air and within the ocean. Now we focus on the very small portion of the water on Earth that is stored just below the surface as groundwater. This form of water may very well be the dominant supply of freshwater where you live.

Franz Marc Frei/LOOK/Getty Images

Groundwater is water that is stored below Earth's surface. Understanding groundwater is important because much of our water supply comes from this source. Groundwater also shapes the ground below you, forming fascinating subterranean chambers such as this cave on the Yucatan Peninsula in Mexico. This chapter focuses on the characteristics of groundwater and the impact that it can have on the landscape, both above and below the Earth's surface.

Movement and Storage of Groundwater

Human Interactions with Groundwater

 Hydrologic Cycle and Groundwater

Karst Landforms and Landscapes

▌ LEARNING OBJECTIVES

1. Describe the way that groundwater is absorbed into the ground and stored within it.

2. Compare and contrast the various components of the groundwater model.

3. Explain the geography and formation of the High Plains Aquifer, and human interactions with it.

4. Understand the concept of groundwater contamination.

5. Discuss the different processes that form caves and how these contribute to the development of karst topography.

Movement and Storage of Groundwater • 391

Movement and Storage of Groundwater

The logical place to begin a discussion of groundwater is with a brief review of the hydrologic cycle presented in Chapter 7. If you recall, the hydrologic cycle refers to the way water is stored in various reservoirs (such as rivers) and how it moves to other places (like the oceans). Earth's groundwater system is also part of the hydrologic cycle because it is a place where water can move to, be stored for a period of time, and flow to some other reservoir.

In this context, it is time to examine the various parts of the groundwater system and establish some fundamental terminology. Start by considering what happens to water after it falls to Earth as some form of precipitation. In most circumstances, this water is absorbed by the soil, where it is stored in the *soil–water belt* (Figure 15.1). Remember from Chapter 11 that water is absorbed into the soil by infiltration, which occurs through pore spaces between sediment grains (Figures 11.3 and 11.4), along pathways associated with the soil structure, or down plant roots. When the soil is very dry, water is held tightly to sediment grains as **hygroscopic water** that is largely unavailable

for plant absorption (Figure 15.2). Recall that the *wilting point* is the threshold at which, for any given soil, water is no longer available for plant uptake. With increasing precipitation, the pore spaces begin to fill with *capillary water*, which is water held to soil particles by molecular attraction. Some of this water is lost to the atmosphere through evaporation, but plants absorb a great deal of it before the water is returned to the air through the process of transpiration (see Figure 15.1). The combined processes of evaporation and transpiration are often referred to as *evapotranspiration*.

If precipitation continues to the point where soil pore spaces become completely filled with water, the soil is considered to be at its **field capacity**. Any excess water that does not run off the surface is then free to move more deeply down into the sediment or rock, under the force of gravity, as gravitational water. As this water moves downward, it first passes through an area called the **unsaturated zone** (see Figure 15.1 again), which is the portion of the rock or sediment mass where pore spaces may be partially filled with water, but also contain air (i.e., oxygen and other gases).

As water continues to percolate downward, it will eventually reach a depth where pore spaces are entirely filled with water and will stop moving. This part of the groundwater sys-

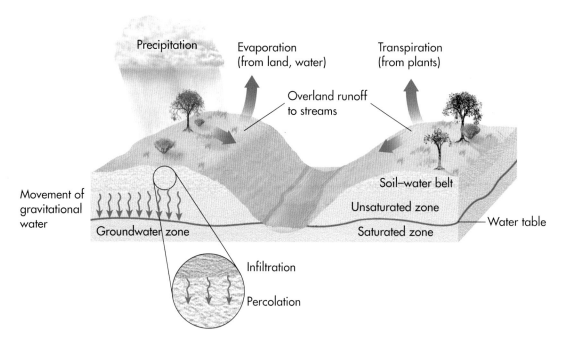

Figure 15.1 The groundwater model. Precipitation is initially stored in the soil–water belt until it becomes saturated. After that occurs, water can either run off the surface in streams or percolate through the unsaturated zone to the saturated zone.

Hygroscopic water *Soil water held so tightly by sediment grains that it is unavailable for plant use.*

Field capacity *The maximum amount of water the soil can hold after gravitational water has moved away.*

Unsaturated zone *The area between the soil–water belt and the water table where pore spaces are not saturated with water.*

Dry periods

Precipitation

Heavy precipitation

Water table

Groundwater zone

Hygroscopic water
(no water available
to plants)

Capillary water
(water available
to plants)

Saturation

Gravitational
water

Figure 15.2 Stages of soil water. During dry periods, soil grains tightly hold onto hygroscopic water. With increasing rainfall, the pores begin to fill with capillary water. Once field capacity is reached, pores are full and excess water can run off the surface and flow downward as gravitational water.

tem is called the **saturated zone** to reflect these conditions. Immediately below the saturated zone is a feature called an **aquiclude** that consists of some kind of impermeable rock mass, such as shale or granite, which keeps water from moving further downward. The top of the saturated zone is called the **water table** and marks the boundary between the ground-water below and the unsaturated zone above. In most places, the water table is not visible because it is beneath the ground. Sometimes, however, topographic depressions or stream channels dip below the level of the water table, as you can see in Figure 15.3. Where such topographic relationships occur, a stream or lake will be present.

Saturated zone   *The zone of rock below and including the water table where pore spaces are completely filled with water.*

Aquiclude   *An impermeable body of rock that may contain water but does not allow transmission of water through it.*

Water table   *The top of the saturated zone.*

NATURAL LAKES

Have you ever visited a region containing many natural lakes? Such places are fascinating ecosystems and are prime vacation spots for people. They are also favored zones of development, specifically for second homes and weekend cabins. Although natural lakes are very attractive places to be, few understand their origin.

An excellent example of a region with numerous natural lakes is the state of Minnesota, which is famous as the "Land of 10,000 Lakes." The majority of these lakes formed in topographic depressions created by glacial ice between about 20,000 and 10,000 years ago. This landscape, as well as other similar places in the upper Midwest in states like Wisconsin and Michigan, looks this way due to an intimate and complex relationship between topography and local and regional water tables. A geographer or hydrologist can tell by looking at the map of Douglas County, for example, that the elevation of the water

table must be high to create so many lakes. Examination of an air photo of a typical Minnesota landscape further reinforces this impression. The next time you are in Minnesota, or a place where lots of natural lakes occur, try to look at the landscape in a more holistic way and think about the local groundwater supply.

(b)

A geographer knows these lakes in Minnesota mean that the regional water table is sufficiently high to fill depressions with water.

(a)

This map of Douglas County, Minnesota, illustrates why the state is known as the Land of 10,000 Lakes. Most of these lakes are associated with topographic depressions created during the most recent glacial period.

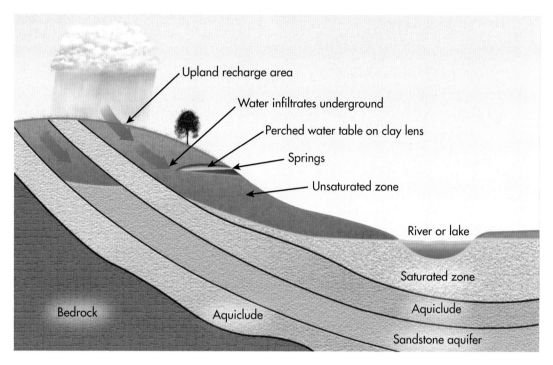

Figure 15.3 Water tables and aquifers. A water table is the top of the saturated zone of rock and soil. Depending on the kind of rock deeper in the ground, water may collect in saturated zones called aquifers, enclosed by layers of impermeable rock called aquicludes. An aquifer is a saturated zone that holds enough water to serve as a source for irrigation or drinking water.

In some places, the groundwater system is simple and consists of a single saturated zone of broad regional extent that overlies an aquiclude. At other places, however, the situation may be more complex with a layered assemblage of saturated zones, aquicludes, and unsaturated zones (Figure 15.3). Where such complexity occurs, saturated zones deep within the system may be slowly *recharged* by water gradually moving from the saturated zones higher in the column or from sources of water far away. In still other settings, a saturated zone may be highly localized because it occurs in association with an isolated aquiclude such as a body of clay-rich sediment deposited by an ancient glacier or stream. Such a localized saturated zone is called a **perched water table** because it is located above the primary groundwater body. Regardless of the setting of a specific saturated zone, or whether it is regionally extensive or more isolated, if it is sufficiently large to be a significant source of water for people, it is called an **aquifer**.

Perched water table *A localized area of saturated sediment that is elevated above the regional water table by a zone of impermeable rock or sediment.*

Aquifer *A geological formation that contains a suitable amount of water to be accessed for human use.*

Human Interactions with Groundwater

The High Plains Aquifer

The most extensive aquifers occur in regions where abundant sandy sediments or sandstone rocks are found across a broad geographical area. Aquifers are one of the most important natural resources on Earth. Perhaps the best-known and most heavily utilized aquifer in North America is the *High Plains Aquifer*. Also known as the *Ogallala Aquifer*, this groundwater reservoir is found in the Great Plains region of the United States and underlies portions of eight states, ranging from South Dakota to Texas (Figure 15.4). The High Plains Aquifer is largely formed in the thick, porous deposits of unlithified sand and gravel of the Ogallala Formation, which consists of sediments that weathered and washed out from the Rocky Mountains between 19 and 5 million years ago. These sediments covered the western part of the Great Plains and slowly filled with water from rainfall and snowmelt during the ensuing millennia. Much of this groundwater accumulated during the ice ages, between about 1.6 million and 10,000 years ago, when the regional climate was much wetter than it is today.

0   100   200 km

0   100   200 mi

As you can see in Figure 15.4 and Table 15.1, the High Plains Aquifer is huge. It covers an area of about 280,000 km² (174,000 mi²) and contains 1,315,000 hectare feet (3,250,000 acre feet) of water. In other words, the aquifer contains the amount of water required to cover over 1,000,000 hectares (3,000,000 acres) to a depth of about 30 cm (12 in.). This volume is approximately equivalent to one of the larger Great Lakes, such as Lake Huron. As you can see from Table 15.1, the thickness of the saturated zone varies dramatically across the region. In Nebraska, for example, the average saturated thickness is currently 104 m (342 ft). In Colorado, on the other hand, the average saturated thickness of the aquifer is only 24 m (79 ft). This variation exists because the sediments are sandy all the way to the surface in Nebraska, resulting in rapid infiltration of water, whereas in Colorado they are covered in most places by relatively impermeable deposits.

As discussed in Chapter 10, the Great Plains is one of the most important agricultural regions on Earth, producing much of the world's wheat, corn, and soybeans. One of the primary reasons for this high productivity is that the soils are mostly Mol-

lisols, which, as we saw in Chapter 11, have thick, organic-rich A horizons and B horizons laden with nutrients such as calcium and magnesium. Despite the semi-arid environment, these fertile soils attracted large numbers of farmers to the region in the late 1800s and early 1900s during the period of western settlement. These farmers quickly learned, however, that the region could be an unforgiving place for agriculture because precipitation could be inconsistent and droughts often led to poor harvests.

The early settlers suspected that a large groundwater supply existed for a couple of reasons. First, they noticed that a number of artesian wells were scattered throughout the region. An **artesian well** occurs when the pressure applied by incoming water elsewhere in the groundwater system causes water to move up to the surface. This pressure can be enhanced by a variety of factors, such as if a saturated zone is sandwiched between two aquicludes, or because of faulting (Figure 15.5a).

Artesian well   *A well in which water from a confined aquifer rises to the surface through natural pressure.*

| TABLE 15.1 | Characteristics of the High Plains Aquifer* |

Characteristic	Unit	Total	CO	KS	NE	NM	OK	SD	TX	WY
Area underlain by aquifer	mi^2	174,050	14,900	30,500	63,650	9,450	7,350	4,750	35,450	8,000
% of total aquifer area	%	100	8.6	17.5	36.6	5.4	4.2	2.7	20.4	4.6
% of each state underlain by aquifer	%	—	14	38	83	8	11	7	13	8
Average area-weighted saturated thickness in 1980	ft	190	79	101	342	51	130	207	110	182
Volume of drainable water in storage in millions in 1980	acre-ft	3250	120	320	2130	50	110	60	390	70

*Modified from E. D. Gutentag et al., 1984, *Geohydrology of the High Plains Aquifer in Parts of Colorado, Kansas, Nebraska, New Mexico, Oklahoma, South Dakota, Texas, and Wyoming*, U.S. Geological Survey Professional Paper 1400-B, p. 63.

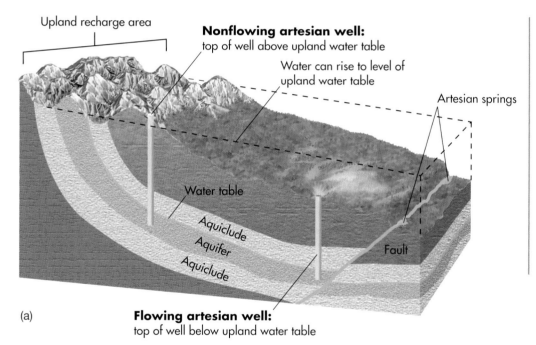

(a)

Upland recharge area

Nonflowing artesian well:
top of well above upland water table

Water can rise to level of upland water table

Artesian springs

Water table

Aquiclude

Aquifer

Aquiclude

Fault

Flowing artesian well:
top of well below upland water table

Figure 15.5 Artesian wells and springs. (a) When aquifers are confined between aquicludes, water may rise to the surface in an artesian well due to pressure applied by incoming water elsewhere in the aquifer. Where depressions or channels intersect the water table—for instance, at a fault—lakes and streams develop. (b) A spring is a place where water naturally seeps out of the ground. This may occur because the groundwater lies between aquicludes, as in an aquifer, or water may flow horizontally along the contact between overlying coarse-textured sediments and an aquiclude.

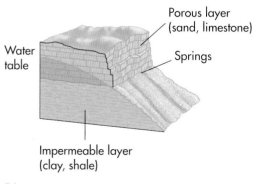

Porous layer (sand, limestone)

Water table

Springs

Impermeable layer (clay, shale)

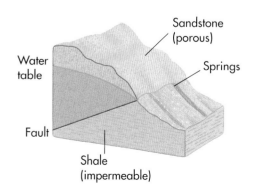

Sandstone (porous)

Water table

Springs

Fault

Shale (impermeable)

(b)

Figure 15.6 Aerial view of center-pivot irrigation. Center-pivot pumping systems such as these in Colorado were developed in the mid-20th century and are almost ubiquitous now in parts of the Great Plains. Each circle is about the size of a 1/4 section in the USPL system discussed in Chapter 2. Look for them when you fly over the country.

Jim Wark/AgStock Images/Corbis

Second, settlers also noticed that many streams had a fairly consistent supply of water in them, even during periods of relative drought. This consistent flow resulted because stream channels intersected with the aquifer's water table, providing a steady source of water. These patterns are common not just to the High Plains Aquifer, but to any place where the water table is high and intersects depressions and streams.

For the first half of the 20th century, access to the High Plains Aquifer was limited and largely confined to artesian and hand-excavated wells. However, the access to water changed dramatically in the early 1950s with the advent of a new technology called *center-pivot irrigation.* Center-pivot systems consist of motorized pumps that tap into the aquifer and suck up water like drinking from a straw. They then distribute the water to farm fields through an elevated sprinkler system that extends to the edge of the field. This sprinkler system slowly rotates around the pump on a set of wheels so that the field is consistently supplied with water. Most such systems cover a circular area approximately 64 hectares (160 acres) in size, although some are larger.

Development of center-pivot systems was a major advancement for the agriculture-based economy of the region, enabling farmers to control the semi-arid environment by providing a reliable supply of water to high-yield crops such as corn that would otherwise not consistently grow there. These systems have been so wildly successful that tens of thousands of them are currently in operation around the region, forming a patchwork of circles that are easily seen from the air (Figure 15.6). The next time you fly across the country, be sure to notice this important part of the nation's cultural landscape and consider how it supplies much of your food. Also consider the massive human footprint on the landscape and how it relates to Chapter 10 and the discussion of the Anthropocene as a potential epoch in the geological timescale.

Although the rapid expansion of center-pivot irrigation has enabled farmers to be successful in marginal regions such as the High Plains, the long-term environmental consequences of extensive groundwater mining in the region are significant. The most obvious impact of extensive groundwater extraction is that the water table has dropped in many places because so much water has been withdrawn. According to the U.S. Geological Survey, 17 billion gallons of water were drawn for irrigation purposes every day in 2000. Another 315 million gallons were withdrawn to supply water for almost 2 million people living in the region. Such a rate of withdrawal far exceeds the rate of recharge, meaning that groundwater levels must inevitably decline. This decline can be seen at the local level with any individual well as a **cone of depression** in which water is drawn down in a circular pattern from the aquifer surrounding the well (Figure 15.7a). If pumping from the well ceases, the cone will fill back in, both by recharge from the surface and by water seeping into the empty pores from the surrounding saturated zone.

In many regions, however, groundwater extraction is increasing because more wells are being installed and water is used year-round for farming and ranching. As a result, the drain on underground water has become serious across the region, as well as in many other parts of the world. Figure 15.7b shows the impact that two wells—a deep well and a shallow, seasonal well—may have on the groundwater supply in a particular area. In the Great Plains of North America, such drawdown has occurred in thousands of places, causing a significant decline in the elevation of the water table in many locations. In areas within southwestern

---

Cone of depression *The cone-shaped depression of the water table that occurs around a well.*

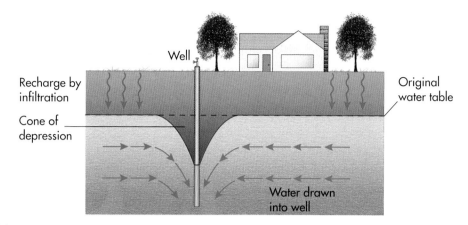

Well

Recharge by infiltration

Original water table

Cone of depression

Water drawn into well

(a)

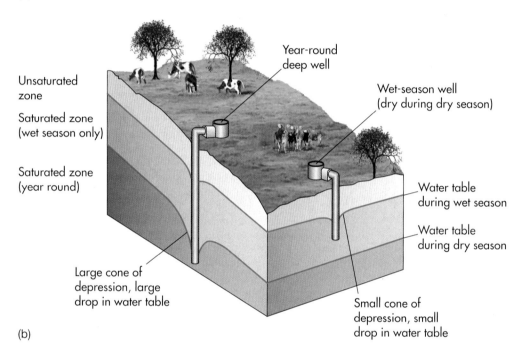

Unsaturated zone

Saturated zone (wet season only)

Saturated zone (year round)

Year-round deep well

Wet-season well (dry during dry season)

Water table during wet season

Water table during dry season

Large cone of depression, large drop in water table

Small cone of depression, small drop in water table

(b)

Figure 15.7 Groundwater depletion. (a) When water is drawn from an aquifer, the water table is depressed in a conical shape. (b) Deep wells draw water from zones that are saturated year round, draining far more water than shallower, seasonal wells.

Kansas and the Texas Panhandle, for example, the average drop in the level of the water table was more than 12 m (40 ft) between 1980 and 1996 (Figure 15.8a). This rapid decline has occurred in these regions because the withdrawal rate is as much as 22 times greater than the recharge rate, which averages only 0.5 cm/year (0.2 in./year). Signs of this drop in the water table can be seen throughout the region. For example, many of the streams have dried up (Figure 15.8b) because the water table is now below the base of the channel.

As a result of the heavy reliance on groundwater mining, the High Plains Aquifer may disappear in many places by the middle of the 21st century. Estimates suggest it will be entirely depleted over a broad area by 2300. This appraisal is based on the amount of groundwater extracted from the aquifer compared to its recharge rate due to infiltration of precipitation. Estimates are similar for the depletion of other

aquifers around the world. In the Sahara Desert, for example, scientists suggest that aquifers there may last 50 years. Put another way, natural resources that took hundreds of thousands of years to evolve will disappear essentially within a timeframe of 200 years (from the mid-1800s to 2050) due to human impact.

In spite of these trends the spatial extent of irrigated land continuous to expand across the region. This expansion is being driven by a complex interaction of several things, including climate change, population growth, production of biofuels, and other factors. Efforts to regulate and manage the system have shown mixed results due to numerous political, cultural, and economic variables. Some farmers recognize the need for conservation and are shifting to dryland agriculture, which is a crop-rotation system designed to conserve soil moisture by leaving fields fallow (or crop-free) every other year. Other farmers

(b)

Figure 15.8 Drawdown of the High Plains Aquifer. (a) Due to excess groundwater mining, the water table has dropped significantly in an area that ranges from southwestern Kansas to the Texas panhandle. (*Source:* U.S. Geological Survey.) (b) The effect of aquifer depletion has been significant on many streams. For example, prior to aggressive groundwater extraction in the region, Rattlesnake Creek in south-central Kansas was a perennial, spring-fed stream. Now, it is dry most of the time, as shown here, because the water table is far below the level of the stream bed. The students in the foreground are studying the sand dunes exposed along the old stream bed.

are attempting to maximize short-term profits by mining the aquifer until it disappears in their region, at which time they will shift to dry-land agriculture or leave farming entirely. It is a complex issue that bears watching in the future.

Subsidence

Human interactions with the High Plains Aquifer demonstrate what happens to the water table elevation when groundwater is tapped excessively. In addition to these changes, which occur deep within the ground, adjustments sometimes take place on the surface when large amounts of groundwater are removed, either to provide irrigation or to supply drinking water in heavily populated regions. One kind of impact that can occur is that sediments become compacted because water is lost from pore spaces. The most obvious surface change occurs when the land sinks due to the removal of water from underlying aquifers. This type of response is called **subsidence**.

Subsidence *The settling or sinking of a surface as a result of the loss of support from underlying water, soils, or strata.*

Subsidence is a particularly significant problem in and around large cities. A great example of this kind of impact can be seen at Venice, Italy. This beautiful city is approximately 1400 years old and was constructed on a landscape consisting of over 100 low-lying islands in a lagoon in the northern part of the Adriatic Sea. Underlying the islands are about 1000 m (about 3300 ft) of weakly cemented deposits of sand, gravel, clay, and silt. The coarse sands and gravels of these deposits contain fresh groundwater resources that have been extensively exploited for human use. The associated subsidence of the landscape has been so great that many Venetian homes are now at the shoreline and flooding is common. In addition, salt water from the Adriatic is beginning to seep into the city's water supply. Similar problems have also recently occurred in China and California. In the San Joaquin Valley of California, for example, groundwater removal for irrigation has caused the ground to subside as much as 4.8 m (16 ft) in places over a 35-year period. This issue will be discussed more thoroughly in Chapter 20.

Groundwater Contamination

Another major environmental problem associated with the relationship between humans and groundwater is contamina-

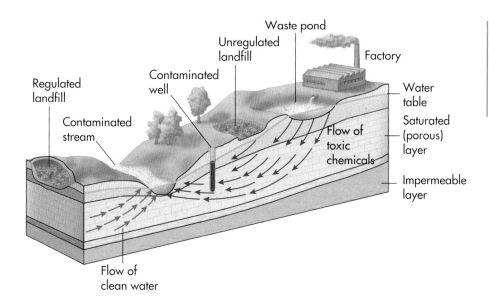

Waste pond
Unregulated landfill
Factory
Contaminated well
Regulated landfill
Water table
Contaminated stream
Flow of toxic chemicals
Saturated (porous) layer
Impermeable layer
Flow of clean water

Figure 15.9 Movement of groundwater contaminants. Infiltrating water at solid-waste disposal sites, such as waste ponds and unregulated landfills, carries toxic chemicals into the groundwater, which flows along natural pathways to pollute wells and streams.

tion by pollutants. Our industrial society produces enormous amounts of solid and liquid waste. Until the latter half of the 20th century, no regulations governed the disposal of this waste. As a result, much of it was simply carted off to large holes in the ground, some excavated and some natural, where it was perhaps burned and buried. Infiltrating rainwater would interact with chemical compounds in the waste and carry pollutants into the saturated zone, where they would then flow along the various groundwater paths to wells and streams (Figure 15.9). Given that drinking water is often obtained from groundwater, contamination can be a significant health risk.

Since the establishment of the Environmental Protection Agency (EPA) in the mid-1970s, a more systematic effort has been made to control the disposal of solid waste in the United States. One important way in which solid-waste disposal is intensely regulated by the EPA occurs in highly engineered landfills. In contrast to the traditional "hole-in-the-ground" approach used previously, modern engineered landfills are designed to eliminate the flow of toxic chemicals and other contaminants into the groundwater supply. For example, landfill engineers can line the base and sides of the landfill with impermeable material so that groundwater contamination is minimized. Any toxic chemicals and contaminants produced in the landfill are pumped to a settling pond where the waste is oxidized and neutralized. In spite of these measures, landfills and other hazardous waste facilities often remain very controversial.

VISUAL CONCEPT CHECK 15.1

This photograph shows the shoreline of a natural lake. Given what you see here, which one of the following statements is accurate?

a) The shoreline represents the place where the ground level meets the local water table.
b) The water table lies below the level of the lake bottom.
c) The local water table is very low.
d) A tremendous amount of groundwater mining has occurred in the area.

Ron Niebrugge/Alamy

Now that the concept of groundwater has been discussed, go to the *Geo Media Library* and open *Hydrologic Cycle and Groundwater*. This animation reviews the processes that have been described so far in this chapter. It begins with the hydrologic cycle and demonstrates how water moves from one reservoir to another. It also animates the relationship among evapo-ration, precipitation, and the way that groundwater is stored and moves. Review how the water table fluctuates and how groundwater depletion impacts the water table. A final exercise tests your understanding of groundwater terminology. Once you complete the animation, be sure to answer the questions at the end to assess your knowledge of this topic.

KEY CONCEPTS TO REMEMBER ABOUT GROUNDWATER

1. Groundwater refers to water that is stored within the ground.

2. Some water is stored in soil pore spaces to form the soil–water belt. When the soil is dry, water is nevertheless held tightly as hygroscopic water. With increasing precipitation, water fills more of the pores as capillary water. If the soil and associated pores are saturated, they reach field capacity and water is free to move downward under the force of gravity.

3. When water percolates downward, it first encounters the unsaturated zone. At some point in its downward flow, this water encounters the aquifer, which is a layer of rock fully saturated with water.

4. The top of the saturated zone is called the water table. In places where the groundwater supply is sufficiently large to be a source of water for people, this zone is called an aquifer.

5. The High Plains Aquifer is the largest aquifer in North America, underlying much of the Great Plains. This aquifer is being extensively mined to irrigate agricultural crops. Consequently, a significant decrease in water table elevation is occurring in many places.

Karst Landforms and Landscapes

Have you ever been to a cool cave system, such as Mammoth Cave in Kentucky or Carlsbad Caverns in New Mexico? Or maybe you explored an unknown cave in southern Indiana, Tennessee, or Kentucky. If so, you explored how groundwater can shape the subterranean landscape over a long period of time. These effects are visible not just beneath the ground, but also at the surface in the form of depressions and sometimes even hills.

A region that contains a lot of caves and other soluble rock features is said to be a landscape with **karst topography**. Karst topography is most closely associated with extensive, thick deposits of limestone. As discussed in Chapter 14, these rocks dissolve easily through the weathering process of carbonation. This erosion produces a complex system of joints that allows water to flow from the surface into the underlying rock. Vegetation also plays a role because it provides organic acids that enhance the solution process. When these conditions occur in humid regions, such as the eastern part of the United States and southeastern China, karst landscapes can be widespread (Figure 15.10). They also occur in some regions that are now arid but were humid at some point in the geologic past, such as New Mexico and northwestern Africa.

Caves and Caverns

Perhaps the best-known karst landforms are **caves**, which consist of underground voids in rock that are sufficiently large for people to enter and explore. Although most caves are small, some cave systems are enormous and contain numerous passageways and large chambers. Such a large cave system is called a *cavern*. To give you an idea of how large chambers can be, consider that *The Big Room* at Carlsbad Caverns is about 110,000 m^2 (1,184,030 ft^2) in area.

People have been aware of caves and caverns for a long time. In Europe and Asia, for example, caves provided shelter for prehistoric people during the Pleistocene Epoch when gargantuan glaciers covered much of the landscape. Evidence for these occupations can be seen in Europe in the number of beautiful paintings (known as *cave art*) that adorn the cave walls.

Karst topography *Terrain that is underlain by soluble rocks, such as limestone and dolomite, where the landscape evolves largely through the dissolution of rock.*

Cave *A cavity in rock, produced by the dissolution of calcium carbonate, that is large enough for someone to enter.*

Figure 15.10 Location of major limestone and karst regions on Earth. Large areas of limestone and karst occur on every continent.

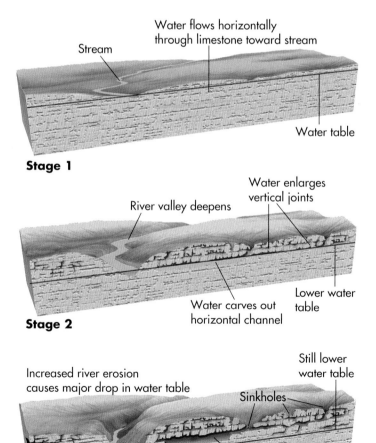

Stage 1

Stream

Water flows horizontally through limestone toward stream

Water table

Stage 2

River valley deepens

Water enlarges vertical joints

Water carves out horizontal channel

Lower water table

Stage 3

Increased river erosion causes major drop in water table

Sinkholes

Still lower water table

New horizontal channel

Caverns now empty

These paintings, as well as burials deep within the caves, have given archaeologists important clues about the way people lived and the animals that they hunted. Today, a lot of people like to explore caves for the fun and challenge of it. Such cave enthusiasts are called *spelunkers*.

Caves and caverns form due to the complex interaction of climate, groundwater, and local streams over time. Figure 15.11 illustrates the evolution of these landscapes. Cave formation begins during a time (stage 1) when the water table is high because the stream has not yet downcut a deep valley by erosion. This concept of downcutting will be discussed in more detail in Chapter 16, but for now, suffice it to say that one of the things that streams do as they evolve is carve out deep valleys. In the context of karst topography, the important relationship is that the water table mirrors the behavior of the stream because groundwater flows to the lowest point on the landscape—in other words, the river channel.

Figure 15.11 Cave and cavern evolution. In Stage 1, limestone below the water table dissolves, forming submerged caverns. As the stream downcuts in Stage 2, the water table drops in response, causing water to carve out channels in the limestone as the water flows to the river. In Stage 3, the water table drops farther, enough so that the formerly submerged caverns are now located in the unsaturated zone. Surface water percolates into these open features, resulting in the formation of stalagmites and stalactites.

Karst Landforms and Landscapes • 403

During stage 1, the process of carbonation is concentrated in the deposits of limestone that occur just below the water table. The resulting dissolved sediments are then carried by underground streams to the surface channel. These kinds of streams evolve because water concentrates along joints in the rocks where it can move freely. These cracks slowly enlarge through dissolution, and the water can thus flow more quickly as an underground stream. Over time, the process of carbonation and associated groundwater flow creates myriad passageways and caverns at the water table boundary. In stage 2, the stream has cut a valley into the rock, lowering the water table. Groundwater cuts large channels and caverns into the rock as it flows through to the stream. At this time, these features are submerged in groundwater. Finally, in stage 3, as the stream continues to downcut and deepens its valley, the water table also lowers still farther. As a result of this lower water table, the formerly submerged passageways of stage 1 are now in the unsaturated zone and thus become open caves and caverns.

Once stage 3 is complete, the cave or cavern has formed and the dominant processes shift to those that occur within the chamber. In order for the cavern to be preserved, some kind of more resistant cap rock must be present to minimize the amount of water percolating into it and keep the roof from collapsing. If such preservation occurs, features such as stalactites (hanging rods), stalagmites (upward-pointing rods), columns, and drip curtains can begin to form within the cave or cavern (Figure 15.12). These unusual formations evolve in caves where large

quantities of surface water percolate downward into the roof of the open cavern. As this water moves through the overlying rock, the solution becomes enriched in dissolved carbonates before it slowly drips into the cave. Stalactites in the roof of a cave form in much the same way that an icicle lengthens downward during a winter thaw. Stalagmites, in contrast, develop when water drips to the cave floor. This water evaporates, leaving the calcium carbonate behind in a deposit that slowly grows upward over time. An easy way to remember the difference between the two forms is that the "c" in stalactites can mean "ceiling," whereas the "g" in stalagmites means "ground."

Karst Topography

Karst landforms can also evolve at the surface. The most common surface landform is a **sinkhole**, which is a depression that occurs in regions of cavernous limestone (Figure 15.11). Sinkholes form when caves enlarge so much that their ceilings collapse due to the force of gravity, causing the surface rocks and sediment to sink, often suddenly. The rapidity of sinkhole formation was tragically exemplified in March 2013 when a sinkhole developed at night beneath the bedroom of a sleeping man in Florida. The hole, which measured 6 m (20 ft) across and about 9 m (30 ft) deep, opened instantaneously and swallowed the victim. His body was never found. Sinkholes are particularly common in parts of Florida because much of the peninsula is underlain by limestone that easily dissolves.

Sinkholes can also form due to excessive groundwater extraction because caves that were previously filled with water become empty. As a result, nothing but air is present below the ceiling of the structure and it collapses under its own weight. Sinkholes frequently enlarge for a period of time after their initial formation because surface water tends to funnel into them. Sometimes this surface flow becomes a **disappearing stream**, which flows down into a sinkhole from the surface and then enters underground channels. Often, this downward-flowing stream causes a pipe to evolve that connects the sinkhole with the caverns and passageways beneath.

At other times sinkholes remain detached from the underlying cavern and simply fill with water, appearing as ponds on the landscape and in topographic maps (Figure 15.13). In regions where karst activity occurs in densely populated regions, sinkholes are a significant geological hazard because they can unexpectedly swallow homes or portions of highways. A spectacular form of karst topography occurs in humid subtropical areas where thick beds of limestone and high water tables exist. This kind of karst is called *tower* (or *haystack*) *karst* and results in dramatic landscapes. In order for tower karst to evolve, karst processes must have operated for a long period of time, during which a variety of subterranean caves and passageways developed. Over time, these features gradually enlarge so that

Figure 15.12 Drip formations at Carlsbad Caverns in New Mexico. Here, dripping water enriched in calcium carbonate has formed numerous stalactites, stalagmites, and columns in the Papoose Room. This form of deposition occurs when percolating groundwater dissolves limestone and drips, leaving calcium carbonate behind.

Rich Reid/National Geographic/Getty Images

Sinkhole   *A topographic depression that forms when underlying rock dissolves, causing the surface to collapse.*

Disappearing stream   *A surface river or stream that flows into a sinkhole and subsequently moves into an underground river system.*

(a)

(b)

Courtesy The Center for Coastal and Watershed Studies, St. Petersburg, FI/USGS

Figure 15.13 Sinkhole topography. (a) Sometimes sinkholes develop that extend below the depth of the water table, forming ponds. (b) Sinkholes show up clearly on topographic maps as ponds, where they are filled with water, and as depressions, where contour lines make concentric circles.

sinkholes and various other collapse structures form. If enough of these karst features exist, the effect is to produce a number of towering hills that are the unaffected remnants of the block of solid limestone that once existed. Streams also play a role in the development of tower karst because they cut steep valleys into the regional rock mass.

An excellent place to see fantastic tower karst is in southeastern China (Figure 15.14), near the city of Guilin in the northern part of Guangxi province. This area lies within the humid subtropical climate region and is affected by the southeastern monsoon. The Guilin area is covered by carbonate rocks that may reach a thickness of 4600 m (~15,000 ft).

Two major surface karst landforms occur in the region: (1) the peak-cluster depressions (Fengcong) and (2) the peak-forest plain (Fenglin). These areas collectively cover an area 2429 km² (938 mi²) in size. The Fengcong landscape consists of a group of peaks separated from each other by collapse depressions, whereas the Fenglin areas are a group of isolated peaks separated by flat ground created by streams. Taken together, these hills literally tower over the landscape, reaching heights of 200 m (660 ft) above the surrounding lowlands and giving the region a beautifully surreal quality. Close to the mainland United States, the only place to see tower karst is the northern part of Puerto Rico.

John Arnold/Alamy Images

(a)

Figure 15.14 Tower karst in China. (a) This surreal landscape of isolated peaks near Guilin, China, formed when thick limestones were dissolved and collapse structures formed on a massive scale. (b) The Guangxi province in southeastern China is especially known for its tower karst.

(b)

Karst Landforms and Landscapes • 405

VISUAL CONCEPT CHECK 15.2

Given what you see in this photograph, which one of the following statements is accurate?

a) The local water table intersects with the ground level at this place.

b) This photograph shows the formation of tower karst.

c) The top of a cave collapsed, forming a sinkhole.

d) The local water table was much higher at some point in the past.

China Tourism Press/Riser/Getty Images, Inc.

SUMMARY OF KEY CONCEPTS

1. Groundwater refers to water that is stored within the ground. Some water is stored in soil pore spaces to form the soil–water belt. If the soil and associated pores are saturated, water is free to move downward under the force of gravity. When water percolates downward, it first encounters the unsaturated zone.

2. At some depth the sediment pore spaces are saturated with water, forming the saturated zone. The sediments in this zone usually consist of sand and gravel that are underlain by an aquiclude of impermeable clay or shale. The top of the saturated zone is called the water table. In places where the groundwater supply is sufficiently large for people to use for irrigation or drinking water, the saturated zone is called an aquifer.

3. Groundwater can become contaminated when chemicals from old landfills or dumps slowly move into the system.

4. Karst landforms and landscapes form in association with groundwater and include features such as caves and sinkholes.

5. Caves begin to form during a period of high water table when flowing groundwater dissolves carbonate-rich limestones through the process of carbonation. Caves become open when the water table falls, usually in association with stream downcutting or, perhaps, climate change.

6. Tower (or haystack) karst topography forms when numerous caverns and subterranean passageways collapse, leaving individual high hills.

1. What is the soil–water belt, and how do the terms *hygroscopic water*, *capillary water,* and *field capacity* relate to this term?

2. What is the unsaturated zone, and where is it located in the groundwater model?

3. What role does an aquiclude play in the formation of a saturated zone?

4. Why do sandy and gravelly deposits make good aquifers?

5. Describe the relationship among topography, the water table, and the development of natural lakes.

6. Describe the physical geography of the High Plains Aquifer.

7. What is a cone of depression, and how is it related to declining water tables in places like the Great Plains?

8. Why are many streams going dry in places where aggressive extraction of groundwater occurs?

9. Why are modern landfills more environmentally safe than unregulated disposal sites of years past?

10. What is karst topography, and how do caves evolve?

11. How does the formation of a sinkhole relate to the process of land subsidence and the formation of tower karst?

ANSWERS TO VISUAL CONCEPT CHECKS

VISUAL CONCEPT CHECK 15.1

The answer to this question is *a*. The shoreline represents the place where the ground level meets the local water table. Such a natural lake exists when depressions dip below the level of the water table.

VISUAL CONCEPT CHECK 15.2

The answer to this question is *d*. The local water table was much higher at some point in the past. This cave formed when carbonate rocks dissolved when the water table was higher. After the water table fell, the cave was left behind.

CHAPTER SIXTEEN | FLUVIAL SYSTEMS AND LANDFORMS

This chapter investigates water that flows across the surface of Earth in streams. Whether these flowing bodies of water occur in the largest rivers or smallest creeks and brooks, they are collectively viewed as *fluvial systems*. This chapter discusses the input of water into these systems, how it behaves within them, and the predictable outputs such as landforms that result.

In many cases, streams are a visible reflection of the more distant geological past. In humid regions, streams have likely been flowing along the same basic course for thousands of years, shaping the landscape gradually through erosion and deposition. Where streams occur in semi-arid to arid regions, however, they may flow only during the wet season, or only every few years when intense rains fall. When they do flow, they may dramatically alter the landscape because moving water has tremendous power. If you like watching streams, or enjoy fishing and canoeing on them, you will like this chapter.

Flowing water is the most important natural geomorphic agent on Earth, shaping the landscape in many interesting ways. Humans also interact with streams on a daily basis, as this image of London's Thames River shows, using them for transportation arteries, supplies of drinking water, and energy production. This chapter focuses on the way that water flows across the Earth's surface in streams, the landforms that are produced by these processes, and some of the ways that people interact with them.

CHAPTER PREVIEW

Overland Flow and Drainage Basins

 The Rhine River

Hydraulic Geometry and Channel Flow

Fluvial Processes and Landforms

 Bedload Transport and Braided Streams

 The Graded Stream

 Stream Meandering

 Fluvial Geomorphology and Stream Processes

Human Interactions with Streams

LEARNING OBJECTIVES

1. Describe the concept of a drainage basin (or watershed) and how such basins are arranged from a hierarchical perspective in the United States.

2. Discuss the various components of hydraulic geometry and the way that water flows in channels.

3. Compare and contrast the concepts of base flow, bankfull discharge, and flood discharge in the context of recurrence interval.

4. Explain how hillslope erosion processes operate and the formation of rills and gullies.

5. Define the concept of a graded stream and how the longitudinal profile of such a stream changes over time.

6. Discuss why base level is an important variable in the behavior of a stream.

7. Compare and contrast the character and development of braided streams with meandering streams.

8. Explain the concept of a stream meander, how water flows within it, and how the landscape is shaped at this location.

9. Describe the various ways that humans impact streams and how they have both positive and negative outcomes.

Overland Flow and Drainage Basins

Streams are important features for a variety of reasons. In the context of the hydrologic cycle, they connect the interior of continents with oceans. Water that falls as precipitation in high mountainous areas, for example, will ultimately flow back to the sea. Flowing water is also the most powerful geomorphic agent on Earth, shaping the landscape in both subtle and profound ways. Streams are important to humans because they are sources of water for both direct consumption and agriculture. In addition, people have long used streams as transportation networks to move economic goods from one place to another. As a result, rivers such as the Rhine in Europe, the Nile in Africa, the Mekong in Southeast Asia, the Amazon in South America, and the Hudson and Mississippi in the United States have played critical roles in the evolution of human societies. Figure 16.1 shows the location of major rivers around the world.

Origin of Streams

A good place to begin considering streams is their origin and the sources of water within them. Some streams flow all year long and are called *perennial streams*, whereas others contain water for only a short time and are thus *ephemeral (or intermittent) streams*. Although you might think stream water comes directly from precipitation that falls into the channel, it is only a negligible contribution. In most places, a more important source of stream water is water that seeps out of the ground where stream channels intersect the water table, as was briefly discussed in Chapter 15 (see Figure 15.3). In some areas, such as high mountain areas, stream water comes from melting snow and glaciers.

Still another important source for streams is water that is stored in lakes and ponds that are fed by both precipitation and groundwater. If the water level in such a lake reaches a sufficient depth, it may spill over at a low spot on the landscape, called an *outlet*, and become the source of a river. Many rivers start at such an outlet. An excellent example is Lake Victoria in east central Africa, which is the source of the Nile River. The lake straddles the borders of Uganda, Kenya, and Tanzania (Figure 16.2a) and lies in a depression created by the East Africa Rift system (Figure 13.9). Approximately 67,850 km² (26,197 mi²) in size, Lake Victoria is the second largest freshwater lake in area on Earth (behind Lake Superior in North America). Excess water spills over an outlet at the north end of Lake Victoria (Figure 16.2b) and marks the beginning of the Nile River. The Nile River is about 10,730 km (6670 mi) long, making it the longest river in the world, and flows northward through Sudan and Egypt to the Mediterranean Sea. The best-known river in North America, the Mississippi, also begins at a lake outlet, Lake Itasca in Minnesota (Figure 16.2c). At this place, you can literally walk across the Mississippi River!

Another significant source of stream water is surface runoff. Runoff occurs most commonly in association with wet periods when soils are saturated and pore spaces can no longer absorb additional precipitation. As discussed in Chapter 15, these periods are times of groundwater recharge because water

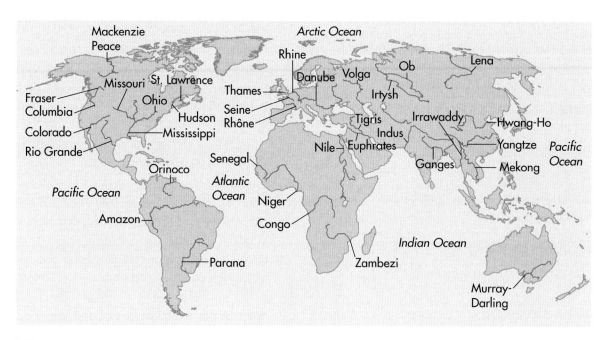

Figure 16.1 Important rivers on Earth. Rivers are channelized bodies of flowing water that connect the continental interior with the ocean. As a result, these rivers, and streams like them, have been used as transportation systems throughout the course of human history.

(a)

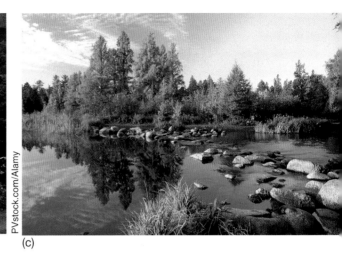

(b)

(c)

Figure 16.2 Sources of major rivers. (a) The Nile River begins at the north end of Lake Victoria and flows northward for about 10,730 km (6670 mi) before it ultimately reaches the Mediterranean Sea in Egypt. (b) This image shows Bujagali Falls in Uganda. This waterfall is located just downstream of Lake Victoria, which is the source of the Nile River. (c) The source of the Mississippi River at Lake Itasca, Minnesota. The boulder bridge marks the point where the Mississippi flows out of the lake (to left of boulder bridge) on its journey south to the Gulf of Mexico, ~4100 km (2550 mi) away.

is free to move from the soil downward through the force of gravity. At the same time that water is moving downward through the soil, water runs off the surface and toward the stream as overland flow (Figure 16.3). You can see the process of runoff clearly in any parking lot during and shortly after a strong storm. The asphalt keeps water from soaking into the soil beneath; as a result, the water flows across the surface in sheets toward a drain of some kind. This type of runoff, called *sheet runoff*, also occurs in areas where slopes are very steep and rainfall simply does not have a chance to soak into the ground.

Drainage Basins

The next step in our discussion of stream systems is to consider how they are spatially organized. A good place to begin is to consider the concept of a **drainage basin**, which is the geographical area that contributes groundwater and runoff to any particular stream. Another term frequently used to define the same area is *watershed*.

To begin this discussion, some basic terminology must first be established. Streams in any two watersheds are separated by a topographic feature called a **drainage divide**, which is an area

Figure 16.3 An example of storm runoff. During periods of heavy rain, soils become saturated and water flows across the surface. This kind of flow is an important source of stream water.

Drainage basin *The geographical area that contributes groundwater and runoff to any particular stream.*

Drainage divide *An area of raised land that forms a bordering rim between two adjacent drainage basins.*

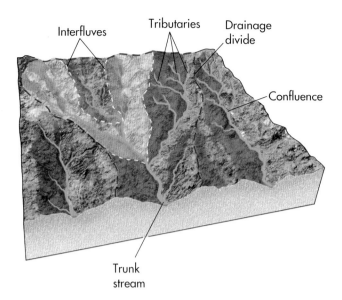

Figure 16.4 The concept of drainage basins. Also called watersheds, drainage basins are separated from one another by high points on the surrounding landscape called divides. Each large drainage basin contains numerous tributaries, which themselves are separated from each other by interfluves.

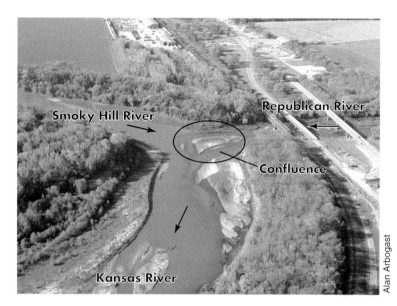

Figure 16.5 A typical river confluence. The Smoky Hill River flows into this image from the upper left, whereas the Republican River enters from the upper right. They join at a confluence in the center of the photograph to form the Kansas River, which is the trunk stream of the Kansas River basin, and flows toward the bottom (east) of the image.

of elevated terrain that forms a kind of high rim around any given basin (Figure 16.4). Below this rim, the topography slopes downward into the core of the basin. As a result, all runoff and groundwater flow into a network of streams that collectively funnel into the **trunk stream**, which is the largest stream in the drainage basin. Each stream that flows into the trunk stream, or a stream that flows into another stream, is called a **tributary**. Tributaries are themselves separated from one another by relatively low topographic rims called **interfluves**. The point where a tributary joins the trunk stream, or any stream for that matter, is referred to as a **confluence** (Figure 16.4).

Drainage basins range from several hectares to thousands of square kilometers in size, with numerous watersheds contained within larger ones in a nested fashion, as you can see in Figure 16.4. Notice, for example, that the small drainage basin portrayed in light orange is nested within the drainage basin highlighted in yellow. In other words, water that drains into the orange basin ultimately flows into the larger stream contained within the yellow basin at a confluence. This stream ultimately flows into the trunk stream at another confluence, like the one pictured in Figure 16.5. This image shows the confluence of the two major tributaries of the Kansas River, which is the trunk stream of the Kansas River basin in the central United States. To see the geographic relationship of these streams, refer back to the map in Figure 15.4. In this map note

that the Republican River drains the northern portion of Kansas and southern Nebraska, whereas the Smoky Hill River drains western Kansas. These two streams meet in eastern Kansas to form the Kansas River.

Now put this discussion of drainage basins into the context of major watersheds in the United States. The map in Figure 16.6 not only shows the geographic position and size of the largest drainage basins, but also indicates into which ocean the trunk stream of any particular basin flows. The largest watershed in North America is the Mississippi River basin (bold outline), which drains most of the United States (and a small part of Canada) and funnels water to the Gulf of Mexico. The Mississippi watershed contains several other large basins, including the Ohio, Arkansas, and Missouri, which are named after their trunk streams. Another way to describe these trunk streams is that they are major tributaries of the Mississippi River. Each of these trunk streams, in turn, contains a number of smaller tributaries. A major tributary of the Missouri River, for example, is the Kansas River shown in Figure 16.5. In other words, one can accurately say that the Smoky Hill and Republican rivers are tributaries not only of the Kansas, but also of the Missouri River and, ultimately, the Mississippi River.

The western divide of the Mississippi basin is the Rocky Mountains. This divide is the famous *continental divide* that separates the Mississippi watershed and its drainage to the

Trunk stream *The primary stream of a drainage basin.*

Tributary *A stream or river that flows into a larger stream or river.*

Interfluves *Topographic high points in a drainage basin that separate one tributary from another.*

Confluence *The place where two streams join together.*

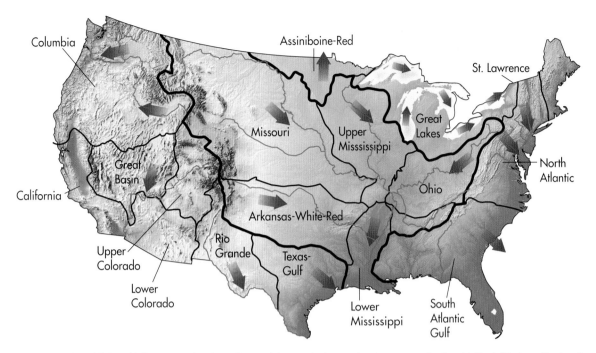

Figure 16.6 Major U.S. watersheds. Several large drainage basins occur in the United States. Each of these basins contains an array of nested drainage basins like those shown in Figure 16.4. Note the direction of flow for each trunk stream. The Mississippi River watershed is the largest drainage basin in the country and is outlined in bold.

Gulf of Mexico from streams that deliver runoff into the Pacific Ocean, such as the Colorado and Columbia rivers. The eastern divide of the Mississippi basin is the Appalachian Mountains, which is also a continental divide because it separates Atlantic-flowing drainages such as the Delaware and Hudson basins from the Gulf-flowing waters of the continental interior.

Drainage Patterns, Density, and Stream Ordering

In the context of individual drainage basins, it is often useful to characterize the spatial configuration of the stream network. Such an assessment sheds light on the evolution of the basin or the nature of the rocks and sediments that lie within it. One method is to look at the drainage pattern, which is the way that the various streams within a drainage basin are spatially organized. The five primary drainage patterns are shown in Figure 16.7.

1. **Dendritic**—A branching, tree-like drainage pattern evolves in areas of uniform rock resistance and structure, with little distortion by folding or faulting. Notice in Figure 16.7a that the streams develop a random branching network similar to a tree.

2. **Rectangular**—A rectangular drainage pattern occurs when joints and faults steer streams at right angles to one another. This pattern occurs because water flows preferentially to these zones of weakened bedrock where the water more freely erodes. Notice in Figure 16.7b the angular nature of the drainage pattern.

3. **Trellis**—A trellis drainage pattern (Figure 16.7c) is a system of streams that develops in areas such as the Ridge and Valley province in the Appalachian Mountains where rocks are folded. In this area, major streams tend to flow parallel to one another in adjacent valleys within the folded mountain belts. Minor tributaries flow into larger streams at right angles.

4. **Radial**—In a radial drainage pattern (Figure 16.7d), streams radiate outward from a central point, forming a spoke-like pattern of rivers. This kind of pattern tends to evolve where streams flow away from rounded upland areas such as a volcano.

5. **Deranged**—A distinctively chaotic drainage pattern (Figure 16.7e), characterized by irregular direction of stream flow, few tributaries, swamps, and many lakes, that develops in recently glaciated terrain.

Another way to characterize a drainage basin is by calculating the drainage density of a watershed. **Drainage density** refers to the relative density of natural drainage channels in a given area. This value is calculated by dividing the total length of all streams in the basin by the area of the basin:

$$\text{Drainage density} = \frac{\text{Total length of all streams}}{\text{Area of drainage basin}}$$

Drainage density *The measure of stream channel length per unit area of drainage basin.*

Overland Flow and Drainage Basins • 413

Figure 16.7 Drainage patterns. Stream networks are configured in five primary drainage patterns: (a) dendritic, (b) rectangular, (c) trellis, (d) radial, and (e) deranged.

Figure 16.8 shows an example of differences in drainage density. This diagram shows two watersheds (A and B) of the same size that clearly contain a variable number of streams. Which area, basin A or B, has the higher drainage density? The answer is basin B. Why might such a difference occur? One reason might be that basin B occurs in a more humid climate, so more streams exist in the area to handle the extra runoff of this region. A second possibility is that basin A occurs on a landscape lying over porous sands that allow water to seep rapidly into the ground, whereas watershed B occurs in a region of relatively impermeable clay that causes water to flow overland. The measurement of drainage density provides a hydrologist or geomorphologist with a useful numerical measure of runoff potential and how much streams have shaped the landscape.

A third way to characterize a drainage network is through the process of *stream ordering*. Streams within

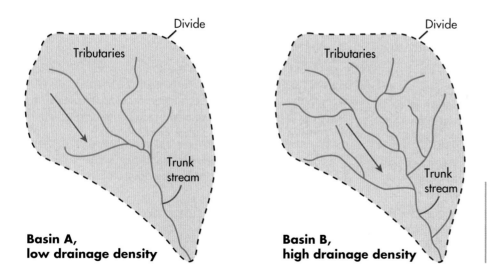

Figure 16.8 Drainage density differences. Basin A has a lower drainage density than basin B. This may result from differences in climate or type of underlying rock or sediment. The arrows indicate the general direction of flow in each watershed.

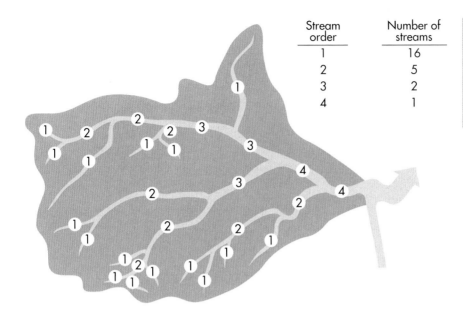

Stream order	Number of streams
1	16
2	5
3	2
4	1

Figure 16.9 Stream ordering. Stream ordering classifies the hierarchy of channels in a drainage basin. Stream order changes only where two streams of the same order merge. A positive relationship exists between stream size and stream order, with streams of progressively higher order portrayed by progressively thicker lines on the diagram.

any drainage network can be arranged into a hierarchy based on their size. Stream ordering is useful because it provides a good relative measure of a stream's place in a basin hierarchy, which is typically a function of how many tributaries occur in any given watershed. Ordering the streams of any basin is a simple task, as shown in Figure 16.9. An important rule to remember when calculating stream order is that it changes whenever two streams of the *same order* join at their confluence. The smallest tributaries of a basin are first-order streams. At the confluence of the first-order tributaries, a second-order stream forms. Subsequently, two second-order streams join to form a third-order stream.

It is tempting to think that the ordering of a stream changes *whenever* it meets another stream, regardless of the second stream's order. For example, Figure 16.9 shows several places where first-order streams flow into streams of a higher order, such as a second- or third-order stream. You can also see a confluence where a second-order stream meets

a third-order stream. Remember that stream order changes *only* at the confluence of streams of the same order. For example, just because a first-order stream meets a second-order stream that does not mean the second-order stream becomes a third-order stream. It does not; rather, it remains a second-order stream. The second-order stream becomes a third-order stream *only* where it joins another second-order stream.

A last point to make about stream ordering is that a positive relationship exists between increased stream order and stream size. In other words, you can expect a second-order stream in a drainage basin to be larger than a first-order stream. Similarly, a third-order stream is bigger than a second-order stream. This increase occurs because a stream of a higher order contains the combined flow of the lower ones. Thus, higher-order streams logically contain more water than lower-order streams. In Figure 16.9, this geographical relationship is portrayed by using progressively thicker lines to represent streams of progressively higher order.

www.wiley.com/college/arbogast

WileyPLUS

The Rhine River

The next section of the chapter focuses on hydraulic geometry and channel flow. You can preview this material by going to the Geo Media Library and accessing *The Rhine River*. This video follows the course of the famous Rhine River in Europe from its origin in the Swiss Alps to where it meets the Atlantic Ocean in Holland. As you watch the video, notice the various landscapes through which the Rhine flows. Listen for the technical terms,

such as *graded stream, rapids,* and *valley,* and try to understand their meaning, as they will be discussed in this text. As you continue through the chapter, keep this video in mind as an excellent example of a stream. Once you complete the exercise, be sure to answer the questions at the end to test your understanding of this concept.

1. Streams consist of channels in which water flows downhill. Runoff refers to water that runs off land surfaces into streams.

2. Streams are fed by runoff or groundwater, or may have a large lake as their source.

3. A drainage basin is the area that contributes runoff or groundwater to any given stream. Drainage basins are separated from one another by divides, which are topographic rims consisting of relatively high ridges and hills.

4. Drainage density refers to the relative density of streams in comparable areas and is influenced by climate, underlying bedrock, or both.

5. Streams are classified by their order, which ranges from low (such as 1 or 2) to high (such as 5 or 6). Stream order changes only at the confluence of two streams of the same order.

Hydraulic Geometry and Channel Flow

Now that the organization of stream networks has been discussed, let's look more closely at the character of flowing water within channels. This assessment falls within the framework of *stream hydrology* and focuses on the geometrical attributes (or *hydraulic geometry*) of river channels, including the measurable width, depth, velocity, slope, and discharge (Figure 16.10). These concepts are respectively defined as:

- w = channel width (how wide the channel is in which the stream is flowing).
- d = depth (how deep the channel is from the water surface to the bed).
- v = velocity (how fast the water is moving in the channel; velocity is greatest in the middle of a channel where water flows freely and is slower on the channel bottom and sides due to frictional forces).
- s = slope (how steep the slope [also called *gradient*] is on which the stream is flowing).
- Q = discharge (how much water is flowing in the channel).

As you might imagine, these variables are intimately related to one another in a holistic way. For example, the velocity of a stream closely correlates to the channel slope, with streams on steep gradients moving more quickly than those on shallower slopes. Increased discharge naturally leads to greater channel width and depth to accommodate higher flow.

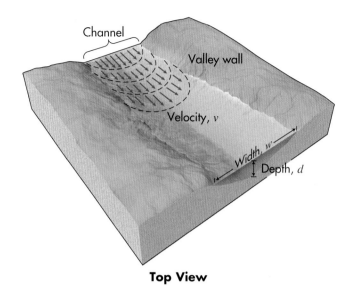

Top View

Side View

Figure 16.10 Hydraulic variables. Streams have a measurable width, depth, velocity, slope (or gradient) down which they flow and discharge. Velocity is lowest on the sides and bottom of the channel due to the effects of friction.

Fluctuations in Stream Discharge

As far as human interactions are concerned, the most important hydraulic variable is stream discharge Q, which refers to the amount of water flowing in the channel. This value is calculated by the simple equation

$$Q = w \times d \times v$$

Discharge is measured with a device called a *stream gauge* that can either be portable or fixed in one place in a gauging station. The United States Geological Survey has a network of stream-gauging stations to monitor discharge in all the major streams in the country (see Figure 1.4 in Chapter 1). Stream discharge is typically expressed in cubic meters per second (m^3/sec) or cubic feet per second (ft^3/sec). With increases in width, depth, and velocity, stream discharge becomes greater.

In most streams, discharge varies over the course of the year, depending on cycles of rainfall and relative drought. The amount of discharge that serves as the baseline level for any stream at any given place and time is **base flow**, which

Base flow *The amount of stream discharge at any given place and time that is solely the product of groundwater seepage.*

Figure 16.11 Rappahannock River hydrograph. This hydrograph illustrates how discharge varied during December 2003 (blue line) in the Rappahannock River near Fredericksburg, Virginia. Base flow (red line) indicates the rate of flow maintained by groundwater influx. Note the peak that occurs in association with increased runoff caused by a storm event. (*Source:* U.S. Geological Survey.)

is the flow rate that is sustained solely by groundwater influx. During periods of heavy precipitation, the amount of runoff to the stream increases, causing stream discharge to increase above the base flow amount. Figure 16.11 is a typical **stream hydrograph** that shows how stream discharge varies over the course of time due to changes in groundwater influx and runoff. This particular hydrograph is from the Rappahannock River in Virginia and was collected over the course of a month from a measuring station near the city of Fredericksburg. Notice the relationship between base flow and the actual amount of discharge. Also observe the fluctuations in discharge that occurred over the month. In particular, note that stream discharge does not immediately increase during a storm event.

This *time lag* between the period of heaviest rainfall and maximum stream discharge is typical of most streams because it takes a period of time for the water to flow overland and reach the channel. This concept will be discussed more thoroughly later in the chapter.

Flooding Although streams are usually well confined within their channels, the amount of water in them can increase dramatically during periods of high rainfall and runoff. During these intervals, the first noteworthy rate of heightened stream flow is **bankfull discharge**, when the channel is literally full of water. If the amount of runoff further increases so that water begins to spill out of the channel onto the adjoining ground,

VISUAL CONCEPT CHECK 16.1

The famous Danube River, shown here, is the second longest river in Europe (the Volga is the longest), flowing from Germany to the Black Sea. Which one of the following statements is correct and can be determined accurately?

a) The Danube is a high-order stream.
b) The discharge of the Danube is low.
c) The watershed of the Danube is small.
d) The slope of the Danube is very steep.

Getty Images

Stream hydrograph *The graphical representation of stream discharge over a period of time.*

Bankfull discharge *The amount of discharge at which the stream channel is full.*

the river is said to be at **flood stage**. In humid regions, most high-order streams are bordered by low, relatively level terrain commonly referred to as the *floodplain*. Although floodplain formation will be discussed in greater detail later in this chapter, it is sufficient here to say that the stream creates this surface, which stands only a few meters above the channel water level at base flow. During the wet season, the combination of significant surface runoff and groundwater influx can cause the floodplain to become inundated by excess discharge. In most humid locations floods are annual to semiannual events and are part of the natural evolution of river systems. In addition, they are important in the maintenance of the flora and fauna that live along a river (which, as you recall from Chapter 10, is called the *riparian zone*).

Occasionally, periods of extremely heavy rains occur in a watershed, causing the streams to rise to spectacular levels that result in great damage to human property and even sometimes significant loss of life. Fortunately, these kinds of floods are rare, with a close statistical relationship between the *return period*—in other words, the frequency of the event—and its magnitude. To put it more simply, more extreme floods happen less frequently than do lower-magnitude floods. Hydrologists who monitor flood frequency may describe a particular large flood as being, for example, a 30-year or 50-year event; in other words, it is expected statistically to occur once every 30 or 50 years. They also refer to the probability of a particular discharge being exceeded by an even larger one.

To see how flood magnitude, return period, and statistical probability can be used to predict stream behavior, examine Figure 16.12, which shows data from the Skykomish River at Goldbar, Washington. The Skykomish has a base flow of about 200 m³/sec (about 7060 ft³/sec), which statistically occurs every year and has a 99% probability of being equaled or exceeded on an annual basis. According to the graph, the 10-year peak discharge in this stream is about 1700 m³/sec (~60,000 ft³/sec), and this flow has a 20% chance of being equaled or exceeded each year. A more unlikely event is the 50-year flood, which has a discharge of approximately 2500 m³/sec (about 88,300 ft³/sec). This discharge has only a 2% probability of being equaled or exceeded on a yearly basis. This is not to say that a 50-year flood or even a larger one could not happen in successive years—it could. It is just not very likely from a statistical perspective.

1993 Mississippi Basin Floods

Perhaps the best example of a major flood in the fairly recent past within the United States is the Mississippi River flood of 1993, which affected a large part of the upper Midwest. This flood was at least a 100-year event and has been classified by some hydrologists

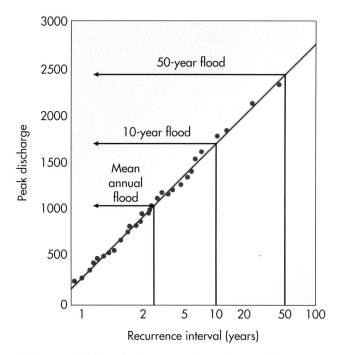

Figure 16.12 Flood frequency data for the Skykomish River at Goldbar, Washington. Each data point is the measured maximum discharge over a 53-year period.

as a 500-year flood. It caused about $30 billion in damage and *easily* remains the costliest flood in U.S. history. The 1993 flood dominated the national media during June and July of that year because of the catastrophic impact it had on people living along the Mississippi River and its tributaries. The causes of this flood, and associated stream response in the basin, nicely demonstrate how environmental variables interact in a holistic way.

The story of the 1993 flood actually begins in the preceding winter when unusually heavy rain and snow saturated the soils in the Midwest. Following this wet winter, an unusual weather pattern developed during June and July of that year, one that resulted in the formation of a strong subtropical high-pressure system that stalled over the southeastern United States (Figure 16.13). Recall from Chapter 8 that atmospheric pressure systems in the midlatitudes rarely stay in one place but instead typically migrate from west to east in association with the westerly winds. The year 1993 was unusual because a high over the southeastern United States remained stationary for about 2 months. This stalled high caused two significant things to occur. First, the midlatitude jet stream persistently flowed over the Midwest because it was blocked from moving east by the southeastern high. Second, water vapor from the Gulf was continuously pumped into the Midwest by the clockwise rotation of the southeastern high. Given that cool/dry cP air was steadily colliding with warm/moist mT air along the stalled jet stream in the upper Midwest, prodigious rains fell during June and July. During July 4 and 5, for example, between 10 cm and 15 cm

---

Flood stage *The level at which stream discharge begins to spill out of the channel into the surrounding area.*

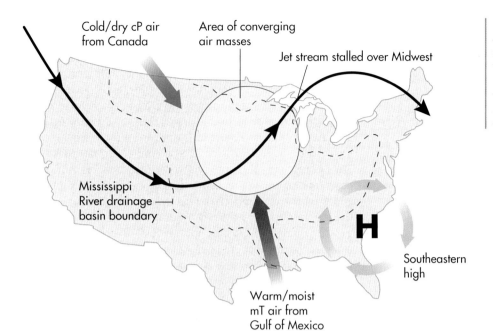

Cold/dry cP air from Canada

Area of converging air masses

Jet stream stalled over Midwest

Mississippi River drainage basin boundary

H

Southeastern high

Warm/moist mT air from Gulf of Mexico

Figure 16.13 Meteorology of 1993 flood. A high-pressure system stalled over the southeastern United States and persistently pumped humid mT air into the upper Midwest. This water vapor condensed and fell as rainfall along the stalled frontal boundary in the region. This intense rain caused widespread flooding.

(4 in. and 6 in.) of rain fell in a broad swath of southeastern Iowa. This kind of rain event was a common occurrence that summer as the region received over *200%* of normal precipitation for that time of year!

Given that the soils of the region were already saturated from the wet winter, most of the rainfall could not be absorbed. As a result, flooding was widespread, affecting most of the major streams and tributaries in Iowa, southeastern Wisconsin, western Illinois, northern Missouri, eastern Nebraska, and northeastern Kansas. Since all these streams progressively merge and ultimately flow into the Mississippi River, flooding along this stream was intense (Figure 16.14).

KEY CONCEPTS TO REMEMBER ABOUT STREAM HYDROLOGY

1. Stream flow in a channel is defined by its hydraulic variables, including width, depth, velocity, slope, and discharge.

2. Hydraulic variables are interrelated with one another. For example, stream discharge is calculated by the equation $Q = w \times d \times v$. Increased slope results in faster-flowing streams.

(a)

USGS

July 4, 1988

July 18, 1993

Mississippi River

Missouri River

St. Louis

(b)

NASA/Science Photo Library/Photo Researchers

Figure 16.14 Flooding along the Mississippi River in 1993. (a) Aerial photograph of flooding along the Mississippi River in 1993. (b) Landsat images of the confluence of the Missouri and Mississippi Rivers north of St. Louis during a normal year (left) and during 1993 (right). In these images, vegetation and urban areas appear in green and pink tones, respectively. The extensive flooding in 1993 appears as blue and black areas, indicating submerged floodplains.

3. Base flow is the average stream discharge in a given stream at any particular time of the year. Bankfull discharge occurs when the channel is full. Floods occur when discharge is sufficiently high that water spills out of the river channel.

4. Prediction of flood discharge is based on statistical probability. The return period refers to the frequency with which a particular discharge is expected to occur. High-magnitude floods occur less frequently than less intense discharges.

Fluvial Processes and Landforms

Now that the nature of stream channel flow has been discussed, let's next examine the landforms that streams produce. Landforms created by fluvial processes are called *fluvial landforms*. Although processes such as glaciation, wind, and ocean waves also act to shape the Earth's surface (and will be covered in subsequent chapters), running water is the single most important geomorphic agent on Earth. This process is driven by gravity, which is the most important geomorphic force on Earth.

Erosion and Deposition

Discussion of stream landforms requires an understanding of erosion and deposition as geomorphic processes because they act in concert to shape the landscape. Simply put, erosion occurs when sediment is removed from one place by a geomorphic process such as slumping, stream flow, or glaciation. Erosion is usually more intense in areas of high relief because the force of gravity enhances the ability of geomorphic processes to do work by increasing their power. Other factors that influence the process of erosion are vegetation and climate. Erosion tends to be less in areas of dense vegetation because the plants protect hillslopes from the effects of running water. Stream erosion is also limited in the driest deserts because very little running water occurs there. In fact, the most intense erosion tends to occur in semi-arid to subhumid zones where plant cover is relatively sparse but heavy rains that can move sediment sometimes occur.

In contrast to erosion, the process of deposition occurs when sediments that are being transported (after being eroded from someplace else) stop moving and are dropped. Deposition occurs when the transporting agent (such as wind or running water) simply loses the power to carry the sediment, which can happen for a variety of reasons. Although erosion and deposition often work in tandem to shape a landscape, one process or the other tends to dominate in any particular place at specific periods of time. Thus, it is possible to classify landforms generally as being either erosional or depositional in their nature. Erosional landforms are created when sediment, soil, or rock is stripped away by some geomorphic process. Depositional landforms, in contrast, form when sediment accumulates after being dropped.

Figure 16.15 shows a simplified example of these categories. Here, the mountain slopes have largely been shaped by erosion due to the high energy created by the steep relief. This

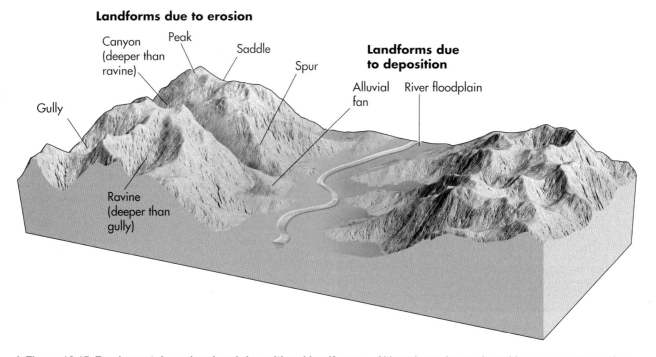

Figure 16.15 Fundamental erosional and depositional landforms. Although erosion or deposition can occur anywhere, one of these processes tends to dominate in any particular place. Erosion tends to be focused on areas of high relief, whereas deposition typically dominates in more level terrain.

Figure 16.16 The process of splash erosion. Splash erosion occurs because large raindrops create a small crater when they impact on bare soil.

USDA

process creates some distinctive landforms. The most prominent feature is a *peak*, which is the highest point on any given mountain. Peaks are typically separated by a lower landform called a *saddle*. As streams cut into the mountain slopes, they first create a shallow *gully*, which can enlarge to become a *ravine* and then, if sufficient time and erosion later occur, a deep and broad *canyon*. These features are separated from one another by a relatively high ridge called a *spur*, which is, in effect, a drainage divide. Over time, the eroded hillslope sediments are transported into the valley below, where they may be deposited on more level terrain within an *alluvial fan* or *river floodplain*. Here the relief lessens and geomorphic processes lose their power. These landforms will be described in more detail later in the chapter.

Fluvial Erosion on Hillslopes The logical place to begin a discussion of stream erosion is by focusing on hillslopes, which are the part of the landscape that is most intensely eroded by running water. Hillslopes are the most active zones of fluvial erosion because, as indicated before, the force of gravity is greatest in areas of high relief, which, in turn, causes running water to flow more quickly and thus with more energy. Sediment transport begins on hillslopes as soil erosion, initiated when the fluid drag associated with overland flow picks up sediment. In humid regions, dense vegetation protects slopes, and erosion occurs sufficiently slowly that the soil maintains all its horizons. However, the process of slope erosion accelerates in more arid regions, deforested areas, and agricultural fields because raindrops fall directly on bare soil. When this pounding occurs, sediment is loosened, lifted, and dropped into new positions through a process called *splash erosion* (Figure 16.16). It subsequently begins to flow down the slope as sheet runoff, mentioned earlier.

As sheet flow moves across the surface, the total water volume and degree of turbulence increase down the slope. This turbulent water loosens sediment and results in the

formation of tiny furrows in the ground called **rills** that look like parallel seams on the slope (Figure 16.17a). These rills, in turn, begin to merge as stream power and erosion intensifies, forming larger and deeper ditch-like features called *gullies* (Figure 16.17b).

Sediments eroded from hillslopes by overland flow are either deposited quickly or carried to a stream where they can be transported great distances. Deposition of eroded slope sediments often takes place at the base of the hill where the slope becomes gentler, the effect of gravity lessens, and less power is available to carry sediment. Recall from Chapter 14 that sediment deposited in this manner is called *colluvium* and can form an apron-like landform that may be difficult to distinguish in humid regions of dense vegetation.

(a)

Sheldan Collins/Corbis

(b)

Richard R. Hansen/Science Source

Figure 16.17 Formation of rills and gullies. (a) Rills are small channels that evolve when heavy rains fall on hillslopes. (b) This gully formed due to the enlargement and merging of rills as a result of rapid runoff.

Rills *Small drainage channels that are cut into hillslopes by running water.*

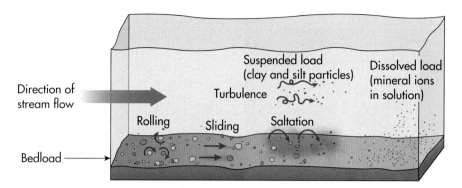

| Figure 16.18 **The sediment load of streams.** Streams carry their sediment load as dissolved, suspended, and bedload.

KEY CONCEPTS TO REMEMBER ABOUT FLUVIAL EROSION AND DEPOSITION

1. Erosional landforms are created where flowing water carries sediment away, as on a steep hillslope.

2. Depositional landforms are constructed through accumulation of sediment when the power to carry sediment lessens, as on a level floodplain.

3. Hillslope erosion is most active on unvegetated slopes and begins with a process called *splash erosion*, caused by raindrops.

4. Flowing water first begins to erode hillslopes in rills. As rills coalesce and enlarge, they form trench-like features called gullies, which eventually broaden into canyons.

Stream Gradation

Let's now examine how landscapes evolve along larger, high-order streams. Although these streams do not vigorously erode like rills and gullies, they nevertheless shape the landscape in subtle ways that have profound effects over a long period of time. In most higher-order streams the combined processes of runoff, erosion, and deposition have been occurring for thousands of years.

Think of a stream in your neighborhood or city. Although its fundamental appearance (e.g., its width, depth, or valley width) probably has not changed much in the past few thousand years, it was likely a very different system tens of thousands (or even millions) of years ago in the early stages of its development. Perhaps this ancestral stream had a much steeper gradient and narrower valley than it currently does. Through time, the system slowly adjusted internally to the environmental setting in which it occurs—specifically, the amount of precipitation that falls, the size of the drainage basin, the density of vegetation on the landscape, and the rock and sediment types found there. An-

other way to look at this change is that streams ideally evolve in a way that allows them ultimately to balance the amount of sediment they carry with the capacity of the stream. This sediment load is derived from hillslopes and tributary channels in the drainage basin and can be carried in the following ways (Figure 16.18):

1. **Dissolved load**—Mineral ions that are carried in solution and are invisible during transport.

2. **Suspended load**—Sediment that floats along in the stream. This load usually consists of clays and silts that are held up by turbulent flow within the water.

3. **Bedload**—Larger particles such as sand and gravel that roll, slide, or bounce along the channel bed in a process called *saltation*. This form of transport dominates in mountain streams where slopes are steep and stream velocity is high.

It is usually possible to determine the dominant kind of sediment a stream is carrying by noting its fundamental channel characteristics. If a stream is transporting mostly suspended sediment, it will typically have a single, deep channel that gradually curves from side to side across the landscape (Figure 16.19a). Such a stream is called a **meandering stream** and often contains muddy-looking water due to the high concentration of suspended sediment that was eroded from deposits containing abundant silt and clay. In the Mississippi River, for example, approximately 90% of the sediment carried within the river is suspended load. No wonder this river is frequently called *Big Muddy*! If a stream's dominant load is bedload, then it will have a **braided stream** pattern (Figure 16.19b), which consists of a maze of interconnected, wide, shallow channels that look like the braids in someone's hair. These streams have this form because the channel banks consist of easily eroded deposits of sand, gravel, and even cobbles that allow the stream to develop a wide, shallow form. In many cases, these coarse-textured deposits are derived from melting glaciers somewhere upstream. This relationship will be examined more thoroughly in Chapter 17.

Meandering stream *A river or small stream that curves back and forth across its valley.*

Braided stream *A network of converging and diverging stream channels within an individual stream system that are separated from each other by deposits of sand and gravel.*

Braided streams are interesting fluvial systems that develop when the sediment they carry is predominantly bedload of sand and gravel. To better visualize the characteristics of these streams, go to the Geo Media Library and select *Bedload Transport and Braided Streams*. This animation contains three videos related to braided streams. The first video shows the nature of bedload transport in flowing water. Subsequently, a second video demonstrates the evolution of a braided system in an experimental setting. The final clip shows the Platte River in the central United States. This stream is one of the best-known braided systems east of the Rocky Mountains. Once you complete the exercise, answer the questions at the end to test your understanding of this concept.

(a)

(b)

Figure 16.19 Stream load and channel characteristics. (a) A meandering stream usually consists of a single channel that winds back and forth across the landscape.(b) Braided streams such as this one in Alaska consist of a maze of wide, shallow channels because the dominant sediment load is sand, gravel, and even cobbles.

The amount and kind of sediment a stream can carry at a given point are determined by a variety of complex factors within the system that are related to the concept of *stream gradation*. This concept is based on the fact that streams adjust internally (by flowing faster, becoming wider, etc.) in a way that allows them to maintain an equilibrium state with the surrounding environmental variables. In this context, it is essential to view streams as holistic systems that can respond in many different ways. Imagine, for example, if a forest fire occurred that removed the vegetation on hillslopes in an area—how would the stream respond? Rill and gully erosion would increase and more sediment would be delivered to streams because vegetation would no longer be present to protect the hillslopes from heavy rain. In an effort to adjust to this higher sediment supply, much of the sediment would initially be deposited on the channel bottom (because the stream could not carry it) in such a way as to increase the overall stream gradient. This kind of deposition, which raises the elevation of the channel bed, is called **aggradation**. The increased gradient caused by aggradation would result in higher flow velocity and power that would, in turn, enable the stream to carry the increased sediment load. If hillslope vegetation was reestablished and erosion slowed through time, the stream would require less energy to carry the reduced sediment load. As a result, the stream would gradually erode its channel bed in a process called **degradation** (or *downcutting*) until a new gradient was established that enabled the stream to be able to carry just the lessened supply of sediment. A stream in this equilibrium condition is referred to as a **graded stream** because a balance exists among erosion, sediment load, deposition, and the capacity of flowing water.

Aggradation *The progressive accumulation of sediment along or within a stream.*

Degradation *The topographic lowering of a stream channel by stream erosion.*

Graded stream *An idealized stream that has achieved a balance between sediment transport and stream capacity.*

(a)

(b)

Figure 16.20 Evolution of a graded stream profile.
(a) Following landscape uplift, the stream profile is very uneven. With time, however, it gradually smooths into an ideal longitudinal profile. From this point forward, the gradient of the channel progressively lowers. (b) A major control of stream behavior is base level, which is the point at which a stream flows into the ocean, lake, or another stream. This example shows the location where Antrim Creek flows into Lake Michigan. In other words, Lake Michigan is base level for Antrim Creek.

Now examine more closely how a stream might evolve to reach a graded condition. Figure 16.20a shows several stages in the development of a hypothetical landscape. The diagram traces the evolution of a stream's **longitudinal profile** (or simply

Longitudinal profile *A graph that illustrates the change in stream gradient in cross section along a stream from its source to its mouth.*

"profile"), which shows the change in the stream's gradient along its length. The beginning point for this developmental sequence could be a period of rapid uplift that elevated the upper part of the basin. Conversely, this sequence could be triggered by a major climate shift or a major environmental change such as the uncovering of the landscape following deglaciation. For simplicity, imagine that the sequence begins with some kind of uplift. Notice in Figure 16.20 that several profiles can be seen

Now that the graded stream concept has been discussed, you can review it by viewing an animation. Go to the Geo Media Library and access *The Graded Stream*. This animation nicely illustrates the idealized process of stream evolution and gradient adjustment. As you watch it, you will first see a longitudinal profile similar to the uppermost one in Figure 16.20a. Notice how this profile smoothes as time progresses through the combined process of knickpoint reduction and filling of lakes and ponds with sediment. Ultimately, the stream will obtain a slope that enables it just to carry the sediment that is provided through hillslope erosion. Once you complete the animation, answer the questions at the end to test your understanding of this concept.

in the side of the illustration. Each one of these profile lines represents the gradient of the channel bed at a distinct period of time in the evolution of the system. For reference, let's say that the entire sequence of events takes about 1 million years to complete.

The top profile line represents the stream gradient following the initial uplift of the landscape above sea level. Soon after the landscape was raised, a stream would begin to develop to handle the runoff associated with heavy rains and groundwater flow. Given the nature of streams, runoff on this landscape would naturally flow downhill toward the sea and would follow the path of least resistance. The gradient would naturally be very steep in some places and relatively shallow in others. The steepest gradients would most likely be associated with resistant rock layers because stream erosion would be slow in these locations compared to the areas of less resistant rock. As a result, the stream gradient is higher where resistant rocks occur and lower in areas of softer rock. A place where the stream gradient is significantly steeper than other places within the stream is called a *knickpoint*. Knickpoints are visible within streams as rapids and, where gradients are especially steep, as waterfalls. Where the gradient is low, overland flow would collect in lakes. Assuming that climate and vegetation cover remained essentially the same, the stream would spend perhaps its first 300,000 to 400,000 years adjusting to the variations in gradient that occur across the landscape. Note, however, that the overall stream gradient becomes smoother between the top two profiles.

One reason longitudinal profiles gradually become smooth is that knickpoints slowly retreat upstream as they wear down by stream erosion. A good example of how this process occurs is at a great waterfall like Niagara Falls at the Ontario–New York border (Figure 16.21). Niagara Falls is on the Niagara River, which flows from Lake Erie into Lake Ontario over the Niagara Dolomite, which is a resistant layer of limestone that overlies a layer of easily eroded shale. This limestone was scraped clean and covered by a continental glacier until about 12,000 years ago, at which time the ice receded into Canada and an ungraded landscape remained. Since the ice melted, the falls have retreated more than 11 km (6.8 mi) at a rate of about 1.3 m/year (4.3 ft/year). The primary reason for this retreat is that the force of the water striking the plunge pool erodes the shale at the base of the falls. As the shale erodes headward over time, the overlying limestone collapses under its own weight and the falls retreat. This cycle has been repeated countless times in the past 12,000 years and continues to the present time. Today about 5522 m³ (195,000 ft³) of water pours over the falls every second. Most of the water flows over the Canadian (or Horseshoe) Falls, which measure about 790 m (2592 ft) in length and are about 49 m (160 ft) high. Although the American Falls are the same height, they are only approximately 305 m (1000 ft) long and thus have less water flowing over them.

Returning to the concept of a smoothing stream profile, let's examine our hypothetical stream about 500,000 years into its history when it becomes graded. In other words, it can carry the average sediment load in the basin. At this time, the stream gradient has the ideal longitudinal profile (see Figure 16.20a), with a steeper slope in the upper reaches of the basin and a shallower slope downstream. From this point forward, this graded profile is maintained and gradually lowered as the landscape is further eroded.

Now let's turn our attention to the place where the stream enters the ocean (or a large lake). This place is called the stream's **base level** because it represents the lowest elevation at which a stream can erode its channel bed. This concept is easy to visualize because the stream cannot erode its bed below the level of the water body in which it flows (Figure 16.20b). If it did, it would be flowing uphill to meet the ocean or lake, which is physically impossible. The concept of base level also applies on a more local level, for example, where one stream flows into another at a confluence. As at the ocean, the elevation of a tributary's stream bed must be the same as the channel bed of the stream in which it flows.

Return to Figure 16.20a to see how base level relates to stream behavior. In this particular example, the elevation of base level did not change throughout the evolution of the

Base level *The lowest level at which a stream can no longer lower its bed, because it flows into the ocean, a lake, or another stream.*

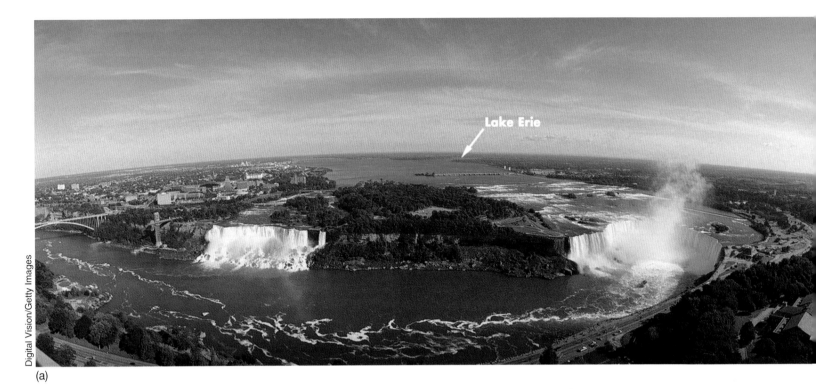

(a)

Lake Erie

(b)

Overlying hard limestone collapses

Retreating escarpments

Plunge pool

Soft shale is eroded by water

Figure 16.21 Niagara Falls. (a) Niagara Falls is a 51-m (167- ft) waterfall on the Niagara River. It actually consists of two waterfalls, the Horseshoe Falls on the Canadian side (to the right) and the American Falls (to the left). The eastern side of Lake Erie lies in the background. (b) The Niagara River flows over a resistant layer of limestone that overlies relatively soft shale. This shale is eroded headward in the plunge pool, which results in collapse of the overlying limestone and retreat of the falls. Niagara Falls has retreated more than 11 km (6.8 mi) in the past 12,000 years.

stream profile. In reality, however, sea level—in other words, the stream's base-level elevation—would probably change frequently over such a period of time and would thus be a major control in the behavior of the stream. If sea level fell, then a wave of erosion would proceed upstream as the stream eroded down to the new base-level elevation. On the other hand, if sea level rose, then the stream would begin to deposit sediment and aggrade its bed as it rose to the new base-level elevation. The stream would also begin to erode if uplift occurred in the source area of the stream, which would cause the channel slope to increase and result in more energy. If such a scenario occurred, the stream system would be said to be *rejuvenated*; that is, it would begin the developmental cycle all over again.

KEY CONCEPTS TO REMEMBER ABOUT STREAM GRADATION

1. Streams carry sediment as dissolved load, suspended load, or bedload.

2. A graded stream is a stream that has evolved to the point where the amount of sediment it carries is balanced with the capacity of the stream.

3. An example of an ungraded landscape is one that was recently uplifted. This uplift creates a very steep gradient. As a result, a stream is able to carry more

(a) Initial uplift

Lake (shallow gradient)

Knickpoint

Waterfall (steep gradient)

(b) Vigorous downcutting

Lake disappears (draining and infilling)

Steep upstream gradient and waterfall

V-shaped valley

(c) Reaching equilibrium

Tributaries form and extend headward

Deeply incised valley

(d) Horizontal expansion

Meanders

Floodplain

Figure 16.22 Evolution of a graded stream and associated valley. During the first stages of valley evolution, the stream downcuts vigorously because it has a steep gradient and excess energy. When the stream reaches a graded state, it begins to migrate horizontally, resulting in a wide valley.

sediment than is provided to it from eroded hillslopes. This stream will downcut its bed through a process called *degradation*.

4. Stream aggradation occurs when the sediment load is greater than the stream's capacity to carry it.

5. Base level is the point where a stream flows into the ocean and is a major control of a stream's behavior. When base level rises, streams deposit sediment to match that elevation; when base level falls, streams erode their channel beds to cut down to sea level.

Evolution of Stream Valleys and Floodplains

The next step in this discussion of stream systems is to investigate the landforms that evolve in association with the long-term process of stream gradation. As you work your way through this section of the chapter, refer to Figure 16.22. This diagram shows a simplified four-stage evolution of a stream and its associated valley.

For simplicity, let's begin by imagining that a landscape was uplifted or deglaciated at some point in the distant past. With this uplift in mind, compare Figure 16.22 with the top

stream profile in Figure 16.20a. Lakes and waterfalls are present in both illustrations. Recall that lakes occur in areas where the slope is shallow, whereas waterfalls are associated with knickpoints where the rocks are resistant and the gradient is steep. In this early stage of gradation, the stream has a capacity to carry more sediment than is being delivered to it. As a result, the stream begins to downcut vigorously in an attempt to reach a state of equilibrium. In so doing, the stream carves out a deep V-shaped valley (Figure 16.22b) that is consistent with the Grand Canyon of the Yellowstone River (Figure 16.23).

As time progresses (Figures 16.22b and c), the lake disappears through a combination of infilling by eroded upstream sediments and draining because the river cut back into it after eroding through its downstream knickpoint. During this interval of time, the gradient becomes smoother, as in the second curve in Figure 16.20a. Turning back to Figures 16.22b and c, the stream begins to extend its valley farther upstream (or *headward*) because the gradient is steepest in this part of the system. Because of this steep gradient, the stream naturally flows at a greater speed and thus has more energy to erode sediments. In Figure 16.22c, you can see that tributaries begin to develop and also extend their valleys upstream by headward erosion. At the same time as this

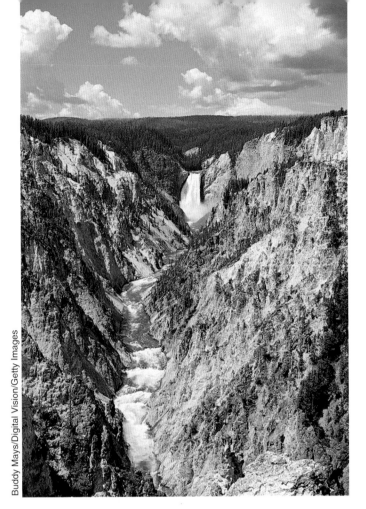

Figure 16.23 The Grand Canyon of the Yellowstone River. Here, the Yellowstone has yet to reach a graded condition, as indicated by the deep V-shaped canyon and waterfall. Compare this photograph with Figure 16.22b.

headward erosion is occurring, the canyon continues to be more deeply incised.

Look at Figure 16.22d. The stream valley looks different from its appearance in the previous three images. Now, the valley is no longer V-shaped but is quite wide. This change is evidence that the stream has reached a graded condition and has the graded longitudinal profile seen in Figure 16.20a. Recall that during the early stages of stream valley evolution the stream erodes vertically because it has proportionately

more energy than sediment load. When a stream becomes graded, however, it no longer needs to downcut throughout the entire system because it has reached a gradient on which it can transport the average load that is provided from the hillslopes in the upper reaches of the system. Although downcutting may continue in the upper reaches of a watershed where the gradient is steep, the stream begins to expend its energy in a horizontal fashion in the lower reaches where the gradient is less.

Stream Meandering As noted previously, streams migrate horizontally through a process called meandering, in which numerous pronounced arcs and curving bends develop in the valley (Figure 16.22d). Stream meanders are fascinating places because both erosion and deposition occur simultaneously. As a result, stream meanders move back and forth across the valley over time. Figure 16.24 shows the processes associated with a typical stream meander. As the river flows around the meander bend, the centrifugal force associated with its current carries the main line of flow of the channel, or *thalweg*, to the outside of the curve. Thus, the channel is deepest and the flow velocity is highest in this part of the stream. Given the high stream power that exists on the outside of the meander, this part of the arc is a zone of aggressive erosion. As a result, the surface leading away from the river is undercut, which causes the overlying sediment to collapse into the stream, forming a steep slope called a *cutbank*. Erosion also scours the channel bottom at these locations, forming deep *pools*. At the same time that erosion shapes the cutbank, deposition occurs on the inside of the meander bend because the channel is shallow in this part and energy is low. As a result, alluvium accumulates as a long, curving deposit of sediment called a *point bar* that is only about 1 m (about 3.3 ft) above the level of the stream.

If you are looking for clues as to whether a stream is actively meandering, a good piece of evidence is the presence of oxbow lakes in a stream valley. An **oxbow lake** is a water-filled

---

Oxbow lake   *A portion of an abandoned stream channel that is cut off from the rest of the stream by the meandering process and filled with stagnant water.*

---

① Maximum current on left side of channel

② Maximum current at center of channel

③ Maximum current on right side of channel

④ Maximum current at center of channel

⑤ Maximum current on left side of channel

Deep channel (thalweg)

Point bar deposits

(a)

Cutbank

Point bar

Alan Arbogast

(b)

Figure 16.24 Features of stream meanders. (a) Deposition of sediment, forming a point bar, occurs on the inside of a meander because stream velocity is low. In contrast, the stream erodes the cutbank on the outside of the meander because velocity is high. (b) Note the distinct point bar and cutbank in this stream meander along the Kansas River.

www.wiley.com/college/arbogast

WileyPLUS

Fluvial Geomorphology and Stream Processes

The chapter thus far has focused on fluvial processes and the nature of stream channels. This point marks a good time to review these concepts with two interactivities. Go to the Geo Media Library and first select *Fluvial Geomorphology and Stream Processes*. This animation reviews the concepts of channel geometry, stream flow, and the way water erodes and transports sediment. As you work your way through this, consider how streams are viewed as holistic systems that shape Earth's surface. A second exercise is *Identification of Stream Landforms,* which is a drag-and-drop exercise that assesses your ability to identify features within streams. Once you complete the exercises, answer the questions at the end to test your understanding of these concepts.

MISSISSIPPI RIVER MEANDERS

You can learn much about a river by examining satellite images of it. This image provides an excellent example. It is a synthetic aperture radar image of the Mississippi River that was acquired on the space shuttle in October 1994. The image is centered at about 32.75° N, 90.5° W and covers an area of about 23 km by 40 km (14.2 mi by 24.8 mi). North is toward the upper right of the image and water appears black. The accompanying diagram shows the location of the oxbow lakes and some of the meander scars. See if you can find some more.

As you can see, the Mississippi River is actively meandering in this area, with numerous oxbow lakes and meander scars visible. By looking at such an image, you can discover where the fastest and slowest velocities are within the river. Note that two such places are identified on the diagram. In addition, one can determine the areas that are prone to flooding. For example, the elevation between the current channel and the modern floodplain boundary is probably low, which means that the area is likely to flood during high discharge. The numerous signs of meandering also mean that the Mississippi River must be close to base level in this area and that the stream is in, or close to, a graded condition.

(a)

(b)

(a) The meandering Mississippi along the borders of Mississippi, Arkansas, and Louisiana. (b) Note the numerous oxbow lakes and meander scars highlighted in the diagram.

(a)

Figure 16.25 Meander evolution and formation of an oxbow lake. (a) Meander in the process of cutting off on the East River in Colorado. (b) Schematic views of the evolution of this meander. Diagram (1) illustrates the possible appearance of the meander before a breach of the neck occurred. Diagram (2) shows the landscape in its present condition. Diagram (3) illustrates how the oxbow lake may look when the meander ends seal with sediment.

(1) Meander forms

(2) Cutoff develops

(3) Oxbow lake forms

(b)

former meander bend that was detached from the main channel in a stream valley. This lake develops when the stream erodes the cutbank as water migrates downstream. At the same time that cutbank erosion is occurring, deposition of sediment takes place at the corresponding point bar. Given the combined impact of these processes, the neck between meander loops gradually becomes narrower. Ultimately, one or more meander loops can be entirely cut through when the stream floods. This neck cutting is part of the stream's continuous process of gradient adjustment and results in a temporary lake (Figure 16.25). With time, an oxbow will usually fill with sediment derived from major floods along the active channel. In the first stages of sediment infilling, the oxbow will become a swamp that contains plants such as reeds and cattails. As sediment continues slowly to accumulate with successive floods, the swamp will ultimately fill completely with alluvium. Nevertheless, evidence that an oxbow lake was once present will remain in the form of a *meander scar*.

Valley Widening and Floodplain Formation In the context of stream meandering, remember that the processes of cutbank erosion and point-bar deposition occur simultaneously. These combined processes cause stream meanders to migrate across the valley over time. This lateral migration, in turn, is important because meanders slowly cut into the side slopes, or *bluffs*, that flank the channel, causing stream valleys gradually to widen, as you can see in Figure 16.22d. Note in this diagram that the stream valley is now broad, essentially flat, and bordered by steep, rocky bluffs.

A stream valley usually contains thick deposits of alluvium that were deposited by the combined processes of point-bar development and overbank deposition during large floods (Figure 16.26). During floods, the coarsest sediments (usually sand grains) are deposited along the channel margin because they cannot be carried far away from the stream due to their size. This near-channel deposition results in a ridge-like landform along the river called a **natural levee** that may be 1 m to 2 m (3.3 ft to 6.6 ft) high. Finer silt and clay-sized particles are carried farther away from the stream during floods and, when deposited, form low **backswamps** because they are poorly drained, for reasons outlined in Chapter 11.

Overall, however, the combined processes of meander development, point-bar formation, and overbank deposition produce the low relief of the stream valley. This part of the landscape is typically called the river *floodplain* because it can be submerged during large floods, even though such an event may be rare. The river floodplain differs slightly from the

---

Natural levee *A small ridge that develops along the channel of a stream through the deposition of relatively coarse sediment when flooding occurs.*

Backswamps *Marshy floodplain landforms that develop behind natural levees in which fine-grained sediments settle after a flood.*

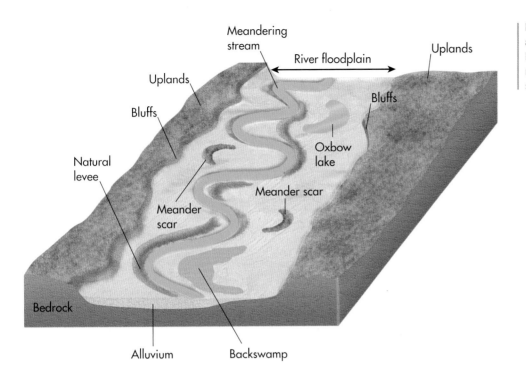

Meandering stream

River floodplain

Uplands

Uplands

Bluffs

Bluffs

Oxbow lake

Natural levee

Meander scar

Meander scar

Bedrock

Alluvium

Backswamp

Figure 16.26 Floodplain features of a graded stream. A graded stream has well developed meanders, natural levees, oxbow lakes, and backswamps.

active floodplain, which is the low, frequently flooded surface that results mostly from point-bar formation on the inside of meander bends.

Entrenched Meanders As previously discussed, a meandering stream develops when a stream reaches a graded condition. These conditions do not last forever, however, as streams periodically adjust to new environmental conditions such as tectonic uplift, a drop in base level, or climate change. The landscape response to these changes is usually preserved in stream valleys as distinct landforms.

KEY CONCEPTS TO REMEMBER ABOUT GRADED STREAMS AND THE EVOLUTION OF STREAM VALLEYS

1. When a landscape is recently uplifted, the resulting ungraded streams degrade by cutting V-shaped valleys that probably contain numerous knickpoints.

2. Waterfalls occur in places where rock is resistant to erosion and progressively erode headward until the gradient is smoothed.

VISUAL CONCEPT CHECK   16.2

This image shows a stream in an alpine setting. Given what you see here, which one of the following statements accurately reflects the stream in this landscape?

a)   The stream has a wide valley.
b)   The stream is actively meandering.
c)   The stream has high discharge.
d)   The stream is not graded.
e)   The stream is a high-order stream.

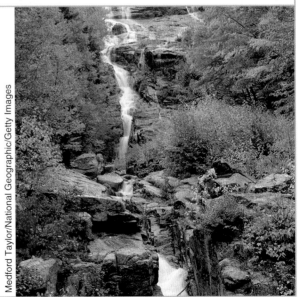

Medford Taylor/National Geographic/Getty Images

3. As streams approach a graded state, they begin to erode horizontally through the process of meandering. This process produces a wide valley because bordering bluffs are eroded.

4. In the process of meandering, erosion occurs on the outside of the meander at the cutbank, whereas deposition is focused at the point bar on the inside of the meander.

5. You can tell that a stream is actively meandering (in present time) if it has well developed oxbow lakes and meander scars.

For example, when a graded, meandering stream responds to a new period of landscape uplift associated with a major tectonic event or base-level drop, the gradient of the stream is dramatically increased. Such a stream once again has a capacity to carry more sediment than is delivered to it from the hillslopes. As a result, the rejuvenated stream will begin a new period of downcutting and channel-bed erosion.

A great place to see the effects of stream rejuvenation is the Colorado Plateau region in the southwestern United States. This part of the country has been affected by broad-scale regional uplift for the past 10 million years. This uplift raised the plateau about 1.5 km (0.9 mi), which caused the stream gradients with the Colorado River system to steepen dramatically. As a result, the Colorado River has downcut extensively, resulting in the famous Grand Canyon discussed in Chapter 12 (Figure 12.20). This wave of erosion also spread to tributaries of the Colorado such as the Green River and San Juan River, creating the widespread canyons that are found throughout the region.

The San Juan River is interesting because it contains well-developed *entrenched meanders*. Prior to uplift of the Colorado Plateau, the San Juan River was clearly a meandering stream and the channel bed was somewhere near what is now the lip of the gorge. In other words, the elevation of the stream channel was much higher at that point in time, and the gorge had not yet formed. As uplift of the plateau progressed and the stream began to downcut, it preserved its meandering form because it was no long eroding horizontally. This process produced the stunning landscape of the famous *Goosenecks of the San Juan River* pictured in Figure 16.27.

Alluvial Terraces Entrenched meanders illustrate the kind of landscape evolution that can occur when extensive steepening of a stream's slope happens due to large-scale regional uplift. Most of the time, however, the environmental adjustments within a drainage basin are more subtle, consisting of relatively minor fluctuations in base level or in the regional climate.

Figure 16.27 Goosenecks of the San Juan River in Utah. These entrenched meanders formed as a result of extensive uplift of the Colorado Plateau in the past 10 million years, which increased the stream gradient of the San Juan River. The resulting gorge is about 370 m (1400 ft) deep.

Although these adjustments do not result in dramatic canyons and gorges, such as the Goosenecks of the San Juan River, they nevertheless create distinctive landforms.

The most prominent landform created by smaller-scale fluvial adjustments is an alluvial terrace. An **alluvial terrace** is a broad, flat surface that occurs within the stream valley and is elevated with respect to the river so that it floods much less frequently than the topographically lower point bars. Although an alluvial terrace is usually considered to be part of the generic floodplain, it is also useful to think of the surface as being an abandoned floodplain because it is no longer frequently flooded.

Floodplain abandonment can occur in one of two ways. One way is that, through repeated flooding, thick deposits of alluvial sediment (or **alluvium**) accumulate on the active floodplain in a process called *alluviation*. As these sediments gradually thicken with repeated flooding, the active floodplain slowly becomes elevated with respect to the stream so that progressively higher discharges are required to flood the surface.

Another way in which floodplain abandonment and terrace formation occur is when a stream downcuts following a period of meandering (Figure 16.28a). This downcutting may be associated with either a subtle period of regional uplift, base-level drop, or regional climate change. When such an event occurs, the stream responds to the new ungraded conditions by downcutting through its own alluvium. This downcutting elevates the former

---

Alluvial terrace *A level, step-like landform that forms when a stream erodes its bed so that a horizontal surface is raised relative to the channel.*

Alluvium *Sediment deposited by a stream.*

(a) **Original floodplain**

(c) **New floodplain**

(b) **River downcutting due to regional uplift, base-level drop, or climate change**

(d)

Courtesy of Keith Richards, Dept. of Geography, University of Cambridge, UK

Figure 16.28 Formation of alluvial terraces. Alluvial terraces form through a combination of meandering, flooding, and downcutting in a river valley. (a) As the stream meanders, it creates an active floodplain that is underlain by alluvial sediment. (b) During a period of downcutting, the formerly active floodplain is elevated with respect to the river; thus, it rarely floods and a terrace forms. (c) A new active floodplain forms at a lower elevation when the stream begins to meander again. (d) Alluvial terraces along the Cave River on the South Island of New Zealand. Notice that two terraces (arrows) occur along this valley, which means that at least two periods of downcutting have occurred.

active floodplain, creating terraces on either side of the river that are at the same elevation (Figure 16.28b). Such terraces are called *paired terraces*. Some of this older alluvial fill (that underlies the terraces) may be removed by cutbank erosion if the stream begins to meander again (Figure 16.28c). If the stream is able to continue meandering *as* it downcuts, it may leave surfaces of unequal elevation called *unpaired terraces* on either side of the river. Many streams have numerous terraces that progressively step down to the river (Figure 16.28d). Whenever you see such a sequence of surfaces, it indicates that the stream has had a very complex history of flooding and downcutting.

A Holistic View of the Graded Stream Model At this point in the chapter, the concepts of stream order, hydraulic variables, stream behavior, and fluvial landform evolution have been discussed. With these concepts in mind, it is now useful to integrate them into a holistic framework that links them within the graded stream model. Remember that once a stream reaches a graded condition, it has the idealized longitudinal profile seen in Figure 16.20a. After this condition is reached, a very close and predictable relationship arises among stream order, hydrau-

lic variables, and the type of landforms you would see along the stream's profile (Figure 16.29). Stream gradient is highest in the upper reaches of the system, which translates to greater velocity in these areas and higher *stream power*. It also has high *stream competence* because it is capable of carrying large particles, perhaps even boulders, at high flows. As a result, the stream expends its energy vertically, and well-developed gullies and canyons appear in this part of the watershed due to extensive erosion. Given that the streams here are low-order systems, however, their width and depth are relatively small because they contain little water compared to higher-order streams. This means that discharge is also relatively low in streams in this part of the drainage basin.

As we move farther downstream into the middle reaches of the profile, notice that slope decreases, which causes stream velocity to drop correspondingly. As a result, the stream has less power and thus begins to meander rather than downcut. This shift results in valley widening in this part of the system. Width, depth, and discharge also increase because streams in this reach have a higher order due to the progressive funneling of low-order streams into the

Upper reaches
High slope gradient
High water velocity
Narrow V-shaped valley
Shallow depth
Low stream order (few tributaries)
Low discharge

Middle reaches
Slope decreases
Reduced water velocity
Wider valley (meanders)
Greater depth (downcutting)
Alluvial deposition (small terraces)
Middle stream order (more tributaries)
Larger discharge (occasional flooding)

Lower reaches
Low slope
Low water velocity
Wide valley (floodplain)
Greater depth
Thick alluvial deposition (broad terraces)
High stream order (many tributaries)
High discharge (major flooding possible)

Pat Behnke/Alamy Images

Martial Colom/Photodisc Green/Getty

Jim Wark/Lonely Planet Images/Getty Images

Figure 16.29 Relationship of stream order, valley form, and hydraulic variables along a graded stream. Streams in the upper reaches of the profile typically have the highest slopes and velocity, but the lowest widths, depths, and discharge. With distance along the profile, stream order and discharge increase, and gradient and velocity correspondingly decrease.

Jacques Descloitres, MODIS Rapid Response Team, NASA/GSFC

Figure 16.30 The Nile River Delta. This true-color image shows the triangular shape of the Nile River Delta where the Nile meets the Mediterranean Sea. The color green represents dense foliage and crops that are being supplied with water and nutrients from the fertile soils of the Nile River Valley and Delta. Note the surrounding desert.

system. With higher discharge the potential of flooding increases, which will result in deposition of alluvium on the active floodplain because the valley is wider than it is in the higher reaches. Deposition of alluvium means that stream terraces may also occur in this part of the profile, especially if some waves of downcutting progressed through the area in response to a drop in base level. The size of the sediment load is usually lower in this part of the system compared to the higher reaches due to a loss in competence.

Moving into the lowest reaches of the profile, the pattern seen in the middle part of the system becomes even more pronounced. Here, the trunk stream is fully developed, which naturally has a relatively high order because it contains virtually all the water that lower-order tributaries can provide. Stream gradient is very low here, resulting in low stream power and the formation of well developed meanders and a correspondingly wide valley. Discharge is high, however, resulting in progressively greater width and depth. Given that all the runoff in the drainage basin now flows through this central conduit, the potential exists for major flooding in this part of the system. Consequently, the alluvium is probably very thick, and well developed terraces are likely present. These landforms may be very prominent if significant base-level fluctuations have occurred through time.

Deltas The graded stream model considers the changes in stream variables and behavior over the course of an idealized longitudunal profile. Now let's examine what happens geomorphically at the location where some streams terminate. In most cases, streams are tributaries of larger streams and simply flow into them at a confluence such as shown in Figure 16.4. In other cases, however, a stream flows directly into a

lake or ocean. Such a place is often called the *river mouth* and is a zone of intense sediment dispersal and deposition. This shift in stream behavior takes places because stream velocity slows dramatically when it enters the relatively still waters of the larger water body. Think of what happens to your own velocity when you run into a lake at the beach. It slows and you then fall. When a stream slows in a similar setting, it drops its sediment.

The landform created by sediment deposition at a river mouth is called a **delta** because it usually has a triangular shape similar to the Greek letter delta (Δ). An example of this shape is the Nile River Delta (Figure 16.30). A delta consists of a level plain of sediment that is sorted geographically by size. In other words, the coarsest sediments, such as sands, are deposited at the point where the stream begins to lose energy. As stream energy further weakens away from the river mouth, the silt-sized fraction of the sediment load is then deposited. The clay fraction is carried the farthest in suspension and ultimately accumulates far from the river mouth.

Another excellent example of a river delta is the Mississippi River Delta in Louisiana, where the Mississippi River meets the Gulf of Mexico. This delta is a classic *birdfoot delta* because it appears to contain a number of individual toes that can easily be seen in the satellite image (Figure 16.31a). Each one of these "toes" is a river channel that is part of a

---

Delta *A low, level plain that develops where a stream flows into a relatively still body of water so that its velocity decreases and alluvial deposition occurs.*

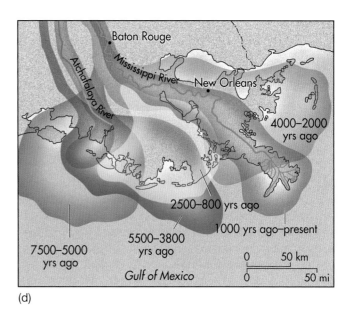

Figure 16.31 The Mississippi River Delta. (a) This satellite image beautifully shows the Mississippi River's birdfoot delta. The delta consists of a distributary network that funnels sediment (whitish hues) into the Gulf of Mexico. (b) The Mississippi River Delta is a complex system of distributary channels that is underlain by copious amounts of sediment derived from the continental interior. (c) The Mississippi Delta is a landscape of very low relief, consisting of numerous swamps and marshes that lie within a network of distributary stream channels. (d) Migration of the delta locations over the past 7500 years.

well-defined distributary network (Figures 16.31b and c). Unlike the tributary system in the upper part of the drainage basin, which funnels water into the trunk stream, the distributary network systematically drains water away from the main channel into the ocean.

Such a network develops because the delta gradient is very low and new channels form wherever a breach occurs in the natural levee along the stream. Wherever these breaks occur freshwater pours into the distributary bay, which subsequently fills with alluvial sediment. When the bay ultimately fills, it seals off and a levee breach occurs at another place in the delta. This cycle occurs simultaneously at several places, which is why the delta has the distinct distributary network. The entire delta is underlain by enormous quantities of sediment (Figure 16.31b) that originated in places ranging from the Appalachian Mountains to the Rocky Mountains. As you can see in Figure 16.31d, the location of delta formation has moved around in the past 7500 years as depositional zones have filled with sediment.

Deltas will be further discussed in Chapter 19 when we investigate how coastal processes shape them.

KEY CONCEPTS TO REMEMBER ABOUT ENTRENCHED MEANDERS, TERRACES, AND DELTAS

1. Entrenched meanders form when the slope of a graded, meandering stream is increased through extensive regional uplift, base-level drop, or climate change.

2. Alluvial terraces are abandoned floodplains that have been elevated through the process of alluviation or stream downcutting so that they are no longer frequently flooded.

This satellite image shows the deltas of the Krishna (bottom) and Godarvi (top) rivers in India. The whitish areas in the Indian Ocean are plumes of sediment from these rivers. Why are they present?

Image Science and Analysis Laboratory, NASA-Johnson Space Center.

3. A delta is a low-lying plain that forms at the mouth of rivers where they meet the ocean. They develop because stream power is lost when the stream flows into the relatively still ocean water. This change in velocity causes massive sedimentation. The sediment contained within a delta is well sorted, with coarse deposits accumulating at the river mouth and progressively finer sediments transported to the delta front.

Human Interactions with Streams

As indicated previously, rivers and streams have long been important for human societies. They function as transportation networks, sources of water for direct human consumption and irrigation, and power that drives factories and even cities. Although this section does not describe every way that people interact with rivers and streams, it provides a good feel for the way we impact streams and how they can impact us. The need to understand these relationships was tragically demonstrated in August 2010 when severe monsoon flooding in Pakistan killed over 1500 people and affected another 10 million due to the low-lying nature of the country.

Urbanization

People have a huge impact on fluvial systems through the process of urbanization. As cities expand, former vegetated soil surfaces are converted into parking lots and other impervious structures. Just consider for a moment the size of the parking lots associated with the malls and shopping centers where you live. These features have the effect of sealing the ground so that water runs directly off the surface rather than soaking into the ground. In many cases, entire drainage basins now lie within urban areas where the ground cover is no longer vegetation and soil, but mostly asphalt and concrete.

The impact of urbanization on stream behavior is dramatic, especially following a heavy rainfall. Figure 16.32 shows a hydrograph that illustrates the theoretical response of a stream to a heavy storm prior to urbanization (green line) and after urbanization (blue line). Before the city was constructed, the lag time between the storm peak (when it was raining hardest) and peak stream discharge was slow. This long lag time occurred because the soil–water belt first had to be saturated before the stream could be supplied with runoff or water from a recharged aquifer.

After urbanization, however, the response of stream discharge is markedly different in two specific ways. Can you see them in the stream hydrograph? First, the stream response is much faster; in other words, the lag time between the storm peak and peak discharge is less. This decreased lag time occurs because rainwater is no longer absorbed first into the soil, but runs off the urban parking lots and other concrete surfaces directly into the stream. Second, the peak discharge in an urbanized setting is *greater* after a given storm event than it was following a storm of similar magnitude during pre-urban times. This increased discharge occurs in the urbanized stream because more water runs off into the stream from the various asphalt and concrete areas rather than being absorbed by the soil. Given these relationships, urban streams are often described as being *flashy* because they react more quickly than their rural counterparts.

These conditions make *flash floods* much more likely in urbanized watersheds than in rural drainage basins if the same

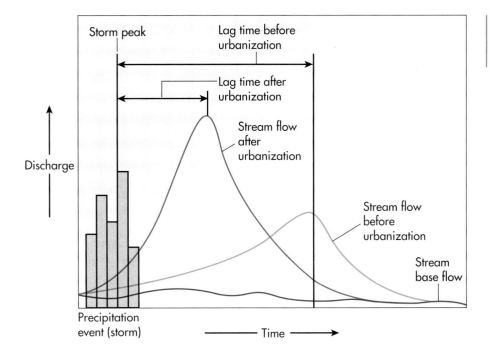

Figure 16.32 Stream hydrograph before and after urbanization. The response of the urbanized stream is much faster and greater than during pre-urban conditions.

type of storm event occurs in both places. In contrast to river floods, which are slow-evolving events such as the 1993 Mississippi basin floods that lasted for weeks, flash floods are brief but intense and can result in both significant property damage and loss of life. Although the scene in Figure 16.33 suggests otherwise, it is not a good idea to drive across a flooded roadway, even if the water appears to be shallow. After all, many automobiles become buoyant and can be carried away in only about 0.6 m (2 ft) of water. Of all the human deaths attributed to flooding, approximately half occur when people are in their vehicles.

Figure 16.33 An example of flash flooding. A flash flood is a brief but intense flood that occurs more frequently in urban areas where stream response is rapid. About half of flood-related fatalities occur when people try to drive across flooded roadways. This particular flash flood occurred in Manila, Philippines, as a result of Typhoon Nesat.

Artificial Levees

Given the increasing number of people living on floodplains around the world, big river floods and flash floods can be economically disastrous. As noted previously, for example, the 1993 floods in the Mississippi basin caused $30 billion in damage. In the context of this kind of cost, humans have engineered a variety of control structures to mitigate the effects of flooding. One kind of flood-control structure is an artificial levee.

Recall that a natural levee is a small ridge that forms along the stream channel where coarse sediments accumulate during a flood. This ridge increases the height of the stream bank so that the next flood must have a greater discharge, and thus be deeper, in order to reach flood stage. An **artificial levee** is based on the same principle as a natural levee, with the difference being that an artificial levee is designed by engineers and is usually built to a height such that a given city or valuable farmland is protected from a major event such as a 50-year or 100-year flood (Figure 16.34). In the United States, most of the large artificial levees along major rivers such as the Mississippi, Ohio, and Missouri are built and maintained by the United States Army Corps of Engineers. For example, in the Memphis district of the Corps of Engineers, levees are built to a height of about 7.6 m (25 ft) above the natural elevation of the Mississippi River bank.

Although levees certainly provide a societal benefit with respect to flood protection, they also have negative side effects. One problem with levees is that they keep the river from replenishing the sensitive wetland ecosystem that borders most rivers. This ecosystem is intimately related to the frequent but

---

Artificial levee *An engineered structure along a river that effectively raises the height of the river bank and thus confines flood discharge.*

Figure 16.34 An example of engineered levees. This artificial levee along the Mississippi River at New Orleans, Louisiana, keeps the Mississippi River (right) from flooding the city. The U.S. Army Corps of Engineers built and maintains this structure, as well as many others like it, along the Mississippi.

low-level floods that naturally occur along the stream and bring a fresh supply of sediment and nutrients into the wetlands. This relationship is lost, however, along a stream that is heavily protected by levees. Another problem with levees is that they may actually increase the intensity of flooding when major floods do occur. Under natural flood conditions, a stream will slowly spill out of its banks and spread across its floodplain. Although this slow spread can certainly be damaging, the gradual nature of the flood decreases the stream power and has a cushioning effect on locations downstream. In a levied section of river, however, the water cannot slowly spread horizontally because it is confined between the levees. As a result, the depth of the water increases beyond what it naturally would, and the associated stream power grows. Much of this energy is transferred to the levee system, which may then fail during especially large floods like the 1993 and 2001 Mississippi River floods. If such a failure occurs, water pours catastrophically through the breach in the levee, resulting in tremendous damage to farmland and human structures where the break occurs.

Dams and Reservoirs

Why Build Dams? Another way that humans interact with rivers is through the construction of dams, which are among the largest human structures in the world. A **dam** is an engineered obstruction that is built across a river to control its flow. When you were young, you may have built dams by piling dirt across small streams to control them. In large dams, this control is obtained through hydraulic gates in the dam that can be adjusted to allow a specific discharge of water to proceed downstream. Dams typically have behind them a large lake or reservoir that forms by impounding stream water that would otherwise flow

Dam *A barrier that blocks or restricts the downstream movement of a stream.*

downstream. A well-known example of a dam is Hoover Dam, which straddles the Colorado River on the border of Arizona and Nevada (Figure 16.35). Formed behind this dam is Lake Mead, which is one of the largest artificially created bodies of water in the world, with an area of 603 km^2 (233 mi^2). Lake Mead will be examined again in Chapter 20.

Historically, dams have been built for three primary reasons: (1) to enhance river navigation, (2) to provide hydroelectric energy, and (3) to provide flood control. The reservoirs behind dams are also extensively used for recreation and as a source of irrigation in many places. Dams improve navigation because the amount of water in the river downstream of a dam can be controlled. For example, Figure 16.36 is a hydrograph from the Chattahoochee River at Norcross, Georgia, which lies downstream of Buford Dam. Note how the discharge of the stream was held much more constant following dam construction than would have occurred naturally before. As a result of this controlling effect, rivers such as the Ohio, Mississippi, and Missouri are important transportation corridors because water depth is generally maintained at a consistent level that is safe for barge traffic.

Dams can also be used to produce hydroelectricity. This form of renewable energy is produced when water from the upstream side of the dam (from the reservoir) flows down and through tunnels in the dam to its outlet into the channel. As the water flows through the dam, it spins turbines within generators that produce electricity. The hydroelectric generators at Hoover Dam, for example, are capable of supplying nearly 1.5 million kW of power and provide electricity to Arizona, Nevada, and Southern California.

The third major function of dams is to provide flood protection for downstream areas because they can be used to store excess runoff from upstream tributaries during periods of heavy rainfall. In theory the gates of a dam can be shut completely during a particularly wet period, which would cause all the

Figure 16.35 Hoover Dam on the Colorado River. This dam was constructed in the early 1930s to provide flood control and hydroelectricity for the region. Lake Mead is the artificial lake upstream of Hoover Dam.

stream discharge upstream of the dam to be trapped in the reservoir. The lake would become larger and progressively fill the trunk and tributary valleys upstream. Below the dam, however, the stream channel would be dry until it began to receive tributary discharge someplace downstream. This kind of extreme scenario rarely occurs, however. Instead, the engineers typically maintain the normal stream discharge downstream of the

dam during a flood event or reduce it slightly to relieve pressure downstream. This approach still causes the size of the impoundment upstream of the dam to grow because more water enters the reservoir than is allowed to leave.

A nice example of flood protection through discharge control can be seen in hydrographs from the 1993 flood at Tescott and Niles, Kansas (Figure 16.37). These small towns lie

Figure 16.37 River hydrographs from Tescott and Niles, Kansas, during the 1993 flood in the Mississippi watershed. Note the difference between simulated peak discharge (blue line) and actual peak discharge (red line) resulting from controlled outflow from the dams upstream of this pair of towns. (*Source:* U.S. Geological Survey.)

downstream of dams on the Saline and Solomon rivers, respectively, within the Kansas River basin. In both graphs the blue line represents the simulated discharge at each town if the dams had not been present, whereas the red line shows the actual discharge. Although actual peak discharge certainly increased during the flood interval, it was about 80% less than what it is simulated to have been. As a result of these two dams, the small towns of Niles and Tescott were protected from the flood.

Controversial Aspects of Dams Although dams have a clear benefit to society because they can be used to control flooding, produce electricity, and enhance shipping, they can also be controversial because of a variety of societal and environmental costs associated with them The most obvious societal cost is that dams are extremely expensive. One example of extraordinary cost to build a dam is the Itaipu Dam on the border of Paraguay and Brazil. This structure is currently the largest hydroelectric dam in the world, producing up to 12,600 mW of electricity, and provides 78% of the electricity used in Paraguay and 26% of that consumed in Brazil. Construction of this dam began in 1975 and was finished in 1983. It cost between US$20 billion and US$25 billion to complete and was such a large financial drain that it contributed to the downturn of the Brazilian economy in the 1990s. Similar costs occurred during construction of the Three Gorges Dam on the Yangtze River in China. This controversial dam cost US$30 billion.

Another negative effect associated with dams is specifically related to the reservoirs they impound. In the course of their creation, reservoirs flood large areas upstream of the dam site. If this flooding occurs in heavily populated areas, many people need to be relocated. Upstream of the Three Gorges Dam, for example, approximately 632 km^2 (244 mi^2) of land will ultimately be inundated. As a result, up to 1 million people have been relocated to avoid being flooded. Such forced relocation has often been controversial, particularly in the United States where individuals have a strong sense of property rights. In a related vein, ancient cultural artifacts are sometimes lost in this way. Upstream of the Aswan Dam in Egypt, which was constructed across the Nile River, impoundment of Lake Nasser resulted in the loss of numerous classical Egyptian sites and artifacts.

In addition to the societal costs of reservoirs, they also have negative environmental impacts. These impacts are most closely related to the loss of ecosystems and spectacular scenery. Prior to construction of the Itaipu Dam, for example, the Parana River flowed through a spectacular gorge that contained a variety of beautiful waterfalls. Many of these waterfalls, along with much of the habitat unique to the plants and animals of the region, are now submerged beneath the reservoir. Similar scenery and habitat loss occurred in the canyon land region of the southwestern United States when dams such as Glen Canyon Dam and Hoover Dam (Figure 16.35) were constructed.

In the context of ecosystem impacts, dams also have negative environmental costs because of the way they regulate the discharge of streams, when, in fact, stream flow is naturally highly variable over the course of the year. The riparian ecosystem is adapted to this variable discharge and can be pressured because stream flow is largely held constant to meet the various societal needs. A variety of animal species, for example, are currently threatened by dams in the United States. In the Pacific Northwest, dams impact salmon because they often form a barrier that stops fish from migrating upstream to spawning sites. Along the Missouri River, the pallid sturgeon is threatened because it requires high spring flows to trigger its reproductive cycle; these flows are moderated significantly by the numerous dams along the river. At this point in time, environmentalists and government agencies are working toward reducing the environmental impact of dams.

The regulation of stream discharge by dams can also be controversial during dry periods when tension arises between the needs of people living near reservoirs and those farther downstream. A good example of such tension occurred in association with the major drought that baked the central United States from 2011 to the early part of 2013. This drought was especially intense in the northwestern part of the Mississippi River drainage basin, particularly in the upper part of the Missouri River watershed. As a result, stream discharge in this portion of the Mississippi drainage basin was dramatically down.

Lower flow in the upper Missouri watershed would naturally have the effect of reducing discharge in the Mississippi River below St. Louis (where the Missouri River and Mississippi River join) because less water would reach the trunk stream. Such a natural reduction in downstream discharge was magnified during the drought, however, because the Army Corps of Engineers reduced the flow out of the large reservoirs in the upper Missouri watershed (Figure 16.38a). This decision was made in an effort to stabilize water levels in those lakes, which were already 20% below capacity because of the drought. At the Gavins Point Dam (Figure 16.38b), for example, flow out of the dam was reduced from 1060 m^3/sec (37,500 ft^3/sec) to 750 m^3/sec (26,500 ft^3/sec) in the latter part of fall 2012.

This lower controlled discharge from the upper Missouri River basin occurred at the same time that the Mississippi River was already low due to the drought in much of the overall watershed. As a result, there was even less water in the Mississippi River below St. Louis (Figure 16.39a) than would be the case if dams were not present in the upper Missouri River basin. Low flows in the Mississippi River are problematic because the stream is a major transportation conduit for agricultural products, with more than 90% of U.S. corn and soybean exports moving to the Gulf of Mexico on barges in this fashion (Figure 16.39b). Mississippi River barges also transport a large amount of coal, oil, and other chemicals. This kind of transport is highly efficient, with the amount of material carried in 15 river barges being equal to 215 rail cars or 1050 large tractor trailers on a highway.

Given lower water levels on the river, barge companies had to reduce efficiency by lowering the load weight on their ships so that they float higher in the river. In an effort to deepen the river, the Corps aggressively dredged in key places. Nevertheless, fears abounded that the river would have to be closed for

MISSOURI RIVER BASIN

FORT PECK
GARRISON
OAHE
BIG BEND
FORT RANDALL
GAVINS POINT

Used with permission, Missouri Department of Natural Resources

(a)

© Corbis

(b)

Figure 16.38 (a) Dams on the upper Missouri River. Six major dams (and associated reservoirs) were constructed along the upper Missouri River to produce hydroelectricity, improve flood control, and provide a source of excess water for irrigation. (b) The Gavins Point Dam. This dam was built in the late 1950s and borders the states of Nebraska and South Dakota. It impounds Lewis and Clark Lake, which has a surface area of 130 km² (50 mi²).

shipping if the drought continued. Such closure would threaten a $7 billion industry and potentially raise prices for food commodities both in the United States and abroad. As a result, barge companies, the governors of Missouri and Illinois, and 15 U.S. senators urged the Corps of Engineers to release more water from reservoirs in the upper Missouri watershed—to no avail. This problem is an excellent example of how a river basin should be viewed holistically and how decisions made far way (in this case, the northern Great Plains) have far-reaching effects that are very much related to climate patterns.

(a) (b)

Figure 16.39 (a) Landsat 7 satellite image of the Mississippi just south of Memphis, Tennessee, in August, 2012. The large sandbars along the river were exposed due to much lower discharge associated with the drought in the central United States. (b) Barge traffic on the Mississippi River. Barges such as this one are used to ship corn and soybeans, as well as coal, oil, and other chemicals, up and down the river. This shipping method is threatened when river levels are low, such as during the recent drought.

KEY CONCEPTS TO REMEMBER ABOUT HUMAN IMPACTS ON STREAMS

1. Most streams in the world have been impacted by humans in some way.

2. Asphalt and concrete structures in cities increase storm runoff. As a result, the lag time of a stream decreases, peak discharge increases, and flash floods are more likely.

3. Artificial levees are ridges built along stream channels to increase the height of the stream bank so that protected areas are rarely flooded. These systems work, but increase the overall power of the stream at high discharge by confining it to a narrow channel rather than allowing gradual flooding.

4. Dams are used to control downstream discharge on rivers by impounding water (in reservoirs) and releasing it slowly. Dams are used for flood control and hydroelectric production, and to facilitate channel navigation.

5. Although dams fill a definite societal need, they have significant environmental impact because the natural rhythm of seasonally variable stream flow is lost. In addition, they are a barrier to migrating river species.

SUMMARY OF KEY CONCEPTS

1. Streams consist of channels in which water flows downhill under the force of gravity. They are fed by runoff or groundwater and may have a large lake as their source. Drainage basins are separated topographically from one another by bordering high ground called divides and interfluves.

2. Channel flow is defined by its hydraulic variables, including width, depth, velocity, slope, and discharge. These variables are interrelated. For example, stream discharge is calculated by the equation $Q = w \times d \times v$.

3. In the course of their evolution, streams gradually sculpt the landscape so that the slope of the stream is sufficiently steep to carry the average discharge in the river.

During this shaping process, the slope of the stream decreases. Once the stream reaches this stage of evolution, it is said to be "graded." The behavior of the stream is controlled in large part by base level and climate.

4. As streams approach a graded state, they begin to erode horizontally through meandering. This process produces a wide valley because bordering bluffs are eroded.

5. Streams produce a variety of erosional and depositional landforms, including gullies, cutbanks, point bars, deltas, and terraces.

6. Given the importance of streams to human societies, they are heavily managed.

1. Describe the concept of runoff.

2. Define the concept of *drainage basin* and describe one of the major watersheds in North America. What are the boundaries of this drainage basin, and what are some of the major tributaries?

3. Imagine that two drainage basins are found in a humid environment. One of the watersheds has high drainage density, whereas the other one does not. Assuming that each drainage basin lies within a region of similar relief, what environmental variable most likely accounts for the difference in drainage density? Why?

4. What is the minimum number of tributaries that a fourth-order stream has? How many stream confluences are there in this watershed?

5. Name four primary hydraulic variables. Provide an example of how one variable influences another.

6. Compare and contrast the concept of *base flow* and *base level*.

7. Which landscape is more likely to contain graded streams: one that has recently been uplifted or one that has been stable for a long period of time with a consistent climate? Why?

8. Name and describe three landforms associated with a stream that is striving to reach a graded condition.

9. Name and describe three landforms associated with a graded stream.

10. How can deposition and erosion both occur along a stream meander?

11. Imagine that you encountered a stream with well-developed meanders that are deeply entrenched. Describe two reasons why this landscape would evolve.

12. Why is a terrace an abandoned floodplain? Describe the two ways that terraces form.

13. Describe the formation of river deltas.

14. Why is peak discharge greater in streams contained within urban areas than those in the surrounding countryside?

15. Describe how a dam provides flood control for a city that lies downstream of the structure.

ANSWERS TO VISUAL CONCEPT CHECKS

VISUAL CONCEPT CHECK 16.1

The correct answer is *a*. The Danube is a high-order stream. You can tell that the Danube is a high-order stream because it is large, with high discharge. This appearance results from the numerous lower-order tributaries that feed into the river.

VISUAL CONCEPT CHECK 16.2

The correct answer is *d*. The stream is not graded. Remember that graded streams have shallow slopes that allow a stream to migrate horizontally. This stream has a very steep slope, which means that most of its energy will be used to erode vertically.

VISUAL CONCEPT CHECK 16.3

The whitish plumes of sediment in the Indian Ocean from the Godarvi and Krishna rivers are present because the ocean is relatively still compared to the flowing streams. As a result, stream velocity slows where they flow into the ocean, causing sediment to be deposited. The whitish plumes are fine-grained sediments such as silts and clays that are floating in suspension.

CHAPTER SEVENTEEN

GLACIAL GEOMORPHOLOGY: PROCESSES AND LANDFORMS

The preceding chapter focused on the processes and landforms associated with water that flows across the surface in rivers and streams. Now it is time to examine the character of flowing *ice* and the landforms that result from this movement. About 77% of the freshwater on land is stored as ice. Although it may not seem possible, this form of water moves as a glacier if it becomes sufficiently thick to flow under its own weight. Like streams, these rivers of ice can also shape the landscape in predictable ways through erosion and deposition.

At the present time, ice covers about 11% of the Earth's land surface. As recently as about 18,000 years ago, however, glaciers covered about 30% of the landmasses, including much of North America and Europe. As the climate warmed in the late Pleistocene and early Holocene, these giant ice sheets melted, and new landscapes created by glacial processes were uncovered. If you live in Canada or any of the northern states such as Wisconsin, Michigan, New York, Washington, or Illinois, your home may very well be built on deposits left by glaciers.

Getty Images

Flowing ice has had a huge impact on Earth throughout history. Although glaciers such as these in the Swiss Alps move slowly, they have tremendous power and are capable of eroding and depositing prodigious quantities of sediment. Much of the landscape seen here has been shaped in this way. This chapter focuses on glacial processes and the landforms that result from flowing ice.

CHAPTER PREVIEW

Development of a Glacier

 The Glacial Mass Budget

Types of Glaciers

 Glaciers in the Cascade Mountains

Glacial Landforms

 Depositional Glacial Landforms

History of Glaciation on Earth

Probable Human Impact on Glaciers

 Glaciers and Climate Change

Periglacial Processes and Landscapes

LEARNING OBJECTIVES

1. Describe how glaciers develop and the environmental conditions that are required.

2. Discuss the various kinds of glaciers that occur and the types of landscapes in which they are found.

3. Compare and contrast the various erosional and depositional glacial landforms.

4. Explain the history of glaciation on Earth, particularly the number of Pleistocene glaciations that occurred.

5. Describe the impact that ongoing climate change is having on glaciers around the world. What is the evidence for that change?

6. Discuss periglacial processes, the environments in which they occur, and the impact they have on the landscape.

Development of a Glacier

A **glacier** is a slowly moving mass of dense ice (Figure 17.1) formed by the gradual thickening, compaction, and recrystallization of snow and water over time. For these processes to take place, glaciers develop in regions where heavy snows fall in the winter and then fail to melt completely during the subsequent summer. For such a scenario to occur, summers are usually short, with temperatures sufficiently cool that some snow remains when the next winter arrives. If the amount of remaining snow gradually grows and thickens with time, it will slowly transform to a river of flowing ice that shapes the landscape.

The Metamorphosis of Snow to Glacial Ice

How is it possible that snow can be transformed into a flowing glacier? Recall the discussion in Chapter 11 about certain kinds of metamorphic rocks and how they form because of high pressure due to compaction from overlying sediments. The process of snow transformation to glacial ice is very similar. The first thing to remember is that the metamorphosis of snow to flowing ice takes a great deal of time, often hundreds or even thousands of years. Over this course of time, snow accumulates annually in a way that slowly increases the overall thickness of the snow mass. These annual accumulations can be seen as distinct layers that are

Figure 17.2 Annual snow layers in a glacier. A key factor in glacial formation is summer temperatures. If snow from the preceding winter fails to melt, it can then accumulate in layers like these in the Quelccaya ice cap in Peru. The average thickness of these layers is about .75 m (2.5 ft). Note the similarity between these layers and those found in nondeformed sedimentary rocks.

Annual Layers, Ice Cliff, Quelccayya Ice Cap, Peru: Photograph by: Lonnie Thompson/Byrd Polar Research Center, The Ohio State University

Figure 17.1 A typical glacier. Valley glaciers like this one in Alaska's Wrangell-St. Elias National Park are rivers of moving ice that slowly flow down the mountains from their snowy source areas.

© Mint Images Limited/Alamy

somewhat similar to sedimentary deposits (Figure 17.2), which is why some glaciologists think of glacial ice as being a form of sedimentary rock.

It is useful to think of snow as being the raw material for a developing glacier. When snow first falls, it is approximately 80% air. This snow is then progressively buried by additional layers of snow in subsequent years. As a result, it gradually compacts and recrystallizes due to the weight of snow above it. In addition, some of the snow might melt and refreeze. Water may percolate downward into the compacting snow from summertime rain that falls on the surface of the snow pack and from the limited snowmelt that takes place within it. As a result of these combined processes, the lowest deposits of snow are slowly transformed into a compact, granular substance called **firn** (Figure 17.3). Firn is somewhat similar to the crunchy substance that forms within the top of snow after it melts a bit during the daytime and

---

Glacier *A slow-moving mass of dense ice that flows under its own weight.*

Firn *The compact, granular substance that is the transition stage between snow and glacial ice.*

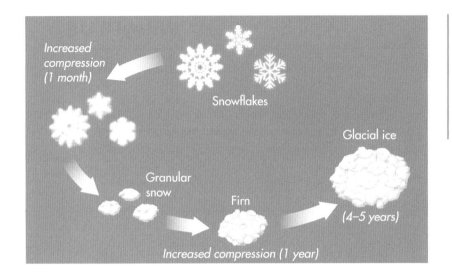

 When snow first falls, it is 80% air. As snow compacts, its density increases to the point that it becomes granular. When these grains of snow become more compact, they collectively form firn. With additional compression, the firn becomes glacial ice. In general, the most dense and compressed layers of ice are at the bottom of a glacier, whereas the top layers reflect relatively recent snowfalls.

then refreezes at night. With additional compaction, pockets of air in the firn further decrease in size or are squeezed out entirely, and the ice crystals become aligned with one another. At this time, the mass is 90% solid and begins to flow slowly under its own weight, much like maple syrup across a pancake. This point in the transformation of snow marks the time when the ice mass is a glacier. In most mountain glaciers, the full transformation to active glacier begins when the snow and ice mass reach a depth of about 40 m (130 ft).

The Glacial Mass Budget

Like streams, glaciers are natural systems that depend on inputs and outputs to function. In the case of streams, the input is liquid water in the form of precipitation, runoff, and groundwater. The outputs are stream discharge and water vapor (through evaporation). The input to the glacial system is snow that accumulates at the upper end of the glacier (Figure 17.4). This area is fittingly called the **zone of accumulation** and is where the addition of annual snows exceeds what is lost through melting, evaporation, and sublimation. Glacial outputs occur primarily down the valley toward the front of the glacier, where the temperatures are higher. This area is called the **zone of ablation** (or *wastage*) and is the place where more annual melting and evaporation of ice take place

(a)

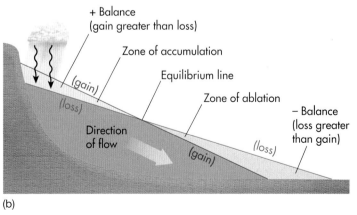

(b)

Figure 17.4 The glacial mass budget. (a) The glacial mass budget is the balance between snowfall in the zone of accumulation and melting and evaporation at the zone of ablation. (b) Side view of the glacial mass budget. Note the geographical relationship of the zones of accumulation and ablation and the way in which ice flows.

Zone of accumulation *The geographical area where snow accumulates and feeds the growth of a glacier.*

Zone of ablation *The part of a glacier where melting exceeds snow accumulation.*

Glacier development depends on the complex interaction of snow accumulation in a source area and melting in the zone of ablation. The advance or retreat of a glacier is governed by the balance between the input of snow or output due to melting. In order to better visualize this process, go to the **Geo Media**

Library and select *The Glacial Mass Budget*. This animation allows you to adjust temperature and the input of snow to influence the advance or retreat of a glacier. It also provides links to studies of glacier flow. After you complete the exercise, answer the questions at the end to test your understanding.

---

than snow accumulation. Somewhere between the ablation and accumulation zones is the **equilibrium line**, which marks the place where system inputs and outputs are in balance.

The overall budget of the glacial system is continually in flux, with variable amounts of input and output over the course of time. During periods when the accumulation of snow in the source area exceeds melting at the zone of ablation, the equilibrium line moves down the valley and the glacier grows and advances. Conversely, when melting in the zone of ablation exceeds the accumulation of snow in the source area, the equilibrium line moves up the valley and the glacier shrinks and retreats.

Glacial Movement

So how does a glacier move? Contrary to what you may think, a glacier is not just a block of ice that happens to slide down a hill. After all, the largest glaciers on Earth moved across fairly level terrain. This type of movement is possible only because these massive bodies of ice are able to flow under their own weight. Most of this movement occurs as a result of *internal deformation* within the ice. This internal movement is related to the fact that the interior of a glacier behaves like a plastic that can be shaped and manipulated to some degree.

Given the relatively high amount of friction that exists between the ice and the rock on which it flows, the ice moves most quickly in the core of the glacier (Figure 17.5). In this way ice-flow dynamics are quite similar to those observed in a stream system. However, in contrast to streams, which flow quickly, average glacial flow is on the order of 10 cm to 30 cm (about 4 in. to 12 in.) per day. This internal deformation continues even if the glacier is melting, shrinking, and retreating. In other words, the interior of the glacier continues to move forward *even though* the front of the glacier may actually be retreating up the mountain valley as a result of fluctuations within the overall mass budget (Figure 17.6).

Not all the ice is plastic, however. The upper part of the ice is quite brittle and breaks easily. When this part of the glacier fractures, prominent **crevasses** can develop that extend from the surface of the ice to some depth within it. Crevasses usually form

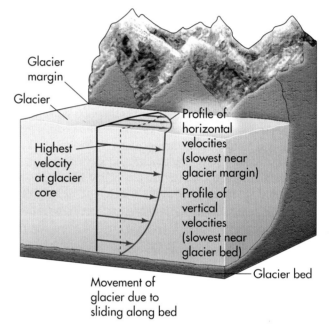

Figure 17.5 Internal flow in a glacier. Glacial ice is a plastic substance that flows under its own weight. Because of friction, the ice flows more slowly on its edge and base where it is in contact with bedrock. Velocity gradually increases toward the core of the glacier because it can move more freely.

when the ice flows over a small ridge in the bedrock below (Figure 17.7). This ridge causes compression to build in the ice on the up-valley side of the ridge. On the downstream side of the ridge, the ice stretches, causing tension to build and crevasses to form. Crevasses are most closely associated with the equilibrium line because that is the place where they become visible as you move down the glacier (see Figure 17.4). These giant cracks in the ice are visible in the zone of ablation because they are not covered by fresh snow. In the zone of accumulation, however, they are not as visible because fresh snow tends to hide them.

Equilibrium line *The place on a glacier where snow accumulation and melting are in balance.*

Crevasse *A deep crack in a glacier.*

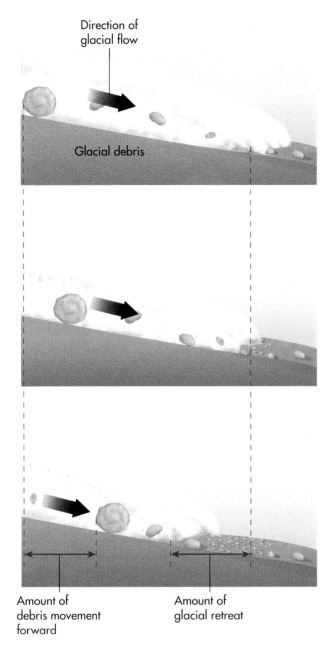

Direction of
glacial flow

Glacial debris

Amount of
debris movement
forward

Amount of
glacial retreat

Figure 17.6 Internal movement of a glacier.   The interior of a glacier continues to flow forward, carrying large rocks and other debris, even though the position of the glacier's front may retreat due to fluctuations in the glacial mass budget.

In addition to the flow associated with internal deformation, glacier movement also takes place through a process called *basal slip*. This process results when high amounts of meltwater accumulate at the base of the glacier, which provides a lubricant between the underlying rock and the ice mass. If conditions are right, this lubricant can suddenly cause a glacier to move rapidly through a process called *surging*. In Alaska and the Himalayas, for example, some glaciers have been observed to have surged up to 100 m (333 ft) in a single day!

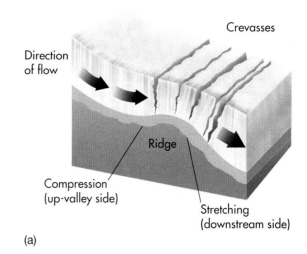

Direction
of flow

Crevasses

Ridge

Compression
(up-valley side)

Stretching
(downstream side)

(a)

(b)

Jerry Dodrill /Aurora /Getty Images

Figure 17.7 Formation of a crevasse.   (a) A crevasse is a deep fissure that extends from the surface to some depth within the ice. Crevasses form when ice is compressed on the upstream side of an underlying bedrock ridge. On the downstream side of the ridge, the ice stretches, tension is high, and crevasses develop. (b) A typical crevasse can be 27 m (~90 ft) deep or more, causing great danger to skiers or hikers attempting to cross a glacier.

Once you understand the glacial mass budget and the appearance of a glacier, it is possible to look at a glacier and interpret where features such as the equilibrium line and zones of ablation and accumulation are located. Look at this image and see if you can identify the zones of accumulation and ablation. Where is the equilibrium line? How can you tell?

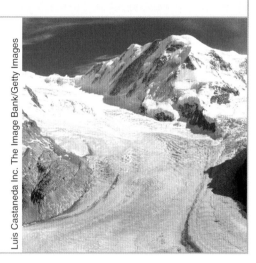

Luis Castaneda Inc. The Image Bank/Getty Images

KEY CONCEPTS TO REMEMBER ABOUT GLACIAL FORMATION, MASS BUDGET, AND FLOW

1. Glaciers form when unmelted annual snows reach a sufficient thickness to promote the metamorphosis of snow to firn, then to glacial ice.

2. Glaciers originate from source areas where snow accumulation is greatest. These source areas are typically located at high altitudes or latitudes where snow does not melt completely in summer because of cool temperatures that season.

3. The glacial mass budget refers to the balance between snow accumulation in the source area and the melting of ice within the zone of ablation.

4. Glaciers behave as a plastic that flows most quickly in the core of the system.

5. The interior of a glacier is always moving forward, even though the actual front of the glacier may be retreating because ablation exceeds accumulation in the source area.

Types of Glaciers

Now that the processes associated with glacier movement have been discussed, let's investigate the different kinds of glaciers by examining the environments in which they form.

Glaciers in Mountainous Regions

Perhaps the most obvious place that glaciers form is in high mountainous environments that have long winters and short, cool summers. Such places are excellent locations for glacier development because heavy orographic snows often accumulate there. Due to the short summers, much of this snow is able to remain into the next winter, forming the foundation of an evolving glacier.

Three fundamental kinds of ice masses occur in such mountainous areas (Figure 17.8). An **ice cap** is a dome-shaped sheet of ice that entirely covers less than 50,000 km² (~19,000 mi²) of land area. In contrast, an **ice field** is more constrained by topography and forms in large basins or on top of plateaus. Many ice fields are located in places where tall mountains, or *nunataks* ("lonely stones"), occasionally rise above the ice. Ice fields are similar in size to ice caps and can also have many glaciers that stream down valleys radiating away from the source area.

Ice caps and ice fields are frequently the source area for the third kind of ice mass found in mountains, **alpine glaciers**, which are glaciers that flow down valleys away from the high country.

Ice cap   *A dome-shaped sheet of ice that covers an area less than 50,000 km² (~19,000 mi²) in size.*

Ice field   *A topographically constrained sheet of ice in mountainous areas that frequently has glaciers streaming away from it.*

Alpine glacier   *A glacier in mountainous regions that flows down preexisting valleys.*

(a)

(b)

CHILE

45° S

Pacific
Ocean

ARGENTINA

50° S

**Patagonia
ice field**

Atlantic
Ocean

70° W

(c)

Figure 17.8 Ice caps and ice fields. (a) The Vatnajökull ice cap in southeastern Iceland. (b) The Patagonia ice field in the Andes Mountains of southern Chile. This long ice mass has numerous glaciers flowing out of it. (c) Nunataks rise above the ice in the Columbia ice field in Alaska. This ice field feeds the Columbia glacier, which flows downslope to the left.

Such glaciers occur in many mountain ranges around the world, including the Southern Alps of New Zealand, the Swiss Alps in Europe, the Canadian Rockies, the Himalayan Mountains, and the Andes Mountains in South America. Several alpine glaciers occur in the United States, with the majority in Washington and Alaska (Figure 17.9). For example, the Columbia ice field pictured in Figure 17.8c is the source for eight such glaciers, including the Athabasca and Columbia, to name two.

In areas where such glaciers flow in steep-sided valleys, they are called *valley glaciers* (Figure 17.10) because they

occupy landscapes originally carved by streams. If a valley glacier extends entirely out of a valley into the lowland beyond the mountain front, it is called a *piedmont glacier*. Some valley glaciers even terminate at the ocean. These glaciers are called *tidewater glaciers* and are particularly common along the coast of Alaska. Such landscapes are attractive tourist destinations because they mark a beautiful interaction of ice and ocean. One of the particularly interesting aspects of tidewater glaciers is that large blocks of ice break off the front of the glacier in a process called *calving*. The calved blocks then fall

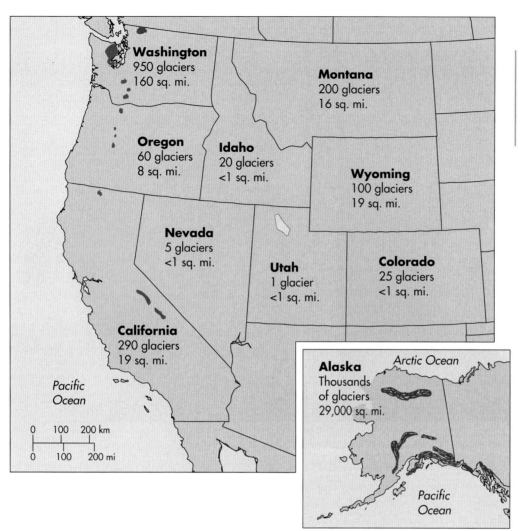

Figure 17.9 Alpine glaciers in the Pacific Northwest and Alaska. Most glaciers in the continental United States are relatively small, as indicated by the total area covered in each state. The largest and most impressive glaciers are in Alaska.

Washington
950 glaciers
160 sq. mi.

Montana
200 glaciers
16 sq. mi.

Oregon
60 glaciers
8 sq. mi.

Idaho
20 glaciers
<1 sq. mi.

Wyoming
100 glaciers
19 sq. mi.

Nevada
5 glaciers
<1 sq. mi.

Utah
1 glacier
<1 sq. mi.

Colorado
25 glaciers
<1 sq. mi.

California
290 glaciers
19 sq. mi.

Pacific Ocean

0 100 200 km
0 100 200 mi

Alaska
Thousands
of glaciers
29,000 sq. mi.

Arctic Ocean

Pacific Ocean

Nunataks

Ice cap or ice field

Meltwater lake

Piedmont glacier

Meltwater stream

Cirque glaciers

Valley glacier

Tidewater glacier

Ocean

Icebergs calved from glacier

(a)

(b)

Bernhard Edmaier/Photo Researchers

Figure 17.10 Valley glaciers. (a) Valley glaciers frequently originate at ice fields and ice caps and flow down terrain initially carved by streams. They become piedmont glaciers if they expand beyond the valley front. (b) Tidewater glaciers occur where glacier ice meets the sea. In these situations, such as seen here at the terminus of the Hubbard glacier in Alaska, ice calves off the front of the glacier and falls into the water to form icebergs. Note the cruise ship in front of the glacier for scale.

Although alpine glaciers are now rare in the contiguous United States, a few nice ones can still be found in the Cascade Mountains in the Pacific Northwest. The best developed of these glaciers occur on the flanks of the large volcanoes in the range, such as Mount Rainier and Glacier Peak. In order to explore these ice masses more thoroughly, go to the Geo Media Library and select *Glaciers in the Cascade Mountains*. After you complete the interactivity, answer the questions at the end to test your understanding of Cascade glaciers.

into the ocean where they become *icebergs*. Such an iceberg, of course, was involved in the tragic sinking of the *Titanic* in 1912.

In addition to ice caps and ice fields, another source area for an alpine glacier is a **cirque**, which is a small, bowl-like depression on a mountain flank formed by glacial erosion. A good example of an environment where numerous cirque glaciers have formed is the Cascade Mountains in the Pacific Northwest. Recall from Chapter 7 that this region receives abundant orographic precipitation from moisture-laden air flowing off the Pacific Ocean to the west. This precipitation is especially high on the numerous volcanoes that arc along the range. Given that the peaks of the volcanoes are usually greater than 3000 m (9800 ft) in elevation, summer temperatures near the mountain crests are typically very cool.

Figure 17.11a shows the effect that these cool temperatures have on snow cover. Note how much snow remains on the flanks of Mount Hood in Oregon even during late summer. Such snow cover is the foundation for several cirque glaciers on this mountain, as well as on other volcanoes in the range. On Mount Hood, a variety of glaciers flank the mountain, extending from near the crest at about 3400 m (about 11,200 ft) down to an elevation of about 1800 m (6000 ft). These glaciers ring the mountain on all sides, as shown in Figure 17.11b.

Continental Glaciers

In addition to mountain glaciers, another form of glacier is called a continental glacier. As the name implies, a **continental glacier** is a gargantuan ice mass that covers a large part of

(a)

Lyn Topinka/USGS

(b)

Figure 17.11 Cascade alpine glaciers. (a) Mount Hood is a composite volcano that reaches an elevation of over 3400 m (11,200 ft) and thus is an excellent place for the development of alpine glaciers. (b) This topographic map illustrates the various glaciers that ring the flanks of Mount Hood.

Cirque *A bowl-like depression that serves as a source area for some alpine glaciers.*

Continental glacier *An enormous body of flowing ice that covers a significant part of a large landmass.*

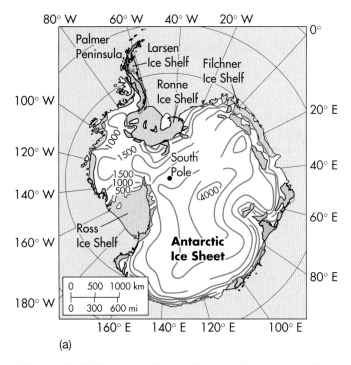

(a)

Figure 17.12 The Antarctic and Greenland ice sheets. These ice sheets collectively contain over 90% of the world's glacial ice. The Antarctic Ice Sheet (a) is over 4000 m (13,000 ft) thick in places, and the Greenland Ice Sheet (b) is over 3000 m (9800 ft) thick in the center.

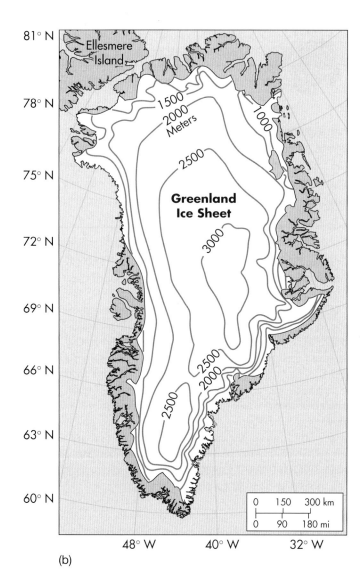

(b)

a continent or large island. Also called *ice sheets*, continental glaciers once covered much of North America and Europe, but are now confined to the island of Greenland and the continent of Antarctica (Figure 17.12) because summer temperatures are cold at these high latitudes.

Modern continental glaciers contain an enormous volume of ice that is more than 3000 m (10,000 ft) deep in places. The Antarctic Ice Sheet covers 90% of the continent and alone contains 91% of all the glacial ice on Earth (amounting to 13.9 million km^3 or 3.3 million mi^3). Similarly, the Greenland Ice Sheet covers 80% of that island and is about 1.8 million km^3 (0.43 million mi^3) in size. In both Greenland and Antarctica, the weight of these extensive ice sheets is so great that the underlying lithosphere has been pushed down into the upper part of the asthenosphere to some degree by a process called *isostatic depression* (see Figure 12.3).

KEY CONCEPTS TO REMEMBER ABOUT TYPES OF GLACIERS

1. Glaciers can be broadly categorized into two classes: alpine and continental glaciers. Alpine glaciers occur in mountainous regions, whereas continental glaciers cover large parts of continents.

2. An ice field is not continuous and has mountains exposed within it. In contrast, an ice cap is a continuous dome of ice that entirely covers the landscape.

3. Ice fields and ice caps are the source areas for valley glaciers. Smaller alpine glaciers form in cirques, which are bowl-like depressions on mountain flanks.

4. The only present-day continental glaciers are located in Greenland and Antarctica and are several thousand meters thick.

Glacial Landforms

Like stream systems, glaciers create landforms through the processes of erosion and deposition. As a glacier advances, it erodes the surface by picking up rocks and other debris that it will later deposit at some other place. In this

Landforms Made by Glacial Erosion

Erosion by glaciers occurs mainly through the processes of abrasion and plucking. **Glacial abrasion** takes place because the moving ice contains particles of sand or rock fragments that are frozen to the bottom of the ice. As the ice moves, these fragments scratch and gouge the underlying bedrock. Such abrasion can result in **glacial striations** or scratches that indicate the direction of glacial movement (Figure 17.13). When the scratches are particularly deep and well expressed, they are called **glacial grooves**.

Another form of glacial erosion is **glacial plucking**, which happens when a glacier pulls large rocks or boulders from the ground as it moves. These boulders are subsequently carried within the ice or are frozen to the bottom of the glacier, where they act as tools for abrasion. When the glacier melts and retreats, it drops the boulders in what appear to be very random locations. These isolated boulders are known as **glacial erratics** because they appear to be out of place since they are now far from their natural outcrops. A great example of glacial plucking and long-distance transport can be seen in the Sioux Quartzite in the midwestern United States. This rock unit is a Precambrian quartzite that crops out in southwestern Minnesota at several places, including Pipestone National Monument. This area was overridden and leveled during the ice ages when a massive continental glacier periodically reached from northeastern Canada to southern Illinois. (This history will be discussed in more detail later in the chapter.) One such advance of the ice plucked rocks from the Pipestone region and carried them to southeastern Nebraska and northeastern Kansas, where they were dropped as glacial erratics (Figure 17.14).

The combined processes of glacial abrasion and plucking scrape and round the hills. The most distinctive landform produced by these combined processes is the roche moutonnée. A **roche moutonnée** is a rounded bedrock hill that has a gradual slope on one side, called the stoss side, and a steep slope on the other, the lee side (Figure 17.15). The stoss side faces the direction from which the glacier came and was smoothed by abrasion as the ice moved over it. The lee side is steeper and more irregular with respect to its surface because the ice plucked rock from this side of the hill.

Although glacial erosion occurs in both alpine and continental environments, its effects are typically more visible in

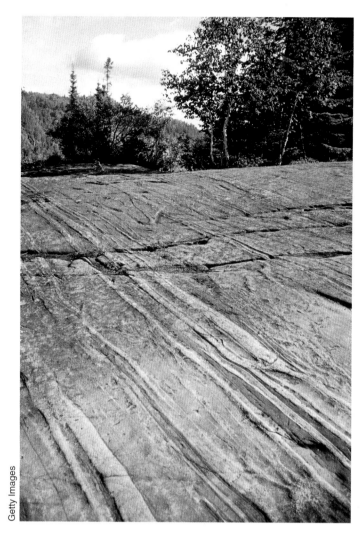

Figure 17.13 Glacial striations. Glacial striations such as these in Minnesota are linear scratches in bedrock produced when rocks at the bottom of a glacier grind across the bedrock.

Getty Images

section, glacial landforms are divided into two groups: (1) landforms created mainly by erosion and (2) landforms produced mainly by deposition. The actual processes associated with glacial erosion and deposition are discussed in some detail.

Glacial abrasion *An erosional process caused by the grinding action of a glacier on rock.*

Glacial striations *Scratches in rock produced by glacial abrasion.*

Glacial grooves *Deep furrows in rock produced by glacial abrasion.*

Glacial plucking *An erosional process by which rocks are pulled out of the ground by a glacier.*

Glacial erratics *Large boulders that have been plucked and transported a great distance before they are deposited.*

Roche moutonnée *A landform produced by glacial abrasion and plucking that has a shallow slope on one side and a steep slope on the other side.*

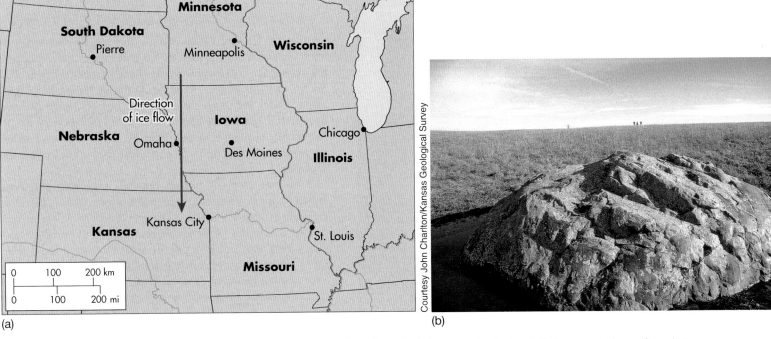

(a)

(b)

Figure 17.14 Plucking and erratics of Sioux Quartzite. (a) The Sioux Quartzite is Precambrian bedrock that crops out in southwestern Minnesota. Rocks from this outcrop were plucked by glacial ice about 600,000 years ago and carried as far south as northeastern Kansas. (b) This Sioux Quartzite erratic lies in northeastern Kansas. It is highly resistant to erosion, which is why it remains a prominent feature even though it has been resting there for about 600,000 years.

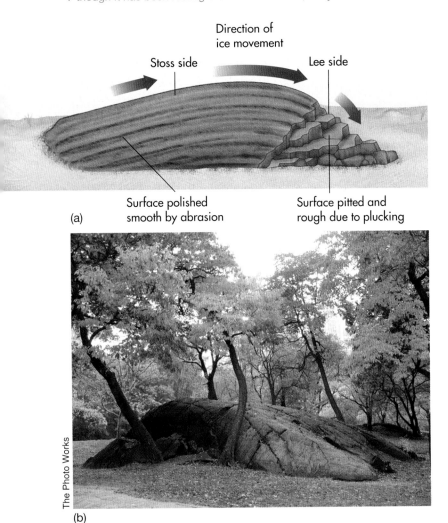

(a)

(b)

mountainous regions, resulting in very dramatic landscapes that are easy to interpret. In general, alpine glaciers modify a landscape by rounding parts of it at the same time that other aspects are sharpened. To see how a preglacial mountainous landscape is transformed by glaciation, refer to Figure 17.16 during the following discussion.

The most obvious landform created by glacial erosion in alpine regions is a cirque, which, as described earlier, is a broad amphitheater that forms on the flanks of a mountain. Cirques typically have very steep side and head walls and a floor that is flat to shallow sloping (Figure 17.17a). Remember that cirques are source areas for alpine glaciation and are enlarged through the combined processes of glacier plucking and mass wasting of the adjacent walls. Sometimes the ice in a cirque melts completely away, leaving water in the quarried depression. Such a small lake is called a **tarn** (Figure 17.17b).

Tarn *A small lake that forms within a glacial cirque.*

Figure 17.15 Roche moutonnée. (a) A roche moutonnée is created through the combined processes of abrasion on the stoss side and plucking on the lee side. (b) A roche moutonnée formed in the Manhattan Schist in New York City's Central Park. The stoss side is on the right-hand side of the image.

(a) Preglacial topography

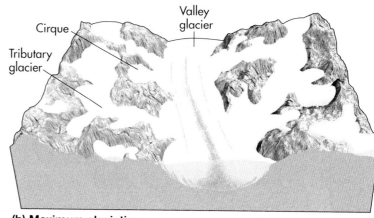

(b) Maximum glaciation

Cirque
Valley glacier
Tributary glacier

U-shaped valley
Hanging valley
Horn
Tarn
Arête

(c) Postglacial topography

Figure 17.16 Modification of mountainous landscapes by glaciation. (a) The preglacial topography of alpine regions consists of rounded mountain slopes and a V-shaped stream valley. (b) During maximum glaciation, well-defined cirque glaciers shape mountain slopes and a thick valley glacier fills the former stream valley. The valley glacier is fed by tributary glaciers flowing into it. (c) The postglacial topography is much sharper on the mountain slopes, and the former V-shaped stream valley is rounded to a U-shaped valley.

In many cases, cirques exist on several sides of an individual mountain. Over time, these cirques may progressively enlarge and steepen their respective side and headwalls so that the ridges between them narrow and become quite steep and thin, with serrated crests (Figure 17.18). Such a thin, steep ridge is called an **arête**, the French word for "fishbone." If a mountain has three or more arêtes resulting from intersecting cirques, then the mountain is called a **horn**. The most famous example of a horn is the Matterhorn in the Swiss Alps.

Glacial erosion not only happens on high alpine peaks, but also within the valleys that slope away from the accumulation zones of snow. Sometimes glaciers never leave their cirque source areas, presumably because snow accu-

Arête *A sharp ridge that forms between two glacial cirques.*

Horn *A mountain with three or more arêtes on its flanks.*

Figure 17.17 Glacial cirques and tarns. (a) Glacial cirques, such as this one in the Canadian Rockies, are bowl-shaped depressions created by the scouring action of glacial ice. (b) A tarn is a meltwater lake that fills the depression within a cirque. This one is located in the North Cascades of Washington State.

(a)

(b)

©Marli Bryant Miller

Alan Arbogast

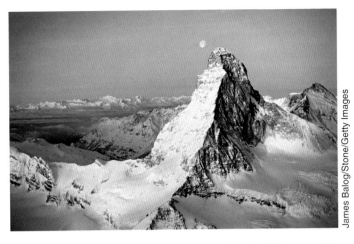

Figure 17.18 The Matterhorn in Switzerland. This mountain is 4478 m (14,692 ft) tall and has been dramatically sculpted by alpine glaciations. Most of this shaping occurred due to erosion on the sides of the mountain.

Figure 17.19 A glacial trough. Extensive alpine glacial erosion creates deep glacial troughs. This well-developed U-shaped trough in Montana was carved by a trunk valley glacier.

mulation is not sufficient to cause glacial advance. If snow accumulation is high, however, a glacier will advance out of the cirque basin and move down the adjoining valley. When this type of advance takes place, the glacier modifies the preexisting V-shaped valley into a U-shaped valley called a **glacial trough** (Figure 17.19). These landforms are *very distinctive* and easy to identify the next time you are in the mountains. The deepest and best-developed glacial troughs are formed by the trunk glacier in a particular mountain system (see Figure 17.16 again). Smaller, tributary glaciers also carve U-shaped troughs, but not to the same depth as the trunk valleys into which they flow because they contain less ice. As a result, a distinctive **hanging valley** is produced at the intersection of the tributary and trunk valleys when the glaciers fully recede. A hanging valley is often a place where a waterfall develops, such as the famous Yosemite Falls, as the tributary stream flows toward the trunk stream.

VISUAL CONCEPT CHECK 17.2

With an understanding of glacial processes, it is possible to visualize how a landscape has evolved through time. Given what we have discussed in this chapter so far, can you describe how glaciation has modified this landscape? What is the evidence for that modification?

Glacial trough   *A deep, U-shaped valley carved by an alpine glacier.*

Hanging valley   *An elevated U-shaped valley (with respect to a glacial trough) formed by a tributary alpine glacier.*

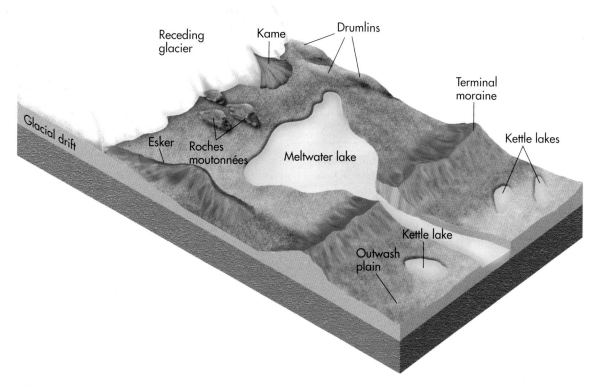

Figure 17.20 **Generalized depositional landforms associated with glaciation.** Note the essential forms and geographic relationships of these landscape features.

KEY CONCEPTS TO REMEMBER ABOUT GLACIAL EROSION AND RESULTING LANDFORMS

1. Glaciers erode through the processes of abrasion (grinding rock) and plucking (pulling rock from the ground).

2. Glacial striations and grooves are produced through abrasion.

3. A roche moutonnée is a streamlined bedrock landform produced by the combined processes of abrasion (on the upstream side) and plucking (on the lee side).

4. Extensive glacial erosion occurs in alpine regions. In general, a preglacial landscape is converted to one that is more angular, with sharper ridges, after glaciation. Resulting landforms include cirques, tarns, arêtes, horns, troughs, hanging valleys, and waterfalls.

Deposition of Glacial Drift and Resulting Landforms

Now that we have examined the landforms created by glacial erosion, let's consider those constructed through glacial deposition. As you study this section of the text, refer frequently to Figure 17.20, which shows most of the landforms discussed as well as their fundamental geographical relationships.

Glacial Till Sediment deposited by a glacier is called **glacial drift** and is found in two general forms: glacial till and glacial outwash. **Glacial till** is sediment that is deposited directly by a glacier. This form of glacial drift typically contains many different particle sizes, ranging from clay to sand and even cobbles and small boulders. As a result, it is considered to be an unsorted deposit because it is a mix of many different sediment sizes.

One form of glacial till is *basal till* (or *lodgement till*), which is deposited at the base of the glacier in way that is somewhat analogous to the way peanut butter is smeared onto bread with a knife. A glacier deposits basal till in somewhat the same way and can do so either as the glacier advances or retreats. Because of the intense pressure at the base of the glacier, due to the weight of the ice, basal till is usually quite dense and relatively fine-grained because it has been ground and crushed.

Glacial drift *Sediment deposited indirectly or directly by a glacier.*

Glacial till *Sediment deposited directly by a glacier.*

Another form of glacial till is *ablation till*, which is sediment that is carried within or on top of the ice and deposited when the glacier melts. Ablation till usually contains coarser fragments such as boulders because they are not crushed at the base of the glacier. As a result, ablation till tends to be even less well sorted than basal till.

Deposition of glacial till can result in very distinctive landforms. A notable type of till landform is a **moraine**, which is a winding ridge-like feature that forms at the front or side of a glacier, or between two glaciers. A glacial moraine is a very poorly sorted deposit, ranging from clays to boulders, that appears when sediments are brought forward by the conveyor action within a glacier until they are dropped out of it (Figure 17.21). A well-developed moraine forms only when the ice front or side is stationary for a relatively long period of time. Only then can the persistent internal motion of the ice continually bring forth enough sediment for a visible ridge to grow.

The specific classification of any given moraine depends on its spatial relationship to the glacier that deposited its till. Figure 17.22 shows examples of the different kinds of moraines. A *lateral moraine* forms on the edge of a glacier and is usually best expressed along the sides of alpine glacial troughs. When a lateral moraine is sandwiched between two glaciers, it is called a *medial moraine*. A *terminal moraine* (also called an *end moraine*) forms at the front of the ice and marks the long-term stationary position at the farthest advance of a glacier. Another type of moraine that forms at the ice front is a *recessional moraine*. Although a recessional moraine may look exactly the same as a terminal moraine, it marks a pause in the

---

Moraine   *A winding ridge-like feature that forms at the front or side of a glacier or between two glaciers.*

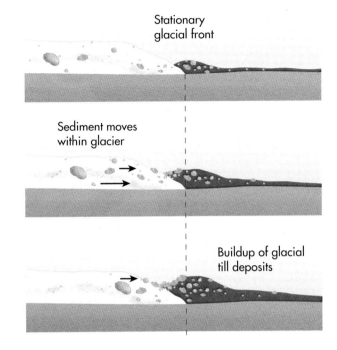

Figure 17.21 Formation of a moraine.   A moraine forms when the position of the ice front is stationary for a long period of time. This stationary position allows till to be deposited in one place due to the conveyor-like action within the interior of a glacier.

overall retreat of the ice, rather than the farthest advance. A final kind of moraine is called *ground moraine*. This kind of moraine forms when the retreat of the ice is slow but steady, allowing an irregular pattern of deposition that creates a hummocky landscape consisting of small hills and depressions (Figure 17.23).

(a)

(b)

Figure 17.22 Types of moraines.   (a) Lateral, medial, and end moraines in an alpine glacier system. (b) The hill in the background is a portion of the Port Huron moraine in northwestern Lower Michigan. This moraine can be traced around much of the northern part of the state's lower peninsula.

Figure 17.23 Ground moraine in the Blackfoot River valley in Montana. This rolling landscape was created because the glacier was retreating slowly at the same time it was depositing sediment in front and beneath it.

After till is deposited, it can be subsequently modified by direct and indirect glacial processes, again leaving distinctive landforms. One such landform is a **drumlin**, which is till or other soft sediment that has been streamlined in the direction of ice flow (Figure 17.24). A drumlin is somewhat similar to a roche moutonnée, but, in this case, the blunt end faces the direction from which the ice advanced. The formation of drumlins is not fully understood. Some may form because a glacier flows over previously deposited till that has uneven resistance, perhaps due to scattered boulder deposits or because some parts of the till have a frozen core. As the glacier flows over the resistant till, it may streamline less resistant sediment downstream of the core. Also, meltwater at the base of the glacier may have a role

in the formation of some drumlins by streamlining sediments when periodic dramatic discharge bursts occur. Drumlins can range in length from 100 m to 5000 m (300 ft to over 3 mi) and can be as high as 200 m (650 ft). They often occur in "swarms" and are particularly common in areas of Wisconsin, Michigan, and New York where continental glaciation was particularly intense during the most recent ice age.

Glacial Outwash In contrast to glacial till, which is sediment that accumulates while in direct contact with an ice mass, other sediments are carried and deposited by water that flows out and underneath a glacier as it melts. This kind of sediment is called **glacial outwash** and results from *glaciofluvial processes*, which are the combined effects of glaciers and streams.

As a deposit, glacial outwash differs fundamentally from glacial till in that it is well sorted, which means that most of the sediments are roughly the same size, namely, sand and gravel. This higher degree of sorting results from the fact that flowing water naturally segregates sediments on the basis of their size (see Chapter 16). For example, flooding stream systems deposit sand near the channel (resulting in natural levees) and clays toward the valley edge (forming wetlands and swamps). When glaciers melt, they produce prodigious amounts of meltwater that flows underneath and out in front of the ice. These flowing bodies behave according to hydraulic principles and therefore sort sediments in the same way as "normal" rivers do. These sediments are also stratified, which means that distinctive layers of sand, silt, or gravel can be traced horizontally for some distance.

Deposition of glacial outwash creates some distinctive landforms. The most common and extensive glaciofluvial landform is an outwash plain. As the name implies, an **outwash plain** is a relatively flat landscape (Figure 17.25) that was created through

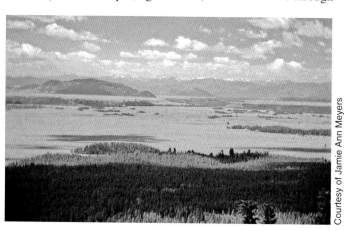

Figure 17.25 Outwash plain at Jackson Hole, Wyoming. Note the broad, horizontal surface that dominates the center of this image. This landform evolved because meltwater streams deposited the sediment in front of a valley glacier.

Figure 17.24 A drumlin in southern Alberta, Canada. These prominent hills were created by the streamlining effect of a glacier on the sediment beneath it.

Drumlin *A streamlined landform created when a glacier deforms previously deposited till.*

Glacial outwash *Sediment deposited by meltwater streams emanating from a glacier.*

Outwash plain *A broad landscape of limited relief created by the deposition of glacial outwash.*

the deposition of sediments carried by meltwater flowing out in front of a glacier. The associated meltwater streams are usually choked with coarse sediment and are therefore braided. Recall that braided streams are associated with aggradation, which, in the case of outwash plains, can result in very thick deposits of water-laid sediment in front of the ice.

Another type of landform associated with glacial meltwater is a **kame**, which is a large mound of sediment deposited by a melting glacier (see Figure 17.20). The formation of a kame begins when sediment accumulates in a depression on the surface of a retreating glacier. The sediment is subsequently deposited on the ground when the ice melts. A kame can form as an alluvial fan at the ice margin or as a deltaic deposit if the ice borders a lake. Kames usually contain stratified deposits of sand and gravel that progressively bury older deposits of a similar kind. Kame landscapes typically consist of rolling and irregular hills that are quite large, so large, in fact, that many ski resorts in the upper Midwest are built on these landforms.

Yet another landform created by flowing water associated with a glacier is an **esker**, which is a winding ridge (Figure 17.26) of coarse sediment deposited by a stream flowing *under* the ice. In this instance, the cross section of a subglacial stream channel is inverted with respect to a normal river channel; that is, the channel looks like an upside-down U in cross section. The flowing water is confined by the ice at the top and sides of the tunnel and by the ground beneath it, and so is the deposition of sediment. When the ice eventually melts, a meandering ridge of well-sorted and stratified glacial sediment is left behind. Most eskers are discontinuous because the conditions required for subglacial fluvial deposition are difficult to maintain over long distances. Nevertheless, eskers can be over 30 m (100 ft) high and tens of kilometers long.

Still another kind of landform associated with glacial outwash is called a **kettle lake**, which forms when a large block of ice falls off the front of a glacier and is subsequently buried by glacial outwash as the ice retreats. At some point, long after the glacier is gone, the buried block of ice melts. In the process of melting, the block creates its own lake, first by forming the depression that results from the subsiding sediments, and second by providing the water to fill the depression. This lake can subsequently be maintained at its original size if the depression intersects the water table. Kettle lakes are very common in heavily glaciated terrain where they are scattered on the landscape (Figure 17.27). As discussed in Chapter 15, the state of Minnesota is known as the "Land of 10,000 Lakes" because it contains thousands of kettle lakes that intersect the water table. The next time you camp at a small lake in this area, or any similar place in the northern tier of the United States, remember that the lake probably evolved because an ancient ice block was buried and then melted. Amazing.

Figure 17.26 An esker. Eskers form when meltwater streams flow in winding tunnels beneath a glacier, depositing sediment that forms a ridge when the ice melts. This esker forms a prominent ridge in Blue Lake, Minnesota.

Louis J. Maher Jr.

Figure 17.27 Kettle lakes. Kettle lakes form when ice blocks break off the front of a glacier, are subsequently buried by glacial sediment, and then melt, forming a depression and a lake if the water table is high.

Getty Images

Kame *A large mound of sediment deposited along the front of a slowly melting or stationary glacier.*

Esker *A winding ridge formed by a stream that flows beneath a glacier.*

Kettle lake *A lake that forms when a block of ice falls off the glacial front, is buried by glacial drift, and then melts, forming a depression that fills with water.*

Depositional Glacial Landforms

Like many other Earth processes, the formation of glacial landforms can be better understood when visualized in animated form. To do so, go to the Geo Media Library and select *Depositional Glacial Landforms*. This animation illustrates the way moving and melting ice shapes the landscape. It also contains a nice video of glaciers in Peru. After you complete the interactivity, be sure to answer the questions at the end to test your understanding.

VISUAL CONCEPT CHECK 17.3

Glaciers construct landforms in many different ways, both in direct contact with the ice and in close proximity to it. Given your understanding of landscape evolution and glaciers, what kind of landform is being constructed in this image? How can you tell? What will the relief of this landform be like when it is fully developed?

©Corbis

KEY CONCEPTS TO REMEMBER ABOUT GLACIAL DEPOSITION AND RESULTING LANDFORMS

1. Two primary kinds of glacial deposits occur: till and outwash. Glacial till is deposited in direct contact with the ice, whereas glacial outwash accumulates through meltwater streams flowing in front of the ice.

2. Glacial till is relatively unsorted and accumulates either as basal (lodgement) till smeared under the bottom of the glacier or as ablation till laid down as the ice melts.

3. A distinctive depositional landform created by glaciers is a moraine (end, lateral, or medial), which is a ridge of till that forms when the ice front or margin is in one place for a relatively long period of time. Drumlins are streamlined till landforms created by the weight and pressure of the overlying flowing ice.

4. Glacial outwash is relatively well sorted because it is deposited by flowing meltwater in front of the ice. The associated streams are typically braided and create a broad, flat surface known as an outwash plain. Sometimes glacial outwash buries a block of ice that broke off the front of a receding glacier. When this ice block subsequently melts, it forms a water-filled depression called a *kettle lake*.

5. Kames and eskers are distinctive meltwater landforms. A kame is an irregularly shaped hill that essentially consists of an alluvial fan or a deltaic deposit that forms in contact with the ice. Eskers are winding ridges that develop when sediment accumulates in a stream that flows in a tunnel at the base of the ice.

History of Glaciation on Earth

The many examples of glacial landforms demonstrate that ice sheets have been an important and widespread feature around the world in the past. The presence of drumlins in places like New York and Michigan, for example, indicates that ice sheets once existed where they no longer do. You might wonder if it is possible for such conditions to return at some point in the future. This is a great question and provides the context for a discussion of Earth's glacial history.

In order to provide the proper context about the glacial past, some fundamental terminology must be established. Periods of glacial advance, when glaciation is a dominant worldwide process, are called *glacials*. During glacials, temperatures are cooler on average, ice sheets expand and become larger, and sea level is lower because more water is stored in the ice sheets. Periods of glacial retreat, such as we are in now and have been for about the past 10,000 years, are called *interglacials*. During interglacials, average global temperatures are warmer, ice sheets and alpine glaciers are smaller, and sea level is higher because meltwater from the glaciers has returned to the sea.

Although the evidence is extremely vague, the first major glacial period on Earth appears to have occurred during Precambrian time about 2.3 billion years ago. This glaciation was in response to a major fluctuation in the composition of the early atmosphere. Prior to this period, the atmosphere was composed mostly of methane, an important greenhouse gas. Sometime approximately 2.3 billion years ago the amount of oxygen in the atmosphere increased dramatically, which in turn cooled the planet so that much of it was covered by ice. Ice was so extensive at this time that the period is sometimes called the *snowball Earth*. During this glacial interval, many of the world's oceans were frozen to a depth of about 0.8 km (0.5 mi). Similar "snowball" glacials occurred later in Precambrian time, about 750 million and 600 million years ago, and are most closely associated with decreases in carbon dioxide, another greenhouse gas.

Although much of Earth was heavily glaciated during these so-called snowball glaciations, the most recent and best-understood glacial periods are those associated with the Pleistocene Epoch between about 1.6 million to 10,000 years ago. During this epoch of time, alpine and continental glaciation occurred on such a large scale that 30% of the Earth's landmasses were covered by glacial ice at its maximum extent. For perspective, remember that contemporary glaciers now extend over only 11% of the Earth's land surface.

Until the latter part of the 20th century, glacial geomorphologists believed that four major advances of glacier ice occurred during the Pleistocene Epoch. This history was suggested through the reconstruction and dating of stratified tills in Europe and North America. In North America, these "classic" advances were named after the locations in which the ice advanced to its farthest point and strong evidence was found in the form of distinctive glacial drift. These glacial periods were given the following names (and approximate ages):

1. Nebraskan—1 million years ago
2. Kansan—600,000 years ago
3. Illinoisan—300,000 years ago
4. Wisconsin—35,000 to 10,000 years ago

In each of these advances, a thick continental glacier extended from the Arctic to a point deep within the American Midwest. East of the Rocky Mountains, this glacier is typically referred to as the **Laurentide Ice Sheet** because it covered the Laurentian Mountains in eastern Canada. Thick ice caps also covered much of the Rockies and extended west to the Pacific coast and south into Washington. Collectively, these ice caps are referred to as the **Cordilleran Ice Sheet**, which, at its maximum extent, merged with the Laurentide Ice Sheet in central Canada to form a single ice mass across much of North America. According to classical theory, the Laurentide Ice Sheet reached its southernmost point during the so-called *Kansan glaciation* about 600,000 years ago (Figure 17.28). This is the glacial advance that stopped at present-day northeastern Kansas and deposited the glacial erratic shown in Figure 17.14.

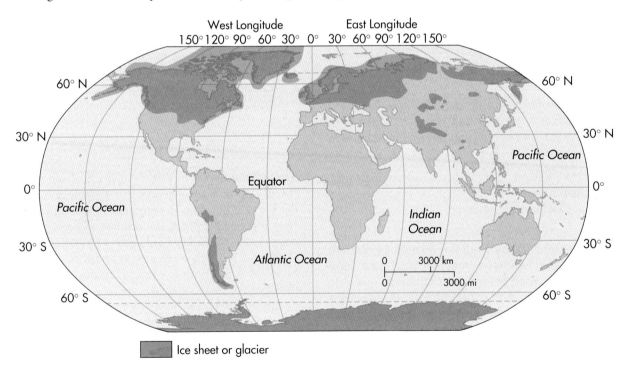

Figure 17.28 Maximum extent of glaciation during the Pleistocene Epoch. During this period of time, glaciers covered 30% of the Earth's land surface, compared to 11% today.

Laurentide Ice Sheet *The continental glacier that covered eastern Canada and parts of the northeastern United States during the Pleistocene Epoch.*

Cordilleran Ice Sheet *The ice cap that covered much of the mountains in the northwestern part of North America during the Pleistocene Epoch.*

As discussed in Chapter 9, however, reconstruction of oxygen-isotope records derived from glaciers and ocean sediments indicates that more than 20 such glacial periods occurred during the Pleistocene Epoch. These numerous cycles are apparently related to Earth's orbital fluctuations associated with the Milankovitch theory, which was also discussed in Chapter 9. Recall that this theory describes how Earth's geometric relationship with the Sun slowly fluctuates over time, with changes in axial tilt, precession of the equinoxes, and orbital shape at distinct cyclic intervals. In this context, the most intense glaciations would be theoretically associated with periods when (1) Earth was farthest away from the Sun in summer, (2) axial tilt is less than present, and (3) equinox occurred during the summer and winter months. As these geometric conditions slowly aligned over tens of thousands of years, the amount of solar radiation striking the high latitudes of the Northern Hemisphere in summer would gradually decrease, resulting in progressively cooler summer temperatures and the growth of ice sheets. Because of these new understandings, the only names maintained from the traditional glacial model in North America are the Illinoisan and Wisconsin advances because the deposits associated with these most recent glaciations are easily recognizable. All other glaciations are lumped into one category called *Pre-Illinoisan*.

The Wisconsin Glaciation and Evolution of the Great Lakes

Given that the Wisconsin glacial advance is the most recent, it is easily the most studied of all Pleistocene glaciations because its effects can still be seen clearly at the surface. Although strong evidence for this glaciation is found in northern Europe and western Canada, a great place to examine its impact on the landscape is the Great Lakes region of North America. Following the post-Illinoisan interglacial period, the Laurentide Ice Sheet began to expand again about 110,000 years ago. For the next approximately 90,000 years ice advanced south, reaching its most southerly position in central Ohio, Indiana, and Illinois by about 20,000 years ago. At this time, *all* of what is now known as the Great Lakes and the state of Michigan were buried beneath a sheet of ice (Figure 17.29a) that was probably hundreds of meters thick. This ice subsequently melted from the Great Lakes region between about 16,000 and 10,000 years ago and was part of the overall collapse of the Laurentide Ice Sheet at this time. The glacier melted in large part because Earth's axial tilt had increased to the point that sufficiently more solar radiation was received at high latitudes in summer to cause rapid melting.

Although the most recent deglaciation of the Great Lakes region left a multitude of glacial landforms, including moraines (such as Figure 17.22b), kames, eskers, outwash plains, and drumlins, the most prominent landscape feature is the Great Lakes themselves (Figure 17.29b). The evolution of these lakes is fascinating and is a testament to the power of glacial ice and how rocks of different resistance can influence the shape of a

(a)

(b)

Figure 17.29 Extent and impact of the Wisconsin glaciation. (a) During the most recent ice age, the Laurentide Ice Sheet and Cordilleran Ice Sheet collectively covered most of northern North America. (b) Satellite image of the Great Lakes acquired in April 2005. These lakes were formed when the advancing ice further enlarged and deepened preexisting glacial troughs that had been scoured out of relatively soft rocks. Note the extent of snow cover, frozen lakes, and the suspended sediment in Lake Erie.

THE CHANNELED SCABLANDS IN EASTERN WASHINGTON

The Channeled Scablands in eastern Washington are known for the numerous dry canyons carved into the landscape. In the early 20th century, the geologist J. Harlan Bretz hypothesized that these landforms were created by catastrophic floods caused by melting glaciers. He was ridiculed for most of his career about this idea because most geologists believed that Earth-changing processes due to glaciers had to be slow and gradual. By the 1970s, however, his hypothesis gained increasing support because more had become known about the way that water flows in open channels. His hypothesis began to be accepted by the scientific community in the 1970s. By the time Bretz died in 1981, however, his hypothesis had been accepted. Indeed, he received geology's highest honor, the Penrose Medal, in 1979.

Geologists now believe that a lobe of the Cordilleran Ice Sheet—the Purcell Trench Lobe—advanced far enough into Montana that it blocked a valley through which glacial meltwater flowed west. This blockage created an ice dam that allowed a deep lake to form behind it, spreading back into the mountain valleys to the east. This lake is known to geomorphologists as *Glacial Lake Missoula* because it formed shorelines that are still visible around Missoula, Montana. These shorelines indicate that the lake was as big as today's Lake Ontario and Lake Erie combined. As the lake deepened, it eventually began to flow over the top of the ice dam, finally causing the dam to break catastrophically.

Torrential floodwaters, up to 300 m (1000 ft) deep, subsequently burst onto the plains of what is now eastern Washington, flowing at a rate 10 times the total of all the rivers on Earth and emptying Lake Missoula in perhaps as little as 48 h. This type of catastrophic discharge may have happened up to 100 times and created the enormous valleys carved into the landscape. The network of now-dry channels is so well defined that it can be seen on satellite imagery.

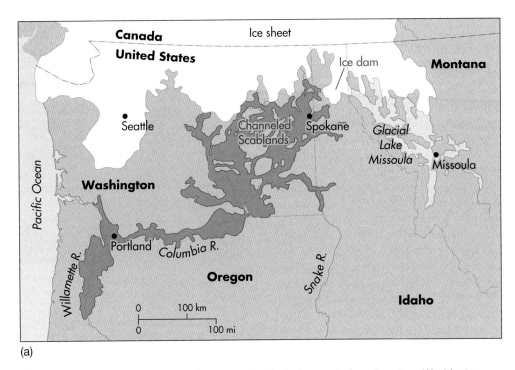

(a)

| The Channeled Scablands cover a large area that includes most of southeastern Washington.

(b)

The horizontal lines visible in this hillside in Missoula, Montana, are old shorelines created by wave action of glacial Lake Missoula during the most recent ice age. To imagine the depth of the water, note the school building in the foreground for scale.

Vincent P. Gutowski

Jeff Schmaltz, MODIS Rapid Response Team, NASA/GSFC

(c)

The Channeled Scablands show up beautifully as light brown channel forms (arrows) in this satellite image that focuses on eastern Washington, northeast Oregon, and western Idaho.

Greg Vaughn / Alamy

A typical view of the Scablands landscape. This deep gorge was cut by catastrophic discharge of glacial meltwater.

(d)

landscape. Prior to Pleistocene glaciation, the areas now occupied by the Great Lakes were large river valleys that were cut into soft shale deposits. Very resistant granites, limestones, and sandstones underlie the areas that are now regional landmasses, such as the upper and lower peninsulas of Michigan.

As the Laurentide Ice Sheet oscillated back and forth across the region during the Pleistocene, the glacier preferentially eroded the softer rocks. In this way, the former river valleys gradually changed into enormous glacial troughs. Because these troughs progressively enlarged and deepened during subsequent glaciations, they became the preferential pathways of the ice front whenever it advanced back into the region. The modern Great Lakes are the legacy of that complex geomorphic history. More resistant rocks were much less eroded and became the foundation for landforms such as the Upper and Lower Peninsulas of Michigan, the Bruce Peninsula in Ontario, the Keweenaw Peninsula in Upper Michigan, and a variety of islands.

KEY CONCEPTS TO REMEMBER ABOUT THE HISTORY OF GLACIATION ON EARTH

1. Several periods of intense glaciation occurred during Precambrian times in which most of the planet was covered by ice. Such a period is referred to as a snowball Earth.

2. In the context of geologic time, the most recent period of major continental glaciations is the Pleistocene Epoch, which lasted from approximately 1.6 million to 10,000 years ago.

3. Until recently, it was believed that four major glaciations occurred during the Pleistocene Epoch. Oxygen isotope records derived from ice cores and marine sediments indicate that more than 20 major glacial periods occurred during the Pleistocene. The best understood of these glaciations are the Illinoisan and Wisconsin periods, which occurred most recently. All other glacial periods are lumped into one category called the Pre-Illinoisan.

4. Most glacial landforms in the upper Midwest, the northeastern United States, and Canada are products of the Wisconsin glaciation, which occurred between about 110,000 and 10,000 years ago. The most distinctive large-scale landforms are the Great Lakes, which are glacial troughs cut into soft bedrock.

Probable Human Impact on Glaciers

With an understanding of past glaciations in mind, a key question to consider now is the status of contemporary glaciers with respect to the issue of ongoing climate change. Although this period of warming may be part of a natural cycle, most climatologists believe that it is associated with increased levels of atmospheric carbon dioxide (and other greenhouse gases) caused by human industrial activity. Regardless of the cause, Earth appears to be in a distinct warming phase that has been linked to the increased incidence of drought, species migration and loss, and rise in sea level. Global warming is also having a major impact on glaciers around the world.

As previously discussed in this chapter, glacial advance and retreat are controlled by fluctuations in the glacial mass budget (see Figure 17.4). When snowfall in the source area exceeds melting in the zone of ablation, glaciers thicken and expand, either down valleys in alpine settings or across landmasses in continental glaciers. Glaciers naturally recede during periods when the rate of melting in the zone of ablation exceeds the accumulation of snow in the source area. During the Pleistocene Epoch, the budget of global ice masses was controlled largely by fluctuations in Earth–Sun geometry as described in the Milankovitch theory. The most recent noteworthy episode of glacial advance occurred during the Little Ice Age, which was a time of global cooling that occurred between about A.D. 1500 and 1850.

Since the end of the Little Ice Age, glaciers have generally been retreating around the world (Figure17.30a). Many scientists have attributed this rapid rate of glacial retreat to human-induced global warming. Although some of the world's glaciers have actually advanced in the past few decades, such as the Perito Moreno Glacier in Argentina, the vast majority are melting at rapid rates. Glacier retreat has been very noticeable in most alpine environments. In the Swiss Alps, for example, the Rhône glacier has retreated about 2.5 km (1.5 mi) in approximately the past 140 years. In Peru, the Ururashraju Glacier retreated 500 m (1640 ft) between 1986 and 1999 alone! At Glacier National Park in Montana, glacier retreat has been dramatic. In 1850 approximately 150 glaciers were present in the area and most were still there in 1910 when the park was established. By 2010, however, only 25 glaciers larger than 10 hectares (25 acres) remained. Repeat photography of the Grinnell Glacier, for example, is representative of the kind of change seen elsewhere in the park. Notice the dramatic difference in that system between 1900 (Figure 17.30b) and 2008 (Figure 17.30c). If such rate of melting continues as environmental models suggest, the Grinnell Glacier, as well as the remaining glaciers in the park, will disappear entirely in the next few decades.

Rapid melting of glaciers has also been observed on the Greenland and Antarctic ice sheets. On the Greenland Ice Sheet, the area on the edge of the glacier affected by melting increased by 16% between 1979 and 2002. You can see how the area affected by melting has changed in the past 20 years in Figure 17.31. On the Antarctic Ice Sheet, the rate of glacial melting is even more impressive. An example of this melting is the collapse of the Larsen B Ice Shelf that occurred over a 35-day period beginning January 31, 2002. An ice shelf is a plate of glacial ice that extends from the landmass into the water; many

Courtesy of United Nations Environment Programme, figure based on
data analysis by R. Prinz, University of Innsbruck, Austria; data from WGMS.

(a)

Figure 17.30 Glacier melting due to climate change. (a) Temporal assessment of glacier length changes around the world since the mid-nineteenth century. The number of advancing glaciers (vertical blue bars) and retreating (vertical red bars) are plotted for each calendar year based on survey observations. Note that the vast majority of the world's glaciers have been retreating during this time span. (b) Photo of the Grinnell Glacier in Glacier National Park (Montana) in 1900. (c) Photo of the Grinnell Glacier in 2008. Note the dramatic change in the size of the glacier. This degree of change is representative of glaciers elsewhere in the park.

F. E. Matthes photo, courtesy of Glacial National Park Archives

(b)

Lisa McKeon photo, USGS

(c)

Probable Human Impact on Glaciers • 471

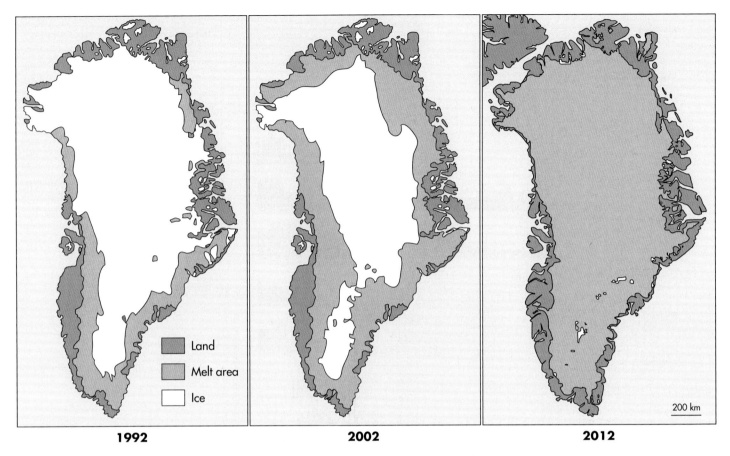

Land
Melt area
Ice

200 km

1992   **2002**   **2012**

Figure 17.31 Recent melting of the Greenland Ice Sheet. Notice the change in the area of active melting in 1992, 2002, and 2012. (*Source*: NASA.)

of these features extend off the Antarctic landmass. During the Antarctic summer of 2002, about 3500 km² of the Larsen B Ice Shelf disintegrated. For scale, consider that the entire state of Rhode Island is 2717 km² (1049 mi²) in size. Figure 17.32 shows the location of the ice shelf and the progression of its breakup in two satellite images. The breakup is believed to be due to the rapid warming (about 0.5°C [0.9°F] per decade) that has occurred since the 1940s. This warming has had a profound impact, causing the extent of seven ice shelves on the Antarctic peninsula to decline by a total of about 13,500 km² (5212 mi²) since 1974.

As evidence that these processes are still active, a chunk of ice about 16 km (10 mi) wide and 80 km (50 mi) long rapidly broke off the Greenland ice cap in late July 2010. This break raised concerns that the ice cap may be reaching an important threshold as far as its future stability is concerned.

www.wiley.com/college/arbogast

WileyPLUS

Glaciers and Climate Change

Evidence strongly indicates that glaciers around the world are melting at rapid rates due to climate change. These changes have been documented at many glaciers through repeat photography and time-lapse imagery. To see examples of such changes, go to the Geo Media Library and select *Glaciers and Climate Change*. This animation contains an excellent discussion of changes at Glacier National Park in Montana. It also presents links to sites where glaciers have been closely monitored in the past decade. Once you complete the exercise, be sure to answer the questions at the end to test your understanding.

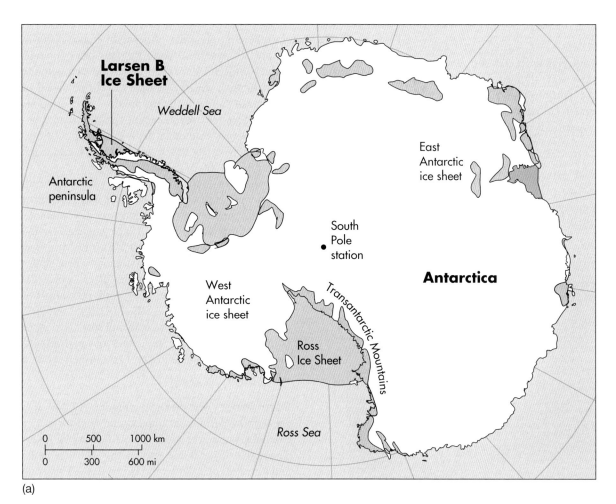

(a)

Figure 17.32 Disintegration of the Larsen B Ice Shelf in Antarctica in 2002. (a) Location of the Larsen B Ice Shelf in Antarctica. (b) The ice shelf when breakup began on January 31, 2002. The dark spots on the shelf are patches of meltwater. (c) Status of breakup on March 5, 2002. The largest ice blocks are hundreds of square kilometers in size.

Periglacial Processes and Landscapes

Although we are clearly living in an interglacial period, one that may even become significantly warmer if global warming models prove accurate, much of Earth remains very cold over much or all of the year. Naturally, these very cold regions are located at high altitudes and latitudes and are associated with polar-type climates. Associated with these climates is a range of specific geomorphic processes referred to as **periglacial processes** that are closely associated with substantial amounts of frost action, specifically the effects that continual freezing and thawing have on the landscape. These processes are closely related to physical weathering, mass movement, and soil modification and affect approximately 20% of the Earth's surface.

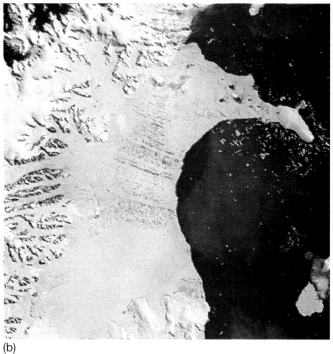

(b)

NASA

Periglacial processes *The suite of processes involving frost action, permafrost, and ground ice that occurs in arctic environments or along the margins of ice sheets.*

Permafrost

The most common periglacial processes are associated with **permafrost**, which, as the name implies, is ground that is permanently frozen. These conditions typically develop when soil or rock temperatures remain below 0°C (32°F) for at least two years. Permafrost may develop because the area is covered by a glacier or it may be in close proximity to the ice margin and thus periglacial.

From a geographical perspective, regions of permafrost can be subdivided into two primary categories: (1) continuous permafrost and (2) discontinuous permafrost. Figure 17.33 shows the geographical distribution of these regions in the Northern Hemisphere. Large areas of permafrost also occur in Antarctica but are not described here. Regions of continuous permafrost in the Northern Hemisphere are associated with the most frigid temperatures, which essentially occur poleward of the −7°C (19°F) mean annual isotherm. In areas of continuous permafrost, all surfaces are frozen except those insulated beneath deep lakes or rivers. Although the average depth of permafrost in these regions is about 400 m (1300 ft), it can reach the impressive depth of 1000 m (3300 ft).

South of the region of continuous and deeply frozen ground, the mean annual temperatures become warmer and permafrost is discontinuous and shallower. This region of discontinuous permafrost generally extends in a broad area bounded by the −7°C (19°F) and the −1°C (30.2°F) isotherms. Permafrost in this region is discontinuous where slopes face south toward the Sun and in places where the ground is insulated by snow. Overall, permafrost of some kind occurs in about 50% of Canada and 80% of Alaska. To see how the depth and spatial extent of permafrost vary across the continuous and discontinuous zones, examine Figure 17.34, which shows a typical cross section across the high latitudes.

Permafrost Processes Figure 17.34 also shows how permafrost impacts the landscape. For example, notice the areas denoted as *active layer*. The active layer is the part of the soil that melts and refreezes on a daily or seasonal basis. This zone lies above the subsurface permafrost layer and predictably varies with latitude, ranging in thickness from up to 2 m (6.6 ft) in areas of discontinuous permafrost to only 10 cm (4 in.) in northern areas of continuous permafrost.

As you can imagine, the active layer is sensitive to climate change. Although the response is slow, colder temperatures gradually cause the thickness of the active layer to decrease, whereas warmer temperatures result in permafrost melting and an increased thickness of the active layer. In this context, abundant evidence indicates that Arctic permafrost is rapidly thawing in Earth's current warming phase. This thawing is releasing tons of CO_2 into the atmosphere, which should contribute to more warming in the future.

In addition to the active layer, another important feature in a permafrost landscape is a talik (see Figure 17.34 again). A *talik* is a body of unfrozen ground that is found within a zone of discontinuous permafrost or beneath a lake or river in the continuous region. Taliks are especially important in discontinuous permafrost zones because they are a link between the active layer and groundwater, forming a kind of conduit through which water can move freely.

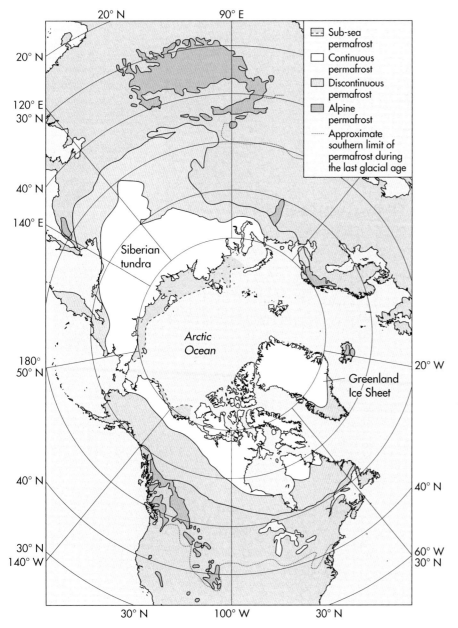

Figure 17.33 Geographical distribution of permafrost in the Northern Hemisphere. An extensive part of the Northern Hemisphere contains permafrost. Note that it can occur in the oceans as well as on land.

Permafrost *Ground that is permanently frozen.*

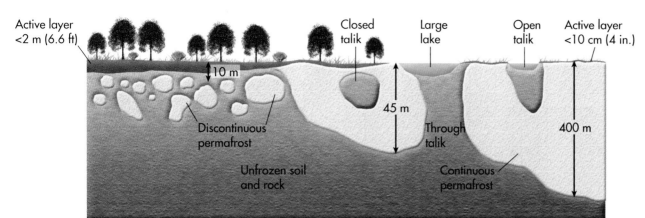

SOUTH

NORTH

Active layer <2 m (6.6 ft)

10 m

Discontinuous permafrost

Unfrozen soil and rock

Closed talik

Large lake

Through talik

45 m

Open talik

Active layer <10 cm (4 in.)

400 m

Continuous permafrost

Figure 17.34 Typical periglacial landscapes. Note the differences in the spatial extent and depth of permafrost across this hypothetical cross section (not to scale).

Ground Ice and Associated Landforms

Within permafrost regions, distinct zones of frozen water occur within the ground. These areas are referred to as **ground ice** and contain highly variable amounts of water, ranging from near zero to completely saturated conditions. Areas of ground ice expand and contract along freezing fronts, which are the boundaries between frozen and unfrozen ground. As these areas expand and contract, they cause physical weathering of the landscape through the frost action processes described in Chapter 14. Remember that when water freezes, it expands by about 9%. This expansion causes rocks to break along joint lines. In addition, water expansion can cause soil to move

vertically or horizontally through the processes of *frost heaving* and *frost thrusting*, respectively.

Ground ice comes in many forms. The most basic kind of ground ice is *pore ice* where water freezes in soil pore spaces. Ground ice also occurs as *horizontal lenses* and veins of ice that extend in random directions. A third kind of ground ice is an *ice wedge*, which results when water enters a crack in the ground (Figure 17.35). *Segregated ice* is ground ice that is buried but grows

(a)

Figure 17.35 Formation of an ice wedge. (a) An ice wedge in northern Canada. (b) Sequence of steps in the formation of an ice wedge. These features form when water penetrates an open crack in the ground and freezes. Over time, repeated freezing and thawing slowly cause the crack to expand, forming a wedge. (*Source:* Modified from Arthur Lachenbruch, 1962, "Mechanics of Thermal Contraction and Ice-Wedge Polygons in Permafrost," Geological Society of America Special Paper 70.)

Courtesy US Fish & Wildlife Service

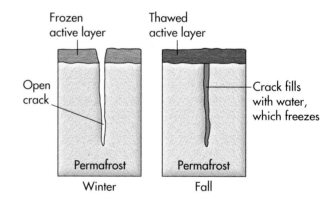

Frozen active layer

Thawed active layer

Open crack

Crack fills with water, which freezes

Permafrost

Permafrost

Winter

Fall

After 500 years

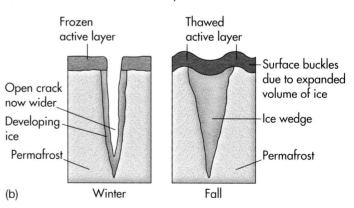

Frozen active layer

Thawed active layer

Open crack now wider

Developing ice

Permafrost

Surface buckles due to expanded volume of ice

Ice wedge

Permafrost

(b)

Winter

Fall

Ground ice *Distinct zones of frozen water that occur in permafrost regions.*

Figure 17.36 A pingo. A pingo, such as this one in the Northwest Territories in Canada, forms when water is injected into the ground under some form of hydraulic pressure. This pressure forces the ground to bulge upward.

Timothy Epp/Alamy

Figure 17.37 Patterned ground. The distinctive polygons on this landscape form when frost action preferentially brings large stones and rocks to the surface.

Jim Wark/Airphoto

when additional water is added. Finally, *intrusive ice* is ground ice that forms when water is injected under pressure, perhaps because it lies between a pair of aquicludes. Where intrusive ice is found, the surface of the ground can bulge up to 60 m (200 ft) upward, forming a landform called a *pingo* (Figure 17.36).

In addition to pingos, frost action can result in a variety of other landforms. A *palsa* (Spanish for "elliptical") is a rounded or elliptical mound of peat that is similar to a pingo in that it

is thrust upward. It differs from a pingo, however, because it is only 1 m to 10 m (3 ft to 30 ft) high and contains ice lenses rather than a solid core. Another distinctive periglacial landform is *patterned ground*. As the name implies, patterned ground is a landscape that has evolved such that it contains distinctive shapes. These shapes are usually polygons (Figure 17.37) that form when frost action preferentially brings coarser materials—stones and boulders—to the surface.

SUMMARY OF KEY CONCEPTS

1. Glaciers form when snow accumulates annually and does not melt during the summer months. In this way, the snow mass gradually thickens with each passing year and gradually changes into glacial ice through compression.

2. A growing ice mass becomes a glacier when it begins to flow under its own weight. Whether a glacier advances or retreats depends on the glacial mass budget, which is the balance between snow accumulation in the source area and melting at the zone of ablation. Melting and accumulation are in balance along an equilibrium line.

3. Ice fields and ice caps are large ice masses that spread horizontally across mountainous landscapes. Valley glaciers flow down preexisting stream valleys in alpine settings. Continental glaciers, such as the Antarctic Ice Sheet, are thousands of meters thick and spread across continental landmasses.

4. Glaciers erode the ground beneath them through the combined processes of abrasion and plucking. In alpine settings, these processes change the landscape from generally rounded to much more angular, with sharp features such as arêtes, horns, and cirques.

5. Glaciers deposit sediment either directly as till or indirectly from meltwater streams as outwash. These processes produce landforms such as moraines, kames, and outwash plains.

6. During the Pleistocene Epoch (between about 1.6 million and 10,000 years ago), glacial ice covered up to 30% of the Earth's land surface including extensive parts of North America, Europe, and Asia. The most recent period of massive glacial advance within the Pleistocene was the Wisconsin glaciation, which ended approximately 10,000 years ago. Glaciers now cover about 11% of Earth's landmasses. These tremendous fluctuations in ice volume are intimately related to gradual changes that occur with respect to Earth–Sun geometry.

7. Evidence for the Pleistocene ice ages is found in many places in the northern part of the United States and Canada. The Great Lakes, for example, were created because large ice lobes enlarged preexisting stream valleys.

8. Permafrost refers to ground that is perpetually frozen, forming features such as ice wedges and pingos. Areas of current permafrost conditions occur at high latitudes and altitudes.

1. Why are summertime temperatures a critical variable with respect to the formation of a glacier?

2. Describe the steps involved in the metamorphosis of snow to glacier ice.

3. How does a glacier flow? Where does the ice move the fastest and slowest? Why?

4. Compare and contrast the various kinds of glaciers on Earth.

5. Describe the difference between glacial abrasion and glacial plucking. What landform is produced by a combination of these two processes?

6. How can you tell if a mountainous landscape has been glaciated? What are some of the characteristic landforms that should be present if the landscape was glaciated?

7. How does a hanging valley form?

8. In what two regions of Earth are large continental glaciers currently found?

9. What is the difference between glacial till and glacial outwash? What are the diagnostic characteristics of each deposit?

10. How does the formation of a recessional moraine differ from that of a terminal moraine?

11. Explain the history of glaciations on Earth.

12. What is the impact of ongoing climate change on most glaciers around the world?

13. What is the difference between continuous and discontinuous permafrost? What is their geographical distribution, and why does it make sense given your understanding of permafrost processes?

14. What is ground ice? Describe two landforms created by this phenomenon.

ANSWERS TO VISUAL CONCEPT CHECKS

VISUAL CONCEPT CHECK 17.1

The glacial mass budget refers to the balance between snow accumulation in the source area and melting in the zone of ablation. The point where these two processes are in balance is known as the equilibrium line. Below the equilibrium line, crevasses are visible, whereas above the equilibrium line they are covered with snow.

VISUAL CONCEPT CHECK 17.2

Glaciers modify alpine landscapes in a very intense way. Prior to glaciation, alpine landscapes are generally rounded, with streams flowing in V-shaped valleys. Through the combined processes of abrasion and plucking, alpine glaciers cause mountainous regions to become more angular, forming features such as cirques, arêtes, and horns. Valley glaciers enlarge and deepen former stream valleys, changing them to U-shaped valleys.

VISUAL CONCEPT CHECK 17.3

This photograph of a stream system in the Southern Alps of New Zealand is a nice example of how outwash plains form. Several lines of evidence point to this conclusion. The snow-capped peaks in the background suggest that alpine glaciers are present, which must be delivering coarse sediment to the stream in glacial meltwater. You can tell that the sediment is coarse because the stream is braided. Given that fluvial deposition is occurring, you can surmise that the landscape will have very limited relief and will thus be consistent with an outwash plain.

CHAPTER EIGHTEEN | ARID LANDSCAPES AND EOLIAN PROCESSES

This chapter focuses on the geomorphology of arid regions and the way that flowing air shapes the landscape. Arid landscapes are fascinating places because landforms are quite visible and the factors that influence their evolution can often be clearly seen. One of the dominant geomorphic processes that shapes landscapes in these regions is the wind. Wind processes are typically referred to as *eolian processes*, which, in turn, produce eolian landforms. (The term *eolian* is derived from Aeolus, the Greek god of the winds.) Although the ability of wind to shape the landscape is low relative to moving water and ice, flowing air can nevertheless produce a fascinating array of erosional and depositional landforms. It is important to understand the role of wind in the evolution of landforms because more than one-third of the Earth's land surface is located within arid or semi-arid climate zones. In addition, many of these landscapes are agriculturally important and can be adversely affected by the wind.

Some of the most beautiful landforms on Earth are in arid regions, such as these sand dunes in Death Valley, California. This chapter focuses on these areas, including how rock structure influences topography and how streams shape the landscape. It also examines the behavior of wind and how it creates landforms such as those pictured here.

Arid Landscapes

Eolian Erosion and Transport

Eolian Deposition and Landforms
Ⓖ *Eolian Processes and Landforms*

Human Interactions with Arid Landscapes and Eolian Processes

▌ LEARNING OBJECTIVES

1. Describe the physical geography of arid landscapes.

2. Discuss how rock structure influences the landscapes of the southwestern United States.

3. Explain the formation of an alluvial fan.

4. Understand the character of eolian processes and how they move sediment.

5. Describe the formation of a sand dune.

6. Compare and contrast the various forms of sand dunes.

7. Describe the character of loess and where it is found on Earth.

8. Discuss human interactions with eolian processes in the African Sahel and Great Plains.

Arid Landscapes • 479

Arid Landscapes

A good way to begin investigating eolian processes is by looking more closely at the geography and character of arid environments because these areas are often most closely associated with the influence of flowing air. This discussion incorporates many of the topics covered earlier in the text—atmospheric circulation, plant geography, soils, geology, and fluvial processes. This part of the chapter seeks to provide you with a good understanding of desert environments and their geography.

Figure 18.1 shows the geography of arid and semi-arid regions on Earth. Such climates occur in both warm and cold settings and collectively form the largest climate region on the planet, one that covers perhaps as much as 35% of Earth's land surface. Within the Köppen climate system (Figure 9.2), these climate regions fall within the arid desert (BW) and semi-arid steppe (BS) climate categories. The spatial distribution of these dry regions is based on one of three factors:

1. *Dominance of Subtropical High-Pressure Systems*— In areas located between about 15° and 35° N and S latitude, strong high-pressure systems dominate the weather for all or significant portions of the year. These descending air masses are components of the Hadley cells (shown in Figures 6.16 and 6.19) that circulate within the atmosphere in the tropical latitudes. The subtropical high-pressure systems are very closely associated with the largest deserts on Earth, such as the Sahara and Kalahari in Africa, the Simpson Desert in Australia, and the Chihuahuan Desert in Mexico (Figure 18.1).

2. *Rain Shadow Effect*—Another environmental factor associated with deserts is the rain shadow effect. If you recall from Chapter 7, a rain shadow occurs on the lee side of mountain ranges where descending air warms adiabatically. This process was outlined in Figure 7.22 and is responsible for the large zones of arid and semi-arid climates in the western United States. It also is related to the presence of the Atacama Desert along the west coast of South America.

3. *Distance from Large Water Bodies*—The last environmental variable that is related to dry regions is the proximity to a water body such as an ocean. Locations deep within continents tend to receive relatively small amounts of moisture-laden air and can therefore be quite dry. An excellent example of this kind of situation is the Taklamakan and Gobi deserts in Asia (Figure 18.1), as well as the other large areas of arid and semi-arid lands that lie deep within that continent.

Desert Geomorphology

A key element that all deserts and semi-arid regions share is a relative lack of vegetation when compared to more humid areas. As a result, descriptive terms such as *sparse* and *barren* are frequently used to describe dry regions. This relationship of climate and vegetation was thoroughly explored in Chapters 9 and 10 and can be seen in numerous photographs of desert environments, such as Figures 9.5 and 10.17.

Although the barren nature of these environments may at first seem unattractive or even foreboding, desert landscapes are beautiful in their own way, with spectacular rock outcrops and panoramic vistas. In the context of geomorphology, deserts are

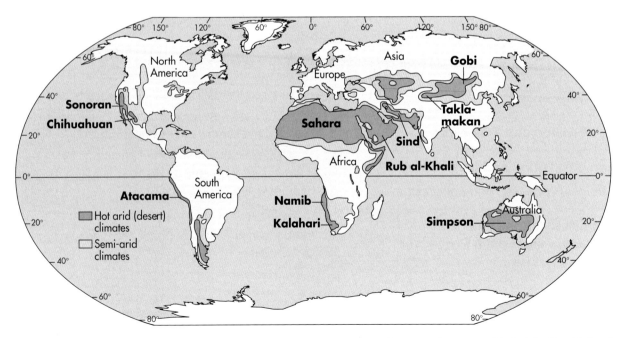

Figure 18.1 Deserts of the world. Arid and semi-arid deserts collectively cover about one-third of the Earth's land surface. The largest deserts are associated with the belt of subtropical high-pressure systems.

Figure 18.2 Uluru in central Australia. Uluru is the best example of an inselberg, or island mountain, in the world. This iconic landform is composed of hard sandstone that was once buried by other rocks that have been removed by erosion. The beauty of Uluru is magnified in the arid environment of central Australia.

(a)

(b)

(c)

Figure 18.3 The Basin and Range Province. (a) The Basin and Range Province extends over much of the American Southwest. (b) Satellite image of the Basin and Range Province. Notice the semilinear network of mountain ranges (darker shades) and intervening basins that lie between the Wasatch Range and Sierra Nevada. (c) Panoramic view of a portion of the Basin and Range Province. Note the distinct mountain ranges and the intervening basins.

great places because you can see landforms very clearly and the important elements such as rock structure and differential weathering that influence their evolution. A superb example of this visual effect can be seen at *Uluru* (formerly known as Ayer's Rock) in central Australia (Figure 18.2). Uluru is the best example of an *inselberg* (island mountain) in the world. Composed of hard sandstone, this prominent landform towers over the surrounding landscape because differential weathering over millions of years removed the rocks that once covered it. In this desert environment, it is possible to visualize this process. Although landforms in humid regions may have similar histories, it can be more difficult to visualize because thick stands of vegetation often mar the view.

The Southwestern United States An excellent place to see desert geomorphology where you may one day visit is the southwestern United States. This region lies in the rain shadow of the Sierra Nevada (Figure 13.23) and encompasses a number of previously discussed arid places such as the Sonoran Desert (Figure 10.16) and the Grand Canyon (Figure 12.20a). Perhaps the most obvious variable, other than climate, that strongly influences landforms in the West is rock structure. A spectacular example of the effects of normal faulting (Figure 13.22), for example, is the geologically famous Basin and Range Province in the southwestern United States. This region extends from the Wasatch Mountains in eastern Utah to the Sierra Nevada along the border of California and Utah (Figure 18.3a). It is characterized by numerous mountain ranges and intervening valleys that are generally aligned from north to south (Figure 18.3b). The relief between the crests of mountain ranges and the adjacent basins can be as much as about 3050 m (10,000 ft), resulting in a distinctive view from the air (Figure 18.3c).

The Basin and Range Province began to form about 20 million years ago when the North American crust started to stretch in this part of the continent due to upward pressure

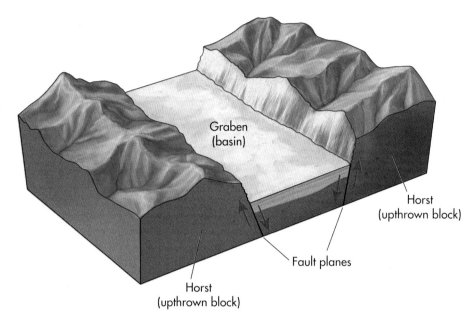

Figure 18.4 Formation of the Basin and Range. The Basin and Range Province consists of a series of normal faults produced by stretching of the continental crust over the past 20 million years. After uplift, erosion attacks the mountain ranges (horsts). The resulting sediments are then deposited in the intervening basins (grabens), slowly filling them with sediment.

Graben (basin)

Horst (upthrown block)

Fault planes

Horst (upthrown block)

applied by an underlying magma plume. As a result of this stretching, the crust thinned, cracked, and pulled apart, creating numerous normal faults that have planes inclined at about a 60° angle (Figure 18.4). This faulting created the series of alternating basins and ranges that characterize the area.

The importance of rock structure can also be seen where the rocks are horizontal and have not been deformed (Figure 18.5). In these areas, which are very common in the West, the rocks usually consist of alternating layers of limestone, sandstone, and shale that contribute to a distinctive pattern of differential weathering. The limestones and sandstones are generally the most resistant rocks to erosion, whereas the shales are relatively soft and thus easily eroded. Where these patterns occur, the hard rocks usually form a resistant caprock that protects the underlying shale to some extent from erosion. Notice in Figure 18.5 that shale tends to form shallow slopes, whereas the more resistant rocks are associated with steep slopes. Shale tends to erode backward gradually due to slope wash when it does rain. This gradual slope retreat ultimately undermines the

caprock above and causes it to collapse due to rockfall, and the cycle begins again.

The net effect of this kind of landscape evolution is that large parts of the western United States are extensively dissected (appear to be cut up), with upright landforms of various sizes that rise above the surrounding lower ground. To the novice these landforms seem to have some relationship to one another, but it is difficult to imagine how they could be related. Actually, this landscape represents a slow progression of erosion over millions of years. Figure 18.5 illustrates the landforms associated with this progressive evolution. It is interesting to notice how each of the rock layers can be visually traced from one landform to another. This linkage occurs because the rock was once a uniform mass with broad regional extent. Through a combination of stream erosion, differential weathering, and mass wasting over millions of years, the landscape was slowly dissected, leaving a succession of flat-topped remnants that tower throughout the area.

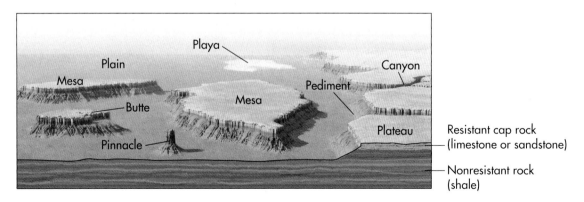

Playa

Plain

Mesa

Butte

Pinnacle

Mesa

Pediment

Canyon

Plateau

Resistant cap rock (limestone or sandstone)

Nonresistant rock (shale)

Figure 18.5 Prominent desert landforms associated with horizontal rock structure. Over time, a plateau is dissected into progressively smaller landforms that give the region a distinctive appearance.

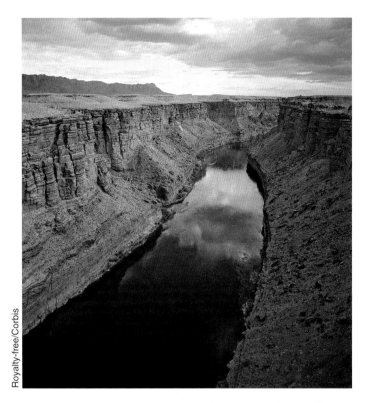

Figure 18.6 A typical canyon in the western United States. Canyons such as this one are very common in the region and evolved as a result of vigorous stream incision. Exposed in the canyon walls are layers of sedimentary rock.

Figure 18.7 Landforms in Monument Valley, Utah. This photograph nicely illustrates the concept that the mesa, buttes, and pinnacles were once part of a uniform rock mass that was extensively eroded over a long time.

The landforms produced by these long-term processes are easy to see if you travel in the Southwest. A **plateau** is a broad platform tens of kilometers across that is elevated up to perhaps 300 m (about 980 ft) above the surrounding terrain. Dissection of a plateau occurs when a stream downcuts vigorously into the underlying rock strata to form a **canyon**, which is a deep, narrow valley with very steep sides (Figure 18.6). The most spectacular canyon, of course, is the Grand Canyon, but numerous smaller canyons carve the landscape. As noted in Chapter 16, this system of canyons is largely associated with the uplift of the Colorado Plateau, which began about 30 million years ago. This uplift increased the slope of streams dramatically, which, in turn, prompted vigorous downcutting since that time (see Figure 16.20a).

With the system of canyons in place, further stream erosion and mass-wasting processes begin to attack the slopes of the dissected plateau. These combined processes gradually leave isolated remnants of the plateau standing alone in progressively smaller and smaller landforms. You can easily visualize this progression in Figure 18.5. The largest of these plateau fragments is a **mesa**, which can be up to several kilometers across. A somewhat smaller remnant is a **butte**, which gradually wears down to form a tower-like landform called a **pinnacle**. A great place to see these kinds of landforms in close proximity to one another is Monument Valley in Utah (Figure 18.7).

The net effect of the progressive dissection and shrinking of plateaus, mesas, buttes, and pinnacles is that the landscape is gradually worn down into a series of rocky surfaces. Where such a surface extends away from an escarpment of the Basin and Range or cliff slope of a mesa or butte, it is called a *pediment*. In other places, the intervening plains and basins are covered with alluvium derived from long-term erosion of hillslopes. A common feature in these areas is a dried lake bed called a **playa** (Figures 18.5 and 18.8). Playas are associated with closed topographic depressions that may temporarily fill with water during wet periods. They subsequently dry out during extended droughts, leaving salty evaporites on the surface. These old lake beds often contain a lengthy record of climate change and sedimentation in the region.

Plateau *A very broad, horizontal surface that is upheld by resistant caprock.*

Canyon *A very steep-sided valley that is cut into bedrock.*

Mesa *A broad horizontal surface, smaller than a plateau, that is upheld by resistant caprock.*

Butte *A steep-sided hill or peak that is often a remnant of a plateau or mesa.*

Pinnacle *A steep-sided, narrow tower that is the final remnant of a plateau, mesa, or butte.*

Playa *A dried lake bed that forms when runoff collects in closed topographic depressions in arid regions.*

Figure 18.8 Playa in Death Valley, California. Lakes form in basins and on plains in the American West because they collect storm runoff. They subsequently dry out during drought periods, leaving this distinctive landscape behind.

Figure 18.9 Arroyos in the Southwest. Arroyos are steep-sided gullies cut into alluvium. Flowing water within this arroyo suggests that a heavy storm recently passed through the area.

In this discussion of southwestern geomorphology, you may wonder how streams can play such an important geomorphic role in such an arid place. Although it might seem that the influence of water would be minimal in dry lands where most streams are ephemeral, streams are actually very important geomorphic agents in these landscapes. One reason for this impact is that when it does rain in arid regions, it often rains very hard for a short period of time, resulting in flash flooding. This high discharge has a disproportionate impact in desert regions because vegetation cover is minimal and hillslopes are not protected by plants. As a result, the landscape can be shaped dramatically when streams do flow. Also remember that these processes have been ongoing for many millions of years and you are seeing the *net effect* of long-term change.

Along with the overall dissected terrain of the Southwest, ephemeral streams also produce some distinctive landforms. One such feature is an **arroyo**, which is a steep-sided gully cut into alluvial sediments (Figure 18.9). Arroyo formation appears to be linked to wet/dry climate cycles related to El Niño and La Niña conditions, respectively, in the Pacific Ocean. Research shows that episodic cutting of arroyos has been especially intense during the past 4000 years.

Perhaps the most interesting landform created by streams in the Southwest and many other arid regions is an alluvial fan. An **alluvial fan** is a depositional landform created where bedload-dominated streams flow out of mountainous or hilly areas onto an adjacent plain. Very simply, as the stream travels down the steep gradient, it can carry the coarse sediment derived from the mountain or hill. When the stream reaches the plain at the base of the hill or mountain, however, it loses energy because the gradient there is much lower. As a result of this reduced gradient, stream power is lost and aggradation occurs. Sediment deposition begins at the fan apex, which is the point where the stream leaves the mountain. Below the apex, the stream sweeps back and forth through time and in so doing creates a semicircular landform that looks much like a hand-held fan from above (Figure 18.10a).

Alluvial fans can occur virtually anywhere that streams flow abruptly from areas of high relief to surfaces that have lower slopes, even in humid areas. However, some of the most impressive fans in the world are those in the southwestern United States. Here, alluvial fans occur along many of the mountain fronts within the region and are sometimes many kilometers wide and have steep gradients. These fans are so large that they often overlap with one another along the mountain front to form a **bajada**. Many of these fans are complex in that they contain interbedded deposits of mud, sand, gravel, cobbles, and even boulders. The sand and gravel layers indicate deposition by a bedload stream, whereas the very coarse rocks indicate prior debris flows. The fine sediments accumulate during major mudflows that periodically occur. Often, the surface of fans consists of a mix of gravels and cobbles (Figure 18.10b).

Arroyo *A deep, steep-sided gully that is cut into alluvium.*

Alluvial fan *A fan-shaped landform of low relief that forms where a stream flows out of an area of high relief into a broad, open plane where the gradient is less and deposition thus occurs.*

Bajada *A landscape consisting of overlapping alluvial fans along a mountain front.*

(a)

(b)

Figure 18.10 Alluvial fans in the southwestern United States. (a) This typical alluvial fan in Death Valley, California, has a fan-like shape where sediment is deposited by water periodically flowing out of the mountains. (b) Surface of an alluvial fan in Death Valley, California. Note the steep slope and coarse texture of this surface.

KEY CONCEPTS TO REMEMBER ABOUT ARID LANDSCAPES

1. Approximately one-third of the Earth's land surface is a mix of arid and semi-arid deserts.

2. Arid lands are associated with subtropical high-pressure systems, rain shadows, and locations that are deep within continents.

3. Deserts and other dry lands are excellent places to study geomorphology because the vegetation cover is relatively sparse, and thus features such as rock structure and landforms are easy to see.

4. In many places of the American West, a distinct progression of landforms is associated with horizontal rock structure. This progression includes plateaus, canyons, mesas, buttes, pinnacles, and plains.

5. A playa is a dried lake bed that occurs in plains and grabens. These lakes temporarily fill with water during wet periods and subsequently dry out during periods of drought.

6. An alluvial fan is a landform created by deposition of stream sediment at the base of steep hillslopes where the gradient drops at the intersection of the basin.

Eolian Erosion and Transport

In addition to the progression of erosional landforms that occurs in many deserts, another prominent geomorphic agent in these areas is **eolian processes**, which are those involving the wind. Although such processes can occur in subhumid and even humid regions, they are most active in desert regions because (1) strong winds are common, (2) a large supply of sand and silt is often available that can be blown, and (3) vegetation cover is minimal and the wind is thus free to erode sediment. This section of the chapter focuses on the way that moving air behaves and the way it shapes the landscape.

The Fluid Behavior of Wind, Erosion, and Sediment Transport

Like water, wind behaves according to specific physical laws associated with fluids. From this perspective, the primary difference between wind and water is that water is much denser than wind. For example, think of how much more buoyant you are in water than in air. Because of this variation in buoyancy, winds must have speeds almost 30 times greater than the speed of water currents to move the same particle.

Both water and wind behave as a fluid with a predictable velocity gradient in the vertical dimension. When air flows over a bare, unvegetated surface, a very thin boundary layer is present where the wind velocity is zero. Above this layer, to a height of about 3 cm, velocity initially increases rapidly, but then slows with additional height.

The ability of wind to erode sediment is similar to that of water in that sediment particles begin to move once a critical threshold velocity is reached. In most circumstances, the mobilization of sand grains begins when wind velocity is greater than about 5 m/sec (~16 ft/sec) at a height of approximately 1 m (3.3 ft) above the ground. This threshold varies depending on the size of any given particle, with small grains more easily moved than larger ones (Figure 18.11). To see how wind moves sediment, consider Figure 18.12 and how a single

---

Eolian processes *Geomorphic processes associated with the way that wind erodes, transports, and deposits sediment.*

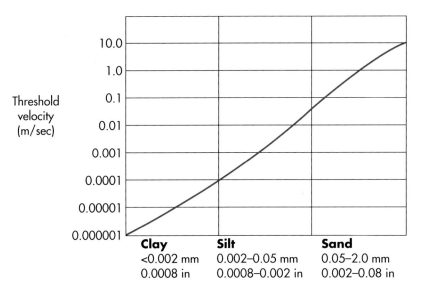

Figure 18.11 Threshold velocities required for wind to carry various-sized particles. The largest sand particles require a wind velocity of 10 m/sec (~33 ft/sec) to be moved by the wind.

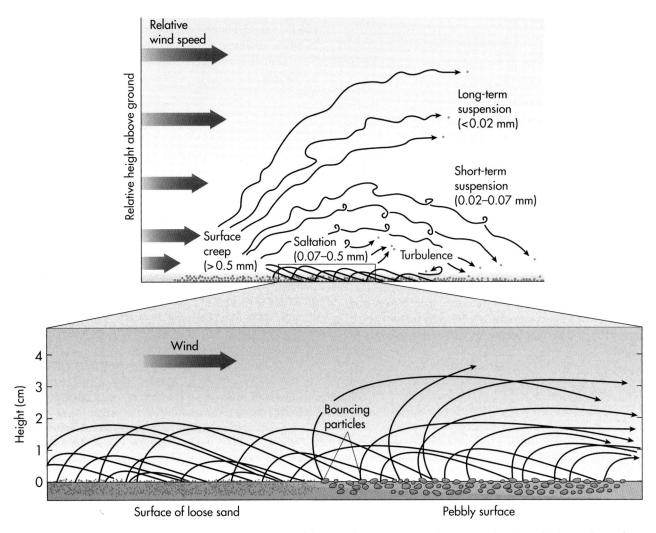

Figure 18.12 Transportation of eolian sediment. Windblown sediment can move by suspension, by saltation, or by surface creep. Note the flow lines to visualize how these processes operate.

Figure 18.13 Eolian dust in suspension. Silt and clay particles can be carried in the air by turbulent flow, as demonstrated here in north central Iowa.

Lynn Betts/NRCS/USDA

particle of sediment begins to move. Initial movement of the particle begins when the threshold velocity is crossed, which causes the sediment to oscillate at the surface beneath the wind. If the particle is less than 0.06 mm—in other words, classified as silt or clay—it will suddenly fly into the air in a nearly vertical fashion. The movement of sediment in this manner is called **deflation**. Although deflation may occur over a very broad area, it may also be localized, forming a bowl-shaped depression called a **deflation hollow**. Regardless, deflated sediments are kept in the air by suspension (Figure 18.13) and may travel hundreds of kilometers before they settle back to Earth.

In contrast to silts and clays, which are carried in suspension, larger particles such as sand either bounce across the ground through the process of *saltation* or roll along the surface in a process called *creep* (Figure 18.12). Saltation begins when a sand grain flies into the air. Because this grain is large relative to silts and clays, it is carried downwind only a short distance and then falls back to the surface under the force of gravity. When the grain strikes the ground, it transfers momentum to other sand grains that, in turn, fly into the air to repeat the process farther downwind. Creeping grains are larger, such as small pebbles, and consistently maintain contact with the surface. Like saltating grains, however, they cause movement in other grains through momentum transfer.

Fine sands carried by the wind can further erode the landscape through the process of **wind abrasion**. In the previous chapter, we investigated the impacts of glacial abrasion, which creates such features as striations in bedrock. In the case of wind abrasion, the process is similar to the sandblasting conducted by work crews to clean buildings and bridges by blowing sand in a compressed stream of air. The process of wind abrasion works much more slowly than mechanical sandblasting and is confined to the space immediately above the ground. Nevertheless, it has the same effect over a long period of time because it smooths and polishes rocks. Many factors influence how rapidly abrasion occurs, including the strength and persistence of the wind, the hardness and angularity of the blowing sand grains, and the resistance of the rock being abraded.

The effects of wind abrasion can be seen in fascinating features called **ventifacts** ("artifacts of the wind"), which are rocks that have clearly been pitted, grooved, or polished (Figures 18.14a and b). In addition, they are typically aerodynamically shaped in the direction of the prevailing winds. Sometimes an entire rock outcrop or series of rock outcrops is streamlined by abrading winds. When this larger-scale abrasion occurs, elongated, wind-sculpted ridges called **yardangs** form (Figure 18.14c). Yardangs are similar to glacial roche moutonée in that they form as a result of two distinct processes, with abrasion dominating on the windward side of the feature and deflation on the leeward side. Yardangs range from a few meters to several kilometers in length and are very prominent in desert regions such as those in Egypt and Iran.

KEY CONCEPTS TO REMEMBER ABOUT EOLIAN PROCESSES

1. Like water, wind is a fluid that behaves in a predictable physical way with respect to flow and velocity.

2. As a fluid, wind is much less dense than water, which means that it must flow at a higher velocity (than that of water) to generate enough force to carry a particle of a given size.

3. Sediment is moved by the wind in three ways, depending on particle size: suspension (silt and clay), saltation (sand), and creep (pebbles and, in strong winds, gravel).

4. Wind erosion works through the processes of deflation and abrasion. Deflation occurs when sediment is physically moved from one place to another. Abrasion is a natural form of sandblasting.

Deflation   *Removal of sediment from a surface by wind action.*

Deflation hollow   *A depression created by wind erosion.*

Wind abrasion   *Erosion that occurs when rock is ground, smoothed, or polished by the wind in a manner similar to sandblasting.*

Ventifact   *An individual rock that is pitted, grooved, or streamlined through wind abrasion.*

Yardangs   *Ridges that are sculpted and streamlined by wind abrasion and deflation.*

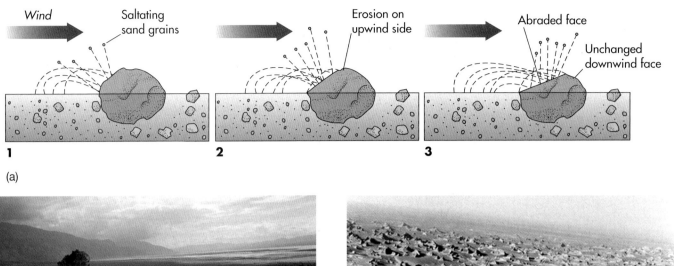

1 Wind — Saltating sand grains

2 Erosion on upwind side

3 Abraded face — Unchanged downwind face

(a)

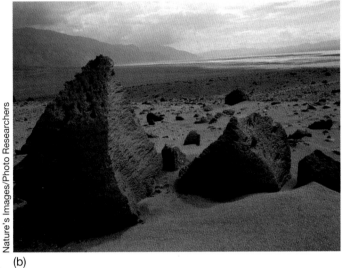

Nature's Images/Photo Researchers

(b)

(c)

NASA

Figure 18.14 Erosional features created by the wind. (a) Ventifacts are individual rocks that have been streamlined by the wind. (b) Ventifacts such as these in Death Valley form when wind sandblasts rocks. (c) These yardangs in southeastern Iran are up to 80 m (~263 ft) high and are some of the largest on Earth.

VISUAL CONCEPT CHECK 18.1

Imagine that you are driving through Death Valley, California, and encounter this strange looking feature. What is it and how did it form?

Alan Arbogast

Eolian Deposition and Landforms

The preceding section discussed how wind behaves as a fluid and how it erodes and transports sediments. These processes produce distinct erosional landforms, such as deflation hollows and yardangs. Once these sediments are eroded they are ultimately deposited someplace else, forming depositional landforms. This part of the chapter focuses on the landforms produced by eolian deposition.

Airflow and the Formation of Sand Dunes

When people are asked to consider the landforms they believe are most associated with wind deposition in desert environments, the typical response is sand dunes. After all, these beautiful and exotic landforms are frequently used in popular culture to epitomize some of the most extreme places on Earth, such as the Sahara Desert. Sand dunes are also fascinating from a scientific perspective because they are great indicators of environmental change. In certain places, they can form or reactivate even when a slight decrease in rainfall or change in wind speed and direction occurs. Given this sensitivity, dunes are fun to study because they can literally change shape overnight, whereas most landforms associated with rivers and glaciers usually require years, decades, or even centuries to change.

The formation of dunes is dependent on the interactions of sand supply, wind speed and direction, and vegetation, as these factors relate to the erosion and deposition of sand. Given that these relationships are complex and not always easily explained, let's look at a simplistic example that is easy to visualize. Imagine first that a bare, sandy patch is exposed to wind flowing over it and is thus a source for a dune that is about to form (Figure 18.15a). This patch might be present because a lot of sand accumulated along a beach or because drought reduced vegetation cover. During the initial phase of sand movement, the airflow over the patch is essentially laminar, which means that air molecules are moving parallel to one another downwind with no turbulence. Now imagine that the moving sand grains encounter some obstruction, such as closely spaced clumps of grass, that results in the deposition of sand and the formation of a small mound. Such accumulation of sand can also be caused by a change in surface texture that slows air speed.

As deposition increases, wind speed at the top of the mound accelerates because the vertical space in which the air is flowing compresses (Figure 18.15b). At the same time, airflow slows and begins to separate in the lee of the mound because the wind can no longer follow the downwind slope. Given this reduced wind speed, sand deposition begins to occur downwind of the growing mound. This separation of airflow increases as the mound grows further, creating an area of erratic air movement in the lee of the mound and even reversal of flow (Figure 18.15c). Although gusts of wind can occur in this area, this zone is relatively calm compared to the flowing air above.

Once the relationships between airflow and topography develop fully, the mound of sand transitions into a fully active sand dune, with distinct patterns of erosion and deposition that result in downwind migration of the landform.

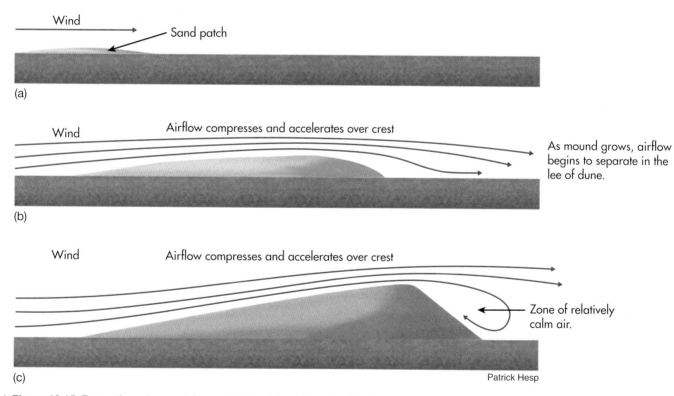

(a)

Wind

Sand patch

(b)

Wind

Airflow compresses and accelerates over crest

As mound grows, airflow begins to separate in the lee of dune.

(c)

Wind

Airflow compresses and accelerates over crest

Zone of relatively calm air.

Patrick Hesp

Figure 18.15 Formation of a sand dune. (a) Sand dune formation begins when a bare patch of sand sediment is eroded by the wind. (b) Moving sand encounters an obstruction or change in surface texture that traps sediment, causing a mound of sand to grow. Airflow is compressed and accelerates on the windward side of the mound. At the same time, airflow on the opposite side of the mound begins to separate and slow. As a result, the mound slowly grows and begins to move downwind. (c) The mound grows further, resulting in full separation of flow and the development of a relatively calm zone in the lee of the dune. Note also that the dune has migrated further downwind.

Figure 18.16 Dune components and migration. (a) When viewed in cross section, a freely active dune has three primary components: the backslope, crest, and slip face. Dunes migrate downwind through the combined processes of sand saltation on the backslope and slumping on the slip face. (b) This large sand dune in central Brazil has a beautiful example of a slip face, which slopes down toward the river. Airflow is left to right across the image. Note the pinnacle in the background and the person (arrow) for scale.

Figure 18.16a illustrates how these processes collectively operate. The best way to visualize these patterns is to begin on the windward slope of the dune where the compressed and accelerated airflow shown in Figure 18.16 causes erosion. Notice in Figure 18.16a that this slope is called the **backslope** and has the shallowest slope of the feature, ranging between about 10° and 15°. Once wind begins to move sand grains on this surface, they bounce up the slope by saltation. When they reach the top of the dune, which is called the **crest**, they fly into the zone of calm air in the lee of the dune. As they fall, they are deposited on an evolving slope that lies at the angle of repose. Remember from Chapter 14 that this angle for sand stands between about 30° and 34°. This slope on a dune is called the **slip face** because sand on this relatively steep slope

sometimes slides downhill under the force of gravity after it is deposited. Due to the combined effects of erosion on the backslope and deposition on the slip face, the dune migrates downwind. This migration is depicted in Figure 18.16a by the dashed red lines that represent the former and future position of the dune. A great example of an actual dune slip face can be seen in Figure 18.16b.

The extent of sand dune migration depends on the balance between erosion on the backslope and deposition on the lee slope. As a result, maintenance of the dune form depends on the forward movement of the entire feature because the amount of backslope erosion is balanced by the amount of deposition on the lee slope. In reality, variations in fluid dynamics and other factors result in complex cross-sectional

Backslope *The gradual slope of a dune that faces the prevailing winds.*

Crest *The highest point of a dune.*

Slip face *The steep slope that lies on the leeward side of a sand dune at the angle of repose.*

Frans Lemmens/Iconica/Getty Images

(a)

Daniel R. Muhs/USGS

(b)

Figure 18.17 Sand seas. (a) The Sahara Desert is best known for the massive sand dunes present in this very dry environment dominated by the Subtropical High pressure system. (b) The Nebraska Sand Hills is the largest sand sea in the Western Hemisphere and contains a wide variety of dune forms. Although isolated areas of blowing sand occur in the area, the dunes are mostly stabilized by an extensive cover of grass. Do you see the similarity in form with the Sahara?

characteristics in most dunes. Overall, the most important factors that influence the shapes of dunes are wind speed, the amount of stabilizing vegetation, and the sand supply—that is, how much sand is present for the wind to move. The amount and kind of vegetation is a critical variable because plant roots hold deposits of eolian sand together. If a sand dune is well vegetated, it tends to remain stable—that is, it stays in one place—unless it is significantly disturbed through human impact, drought, or a major storm. If such an event occurs, bare sand can be exposed and the dune becomes active and begins to move.

Classification of Sand Dunes and Related Landforms

Although dunes acquire a distinctive cross-sectional shape through the combined processes of erosion and deposition, their overall form can vary in a way that promotes classification. On a very broad scale, the largest depositional features related to eolian sand are *sand sheets* and *sand seas*. Sand sheets are horizontal to semi-horizontal bodies of sand that exhibit little or no surface topography. In contrast, sand seas are vast regions where enormous quantities of sand result in a wide variety of dune types. The best-known sand sea is in the Sahara Desert in Africa (Figure 18.17a). Sand seas occur elsewhere in the world, however, such as the Namib sand sea in the Namib Desert on the southwest coast of Africa (Figure 18.1). The largest sand sea in North America is the Nebraska Sand Hills, which is over 32,000 km² (about 12,350 mi²) in size. This sand sea is largely stabilized at the present time due to an extensive cover of grass (Figure 18.17b). Research indicates, however, that it has been very active at various times within the Holocene Epoch (the past 10,000 years) when intensive droughts reduced the vegetation cover.

On a much more local level, individual dunes can be classified based on a variety of factors that depend on the complex interaction of sand supply, the amount of vegetation present, and wind speed and prevailing direction. The variability of these interactions results in at least eight major classes of dunes that can be seen in Figure 18.18. For simplicity, this discussion will be initially subdivided on the basis of dunes that develop in landscapes that are poorly vegetated as opposed to those where plants are common. Another way to consider this relationship is that dunes in poorly vegetated areas are "free," whereas those in vegetated areas are "anchored." As you work your way through this classification system, refer to Figure 18.18.

Beginning with free dunes, let's first examine those that form where sand supply is limited. The first of such dunes to consider is a *dome dune*, which is an individual mound of sand that has an oval form (Figure 18.19a) with no slip face. Dome dunes are temporary features with low relief and are thought to form because the wind has not blown from one direction long enough for a slip face to form, or because wind direction frequently changes. If sufficient time elapses, or winds become unidirectional, dome dunes can be modified into *barchan dunes*. These dunes look like a crescent from above and have a gently inclined windward slope with steep lee side, around which tapering cusps of the dune project downwind (Figure 18.19b). Usually, barchans are isolated dunes that form in areas of strong wind and relatively small amounts of sand. Where winds are weak, however, and massive quantities of sand are present, *barchanoid ridges* may develop (Figure 18.18). These dunes are sinuous, asymmetrical landforms that are transitional to yet another kind of dune called a *transverse ridge*. Transverse ridges are primarily a consequence of unidirectional winds that result in a linear deposit of sand perpendicular to the prevailing wind direction.

Eolian Deposition and Landforms • 491

Dome
Small oval dunes with no
slip face

Barchan
Strong wind in one direction,
limited amounts of sand

Barchanoid ridge
Formed from rows of merged barchans

Transverse
Weak wind in one direction,
large amounts of sand
deposited perpendicular to wind

Longitudinal
Opposing winds flowing in similar
directions, two slip faces, aligned with
wind

Star
Multidirectional winds, largest dunes
grow tall instead of moving

Reversing
Two winds in opposite directions,
changes shape periodically

Blowouts
Localized deflation basins in partially
vegetated landscape

Parabolic
Arms stabilized upwind by vegetation,
often start as semicircular blowouts
that elongate downwind

Figure 18.18 Sand dune classification. Eight major varieties of sand dunes occur, with form depending on wind direction, sand supply, and amount of vegetation cover. Arrows represent the direction of prevailing winds.

(a)

(b)

Figure 18.19 Dome dunes and barchan dunes. (a) A typical dome dune (foreground). Note the oval shape, lack of a slip face, and low relief of this dune. (b) A well-developed barchan dune. Airflow is from left to right across the image. Note the crescent form, lee slope, and tapering cusps.

(a)

(b)

Figure 18.20 Longitudinal dunes and star dunes. (a) A nicely developed longitudinal dune. Such a dune forms as a result of two dominant wind directions. (b) Star dunes such as these in the Sahara desert form when winds are multidirectional.

Another kind of dune that forms in barren landscapes is a *longitudinal dune* (Figure 18.20a). These dunes have the same linear form as transverse ridges, but differ because they develop along the axis of two prevailing wind directions. In areas of multidirectional winds with very little vegetation, *star dunes* may form (Figure 18.20b). Star dunes can grow to be very large. In the Sahara, for example, star dunes may be several hundred meters in height and many kilometers in diameter. Rather than migrate, star dunes simply grow in height. A transitional dune between star dunes and transverse ridges is the *reversing dune*, which forms where two winds from nearly opposite directions are balanced with respect to strength and duration. As a result, a second slip face periodically develops.

In contrast to free dunes, which actively migrate, anchored dunes are somewhat more stable landforms. The best known anchored dunes are *blowout* and *parabolic* dunes (Figure 18.18). Both of these dune types are essentially deflation features with slip faces that slope in many directions. A typical blowout is a semicircular to circular deflation basin with lateral erosional walls (Figure 18.21). Sand eroded from the deflation basin is moved immediately downwind to form a depositional lobe. If the blowout continues to enlarge and elongate downwind, a parabolic dune will form with U- or V-shaped arms that point upwind (Figure 18.18). These arms are usually stabilized by vegetation, which causes the dune to further elongate as it migrates downwind.

Deposition of Windblown Dust

Although sand dunes are the most common feature associated with eolian processes, deposition of windblown dust can also occur in ways that shape landscapes in distinctive, yet subtle ways. Such sediment consists of fine-grained clays and silts that are deflated from source areas and subsequently transported in suspension before falling back to Earth's surface. These dust deposits may travel a short distance before they are deposited, or they may be carried hundreds of kilometers.

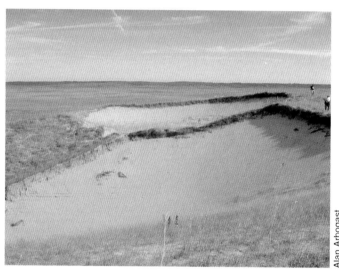

Figure 18.21 A typical blowout. Blowouts like these in the Nebraska Sand Hills form when vegetation thins and strong winds deflate the core of a sandy deposit.

One of the most interesting landforms created by dust deposition in arid environments is **desert pavement**, which is a flat surface mantled with coarse pebbles, gravel, and even larger rock clasts (Figure 18.22). At first glance it defies logic that desert pavement could be the result of eolian deposition because it is impossible for wind to carry the coarse-textured sediment at the surface. For some time, it was believed that these surfaces develop due to deflation of fine sediments away from host deposits that also contain coarse particles. In this traditional model, the fine particles are blown out of the sediment matrix,

Desert pavement *A resistant, pavement-like surface created when fine particles blow away and coarse sediment such as pebbles and gravel is left behind.*

Figure 18.22 Desert pavement. (a) A typical desert pavement in the southwestern United States. Note the very coarse texture of this surface.

leaving a coarse lag at the surface that could not be moved by the wind.

As a result of detailed research, it is now believed that desert pavement forms in association with dust deposition rather than deflation (Figure 18.23). This development occurs because dust accumulates on the coarse-textured desert surfaces and settles in the gaps between the rock clasts when the sporadic rains move it downward. In the early phase of pavement development, the eolian sediment forms a crust that adheres to the bottom of the clasts. Clays within the sediment subsequently swell and shrink with wetting and drying over time, respectively, causing the rock clasts to ride upward. Salt crystallization within the sediment probably also plays a role in this lifting process. With the addition of progressively more dust, the zone of accumulation below the clasts thickens further, forcing the pavement zone even higher and producing a smoothing effect on the landscape. This thickened dust zone, which is classified as a form of soil A horizon, is interesting in and of itself because it contains microscopic pores called vesicles. Desert pavements are usually a mix of dark colors, collectively referred to as *desert varnish*, which consists of fine clays and bacteria that coat rock surfaces over many decades.

Another form of eolian deposit is called **loess** (a German word that, in the United States, is usually pronounced "luss"), which consists of fine-grained, windblown silt. Loess is typically highly calcareous (as much as ~40% $CaCO_3$) and yellowish-tan in color (buff-colored). Loess often originates within the same region as the sand that forms dune fields. Given that silt can be carried far in the air by suspension, however, it is sorted out from the sand and subsequently deposited great distances from its original source, leaving the dune fields behind (Figure 18.24). Loess also originates in stream valleys after silt is deposited on floodplains during floods. A great deal of loess

---

Loess *Windblown silt.*

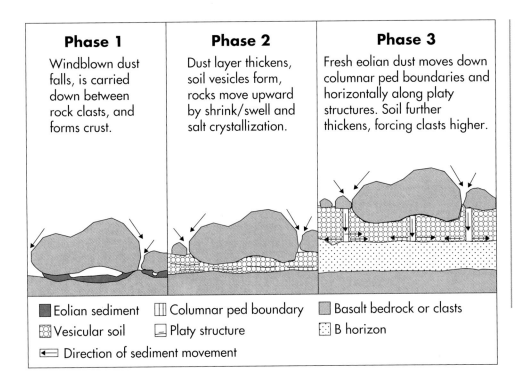

Phase 1	Phase 2	Phase 3
Windblown dust falls, is carried down between rock clasts, and forms crust.	Dust layer thickens, soil vesicles form, rocks move upward by shrink/swell and salt crystallization.	Fresh eolian dust moves down columnar ped boundaries and horizontally along platy structures. Soil further thickens, forcing clasts higher.

■ Eolian sediment ⫼ Columnar ped boundary ▨ Basalt bedrock or clasts
▧ Vesicular soil ⬚ Platy structure ⬚ B horizon
⬅ Direction of sediment movement

Figure 18.23 Formation of desert pavement. Desert pavement forms through the process of dust deposition. In Phase 1, fine dust falls between coarse rock clasts and forms a crust. As time passes (Phase 2) and more dust accumulates, the crust thickens, soil vesicles form, and rocks move upward by a combination of shrink/swell and salt crystallization. In Phase 3, fines move down columnar ped boundaries and horizontally along platy structures. The soil thickens further and the clasts move further upward. (*Source*: K. Anderson, S. Wells, R. Graham. "Pedogensis of Vesicular Horizons, Cima Volcanic Field, Mojave Desert, California." *Soil Science Society of America Journal* 2002; 66:878–887. Reprinted by permission, ASA, CSSA, SSA.)

Figure 18.24 The process of eolian sorting. After sediments are deflated from a source area, silts and clays are transported great distances in suspension. Coarse sands, in contrast, travel a relatively short distance, forming dune fields.

also owes its origins to the thick deposits of glacial till and outwash deposited during the Ice Age. Following the deposition of these sediments by the glacier, the ice retreated, exposing surfaces that were unvegetated and easily deflated by strong winds. These winds picked up the silts contained within these deposits and carried them downwind.

Loess is a distinctive deposit for two reasons. First, it is often found over very broad areas as a mantle of windblown sediment. Approximately 10% of the Earth's land surfaces are covered in this way, including much of the central United States and parts of Argentina, Russia, and China (Figure 18.25). These areas are some of the most productive agricultural regions on

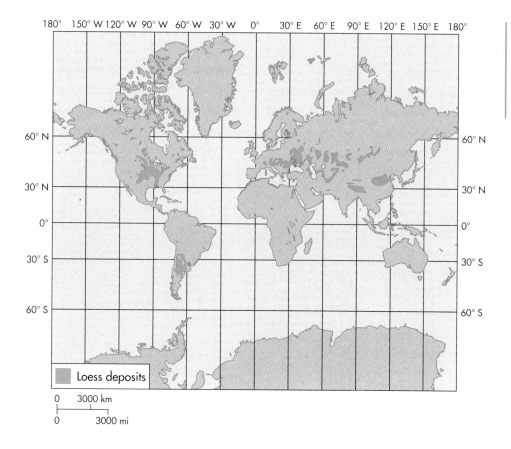

Figure 18.25 Significant loess deposits in the world. Loess covers about 10% of the Earth's landmass and about 30% of the United States. Much of the loess in the United States originated from glacial sediments that were subsequently deflated.

Since north is at the top of the photograph, the orientation of these barchanoidal transverse dunes indicates that prevailing winds are southwesterly. These winds are moving the dunes in a northeasterly direction toward the upper right of the photograph. The backslopes face the direction from which the winds are coming, whereas the slip faces lie in the shadows. The presence of these dunes indicates that the available sand supply is relatively low, causing distinct dunes to form rather than a sand sea.

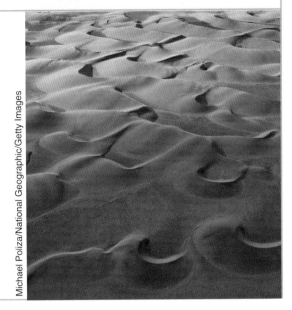

Michael Poliza/National Geographic/Getty Images

● DISCOVER...

SAND DUNES ON MARS

You may know that a very active research program exists to learn about the environment and geologic history of Mars. The reason for this interest is that Mars and Earth are about the same size, are relatively close to one another, and are about the same age. Evidence suggests that flowing water once existed on Mars, which means that the planet may have had an atmosphere somewhat similar to Earth's at one time. It also means that some form of primitive life may have once existed there, or may still exist in some obscure place.

One of the ways that Mars is being studied is with satellite remote sensing. This program is revealing much about the character and history of Mars, including what you can see in this image. Do you recognize these features? They are barchan dunes. These dunes mean that wind blows sand on the planet, just as it does on Earth. It also means that the amount of sand available to blow is somewhat limited. Of particular interest regarding these dunes are the randomly spaced dark patches. If these dunes were on Earth, you might assume that these spots are patches of vegetation. But, of

course, vegetation does not exist on Mars. So what are they? The best current hypothesis is that the dunes are covered with a light frost in the Martian winter and that the spots represent places where the frost sublimates (turns to vapor) as summer approaches. Fascinating, yes?

NASA

Now that we have discussed the basic components of eolian processes and associated landforms, let's visualize these concepts in an animated format. To do so, go to the Geo Media Library and open *Eolian Processes and Landforms*. This animation describes the fluid flow of air and how it shapes the landscape. In this animation, you will see how abrasion shapes rock outcrops; how the variables of sand supply, wind direction, and vegetation influence the formation of dunes; and how loess develops. After you complete the animation, be sure to answer the questions at the end to test your understanding of this concept.

Earth because loess is calcareous and has high water-retention capacity. Second, loess maintains nearly vertical faces upon exposure (Figure 18.26). This durability occurs because the silt grains are strongly attracted to one another and are bound together by the calcium carbonate contained within it.

KEY CONCEPTS TO REMEMBER ABOUT EOLIAN LANDFORMS

1. The best-developed depositional eolian landforms are sand dunes. Sand dunes occur in a variety of forms, determined by wind direction and strength, sand supply, and the amount of stabilizing vegetation.

Figure 18.26 Loess landscape in south central Nebraska. This roadcut exposes a thick sequence of loess that largely accumulated during the Wisconsin glaciation when the climate in the region was drier and strong winds blew. The dark band in the exposure is a buried soil that formed at the end of the ice age when the landscape stabilized. Above the soil is loess that accumulated during the early Holocene when the climate was warmer and drier. The Platte River valley lies in the background. The hills between the exposure and river valley are also underlain by loess.

2. The wind is an efficient sorting mechanism, which means that sediments are separated by size. Clay-sized particles travel the farthest from the source area. Silts move an intermediate distance, and pebbles and gravels remain in place because they are too large to move by wind.

3. Another eolian deposit is loess, which is windblown silt that is carried great distances in suspension before it settles to the ground. Loess covers much of the Earth and is very fertile.

Human Interactions with Arid Landscapes and Eolian Processes

In Chapter 16 we examined how humans interact with stream systems, such as in urban environments or by building dams. Humans can also impact the eolian landscape in many ways. In several examples around the world, people have contributed to the degradation of a landscape through the process of **desertification**, that is, transforming a formerly vegetated landscape into one that is relatively barren and highly susceptible to wind erosion. This process is enhanced in marginal, semi-arid landscapes where the vegetation density is already low. Figure 18.27 shows extensive regions of Earth that are susceptible to desertification. Until recent times, people have generally avoided these kinds of landscapes because rainfall to sustain crops and animal herds is unpredictable. However, with increasing population pressure, more and more people have been moving into these formerly uninhabited places in search of land that can be cleared for farmland. In the process of this development, the impact on the landscape has been enormous. As you will see in the following discussion, human-induced desertification can have catastrophic consequences.

Desertification *The process through which a formerly vegetated landscape gradually becomes desert-like.*

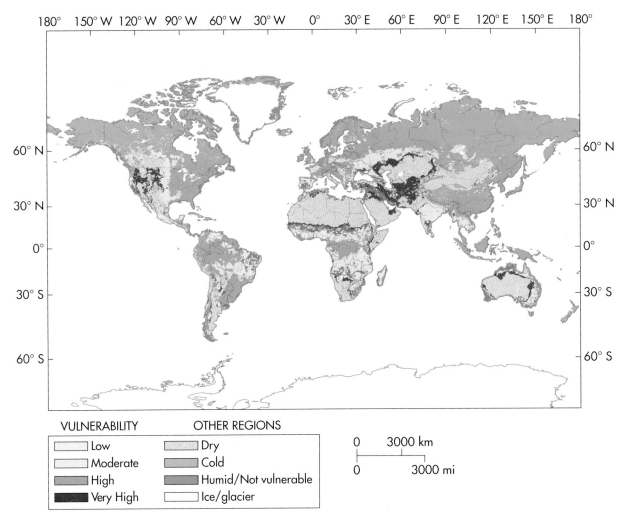

Figure 18.27 Regions on Earth prone to desertification. Much of the Earth's land surface lies in semi-arid to arid climate zones where desertification is a potential hazard.

Desertification in the African Sahel

One place where human-induced or human-enhanced desertification has occurred is the Sahel region of Africa. The Sahel is a narrow band of latitude that lies between about 15° and 18° N and extends from the country of Senegal on the west coast to the country of Sudan on the east coast (Figure 18.28a). Recall from Chapter 9 that this part of Africa has a distinct wet/dry precipitation cycle that depends on the seasonal passage of the Intertropical Convergence Zone (ITCZ). Consequently, the wet season occurs from late June to mid-September when the ITCZ is in the region. In contrast, the remainder of the year is quite dry because of the dominance of the Subtropical High (STH) Pressure System. In the context of vegetation, the region lies between the Sahara Desert to the north and the tropical savanna (grass and open forest) to the south. In other words, this region is an **ecotone** because it lies between two distinct ecosystems and serves as the transition from one to the other. The

Ecotone *The transition between two distinct ecosystems that contains species of each area.*

photograph in Figure 18.28b gives you a feel for the character of the place.

Prior to the 19th century, the Sahel was largely home to nomadic pastoralists and small-scale sedentary farmers. The nomads lived mostly in the driest (northern) part of the region, which bounds the Sahara Desert to the north, whereas the farmers lived in the humid southern part of the region. The nomads maintained small herds of cattle and goats and followed a practice of moving their herds to find grass that their cattle could graze. Given the seasonal movement of the ITCZ and associated rainfall, their lifestyle evolved into a pattern of annual north–south migrations across the region; in other words, they moved their cattle to follow the rain. The farmers, however, instead tended to stay in one place and had evolved a subsistence strategy that allowed them to be successful in this marginal environment. These strategies included staggering their plantings to have a continuous source of food, growing grains that mature quickly, and farming near a source of water such as river valleys.

Overall, the people in the Sahel had evolved a lifestyle that was compatible with the carrying capacity of the landscape;

(a)

(b)

Figure 18.28 The Sahel region in Africa. (a) This region is an ecotone that lies between the grass and forested regions to the south and the Sahara Desert to the north. (b) The Sahel landscape. Note the widely-spaced trees and intervening grass that makes good pasture during the wet season. It can easily be overgrazed, however, leading to landscape disturbance.

that is, the number of people living there did not exceed what the landscape could support without degradation. This sensitive relationship began to change in the late 1800s when European nations (primarily Great Britain and France) parceled the region into distinct colonial entities. With this organization, artificial boundaries arose that became barriers to the annual cycle of nomadic migration. In addition, the economic focus changed from living within the regional carrying capacity to exploiting raw materials for consumption in Europe. Although the colonial era ended in the mid-20th century with the advent of African nationalism, the political boundaries, and their effect on human behavior, remained in place. People became increasingly sedentary, which resulted in the cultivation of fragile soils, overgrazing, and the removal of trees for firewood. In addition,

overall population increased because people from more humid regions in the south migrated to the Sahel in search of places to cultivate.

As a result of the combined factors of boundary establishment and increasing population, the Sahel landscape began to degrade in the early 20th century because the carrying capacity was greatly exceeded. Landscape degradation and desertification of the Sahel reached a catastrophic level during the late 1960s and early 1970s. As you can see in Figure 18.29, this period coincides with a major period of drought that has basically lasted to the present time. Coupled with the preexisting human-induced degradation that had already occurred on the landscape, this drought resulted in widespread desertification in the region.

Figure 18.29 Annual rainfall from June through October in the Sahel, 1898–2010. In this graph, 0 represents the mean amount of rainfall during this period. Positive numbers above the mean are annual amounts of rainfall that are above the mean. Conversely, negative values below the mean represent below-average amounts of precipitation. Until the late 1960s, the region was relatively wet. Drought conditions have dominated since that time. (*Source*: NOAA.)

Human Interactions with Arid Landscapes and Eolian Processes • 499

Figure 18.30 Atmospheric dust off the northwest coast of Africa. The combination of drought and human environmental impacts in the Sahel may have increased the incidence of dust blowing off of the African continent.

In association with this drought, hundreds of thousands of people and countless numbers of livestock died of starvation; it has been estimated that approximately 100,000 people died in 1973 alone. The amount of wind erosion has also increased over the region, frequently resulting in the transportation of large amounts of atmospheric dust (Figure 18.30). Some researchers believe that this increased eolian dust is a direct result of the human impact on the landscape. Others believe, however, that the dust is a result of the increased occurrence and intensity of tropical waves (low-pressure systems) in the region.

Desertification in the Great Plains of the United States: The Dust Bowl

People in the United States sometimes think that catastrophic environmental impacts such as those in the Sahel occur only in poor countries where most people have relatively little education and economic opportunity. In fact, within the so-called developed world examples abound of human impacts on landscapes that resulted in dire environmental consequences. In the United States an excellent such example is the famous *Dust Bowl* in the 1930s.

The Dust Bowl region is located in the western Great Plains and includes parts of Colorado, Kansas, Nebraska, New Mexico, Oklahoma, and Texas (Figure 18.31). As you know by now, the climate of this region is semi-arid due to the rain shadow of the Rocky Mountains. In response to this climate, the natural vegetation is predominantly short-grass prairie.

Until the latter part of the 1800s, this region was largely uninhabited, except for small bands of Comanche and Pawnee tribes. Given the hunting and gathering subsistence lifestyle of these people, they had very little impact on the landscape because they moved frequently from one place to another. Beginning in the 1870s, however, large numbers of European Americans from the eastern United States began to settle in the region in association with the overall pattern of westward expansion and the Homestead Act of 1862. In this context, each family of "homesteaders" was given a free quarter section (see Figure 2.9b), or about 65 hectares (160 acres), of land with the understanding that it would be developed in some way. Given the high fertility of the Mollisols in the region, the area was an attractive place to grow wheat. Although drought was a concern, many early settlers believed the claim by land speculators that *rain follows the plow*; in other words, if the land was plowed,

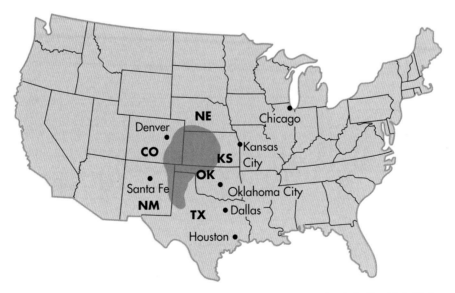

Figure 18.31 The American Dust Bowl. Most native grasslands in the High Plains were plowed in the late 19th and early 20th centuries during a settlement boom. Subsequently, a severe drought in the 1930s impacted a very large area that included parts of Nebraska, Kansas, Colorado, Oklahoma, New Mexico, and Texas.

it would rain more in this semi-arid land because water stored within the soil would be released and thus fall immediately back to Earth. Fueled in part by this myth, as well as a strong sense of rugged individualism, the settlers gradually converted progressively larger tracts of native grassland to farmland by plowing it up (Figure 10.32).

As settlement in the area progressed, the first few decades were sufficiently wet to give farmers reason for hope in their economic future. Many even believed that it was indeed raining more because much of the grassland had been plowed. A drought in the early 1890s caused some concern, but was not overly extreme because favorable environmental conditions quickly returned and wheat harvests rebounded. Confidence grew, which promoted further plowing and planting. With the advent of World War I in the 1910s, wheat prices surged due to crop shortages in Europe, and even more land was plowed and converted to wheat acreage. After the war ended, however, wheat prices crashed. In response, farmers converted even the marginal lands to wheat in an effort to make up the price differential. The net effect, however, was to reduce wheat prices even more because supply drastically exceeded demand. By this point, virtually the entire prairie had been plowed, and the land lay bare of the grassland that had protected it for thousands of years.

Disaster struck in the early 1930s when the region was crippled by a searing drought that lasted for several years. Precipitation in many parts of the region was less than half of the ~50 cm (~20 in.) per year that normally fell and crops failed everywhere. Such an intense drought would have impacted the landscape, perhaps through the remobilization of small dune fields, even if it had occurred during the pre-settlement period when the ground was covered with grass. On the other hand, the deep-rooted native grasslands would have protected the soil in most places. Given that most of the grass had been plowed under in the preceding few decades, however, the soils were utterly exposed and thus easily deflated by the very high winds that accompanied the drought.

Given the exposed soils and high winds, the region was pounded by crippling dust storms (Figure 18.32) throughout much of the 1930s. Such storms repeatedly blasted an area about 400,000 km^2 (~155,000 mi^2) in size and were so intense that they were called *black blizzards*. Perhaps 800 million tons of dust was blown from the Plains and carried as far as Chicago and Washington, D.C. The dust was so thick in Kansas City that streetlights were often left on during daylight hours. Perhaps the worst of it occurred on April 14, 1935 when 20 such storms occurred throughout the region, turning day into night and reducing visibility to less than 2 m (~6 ft). No wonder this day is now referred to as Black Sunday. Overall, about 350 such storms occurred from 1937 and the first nine months of 1937. Can you imagine what it must have been like to live there?

The overall result of the drought was to desertify the landscape in much of the region. From a cultural and economic standpoint, the impact was devastating, with many families left literally on the verge of starvation. Many hundreds of people also contracted *dust pneumonia* from inhaling too much fine

Figure 18.32 A black blizzard during the Dust Bowl. Dust storms in the Great Plains, such as this one in Oklahoma, were common during the 1930s. These massive storms resulted from the combined impacts of drought, high winds, and plowed soils.

sediment, with young and old alike succumbing to the condition. Although many people chose to stay and hope for rain, large numbers moved out of the region in search of a better life. Over 200,000 of these refugees migrated to California, as depicted in John Steinbeck's famous novel, *The Grapes of Wrath*. By 1940 over two million people had left the Plains states, with some counties losing over half the population.

The environmental catastrophe associated with the Dust Bowl prompted the federal government's first systematic effort to promote environmental conservation. As discussed in Chapter 11, the U.S. government established the Soil Erosion Service in 1933 (now the NRCS) to promote soil conservation in the region. Farmers were taught to plow and terrace their fields in ways to control soil erosion, as well as to plant *windbreaks* on the edge of fields to block the wind (Figure 18.33). They also

Figure 18.33 A conservation windbreak. Windbreaks such as this one are rows of tall trees planted along fields that block the wind from blowing directly across the ground.

Human Interactions with Arid Landscapes and Eolian Processes • 501 began systematically to employ *conservation tillage*, which is a shallow plowing method that leaves about 30% of the previous year's crop residue on the ground. This residue increases the surface roughness of the ground, which helps protect the soil from wind erosion and conserve water in the soil.

Although the Dust Bowl occurred about 80 years ago, its effects echo to the present day. These echoes are particularly strong as of this writing because the region is experiencing a severe drought, one that may ultimately rival the drought years in the 1930s. Current levels of annual precipitation are about half the norm, and agriculture has been hard hit. One major advantage that modern farmers have over their counterparts in the *Dirty '30s* is the soil conservation measures now employed to protect fields. In addition, many modern farmers use central pivot irrigation, which, as discussed in Chapter 15, can provide a steady flow of water to crops, even during a drought. On the other hand, climate models discussed in Chapter 9 suggest that the region will likely become even more drought prone in the future as global warming intensifies. In this context, how will farmers respond if and when groundwater supplies disappear, as they very well may this century? Is it possible that another Dust Bowl will occur in the future? Only time will tell.

KEY CONCEPTS TO REMEMBER ABOUT DESERTIFICATION OF SEMI-ARID LANDS AND SAND DUNES AS INDICATORS OF CLIMATE CHANGE

1. Large areas of Earth's surface are prone to desertification, which is the transformation of formerly stable (by vegetation) landscapes to desert-like conditions.

2. The Sahel region of Africa is an ecotone that lies between the Sahara Desert and the tropical rainforest. This region has undergone extensive desertification in the past 30 years due to human impacts and extensive drought.

3. The American Dust Bowl is a large area in the central United States that was extensively desertified during the 1930s due to a combination of human settlement patterns and extensive drought.

4. Soils in semi-arid lands can be protected to some extent through such measures as windbreaks and conservation tillage.

SUMMARY OF KEY CONCEPTS

1. Arid and semi-arid environments collectively comprise over 30% of the Earth's land surface. Arid landscapes are excellent places to study geomorphology because elements such as rock structure are easily seen. In places where the structure is horizontal, desert landscapes often contain a progression of erosional landforms, including plateaus, mesas, buttes, and pinnacles.

2. Eolian processes involve the shaping of the Earth due to the wind. The wind moves sediment in one of three ways: (1) fine grains such as silts and clays are carried great distances in suspension; (2) sand grains bounce along the ground in the process of saltation; and (3) heavier particles roll along the ground by creep.

3. The wind shapes the landscape through the combined processes of erosion and deposition. The formation of yardangs is an example of wind erosion. In contrast, sand dunes form when windblown sand is deposited and shaped.

4. Sand dunes come in many different shapes and sizes. The shapes of dunes depend largely on the supply of sand, the amount of vegetation, and the prevailing wind speed and direction.

5. Loess is windblown silt carried a great distance by suspension. Thick deposits of loess occur in many places around the world, with most accumulating during glacial cycles.

6. Human-induced desertification is a major environmental issue facing the world today. Desertification occurs when the stabilizing vegetation on the landscape is reduced through human activity. This process exposes soils to the wind, which causes severe soil erosion and the transport of eolian sediment.

CHECK YOUR UNDERSTANDING

1. What are the three large-scale environmental variables that contribute to the presence of deserts? What are the processes associated with each of these factors, and why do they produce arid environments?

2. Discuss how normal faulting is responsible for the dominant landforms of the Basin and Range Province.

3. Beginning with a plateau, describe the sequence of landform evolution where horizontal rocks of differing resistance influence the regional geomorphology.

4. What is a playa, and where is a good place to find one?

5. Describe the formation of an alluvial fan.

6. How does flowing air compare to flowing water? How is it similar and how is it different?

7. In the context of eolian processes, describe suspension, saltation, and creep. How do these various processes function, and what kind of sediment is associated with each one?

8. What is the difference between deflation and abrasion?

9. Describe how ventifacts and yardangs form.

10. Discuss the early stages of dune formation and the nature of airflow during this time.

11. Sketch and describe a typical dune cross section and identify the three important components. How does sand move in these places, and why do these processes collectively result in dune migration?

12. How do the factors of sand supply, wind (speed and direction), and vegetation influence the behavior and form of sand dunes? Provide at least three examples.

13. What is loess, what is its source, and how is it transported?

14. Describe the process of desertification. Why are marginal landscapes most likely to experience human-induced desertification?

15. Name and describe two conservation techniques used to protect soils from wind erosion.

ANSWERS TO VISUAL CONCEPT CHECKS

VISUAL CONCEPT CHECK 18.1

This bizarre looking rock feature is a great example of a ventifact. It formed due to repeated sandblasting and abrasion caused by windblown sediment striking it.

VISUAL CONCEPT CHECK 18.2

Since north is at the top of the photograph, the orientation of these barchanoidal transverse dunes indicates that prevailing winds are northeasterly, that is, from the north and east. These winds are moving the dunes in a somewhat southwesterly direction toward the lower left of the photograph. The backslopes face the direction from which the winds are coming, whereas the slip faces lie in the shadows. The presence of these dunes indicates that the available sand supply is relatively low, causing distinct dunes to form rather than a sand sea.

CHAPTER NINETEEN

COASTAL PROCESSES AND LANDFORMS

The previous three chapters investigated fluvial, glacial, and eolian processes and their distinctive landforms. Now we turn to the various processes and landforms that occur along coastlines. Although these processes affect a very small part of the Earth's surface, they create some of the most distinctive and popular landscapes you see.

Why should you be interested in coastal processes and landforms? One important reason is that you may very well live along or near a coastline, as do a majority of people in the United States (and indeed, the world). People are attracted to coastlines for a variety of economic, recreational, and lifestyle reasons and are affected by the geomorphic processes that occur along these narrow strips of land. In turn, people influence the behavior of coastlines in dramatic ways through development. Thus, understanding the geomorphology of coastlines is important, not only for gaining a better appreciation of this landscape, but also for becoming a good steward for these areas.

James Morgan/Robert Harding World Imagery/Getty Images

Are you attracted to this landscape? If so, you are certainly not alone. Coastlines are perhaps the most exotic landscapes on Earth, and people love to live near or visit them. This chapter focuses on the oceans of the world and the way they shape the landscape along coastlines.

CHAPTER PREVIEW

Oceans and Seas on Earth

The Nature of Coastlines: Intersection of Earth's Spheres

 Tides

Coastal Landforms

 Waves and Coastal Erosion

 Longshore Processes and Depositional Coastlines

Human Interactions with Coastlines

 Evolution of the Louisiana Coastline

LEARNING OBJECTIVES

1. Distinguish among oceans, seas, and gulfs on Earth.

2. Describe the nature of eustatic sea-level changes during the past 150,000 years.

3. Discuss how waves form and the ways in which they interact with coastlines.

4. Explain how tides form and the variables that influence them.

5. Compare and contrast the various erosional coastal landforms.

6. Describe the character of coastal depositional landforms and how they form.

7. Discuss the various ways that humans interact with (and manage) coastlines.

Oceans and Seas on Earth

Before we begin discussing various coastal processes, let's take a closer look at the character of the world's oceans and seas because it provides the context in which to view the shore. Figure 19.1 shows the names and locations of most of the notable large water bodies on Earth. The term **ocean** refers to the enormous body of salty water that covers about 71% of Earth. Five geographic divisions of the world's oceans are recognized. The largest is the Pacific Ocean, which encompasses a third of Earth's surface. This ocean is significantly larger than Earth's entire landmass, having an area of 179.7 million km^2 (69.4 million mi^2). The next largest ocean is the Atlantic, followed by the Indian Ocean, then the recently recognized (in 2000) Southern Ocean, and the Arctic Ocean. The Arctic Ocean covers an area of about 14,090,000 km^2 (5,440,000 mi^2), which is slightly less than 1.5 times the size of the United States. The Southern Ocean was recognized because it encircles Antarctica and has distinctive circulatory patterns.

We use the term *ocean* to refer to these large water bodies. The terminology for smaller areas of water more closely associated with land is inconsistently applied. In general, the next largest bodies of water are considered to be **seas**, which are subdivisions of oceans and partially enclosed by land. For example, the Mediterranean Sea (Figure 19.1) is mostly surrounded by the European, Middle Eastern, and African landmasses. In fact, it opens to the Atlantic only in its far western end.

After seas the next largest body of water is considered to be a **gulf**, which is a smaller arm of an ocean or sea that is also partially enclosed by land. Unfortunately, use of the terms *gulf* or *sea* can be misleading because some named gulfs are actually larger than some seas. The Gulf of Mexico, for example, is much larger at 1.6 million km^2 (615,000 mi^2) than the Black Sea in Eastern Europe, which is 422,000 km^2 (163,000 mi^2).

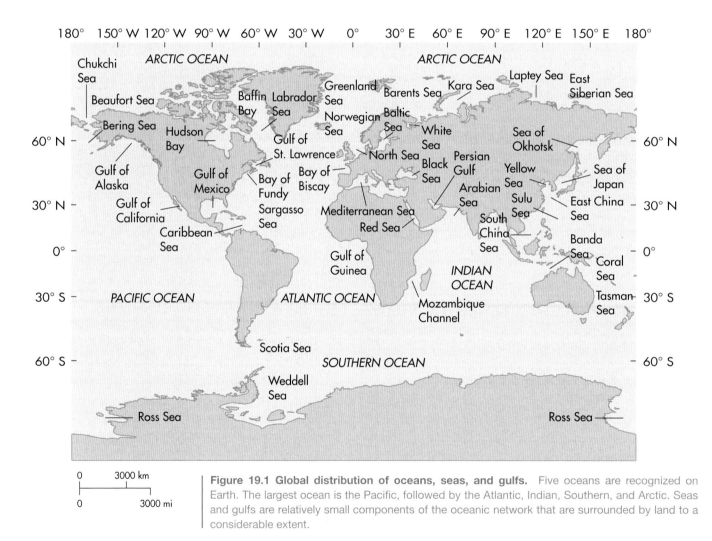

Figure 19.1 Global distribution of oceans, seas, and gulfs. Five oceans are recognized on Earth. The largest ocean is the Pacific, followed by the Atlantic, Indian, Southern, and Arctic. Seas and gulfs are relatively small components of the oceanic network that are surrounded by land to a considerable extent.

Ocean *The entire body of salt water that covers about 71% of Earth.*

Sea *A subdivision of an ocean that is partially enclosed by land.*

Gulf *A relatively small body of salt water that is surrounded by land on three sides and opens to a sea or ocean.*

A distinctive characteristic of a gulf is that it usually has a recessed shoreline that opens outward to a larger body of water. Notice in Figure 19.1 how the shoreline of the Gulf of Mexico curves northward and how the Gulf opens to the Caribbean Sea east of Central America. On a still smaller level, a **bay** is a relatively small indentation on the coast that is directly connected to an ocean, sea, or gulf.

Although oceans, seas, and gulfs differ by size and their relationship to land, they are essentially interconnected and therefore have the same basic chemical composition. This chemistry is a result of complex interactions among several factors, including the atmosphere, seawater, minerals, sediments, and the myriad of organisms living in the water. Given that water is an excellent solvent, the world's oceans and seas contain a wide variety of dissolved solids, such as chlorine, sodium, and magnesium ions, among others. The term **salinity** refers to the concentration of dissolved solids in seawater and is most commonly expressed in parts per thousand. Overall, global salinity varies between 34‰ and 37‰. Water that exceeds a salinity of 35‰ is considered **brine**, whereas water with salinity less than 35‰ is called **brackish**.

Figure 19.2 A typical coastline in northern California. Coastlines such as this one at Big Sur are where the hydrosphere, lithosphere, and atmosphere meet in a very big way.

The Nature of Coastlines: Intersection of Earth's Spheres

The primary distinguishing characteristic of coastlines is that they are places where landmasses and large water bodies intersect (Figure 19.2), often for hundreds or even thousands of kilometers. Although coastlines are usually associated with oceans and seas, they also occur around large lakes such as the Great Lakes of North America. Coastlines are especially noteworthy because they represent the intersection of three major Earth spheres, namely the hydrosphere, lithosphere, and atmosphere.

Coastlines come in all shapes and sizes. Some coasts have wide, sandy beaches, whereas others feature rocky bluffs and cliffs. Along some coastlines, mountains extend straight down to the water. Because of the wide range of environments associated with coastlines, some are passive and change slowly, whereas others erode vigorously and are in a near-constant state of adjustment.

Processes That Shape the Coastline

A variety of physical processes influence the shape of a coastline, with most directly related to how water moves along the shore. This movement occurs at several different temporal and spatial scales—some movements occur very slowly over centuries, whereas others are related to daily oscillations and flow patterns. This section discusses the primary ways in which water moves along a coastline.

Fluctuations in Water Level Over long periods of time, water levels in oceans and large lakes change for a variety of reasons. For instance, changes can occur when tectonic forces cause a landmass to be uplifted or sink. Water levels also fluctuate when the amount of water in the ocean or lake varies as a result of adjustments in the hydrologic cycle. This form of water-level fluctuation is called **eustatic change**.

The best example of eustatic change in recent Earth history is the great change in sea level that occurred at the end of the Pleistocene Epoch, when the massive continental glaciers melted and returned water to the oceans. Figure 19.3 shows that sea level changed dramatically between the Illinoisan glaciation (called the *Riss glaciation* in Europe) around 130,000 years ago and the Holocene Epoch (the past 10,000 years). Major periods of low ocean water level occurred during both the Illinoisan/Riss glaciation and Last Glacial Maximum (the Wisconsin glaciation discussed in Chapter 17), when enormous volumes of water were stored in the immense ice sheets that covered much of North America and Europe. When the ice subsequently melted, the sea level rose rapidly because water poured back into the ocean.

Bay *An indentation in the shoreline that is generally associated with an ocean, sea, or gulf.*

Salinity *Concentration of dissolved solids in water that is measured in parts per thousand (‰).*

Brine *Water that has salinity greater than 35‰.*

Brackish *Water that has salinity less than 35‰.*

Eustatic change *Fluctuations in sea level associated with adjustments in the hydrologic cycle.*

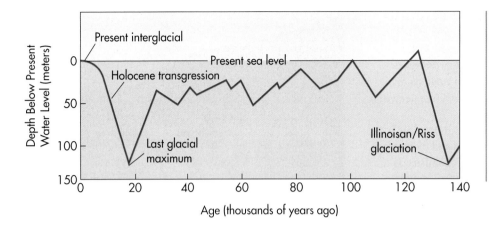

Figure 19.3 Eustatic sea-level changes during the late Quaternary Period at New Guinea. Dramatic low sea levels occurred during the Illinoisan glaciation (called the *Riss glaciation* in Europe) and Last Glacial Maximum, with increases following these glaciations. Sea-level rise the past 10,000 years is often called the *Holocene Transgression*. (*Source*: Reprinted by permission from Macmillan Publishers Ltd: J. Chappell and N. J. Shackelton, 1986, "Oxygen Isotopes and Sea Level," *Nature* 324: 137–140.)

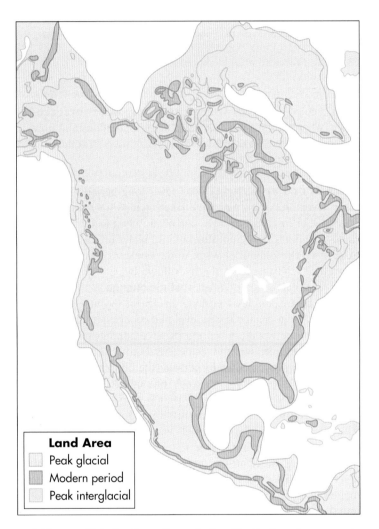

Figure 19.4 Configuration of the North American coastline during peak glacial, peak interglacial, and modern periods. Over time sea level has fluctuated greatly due to the amount of water tied up in ice sheets. Note the configuration of the coastline if all ice sheets melted.

One of the important outcomes of the Pleistocene eustatic changes was that the amount of continental land subaerially exposed—that is, above water—varied a great deal over time. This variation can be seen in Figure 19.4, which shows the configuration of the North American coastline today, at peak glacial time, and at a peak interglacial if all global ice sheets melted. As you can see, the amount of land above sea level during the peak glacial period was significantly greater than it is today.

Referring back to Figure 19.3, note that water levels during the most recent glacial periods were about 125 m (410 ft) lower than they are today. At these times, large parts of the now-submerged continental shelf were subaerially exposed (see Figure 19.4 again) and rivers extended onto that platform. The evidence for these rivers can be seen in the form of canyons cut into the shelf deposits during the major low ocean levels. An excellent example of such a canyon is the Hudson Canyon, which was eroded into the continental shelf by the ancestral Hudson River. Similar features also exist on the floor of Chesapeake Bay (Figure 19.5a), which is essentially a large river valley, or **ria**, that was flooded when the sea level rose during the Holocene. Rias are common along many coastlines, with some of the most beautiful found in New Zealand (Figure 19.5b). The glacial counterpart of a ria is a **fjord**, which is an ice-formed (u-shaped) trough that becomes flooded when sea level rises. Excellent examples of fjords exist in places such as Alaska, Norway, and the south island of New Zealand.

Tides Tides are regular and predictable oscillations that occur with respect to the level of the world's oceans (Figure 19.6). These oscillations are related to Newton's law of gravitation, which states that every particle in the universe attracts every other particle. According to this law, both the Moon and Sun exert a gravitational force on the Earth. Although it would seem that the Sun has the biggest influence on tides due to its immense size, the Moon's gravitational effect is actually greater. This is because the strength of the force is inversely proportional to the distance: the closer the two bodies, the stronger the pull. Recall from Chapter 3 that the average distance from Earth to the Sun

---

Ria *A former river valley along the coast that is flooded by rising sea level.*

Fjord *A former glaciated valley along the coast that is flooded by rising sea level.*

(a)

Figure 19.6 Tidal fluctuations in Chesapeake Bay. This image shows a boat dock at low tide. At high tide, the water reaches approximately halfway up the support beams of the dock (white arrow).

(b)

Figure 19.5 Flooded valleys on the continental margin. (a) Chesapeake Bay is really a large river valley, created by joining rivers, including the Susquehanna, Potomac, and James. The valley was later flooded by postglacial rising sea levels. Delaware Bay, shown in the upper right of the image, is also a ria that was created when a portion of the ancestral Delaware River valley was flooded because of eustatic change. (b) Queen Charlotte Sound, New Zealand, evolved when rising sea levels flooded this valley on the northern tip of New Zealand's south island.

is approximately 149,000,000 km (93,000,000 mi). In contrast, the Moon is only about 386,000 km (240,000 mi) away from Earth. As a result, the Moon's gravitational pull is responsible

for about 56% of the daily tides, whereas about 44% is related to the Sun's effect. Other factors influencing the nature and timing of tides are related to the geometric relationship of Earth to the Moon and Sun and how it varies on a daily and monthly basis.

The simplest tide to explain is the one created on the side of Earth that faces the Moon. Think of this tide as a bulge of water that forms when ocean water is drawn up and along by gravity as the Moon passes over. A similar bulge also forms on the opposite side of Earth, however, which may seem odd because this side of the planet faces away from the Moon. How can this tide form? Tidal bulges appear on both sides of Earth because of the combined effects of the Moon's gravitational pull *and* the fact that the Earth and Moon rotate around each other on a common axis.

Examine Figure 19.7 to see how these relationships work. If the Earth and Moon did not rotate around each other, they would fly through space side by side and the tidal bulge would always be on the side of Earth facing the Moon. Because they rotate around a common axis, however, a centrifugal force exists that is equal to the gravitational forces between the two objects. As a result, the Earth and Moon are always the same distance apart and each particle of Earth experiences both forces. A tidal bulge forms on the side of Earth most directly facing the Moon because the Moon's gravitational pull there is slightly greater than the centrifugal force toward the other side of Earth (Figure 19.7a). The high tide associated with this gravitational force is called a *direct tide*. A tidal bulge also forms on the other side of Earth because the centrifugal force (of the rotating bodies) on this side of the planet is slightly greater than the gravitational pull of the Moon (Figure 19.7b). The high tide associated with this centrifugal force is called an *indirect tide*. Each of these tides has a corresponding low tide.

To fully understand the timing of the direct and indirect tides at any given point, you must first consider the daily (24-h) rotation of Earth on its axis. If the Moon were stationary relative to Earth, any coastal location on the planet would rotate

(a)

(b)

Figure 19.7 Formation of tides. (a) Tide-raising force on the side of Earth facing the Moon. On this side of Earth, a tidal bulge forms because the Moon's attractive forces exceed Earth's centrifugal force. The bulge is most pronounced where Earth is closest to the Moon due to the curvature of both bodies. (b) Tide-raising force on the side of Earth opposite the Moon. A tidal bulge forms on this side of Earth because the centrifugal force associated with the combined rotation of Earth and Moon is greater than the Moon's gravitational force. The height of the observed tides in the figure is significantly exaggerated to illustrate the forcing variables and effects.

under the two tidal bulges (the direct and indirect) each day. Similarly, each coastal locality would also experience two low tides between the bulges. If the Moon were stationary, the timing of these high and low tides each day would always be the same. The Moon is not stationary, of course, but instead slowly orbits Earth approximately every 28 days. As a result, the timing of the high tides is about an hour later each day.

Tides are even more complicated when the role of the Sun is factored into their timing and character (Figure 19.8). Although the gravitational force of the Sun is less than that of the

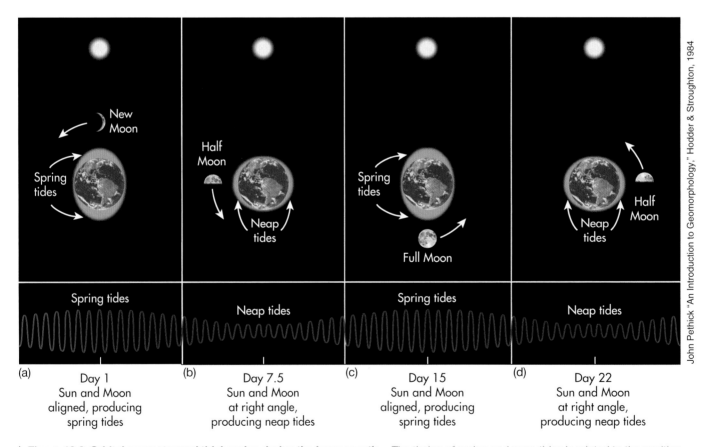

(a) Day 1
Sun and Moon
aligned, producing
spring tides

(b) Day 7.5
Sun and Moon
at right angle,
producing neap tides

(c) Day 15
Sun and Moon
aligned, producing
spring tides

(d) Day 22
Sun and Moon
at right angle,
producing neap tides

John Pethick "An Introduction to Geomorphology," Hodder & Stroughton, 1984

Figure 19.8 Orbital geometry and tidal cycles during the lunar month. The timing of spring and neap tides is related to the position of the Moon relative to the Earth and Sun. Spring tides occur when the Earth, Moon, and Sun are aligned, whereas neap tides happen when the Moon is at a right angle to Earth relative to the Sun. There is usually a short (~1 day) lag between the celestial and tidal cycles.

Moon, it is nevertheless important. When the Sun, Moon, and Earth are aligned, the tide-raising force is maximized, resulting in the very highest tides called *spring tides*. In this case, "spring" refers to upwelling, rather than the season. When the Sun and Moon are at right angles to Earth, however, somewhat lower high tides result; these tides are called *neap tides*.

In addition to the tidal variations that occur with respect to the geometric relationship of the Earth, Moon, and Sun, the overall tidal range can vary dramatically between places. For example, as you can see in Figure 19.9, the average tidal range at San Francisco, California, is about 2.5 m (8.2 ft). The largest tidal range on Earth occurs in the Bay of Fundy, which lies between the Canadian provinces of New Brunswick and Nova Scotia in eastern North America (Figure 19.1). The shape of the coastline here has a funneling effect that produces a tidal range varying from about 6 m (~20 ft) at the mouth of the bay to 16 m (53 ft) at the head of the bay!

Waves The most dynamic form of water movement along coastlines is associated with **waves**, which are oscillations in a body of water resulting mostly from the force of friction generated by wind blowing across the water surface. The specific wind factors involved in the formation of waves are (1) wind strength, (2) wind duration, and (3) wind *fetch*—that is, the

---

Waves *Oscillations in a body of water that form mostly due to the frictional force generated by wind blowing across the surface of the water.*

distance of unimpeded airflow across the ocean or large water body. Although waves appear to be bodies of water that are moving horizontally, they are, in fact, nothing more than rising and falling disturbances of water masses. Very little water moves forward in a wave; instead, it is the wave energy that is transmitted forward through the elastic medium of the water. If you want to test this concept yourself, place a floatable object in the water the next time you are at the ocean and watch as the undulations pass beneath it. You will see that, although many different waves pass by, the object moves only slightly.

Water waves are similar to the electromagnetic waves discussed in Chapter 4 because they consist of various measurable components. The *wave crest* is the high spot of an individual wave and forms when a vertical column of circularly rotating water particles passes through the water and causes the water to rise. Although the column of rotating water continues to some depth in the water, the diameter of individual oscillations progressively becomes less until they disappear at the *wave base*. After the wave crest passes, the water lowers into the *wave trough*. The horizontal distance between successive wave crests is the *wavelength*. *Wave amplitude* refers to the vertical distance between the wave crest and the level of the water if it were still. *Wave height* is somewhat different from wave amplitude in that it measures the vertical distance between the wave crest and the base of the wave trough. In general, stronger winds result in greater wave heights and amplitudes. When waves move out of the area where the generating winds occur, they continue to move forward and are then known as *swells*.

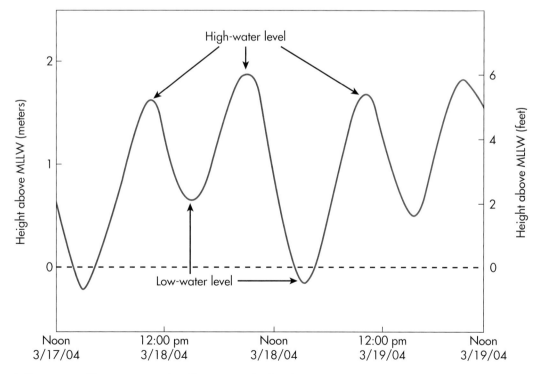

Figure 19.9 Tidal range at San Francisco, California, over a two-day period. The point of reference used here is the Mean Lower Low Water (MLLW), which is the average of the lower low-water height of each day and is represented by the 0 baseline in the graph. Relative to this level, the average tidal range is about 2.0 m (about 6.5 ft). (*Source*: NOAA.)

As waves approach the coastline, they begin to encounter shallower water because the seafloor gradually slopes upward to the shore (Figure 19.10). At some point, the water depth becomes sufficiently shallow that the wave base and the ocean bottom intersect. From this point forward, the vertical space in which the rotating water has to move becomes progressively restricted, causing the overall wave height to increase and the overall speed of the waves to slow. At the same time that the forward speed of the waves is decreasing, from the point where the interaction of the wave base and ocean bottom begins, additional waves continue to advance into the area, resulting in compressed wavelengths as more waves crowd into a smaller space. Ultimately, the height of each individual wave exceeds its vertical stability, and the wave crashes in a feature called a **breaker**. This point typically occurs when the wave height is about seven times greater than the wavelength. From the time the wave breaks onward, the forward momentum of the wave causes water to rush onto the beach as *surf*, which is the only part of the wave system where water is actually moving forward, instead of up and down. In places that are typically sheltered from waves in

some fashion, or have a short fetch, a sudden increase in wave energy can cause erosion and significant modification of the coast. On the other hand, many other coastlines are consistently influenced by strong waves and are thus in equilibrium with no net erosion or deposition of sediment due to high wave energy.

Littoral Processes and Sediment Transport   Let's now take a closer look at wave/shore interactions and how they affect the way that sediment is transported along the coast. These interactions depend on many factors, including the shape of the coastline, the direction and seasonal variation of prevailing winds, intensity of waves, tidal range, and the nature of coastal sediment, to name but a few. In some places sediment moves freely along straight coasts for great distances, whereas in others it can be trapped in nooks and crannies along the shore after moving only a short way. As you might imagine, the various combinations of these factors can result in myriad interactions, most of which are far beyond the scope of this text.

Perhaps the most simplistic and commonly considered way to view these interactions is to observe what happens on a relatively straight coast when waves approach at some kind of angle (Figure 19.11), as they do in many places. Because of this oblique approach, the force of the water

---

Breaker   *A wave that rises and crashes when the forward momentum of oscillations cannot be maintained.*

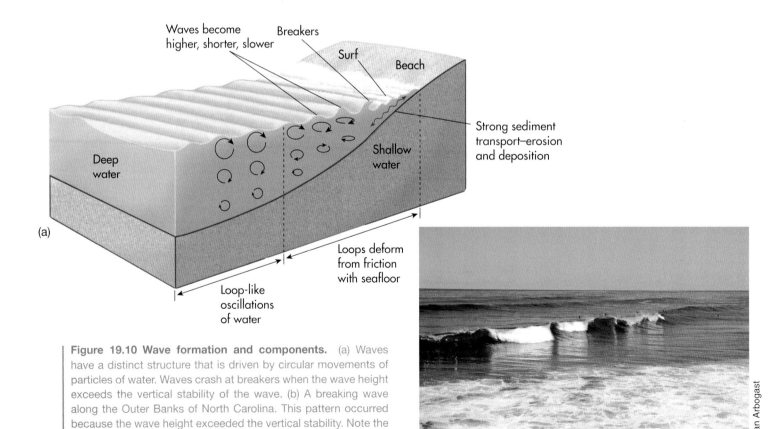

(a)

(b)

Figure 19.10 Wave formation and components. (a) Waves have a distinct structure that is driven by circular movements of particles of water. Waves crash at breakers when the wave height exceeds the vertical stability of the wave. (b) A breaking wave along the Outer Banks of North Carolina. This pattern occurred because the wave height exceeded the vertical stability. Note the pair of incoming swells in the background and the surfers waiting to catch a great wave.

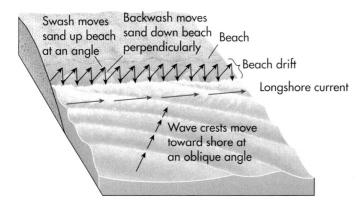

Figure 19.11 Littoral drift. A longshore current develops in places where waves strike the coast obliquely. Sediment subsequently moves down the shore as littoral drift. Littoral drift includes beach drift, which is the transport of sediment down shore due to the zig-zag pattern of swash and backwash, and longshore drift, which is the movement of sediment by the longshore current.

is deflected downwind when it interacts with the shore. This deflection causes the formation of a **longshore current** that flows parallel to the shore, one that is capable of moving sediment down the coast in a process called **longshore drift**.

Obliquely approaching waves also cause an interesting pattern of sediment transport on the beach proper. In these situations the surf also flows onto the beach (known as *swash*) at an angle. After the swash reaches its apex on the beach, it flows back to the ocean at a right angle to shore as *backwash*. Sediment is picked up in this fashion and transported down the coast in a zig-zag pattern known as **beach drift**. Taken together, the combined processes of longshore drift and beach drift are called **littoral drift**.

As noted previously, the various combinations of wave/beach interactions are too numerous to describe in this text. The simplistic pattern just described nicely explains the pattern of littoral drift along the coast of Southern California where the strong, persistent longshore current freely moves sediment from northwest to southeast. Along the eastern shore of Lake Michigan, however, seasonal prevailing winds shift from northwesterly in winter to southwesterly in summer. This shift in wind direction causes sediment to move up and down the coast to some degree. Still another pattern is observed along the concave southeastern coast of Texas where waves approach from the southeast (Figure 19.12). In the central part of this portion of the Texas coast, the waves approach in a perpendicular fashion. On the flanks of this section of coast, however, the curvature of the shore causes the waves to approach obliquely. This interaction results in opposing longshore currents that converge in the central part of the coast.

KEY CONCEPTS TO REMEMBER ABOUT WATER MOVEMENT ALONG COASTLINES

1. Coastlines are the places on Earth where the hydrosphere, lithosphere, and atmosphere intersect over a large spatial extent.

2. The amount of water in the oceans fluctuates with time due to adjustments in the hydrologic cycle. These changes are called eustatic changes. Many coastlines show evidence of long-term eustatic change in the form of rias and fjords.

www.wiley.com/college/arbogast

WileyPLUS

Tides

The formation of tides involves the complex geometric interaction of the Earth, Moon, and Sun. These interactions cause gravitational forces that pull on the Earth's oceans in such a way as to cause predictable oscillations in local sea level. To see how these forces appear in animated form, go to the *Geo*

Media Library and select *Tides*. This animation will allow you to better visualize the various forces acting on the Earth's oceans and how they move water. Once you complete it, be sure to answer the questions at the end to test your understanding of this concept.

Longshore current *The current that develops parallel to a coast when waves approach the coast obliquely and forward momentum is deflected.*

Longshore drift *Transport of sediment by the longshore current.*

Beach drift *Sediment that is transported in the surf zone by swash and backwash, which form due to the oblique approach of waves.*

Littoral drift *Sediment that is transported through the combined processes of longshore drift and beach drift.*

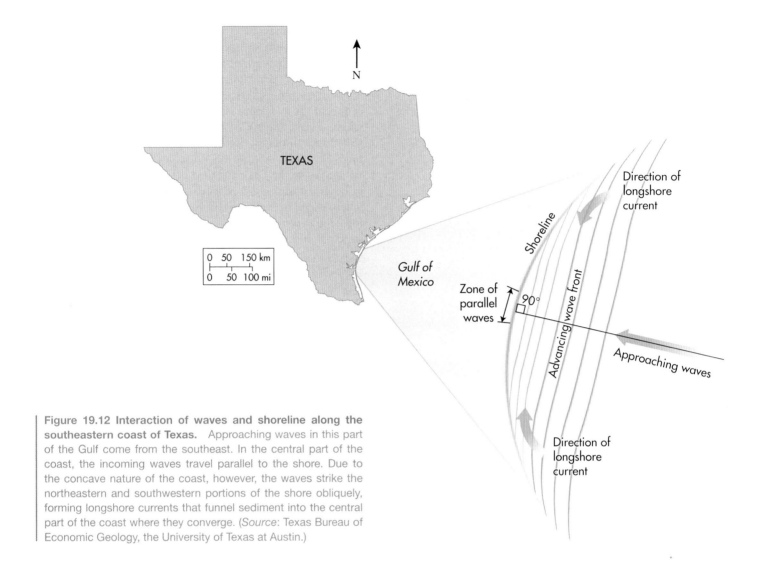

Figure 19.12 Interaction of waves and shoreline along the southeastern coast of Texas. Approaching waves in this part of the Gulf come from the southeast. In the central part of the coast, the incoming waves travel parallel to the shore. Due to the concave nature of the coast, however, the waves strike the northeastern and southwestern portions of the shore obliquely, forming longshore currents that funnel sediment into the central part of the coast where they converge. (*Source*: Texas Bureau of Economic Geology, the University of Texas at Austin.)

VISUAL CONCEPT CHECK   19.1

With your understanding of littoral processes in mind, examine this image from Queensland, Australia. Assume the axis of the coast is aligned north and south and see if you can determine:

   a)   The direction of wave-forming winds
   b)   Which way the longshore current is flowing
   c)   Which way the beach drift is moving

Hint: North is at the bottom of the photo.

David Wall/Alamy

3. Tides are daily adjustments that occur with respect to the water level along any given coastline. Tidal processes are driven largely by the gravitational pull of the Moon and, to a lesser extent, the Sun. Other important factors are the centrifugal force of the combined Earth–Moon rotation and the geometric relationship of the Sun and Moon.

4. Waves are the dominant factor that shapes coastlines. Waves are disturbances in bodies of water and form as a result of frictional forces related to wind. They generate rotating bodies of water that migrate obliquely toward the shore. At the point where the wave base interacts with the ocean bottom in shallow water, the oscillations become deformed, ultimately causing waves to crash.

5. Littoral processes refer to the way that sediment is transported and deposited in the shore zone. An important feature is a longshore current that develops along a coastline due to the deflected force of oblique waves striking the shore. In a related process, water runs up the beach and back into the ocean or sea in processes called swash and backwash, respectively.

Figure 19.13 Headlands along the southern coast of Tasmania in Australia. Headlands are prominent coastal features that project out into the water. Note that at least three headlands appear in this image.

Coastal Landforms

Let's now turn to the various landforms created along shorelines. Recall that fluvial, glacial, and eolian landforms can be subdivided into erosional and depositional categories. Coastal landforms can be viewed in the same way. This discussion begins with erosional coastlines.

Erosional Coastlines

Most coastal erosion is accomplished through the unceasing pounding of the shoreline by waves. Waves have tremendous power after they break, with the associated spray moving as fast as 113 km/h (70 mi/h). Given that water has a very high density, the rapidly moving water has a lot of power and is responsible for the majority of coastal erosional landforms.

Although wave erosion can occur along any part of the coast, it tends to be focused on a particular part of the landscape called a headland. A **headland** is a promontory that juts into the ocean or sea and thus is surrounded on three sides by water (Figure 19.13). Headlands tend to form along shorelines where bands of rock with alternating resistance run perpendicular to the coast. The headlands are associated with the more resistant

rock such as limestone or granite. Where the rock is relatively soft, perhaps shale or sandstone, waves can erode sediment more effectively and a bay forms.

Once the pattern of headlands and intervening bays develops, the erosive power of waves begins to concentrate on the headlands. This change occurs because approaching waves initially begin to slow down in front of the headlands, where they first encounter shallow water. This friction causes the waves to pivot around the headlands in a process called **wave refraction**. As you can see in Figures 19.14a and b, waves bend around a headland, which results in the energy of the waves being expended on three sides of the landform. In this way, headlands are more vigorously eroded than the intervening bays. The bays become zones of deposition, forming features called *pocket beaches*, because sediment is funneled into them from the eroded headlands (Figure 19.14c).

The basic coastal erosional landform is the wave-cut bluff, which is a vertical face that commonly forms along rocky shorelines. Excellent examples of large wave-cut bluffs occur on many parts of the coast of Scotland (Figure 19.15), as well as the coasts of England, Australia, and California. Wave-cut bluffs evolve in a predictable sequence of events as shown in Figure 19.16. This sequence begins with waves that are pounding against a noneroded shoreline (Figure 19.16a). With time (Figure 19.16b), the waves cut a vertical face into the rock outcrop, and at the same time they plane a more horizontal surface

---

Headland *A portion of the coast that extends outward into a large body of water.*

Wave refraction *The process through which waves are focused and bent around headlands.*

Pocket beach

Headland

Bay

Wave energy dissipated in bay

Wave energy focused on headland by refraction

(a)

©Marli Bryant Miller

(b)

Alan Arbogast

(c)

Figure 19.14 Headland erosion. (a) A model showing how waves refract around headlands, causing the formation of a pocket beach in the intervening bay. (b) Wave refraction along the coast of California. Notice how the waves are bending around this headland? (c) Pocket beach on the Pacific Coast north of San Francisco. This beach formed when sandy sediment eroded from headland in the background (arrow) and behind (south) the view was funneled into this small embayment.

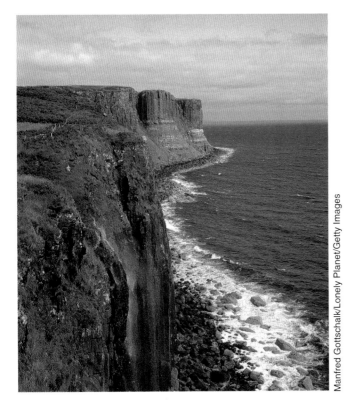

Manfred Gottschalk/Lonely Planet/Getty Images

Figure 19.15 Wave-cut cliff of Kilt Rock, Scotland. This massive wave-cut bluff was eroded into basalt cliffs over time. Note the waterfall in the foreground.

(a) Noneroded rocky shoreline

(b) Formation of vertical bluff

Eroded vertical face
Wave-cut platform

(c) Waves cut notch into base of cliff

Rock overhang
Notch

(d) Overhang collapses, bluff retreats

Direction of retrogradation

New wave-cut platform

Figure 19.16 Bluff formation and retreat. (a) Waves attack a noneroded rock outcrop. (b) With time, the waves cut a vertical bluff into the rock. (c) During strong storms or relatively high water, the waves cut a notch into the lowermost rocks of the bluff. (d) If the notch is cut sufficiently deep, the overhanging rocks in the bluff collapse, causing bluff retreat.

(a)

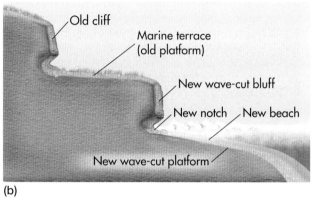

(b)

Figure 19.17 Evolution of a marine terrace. (a) Waves cut a bluff and platform. (b) Tectonic uplift or sea-level fall raises the previous wave-cut features relative to the position of current erosion, resulting in a marine terrace.

called a *wave-cut platform* into the rocks in front of it. During higher water stages or strong storms, waves run up and cut a notch into the lower rocks of the bluff outcrop and further plane the nearshore platform (Figure 19.16c). If the notch is cut sufficiently deep, the rock overhang falls under the influence of gravity and the bluff retreats farther inland, leaving a longer wave-cut platform in front of it (Figure 19.16d). As the coastline retreats due to erosion, it is said to be undergoing **retrogradation**.

As described earlier, one way that water moves along coastlines is through eustatic changes in sea level. In addition to the submerged river channels that occur on the continental shelf, one of the best lines of evidence for prehistoric eustatic sea-level changes is the presence of abandoned bluffs and marine terraces. Similar to a fluvial terrace, a marine terrace represents the former position of the active shoreline on the landscape. Figure 19.17 shows how this kind of landscape evolves. In Figure 19.17a, waves create a wave-cut bluff and platform similar to that shown in Figure 19.16. Imagine what would happen if the landscape were subsequently uplifted or if sea level fell. If such a eustatic change occurred, a new bluff and platform would form below the abandoned one (Figure 19.17b), leaving a stair-stepped coastal landscape. Such landscapes are

Retrogradation *The process through which a shoreline retreats through erosion.*

Figure 19.18 Marine terraces on San Clemente Island, California. Each of these terraces formed during a high sea level associated with an interglacial period, with the highest surface representing the oldest high-level sea stand. Sea level subsequently dropped during glacial periods when enormous volumes of water were stored as ice. At the same time, the old wave-cut bluff was slowly uplifted due to tectonic activity. With each successive interglacial, sea level rose again, creating new wave-cut bluffs that were progressively lower than the prior ones.

common along the active margin of the west coast of the United States (Figure 19.18).

In addition to wave-cut bluffs and terraces, additional evidence for coastal retrogradation comes in a variety of forms, which evolve in a continuum as the shoreline recedes. *Sea caves* are semicircular notches cut into the base of rock bluffs. If the rock is entirely cut through so that water passes through the notch, a *sea arch* results (Figure 19.19). With

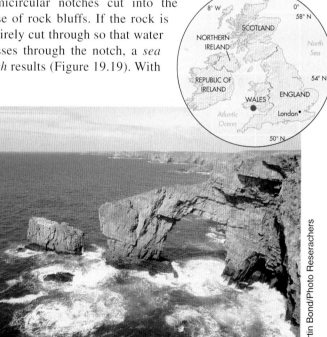

Figure 19.19 Erosional landforms along the coast of Wales. Sea arches develop when sea caves are entirely cut through, allowing free water passage beneath the arch. If the top of the arch collapses, an isolated sea stack remains, such as the one to the left. Note the tilted rock strata.

Land

Bay

Sea

Headland

Bay fills in
as deposits
continue

Bluff retreats
as erosion
continues

Sea
stacks

Beach forms
by deposition

Bluff forms from
wave erosion

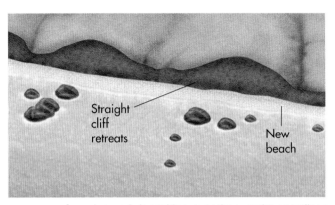

Straight
cliff
retreats

New
beach

Figure 19.20 Evolution of a rocky coastline Through the combined processes of erosion and deposition over time, rocky coastlines evolve in a predictable way that wears down headlands to form relatively straight beaches.

time and additional erosion, the top of the arch may collapse, leaving a detached fragment of the original bluff, called a *sea stack*, isolated in the ocean. These features are particularly common along the rocky coast of the Pacific Northwest. The net effect of shoreline retrogradation in places like this is that the coastline slowly transforms from one that has a number of prominent headlands to one that is relatively straight (Figure 19.20).

KEY CONCEPTS TO REMEMBER ABOUT EROSIONAL COASTLINES

1. A coastline that is essentially retreating through erosion is considered to be retrograding.

2. Erosional coastlines are shaped primarily by strong waves. These coastlines tend to be rocky with high bluffs.

VISUAL CONCEPT CHECK 19.2

Imagine that you are driving down the Pacific Coast in Oregon and see the beautiful landscape pictured here. What conclusions can you reach about the history of landscape evolution at this location? What were the processes involved? In short, how did this landscape form?

Danita Delimont/Alamy

Waves are fascinating features that often shape the coastline in dramatic ways. Aside from witnessing these processes yourself, the best way to visualize them is to see video of them in action. To do so, go to the *Geo Media Library* and open *Waves and Coastal Erosion*. The first part of this animation is a video that demonstrates wave processes and their interaction with the

shore. Following this video is another film clip that shows coastal erosion along the California coast in 2008. This video was filmed during a severe storm that produced strong waves that battered coastal bluffs. Watch how the waves crashed against the base of the bluff and how the upper part of the cliff was destabilized in the manner discussed in this chapter.

---

3. Headlands are prominent landforms that protrude into the ocean. These landforms are eroded more vigorously than other places along the shore because waves are refracted around them and erode on three sides.

4. Marine terraces represent ancient shorelines. The landforms indicate either variable water levels or uplift due to tectonic activity.

5. Other indicators of coastline recession include caves, sea arches, and sea stacks. These erosional features evolve in a continuum as a coastline erodes.

Depositional Coastlines

We have just seen how coastlines are shaped by erosion and some of the diagnostic landforms that result from this process. Now, let's turn to the coastal landforms created when sediment is deposited in various ways and places. You were briefly introduced to this concept with the formation of a pocket beach between two eroding headlands (Figure 19.14c). Coastlines in the process of extending outward into the water through deposition are said to be undergoing **progradation.**

Beaches   The part of the coastal landscape people are probably most familiar with is the beach. Beaches are dynamic places where sediment is deposited through the combination of waves, beach drift, and wind. In some places, beaches are supplied with alluvial sands derived from land far inland. The beach is a transition between the water and the landmass and consists of exposed, unconsolidated sediments that usually range from sand to cobbles. Although finer silt and clay-sized particles are

sometimes contained within a beach, they are usually carried away in suspension by the longshore current. Most beach growth occurs during the summer months in the Northern Hemisphere when the weather is relatively calm. In winter, however, beaches can be significantly eroded due to large waves created by strong storms. If you happen to live on or near a beach, notice how the shape of the beach can change between seasons.

Figure 19.21 shows how beach components are differentiated on the basis of their relative position to the water. The **offshore** is permanently submerged and is the zone where waves break and the surf is most active. Moving landward, the **foreshore** is the part of the beach that is influenced by the rise and fall of the tides; that is, it is regularly exposed and submerged. Separating the offshore and foreshore is a submerged ridge called an **offshore bar**, which forms as a result of the interaction between breakers and the ocean bottom in nearshore environments, where wave velocity is relatively low. Another noticeable feature on many coasts is a **berm scarp**, which is cut at the high water line of the beach. Landward of the high water line is the **backshore**, which is the relatively flat part of the beach that is covered by water only during severe storms.

Spits and Baymouth Bars   As discussed earlier, one of the ways that water moves along the coast is in association with the longshore current (see Figure 19.11). This current is capable of eroding a great deal of fine sediment from the foreshore and offshore environments and transporting it down the beach. As long as the beach continues in an uninterrupted fashion along the shore, the current and its sediment load will flow progressively down the coast because of the interaction of waves and

---

Progradation *Outward extension of the shoreline through deposition of sediment.*

Offshore *The nearshore zone that is permanently submerged and where waves break.*

Foreshore *The nearshore zone that is regularly exposed and submerged through the tidal fluctuations and movement of surf.*

Offshore bar *A small ridge on the bottom of the ocean that separates the offshore and foreshore.*

Berm scarp *A miniature cliff created by wave erosion that fronts a beach berm.*

Backshore *The part of the beach that lies between the berm scarp and foredune and is covered by water only during strong storms.*

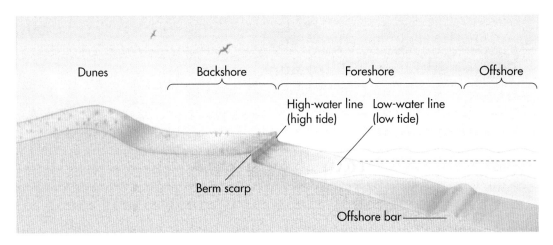

Figure 19.21 Generalized beach cross section. Beaches are divided into several regions, with each associated with a different water level or the action of wind.

the landmass. Once the current encounters the mouth of a deep bay, however, it loses velocity in a manner consistent with a stream's loss of velocity when it meets the ocean. As with river deltas, sediment carried by the longshore current is deposited in a predictable fashion that produces distinctive landforms in bay areas (Figure 19.22).

The variety of landforms created by the deposition of longshore drift in bays forms a geomorphic continuum that evolves with time. This continuum begins with the formation of a **spit**,

Spit   *A linear bank of land that extends into a bay made by the deposition of longshore sediment.*

which is a linear bank of land that extends into a bay when longshore sediment is deposited (Figure 19.22). An important distinction to make here is that a spit is connected to land on the upcurrent side of the bay and extends into open water on the other side. A spit is usually straight for most of its length, but typically curves landward at its tip, forming a hook when viewed from the air above. This form develops because at the same time the longshore sediment is being deposited as the spit extends, incoming waves are moving the tip shoreward. Cape Cod, Massachusetts is an excellent example of a curved spit (Figure 19.23). With sufficient time, the spit could extend entirely across the bay, closing it off with a landform called a

Figure 19.22 Depositional coastal landforms associated with littoral processes. On a depositional coast landforms such as spits, tombolos, and baymouth bars form where sediment accumulates due to littoral processes.

Figure 19.23 Cape Cod, Massachusetts. This satellite image shows a beautiful example of a curved spit, one that formed because the prevailing longshore drift flows to the north.

and are parallel to the shore (Figure 19.25a). These dunes develop due to shore vegetation trapping sand blown by the wind off the beach and gradually forming a linear (along-shore) ridge or dune. Other kinds of dunes commonly seen in coastal settings are blowouts and parabolic dunes, discussed in Chapter 18 (Figure 18.18). In places where the supply of sand is especially large and the winds are consistently strong, massive *transgressive dune fields* form. These dune fields are

(a)

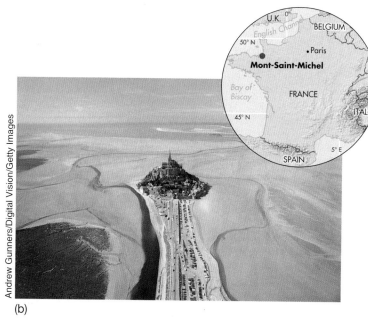

(b)

Figure 19.24 Tombolos. (a) Tombolos form when longshore drift extends to a preexisting island. (b) A beautiful example of a tombolo is Mont-Saint-Michel in France.

baymouth bar. If and when this closing occurs, the bay will no longer be directly connected to the ocean and will then be called a **lagoon**. The water in a lagoon is typically brackish relative to the open ocean and thus is an important ecosystem. Sometimes longshore sediments converge from two different directions, causing a spit to grow perpendicular to the shore. If this kind of spit grows so that it reaches a large sea stack or island, it is called a **tombolo** (Figure 19.24).

Coastal Dunes   Another type of depositional landform associated with sandy coasts is sand dunes. In contrast to desert environments where dunes form in poorly vegetated, sandy areas, coastal dunes develop because the supply of sand along beaches can be very high. A common type of dune form associated with beaches is a **foredune**. Foredunes range in height from 1 m to perhaps 10 m (3.3 ft to 33 ft)

Baymouth bar   *A spit that entirely encloses a bay.*

Lagoon   *A brackish body of water that lies behind a baymouth bar.*

Tombolo   *A spit or sandbar that connects an island to the mainland.*

Foredune   *A dune that forms parallel to the shore when sand blows inland from the beach.*

Coastal Landforms • 521

(a)

(b)

(c)

Alan Arbogast

Dennis Frates/Alamy

Figure 19.25 Coastal sand dunes. (a) Foredune (arrow) along the east coast of Lake Michigan. The dune formed because sand blew up from the beach, which can be seen to the left. Note the person for scale standing in a swale in the lee of the dune and the way the dune axis mirrors the beach. (b) Transgressive dune field on North Head, New Zealand. Here, eolian sand has spread inland due to very high sand supply and strong winds blowing off the Tasman Sea (left). (c) Coastal dunes on the Oregon coast. Westerly winds here have produced this beautiful dune field. Do you see the sea stacks in the background?

so named because they transgress (advance) inland and bury older surfaces (Figure 19.25b).

Coastal dunes are found all over the world. Spectacular dune fields occur along the coast of Namibia in Africa, Australia, Brazil, India, many shores in Europe, Mexico, and numerous other places. Coastal dune fields also line much of the U.S. shore. Perhaps the best example of such dunes in a marine environment along a U.S. coast is seen in the fantastic dune fields in Oregon and Washington (Figure 19.25c). These dune fields have formed in areas where the high sand supply was remobilized by strong westerly winds off the Pacific Ocean.

Barrier Islands Another type of depositional landform that occurs along coastlines is a **barrier island,** which is an accumulation of sediment (sand, gravel, boulders, and/or shells) parallel to the shore in the ocean that is formed by waves, tides, and winds (Figure 19.26). These islands tend to form on broad, gently sloping continental shelves. Although no firm agreement exists about how these islands develop, a leading hypothesis is that they are linked to rising sea levels following the most recent ice age. In this model, sediment deposited on the continental shelf by rivers during the Wisconsin glacial period was reworked and molded when sea level began to rise as the ice sheets melted. As the water level slowly rose, the shelf sediments were shaped

Thomas R. Fletcher/Alamy

Figure 19.26 A typical barrier island Barrier islands are essentially long sand bars, with some capping sand dunes, that form parallel to the coast. These landforms may have formed when rising sea levels at the end of the Wisconsin glacial period reworked sediments on the continental shelf into these linear features. Note the dunes on this island due to high sand supply.

Barrier island *An elongated bar of sand that forms parallel to the shore for some distance.*

COASTAL DUNES ALONG THE EASTERN SHORE OF LAKE MICHIGAN

Although coastal dunes are typically associated with marine environments, some of the most impressive dune fields in North America occur along the many shores of the Great Lakes in Wisconsin, Michigan, and Ontario, Canada. Of these freshwater dune systems, perhaps the best are the fantastic dune fields that line much of the eastern shore of Lake Michigan, ranging from northern Indiana to the northern end of Lower Michigan. These dune fields contain a variety of foredunes (see Figure 19.25a) and are often large transgressive systems with numerous parabolic dunes embedded within them. An air photo from Warren Dunes along the southeastern shore of Lake Michigan nicely shows an example of these landfroms. Some of these parabolic dunes are up to 60 m (~195 ft) high and contain enormous quantities of sand. Along the northwestern shore of Lower Michigan, dune fields often mantle high bluffs and are thus called *perched dunes*. These landscapes collectively are some of the most beautiful in the Midwest, if not the country.

Aside from their scenic character, I am particularly attracted to Lake Michigan coastal dunes because I am a geomorphologist and study them in my own research. Earth scientists have long known that the eastern shore of Lake Michigan is a perfect place for the development of large coastal dune fields for three reasons: (1) prevailing westerly winds blow across the lake, (2) Lake Michigan has a long fetch, which means that the air flows cleanly across the lake before it strikes the shore, and (3) sand supply is very high from deposits left behind by Pleistocene glaciations. Working with some excellent colleagues, my research focuses on the history of sand mobilization and dune growth as it relates to climate change and fluctuations in the level of Lake Michigan over time. With a variety of dating techniques, we have been able to show that the dunes began to form about 5000 years ago in many places and have grown in distinct stages since that time. These periods of dune growth seem to have most often occurred when the lake was rising, which apparently pushed sand on to the beach where it could be blown by the wind inland. These findings are important for environmental management in Michigan because the dune landscapes are heavily utilized for real estate development and sand mining. In addition, about 3 million people a year visit the numerous scenic parks, such as Warren Dunes, that are scattered along the coast. If you have a chance to visit them someday, I highly recommend it because they rival many of the coastal dune systems anywhere else in the world!

NRCS

Aerial image of Warren Dunes State Park along the southeastern shore of Lake Michigan. Note the big parabolic dunes and blowouts that dominate the park. The orientation of the dunes indicates the prevailing winds are westerly, although some dunes were shaped by southwesterly winds.

Alan Arbogast

View of dunes along the southeastern shore of Lake Michigan from the beach. Lake Michigan is to the left (west). Dunes like this line much of the shore in this part of Michigan and are up to 60 m (195 ft) high.

by waves, tides, and the wind into the linear islands. At the same time as these islands began to form, they were gradually nudged toward their present position by the slowly rising water and storms crashing into them.

A key factor in the overall topography of a barrier island is the supply of sand fed to the system. In barriers with low sand supply, the landform will have very low relief and lie only a meter or two above sea level. Where barriers have high sand supply, dunes typically form and provide additional relief to the landscape. Barrier islands lie in various proximities to the shoreline, with some being only several hundred meters away and others many kilometers from the coast. An excel-

lent place to see barrier islands is along the Atlantic and Gulf coasts of the southeastern United States, where approximately 280 barrier islands are found. Figure 19.27a shows a satellite image of a reach of barrier islands along the Gulf Coast from Louisiana (left) to Alabama (right). The Chandeleur Islands are part of the complex, lying east of the Mississippi Delta. These islands have very low relief due to limited sand supply and are thus pounded when large waves from strong storms and hurricanes wash over. As you can see in Figure 19.27b, hurricanes Rita (2004) and Katrina (2005) dramatically altered the shape of these islands, producing numerous gaps and reducing island size.

(a)

(b)

Figure 19.27 Barrier islands along the Gulf Coast. (a) Satellite image of a complex of barrier islands east of the Mississippi Delta along the coast of Mississippi and Alabama. (b) Time sequence of the Chandeleur Islands from 2001, 2004, and 2005. Notice the profound impact hurricanes Rita and Katrina had on these inlands.

Look at this satellite image of a landform on the coast of Japan. Assuming that north is at the top of the image, in what direction is the long-shore current flowing? Why is the curve present at the end of the feature?

Japan Space Systems

www.wiley.com/college/arbogast

WileyPLUS

Longshore Processes and Depositional Coastlines

Now that we have discussed the basic processes associated with longshore drift, we can view them in an animated format. To do so, go to the *Geo Media Library* and access *Longshore Processes and Depositional Coastlines*. This animation describes the way that water moves along a shoreline and the landforms that result from this flow. Watch how oblique-approaching waves cause a current to develop parallel to the shore. Also see how this current moves sediment until it reaches bays where it is deposited first to form spits and then baymouth bars. Most importantly, watch how the coastal landscape is transformed through time and process. After you complete the animation, be sure to answer the questions at the end to test your understanding of this concept.

Like baymouth bars, brackish lagoons form behind barrier islands that are close to the shore because they are not fully in contact with the ocean and thus have lower salinity. Once these lagoons become established, they become the focus of sediment deposition from three sources: (1) streams flowing off the mainland, (2) eolian sand blowing landward from the island, and (3) tides. Taken together, this deposition produces new land bodies called mudflats (Figure 19.28a), which, if vegetated, become salt marshes (Figure 19.28b). If plant colonization continues unchecked, due to the lack of strong storms or an outlet to the ocean, the lagoon is slowly transformed into a continuous landmass connected to the barrier island.

(a)

(b)

Figure 19.28 Infilling of a lagoon. (a) Because lagoons are water bodies of low energy, they are the focus of sediment deposition, forming mudflats. (b) With time, the lagoon will be transformed into a vegetated salt marsh that connects the mainland with the barrier island.

Impact of Coastal Processes on Delta Form As discussed in Chapter 16, one of the most interesting landscapes on Earth is the place where a river meets the ocean to form a delta. Deltas are noteworthy purely in the context of fluvial processes because they mark the location where sediment from the continental interior is deposited due to very low stream gradients and reduced velocity. These places are also fascinating from the standpoint of various coastal processes because their overall form is directly related to the influence (or lack thereof) they have on the delta as it develops.

The most common way to examine delta form is to consider whether waves, tides, or fluvial processes dominate the coastline at any given place. Although many deltas are influenced equally by both waves and tides, or fluvial processes and waves, it is instructive to consider their shape as it relates to the singular dominance of any particular process. This triangular relationship is shown in Figure 19.29. As you examine the corners of this diagram, remember that a continuum of delta forms actually exists and that you are observing the extreme examples of each kind at this level.

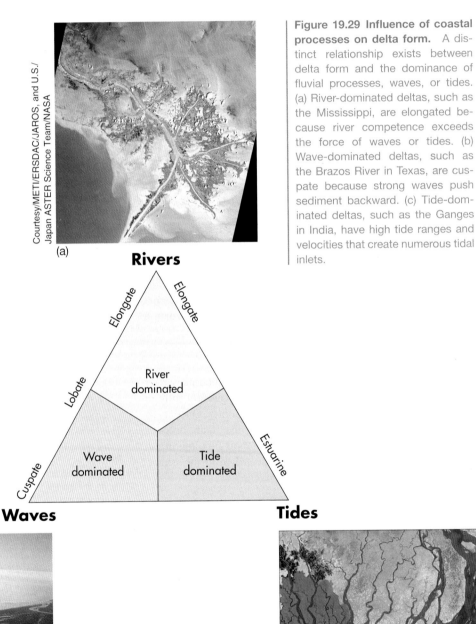

Figure 19.29 Influence of coastal processes on delta form. A distinct relationship exists between delta form and the dominance of fluvial processes, waves, or tides. (a) River-dominated deltas, such as the Mississippi, are elongated because river competence exceeds the force of waves or tides. (b) Wave-dominated deltas, such as the Brazos River in Texas, are cuspate because strong waves push sediment backward. (c) Tide-dominated deltas, such as the Ganges in India, have high tide ranges and velocities that create numerous tidal inlets.

Let's begin this investigation by considering the shape of deltas dominated by rivers (Figure 19.29a). This kind of delta occurs along a coastline where wave energy and tidal range are typically low and stream energy is relatively high. In these situations, such as the Mississippi Delta, the delta is elongated because the stream is able to extend into the ocean through progradation because waves and tides are unable to blunt this growth. Wave-dominated deltas are those that form where the energy of incoming waves is high. These deltas tend to be blunted and even cuspate because strong waves push sediments back toward the mainland. A good example of such a cuspate delta can be seen where the Brazos River meets the Gulf of Mexico along the Texas coast (Figure 19.29b). Finally, tide-dominated deltas form along coasts that have a high tidal range. In these settings, such as the Ganges River Delta in India (Figure 19.29c), the velocity of incoming and outgoing tides exceeds river velocity except during floods. As a result, the delta takes on the form of an estuary with numerous tidal inlets.

Coral Reefs

Thus far our discussion of depositional coastlines has focused on landforms that develop due to the accumulation of sediment. Another way that coastal landforms evolve is through the growth of **coral reefs** (Figure 19.30). Coral reefs form due to the symbiotic interaction between coral polyps and microscopic algae (called *zooxanthellae*) that give the corals their color. When the polyps die, they leave an exoskeleton behind composed of calcium carbonate. These deposits grow larger with time because, when one generation of corals dies,

---

Coral reefs *Resistant marine ridges or mounds consisting largely of compacted coral together with algal material and biochemically deposited calcium carbonates.*

Stephen Frink/Corbis

Figure 19.30 Pillar coral in the Cayman Islands. Coral comes in a wide variety of colors, shapes, and species, but they all produce calcium carbonate external skeletons.

a new one grows on top of it. Coral reefs grow best in tropical waters between about 30° N and 25° S that are warmer than 20°C (68°F), free of suspended sediment, and well aerated. Figure 19.31 shows the distribution of coral reefs on Earth.

Coral reefs form in three distinct types of settings. A *fringing reef* forms on a shallow platform attached to the shore or

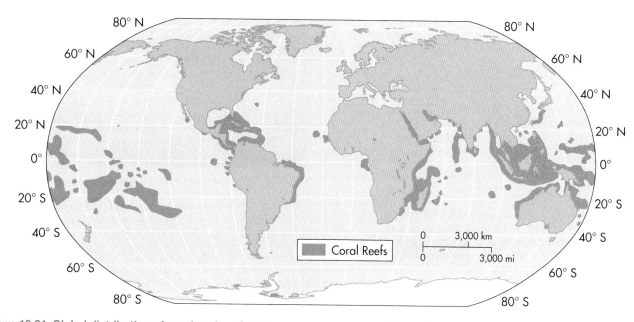

| Figure 19.31 Global distribution of coral reefs. Coral reefs are very common in tropical areas.

Sylvain Grandadam/Stone/Getty Images

Figure 19.32 Coral reefs. A fringing reef along the island of Moorea in Tahiti, South Pacific Ocean.

island, such as the one shown in Figure 19.32. A second kind of reef is a *barrier reef*, which is a line of coral that is roughly parallel to the shore. Such reefs are separated from the coast either by a deep and wide lagoon or, in some cases, by several kilometers of open ocean that is much deeper than the water above the coral. The best-known barrier reef is the Great Barrier Reef in Australia. This spectacular reef is actually a complex of more than 2000 individual reefs that lie about 50 km (30 mi) from the northeast Australian coast. This complex of reefs extends for about 2000 km (1250 mi), making it the longest reef system in the world.

A third type of coral reef is an atoll. *Atolls* are semicircular reefs that are most closely associated with degradation of volcanic islands. These reefs evolve in a predictable pattern. In the first stage of development (Figure 19.33a), a fringing reef grows on the nearshore platform surrounding a relatively new volcanic island. In the second stage of development (Figure 19.33b), the volcano is dormant and begins to wear down and become somewhat submerged due to erosion and subsidence. During this phase of reef development, the reef becomes more like a barrier reef because it is separated from

Guido Alberto Rossi/Age Fotostock

Figure 19.33 Three-stage evolution of an atoll. (a) During the first stage, a fringing reef grows on the shallow platform surrounding a volcanic island. (b) In the second stage, the volcano begins to collapse on itself, which separates the reef from the landform, resulting in a barrier reef. (c) With time, the volcanic island is entirely submerged, leaving only the coral above water as an atoll. (d) This photograph of the Maldive Islands in the Indian Ocean illustrates the evolution of atolls, beginning with the fringing reef in the foreground.

the volcanic island. With the passage of additional time, the volcano disappears entirely (Figures 19.33c, d), leaving an atoll reef ring isolated in the ocean. Atolls are particularly common in the South Pacific Ocean and, like reefs everywhere, are home to a fascinating and colorful variety of ocean animal species (Figure 19.34).

KEY CONCEPTS TO REMEMBER ABOUT DEPOSITIONAL COASTLINES

1. Depositional coastlines form through the process of sediment accumulation. A coastline that is extending outward into a body of water is said to be prograding.

2. The most familiar coastal depositional feature is the beach. Sediment is supplied to beaches through longshore deposition, swash/backwash, and wind. This landform is subdivided into three regions (offshore, foreshore, backshore) on the basis of water depth and frequency and nature of submergence.

3. Sand dunes commonly occur in the lee of beaches where the supply of sand is high and strong winds frequently blow.

4. A range of depositional features is associated with the decreased velocity of the longshore current, including spits, baymouth bars, and tombolos.

5. A coral reef evolves through the growth of small marine organisms and can be categorized as a fringing reef, barrier reef, or atoll.

Human Interactions with Coastlines

As noted previously, coastlines are a favored place for human interactions related to economic development, trade, housing, and recreation. Of the 6 billion people who live on Earth, approximately 37% (2 billion) live within 100 km (60 mi) of a major coastline. Nearly 50% live within 200 km (120 mi) of a shore. To see how these figures compare to the United States, examine Table 19.1. Overall, nearly 144 million people (~53% of the population) live near a coastline. Almost 63 million people live along the Atlantic Coast in cities such as New York, Boston, Philadelphia, and Miami. Another 17.3 million people live along the Gulf Coast in cities such as Tampa, New Orleans, and Houston. The Great Lakes region is home to almost 27 million people who live in cities such as Cleveland, Detroit, and Chicago. Still another 37 million people reside along the Pacific Coast, with most living in cities such as Seattle, Portland, San Francisco, and Los Angeles.

Given that most people in the United States (and the world) live fairly close to coastlines, it is important to understand some of the human impacts in these geomorphically

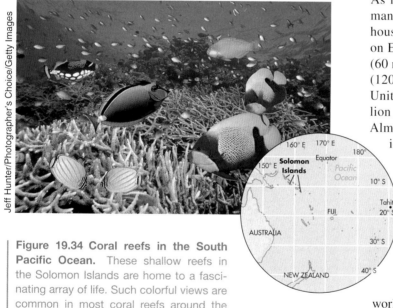

Figure 19.34 Coral reefs in the South Pacific Ocean. These shallow reefs in the Solomon Islands are home to a fascinating array of life. Such colorful views are common in most coral reefs around the world.

Jeff Hunter/Photographer's Choice/Getty Images

TABLE 19.1	U.S. Coastal Population Statistics Reported by the U.S. Department of Commerce					
Total U.S.	Atlantic Coast	Gulf of Mexico	Great Lakes	Pacific Coast	Total Coastal	Balance of U.S.
Population in Millions						
272.7	62.7	17.3	26.8	37.0	143.9	128.8
Percent of Total U.S. Population						
100.0	23.0	6.3	9.8	13.6	52.8	47.2

Source: U.S. Department of Commerce, Bureau of the Census.

sensitive areas. Many of these impacts involve stabilizing beaches in order to maintain attractive tourist destinations for economic purposes. Others are potential future outcomes related to global climate change and the effects rising sea level will have on people living near the shore. Still other human interactions are based on the desire to keep coastlines as environmentally healthy as possible.

In this latter context, the nation was focused in July 2010 on the catastrophic oil leak from a ruptured well in the Gulf of Mexico. Estimates indicated that over 757,000 L (200,000 gal) of oil were streaming into the Gulf every day, causing grave concerns about the impact on beaches and marine life in the region. Geographers were involved in mapping the extent of the spill and the coastal zones that were potentially harmed. This section will discuss some of the ways humans impact coastlines.

Coastal Engineering

One of the most obvious ways that people impact the coastline is through engineered structures. Although these structures come in a wide variety of forms, they are generally built with the idea of (1) protecting the shore and shorefront homes from coastal hazards such as erosion and storm surges, (2) stabilizing and nourishing beaches, or (3) maintaining or improving the flow of trade or traffic into ports.

Shoreline Protection   Of the many people who live near U.S. coastlines, a large number choose to reside or work directly on the shore (Figure 19.35). The reasons for this choice are probably obvious, as the shore is a pleasurable landscape in which to live and work, with spectacular views and numerous recreational opportunities. However, coastlines are one of the more sensitive landscapes on Earth, with changes occurring on a daily basis and major adjustments taking place when strong storms such as hurricanes strike.

Although extreme events such as hurricanes are indeed dangerous, they are statistically rare along any particular stretch of coast. As a result, people tend to be much more concerned about the erosion associated with less severe storms, waves, and high-water level. Unchecked, the effects of shoreline erosion on coastal communities can be disastrous (Figure 19.36). In an effort to control these effects, people have devised a variety of protective structures.

Perhaps the most common way people attempt to mitigate coastal erosion is to build *sea walls* and *revetments* that armor the shore. As the name implies, a sea wall is a vertical, wall-like structure built along the coast, usually at the landward edge of a beach, where a bluff or dune may occur. Figure 19.37 shows a sea wall at Palm Beach, Florida. Many protective structures are less elaborate, consisting of large rocks and sometimes even cement blocks from old roads. This kind of protection is called *rip-rap* and forms a revetment because it is usually applied to a preexisting slope, rather than being built vertically. Although sea walls and revetments protect the property they front from erosion, they cause more extensive erosion up and down shore than would otherwise occur naturally because the erosive power of waves is transferred to and concentrated on those locations (Figure 19.38).

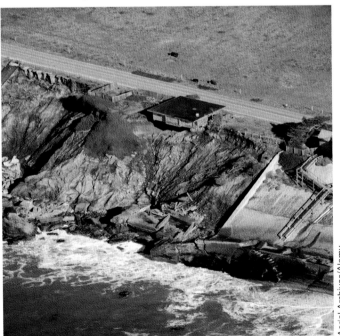

Figure 19.35 Coastal development on Padre Island, Texas. Much of this barrier island, which is shown in the satellite image in Figure 19.26, is heavily developed. This pattern of development is very common along the Gulf and Atlantic coasts of the United States.

Figure 19.36 Effects of coastal erosion on personal property. Severe bluff erosion along this stretch of the California coastline can be a problem for those who choose to live really close to shore.

Figure 19.37 Sea walls. This sea wall (arrow) at Palm Beach, Florida, protects a hotel complex from coastal erosion. Structures like this armor much of the shore in southeastern Florida.

Beach Nourishment One of the primary reasons so many people like coastlines is that they are attracted to the beach and all the recreational and leisure opportunities it provides. Much economic activity is directly tied to beaches, through either tourism or the many people who live there. Given these important relationships, people have a strong desire to maintain the physical integrity of their particular beach, whether it fronts a luxury hotel, condominiums, or a row of homes.

However, this desire to maintain an attractive beach is frequently at odds with natural coastal processes because beaches often consist of unconsolidated sands that are easily eroded. In an effort to offset coastal erosion, many communities have organized beach nourishment programs that bring in more sand to replace sand that is lost.

Figure 19.38 Beach walls and coastal erosion. Although sea walls protect the property they front from erosion, they cause extensive erosion up and down shore because wave power is deflected.

Though effective, these programs are very costly. In the very small state of Delaware, for example, over 2,300,000 m³ (3,008,286 yd³) of sand has been used to nourish beaches since the early 1960s at a cost in 2010 dollars of over $60,000,000. In the tourist-dominated state of Florida, the cost to nourish beaches over the same length of time has been over $432,000,000 in 2010 dollars.

One way that people nourish beaches is simply to bring fresh deposits of sand into a beach system to replenish sand lost through erosion. In this manner of beach maintenance, which usually occurs at the end of the winter storm season, large earth-moving equipment transports sand from another source to a heavily eroded beach. Another way that people try to maintain a functional beach is to limit the loss of sand to erosion. The most common method used to stabilize sand supply is building a simple structure called a **groin**, which is a low wall built at a right angle from the shore a short distance into the water. This wall intercepts and slows the longshore current on the upcurrent side of the structure, causing deposition of sand at that locality (Figure 19.39).

Groin *A short, low wall built at a right angle to the coast that intercepts the longshore current.*

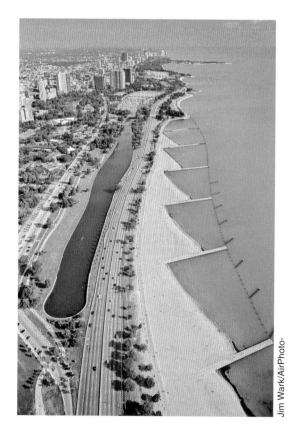

Figure 19.39 Groins and shoreline processes. Groins such as these along Lake Michigan in Chicago are walls built perpendicular to the shore. These walls intercept the longshore current, causing deposition of sand on the upcurrent side and erosion on the downcurrent side of the structure. In Lake Michigan, the prevailing longshore current flows from north (at the top of the photo) to south. Note how sand has accumulated on the north side of these groins and has washed away on the south side.

Human Interactions with Coastlines • 531

Although this process results in beach widening on one side of the groin, the portion of the beach immediately downcurrent of the structure becomes a zone of erosion because the longshore current is regenerated there due to unaltered waves striking the shore obliquely. Thus, it is possible for one landowner, say, landowner A, to build a wide beach in front of her property by installing a groin that, in turn, causes erosion and beach loss at the adjacent property of landowner B down the coast. To avoid this loss, landowner B might also install a groin to intercept the sand eroded on his property caused by the groin on the property immediately up the coast. Given this downshore ripple effect, it is quite common to see a series of groins along a coast in distinct *groin fields*.

Jetties   Another purpose for protecting beaches and coastlines is to maintain a channel between the open water and interior water bodies such as lagoons, lakes, and rivers. This access is maintained with **jetties**, which are long stone or concrete structures used to create a permanent channel by reinforcing both sides of the passageway (Figure 19.40). From the shore, jetties project several hundred meters out into the ocean and provide a zone of relatively quiet water for a boat to approach

Courtesy of Lanmon Aerial Photography, Corpus Christi, TX.

Figure 19.40 A typical jetty.   This jetty on the Texas coast provides a reliable passageway into the ocean for boats. Given the variable beach thickness on either side of the jetty, can you tell which way the longshore current is flowing? Note the sequence of foredunes on the updrift (arrow) side of the jetty. These dunes formed due to high sand supply because sand was trapped.

Jetties   *Long stone or concrete structures used to create a permanent opening for a channel by reinforcing both sides of the passageway.*

the channel safely. Like groins, jetties cause deposition on the upcurrent side of the structure and erosion on the downcurrent side. To offset beach loss on the downcurrent side of jetties, beaches are often nourished either through direct deposition of sand or with groins.

Global Climate Change and the Impact on Coastlines

Recall from Chapter 9 that one of the major environmental issues today is global climate change. As noted in that chapter, a strong scientific consensus now exists that average global temperature will likely increase between 2°C and 4°C (4°F to 7°F) by the end of this century. Among other negative effects, this increased temperature will cause further melting of polar ice caps and alpine glaciers, as we discussed in Chapter 17. This meltwater will return to oceans, causing sea level to rise. The rising sea level caused by melting glaciers will be magnified by the thermal expansion of water due to increased global temperature. According to some of the nation's top Earth scientists, sea level appears to be rising faster than was predicted just a few years ago. More importantly, current projections indicate that sea level will rise somewhere between 20 cm and 2 m (8 in. and 6 ft) by 2100 (Figure 19.41a).

Assuming that the amount of sea level rise by 2100 falls somewhere in the middle between the two extremes shown in Figure 19.41a, it still means an increase of about 1 m (3 ft) this century. The impact of such a rise in sea level on coastal zones around the world will be significant. On a global scale, it is estimated that over 100 million people are at risk of coastal flooding, with governments in countries like Bangladesh, Thailand, Indonesia, China, and the Netherlands particularly concerned. In the United States, significant regions are susceptible to rising water levels, with up to 8 million people at risk. For example, Figure 19.41b shows the areas vulnerable to these changes along the coast of North Carolina. This map is color-coded: areas in red are coastal zones less than 1.5 m (5 ft) above sea level, whereas those in blue lie between 1.5 m (5 ft) and 3.5 m (11.5 ft) above sea level. As you can see, much of the eastern part of the state is very close to sea level. Of these regions, the extensive zone less than 1.5 m (5 ft) will likely be impacted somehow from coastal flooding associated with rising sea level. The likely response of environmental planners and government officials to this threat will be more funds devoted to coastal engineering and beach stabilization.

Another reason rising sea level is a cause for concern is that it will likely increase the amount of damage associated with strong storms and hurricanes in the future. As discussed in Chapter 8, one reason why Hurricane Sandy was such a catastrophic storm was that it made landfall during the highest tide of the month. In addition, sea level was a few cm higher due to thermal expansion than during the winter months when water temperature is cooler. Given these combined factors, the storm surge in northern New Jersey and Staten Island, New York, was particularly intense and the impact on coastal communities disastrous (Figure 8.27). It is possible that this storm was a harbinger of future events should the sea levels predicted

(a)

(b)

Figure 19.41 (a) Potential sea-level rise scenarios by 2100. Present research suggests that sea level will rise somewhere between 20 cm (8 in.) and 2 m (6 ft) this century. (b) Predicted extent of coastal flooding in North Carolina. Much of the coastline lies less than 1.5 m (5 ft) above sea level and thus will likely be impacted to some extent by coastal flooding associated with global warming. (*Source:* a: NOAA.)

in Figure 19.41a transpire. As potentially bad as the impact of rising sea levels may be on coastal communities in the United States and other developed countries, at least the very security of the nation will not be threatened as it is in some places. An area that is particularly threatened by rising sea levels is the Pacific Island region, which contains hundreds of low-lying atolls and volcanic islands. Countries such as the Marshall Islands, which consists of an archipelago of 34 atolls and coral reefs with a maximum elevation of less than 3 m (10 ft) above sea level, are already being affected by rising sea level. Several islands have been lost to erosion throughout the Pacific Island region, and plans are being developed in many places to relocate coastal inhabitants to more interior locations within the islands. In response to this regional crisis, people of the various island nations have formed the *Alliance of Small Island States* to lobby the United Nations about their particular plight, which, they argue, is impacting them disproportionately given that they produce only 0.06% of all carbon emissions.

Coral-Reef Bleaching Another by-product of global climate change is the impact that warmer ocean waters are having on coral-reef systems around the world. These systems are being affected by a process called *coral bleaching*, which occurs when coral reefs are stressed by warmer-than-normal waters. Coral bleaching can occur naturally as a result of short-term temperature changes, such as those that occurred in association with the El Niño in the early 1980s. Coral bleaching also occurs due to overexploitation of coral resources by people and other forms of local environmental degradation. Since the 1980s, however, growing stress caused by warmer oceans has been increasingly linked to coral bleaching. This stress loosens the algae that help feed the coral-building organisms from the reef. Because the algae give the corals their color, the starved corals look pale or bleached. Ultimately, continued bleaching kills corals.

In the past decade, significant amounts of bleaching associated with above-normal ocean temperatures have been reported in a variety of Pacific islands, the Caribbean Sea, Indian Ocean, and along the coast of Australia. In August 2010, for example, a major bleaching event occurred in the corals of Sumatra in Indonesia. Over 60% of corals in the area were bleached, with over 80% of some species lost. This event occurred because ocean temperatures were over 32°C (90°F).

KEY CONCEPTS TO REMEMBER ABOUT HUMAN IMPACTS ON COASTLINES

1. Coastlines are favored places for people to work, live, and play. Given the inherent sensitivity of the coastal landscape, human impacts are significant and growing. Most human impacts along coastlines are associated with beach stabilization, protection, and access (to port) maintenance.

2. Beaches are stabilized predominantly in two ways: beach nourishment and groin construction. Beach

Rising sea levels threaten many coastlines around the world. Perhaps the most sensitive coastal zone in the United States is the Gulf shore from Texas to Florida. The coast of Louisiana is particularly susceptible to rising sea levels because of the close relationship it has with the Mississippi River and its delta. Large areas of coastal loss have already occurred in the past few decades. To see the complex history of coastal zone, and the rates of projected sea-level change in the future, go to the *Geo Media Library* and select *Evolution of the Louisiana Coastline*. Once you complete the animation, be sure to answer the questions at the end to test your understanding.

nourishment is an active process involving the physical transport of sand to the beach, whereas groins take advantage of the natural behavior of the longshore current.

3. Beach and bluff protection is mostly accomplished through the installation of sea walls and revetments, which armor the shore. Although these features work in the sense that they protect the portion of shore they front, they transfer wave energy to adjacent localities where it is concentrated as an erosive force.

4. Jetties are long walls built along river channels and inlets to maintain a passageway from the open ocean to the interior. These features project some distance into the water to provide a safe coastal approach for ships.

5. Global warming appears to be causing a rise in sea level. In this context, significant parts of global coastlines are susceptible to increased erosion and, in some extreme cases, ultimate submergence. Warmer oceans are also contributing to coral bleaching.

SUMMARY OF KEY CONCEPTS

1. A coastline is a narrow zone where the hydrosphere, lithosphere, and atmosphere interact on a very large scale. These interactions result in very distinctive processes and landforms. Most coastlines are associated with the world's oceans, seas, and gulfs, but some prominent coastlines occur along very large lakes such as the North American Great Lakes.

2. The world's oceans collectively constitute the largest component of the hydrological cycle. Smaller bodies of water associated with oceans are seas and gulfs.

3. The most important agents of coastal change include water fluctuations, tides, waves, and littoral processes. The most extensive water fluctuations occurred when continental glaciers advanced and retreated during the Pleistocene. Tides are daily fluctuations caused by the gravitational pull of the Moon (mostly), and to a lesser degree the Sun. Waves form when wind blows across the water, causing energy within the water to roll forward due to friction. Waves often approach coastlines obliquely, resulting in the littoral processes

of (1) longshore current in the water along the shore and (2) swash/backwash of water moving up and down, respectively, on the beach.

4. Erosional coastlines evolve when sediment is removed by coastal processes. Prominent erosional landforms along coastlines are headlands, bluffs, sea stacks, and marine terraces. Depositional coastlines develop where sediment accumulates. These landforms include beaches, spits, baymouth bars, tombolos, and barrier islands.

5. A slight majority of people on Earth live near coastlines. Given this population distribution, coastlines are impacted greatly by human behavior. Most of this impact occurs in the form of engineered structures, such as seawalls and jetties, which protect the shore and maintain the location of shipping channels.

6. It appears that global warming is having a major impact on coastlines around the world. These impacts include bleaching of coral reefs and rising sea levels, which are threatening coastal communities.

CHECK YOUR UNDERSTANDING

1. What is the difference between an ocean and a sea? Name the oceans of the world and at least five major seas.

2. What does the term *salinity* mean, and how is it related to the chemical composition of ocean water?

3. In the past approximately 150,000 years, what has been the primary cause of eustatic sea-level change? Why have these changes occurred, and what are some prominent landscapes that indicate such fluctuations?

4. With respect to tides, what is the difference between centrifugal and gravitational forces, and how do they produce tidal bulges?

5. Compare and contrast spring and neap tides.

6. Describe the formation and movement of waves. What happens to waves when they approach the shore?

7. Describe the processes associated with littoral drift. How are they related to the approach of waves to the shore?

8. How does a marine terrace indicate that (1) a period of coastal erosion occurred, and (2) sea level fell or tectonic activity transpired?

9. How are sea caves, sea arches, and sea stacks part of a geomorphic continuum that indicates shoreline erosion is occurring?

10. What is the difference between an erosional and a depositional coastline?

11. Name and describe the depositional landforms associated with the longshore current.

12. Why are coastlines a good place for dune fields to evolve?

13. How is the shape of river deltas influenced by coastal processes?

14. How do coral reefs form, and why is this process fundamentally different from other geomorphic processes?

15. What are the various ways that people attempt to protect beaches and bluffs from erosion? Why do these approaches also cause problems in nearby parts of the coast?

16. How does a groin take advantage of the longshore current to cause beach growth? Why are groin fields common?

17. Name and describe the two ways in which global warming may cause sea level to rise.

ANSWERS TO VISUAL CONCEPT CHECKS

VISUAL CONCEPT CHECK 19.1

Assuming that the axis of this part of Australia's coast aligns north and south, the wave-forming winds must be out of the southeast. These winds are causing swells to move slightly toward the northwest before they break as waves and strike the coast. This oblique approach causes the force of the water to be deflected toward the north, resulting in the formation of a longshore current flowing in that direction. In association with this current and swash/backwash of the surf, littoral drift is toward the north along the shore.

VISUAL CONCEPT CHECK 19.2

Sea stacks and arches are indicative of a retrograding coastline. These landforms are bodies of relatively resistant rock that have, until now, withstood the force of erosion caused by waves.

VISUAL CONCEPT CHECK 19.3

Assuming north is at the top of the photograph, this landform indicates that the longshore current is flowing slightly from northwest to southeast. The spit has formed because the longshore current encountered the relatively still water in the open ocean, causing deposition of sediment and progradation of the landform. The spit is curved because waves are moving sediment landward at the same time that it is being deposited at the tip of the feature.

CHAPTER TWENTY

RELEVANCE OF PHYSICAL GEOGRAPHY TO ENVIRONMENTAL ISSUES

This text has focused on the physical geography of Earth. The vast majority of topics dealt with the natural geographic patterns of the planet, such as seasonal variations in solar radiation, climate variability, distribution of vegetation and soils, location of tectonic plates, behavior of streams, and the configuration of coastlines, to name a few. In addition, an underlying theme has been human interactions with the physical environment. In Chapter 9, for example, we investigated the apparent human impact on Earth's climate system. We also addressed some of the ways that humans interact with streams, soils, eolian processes, and coastlines, and the ways that humans harness energy from the sun, wind, and fossil fuels. This closing chapter will sharpen the focus on this theme of human interactions by looking at some key issues that you will likely hear about in the future—issues that relate to the natural environment and the way that humans interact with it. The purpose of this discussion is to demonstrate how an understanding of physical geography is extremely relevant to many societal issues.

Neil W Zoglauer/Flickr/Getty Images

Harvesting wheat in the Great Plains. Wheat production requires an excellent understanding of the holistic relationship among soils, climate, and vegetation. In a world of rapidly growing human population, good knowledge of physical geography is essential to create a sustainable food supply. This chapter focuses on some of the various ways that physical geography relates to important human/environmental interactions on Earth.

CHAPTER PREVIEW

A Short History of Human Population

Case 1: Water Issues in the Arid American Southwest

Case 2: Soil Salinization in Arid and Semi-Arid Lands

Case 3: Developing Unconventional Oil Supplies in North America

Case 4: Giant Panda Conservation in China

Summary

▌LEARNING OBJECTIVES

1. Describe how an understanding of physical geography is essential to a consideration of contemporary human/environmental issues.

2. Discuss the history of human population and the factors that have fueled its growth over time.

3. Compare and contrast the physical setting of Las Vegas, Nevada, its history of population growth, and ongoing water issues in the region.

4. Describe the various ways that soil salinization occurs and why it is a major problem in California and Australia.

5. Discuss the general history of oil production in the United States, our effort to develop unconventional sources, and potential environmental impacts.

6. Discuss the nature of the panda habitat and how it has been progressively fragmented over time. Explain how geographical principles are being used in an effort to stabilize panda populations in the wild.

A Short History of Human Population

Before beginning the investigation of human/environmental interactions, we should first consider the general history of human population on Earth and how it is expected to change over the course of your lifetime. This discussion provides important context because it is directly related to resource consumption and the ability of people to live in a particular place or consume resources in a *sustainable* way—in other words, in a fashion that can continue for future generations. The **carrying capacity** of any given landscape refers to the number of individuals who can live in it without degrading the natural, social, cultural, and economic environments beyond sustainable limits. A number of variables directly associated with physical geography affect the carrying capacity of the land, including climate, vegetation, and soils.

Technological Development and Population Growth

A good place to begin a discussion of human population is the end of the most recent ice age approximately 10,000 years ago. As discussed in Chapter 9, the beginning of the Holocene Epoch marks the onset of what climatologists generally consider to be "modern climate" (Figure 20.1). Warming in the early Holocene was rapid, and by 9000 years ago the average Earth temperature was about 4°C (~7°F) warmer than it had been just a thousand years before. Although distinct warmer/drier or cooler/wetter periods have occurred within the Holocene, as shown in Chapter 9, the overall climate has generally been much more hospitable to humans in the past 10,000 years than it was at the end of the Pleistocene when glacial ice covered much of the northern latitudes. Partially in response to this more favorable climate, human societies gradually became more complex. People moved into newer areas and technological innovation increased. A key result of this cultural evolution was the beginning of agriculture about 8000 years ago in several widely separated places on Earth. Prior to that time, people were essentially hunters and gatherers who spent much of their lives foraging for food.

The onset of agriculture was critical in human history because people could now grow their own food rather than search for it. In so doing, humans gradually raised the carrying capacity of many landscapes in such a way that larger numbers of people could congregate in centralized locations that ultimately became villages, towns, and even cities. With food more easily accessible, people were freer to devote time to improving agricultural techniques and inventing other things that made life easier. Ultimately, these efforts led to the Industrial Revolution in the late 18th and early 19th centuries, during which numerous technological advances led to mechanized production and increased supply

---

Carrying capacity *The maximum number of organisms that can live in a given area of habitat without degrading it and causing social stresses that result in population decline.*

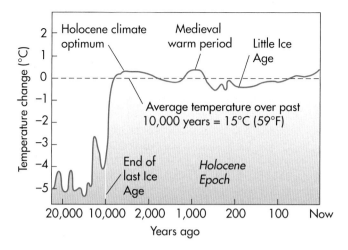

Figure 20.1 Global temperature changes over the past 20,000 years. Note the rapid warming that occurred at the end of the Pleistocene and the beginning of the Holocene. Average temperatures for this time span are compared to those of the present.

of goods. The social and economic changes brought on by the Industrial Revolution ushered in the modern world as we know it.

Historical growth in human population during the past 10,000 years closely mirrors the evolution of human technology and society. At the end of the most recent ice age, probably fewer than 500,000 people lived on Earth (Figure 20.2) and average life expectancy may have been as short as 20 years. With the advent of agriculture and enhanced food production, human population slowly increased to approximately 500 million about 2000 years ago. The population essentially held steady for another 1000 years and then gradually increased to about 600 million around the year 1700, with an average life expectancy of perhaps 40 years at that time.

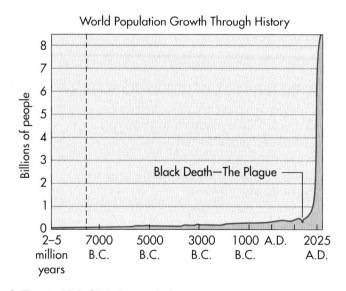

Figure 20.2 Global population change throughout human history. Note the dramatic increase during the past 200 years. It is estimated that human population will be 9 billion by 2050. (*Source*: Population Reference Bureau, *World Population: Toward the Next Century*, 1994.)

The impact of early industrialization on human population was dramatic. By 1800 the global population had reached about 1 billion. Much of this increase was driven by further technological developments, many of which were associated with sanitation, as well as improvements in medicine that dramatically enhanced human health. In response to these advances, average life expectancy in the United States approached 50 by 1900. By 2000 the human population on Earth reached over 6 billion and average life expectancy for those born in the United States was over 77 years. As of July 2013, the global population was about 7.1 billion. According to the United Nations, population is projected at 8 billion by 2028 and 9 billion by 2050. In other words, by the time a current college senior is about 60 years old, 2 billion more people will probably inhabit Earth than do now.

The Impact of Growing Human Population on the Natural Environment

The impact of this rapidly growing human population on the Earth's environment is important in many ways. At the most fundamental level a growing population means that more food is required. The need for more food, in turn, means that more forests are cleared to increase the amount of farmland. As discussed in Chapter 11, soils will also be more heavily utilized and thus degraded as farmers attempt to sustain or improve crop yields. A growing population also means that more environmental pollution occurs, often in ways that the average person cannot imagine.

An excellent example of a growing but little known environmental impact is the huge expanse of trash in the northern Pacific Ocean where the North Pacific gyre circulates (Figure 20.3a). Many of the coastal countries along the Pacific Rim, such as Japan, the Philippines, and the United States, are

Figure 20.3 (a) The North Pacific gyre. Note the counterclockwise circulation in the northern Pacific Ocean. This current moves trash in a circular motion around the Pacific Ocean.

Rosanne Tackaberry / Alamy

Figure 20.3 (b) Coastal trash on the Pacific Rim. This garbage circulated in the North Pacific gyre for an unknown period of time before it washed ashore on this beach.

Linny Morris/The New York Times/Redux Pictures

Figure 20.3 (c) Dead albatross on Midway Atoll, Hawaii. This bird died because it consumed plastic from the Pacific garbage patch. Numerous albatross chicks have died in this manner because plastic was brought back to them to eat by their parents.

densely populated and produce prodigious amounts of garbage. Some of this trash, unfortunately, winds up in the ocean and circulates around the basin in the gyre before it washes up on shore (Figure 20.3b). The amount of trash in the northern Pacific is now so high that the area has been referred to as the *North Pacific garbage patch*. To make matters worse, a huge amount of debris wound up in the northern Pacific following the recent Japanese tsunami discussed in Chapter 13. Most of this trash is plastic, some of which is being consumed by a variety of animals with harmful (and even fatal) results (Figure 20.3c). Countless examples of this kind of impact are occurring around Earth at the present time and will continue into the future.

The environmental impacts of a growing population are enormous. Some parts of the world have industrialized and reached relatively stable population levels, but others are in the process of developing economically and are rapidly growing. An example of an industrialized (or "developed") country is the United States, which has long been a global leader in technological innovation. As a result, the quality of life for the vast majority of Americans is generally high, certainly at least with respect to most other countries. Developing countries such as India and China, in contrast, are in the process of industrializing, which means that large numbers of people are reaching the middle class for the first time and thus now have the income to purchase consumer goods. Still other countries, such as a number in Africa, have just begun to develop economically and are populated mostly by people living on less than $1 per day. Many people in these countries live at a level of poverty that the average American simply cannot imagine.

The distinction between the developed and developing countries is important because their population growth rates are very different. Population in the industrialized world reached about 1 billion in 2000 and has essentially stabilized since that time (Figure 20.4). Future projections suggest that it may even slowly decrease for the next century due to lower birth rates. In stark contrast, population in the developing world began to grow rapidly in the late 20th century when technological and health advances reached those areas. Leading the way have been China and India, which now each contain more than 1 billion people. According to current projections, population in the developing world will continue to increase rapidly for the foreseeable future, with perhaps 7 billion people in these areas by 2050.

The rapid population growth rates in the developing world are very important as far as sustainable development and environmental change are concerned. For one thing, increased population pressure in regions with a moderate climate means that more people will move out of those places into marginal areas with limited carrying capacity, such as the African Sahel or the Middle East. As far as future climate change is concerned, the Intergovernmental Panel on Climate Change (IPCC) believes that the combination of increasing population and rapid economic growth in the developing world makes further temperature increases more likely because this growth will largely depend on fossil fuels that produce greenhouse gases. For example, to meet the rapidly increasing demand for energy in the growing Chinese economy, the government there is in the midst of building 160 new coal power plants in the next four years. In a similar vein, an Indian auto maker is now producing an automobile for the country's growing middle class that sells for less than $3000. Although this car is highly efficient (>40 mi/gal or mpg), its development and potential high sales make increased greenhouse gas production a certainty because tens of thousands of Indians will now own a car for the very first time.

These issues of economic development, growing human population, and environmental impacts are happening all around the world. In dealing with current and future environmental concerns, an understanding of physical geography is important to identify, manage, and, where possible, solve issues associated with sustainable development. The remainder of this chapter presents four case studies that examine the impact that humans have on the environment in a particular place, and vice versa. As you read about them, be sure to notice how these issues are distinctly related to physical geography and look for the topics that were covered in earlier chapters.

Figure 20.4 World population growth in the industrialized and nonindustrialized world from 1750 to 2100. Note the rapid growth in the developing world that is currently taking place, whereas population in the developed world has been stable for some time. (*Source*: The United Nations.)

KEY CONCEPTS TO REMEMBER ABOUT HUMAN POPULATION

1. Carrying capacity refers to the number of individuals who can be supported within a landscape in a way that does not degrade the natural, social, cultural, and economic environment beyond sustainable limits.

2. The onset of the Holocene Epoch is an important time marker in the development of human societies because the climate began warming, making life easier.

3. The development of agriculture about 8000 years ago set the stage for later population growth because it raised the carrying capacity of landscapes and allowed more people to live in smaller geographical areas.

URBAN SPRAWL

Urban sprawl is the expansion of cities outward from the original central core into formerly rural land. It is usually associated with increased suburbanization and expanding traffic networks, such as the example shown here at Phoenix, Arizona, in the lower right. Geographers like to visualize the spatial patterns of sprawl by mapping changes at specific intervals over time. For example, look at this map to see the patterns of urban sprawl in the Detroit metropolitan area in southeast Michigan since 1890. It is color-coded to illustrate the progression of urban expansion over time. The impacts of suburbanization on the physical environment are extensive, including deforestation, the loss of farmland, increased storm runoff because more surfaces are covered with concrete, a greater possibility for the urban heat island effect, and enhanced atmospheric CO_2 concentrations because more people are commuting farther to and from work.

This map shows the expansion of the Detroit metropolitan area over time. Growth patterns like those shown here have occurred in a number of American cities during the past century.

A typical suburban landscape in Phoenix, Arizona. This landscape was formerly desert that has been converted to a neighborhood setting.

4. The Industrial Revolution in the 19th century allowed human population to grow rapidly because mechanization, increased sanitation, and developments in medicine led to longer and better lives. In 1800 the global population was about 1 billion, whereas in 1900 it was approximately 1.7 billion.

5. Global population is now about 6.8 billion and is projected to reach approximately 9 billion by 2050. This large number of people is having huge impacts on the natural environment.

Case 1: Water Issues in the Arid American Southwest

The first case study examines water use in the American Southwest. It focuses on the Las Vegas metropolitan area, which has been one of the fastest-growing urban zones in America since the mid-20th century. According to the U.S. Census Bureau, the population of the Las Vegas metropolitan area increased from about 1.4 million in 2000 to about 2 million today. Ironically,

Las Vegas is located within the low-latitude desert climate (*BWh*) in the Köppen system, which is extremely arid. In other words, hundreds of thousands of people have recently moved into a region where it rarely rains and natural water supplies are scarce. Although the current economic downturn has slowed population growth in the area, the effect at this time has been essentially to stabilize the number of people rather than to cause a steep decline.

The competing pressures of high population and the potential lack of freshwater pose challenging questions for the current and future residents of Las Vegas, as well as other large cities in the Southwest such as Phoenix and Tucson, Arizona, and Albuquerque, New Mexico. Are these cities sustainable places that can continue to grow, or will they one day decline if and when water supplies give out? Regardless of the ultimate outcome, to understand the issues and appreciate the various solutions proposed require a deep understanding of physical geography.

As you probably know, Las Vegas is a world-famous resort city that draws millions of tourists every year. Located in the southern part of Nevada in Clark County (Figure 20.5), the metropolitan area lies within the Las Vegas Valley, which is part of the Basin and Range physiographic province discussed in Chapter 13. The Las Vegas Valley is bordered to the immediate

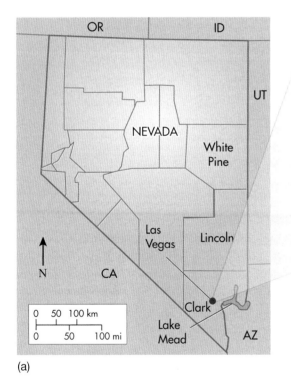

Figure 20.5 Las Vegas, Nevada. (a) Las Vegas is located within Clark County in the southern part of Nevada. (b) Skyline of Las Vegas. Las Vegas is a world-famous resort city that lies within the arid Mojave Desert.

(b)

Dennis Flaherty/Getty Images, Inc.

west by the Spring Mountains and to the east by Frenchman Mountain. Las Vegas is located about 48 km (30 mi) northwest of Hoover Dam, which impounds Lake Mead.

The Las Vegas metropolitan area is located in the central part of the Mojave Desert and is thus one of the hottest cities in the United States. The average daily high temperature is >38°C (100°F) approximately 90 days out of the year (Figure 20.6). The record high temperature is 47°C (117°F), which occurred in July 2005. Winter months are typically mild, with an average daily high temperature of 13°C (57°F) in January. The Las Vegas climate is influenced strongly by the rain shadow effect associated with the Sierra Nevada in California to the west. Average annual precipitation is only about 10 cm (4 in.), and relative humidity is usually very low. As a result, Las Vegas is one of the sunniest cities in the country, with over 300 days of clear skies per year.

The development of Las Vegas is directly tied to its distinctive physical geography. The Las Vegas Valley was "discovered" in 1829 when a scouting party associated with a Mexican trading caravan traveling to Los Angeles left the main trail in search of water. While doing so, the scouts entered an unnamed valley and found an oasis fed by numerous artesian springs. As

a result of this discovery, the site was named *Las Vegas*, which is Spanish for "The Meadows," and became an important place to replenish water supplies, first for Mexican traders and later for American explorers. The area was subsequently annexed by the United States following the Mexican-American War in 1848 and was soon occupied by Mormon missionaries, who built a fort near present-day downtown Las Vegas in the mid-1850s. The fort was soon abandoned and later reoccupied in 1865 by Octavius Gass. After constructing an irrigation system drawing water from the springs, Gass developed a farm complex that was about 243 hectares (600 acres) in size and renamed the area *Las Vegas Rancho*. Gass's farm became an important rest stop for people traveling on the nearby Mormon trail from Salt Lake City, Utah, to Los Angeles.

The population of Clark County began to grow very slowly after Nevada became a state in 1864. A major factor in the early growth of the area was the State Land Act of 1885, which allowed homesteaders to purchase land for $1.25/acre ($309/km²). The lure of cheap land drew many people to the area, most of whom were farmers. As a result, agriculture dominated the local economy until the early 1900s and was supported by an increasingly elaborate irrigation system that drew water from the wells. Some of this water was directed into the town proper, and Las Vegas became firmly established as a stop for travelers moving to and from Los Angeles and places to the east such as Albuquerque, New Mexico.

Despite the irrigation improvements in the Las Vegas Valley in the late 19th century, only about 1100 people lived in Clark County in 1900. The first surge in population occurred after the San Pedro, Los Angeles & Salt Lake Railroad linked the area to other places in the West in 1905. The city of Las Vegas was founded that same year, and the population of Clark County grew to about 3300 in 1910. Las Vegas was subsequently connected to Southern California by a highway in 1926, and the population of Clark County reached about 8500 in 1930. Approximately 60% of those residents lived in Las Vegas at that time.

Establishing Water Rights

The next major population surge in Clark County occurred in the 1930s and is directly related to the establishment of the *Colorado River Compact*, which was signed in 1922 by representatives of the seven states that lie within the Colorado River basin. This agreement was part of an overall effort by the U.S. federal government to draw people to the region and "reclaim" the arid lands for economic development. The purpose of the compact was (and remains) to sustainably allocate water to users in the basin in a fair and equitable manner by establishing a system of **water rights** defined by law. It does so by dividing the watershed into an upper basin (including Wyoming, Colorado,

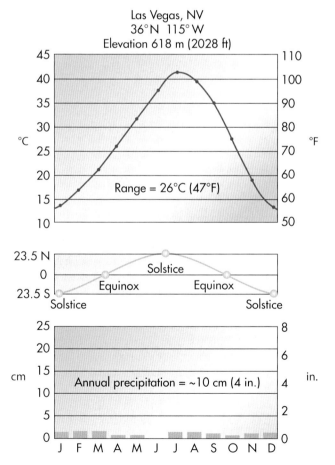

Figure 20.6 Climograph for Las Vegas, Nevada. Average monthly temperatures range from mild in the winter to blistering hot in the summer. Total annual precipitation is about 10 cm (~4 in.).

Water rights *Legally protected rights to take possession of water occurring in a natural waterway and to divert that water for beneficial purposes.*

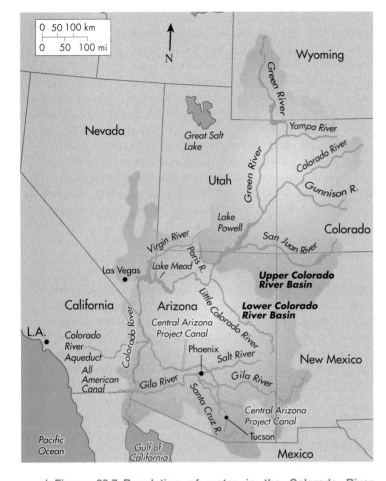

Figure 20.7 Regulation of water in the Colorado River basin. The Colorado River Compact divided the watershed into an upper and lower basin. Water in each of these basins is strictly regulated, with a set amount allowed for each region. The Central Arizona Project and Colorado River Aqueduct divert water from the Colorado River to southern Arizona and Southern California, respectively, for human purposes. Note the locations of Los Angeles, Las Vegas, Phoenix, and Tucson.

conservation and (2) producing hydroelectricity for southwestern states. During the course of construction, the population of Clark County swelled to about 25,000. It was at this time that the first of the now-famous casinos were established in Las Vegas to entertain the construction workers. After the dam was completed in 1935, the structure and Lake Mead became tourist attractions that further propelled the growth of Las Vegas, along with an influx of organized crime. By 1950 the population of Clark County was nearly 50,000.

Since the advent of air conditioning in the 1950s, the population in the Las Vegas metropolitan area has grown at a rapid rate to about 2 million (Figure 20.8a) and was expected to reach 3 million by 2020 prior to the current economic downturn. It may still reach that figure, but such projections are a bit uncertain as of early 2013. Regardless, the city is visited by about *40 million* tourists every year who are attracted to the casinos, entertainment, and area's natural beauty. This rapid population increase mirrors the rate of change in other hot southwestern cities such as Phoenix, Tucson, and Albuquerque, and is part of a general migration of Americans during the past 50 years to the southern part of the country or *Sun Belt*. Over 6 million people have moved into the region in that time, with urban areas sprawling rapidly into the surrounding deserts. A comparison of satellite images between 1973 and 2006 (Figure 20.8b) illustrates the impact that rapid population growth in Las Vegas has had on the size of the city. Las Vegas is now a major American city with wide boulevards lined with luxury hotels, waterfalls, and fountains that spray millions of gallons of water into the dry desert air.

Will Water Supplies Disappear?

As the growth of Las Vegas accelerated in the latter part of the 20th century, many people began to become concerned about the supply of water in the arid desert. Was there enough water available for sustainable growth? Was it possible that water supplies would disappear? These concerns resulted in the establishment of the *Southern Nevada Water Authority (SNWA)* in 1991. The purpose of this organization has been to develop and manage local water resources, theoretically in a sustainable way. At the present time 90% of water used in Las Vegas comes from Lake Mead on the Colorado, whereas the remaining 10% is derived from groundwater. In an effort to manage these water supplies, the SNWA established a four-phase hierarchy of drought status, including (1) no drought, (2) drought watch, (3) drought alert, and (4) drought critical, to keep people informed of current environmental conditions and the various kinds of water use that are permitted at each level.

The concerns about water have magnified since the late 1990s, when the southwestern United States entered into a major drought that has only recently eased somewhat. Average annual precipitation in the region has been about 70% to 75% of normal, with a significant decrease in the amount of snowfall each winter in the Rocky Mountains. The amount of snow is important because it provides spring runoff to the Colorado River and its tributaries, which is the major source of water for

New Mexico, and Utah) and a lower basin (including Nevada, Arizona, and California; see Figure 20.7). The key stipulation of the compact is that both the upper basin and lower basin are allowed 7.5 million acre-feet of water per year, in other words, the amount of water required to cover 7.5 million acres in 1 foot of water. Given that the Colorado River flows for a short distance in Mexico before it reaches the Gulf of California, 1.5 million acre-feet of water is allocated to Mexico per year. This stipulation means that the upper basin states cannot use all the water they desire; instead, they must allow a sufficient amount of water to flow downstream so that both the lower basin and Mexico receive their share. The lower basin, in turn, must assure that Mexico receives the amount of water required by the agreement.

The Colorado River Compact set the stage for construction of the Boulder Dam (later renamed the Hoover Dam), which began near Las Vegas in 1931. In association with the Colorado River Compact, Hoover Dam was designed for two purposes: (1) impounding water (in Lake Mead) for irrigation and water

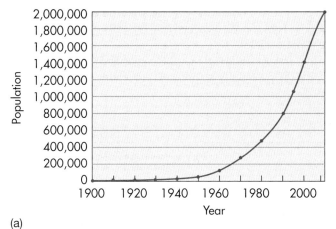

(a)

Figure 20.8 **Growth in Las Vegas population.** (a) Graph showing the growth of Las Vegas since 1900. Note the rapid increase after 1950. (b) Satellite images showing the expansion of Las Vegas between 1973 (left) and 2006 (right). The 1973 image shows an area about 25 km (~16 mi) across.

NASA Goddard Space Flight Center

(b)

VISUAL CONCEPT CHECK 20.1

What is the most likely source of water used in this fountain in Las Vegas?

a) Groundwater

b) Snowmelt from the Sierra Nevada Mountains

c) The Colorado River

d) Rainfall

Ethan Miller/Getty Images, Inc.

Las Vegas and many other southwestern cities. As you can see in Figure 20.9a, the average spring runoff in the Colorado was generally almost half of the normal amount between 2000 and 2007. This decrease in runoff has caused water levels in Lake Mead to drop to the point where it now contains only about 50% of its total capacity (Figure 20.9b). The extent of this decrease is visible on the bleached canyon walls on the edge of the lake (Figure 20.9c).

The impact of drought on Las Vegas and other southwestern cities is potentially significant, especially given the recent

(a)

(b)

(c)

Figure 20.9 Colorado River runoff and Lake Mead. (a) Average spring runoff in the Colorado River since the 1980s has been highly variable, with the first decade of the 21st century particularly low. (*Source*: U.S. Bureau of Reclamation.) (b) Due to reduced spring runoff, water levels in Lake Mead have dropped significantly in the past two decades (*Source*: U.S. Bureau of Reclamation.) (c) Current water level in Lake Mead. Since 2000, the lake level has dropped about 30 m (100 ft), exposing a ring of bleached rock around the edge of the lake.

Alan Arbogast

influx of people into the region. As a result of the drought, the SNWA moved the drought status from *drought watch* to *drought alert* and imposed a series of tougher water restrictions. For example, no new grass turf is allowed in residential yards and is limited in rear and side yards to 50% of the area, with a 3 m (10 ft) minimum dimension and a 464 m² (5000 ft²) maximum. In addition, lawn watering is allowed on only three assigned days per week and golf courses are governed by strict water budgets. If violations of these and other restrictions are observed, customers are first warned, and then subsequently fined if further waste occurs. As of October 2013, these restrictions are still in place, and, in fact, the SNWA is now subsidizing people to tear out their lawns to save water.

A Drought-Prone Future?

In the context of the recent drought, one cannot help but wonder what future climate conditions may be like in the Southwest. To answer this question it is useful to look at the past. Recent tree-ring studies, like those discussed in Chapter 9, indicate that eight droughts with intensities similar to the current one have occurred in the region within the past 500 years, with perhaps two or three being even more severe. These studies show that arid conditions are the norm and that extreme droughts are fairly common in the American Southwest. These studies also reveal that the Colorado River Compact was signed during a relatively wet phase in the regional climate, one resulting in more runoff into the Colorado than normal. In other words, the current distribution of water in the upper basin, lower basin, and Mexico is based on ideal conditions that may, in fact, be quite rare.

Unfortunately, it appears that ideal climate conditions in the Southwest may become even more rare in the future. This assessment is based on climate models that predict future environmental conditions. These models demonstrate that the southwestern climate indeed became more arid in the late 20th century and is continuing the trend into this century. The prevailing belief is that ongoing global climate change is altering the movement of storms and moisture in the atmosphere and that the drying trend will continue at least until 2100. In short, climate researchers are concerned that the drought conditions of the past decade will become the new norm for the region.

If this assessment of future climate conditions is true, it has important implications for water resources in southwestern cities such as Las Vegas and Phoenix, especially if recent growth rates pick up again as the national economy improves. An increasing number of scientists argue that planners should acknowledge that water resources are limited and that shortfalls will occur more frequently in the future. Some even suggest that recent growth rates are unsustainable and that limits should actually be placed on the number of people who can live in the region. In response to these concerns, a 2007 planning study argued that limiting growth in Clark County was an unrealistic method of drought control because the economy of Nevada would suffer. Instead, the focus should be to increase the efficiency of water usage and to continue an aggressive search for other water sources to supplement the current supply from the Colorado River and local groundwater. Efficiency has indeed improved dramatically in the recent past, as Las Vegas residents now stretch their water supplies further through the restrictions described earlier. Unfortunately, these gains are offset by the increased overall demand due to growing population. As a result, a systematic effort is currently ongoing to acquire water from groundwater supplies in Lincoln and White Pine counties, which lie north of Las Vegas (Figure 20.5a). The extent of these supplies is currently being assessed, with the goal of building a pipeline that will deliver water to Las Vegas in the near future.

Although these counties are lightly populated (~14,000 total people), many residents are fighting to keep what they consider to be *their* water. They believe that groundwater extraction from the counties will degrade the environment and destroy their traditional ranching lifestyle. They recall the negative impacts that water diversion from Owens Valley, California, to Los Angeles in the early 20th century had on the environment. Much of this water was diverted from Mono Lake, which is about 385 km (240 mi) northeast of Los Angeles, to fuel the rapid growth of southern California. As a result of this diversion, the level of Mono Lake fell about 14 m (45 ft), approximately 7500 hectares (18,500 acres) of shoreline were consequently exposed to blowing wind, and the salinity of the lake doubled by the early 1980s. The concerns of residents in Lincoln and White Pine counties are shared by officials in neighboring Utah because groundwater extraction in Nevada may impact groundwater resources there. They are particularly concerned about the viability of Fish Springs, which is a lake that is home to a fish called the least chub. Environmentalists are concerned that the least chub will become endangered if the lake level declines as a result of groundwater extraction in Nevada. Tapping groundwater resources may also be problematic because it can cause surface subsidence and fracturing. Such problems have already been identified in Tucson basin in Arizona, for example, where as much as 13 cm (~5 in.) of subsidence has occurred in some areas due to compaction of the aquifer system.

▌ **KEY CONCEPTS TO REMEMBER ABOUT LAS VEGAS WATER ISSUES**

1. The Las Vegas Valley lies within the rain shadow of the Sierra Nevada Mountains and thus receives about 10 cm (4 in.) of precipitation per year. Average annual temperatures range from warm in the winter to exceptionally hot in summer.

2. A major factor in the growth of Las Vegas was the Colorado River Compact, which was signed in 1922. This document legally subdivided the Colorado River basin into separate basins with distinct water rights.

3. Rapid growth of Las Vegas began in the 1950s, in part due to the advent of air conditioning. Population has grown from about 50,000 in 1950 to over 2 million today.

4. Increased population has raised serious concerns about the sustainability of water supplies in the region.

As a result, water managers in Las Vegas can employ strict use restrictions and are aggressively searching for supplemental sources of water from neighboring areas.

5. Regional climate models suggest that the southwestern United States will become more drought-prone in the coming decades, potentially making water concerns more acute.

As you can see, a full understanding of population growth and water use in marginal environments such as the southwestern United States requires a holistic assessment of many factors directly related to physical geography. This particular case study has included several topics discussed in earlier chapters, such as Köppen climates, the Basin and Range province, the rain shadow effect, groundwater, drainage basins, runoff, and global climate change. All these factors are intimately related to create a distinctive environment in the southwestern United States, one that happens to be the center of a recent population boom.

The societal tensions resulting from issues like this one are loosely referred to as *water wars* and are particularly intense in places where different groups of people compete for the same water supply. According to current estimates, about 1 billion people on Earth lack access to reliable freshwater supplies. This number will undoubtedly increase as global population continues to grow. In many other places, access to water is being used as a political weapon to influence the behavior of neighboring countries. An excellent example is Turkey, where dams have been built across the Tigris and Euphrates rivers that reduce stream flow into neighboring Iraq. Examples such as these abound around the world and will be political flashpoints for the rest of your life. To fully understand them, and their potential solutions, requires an excellent understanding of physical geography.

Case 2: Soil Salinization in Arid and Semi-Arid Lands

This next case study expands on the theme of human interactions in marginal lands by focusing on some of the problems that can develop in association with agriculture in these regions. As we discussed earlier in this chapter, the development of agriculture increased the carrying capacity of landscapes and gradually led to increased population. During the previous hunter/gatherer phase of human evolution, the global population was probably less than 5 million people. Today, modern agricultural techniques support a global population of about 6.8 billion, with 9 billion (perhaps more) projected by 2050.

The ability of agriculture to feed everyone on a sustainable basis poses enormous challenges. The production of cereal grains such as wheat and sorghum, for example, will have to *double* between now and 2050, not only to meet the basic needs of so many people, but also because more people will consume grain-fed beef as their standards of living improve. To meet this growing demand, further increases in agricultural output are essential for global political and social stability and equity. An even greater challenge is to meet this demand in a sustainable way that maintains the health of environmental systems.

One of the primary factors complicating the task of increasing global food production is simply the limited supply of **arable land** on Earth (Figure 20.10). Recall from the discussion of the hydrological cycle in Chapter 7 that about 71% of

---

Arable land *Land that has the high potential to be cultivated for crop production.*

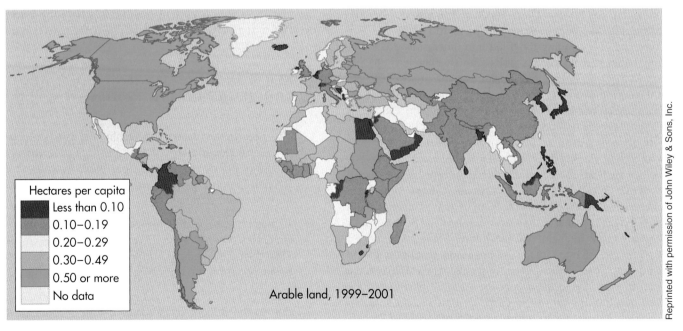

Hectares per capita
Less than 0.10
0.10–0.19
0.20–0.29
0.30–0.49
0.50 or more
No data

Arable land, 1999–2001

Reprinted with permission of John Wiley & Sons, Inc.

Figure 20.10 Map of global arable lands. This map shows the countries that contain arable land; it is somewhat misleading. Much of the western United States, for example, as well as most of Australia, contains very little arable land. Regardless, note that this map indicates that extensive parts of the world contain few lands suitable for agriculture.

the Earth's surface is covered with water, mostly contained in oceans. Of the remaining 29% of the Earth's surface, significant parts of the land area are simply unsuitable for large-scale agriculture on a sustainable basis. Such places include the world's arctic and subarctic climate zones, the major desert regions, and large parts of the tropical rainforests. In fact, only 22% of the Earth's land area is potentially productive for agriculture. Of these areas, only a few such as the American Midwest, southeastern Asia, parts of South America, and eastern Europe are ideal for intensive agriculture because they have fertile soils and abundant precipitation.

With the increasing demand for food, more and more agriculture is being practiced in marginal lands where the potential for major environmental degradation is high. In Chapter 18, two such examples of human impact were the overplowing of soils that created the American Dust Bowl of the 1930s and changing land-use policies that led to massive desertification in the Sahel region of Africa. This second case study focuses on another environmental impact of expanding agriculture in arid and semi-arid regions. This impact is **soil salinization**, which is the accumulation of soluble salts such as sulfates of calcium and potassium in soil. Salt buildup in soils is toxic to most plants because salts decrease the osmotic potential of the soil, that is, the ability of soil water to enter permeable plant roots. As soils become increasingly salty, they have greater concentrations of solute than the plant roots do, so plants cannot absorb water from the soil.

Salinization of soils can occur under natural conditions and is most commonly associated with topographic depressions (Figure 20.11). In these microenvironments dissolved salts are washed by overland flow into the low area, or alternatively, the base of the depression lies just above the water table. The source of these salts can either be the weathering of minerals on the slopes above the depression, ions precipitated in the rainwater, or solutes leached from saline rocks such as gypsum. If the salts are washed into the depression from the nearby slopes, they will recrystallize at the surface when the water evaporates. If the water table is shallow, salts can be brought up from deeper horizons by the process of capillary action (or "wicking") described in Chapter 15. When excess salts accumulate, soil structure breaks down because clays no longer bind together. Recall from Chapter 11 that soils with a moderate amount of clay typically have good structure, which allows water to flow through the soil rapidly. When this binding ceases, the clay particles act individually to plug soil pores. As a result, the soil becomes increasingly impermeable to water.

Soil salinization can also occur when agricultural practices change the hydrological and chemical aspects of soils. This change can happen in two ways. One way is related to

---

Soil salinization *The process in which soils become enriched in soluble salts such as sulfates from calcium and potassium.*

(a)

(b)

Figure 20.11 Natural salinization of soils. (a) Salt crusts naturally form in topographic depressions within arid regions because slopewash evaporates and leaves a salt crust behind. (b) Salt crust at Bad Water in Death Valley, California. This playa contained water during the Wisconsin ice age that evaporated during the Holocene, leaving mineral salts behind. Today, occasional rains promote runoff into the playa. Some salts dissolve and then recrystallize when the water evaporates. The hexagonal pattern results from repeated freeze/thaw and evaporation cycles that push the crust into these forms.

rising water tables associated with the clearing of native vegetation (Figure 20.12). In arid and semi-arid areas where natural vegetation occurs, rainwater typically leaches salt from the permeable soil deep into the ground and into the groundwater. Even though the salinity of the groundwater is high in these circumstances, native plants do fine as long as the water table is about 4 m (13 ft) below the surface. In places where the water table is shallower, some native plants such as salt grass can cope with higher levels of groundwater salinity. When native vegetation is cleared in these areas, the water table typically rises because less precipitation is intercepted by plants and less evapotranspiration occurs. If the water table rises to within about 2 m (~6.5 ft) of the ground, saline water can begin to seep into topographic depressions and salt crusts can form through evaporation. In addition, saline water can begin to rise to the surface by capillary action. In this way, soils become laden with salts.

Before land clearing, transpiration from
trees helps to keep the water table down

(a)

After land clearing, discharge through
transpiration decreases, causing the
water table to rise

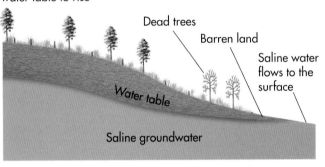

Dead trees

Barren land

Saline water
flows to the
surface

(b)

**Figure 20.12 Salt crust formation after native vegetation is
cleared.** (a) Native vegetation keeps the water table down due
to transpiration. (b) Once the native plants are removed, more
groundwater recharge occurs because agricultural crops absorb
less water. The water table then rises, causing salinization in topographic depressions.

Another way that agriculture causes soil salinization is
through irrigation (Figure 20.13). In irrigated fields the water
applied to the soil can raise the water table. Evidence from
around the world indicates that irrigation can cause the local
water table to rise rapidly, with increases up to 3 m (~10 ft) reported in some places. As with the clearing of native vegetation,
such a rise can lead to wicking of salt in the soil and formation
of salt crusts as the water evaporates. Especially in warm and
dry areas, if the applied water is already concentrated with salts,
the irrigated water will rapidly evaporate from the surface, leaving higher levels of salt in the soil. Human-induced soil salinization has been a problem since the dawn of civilization. One of
the best examples of the impact that increased soil salinity has
on people is ancient Sumeria in Mesopotamia.

Ancient Sumeria

The Sumerians lived between about 4000 and 2300 B.C. in the
fertile alluvial plain of the lower Tigris and Euphrates river valleys in what is now southern Iraq. This region lies within the
southeastern part of the famed *Fertile Crescent* (Figure 20.14a),
which is one of the key places where early human civilization
developed. Although this region lies within the hot low-latitude

(a)

Plants transpire water

Water
evaporates

Water plus salt
goes into soil

Salts remain behind

(b)

Figure 20.13 Soil salinization due to irrigation. (a) Soil salinization can occur in irrigated fields because application of water
causes the water table to rise close to the root zone. Once this
occurs, water and salts move to the surface by wicking. The water dissolves, leaving salt behind. (b) Soil salinization in farm fields
can also occur if saline water is applied to fields. Once the water
evaporates, salt crusts form at the surface.

desert climate region (*BWh*), the Sumerians thrived for centuries because they developed a complex irrigation network of
canals and aqueducts to divert water from the rivers to outlying
farm fields. It is believed that wheat was first grown as a domes-
tic crop during this time. As carrying capacity increased, the
first city-states (Figure 20.14b) evolved in this part of the world
and the first form of systematic writing began (Figure 20.14c).

At the same time that Sumerian society became increasingly
sophisticated, its foundation was apparently slowly collapsing
as soil salinity increased due to rising water tables and wicking.
Indirect evidence for increased salinity can be seen in the counts
of grain impressions in excavated pottery, which indicate that by

(a)

(b)

(c)

Figure 20.14 The ancient Sumerians. (a) The Sumerians lived in the lower Tigris and Euphrates valleys in what is now southern Iraq. This area was part of the Fertile Crescent. (b) Artist's rendering of a Sumerian temple. The development of agriculture raised the carrying capacity of the landscape and allowed people to congregate in smaller places and diversify tasks. (c) The Sumerians were the first people to use writing, such as seen on this tablet.

arable land and to experience a near-constant state of conflict that ultimately caused Sumerian society to collapse. The legacy of salinity is still keenly felt in modern Iraq today, where at least 20% of the potentially arable land is permanently destroyed.

The Current Global Extent of Salinization

Soil salinization is currently a problem in many parts of the world (Table 20.1). Today, it affects between 20 and 30 million hectares (50 and 75 million acres) of the world's current

2500 B.C. wheat production had dropped by 83% and been replaced by barley. This switch was apparently made because wheat is one of the least salt-tolerant crops, whereas barley does somewhat better. Between 2400 and 1700 B.C., however, barley yields dropped substantially as salinity increased further and massive crop failures began soon thereafter. Archaeologists believe that these stresses led the various city-states in the region to compete for the remaining

TABLE 20.1	Extent of Salinized Soils Throughout the World	
Region	Area in Millions of Hectares	Area in Millions of Acres
Australia	84.7	209.3
Africa	69.5	171.7
Latin America	59.4	146.8
Near and Middle East	53.1	131.2
Europe	20.7	51.2
Asia and Far East	19.5	48.2
North America	16.0	39.5

Source: FAO/UNESCO.

260 million hectares (642 million acres) of irrigated land and has the potential to severely limit world food production. Every inhabited continent has areas that are being affected by salinization, with Australia leading the way at over 84 million hectares (209 million acres) affected. In the United States, salinization may be lowering crop yields on as much as 25% to 30% of the nation's irrigated lands. In Mexico, salinization is estimated to be reducing grain yields by about 1 million tons/year, which is enough to feed approximately 1 million people. The remainder of this case study examines two examples of salinized soils—the San Joaquin Valley of California and Australia.

The San Joaquin Valley in California

Perhaps the best example of the impact of soil salinity in the United States today is the San Joaquin Valley of California, which encompasses the southern half of California's Central Valley (Figure 20.15). In contrast to much of the rest of California, the valley terrain has very limited relief due to the complex geological history of the area. The geological foundation of the valley was created when subduction along the North American coastline during the Mesozoic Era established a structural basin that lay below sea level. This basin was later filled with about 7600 m (25,000 ft) of marine sediments. These deposits were then buried by as much as 730 m (2400 ft) of alluvial sands and gravels derived from the rising Coast Ranges to the west and the Sierra Nevada to the east. Taken together, this depositional sequence created an extremely flat landscape that

Figure 20.15 The San Joaquin Valley. Map of annual precipitation within the San Joaquin Valley and surrounding area. The dry area on the west side of the valley lies within the rain shadow of the Coast Ranges.

generally lies less than 122 m (400 ft) above sea level. Given the higher elevations of the Coast Ranges to the west, the San Joaquin Valley lies within a rain shadow and thus receives less than 38 cm (15 in.) of rainfall per year. The climate of the region is generally classified as hot low-latitude steppe (*BShs*) and even low-latitude desert (*BWh*) in places, with hot and dry summers and cool, damp winters. The northern part of the region is drained by the San Joaquin River. Streams in the southern end of the valley flow into Tulare Lake in the Tulare Basin.

In spite of the semi-arid climate of the region, the San Joaquin Valley is one of the most important agriculture zones in the United States, with 5 of the top 10 farming counties in the country. The valley has often been referred to as the *nation's salad bowl* for the amazing array of fruits and vegetables grown here. If you happen to like asparagus, grapes, raisins, almonds, and pistachios, they are likely grown in the San Joaquin Valley. In addition, the largest single cotton farm in the country is found here. Agriculture has been so successful in the valley because a complex irrigation network was developed that delivers water to fields whenever it is needed (Figure 20.16).

Large-scale diversion of water began in 1935 when the Bureau of Reclamation launched the *Central Valley Project (CVP)*. Like the hydrological projects we viewed previously around Las Vegas, the purpose of the CVP (and later, the State Water Project) has been to manage water resources in central California and, among other things, relocate water with a system of dams and canals from the humid northern part of the Central Valley to the more arid south. In addition, the underlying Central Valley aquifer has been systematically tapped, much like the High Plains aquifer discussed in Chapter 15. Taken together, these water projects have turned what was previously a semi-arid desert into productive farmland. As a result, the area is now considered by some to be the engine of the California economy. This significance is not to be underestimated, as the California economy provides 13% of all goods and services in the United States and would be the 10th largest in the world if the state was its own country.

Consequences of Irrigation   Nevertheless, the impact of irrigation on the natural environment in the valley has been extensive. In some places, withdrawal of water from the Central Valley aquifer has caused the land surface to subside, with millions of dollars of damage to buildings, aqueducts, bridges, and highways. Irrigation has also caused extensive salinization of the soils in the valley (Figure 20.17), the effects of which have been magnified by the hydro-geologic relationships in the area. To water their fields, farmers divert water from irrigation canals or draw fresh groundwater from the Central Valley aquifer (Figure 20.16). Early on, these irrigation strategies were beneficial to the soil because they flushed salts from the sediment, which naturally occurred given the marine geologic history of the region. Unfortunately, a fine-grained marine deposit, called Corcoran clay, lies at a shallow depth below the surface in the central and western parts of the valley. This clay layer is impermeable, which causes percolating irrigation water to perch on top of it and the water table to rise. Because this part of the groundwater system does not reach the freshwater in the main aquifer, the groundwater perched on

(a)

(b)

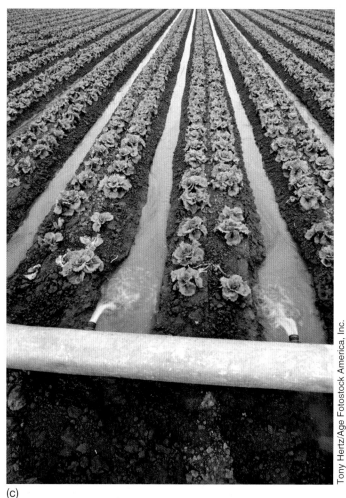

(c)

Figure 20.16 Irrigation in the San Joaquin Valley. (a) Water used in spray irrigation is derived from the Central Valley aquifer. (b) Irrigation canals such as this one divert water from the streams to specific farm areas. (c) Row irrigation of lettuce. Water from irrigation canals is delivered to a key part of your next salad in this way.

the clay becomes perpetually more saline, which, in turn, causes the salinization of soils in a manner entirely consistent with what the Sumerians experienced thousands of years ago.

Figure 20.17 Soil salinization in the San Joaquin Valley. This photo nicely illustrates the effects of salinization in the San Joaquin Valley. Notice how the salty patch (large white area) is very poorly vegetated.

In an effort to mitigate the impact of increased soil salinity, the Bureau of Reclamation devised a way in the 1960s to drain away the shallow groundwater to stabilize the depth of the water table below the root zone of plants and to facilitate leaching of surface salts. This diversion was accomplished by the construction of the San Luis Drain, which flowed 134 km from the southern end of the valley to Kesterson Reservoir (Figure 20.18), a manmade impoundment containing 12 evaporation ponds within the Kesterson National Wildlife Refuge. The drainage system was completed in 1971 and appeared to work well until it was discovered in the early 1980s that water in the impoundment contained elevated levels of selenium, which is toxic to wildlife in high concentrations. Although selenium occurs naturally in the area soils, especially the eastern side of the valley near outcrops of marine rocks within the Coast Ranges, it became concentrated in the groundwater because percolating irrigation water flushed the selenium from the surface and it ultimately drained into Kesterson Reservoir. The impact of this selenium enrichment in the groundwater became apparent when many overwintering migratory birds and newborn chicks at the reservoir developed deformities or became so sick that they drowned. In addition, ranchers near the drain and reservoir reported numerous livestock deformities

(a)

(b)

(c)

Figure 20.18 Drain system in the San Joaquin Valley. (a) Map of the San Luis Drain and Kesterson Reservoir. This system was used to drain excess groundwater from farm fields in the valley, some of which contained high levels of selenium derived from the Coast Ranges. (b) Photograph of Kesterson Reservoir. (c) Deformities in a bird chick linked to high levels of selenium in the water at Kesterson Reservoir.

and deaths. As a result of this environmental problem, the San Luis Drain was closed in 1985. Amid much public outcry, it was reopened in 1996 to receive limited amounts of drain water.

Possible Solutions Given the environmental problems associated with draining saline groundwater from the San Joaquin Valley, environmental managers continue to look for a way to reduce soil salinity in the region. In the mid-1980s the United States Geological Survey (USGS) took the lead in the search for new and better ways to rid the valley of excess groundwater. A number of potential solutions continue to be discussed. One idea is to

pay farmers hundreds of millions of dollars to take about 79,000 hectares (~195,000 acres) of the most sensitive farmland out of production. Another idea is to build more than 2000 artificial evaporation ponds into which groundwater would drain and evaporate. This method would concentrate salts at specific locations, which could then perhaps be transported to the ocean and dumped. It would require removing land from productivity, however, to create room for the ponds. In the western part of the valley, for example, it is believed that one pond is required for every nine fields. Still other ways of coping are to treat the water with expensive technology and to grow more salt-tolerant crops such as cotton.

Imagine you are flying over the San Joaquin Valley in California and see this landscape when you look out the window. What is the best explanation for the whitish area visible in the center of the image?

Scott Bauer, courtesy of the U.S. Department of Agriculture

At this point in time, no systematic plan has been implemented to deal with soil salinity in the San Joaquin Valley. Regardless of the method(s) employed, some kind of soil salt balance must be achieved for irrigated agriculture to be sustainable in the long term in this critical agricultural zone. Recent modeling of groundwater salt concentrations suggests that in the near term some kind of soil salt balance can be attained. The long-term prospects are not as favorable, however, primarily because it appears that salt concentrations in the deep Central Valley aquifer are slowly rising. These rising concentrations will be a problem in the coming decades because this water supply is used extensively for irrigation. Increasing salt levels in this aquifer mean that increasingly salty water will be applied to crops, thereby damaging them and raising soil salinity levels. Given the need to produce an ever-increasing amount of food in the future, this potential outcome has serious implications.

Australia

Australia is a fascinating country in many ways and has long been a draw to people throughout the world. It is the largest island on Earth, while at the same time it is a continent about the size of the contiguous United States (its lower 48 states). Some of the oldest rocks in the world, dating to about 3.5 billion years ago, are found in Australia. The Australian continent has been essentially inert from a geologic perspective for some 60 million years, which means that some of the oldest and most weathered soils on the planet occur there. Given its long-term isolation, the flora and fauna are exotic, with 80% of Australian plants and animals like the kangaroo, platypus, and koala found nowhere else on Earth. Some of the most lethal animals on Earth are indigenous to Australia, including the salt-water crocodile, trap-door spider, and box jellyfish. The fantastic Great Barrier Reef fringes the northeast coast. Among many other unusual things, Australia has the most salinized soils in the world. Most of the country is semi-arid to arid, receiving less than 50 cm (20 in.) of precipitation a year (Figure 20.19). The combined effect of low precipitation and ancient soils poor in nutrients produced one of the most desolate places on Earth, the famed Australian outback.

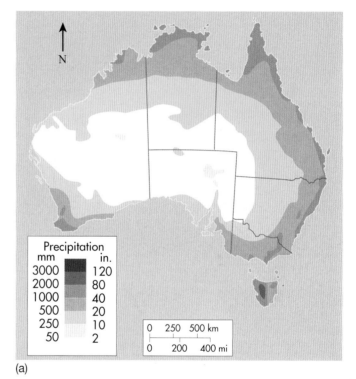

Precipitation
mm		in.
3000		120
2000		80
1000		40
500		20
250		10
50		2

0 250 500 km
0 200 400 mi

(a)

Panoramic Images/Getty Images, Inc.

(b)

Figure 20.19 Australia precipitation. (a) Map of average annual precipitation in Australia. Note that most of the country is semi-arid to arid. (b) Typical scene in the Australian outback. This landscape is the result of sparse precipitation and ancient soils that are low in nutrients. Notice the redness of the soils, which indicates deep weathering of the host sediment.

Understanding the complexity of Australia's physical geography is essential to appreciating the impact humans are having there. The environment is extremely sensitive, and the potential environmental issues in the future are thus critical. Consider first the population of the country. As of 2012, the population of Australia was a little over 22 million, with most people living on the relatively humid margins of the country, especially the continent's eastern side (Figure 20.20a). To put that number in perspective, consider that the population of the United States, which is fundamentally the same physical size as Australia, is now about 314 million. Certainly, some of Australia's relatively

low population is related to its isolation and a very different immigrant history compared to the United States. Another more important factor is that the comparatively sterile physical environment of the country simply cannot support more people on a sustainable basis. In fact, some believe that the country is *already* overpopulated and that the sustainable population may be as low as 8 million people.

Given potential population issues in the country, it is essential that Australian farmers maximize food production. Unfortunately, a key variable limiting this goal is the simple fact that the vast majority of the country is unsuitable for agriculture, as you can see in the land-use map in Figure 20.20b. As a result of these opposing realities, many farmers have focused on marginal drylands on the fringe of the outback to produce food, particularly wheat. The primary wheat areas in Australia are the Murray-Darling River basin in the southeastern part of the country and the wheat belt region in the extreme southwest (Figure 20.20c). The wheat belt region, for example, produces about 20 to 25 million tons of wheat per year.

As you have seen in ancient Sumeria and the San Joaquin Valley, dryland farming can increase soil salinity. Such soil salinization is a serious problem in the wheat areas of Australia. These regions are particularly susceptible to salt buildup because Australia is naturally a very salty place. This natural salinization of the continent has occurred due to the interaction of several factors related to physical geography. First, much of the continent is essentially a closed basin where very little water flows to the sea. As a result, when rain does fall, it tends to settle in specific places rather than flow away. Another factor is that any rain which does fall contains microscopic salt particles derived from the saline marine waters that surround the continent. Over millions of years these salts have gradually built up in Australian

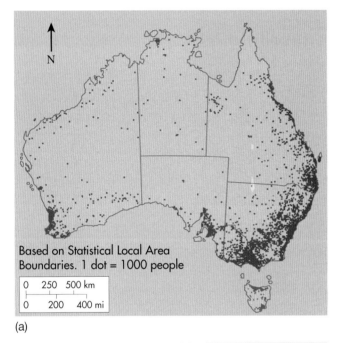

Based on Statistical Local Area Boundaries. 1 dot = 1000 people

| 0 | 250 | 500 km |
| 0 | 200 | 400 mi |

(a)

ABS data used with permission from the Australian Bureau of Statistics

—— Approximate Outback boundary

Tropical fruits
Forestry
Natural vegetation
Wheat
Extensive grazing
Barren land

| 0 | 250 | 500 km |
| 0 | 200 | 400 mi |

(b)

Bill Bachman/Alamy

(c)

Figure 20.20 Australian population and land use. (a) Population geography of Australia. Note the relationship of dense population clusters and the precipitation map in Figure 20.19. (b) Map of land use in Australia. Again, note the geographical relationship of land-use patterns and precipitation shown in Figure 20.19. (c) Photograph of the wheat belt region in southwestern Australia.

soils, especially since water rapidly evaporates in the semi-arid to arid climate. A third factor is that much of the groundwater in Australia is saline—in some cases, as much as the ocean that surrounds the continent. This natural groundwater salinity is a result of the gradual salinization of Australia over tens of millions of years and the slow, persistent downward movement of salt into underlying aquifers.

Given the natural salinity of Australia, the potential for soil salinization due to agriculture is extremely high. Significant areas of the country are already being affected by the problem (Figure 20.21). In the Murray-Darling River basin, salinization occurs because irrigation is increasing soil salinity in a manner very similar to what we discussed in the San Joaquin Valley. The largest areas of dryland salinity are in the wheat belt of Western Australia, where salinization is largely related to the clearing of native vegetation and the associated rise of the water table. In this context, salts are either accumulating in depressions or are wicked up from the shallow water table.

The real and potential costs of soil salinization in Australia are staggering. About 5.7 million hectares (14 million acres), or approximately 16% of the agricultural area, are currently at risk. The reduced or lost crop yields from this large area have cost farmers about $1.3 billion Australian dollars (~$800 million U.S.) in lost income. Soil salinization is also posing risks to urban and rural infrastructure (because it deteriorates iron and concrete), as well as the health of stream ecosystems and wetlands.

The problem of soil salinization has a very high profile in Australia and is a major source of concern. Unfortunately, the situation is expected to become even worse in the future. A good place to see the likely future outcome is in the wheat belt in Western Australia. At present, about 3.5 million hectares (~8.6 million acres) of the region are at risk of eventual soil salinization. At current rates of salinization, the amount of land that will be affected is expected to double by 2050 (Figure 20.22). In the country as a whole, the amount of land salinized may reach 17 million hectares (42 million hectares).

(a)

(b)

Figure 20.22 Dryland salinity in southwestern Australia. (a) Map showing the extent of dryland salinity in 2000. (b) Map showing area expected to be at risk for salinization by 2050. (*Source*: Commonwealth of Australia.)

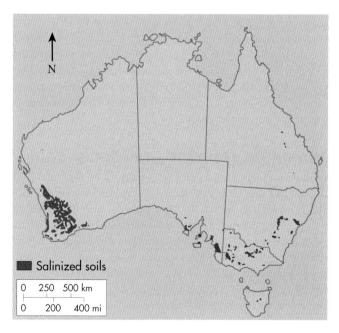

Figure 20.21 Australia soil salinization. Map of areas in Australia where soil salinization is a problem. (*Source*: Commonwealth of Australia.)

Clearly, the Australian people face a critical issue, especially in the context of a growing population. Australians are attempting to deal with the problem, specifically with efforts to restore water and salt balances by controlling groundwater recharge rates in some areas. In other places, farmers and ranchers are attempting to reduce the effects of salinization by planting salt-tolerant trees, such as Saltgrow (a hybrid eucalyptus tree). These efforts will take a substantial amount of time, however, which means that the problem of salinization in Australia is going to be chronic for the foreseeable future and will likely get worse before it gets better. The potential negative impact that salinization has on wheat production is magnified if global climate change scenarios are included in the equation, because most models suggest that Australia will become even warmer and drier in the future. If this is true, then the challenge of producing enough wheat in Australia is obvious. Stay tuned.

The Challenge of Sustainable Agriculture in Semi-Arid and Arid Regions

This case study focused on the issue of soil salinization in semi-arid and arid regions of the world. The factors that influence the salinization of soils are clearly related to physical geography and the ways that humans interact with the environment. Salinization is prone to occur in marginal lands where people are attempting to farm in an effort to produce more food. These environments are extremely sensitive to disturbance and modification of the landscape to fit agricultural needs. Understanding the subsequent salinization of soils that follows such disturbance requires an appreciation of soils, groundwater, and evaporation rates. Archaeological evidence indicates that soil salinization has been a problem since the dawn of human civilization, resulting in the collapse of past human societies. The problem remains with us today and will likely intensify in the future, barring some yet undeveloped way to deal with soil salts systematically. Given the expected growth in human population in the coming decades, the likely loss of farmland or reduced crop yields in salinized areas will increase the challenge of producing sustainable levels of food for the world.

KEY CONCEPTS TO REMEMBER ABOUT SOIL SALINIZATION

1. Soil salinization refers to the buildup of salts in soil and usually occurs in semi-arid and arid regions. It can occur naturally when salty slopewash accumulates in a depression and evaporates, leaving a salt crust behind.

2. Soil salinization can also occur due to agricultural practices in marginal lands. It can happen when the clearing of native vegetation causes local water tables to rise sufficiently close to the surface that wicking can occur. This form of salinization is magnified when irrigation causes the water table to rise further. Soil salinization can also occur when salty irrigated water evaporates after it is applied to fields, leaving behind salt deposits.

3. A major area of soil salinization in the United States is the San Joaquin Valley in California, which is one of the most important agricultural regions in the nation.

4. Extensive soil salinization is also occurring in the agricultural belts of Australia. Soil degradation in these areas has the high potential to significantly reduce crop yields such as wheat.

Case 3: Developing Unconventional Oil Supplies in North America

The third case study moves away from marginal lands by focusing on some of the challenges we face with regard to petroleum supplies. As discussed in Chapter 12, the vast majority of energy required to power our economy is derived from fossil fuels (coal, gas, petroleum). Of these fuels, people are generally most familiar with petroleum because they routinely interact with its refined by-product at the gas pump. Given the importance we place on freely moving about in our vehicles, people are keenly aware of price fluctuations, especially if the cost of gas spikes for a period of time. How many of us understand, however, the factors involved in maintaining a steady supply of gas to the American consumer? Certainly, such a discussion could focus solely on the economic or political factors that influence the supply and cost of oil. On the other hand, it can be viewed through the prism of physical geography, which this case study does, to better understand some of the new issues we face just to extract oil from the ground.

Oil Supply, Demand, and Geography

As indicated in Chapter 12, the beginning of the modern petroleum era can be traced to the Industrial Revolution in the mid-1800s. Through the first few decades of industrialization in the United States, supply easily kept pace with demand. By 1906, for example, U.S. oil production was about 126.5 million barrels per year, and extensive fields were discovered in Texas (Figure 20.23), Oklahoma, and California that expanded our national supply. During this phase of development and extending to the recent past, extraction methods can be viewed simply as an elaborate straw system that sucked oil from reservoirs deep within the ground. During the 1950s petroleum supplanted coal as the world's dominant fuel source, and the United States was the world's leading oil producer, yielding

Figure 20.23 The Lucas gusher. In 1901 the Corsican oil field in southeastern Texas was tapped by Patillo Higgins, a self-taught geologist. The pressure in the well was so great that the oil (or "black gold") erupted to a height of about 50 m (150 ft) for 9 days after the initial strike. Approximately 1,000,000 barrels of oil were lost, which would be worth about $17,000,000 today.

approximately 4 billion barrels per year. At about this same time, oil companies such as Shell and British Petroleum discovered huge reserves in the Middle East and Venezuela that dwarfed our own. These reserves were ultimately nationalized by the host nations, which then formed the *Organization of the Petroleum Exporting Countries* (OPEC) to coordinate petroleum policies.

The economic geography of petroleum began to change in the late 1960s as the United States slowly reached peak oil production, which occurred in 1971. Our demand had grown to the point where we had to import another 23% from foreign sources to meet our needs. Our vulnerable dependence on foreign supplies first became apparent in 1973 when OPEC organized an oil embargo to flex political muscle. As a result of this disruption in supply, prices in the United States jumped and long gas lines were commonplace (Figure 20.24). In an effort to conserve gasoline, the speed limit on the nation's highways was lowered to 55 mph, and automobile manufacturers focused on building fuel-efficient cars. Yet another embargo occurred in 1979, further disrupting supplies and increasing gas prices. Throughout most of the 1980s, global oil supplies were tight and prices at the pump remained high.

In the late 1980s and 1990s, global oil supply began to increase when better extraction techniques were developed and new reserves, such as those in Alaska's North Slope and in the North Sea of Europe, were systematically tapped. As a result, prices fell dramatically to about $15 per barrel and Americans generally moved away from fuel-efficient cars to sport utility vehicles (SUVs), which consumed far more fuel. At the same time, suburban development in the United States exploded in association with urban sprawl (see Discover—*Urban Sprawl* above) and the length of daily work commutes increased. Global oil prices slowly began to rise again in 2004 and reached about $88 per barrel in 2006, in large part because of rising demand in China and India, which have rapidly growing populations. Prices rose quickly in late 2007 and early 2008 when future supply fears, speculation, and a weak dollar pushed the price of a barrel of oil to over $140. This rapid increase caused prices at the pump to exceed $4 per gallon in many places within the United States. Just as consumers were adjusting to this new reality, the world experienced a massive financial crisis in late 2008 and supply fears decreased due to a deepening global recession. At that time, the average price for a gallon of gas in the U.S. midwest was about $2.70. Since then, the economy has gradually expanded and the demand for petroleum thus increased. As a result, the average price for a gallon of gas was about $3.40 in October, 2013.

At the time of the most recent economic downturn, concerns abounded that global oil supplies were reaching a critical level that could dampen economic growth for the foreseeable future. The U.S. economy has long been vulnerable to tight supplies because we consume about 25% of the world's oil (even though we have only 4% of the population) to keep our industrial society humming. Given our own low supplies of easily accessible oil, we now import about 60% of what

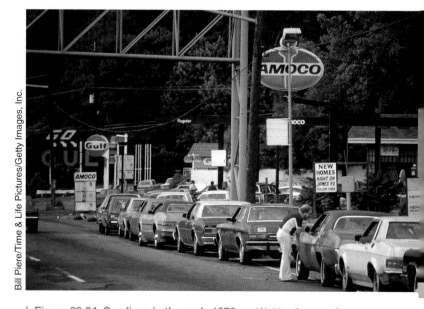

Figure 20.24 Gas lines in the early 1970s. Waiting for gasoline like this was a common occurrence during the first oil embargo because service stations often ran out of gas.

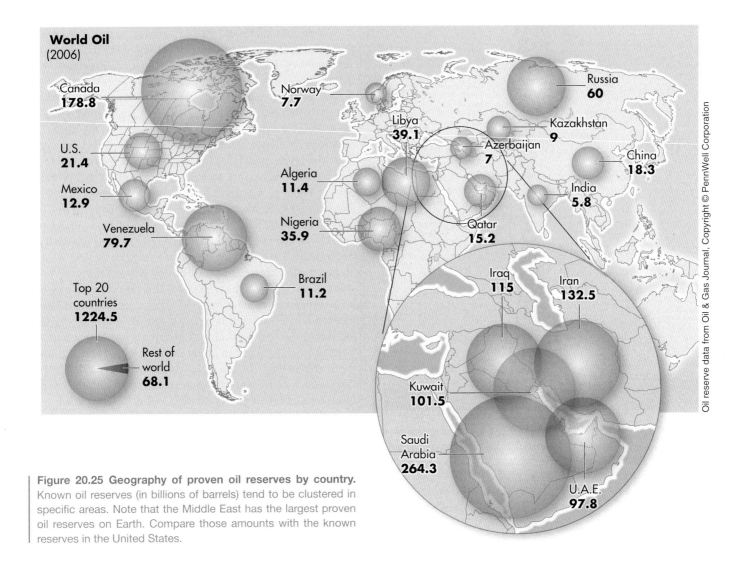

Figure 20.25 Geography of proven oil reserves by country. Known oil reserves (in billions of barrels) tend to be clustered in specific areas. Note that the Middle East has the largest proven oil reserves on Earth. Compare those amounts with the known reserves in the United States.

we need. Although we import our oil from a variety of places, prices are set by OPEC, which is dominated by politically unstable countries in the Middle East where most conventional oil reserves occur (Figure 20.25). With this backdrop in mind, many geoscientists in the early 21st century believed that we have reached peak global oil production (Figure 20.26). Although the peak oil scenario is a controversial one, signs of it were beginning to emerge in early 2008 when global oil production (88 million barrels per day) barely exceeded global demand (86 million barrels of oil per day). This tight relationship between supply and demand is of particular concern given that global oil demand is expected to reach *130 million barrels per day* in 2030.

Developing Unconventional Sources of Oil

Although the peak oil concept has been controversial, it is generally agreed that a lot (most?) of the easily accessible oil that can be mined with the conventional "straw" system has

been consumed. As a result, future economic growth may be threatened if the cost of petroleum spikes It has been these concerns—along with worries about fossil fuel use and climate change (Chapter 9)—that have spawned the major push toward renewable fuels such as solar power (Chapter 4), wind power (Chapter 6), and hydroelectric power (Chapter 16). Nevertheless, fossil fuels are still used to generate over 85% of the energy in the United States, with 40% produced by petroleum alone. Clearly, the global economy still needs a whole lot of oil for the foreseeable future.

In this context, recent work within the oil industry has focused on locating unconventional reserves of oil and developing technology to extract them. Such work requires an excellent understanding of physical geography and geology. The largest reserve of unconventional oil in North America is the Athabasca oil sands (or tar sands) in Canada (Figure 20.27). These deposits consist of loose sand or partially cemented sandstones that contain a very dense and extremely viscous form of petroleum called *bitumen*. Estimates indicate that as many as *178 billion barrels* of oil occur in these deposits (Figure 20.25).

Figure 20.26 Peak oil curve. The concept of peak oil revolves around the notion that we may be reaching global peak oil production. The subdivisions of the graph represent the contributions of specific geographical regions to the overall global oil supply over time. NGL refers to natural gas liquids, whereas heavy supplies are those that do not flow easily and have thus been previously ignored. If this scenario is accurate, note the progressive drop in global oil supplies in the coming decades.

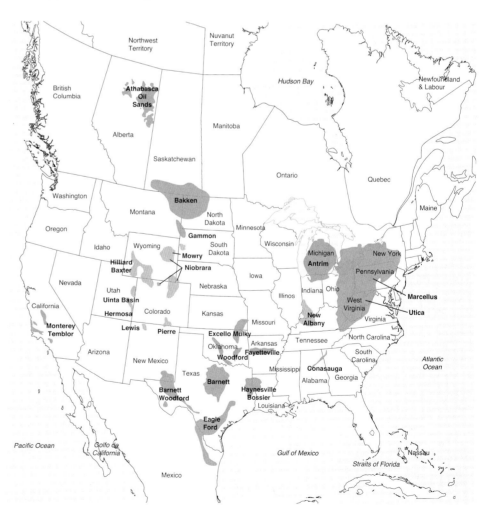

Figure 20.27 Geography of unconventional oil and gas deposits in North America. The Athabasca oil sands are believed to contain 178 billion barrels of oil, which is currently in a semi-solid form called bitumen. Unconventional deposits in the United States are mostly associated with extensive shale deposits scattered around the country. Although most of these plays are largely known for their gas deposits, several contain abundant supplies of shale oil.

(a)

NASA image created by Jesse Allen, using data provided courtesy of NASA/GSFC/METI/ERSDAC/JAROS, and U.S./Japan ASTER Science Team

Associated Press

(b)

Figure 20.28 Millennium pit mine in Alberta, Canada. (a) The Millennium mine is a gargantuan open-pit mine that is used to access oil sands. This mine produced over 300,000 barrels of oil per day in 2012. (b) ASTER image of the Millennium mine from NASA's Terra satellite in 2007. Note the large size of the mine, the tailing pond where waste is stored, and its close proximity to the Athabasca River.

The United States currently imports about 2 million barrels a day from this source. The semi-solid oil in Canada is extracted either by open-pit mining at the Earth's surface (Figure 20.28) or by injecting steam into the oil-bearing sands at depth. This infusion of hot water through the wellbore liquefies the semi-solid bitumen so that it can be brought to the surface in the well. Several other methods have also been used to extract oil from these deposits, including injection of a hydrocarbon solvent to separate the oil from the sand.

In the United States unconventional oil resources are typically associated with shale rock units generally referred to as *oil shales*. Recall from Chapter 12 that shale consists dominantly of lithified clays that accumulated in calm coastal bays and lagoons (Figure 12.12). According to the United States Geological Survey, the contiguous United States contains abundant shale deposits that hold vast quantities of oil and natural gas. Each of these areas is called a **shale play** that is controlled by a unique set of geological circumstances within the area and is a favorable site for exploration. Such plays are named after the host shale in which the fossil-fuel reserves are contained, including the Marcellus in the northeast, the Antrim in Michigan, the Bakken in the northern Great Plains, the Mancos in the Rocky Mountains, and the Monterey/Santos in California, to name a few (Figure 20.27). Although these shale plays are known more for the amount of natural gas they contain (the Marcellus Shale holds approximately 410 trillion cubic feet of natural gas, for example), many hold significant oil reserves. For example, the Bakken Shale is believed to

contain about 4 billion barrels of oil, but may contain more. On the Pacific coast, the Monterey/Santos play is estimated to hold about 15 billion barrels of oil.

Oil from shale plays is extracted with a method called **hydraulic fracturing** (or *fracking*). In a fracking operation, water, sand, and various chemicals are injected down the well bore into the oil-bearing strata. When it reaches that level, it creates fractures in the rock that permit the oil (as well as natural gas) to escape and flow out of the well (Figure 20.29a). In contrast to conventional wells, fracking permits horizontal drilling into rocks, which extends the reach of the mining operation. Fracking is widely popular at the present time, with over 40,000 wells installed nationwide between 2011 and 2013. The Bakken play in North Dakota has experienced such explosive growth that the artificial lights associated with operations there can now be seen from space (Figure 20.29b). As a result, production of U.S. oil rose from about 100,000 barrels of oil per day in 2003 to 2 million barrels per day in 2012.

Is the Environment Threatened by Fracking?

Given our nation's current dependence on oil to fuel large parts of our economy, many people are understandably excited about the potential these unconventional sources have in enhancing our supply needs. The development of these resources also shows much promise for reducing our dependence on oil from unstable countries and enhancing our overall energy security.

Shale play *A geographic area, dominated by shale bedrock, that is targeted for exploration due to favorable geological conditions that suggest high potential to yield an economic quantity of oil or natural gas.*

Hydraulic fracturing *Fracturing of rock at depth by the injection of fluids to release oil and natural gas contained within shale beds.*

Believe it or not, this glob of Earth held in this gloved hand can be refined to power your automobile. What is this material called, where is it found, and what are two ways it can be extracted from the ground?

Ashley Cooper/Alamy

Enhanced energy security is desirable for many reasons; in particular, it may keep us out of costly conflicts in unfriendly places. On the other hand, concerns abound that extracting these unconventional oil deposits may have a huge negative impact on the environment. You may have heard about some of these issues.

The environmental impacts of open-pit mining in the Athabasca oil sands are pretty straightforward. As you can clearly see in Figure 20.28, such mining operations play havoc with the landscape. Giant earthmovers excavate hundreds of thousands of metric tonnes of Earth each day, which is then hauled to ore crushers where the extraction of bitumen begins. A number of roads have been carved out of the wilderness to facilitate the process, which has certainly impacted wildlife in the area. The visible tailing pond may also leach pollutants into the nearby Athabasca River, such as mercury and naphthenic acid, that are harmful to fish and wildlife. In the places where bitumen at depth is liquefied by steam, a tremendous amount of water from the river is required in the process; estimates are that 2 to 4.5 m^3 of water is needed to produce 1 m^3 of oil. Yet another environmental impact is related to the fact that extraction of bitumen requires energy generated by burning natural gas. It is estimated that mining these unconventional resources contributes 5% to 15% more carbon emissions into that atmosphere than conventional mining operations. These higher carbon emissions make future climate change more likely.

Several of the environmental problems associated with fracking are similar to those observed in the Canadian oil sands. For example, some have argued that the fracking process releases greenhouse gases into the atmosphere in the form of methane from equipment and from the well through leaks and venting. Fracking also requires a lot of water. A typical well requires 11,000 to 19,000 m^3 (3 to 5 million gallons) of water for the initial injection and potential restimulation at a later date. Such usage can be a problem for streams in semi-arid to arid zones because it reduces discharge.

The primary environmental issue associated with fracking revolves around the possibility of groundwater contamination. As you can see in the geological layers shown in Figure 20.29a, a stratum containing groundwater overlies the oil-bearing shale below. This groundwater zone was penetrated by the well as it proceeded through to the underlying shale bed. Mining firms are required to line the well hole with steel casing, which is cemented into place, to protect groundwater layers from contamination by methane released in the fracking process or the chemicals (such as sodium chloride, ethylene glycol, borate salts) used to improve flow. It is conceivable, however, that such casings can leak, causing groundwater contamination. Yet another way that groundwater can theoretically be contaminated is by the release of methane (or chemicals) from the fracked shale into an overlying groundwater zone.

At this point, it seems safe to say that uncertainty prevails with respect to the degree of groundwater contamination caused by fracking. Environmentalists argue that the process is woefully unsafe and that groundwater contamination is common. Unacceptable levels of methane and chemicals such as arsenic, barium, and glycol compounds have certainly been measured in some areas where fracking is commonplace. In a similar vein, people living in and around places heavily fracked have reported that water tastes odd and that their livestock have become ill by drinking the local water. In 2011 the U.S. Environmental Protection Agency reported for the first time that groundwater contamination from fracking likely occurred in Wyoming. The industry maintains, however, that the process is safe and that pollutants can come from a number of places unrelated to fracking. For example, the industry argues that it is possible that personal water wells and municipal wells, which may be less well designed than industrial systems, can penetrate methane-rich deposits that contaminate groundwater. As a result of these conflicting points of view, states are grappling with how to regulate the industry to ensure environmental safety. Meanwhile, the growth of fracking operations continues to grow at a frenetic pace.

This case study focused on an issue that is important to everyone—oil resources. Given that our industrial society currently depends so much on reliable oil supplies, it is important to understand the factors involved in maintaining them. As you can see, this issue is directly related to geology and physical geography. The ongoing search for unconventional oil supplies depends on an understanding of sediment, rock strata, and mapping to isolate the various plays that contain mineable petroleum. Dealing with the potential environmental issues requires an understanding of streams, the carbon cycle, and groundwater. You will likely hear about these issues at various times in your life, and perhaps you may think about them if the price of gas rises dramatically at some point in the future. Your understanding of physical geography will enable you to better understand the choices that need to be made.

KEY CONCEPTS TO REMEMBER ABOUT UNCONVENTIONAL OIL SUPPLIES IN NORTH AMERICA

1. The United States was the world's leading oil producer from 1900 until 1970.

2. U.S. oil production peaked in the early 1970s. In the face of a greater demand, we began to import increasing amounts of oil. Much of this imported supply came from politically unstable OPEC nations, raising questions about national security.

3. Oil prices have varied dramatically in the past 40 years due to fluctuations in supply, political instability, population growth, economic conditions, and demand.

4. With conventional oil supplies waning, recent efforts have focused on developing unconventional resources. The largest of these sources in North America are found in the Canadian oil sands and various shale plays in the United States.

5. Extraction of unconventional oil is conducted by nontraditional methods such as open-pit mining and fracking. These methods may have serious environmental impacts.

(a)

(b)

Figure 20.29 Hydraulic fracturing (a) Schematic illustration of the fracking process. Fracking is the method used to extract oil and natural gas from shale deposits buried deep within the ground. The wellbore passes through a series of geological strata to reach the host shale. At that point the drill turns horizontally to access as much shale as possible. Water, sand, and various chemicals are then injected into the shale so that it fractures, releasing the trapped oil and natural gas. Note the infrastructure at ground level required to work the well. (b) Satellite view of western North Dakota and eastern Montana at night. The cluster of lights within the circle results from equipment, buildings, and gas flares associated with work in the Bakken shale play.

Case 4: Giant Panda Conservation in China

In the fourth and final case study, we will investigate the effort to conserve giant panda habitat in China. Giant pandas (*Ailuropoda melanoleuca*) are among the most popular animals on Earth because of their distinctive markings and docile demeanor (Figure 20.30a). These animals are also an iconic cultural symbol in China. Nevertheless, pandas are highly threatened due to human land-use practices and thus are officially an endangered species. Only about 1600 of them live in the wild. Given these very small numbers and the potential for extinction, aggressive efforts are ongoing by the Chinese government, international conservation agencies, and leading scientists to protect remaining habitat so pandas can maintain a sustainable population in the wild. Physical geography helps us understand the panda's landscape, the impact that growing human population has on the panda environment, and the con-

servation efforts required to save them in the wild. In contrast to the potentially bleak case studies investigated in Las Vegas, California, and Australia, the panda case demonstrates how a potentially disastrous environmental issue may be stabilized and perhaps reversed in the future by using holistic geographical principles.

Giant pandas are solitary members of the bear family (*Ursidae*). They live in old-growth forests that have a healthy bamboo understory and forest clearings (Figure 20.30b). Although pandas are classified as carnivores and have been known to eat honey, bulbs, fish, and eggs, their diet consists almost entirely (>95%) of bamboo shoots. Given the low nutrient composition of bamboo, and the pandas' inability to digest cellulose efficiently, they derive little energy from their food and thus must consume up to 13.5 kg (~30 lb) of shoots daily. To meet this need, they spend 10 to 12 h a day feeding and conserve energy by moving slowly. Pandas neither hibernate nor make permanent dens like many other bears and use hollow trees and caves for shelter.

Karen Su/China Span/Getty Images, Inc.

(a)

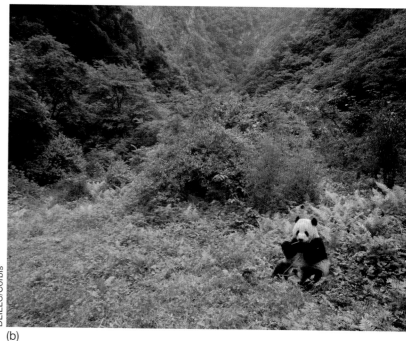

DLILLC/Corbis

(b)

Figure 20.30 Giant pandas. (a) Pandas are members of the bear family and are known for their distinctive markings and docile demeanor. (b) Panda habitat consists of old-growth forests with a prominent bamboo understory and forest clearings.

Figure 20.31 Prehistoric and current range of giant pandas in southeastern Asia. Note the extensive decline in panda range that has occurred over time. This loss of habitat has generally coincided with two major periods of population growth in China.

In prehistoric times, the pandas' natural range once extended throughout the forests of southeastern China, Myanmar (formerly Burma), and even into India (Figure 20.31). This area lies largely within the humid subtropical hot-summer climate (*Cfa, Cwa*) region of southeastern Asia, which, if you recall from Chapter 9, is dominated by monsoon atmospheric circulation. During the 18th century the extent of panda habitat progressively shrank because lowland subtropical forests were cleared for agriculture and wood fuel for cooking and home heating. This expansion of agriculture and loss of panda habitat occurred during a major growth phase in Chinese population from about 140 million people in 1700 to approximately 400 million in 1900 (Figure 20.25). As a result, by 1900, panda habitat was confined to the Qinling Mountains in the north-central part of their historical range and the Minshan, Xiangling, Qionglai, and the Liangshan Mountains in the west-central part of their ancestral range (Figure 20.31).

During the 20th century panda habitat was further reduced as agriculture expanded along the major river valleys within the region. This growth of farming occurred at the same time that Chinese population grew rapidly from about 500 million people in 1950 to over 1.3 billion presently (Figure 20.32). Pandas were also hunted for pelts and because it was believed that their body parts had medicinal qualities valued in traditional Chinese culture. Today, pandas are found only in about 20 isolated parts of the six mountain ranges mentioned previously and live in dense stands of bamboo that lie between about 1500 m and 3500 m (~4900 ft and 11,500 ft) above sea level. Their optimal altitudinal range is between 2500 m and 3000 m (~8200 ft to 9800 ft) above sea level. Given their need to conserve energy, they prefer level terrain (<15° slopes) within the mountains so they can move about easily.

As population in China increased in the latter half of the 20th century, concerns about the viability of wild panda populations began to grow. In the early 1960s the first four panda reserves were established by the Chinese government and hunting was outlawed. Additional reserves were established in the 1970s, including a permanent research station at the Wolong Nature Reserve. In the 1980s pandas were given even more legal protection and hunters began to be prosecuted vigorously for killing them. At about this same time, however, geographical surveys indicated that the extent of panda habitat dropped from 29,500 km²

Heilig, G.K. (1999): Can China Feed Itself? A System for Evaluation of Policy Options. International Institute for Applied Systems Analysis (CD-ROM) (http://www.iiasa.ac.at/Research/SRD/ChinaFood/index_m.htm)

Figure 20.32 Graph of Chinese population the past 2000 years. Note the rapid growth phases from 1650 to 1900 and between 1950 and the present. Over 1 billion people now live in the country.

(11,390 mi^2) in 1975 to only 13,000 km^2 (5,019 mi^2) by 1985 due to extensive logging. Results of the third national panda survey in 2004 indicated that China's wild giant panda population included approximately 1600 animals, which is about 40% more than previous estimates. Although hailed as good news, it was believed that this increase is probably related to more accurate survey methods rather than an actual increase in the number of pandas.

The Wolong Nature Reserve

In the past decade, a systematic effort has been made to fully document the extent of panda habitat and to monitor its health and distribution. Much of this effort has focused on the Wolong Nature Reserve, which is located in the Qionglai Mountains in the west-central part of the modern panda range (Figure 20.31 and 20.33a). The reserve was established in 1963 and is about 200,000 hectares (~494,200 acres) in size. It contains numerous high mountains and deep valleys, with overall elevation ranging from 1200 m to 6250 m (~3900 ft to 20,500 ft) above sea level. The vegetation in the reserve has a distinct vertical zonation, such as that discussed in Chapter 10, ranging from evergreen and deciduous broad leaf trees at the lower elevations to subalpine coniferous forests at higher elevations. Given this broad ecological range, the reserve contains over 6000 different plant and animal species, including more than 150 giant pandas. The reserve also includes the Wolong Giant Panda Reserve Center, which was established in the 1980s and is thus one of the oldest research centers devoted to pandas in China. In an effort to support the wild panda population, this research center focuses largely on giant panda breeding and the study of bamboo ecology.

KEY CONCEPTS TO REMEMBER ABOUT PANDA HABITAT AND CONSERVATION

1. Giant pandas are members of the bear family who live in the bamboo understory of old-growth forests in China. They mostly eat bamboo shoots.

2. The range of pandas once included much of southeastern Asia. As a result of increased human population and logging, they are now confined to several small areas in south-central China, including the Wolong Nature Reserve. Approximately 1600 pandas live in the wild, and they are severely threatened with extinction.

3. Analysis of land-cover change at the Wolong Nature Reserve shows the patterns of forest loss and fragmentation of panda habitat in the late 20th century.

4. Understanding the patterns and causes of panda habitat loss, coupled with concerns over their status, has led to reforms aimed at protecting them in the wild.

5. The loss of panda habitat may be stabilizing, potentially resulting in the increased viability of pandas in the future.

A great deal of scientific research has been conducted in the Wolong Nature Reserve to monitor the health of the panda habitat and the human impact on it. Much of this research uses the remote sensing technology discussed in Chapter 2. Figure 20.33b is a good example of a satellite image of the Wolong Nature Reserve. This image is a composite digital el-

A. S. Vin et al., "Temporal Changes in Giant Panda Habitat Connectivity across Boundaries of Wolong Nature Reserve, China," *Ecological Applications* 17(4), 2007, pp. 1019–1030.

(a)

Elevation

1200-2500 m
2500-3000 m
3000-3500 m
3500-4000 m
4000-4500 m
>4500 m

10   0   10 km

(b)

N

Reserve Boundary

0   6 mi
0   10 km

NASA

(c)

Images Limited/Getty Images

Figure 20.33 The Wolong Nature Reserve. (a) Map showing the location of the Wolong Nature Reserve in Sichuan Province and elevation in the area. (b) Satellite image of the Wolong Nature Reserve in China. Note the rugged terrain in the western part of the reserve and how it correlates with the elevation map in (a). The best panda habitat is in the less rugged terrain on the east side of the reserve. (c) Qionglai Mountains in the Wolong Nature Reserve. This oblique image provides a great sense of the terrain in this part of China.

evation model (DEM) that provides a sense of the topographic relationships in the reserve. This image reveals that the most rugged terrain is on the western side of the reserve and is unsuitable for pandas (Figure 20.33c), whereas the best panda habitat is to the east.

A major effort at the Wolong Reserve is to document changes in the distribution of panda habitat over the past few decades and to predict future patterns. Much of this research assesses the changes in **land cover**, such as roads and forests, on the landscape over time. Satellite imagery from the mid-1960s to 2001 at the Wolong Nature Reserve, for example, reveals that a significant amount of deforestation occurred throughout the reserve during that period of time (Figure 20.34). In 1965, about 43% of the nature reserve was forested, whereas by 2001 forests had decreased to approximately 36% of the area. This modification in vegetation was most closely associated with the change from forest to cropland, grassland, and shrub land, particularly along the rivers and the main roads. Although the area is technically a nature reserve, ~5000 people live within the boundaries of the park. About 90% of them are farmers who engage in agriculture or livestock breeding in cleared areas. Perhaps the most important cause of forest loss was the collection of wood fuel by peasant farmers who lacked the income to purchase other sources of fuel for cooking and home heating.

Aside from the extensive loss of forest cover, a major outcome of deforestation between 1965 and 2001 was the intensive fragmentation of the remaining panda habitat. In other words, patches of healthy bamboo in the understory were cut off from one another. You can see the nature and extent of habitat fragmentation by examining Figure 20.35, which again shows a time series of forest loss, this time between 1965 and 1997. This degree of fragmentation is potentially a serious problem because it reduces the quality and quantity of panda habitat and decreases the carrying capacity for the animals. It also limits the ability of pandas to move from one area of suitable habitat to another, which increases the potential for the animals to become genetically isolated from one another. If this trend were to continue, it would dramatically increase the probability of extinction through inbreeding. Unchecked, this type of human impact is projected to result in the loss of up to 16% of all existing panda habitat by 2035.

As the geography of forest loss in the Wolong Nature Reserve and other panda habitats became more apparent in the latter part of the 20th century, the Chinese government responded by outlawing logging in protected panda areas in 1998. In return, farmers were offered government payments to allow cropland to convert back to forest. At about the same time, conservation efforts by international organizations such as the World Wildlife Fund (WWF) began to have significant positive impacts. In 2002, for example, an agreement between

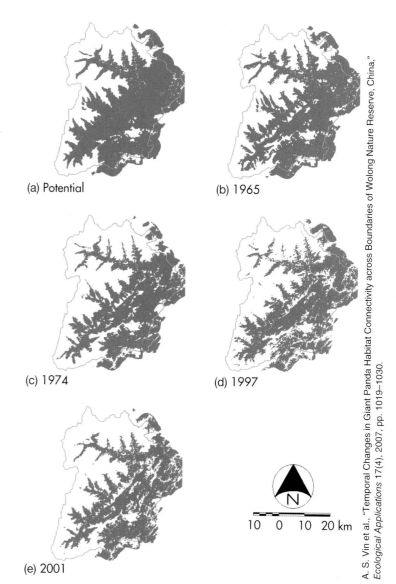

(a) Potential   (b) 1965

(c) 1974   (d) 1997

(e) 2001

A. S. Vin et al., "Temporal Changes in Giant Panda Habitat Connectivity across Boundaries of Wolong Nature Reserve, China," *Ecological Applications* 17(4), 2007, pp. 1019–1030.

Figure 20.34 Deforestation of the Wolong Nature Reserve between 1965 and 2001. Note the change in forest cover that occurred during this interval of time. The "potential" map refers to the expected amount of forest if no deforestation had occurred in the area.

the WWF and the Shaanxi Forestry Department was signed to create 13 new panda reserves and develop the first corridors in the Qinling Mountains to connect fragmented habitats. A further positive development was that rising wages associated with regional economic development reduced pressure on forests because more people gained access to electricity and thus did not need wood fuel for heating. These combined factors, among others, have the potential to stabilize the ongoing loss of panda habitat and to slowly turn things around in the coming decades. Panda habitat may begin to recover after about 40 years of forest regeneration. If the logging ban is maintained, previously forested tracts may regenerate entirely and the odds for long-term survival of pandas will increase.

Land cover *The various things that cover the landscape, such as forests, roads, and water bodies, at any point in time.*

Used with permission of the American Association for the Advancement of Science; Liu et al., "Ecological Degradation in Protected Areas: The Case of Wolong Nature Reserve for Giant Pandas," Science 292, 2001, pp. 98–101.

1965

1974

1997

1997

0   5   10 Kms

Forest/Non-forest   Panda Habitat

(a)   (b)

AP/Wide World Photos

Figure 20.35 Fragmentation of panda habitat from 1965 to 1997. Map on left shows forest extent in Wolong Nature Reserve in 1997. Sequence of gray and white larger-scale maps in the center illustrates change from forest (gray) to nonforest (white) during study interval. Sequence of three color-coded maps on right shows changes in areas of highly suitable (red), suitable (yellow), marginally suitable (green), and unsuitable (black) panda habitats from 1965 to 1997. (b) Deforested hillslopes in the Wolong Nature Reserve. Such a deforested area fragments the panda habitat and makes it difficult for pandas to cross to more suitable areas.

SUMMARY

Our first two case studies dealt with water use in the American Southwest and soil salinization in marginal lands. These case studies showed that it is necessary to understand such variables as the nature and cause of regional climate characteristics, groundwater, runoff, and soils in order to place human/environment interactions in a proper context. In each case, the current trends in water and land use are probably unsustainable in the long term and the potential for significant environmental and social impacts is very high. One cannot help but wonder what people will do in Las Vegas and other southwestern cities if and when water supplies decrease drastically should global warming scenarios actually come to pass in the coming decades. Similarly, what does the future hold for farmers in California and Australia if soil salinization continues? More important for you, per-

haps, is what happens to the cost of food in the coming decades as global population continues to grow and more farmland is potentially lost to salinization. These are tough questions indeed, and the ultimate outcome is uncertain.

The third case study revolved around oil supplies, which are important to everyone given our reliance on petroleum to fuel the economy and our personal vehicles. The trends seem to show that we are reaching the limit of the conventional oil supplies that can be easily extracted with traditional methods. Given ongoing demand for oil, and the likely increase in future demand with increasing population and associated economic expansion around the world, the oil industry is focusing more and more on extracting unconventional oil resources. In North America, these efforts have centered in the Athabasca oil sands in Canada and the various shale plays in the United States that collectively contain about 200 billion barrels of oil. Many regard successful extraction of these fossil-fuel resources as essential to a healthy and expanding U.S. economy. On the other hand, extracting these resources may very well have significant negative environmental impacts that could even harm human health. Is this a deal that we are willing to accept? Only time will tell.

Our last case on panda habitat also requires good knowledge of physical geography to fully comprehend the issue and possible outcomes. Understanding panda conservation requires knowledge of the geographical charac-

teristics of their habitat, such as the distribution of bamboo, elevation, and relief. It also requires an understanding of human population trends in the area and how expansion of agriculture and wood fuel collection has impacted the habitat. This case study also demonstrates how scientists have used geographical techniques to reconstruct the changes in habitat over time. Their studies have helped shape conservation efforts within China and by various international organizations in a positive way that may very well have stabilized the giant panda population in the wild and saved them from extinction.

SUMMARY OF KEY CONCEPTS

1. An understanding of physical geography is essential to understanding human environmental interactions around the world. Many of the topics covered in this text, such as solar radiation, atmospheric circulation, soils, and plant geography, are directly related to major societal issues.

2. Human population has grown dramatically in the past 200 years, from about 1 billion people in 1800 to approximately 6.7 billion today. It is projected that the population will reach over 9 billion by 2050. This rapid growth is tied directly to industrialization and improvements in human health. It poses profound questions regarding the impact on environmental systems around the world.

3. An important ongoing and future issue is the availability of freshwater for human consumption in many places around the world. As population grows, more people are moving to marginal lands where water supplies are scarce. This scarcity in places such as Las Vegas, Nevada, poses major challenges regarding sustainable economic development in semi-arid and arid regions in the future.

4. With the growing need for food, agriculture is being practiced on a massive scale in many places, such as the San Joaquin Valley in California and in several parts of Australia. The clearing of native vegetation and the use of irrigation have caused intensive soil salinization to occur in many parts of these landscapes. This process significantly harms soils and reduces crop yields at a time when more food is required.

5. As a result of rising demand for oil and decreasing supply, the oil industry is systematically developing unconventional supplies such as those found in the Athabasca oil sands in Canada and various shale plays in the United States. Although these oil sources show great promise, extraction techniques such as open-pit mining and fracking may have serious environmental impacts.

6. A growing population has resulted in massive deforestation around the world. Such forest loss has threatened the habitat of countless plant and animal species, such as the giant panda in China. Conservation efforts have the potential to stabilize these losses and even reverse them in the coming decades.

CHECK YOUR UNDERSTANDING

1. Describe the overall history of human population on Earth. When were there major growth spurts, and why did they occur? What is the projected growth of population in the future?

2. What is the difference between the developed and developing world, and what are the population trends in each of these areas?

3. Why does population growth in the developing world make it more likely that concentrations of atmospheric CO_2 will increase in the future?

4. Characterize the climate of Las Vegas and describe why this pattern occurs.

5. What is the Colorado River Compact, and how did it influence the population of Las Vegas?

6. How do the concepts of carrying capacity and sustainability relate to the issue of water use and population in Las Vegas?

7. Describe the process of soil salinization and how it can occur naturally.

8. How can agricultural practices cause soil salinization to occur? What impact does increased soil salinity have on crops?

9. In the context of soil salinization, how are the San Joaquin Valley and ancient Sumeria similar? Why are these places particularly susceptible to soil salinization?

10. In the context of carrying capacity, why is soil salinization in Australia an important environmental issue?

11. Describe the general history of oil supply and demand in the 20th century. What factors influenced gas prices, and how has oil been traditionally extracted from the ground?

12. Discuss the unconventional oil deposits in North America. Where are they located, and how is oil extracted from the ground at these sites? What are the various environmental implications of oil recovery in these areas?

13. Describe the nature of the giant panda habitat, and how and/or why its geographical distribution has changed over time.

14. What do studies of land-cover change in the late 20th century indicate about panda habitat at the Wolong Nature Reserve? How have these findings influenced Chinese environmental policy?

ANSWERS TO VISUAL CONCEPT CHECKS

VISUAL CONCEPT CHECK 20.1

Water in this fountain is most likely derived from Lake Mead, which is a reservoir upstream of Hoover Dam on the Colorado River. Water in Lake Mead originates mostly as runoff from melting snow in the Rocky Mountains, which are the headwaters of the Colorado River.

VISUAL CONCEPT CHECK 20.2

The salt crust in the San Joaquin Valley formed because water tables in the basin have increased in the past few decades from irrigation. As a result, water and salts have moved to the surface by wicking. After the water evaporates, salts are left behind on the surface.

VISUAL CONCEPT CHECK 20.3

The material shown in the photograph is bitumen, which is a semi-solid form of petroleum. It occurs in sandy deposits, known as the Athabasca oil sands, in central Canada. Bitumen can be extracted directly from the ground in blobs like this in open-pit mines. It can also be converted to liquid oil at depth by injecting steam into the well.

A

A horizon   The soil horizon that is enriched with humus.   281

Absorption   The assimilation and conversion of solar radiation into another form of energy by a medium such as water vapor. In this process, the temperature of the absorbing medium is raised.   71

Acid rain   The precipitation by rain, fog, or snow of strong mineral acids, primarily sulfur dioxide and nitrogen oxides, that originated in factories.   379

Actual evapotranspiration (actual ET)   The quantity of water actually removed from a given land area by evaporation and transpiration.   203

Adiabatic processes   Changes in temperature that occur due to variations in air pressure.   149

Adret slope   The slope that faces the Sun most directly.   252

Advection   The horizontal transfer of air.   108

Advection fog   Fog that develops when warm air flows over cooler air.   153

Aerial photographs   Photographs taken of the Earth's surface from the air.   32

Aggradation   The progressive accumulation of sediment along or within a stream.   423

Air mass   A large body of air in the lower atmosphere that has distinct temperature and humidity characteristics.   170

Air pressure   The force that air molecules exert on a surface due to their weight.   106

Albedo   The reflectivity of features on the Earth's surface or in the atmosphere.   73

Albic horizon   A diagnostic horizon of podzolization from which clay and free iron oxides have been removed, resulting in a light-colored E horizon.   290

Alfisols   Soils generally found in seasonal midlatitude regions that formed through podzolization and have an alkaline argillic horizon.   289

Alluvial fan   A fan-shaped landform of low relief that forms where a stream flows out of an area of high relief into a broad, open plane where the gradient is less and deposition thus occurs.   484

Alluvial terrace   A level, step-like landform that forms when a stream erodes its bed so that a horizontal surface is raised relative to the channel.   433

Alluvium   Sediment deposited by a stream.   433

Alpine glacier   A glacier in mountainous regions that flows down preexisting valleys.   452

Andisols   Soils formed in parent material that is at least 50% volcanic ash.   299

Angle of incidence   The angle at which the Sun strikes Earth at any given place and time.   74

Anthropocene   A proposed new epoch in the geological timescale based on the recognition that humans are now the foremost agent of change on the Earth's surface.   328

Anthropogenic   Caused by humans.   226

Anticlinal valley   A topographic valley that occurs along the axis of a structural anticline.   345

Anticline   A convex fold in rock in which rock layers are bent upward into an arch.   342

Anticyclones   High-pressure systems.   108

Aphelion   The point of the Earth's orbit where the distance between the Earth and Sun is greatest (~152 million km or 94.5 million mi).   47

Aquiclude   An impermeable body of rock that may contain water but does not allow transmission of water through it.   393

Aquifer   A geological formation that contains a suitable amount of water to be accessed for human use.   395

Arable land   Land that has the high potential to be cultivated for crop production.   548

Arête   A sharp ridge that forms between two glacial cirques.   459

Argillic horizon   A B horizon that is enriched in eluviated clay.   283

Aridisols   Mineral soils that form in arid environments and thus are poorly developed.   295

Arroyo   A deep, steep-sided gully that is cut into alluvium.   484

Artesian well   A well in which water from a confined aquifer rises to the surface through natural pressure.   396

Artificial levee   An engineered structure along a river that effectively raises the height of the river bank and thus confines flood discharge.   439

Asthenosphere   The layer of very soft rock that occurs in the upper part of the upper mantle. This region is about 40 km to 250 km (25 mi to 105 mi) below the surface of Earth. The soft character of this rock allows isostatic adjustments to occur.   308

Atmosphere   The gaseous shell that surrounds Earth.   7

Autotrophs   Organisms that synthesize their own food using heat or light as the source of energy.   236

Avalanche   A large mass of snow or rock that suddenly slides down a mountainside.   386

Axis   The line around which the Earth rotates, extending through the poles.   47

B

B horizon   The soil horizon that forms below the E horizon because translocated minerals recrystallize.   281

Backshore   The part of the beach that lies between the berm scarp and foredune and is covered by water only during strong storms.   519

Backslope   The gradual slope of a dune that faces the prevailing winds.   490

Backswamps   Marshy floodplain landforms that develop behind natural levees in which fine-grained sediments settle after a flood.   431

Bajada   A landscape consisting of overlapping alluvial fans along a mountain front.   484

Bankfull discharge   The amount of discharge at which the stream channel is full.   417

Barrier island   An elongated bar of sand that forms parallel to the shore for some distance.   522

Base flow   The amount of stream discharge at any given place and time that is solely the product of groundwater seepage.   416

Base level   The lowest level at which a stream can no longer lower its bed, because it flows into the ocean, a lake, or another stream.   425

Bay   An indentation in the shoreline that is generally associated with an ocean, sea, or gulf   507

Baymouth bar   A spit that entirely encloses a bay.   521

Beach drift   Sediment that is transported in the surf zone by swash and backwash, which form due to the oblique approach of waves.   513

Berm scarp   A miniature cliff created by wave erosion that fronts a beach berm.   519

Big Bang theory   The theory that the Universe originated about 14 billion years ago when all matter and energy erupted from a singular mass of extremely high density and temperature.   44

Biomass   The amount of living matter in an area, including plants, animals, and insects.   239

Biome   The complex of living communities maintained by the climate of a region and characterized by a distinctive type of vegetation.   241

Biosphere   The portion of Earth and its atmosphere that supports life.   8

Bioturbation   The mixing of soil by plants or animals.   272

Brackish   Water that has salinity less than 35‰.   507

Braided stream   A network of converging and diverging stream channels within an individual stream system that are separated from each other by deposits of sand and gravel.   422

Breaker   A wave that rises and crashes when the forward momentum of oscillations cannot be maintained.   512

Brine   Water that has salinity greater than 35‰.   507

Butte   A steep-sided hill or peak that is often a remnant of a plateau or mesa.   483

C

C horizon   Unaltered soil parent material.   281

Calcic horizon   A diagnostic soil horizon of calcification enriched in illuviated calcium carbonate.   293

Calcification   A regional soil-forming process in which calcium carbonate is cycled within the soil.   292

Canyon   A very steep-sided valley that is cut into bedrock.   483

Capillary action   The process through which water is able to move upward against the force of gravity.   138

Caprock   A unit of relatively resistant rock that caps the top of a landform and thus protects underlying rocks from erosion.   372

Carbonation   A type of chemical weathering caused by rainwater that has absorbed atmospheric carbon dioxide and formed a weak carbonic acid that slowly dissolves rock.   378

Carrying capacity   The maximum number of organisms that can live in a given area of habitat without degrading it and causing social stresses that result in population decline.   538

Cartography   The design and production of maps.   22

Cation exchange capacity (CEC)   The total amount of exchangeable cations that a soil can absorb.   279

Cations   Positively charged ions, such as sodium, potassium, calcium, and magnesium.   279

Cave   A cavity in rock, produced by the dissolution of calcium carbonate, that is large enough for someone to enter.   402

Celestial dome   A sphere that shows the Sun's arc, relative to the Earth, in the sky.   54

Chemical weathering   The decomposition and alteration of rocks due to chemical actions of natural physical and biological processes.   375

Chinook winds   Downslope airflow that results when a zone of high air pressure exists on one side of a mountain range and a zone of low pressure exists on the other.   129

Cinder-cone volcano   A small, steep-sided volcano that consists of solidified magma fragments and rock debris that may form in only one eruption.   361

Circle of illumination   The great circle on Earth that is the border between night and day.   47

Cirque   A bowl-like depression that serves as a source area for some alpine glaciers.   455

Cirrus clouds   Thin, wispy clouds that develop high in the troposphere.   153

Climate   Average precipitation and temperature characteristics for a region that are based on long-term records.   202

Climax vegetation   The vegetation within an area that has reached its ultimate complexity.   254

Coal   A solid fossil fuel that consists of carbonized plants and animals.   318

Cold front   A frontal boundary where cold air is advancing into relatively warm air. This front is typically associated with intense rain of short duration.   172

Colloids   Very small (10 nanometers to 1 micrometer), evenly divided solids that do not settle in solution.   279

Colluvium   Unconsolidated sediment that accumulates at the base of a slope.   381

Composite volcano   A large, steep-sided volcano that grows through progressive volcanic eruptions, which are usually explosive, and consists of layers of volcanic debris.   361

Condensation   The process through which water changes from the vapor to liquid phase.   140

Condensation nuclei   Microscopic dust particles around which atmospheric water coalesces to form raindrops.   152

Conduction   The transfer of heat energy from one substance to another by direct physical contact.   71

Cone of depression   The cone-shaped depression of the water table that occurs around a well.   398

Confluence   The place where two streams join together.   412

Conformal projection   A map that maintains the correct shape of features on the Earth but distorts their relative size to one another.   24

Conglomerate   Sandstone that contains a wide variety of particle sizes.   315

Constant gases   Atmospheric gases such as nitrogen, oxygen, and argon that maintain relatively consistent levels in space and time.   63

Continental   A place that is surrounded by a large body of land and that experiences a large annual range of temperature.   94

Continental crust   Granitic part of the Earth's crust that makes up the continents. Continental crust averages about 40 km (25 mi) in thickness and is also called sial because it consists largely of silica and aluminum.   308

Continental drift   The theory that the continents move relative to one another in association with plate tectonics.   335

Continental glacier   An enormous body of flowing ice that covers a significant part of a large landmass.   455

Contours   Isolines that connect points of equal elevation.   30

Convection   A circular cell of moving matter that contains warm material moving up and cooler matter moving down.   71

Convectional uplift   Uplift of air that occurs when bubbles of warm air rise within an unstable body of air.   157

Convergent uplift   Uplift of air that occurs when large bodies of air meet in a central location.   157

Coral reefs   Resistant marine ridges or mounds consisting largely of compacted coral together with algal material and biochemically deposited calcium carbonates.   527

Cordilleran Ice Sheet   The ice cap that covered much of the mountains in the northwestern part of North America during the Pleistocene Epoch.   466

Coriolis force   The force created by the Earth's rotation that causes winds to be deflected to the right in the Northern Hemisphere and to the left in the Southern Hemisphere.   112

Counterradiation   Longwave radiation that is emitted toward the Earth's surface from the atmosphere.   64

Crest   The highest point of a dune.   490

Crevasse   A deep crack in a glacier.   450

Cumulus clouds   Individual puffy clouds that develop due to convection.   153

Cyclogenesis   The sequence of atmospheric events along the polar jet stream that produces midlatitude cyclones.   174

Cyclones   Low-pressure systems.   108

D

Dam   A barrier that blocks or restricts the downstream movement of a stream.   440

Debris flow   A rapidly flowing and extremely powerful mass of water, rocks, sediment, boulders, and trees.   386

Debris slide   A mass-wasting process in which slope failure occurs along a plane that is roughly parallel to the surface.   384

Decomposers   Organisms that consume dead or decaying organic substances for nutrition.   236

Deflation   Removal of sediment from a surface by wind action.   487

Deflation hollow   A depression created by wind erosion.   487

Deforestation   The removal of trees for economic or agricultural purposes.   256

Degradation   The topographic lowering of a stream channel by stream erosion.   423

Delta   A low, level plain that develops where a stream flows into a relatively still body of water so that its velocity decreases and alluvial deposition occurs.   436

Dendrochronology   The dating of past events and variations in the environment and climate by studying the annual growth rates of trees.   219

Deposition   The process by which water vapor changes directly to ice.   140

Desert pavement   A resistant, pavement-like surface created when fine particles blow away and coarse sediment such as pebbles and gravel is left behind.   493

Desertification   The process through which a formerly vegetated landscape gradually becomes desert-like.   497

Dew-point temperature   The temperature at which condensation occurs in a definable body of air.   145

Direct radiation   Solar radiation that flows directly to the surface of Earth and is absorbed.   71

Disappearing stream   A surface river or stream that flows into a sinkhole and subsequently moves into an underground river system.   404

Diurnal cycle   A 24-hour cycle   54

Dolomite   Sedimentary rock consisting of over 50% calcium-magnesium carbonate ($CaMg[CO_3]_2$).   317

Downdraft   A rapidly moving current of cool air that flows downward in a thunderstorm.   180

Downwelling current   A current that sinks to great depths within the ocean because water temperature drops and salinity increases.   132

Drainage basin   The geographical area that contributes groundwater and runoff to any particular stream.   411

Drainage density   The measure of stream channel length per unit area o$_f$ drainage basin.   413

Drainage divide   An area of raised land that forms a bordering rim between two adjacent drainage basins.   411

Drumlin   A streamlined landform created when a glacier deforms previously deposited till.   463

Dry adiabatic lapse rate (DAR)   The rate at which an unsaturated body of air cools while lifting or warms while descending. This rate is 10°C/1000 m (5.5°F/1000 ft).   149

E

E horizon   The soil horizon that is progressively depleted of minerals through dissolution and translocation.   281

Earthflow   A slow-to-rapid type of mass movement that involves soil and other loose sediments, some of which may be coarse.   385

Earthquake   Shaking of the Earth's surface due to the instantaneous release of accumulated stress along a fault plane or from underground movements within a volcano.   348

Easterly wave   A slow-moving trough of low pressure that develops within the tropical latitudes.   190

Ecosystems   A community of plants, animals, and microorganisms linked by energy and nutrient flows.   236

Ecotone   The transition area where two or more ecosystems merge.   243

Electromagnetic spectrum   The radiant energy produced by the Sun that is measured in progressive wavelengths.   60

Eluviation   The dissolution and downward mobilization of minerals by water in the soil.   270

Emissivity   The amount of electromagnetic energy released by some aspect of the Earth's surface.   35

Entisols   Soils that are very weakly developed and thus have no distinct horizonation.   296

Environmental lapse rate   The decrease in temperature that generally occurs with respect to altitude in the troposphere. This rate is 6.4°C per kilometer or 3.5°F per 1000 ft (a positive lapse rate).   86

Eolian processes   Geomorphic processes associated with the way that wind erodes, transports, and deposits sediment.   485

Epicenter   The point on the Earth's surface that lies directly over the focus of an earthquake.   349

Epipedon   The uppermost horizon or horizons of a soil that is used for diagnostic purposes.   283

Equator   The great circle that lies halfway between the North and South Poles.   15

Equatorial trough   Core of low-pressure zone associated with the Intertropical Convergence Zone.   116

Equilibrium line   The place on a glacier where snow accumulation and melting are in balance.   450

Equivalent projection   A map projection that accurately portrays size features throughout the map.   26

Esker   A winding ridge formed by a stream that flows beneath a glacier.   464

Eustatic change   Fluctuations in sea level associated with adjustments in the hydrologic cycle.   507

Evaporation   The process through which water changes from the liquid to vapor phase.   140

Evaporites   Surface salt residue that collects through the evaporation of water and the crystallization of sodium.   317

Evapotranspiration   The combined processes of evaporation and transpiration.   148

Exfoliation   Form of physical weathering in which sheets of rock flake away due to seasonal temperature changes or by the expansion of the rock due to unloading.   375

Extrusive igneous rocks   Rocks that form when magma cools on the Earth's surface.   311

F

Fall Equinox   Assuming a Northern Hemisphere seasonal reference, the Fall (or autumnal) Equinox occurs on September 22 or 23, when the subsolar point is located at the Equator (0°).   53

Fault   A crack in the Earth's crust that results in the displacement of one lithospheric plate or rock body relative to another.   348

Fault escarpment   A step-like feature on the Earth's surface created by fault slippage.   352

Field capacity   The amount of water remaining in the soil after the soil is completely drained of gravitational water.   268

Firn   The compact, granular substance that is the transition stage between snow and glacial ice.   448

Fjord   A former glaciated valley along the coast that is flooded by rising sea level.   508

Flood stage   The level at which stream discharge begins to spill out of the channel into the surrounding area.   418

Foredune   A dune that forms parallel to the shore when sand blows inland from the beach.   521

Foreshore   The nearshore zone that is regularly exposed and submerged through the tidal fluctuations and movement of surf.   519

Fossil fuels   Carbon-based energy sources, such as gasoline and coal, that are derived from ancient organisms.   227

Freezing   The process through which water changes from the liquid to solid phase.   139

Frontal uplift   Uplift of air that occurs along the boundary of contrasting bodies of air.   157

Frost heaving   Upthrust of sediment or soil due to the freezing of wet soil beneath.   374

Frost wedging   Expansion and contraction of water in rock cracks due to freezing and thawing.   373

Fumerole   A steam vent that results because underlying groundwater is boiled away before reaching the surface.   368

G

Gelisols   Soils in subarctic and arctic environments that contain permafrost within 2 m of the surface.   291

Geologic time   The period of time that encompasses all of Earth history, from its formation to the present.   322

Geomorphology   The branch of physical geography that investigates the form and evolution of the Earth's surface.   342

Geostationary orbit   An orbit where satellites remain over the same place on Earth every day. This orbit is achieved because the satellite is placed very high above Earth and travels at the same speed as the Earth's rotation.   34

Geostrophic winds   Airflow that moves parallel to isobars because of the combined effect of the pressure gradient force and Coriolis force.   114

Geyser   A superheated fountain of water that suddenly sprays into the air on a periodic basis.   367

Glacial abrasion   An erosional process caused by the grinding action of a glacier on rock.   457

Glacial drift   Sediment deposited indirectly or directly by a glacier.   461

Glacial erratics   Large boulders that have been plucked and transported a great distance before they are deposited.   457

Glacial grooves   Deep furrows in rock produced by glacial abrasion.   457

Glacial outwash   Sediment deposited by meltwater streams emanating from a glacier.   463

Glacial plucking   An erosional process by which rocks are pulled out of the ground by a glacier.   457

Glacial striations   Scratches in rock produced by glacial abrasion.   457

Glacial till   Sediment deposited directly by a glacier.   461

Glacial trough   A deep, U-shaped valley carved by an alpine glacier.   460

Glacier   A slow-moving mass of dense ice that flows under its own weight.   448

Graben   A downthrown block of rock that lies between two steeply inclined fault blocks.   352

Graded stream   An idealized stream that has achieved a balance between sediment transport and stream capacity.   423

Great circles   Circles that pass through the center of the Earth and that divide the planet into equal halves.   14

Greenhouse effect   The process through which the lower part of the atmosphere is warmed because longwave radiation from Earth is trapped by carbon dioxide (CO2) and other greenhouse gases.   64

Groin   A short, low wall built at a right angle to the coast that intercepts the longshore current.   531

Ground ice   Distinct zones of frozen water that occur in permafrost regions.   475

Gulf   A relatively small body of salt water that is surrounded by land on three sides and opens to a sea or ocean.   506

Gyres   Large oceanic circulatory systems that form because currents are deflected by landmasses.   132

H

Hadley cell   Large-scale convection loop in the tropical latitudes that connects the Intertropical Convergence Zone (ITCZ) and the Subtropical High (STH).   118

Hanging valley   An elevated U-shaped valley (with respect to a glacial trough) formed by a tributary alpine glacier.   460

Headland   A portion of the coast that extends outward into a large body of water.   515

Heterotrophs   Organisms that consume complex organic substances for food.   236

High-pressure ridge   An elongated area of elevated air pressure in the upper atmosphere that is typically associated with sunny skies and calm winds.   176

High-pressure system   A rotating column of air that descends toward the surface of Earth where it diverges.   108

High latitudes   The zone of latitude that lies from about 55° to 90° in both hemispheres.   17

Histosols   Organic soils that form in cool, wet environments where organic carbon decomposes very slowly.   291

Hogback ridge   A ridge underlain by gently tipped rock strata with a long, gradual slope on one side and a relatively steep scarp or cliff on the other.   345

Hook echo   The diagnostic feature in Doppler radar indicating that strong rotation is occurring within a thunderstorm and that tornado development is thus possible.   189

Horn   A mountain with three or more arêtes on its flanks.   459

Horst   An upthrown block of rock that lies between two steeply inclined fault blocks.   352

Hotspot   A stationary zone of magma upwelling that is associated with volcanism within the interior of a crustal plate.   365

Humidity   A measure of how much water vapor is in the air. The ability of air to hold water vapor is dependent on temperature.   142

Humus   Decomposed organic matter, typically dark, that is contained within the soil.   266

Hurricane   A tropical circulatory system with maximum sustained winds greater than 63 knots (73 mph).   190

Hydraulic fracturing   Fracturing of rock at depth by the injection of fluids to release oil and natural gas contained within shale beds.   562

Hydrologic cycle   A model that illustrates the way that water is stored and moves on Earth from one reservoir to another.   141

Hydrolysis   A chemical weathering process that results in the decomposition of silicate molecules within rock through the reaction of hydrogen and hydroxyl ions in water.   377

Hydrosphere   The part of Earth where water, in all its forms, flows and is stored.   8

Hygroscopic water   Soil water held so tightly by sediment grains that it is unavailable for plant use.   392

I

Ice cap   A dome-shaped sheet of ice that covers an area less than 50,000 km^2 (~19,000 mi^2) in size.   452

Ice field   A topographically constrained sheet of ice in mountainous areas that frequently has glaciers streaming away from it.   452

Igneous rocks   Rocks that form when magma rises from the mantle and cools, either within the Earth's crust or on the surface.   311

Illuviation   The recrystallization of minerals that occurs directly below the zone of eluviation.   270

Inceptisols   Soils that have one or more weakly developed horizons due to some alteration and removal of soluble minerals.   298

Indirect radiation   Radiation that reaches Earth after it has been scattered or reflected.   74

Inner core   The inner part of the Earth's core. This area is about 1220 km (760 mi) thick and consists of solid iron and nickel.   306

Insolation   Amount of solar radiation measured in watts per square meter (W/m^2) that strikes a surface perpendicular to the Sun's incoming rays.   70

Interfluves   Topographic high points in a drainage basin that separate one tributary from another.   412

International Date Line   This line generally occurs at 180° longitude, with some variations due to political boundaries, and marks the transition from one day to another on Earth.   48

Intertropical Convergence Zone (ITCZ)   Band of low pressure, calm winds, and clouds in tropical latitudes where air converges from the Southern and Northern Hemispheres.   116

Intrusive igneous rocks   Rocks that form when magma cools within the Earth's crust.   311

Isobars   Isolines that connect points of equal atmospheric pressure.   30

Isohyets   Isolines that connect points of equal precipitation.   30

Isolines   Lines on a map that connect data points of equal value.   29

Isopachs   Isolines that connect points of equal sediment or rock thickness.   30

Isotherms   Isolines that connect points of equal temperature.   30

J

Jetties   Long stone or concrete structures used to create a permanent opening for a channel by reinforcing both sides of the passageway.   532

Joint   A crack or fissure along horizontal or vertical planes in a rock mass that divides the rock into large blocks.   373

K

Kame   A large mound of sediment deposited along the front of a slowly melting or stationary glacier.   464

Karst topography   Terrain that is underlain by soluble rocks, such as limestone and dolomite, where the landscape evolves largely through the dissolution of rock.   402

Katabatic winds   Downslope airflow that evolves when pools of cold air develop over ice caps and subsequently descend into valleys.   129

Kettle lake   A lake that forms when a block of ice falls off the glacial front, is buried by glacial drift, and then melts, forming a depression that fills with water.   464

Kinetic energy   The energy of motion in a body, measured as temperature, that is derived from movement of molecules within the body.   89

L

Lagoon   A brackish body of water that lies behind a bay-mouth bar.   521

Land breeze   Nighttime circulatory system along coasts where winds from a zone of high pressure over land flow to a zone of relatively low pressure over water.   129

Land cover   The various things that cover the landscape, such as forests, roads, and water bodies, at any point in time.   569

Landform   A natural feature, such as a hill or valley, on the surface of Earth.   342

Landslide   An instantaneous movement of soil and bedrock down a steep slope in response to gravity, triggered by an event such as an earthquake.   384

Large-scale map   A map that shows a relatively small geographic area with a relatively high level of detail.   27

Latent heat   Heat stored in molecular bonds that cannot be measured.   74

Laterization   A regional soil-forming process in tropical and subtropical environments that results in extensive eluviation of minerals except for iron and aluminum.   286

Latitude   The part of the Earth's grid system that determines location north and south of the Equator.   15

Laurentide Ice Sheet   The continental glacier that covered eastern Canada and parts of the northeastern United States during the Pleistocene Epoch.   466

Lava dome   A steep-sided volcanic landform consisting of highly viscous lava that does not flow far from its point of origin before it solidifies.   361

Leeward side   The side of a mountain range that faces away from prevailing winds.   129

Level of condensation   The altitude at which water changes from the vapor to liquid phase.   151

Limestone   Sedimentary rock consisting of over 50% calcium carbonate ($CaCO_3$).   316

Liquid outer core   The outer part of the Earth's core. This area is about 2250 km (1400 mi) thick and consists of molten iron and nickel.   307

Lithification   The process whereby sediments are cemented through compaction to form rock.   313

Lithosphere   The outer, solid part of Earth that is about 70 km (44 mi) thick and includes the uppermost part of the asthenosphere and the crust.   308

Littoral drift   Sediment that is transported through the combined processes of longshore drift and beach drift.   513

Loess   Windblown silt.   494

Longitude   The part of the Earth's grid system that determines location east and west of the Prime Meridian.   17

Longitudinal profile   A graph that illustrates the change in stream gradient in cross section along a stream from its source to its mouth.   424

Longshore current   The current that develops parallel to a coast when waves approach the coast obliquely and forward momentum is deflected.   513

Longshore drift   Transport of sediment by the longshore current.   513

Longwave radiation   The portion of the electromagnetic spectrum that includes thermal infrared radiation.   60

Low-pressure system   A rotating column of air where air converges at the surface and subsequently lifts.   108

Low-pressure trough   An elongated area of depressed air pressure in the upper atmosphere that is typically associated with cloudy skies and rain.   176

Low latitudes   The zone of latitude that lies between about 35° N and 35° S.   17

M

Magma   Molten rock beneath the surface of Earth.   311

Mantle   The layer of the Earth's interior that lies between the liquid outer core and the crust. This area is about 2900 km (1800 mi) thick and consists largely of silicate rock.   307

Map projection   The representation of the three-dimensional Earth on a two-dimensional surface.   23

Map scale   The distance ratio that exists between features on a map and the real world.   27

Maritime   A place that is close to a large body of water that moderates temperature.   94

Maritime vs. continental effect   The difference in annual and daily temperature that exists between coastal locations and those that are surrounded by large bodies of land.   94

Mass wasting   Movement of rock, sediment, and soil downslope due to the force of gravity.   381

Maximum humidity   The maximum amount of water vapor that a definable body of air can hold at a given temperature.   143

Meandering stream   A river or small stream that curves back and forth across its valley.   422

Mechanical weathering   The breakup of a body of rock into smaller rocks of the same type.   373

Meridians   Lines of longitude.   17

Meridional flow   Jet stream pattern that develops when strong Rossby waves exist and the polar front jet stream flows parallel to the meridians in many places.   121

Mesa   A broad horizontal surface, smaller than a plateau, that is upheld by resistant caprock.   483

Mesocyclones   Strong updrafts that rotate within a supercell thunderstorm.   185

Mesopause   The upper boundary of the mesosphere where temperature reaches its lowest point.   88

Mesosphere   A layer of decreasing temperature in the atmosphere that occurs from about 50 km to 80 km (~30 mi to 50 mi) in altitude.   88

Metamorphic rocks   Rocks that form when igneous, sedimentary, or other metamorphic rocks are subjected to intense heat and pressure.   318

Microclimate   Average temperature and precipitation characteristics of a small area within a larger climate region.   252

Mid-oceanic ridge   A ridge-like feature that develops along a rift zone in the ocean due to magma upwelling.   339

Midlatitude cyclone   A well-organized low-pressure system in the midlatitudes that contains warm and cold fronts.   173

Midlatitudes   The zone of latitude that lies between about 35° and 55° in both hemispheres.   17

Milankovitch theory   The theory that best explains Quaternary glacial/interglacial cycles through long-term variations in Earth's orbital eccentricity, tilt, and axial precession.   224

Minerals   Naturally occurring substances with distinctive chemical configurations that usually manifest themselves in some kind of crystalline form.   310

Mohorovicic discontinuity   The boundary between the Earth's crust and the upper part of the asthenosphere; seismic waves change speed at this boundary.   308

Mollisols   Soils that form through calcification and have a mollic epipedon that overlies mineral matter, which is more than 50% saturated with base ions.   292

Monocline   A geologic landform in which rock beds are inclined in a single direction over a large distance.   341

Monsoon   The seasonal change in wind direction that occurs in subtropical locations due to the migration of the Intertropical Convergence Zone (ITCZ) and Subtropical High (STH) Pressure System.   124

Moraine   A winding ridge-like feature that forms at the front or side of a glacier or between two glaciers.   462

Mountain breeze   Downslope airflow that develops when mountain slopes cool off at night and relatively low pressure exists in valleys.   129

Mudflow   A well-saturated and highly fluid mass of fine-textured sediment.   385

Mudpot   A bubbling mixture of gaseous mud and water at the Earth's surface that is associated with geothermal activity.   368

Multipath error   Error that results when a satellite signal reaches the GPS receiver directly from the satellite as well as others reflected from nearby buildings or other surfaces.   39

N

Natural gas   Naturally occurring mixture of hydrocarbons that often occurs in association with petroleum and is found in porous geologic formations.   318

Natural hazard   An actual or potentially occurring natural event, such as an earthquake or volcanic eruption, that has a negative effect on people or the environment.   353

Natural levee   A small ridge that develops along the channel of a stream through the deposition of relatively coarse sediment when flooding occurs.   431

Net radiation   The difference between incoming and outgoing flows of radiation.   76

Normal fault   A steeply inclined fault in which the hanging rock block moves relatively downward.   350

Northern Hemisphere   The half of Earth that lies north of the Equator.   15

O

O horizon   The uppermost soil horizon that consists of undecomposed plant litter.   281

Occluded front   The area where a cold front begins to overtake a warm front and thus lift warm surface air aloft.   178

Ocean   The entire body of salt water that covers about 71% of Earth.   506

Oceanic crust   Basaltic part of the Earth's crust that makes up the ocean basins. Oceanic crust is about 8 km (5 mi) thick and is also called sima because it consists largely of silica and magnesium.   308

Offshore   The nearshore zone that is permanently submerged and where waves break.   519

Offshore bar   A small ridge on the bottom of the ocean that separates the offshore and foreshore.   519

Orogeny   A period of mountain building, such as the Allegheny Orogeny.   344

Orographic uplift   Uplift that occurs when a flowing body of air encounters a mountain range.   157

Outwash plain   A broad landscape of limited relief created by the deposition of glacial outwash.   463

Overthrust fold   A structural feature where one part of the rock mass is shoved up and over the other.   342

Overturned fold   A structural feature in which the fold limb is tilted beyond vertical, which results in both limbs inclined in the same direction, but not at the same angle.   342

Oxbow lake   A portion of an abandoned stream channel that is cut off from the rest of the stream by the meandering process and filled with stagnant water.   428

Oxic horizon   A diagnostic soil horizon in tropical and subtropical environments rich in iron oxides.   287

Oxidation   A form of chemical weathering in which oxygen chemically combines with metallic iron to form iron oxides, resulting in the loss of electrons.   378

Oxisols   Mineral soils in tropical and subtropical environments that form through laterization and thus have an oxic horizon within 2 m of the surface.   286

Oxygen isotope stages   Periods of time that have distinct O-18/O-16 ratios, which are used to reconstruct prehistoric climate change.   222

Ozone hole   The decrease in stratospheric ozone observed on a seasonal basis over Antarctica, and to a lesser extent over the Arctic.   67

Ozone layer   The layer of the atmosphere that contains high concentrations of ozone, which protect the Earth from ultraviolet (UV) radiation.   66

P

Pacific Ring of Fire   The chain of volcanoes that occurs along the edge of the Pacific lithospheric plate.   362

Pangaea   The hypothetical supercontinent, composed of all the present continents, that existed between 300 and 200 million years ago.   334

Parallels   Lines of latitude.   17

Parent material   The mineral or organic material in which soil forms.   271

Passive margin   A place where the continental crust and the oceanic crust are on the same tectonic plate and thus do not move relative to each other.   338

Pedogenic processes   The natural processes of soil formation that involve additions, translocations, transformations, and losses.   268

Perched water table   A localized area of saturated sediment that is elevated above the regional water table by a zone of impermeable rock or sediment.   395

Periglacial processes   The suite of processes involving frost action, permafrost, and ground ice that occurs in arctic environments or along the margins of ice sheets.   473

Perihelion   The point of the Earth's orbit where the distance between the Earth and Sun is least (~147 million km or 91.5 million mi).   47

Permafrost   Ground that is permanently frozen.   474

Petroleum   Naturally occurring oily liquid that consists of ancient hydrocarbons.   318

Photosynthesis   The conversion of solar radiation into chemical energy. Sugars and starches are produced from carbon dioxide and water through the interaction of light and chlorophyll in green plants. The process releases oxygen into the atmosphere.   237

pH   The measure of acidity or alkalinity of a solution, ranging from 0 to 14, based on the activity of hydrogen ions (H^+).   278

Physical geography   Spatial analysis of the physical components and natural processes that combine to form the environment.   6

Pinnacle   A steep-sided, narrow tower that is the final remnant of a plateau, mesa, or butte.   483

Pixel   The smallest definable area of detail on an image; short for pixel element.   34

Plane of the ecliptic   The flat plane on which the Earth travels as it revolves around the Sun.   45

Plant succession   The predictable vegetation transition that occurs within a landscape over a period of time.   254

Plate tectonics   The theory that the Earth's crust is divided into a number of plates that move because they float on the asthenosphere.   334

Plateau   A very broad, horizontal surface that is upheld by resistant caprock.   483

Playa   A dried lake bed that forms when runoff collects in closed topographic depressions in arid regions.   483

Pluton   An extremely large mass of intrusive igneous rock that forms within the Earth's crust.   311

Podzolization   A regional soil-forming process in cool, humid environments that results in the eluviation of iron, aluminum, and organic acids to form well-developed E and Bs horizons.   290

Polar easterlies   Band of easterly winds at high latitudes.   123

Polar front   The contact in the midlatitudes between warm tropical air and colder polar air.   119

Polar front jet stream   River of high-speed air in the upper atmosphere that flows along the polar front.   120

Polar High   Zone of high atmospheric pressure at high latitudes.   122

Potential evapotranspiration (potential ET)   A measure of the maximum possible water loss from a given land area assuming sufficient water is available.   203

Potential natural vegetation   The vegetation that would occur naturally within a specific area if no human influence occurred.   241

Pressure gradient force   The difference in barometric pressure that exists between adjacent zones of low and high pressure that results in airflow.   111

Prime Meridian   The arbitrary reference point for longitude that passes through Greenwich, England.   17

Process   A naturally occurring series of events or reactions that can be measured and that result in predictable outcomes.   6

Progradation   Outward extension of the shoreline through deposition of sediment.   519

Proxy data   Indirect evidence of an event. For example, fossil pollen is a proxy indicator of climate change because vegetation reflects climate.   218

Pyroclastic material Fragmented rock materials resulting from a volcanic explosion or ejection from a volcanic vent. 361

R

R horizon Unweathered bedrock that underlies soil. 281

Radial dikes A long, wall-like feature of intrusive igneous rock that forms when magma is injected into thin cracks within older rock and cools. 328

Radiation Energy that is transmitted in the form of rays or waves. 70

Radiation budget The overall balance between incoming and outgoing radiation on Earth. 76

Radiation fog Fog that develops at night when a temperature inversion exists. 153

Radioactive isotopes Unstable isotopes that emit radioactivity as they decay from one element to another. 324

Rain shadow The body of land on the leeward side of a mountain range that is relatively dry and hot (compared to the windward side) due to adiabatic warming and drying. 160

Reflection The process through which solar radiation is returned directly to space without being absorbed by Earth. 73

Regolith The fragmented and weathered rock material that overlies solid bedrock. 271

Relative humidity The ratio between the specific and maximum humidity of a definable body of air. 143

Relief The difference between the high and low elevation of an area. 274

Remote sensing The method through which information is gathered about the Earth from a distance. 32

Residual parent material Parent material that forms by the weathering of bedrock directly beneath it. 271

Retrogradation The process through which a shoreline retreats through erosion. 517

Reverse fault A steeply inclined fault in which the hanging rock block moves relatively upward. 352

Ria A former river valley along the coast that is flooded by rising sea level. 508

Richter scale The logarithmic scale used to measure the strength of an earthquake. 350

Rifting The spreading apart of the Earth's crust by magma rising between fractures in the Earth's plates. 339

Rills Small drainage channels that are cut into hillslopes by running water. 421

Riparian zone The strip of land bordering a body of water that supports plants and animals adapted to water systems. 254

Roche moutonnée A landform produced by glacial abrasion and plucking that has a shallow slope on one side and a steep slope on the other side. 457

Rock An amorphous mass of consolidated mineral matter. 310

Rockfall A mass-wasting process in which rocks break free from cliff faces and rapidly tumble into the valley below. 382

Rock outcrop A place where rocks are exposed at the surface of Earth. 311

Rock structure The internal arrangement of rock layers. 341

Rossby waves Undulations that develop in the polar front jet stream when significant temperature differences exist between tropical and polar air masses. 120

S

Salic horizon The diagnostic horizon of salinization that forms due to the recrystallization of secondary salts. 296

Salinity Concentration of dissolved solids in water that is measured in parts per thousand (‰). 507

Salinization A regional soil-forming process in which soluble salts are cycled within the soil. 295

Salt-crystal growth A weathering process involving the buildup of salts on rock surfaces through evaporation. These salts weaken the cement that bonds rock particles, allowing them later to be washed or blown away. 375

Sandstone A sedimentary rock created when individual sand grains are deposited in thick layers by wind or water and lithify. 315

Saturated zone The zone of rock below and including the water table where pore spaces are completely filled with water. 393

Scattering The redirection and deflection of solar radiation by atmospheric gases or particulates. 73

Sea A subdivision of an ocean that is partially enclosed by land. 506

Sea breeze Daytime circulatory system along coasts where winds flow from a zone of high pressure over water to a zone of relatively low pressure over land. 128

Sea fog Fog that develops when cool, marine air comes into direct contact with colder ocean water. 153

Sediment Solid fragments of rocks that are transported to some location and deposited by wind, water, or ice. 313

Sedimentary rocks Rocks that form through the deposition and lithification of small fragments or dissolved substances from other rocks or, in some cases, marine animals. 313

Seismic waves Vibrations that travel through the Earth when stress is released in an earthquake. 306

Sensible heat Heat that can be felt and measured with a thermometer. 74

Shale Sedimentary rock consisting of lithified clay-sized sediments. 315

Shale play   A geographic area, dominated by shale bedrock, that is targeted for exploration due to favorable geological conditions that suggest high potential to yield an economic quantity of oil or natural gas.   562

Shield volcano   A very broad volcano with shallow slopes that forms in association with nonviscous lava flows.   364

Shortwave radiation   The portion of the electromagnetic spectrum that includes gamma rays, X-rays, ultraviolet radiation, visible light, and near-infrared radiation.   60

Sinkhole   A topographic depression that forms when underlying rock dissolves, causing the surface to collapse.   404

Slip face   The steep slope that lies on the leeward side of a sand dune at the angle of repose.   490

Slope   The degree of steepness of a portion of the landscape.   251

Slump   A mass-wasting process in which rock and sediment rotate and move down the slope along a concave plane relative to the surface.   384

Small circles   Circles that intersect the Earth's surface and that do not pass through the center of the planet.   15

Small-scale map   A map that shows a relatively large geographic area with a relatively low level of detail.   27

Soil   The uppermost layer of the Earth's surface that forms by the influence of parent material, climate, relief, and chemical and biological agents.   266

Soil creep   The gradual downhill movement of soil, trees, and rocks due to the force of gravity.   383

Soil-forming factors   The variables of climate, organisms, relief, parent material, and time that collectively influence the development of soil.   271

Soil horizons   The distinct layers within a soil that result from pedogenesis.   280

Soil order   A group of 12 distinctive soils differentiated at the most general level.   283

Soil profile   A vertical exposure in which all soil components can be seen.   280

Soil salinization   The process in which soils become enriched in soluble salts such as sulfates from calcium and potassium.   549

Soil science   The study of soil as a natural resource through understanding of its physical, chemical, and biological properties.   283

Soil structure   The way soil aggregates clump to form distinct physical characteristics.   277

Soil taxonomy   The method of soil classification that is based on the physical, chemical, and biological properties of the soil.   283

Soil–water budget   The balance of soil water that involves the amount of precipitation, evapotranspiration, and water storage and loss.   268

Solar constant   The average amount of solar radiation (~1370 W/m2) received at the top of the atmosphere.   61

Solar noon   The time of day when the Sun angle reaches its highest point as the Sun arcs across the sky.   53

Solifluction   A form of soil creep that occurs in arctic environments where freeze-thaw processes result in lobes of soil moving gradually downslope.   383

Solum   The A, E, and B horizons of a soil, which form through pedogenic processes.   281

Southern Hemisphere   The half of Earth that lies south of the Equator.   15

Spatial analysis   A method of analyzing data that specifically includes information about the location of places and their defining characteristics.   5

Spatial resolution   The area on the ground that can be viewed in detail from the air or space.   34

Specific humidity   The measurable amount of water vapor that is in a definable body of air.   143

Spit   A linear bank of land that extends into a bay made by the deposition of longshore sediment.   520

Spodic horizon   A mineral soil horizon characterized by the illuvial accumulation of aluminum, iron, and organic carbon.   290

Spodosols   Soils in cool, humid regions that form through podzolization and contain a spodic horizon enriched in eluviated iron, aluminum, and organic carbon.   290

Spring Equinox   Assuming a Northern Hemisphere seasonal reference, the Spring (or vernal) Equinox occurs on March 20 or 21, when the subsolar point is located at the Equator (0°).   51

Stable air   A body of air that has a relatively low environmental lapse rate compared to potential uplifting air; thus, strong convection cannot occur.   162

Stationary front   A boundary where contrasting air masses are flowing parallel to one another.   171

Stratopause   The upper boundary of the stratosphere where temperature reaches its highest point.   88

Stratosphere   The layer of the atmosphere, between the troposphere and mesosphere, that ranges between about 12 km and 50 km (between ~7.5 mi and 31 mi) in altitude.   88

Stratus clouds   Layered sheets of clouds that have a thick and dark appearance.   153

Stream hydrograph   The graphical representation of stream discharge over a period of time.   417

Strike-slip fault   A structural fault along which two lithospheric plates or rock blocks move horizontally in opposite directions and parallel to the fault line.   352

Subduction   The process by which one lithospheric plate is forced beneath another. This usually happens when oceanic crust descends beneath continental crust, but it can occur where two plates of oceanic crust meet.   341

Sublimation The process through which water changes directly from ice to the vapor phase.   140

Subsidence The settling or sinking of a surface as a result of the loss of support from underlying water, soils, or strata.   400

Subsolar point The point on Earth where the Sun angle is 90° and solar radiation strikes the surface most directly at any given point in time.   45

Subtropical High (STH) Pressure System Band of high air pressure, calm winds, and clear skies that exists at about 25° to 30° N and S latitude.   118

Summer Solstice Assuming a Northern Hemisphere seasonal reference, the Summer Solstice occurs on June 20 or 21, when the subsolar point is located at the Tropic of Cancer (23.5° N).   51

Sun angle The angle at which the Sun's rays strike the Earth's surface at any given point and time. This angle is high at low latitudes and is progressively less at higher latitudes.   44

Sun-synchronous orbit A slightly inclined polar orbit that keeps pace with the Sun's westward progress as Earth rotates, resulting in regular return intervals over every location on Earth.   34

Supercell thunderstorms Large thunderstorms that contain winds moving in opposing directions and that are associated with strong winds, lightning, thunder, and sometimes hail and tornadoes.   184

Surface tension The contracting force that occurs when the water surface meets the air and acts like an elastic skin.   267

Synclinal valley A topographic valley that occurs along the axis of a structural syncline.   346

Syncline A concave fold in rock in which rock layers are bent downward to form a trough.   342

Systems theory The examination of interactions involving energy inputs and outputs that result in predictable outcomes.   6

T

Talus A pile of rock fragments and boulders that accumulates below a cliff due to rockfall.   382

Tarn A small lake that forms within a glacial cirque.   458

Temperature inversion A layer of the troposphere in which the air temperature increases, rather than cools, with altitude.   87

Temporal lag The difference in time between two events, such as when peak insolation and peak temperature occur.   93

Thermohaline circulation The global oceanic circulatory system that is driven by differences in salinity.   132

Thermosphere The upper layer of the atmosphere, which occurs between about 80 km and 480 km (~50 mi and 300 mi) in altitude.   88

Thunderstorm A brief but intense storm that contains strong winds, lightning, thunder, and perhaps hail.   172

Tombolo A spit or sandbar that connects an island to the mainland.   521

Topographic map A map that displays elevation data regarding the Earth's surface.   30

Topography The shape and configuration of the Earth's surface.   30

Trade winds The primary wind system in the tropics that flows toward the Intertropical Convergence Zone on the equatorial side of the Subtropical High-Pressure System. These winds flow to the southwest in the Northern Hemisphere and to the northwest in the Southern Hemisphere.   118

Transform plate margin A plate boundary where opposing plates move horizontally relative to each other.   338

Transpiration The passage of water from leaf pores to the atmosphere.   148

Transported parent material Parent material such as glacial or stream sediments that has recently been deposited and in which soil forms.   272

Tree line The line that represents the upper limit in mountains and high latitudes where environmental conditions support the growth of trees.   253

Tributary A stream or river that flows into a larger stream or river.   412

Tropic of Cancer The line of latitude at 23.5° N where the subsolar point is located (and Sun angle is thus 90°) on the Summer Solstice in the Northern Hemisphere.   53

Tropic of Capricorn The line of latitude at 23.5° S where the subsolar point is located on the Winter Solstice (and Sun angle is thus 90°) in the Northern Hemisphere.   53

Tropical depression A tropical low-pressure system with central sustained winds ranging between 20 knots and 34 knots (23 mph and 39 mph).   190

Tropical easterlies Band of easterly winds that exists where northern and southern trade winds converge.   118

Tropical storm A tropical low-pressure system with maximum sustained winds between 35 knots and 63 knots (39 mph and 73 mph).   190

Tropopause The top part of the troposphere, which is identified by where the air temperature is –57°C (–70°F).   87

Troposphere The lowermost layer of the atmosphere, which lies between the Earth's surface and an altitude of about 12 km (~7.5 mi).   86

Trunk stream The primary stream of a drainage basin.   412

Tsunami A destructive sea wave caused by a disturbance within the ocean, such as an earthquake, volcanic eruption, or landslide.   355

U

Ubac slope The slope that faces away from the Sun.   252

Ultisols Mineral soils in subtropical environments that formed through laterization and thus are depleted of calcium and have an argillic horizon.   287

United States Public Land Survey The primary way that land is subdivided in the United States by using a grid system based on the four cardinal directions.   19

Unsaturated zone The area between the soil–water belt and the water table where pore spaces are not saturated with water.   392

Unstable air A body of air that has a relatively high environmental lapse rate compared to uplifting air within it; thus, strong convection can occur.   162

Updrafts An area of rapidly flowing air that is moving upward within a thunderstorm.   184

Upwelling current A current that ascends to the surface of the ocean because water temperature increases and salinity decreases.   132

Urban heat island The relatively warm temperatures associated with cities that occur because paved surfaces and urban structures absorb and release radiation differently from the surrounding countryside.   101

V

Valley breeze Upslope airflow that develops when mountain slopes heat up due to re-radiation and conduction over the course of the day.   129

Variable gases Atmospheric gases such as carbon dioxide, water vapor, and ozone that vary in concentration in space and time.   63

Ventifact An individual rock that is pitted, grooved, or streamlined through wind abrasion.   487

Vertical zonation The change in environmental characteristics that occurs with respect to altitude.   252

Vertisols Soils that contain an abundance of expandable clay and thus swell and shrink during wet and dry cycles, respectively.   288

Volcanic arc A chain of volcanoes created by rising magma derived from a subducting tectonic plate.   361

Volcano A mountain or large hill containing a conduit that extends down into the upper mantle, through which magma, ash, and gases are periodically ejected onto the surface of Earth or into the atmosphere.   360

W

Warm front A frontal boundary where warm air is advancing into relatively cool air. This front is typically associated with slow, steady precipitation.   171

Water rights Legally protected rights to take possession of water occurring in a natural waterway and to divert that water for beneficial purposes.   543

Water table The top of the saturated zone.   393

Wave amplitude The overall height of any given wave as measured from the wave trough to the wave crest.   60

Wavelength The distance between adjacent wave crests or wave troughs.   60

Wave refraction The process through which waves are focused and bent around headlands.   515

Waves Oscillations in a body of water that form mostly due to the frictional force generated by wind blowing across the surface of the water.   511

Weather Day-to-day changes that occur in temperature and precipitation.   202

Weathering Physical or chemical modification of rock or sediment that occurs over time.   372

Westerlies Midlatitude winds that generally flow from west to east.   119

Wet adiabatic lapse rate (WAR) The rate at which a saturated body of air cools as it lifts. The average rate is about 5°C/1000 m (2.7°F/1000 ft).   151

Wilting point The threshold amount of soil water below which plants can no longer transpire water.   268

Wind abrasion Erosion that occurs when rock is ground, smoothed, or polished by the wind in a manner similar to sandblasting.   487

Windward side The side of a mountain range that faces oncoming winds.   129

Winter Solstice Assuming a Northern Hemisphere seasonal reference, the Winter Solstice occurs on December 21 or 22, when the subsolar point is at the Tropic of Capricorn (23.5° S).   53

X

Xerophytic plants Plants in very dry places that have a number of survival mechanisms in response to prolonged periods of drought.   249

Y

Yardangs Ridges that are sculpted and streamlined by wind abrasion and deflation.   487

Z

Zonal flow Jet stream pattern that is tightly confined to the high latitudes and is thus circular to semicircular in polar view.   121

Zone of ablation The part of a glacier where melting exceeds snow accumulation.   449

Zone of accumulation The geographical area where snow accumulates and feeds the growth of a glacier.   449

INDEX

A

Ablation till, 462
Abrasion:
  glacial, 457
  wind, 487
Absorption:
  defined, 71
  of solar radiation, 74
Acadian Orogeny, 344
Acid rain, 379–380
Active floodplain, 432
Active layer (permafrost), 474
Active solar heating, 80–8
Actual evapotranspiration (actual ET), 203
Additions, soil, 268–270
Adiabatic processes, 149–151
  dry adiabatic lapse rate, 149–150
  wet adiabatic lapse rate, 151
Adret slope, 252
Adriatic Sea, 400
Advanced Very High Resolution Radiometer
  (AVHRR), 34, 240
Advection, 108, 110
Advection fog, 153
Aerial photographs, 32, 33
Afforestation, 256
Africa:
  Alfisols in, 287
  biomes in, 244, 247
  climate of, 208
  deforestation in, 256
  deserts of, 119
  economic development in, 540
  Great Rift system, 340
African Congo, climate of, 207
Aggradation, 423
Agriculture:
  arable land supply, 548–549
  in Australia, 556
  conservation practices in, 501–502
  and Dust Bowl, 500–501
  and Great Plains drought, 164, 165
  large-scale, 259, 260
  onset of, 538
  salinization due to, 556–557
  slash and burn, 256–258
  and soil quality, 299–302
  sustainable, 558
  vegetation impacted by, 259, 260
A horizon, 281
Air, in soil, 268
Airflow direction. *See also* Atmospheric circulation
  and Coriolis force, 112–114
  and frictional forces, 114–115
  and pressure gradient force, 111–113
  and unequal heating of land surfaces, 110–111
Air mass, 170
Air mass circulation, climate and, 202, 203
Air pressure:
  and 500-mb pressure surface, 174–176
  defined, 106
  and global circulation, 116
  measuring and mapping, 107–108
Air temperature, 90
  adiabatic heating and cooling, 149–151
  and atmospheric pressure, 107
  dew-point temperature, 145–146
  and evapotranspiration, 148
  and humidity, 143, 145
  and large-scale geographic factors, 90, 93, 94
  local factors influencing, 93–98
Alabama, biomes in, 246
Alaska:
  biomes in, 246, 247
  braided stream in, 423

climate of, 215
earthquake of 1964 in, 359
fjords in, 508
Gelisols in, 292, 293
glaciers in, 453, 454
permafrost in, 474
Albedo, 73, 75
Albers, H. C., 26
Albers equal-area projections, 26
Alberta, Canada, drumlin in, 463
Albic horizon, 290
Alfisols, 284, 285, 289, 290
Allegheny Orogeny, 344, 345
Alliance of Small Island States, 533
Alluvial fans, 421, 484–485
Alluvial terraces, 433–434
Alluviation, 433
Alluvium, 433
Alpine and arctic ice biome, 241, 242, 250, 251
Alpine glaciers, 452–455
Alpine tree line, 253
Alpine tundra, 250, 251
Altitude:
  and atmospheric pressure, 107–108
  and temperature, 97
Altocumulus clouds, 154
Altostratus clouds, 154
Amazon rainforest, 258, 261
American Clean Energy and Security Act (2009), 232, 426
American Falls, 425
Amphibole, 313
Anasazi Indians, 80
Andes Mountains, 99
  alpine tundra of, 250
  and Atacama Desert, 210
  climate of, 216
  Patagonia ice field, 453
  and surface temperatures, 99
  vertical zonation in, 252
Andisols, 284, 285, 299, 300
Angle of incidence, 74, 75
Angle of repose, 381
Antarctica:
  climate of, 216
  on maps, 25
  ozone hole over, 67–69
  permafrost in, 474
Antarctic Circle, 16, 53
Antarctic ice cap, 9, 216
Antarctic Ice Sheet, 456, 470, 472, 473
Anthropocene, 328–329
Anthropogenic (term), 226
Anthropogenic influences, on climate change,
  226–232. *See also* Human/environment
  interactions
Anticlinal valley, 345
Anticline, 342
Anticyclones, 108
Antrim Creek, 424
Anvil head, 181
Aphelion, 47
Appalachian Mountains, 334, 344–346
  acidified soils in, 379
  as continental divide, 413
  evolution of, 342–346
  Fraser firs in, 379
  Inceptisols in, 298, 299
Aquiclude, 393
Aquifers, 395–400
Arabian Peninsula, 119
Arable land:
  defined, 548
  and food production, 548–549
Archean Eon, 322
Arches National Park (Utah), 371, 372, 378

Arctic Circle, 16, 53
Arctic Ocean, 506
Arctic outbreak, 122
Arctic tundra, 250, 251
Arête, 459
Argentina:
  climate of, 209
  loess deposits in, 495
Argillic horizon, 283
Argon, in atmosphere, 63
Arid and semi-arid (*B*) climates, 206, 208–210
Arid climates:
  soil formation in, 273
  soil salinization in, 548–558
  water issues in, 542–548
Arid desert (BW) climates, 480
  cold midlatitude desert climate (*BWk*), 209–210
  hot low-latitude desert climate (*BWh*), 208, 209
Aridisols, 284, 285, 295–296
Arid landscapes, 478–485
Arroyos, 484
Artesian wells, 396–398
Artificial levees, 439–440
Asia:
  Alfisols in, 287
  arable land in, 549
  Aridisols in, 295
  biomes in, 244, 245, 249
  climate of, 207–209, 214
  forest in, 256
  monsoon in, 124, 125
  surface temperatures in, 98, 99
Aspect, vegetation distribution and, 251, 252
Asteroid impacts, climate change caused by, 224
Asthenosphere, 306, 308, 309
Aswan Dam (Egypt), 442
Atacama Desert (South America), 210, 480
Athabasca oil sands (Canada), 560–563
Atlanta, Georgia, heat island, 102
Atlantic Ocean, 506
  air masses formed over, 170
  hurricanes in, 190–193
  ocean currents in, 131, 132
  oceanic crust in, 336
  sea-surface temperatures in, 132
Atlantic Rift Zone, 339
Atmosphere, 7, 8
  carbon dioxide in, 228
  composition of, 62–70
  constant gases in, 63
  defined, 7
  energy flows in, 71–74
  mesosphere, 86, 88
  particulates in, 69–70
  stratosphere, 86, 88
  thermosphere, 86, 88–89
  troposphere, 86–88
  variable gases in, 63–69
Atmospheric circulation, 104–134
  atmospheric pressure, 106–108
  atmospheric pressure systems, 108–110
  direction of, 110–116
  and global pressure, 116–125
  local wind systems, 128–130
  midlatitude, 119–122
  and oceanic circulation, 130–134
  polar, 122–123
  and seasonal migration of pressure systems,
    123–125
  sub-tropical, 118
  tropical, 116, 118
  and wind energy, 125–128
Atmospheric moisture, 136–167
  adiabatic processes, 149–151
  and biomes, 242. *See also individual biomes*
  in climate classification, 205

cloud classification, 153–155
cloud formation, 152
and drought, 164–166
evaporation, 148–149
fog, 153, 156
humidity, 142–148
hydrologic cycle, 141–142
hydrosphere, 138–140
and physical properties of water, 138–140
precipitation, 156–164
Atmospheric pressure, 106–108
Atmospheric pressure systems, 108–110
seasonal migration of, 123–125
sub-tropical high-pressure, 118
Atolls, 528, 529
Augustine Volcano (Alaska), 39, 360
Australia:
Alfisols in, 287
Aridisols in, 295
biomes in, 244, 245, 247
bluffs in, 515
climate of, 207, 208, 212
coastal dune fields in, 522
coral bleaching in, 533
physical geography of, 556
precipitation in, 555
soil salinization in, 552, 555–558
surface temperatures in, 99
Vertisols in, 289
Autotrophs, 236
Autumnal equinox, 53
Avalanches, 386–387
AVHRR (Advanced Very High Resolution Radiometer), 34, 240
Axial tilt (Earth), 48, 49
climate change and, 225
and glacial cycles, 225–226
and seasons, 47, 50
as viewed from Sun, 54
Axis, 47

B

Backshore, 519
Backslope (sand dunes), 490
Backswamps, 431
Backwash, 513
Bahama Islands, 316, 317, 322
Bajada, 484
Bakken Shale, 562
Baltimore, Maryland, 36
Banda Aceh, Indonesia, 357, 359
Bandalier National Monument (New Mexico), 377
Banff National Park (Canada), 383
Bangladesh, risk of coastal flooding in, 532
Bankfull discharge, 417, 418
Barchan dunes, 491, 492
Barchanoid ridges, 491, 492
Barometer, 107
Barrier islands, 522, 524–525
Barrier reefs, 528
Bar scale, 27
Basal slip (glaciers), 451
Basalt, 313
Basal till, 461
Base flow, 416–417
Base level (streams), 425–426
Baseline, in USPLS, 20
Basin and Range Province, 481–482, 542
Batholiths, 311
Bays, 507, 520, 521
Baymouth bar, 521
Bay of Fundy, 511
Beaches:
depositional, 519, 520
pocket, 515, 516
shoreline protection, 530–532
Beach drift, 513
Beach walls, 531
Bedload, 422, 423
Benthic foraminifera, 222
Berm scarp, 519

Bermuda High, 211
B horizon, 281
Big Bang theory, 44
Big Sur, California, 5, 507
Biogeography, 236. *See also* Plant geography
Biomass, 239–240
Biome(s), 241–251
defined, 241
desert, 242, 249–250
forest, 242–247
grassland, 242, 247–249
human footprint on, 255–256
tundra, 242, 250–251
Biosphere, 8
defined, 8
global, 240
and remote sensing, 261
Biotite, 313
Bioturbation, 272–273
Birdfoot delta, 436
Bitumen, 560
Black blizzards, 501
Blackfoot River valley moraine (Montana), 463
Black Sea, 506
Black Sunday, 501
Blocky soil structure, 278
Blowout dunes, 492, 493
Blue Lake, Minnesota, 464
Bluffs, wave-cut, 515–516
Boreal forest biome, 241–243, 246, 247
Boulder Dam, 544
Brackish lagoons, 525
Brackish water, 507
Braided streams, 422–423
Brazil:
biomes in, 247
coastal dune fields in, 522
deforestation in, 257, 258
and Itaipu Dam, 442
termite mounds in, 273
Brazos River (Texas), 526, 527
Breakers (waves), 512
Bretz, J. Harlan, 468
Brine, 507
British Isles, climate of, 212
British Petroleum, 559
Buford Dam, 440, 441
Bujagali Falls (Uganda), 411
Bureau of Land Management, 20
Business, GIS used in, 40
Buttes, 483

C

Calcic horizon, 293
Calcification:
defined, 292
in Mollisol formation, 292, 293
California:
airflow and precipitation patterns in, 160
average focal depth in, 349
biomes in, 246
bluff erosion in, 530
bluffs in, 515
bristlecone pine in, 220, 221, 252
climate of, 212
earthquakes/faults in, 352–354
fault plane in, 351
flows in, 385
La Conchita debris flow in, 386
littoral drift along coast of, 512
oil fields in, 558
orographic patterns in, 160
Santa Ana winds in, 130
sea fog in, 153
shale play in, 562
water diversion in, 547
wave refraction along, 516
wildfire in, 248
wind farms in, 126
Calving, 453
Cambrian Period, 322

Canada, 256
acid rain in, 379
air masses over, 170
Alfisols in, 287
Athabasca oil sands, 560–563
biomes in, 245, 246
climate of, 215
ice wedge in, 475
permafrost in, 474
pingo in, 476
Polar High over, 123
surface temperatures in, 98, 99
time zones in, 49
Canadian (Horseshoe) Falls, 425, 426
Canadian Rockies, 459
Canyons, 421, 483
Canyon de Chelly (Arizona), 377
Cape Cod, Massachusetts:
Cape Wind project, 126
false-color image of, 35
spits in, 520, 521
Cape Wind project, 126
Capillary action, 138, 267
Capillary water, 392
Caprock, 372
Carbonate disposition, 316, 317
Carbonation, 378, 379, 404
Carbon cycle, climate change and, 227–229
Carbon dioxide:
in atmosphere, 64–65, 228
in photosynthesis, 237–239
Caribbean Sea, 533
Carlsbad Caverns (New Mexico), 402, 404
Carotenoids, in photosynthesis, 239
Carrying capacity, 538
Cartography, 22–23
Cascade Mountains, 455
Cascade Volcanic Arc (United States), 363–364
Cations, 279
Cation exchange capacity (CEC), 279
Cause-and-effect relationships, holistic exploration of, 9–10
Cave(s), 402–404
Cave art, 402, 403
Cave River (New Zealand), 434
Caverns, 402–404
CEC (cation exchange capacity), 279
Celestial dome, 54–56
Celsius, Anders, 90
Celsius scale, 90, 91
Cenozoic Era, 322–324, 337
Center of the Earth, 306–307
Center-pivot irrigation, 398
Central America, Alfisols in, 287
Central Arizona Project, 544
Central Valley (California), 300, 301
groundwater contamination in, 301
soil fertility in, 300
Central Valley Project (CVP), 552
CFCs (chlorofluorocarbons), 67, 69
Chandeleur Islands, 524
Channeled Scablands (eastern Washington), 468–469
Channel flow, 416–419
Chattahoochee River (Georgia), 440, 441
Chemically-precipitated sedimentary rocks, 314, 316–317
Chemical weathering, 375–379
Chert, 317
Chesapeake Bay, 508–509
Chihuahuan Desert (Mexico), 480
Chile, climate of, 212
China:
Alfisols in, 287
climate of, 209
earthquake of 2008 in, 348
economic development in, 540
forest in, 256
giant panda conservation in, 565–570
human footprints in biomes, 256
industrialization in, 540
karst landscapes in, 402, 405

China *(cont.)*
    loess deposits in, 495
    oil demand in, 559
    rice paddies in, 277
    risk of coastal flooding in, 532
    subsidence in, 400
    Utisols in, 287
Chinook winds, 129, 130
Chlorofluorocarbons (CFCs), 67, 69
Chlorophylls, 239
C horizon, 281
Chroma (soil color), 276
Cinder-cone volcanoes, 361
Circle of illumination, 47
Circumpolar tree line, 253
Cirques, 455, 459
Cirrocumulus clouds, 154
Cirrostratus clouds, 155
Cirrus clouds, 153, 154
Clark County, Nevada, 542–548
Clastic sedimentary rocks, 314, 315
Clays, 315
Clay loam, 277
Claystone, 315
Clean Air Act and amendments, 379, 380
Cliff dwellings, 376–377
Climate(s), 200–224
    arid and semi-arid, 205, 206, 208–210
    defined, 202
    factors affecting, 202–203
    highland, 205, 206, 216–217
    Köppen classification system for, 203–217
    of Las Vegas, 543
    mesothermal, 205, 206, 211–214
    microthermal, 205, 206, 214–215
    polar, 205, 206, 215–216
    reconstructing past climates, 217–224
    and soil formation, 272, 273
    tropical, 205–208
    vegetation and, 241. *See also* Biome(s)
Climate change, 10, 224–232
    and carbon cycle, 227–229
    coastlines impacted by, 532–534
    and glacier fluctuations, 470, 472
    and human/environment interactions, 226–232
    past, causes of, 224–226
    predicting, 230–232
Climate feedbacks, 230–231
Climate reconstruction, 217–224
    with ice core analysis, 221–223
    with pollen records, 218–219
    with tree ring patterns, 218–221
Climatology, 202
Climax vegetation, 254
CLORPT, 271. *See also* Soil-forming factors
Clouds:
    associated with thunderstorms, 185
    classification of, 153–155
    formation of, 152
    orographic, 159
Coal, 318, 319
Coastal dunes, 521–523
Coastal engineering, 530–532
Coastlines, 504–534
    depositional, 519–529
    erosional, 515–519
    human interactions with, 529–534
    landforms of, 515–529
    littoral processes and sediment transport, 512–514
    and oceans/seas, 506–507
    processes shaping, 507–515
    tides, 508–511
    water-level fluctuation along, 507–508
    waves, 511–512
Coast Ranges:
    deposition along, 552
    marine rock outcrops in, 553
    precipitation along, 160
Cold fronts, 172, 173
Cold midlatitude desert climate (*BWk*), 209–210
Cold midlatitude steppe climate (*BSk*), 208, 209

Colloids, 279–280
Colluvium, 381, 421
Color, soil, 275–276
Colorado:
    aquifer saturated thickness in, 396
    climate of, 209
Colorado Plateau, 433, 483
Colorado River, 544, 547
Colorado River Aqueduct, 544
Colorado River Compact, 433, 440, 543, 544, 547
Columbia ice field (Alaska), 453
Columnar soil structure, 278
Comb Ridge Monocline (Utah), 343
Composite volcanoes, 361
Condensation, 140
Condensation level, 151
Condensation nuclei, 152
Conduction, 70, 71
Cone of depression, 398
Confluence (streams), 412
Conformal projections (true shape), 24–25
Conglomerate, 315
Conservation tillage, 502
Constant gases, 63
Contact metamorphism, 319
Contamination, groundwater, 301, 400–401, 563
Continents, Pangaea and, 334, 335
Continental (term), 94
Continental (dry) air masses, 170
Continental Arctic air mass, 170
Continental crust, 306, 308, 340–341
Continental divide, 412–413
Continental drift, 334–337
Continental glaciers, 455–456
Continental locations, air temperature for, 94, 96–98
Continental Polar air mass, 170
Continental Tropical air mass, 170
Continent-to-continent collision, 341–342
Contours, 30
Convection, 70, 71
    atmospheric, 110–111
    defined, 71
    and plate movement, 335–336
Convectional uplift, 157, 158, 160, 162–164
Convection loops, 113
Convergent uplift, 157, 158
Coral bleaching, 533
Coral reefs, 527–529
Cordilleran Ice Sheet, 466, 468
Coriolis force:
    and atmospheric circulation, 112–114
    and oceanic currents, 130
Counterradiation, 64
Creep, 486, 487
Crest (sand dunes), 490
Cretaceous Period, 322
Crevasses, 450, 451
Crop residue recycling, 301
Crop rotation, 301
Cross dating, 221
Crust (Earth), 308–309
    and isostatic adjustment, 308, 309
    rocks and minerals in, 310–322
Cumulonimbus clouds, 153, 155, 162
Cumulus clouds, 153, 154, 162
Cumulus stage (thunderstorms), 180, 181
Cutbanks, 428, 431
CVP (Central Valley Project), 552
Cyclogenesis, 174, 177–179
Cyclones, 121
    defined, 108
    midlatitude, 173–180
    tornadoes vs., 174
    tropical, 190–197

D

Dallas, Texas, temperatures in, 165
Dams, 440–444
    controversial aspects of, 442–443
    defined, 440
    for flood protection, 440–442

Danube River (Europe), 417
DAR (dry adiabatic lapse rate), 149–150
Day, 47, 53–54
Day length, 55
    and air temperature, 90, 93
    at poles vs. at Equator, 78–80
Dead Sea Fault, 338, 339
Death Valley, California, 161
    alluvial fan in, 485
    Bad Water salt crust, 549
    mineral deposits in, 317
    playa in, 484
    sand dunes in, 479
    ventifacts in, 488
Debris flows, 386
Debris slide, 384
Deccan Plateau (India), 289
Decomposers, 236, 237
Decomposition:
    in photosynthesis, 239
    and photosynthesis process, 238
Deep time, 322. *See also* Geologic time
Deflation (sediment), 487
Deflation hollow, 487
Deforestation, 10
    in China, 569, 570
    defined, 256
    vegetation impacted by, 256–259
Degradation, 423
Delaware beaches, 531
Deltas, 436–437, 526–527
Denali National Park (Alaska), 107
Dendritic drainage pattern, 413, 414
Dendrochronology, 218–221
Denver, Colorado, air pressure in, 108
Depletion, soil, 268, 269
Deposition, 140
    coastal, 519–529
    defined, 140
    eolian, 488–497
    as fluvial process, 420–422
    of glacial drift, 461–465
Deranged drainage pattern, 413, 414
Deserts:
    arid and semi-arid, 480
    arid landscapes, 480–485
    desertification, 497–502
    eolian deposition and landforms, 488–497
    eolian erosion and transport in, 485–488
    geomorphology of, 480–485
    and Subtropical High Pressure System, 118, 119, 480
Desert biomes, 242, 249–250
Desertification, 497–502
    in African Sahel, 498–500
    in U.S. Great Plains, 500–502
Desert pavement, 493–494
Developed countries, 540
Developing countries, 540
Developing stage (thunderstorms), 180, 181
Devil's Postpile National Monument (California), 373
Devonian Period, 322
Dew point, 146
Dew-point temperature, 145–146
Differential weathering, 372
Digital technology, 32–40
    Geographic Information Systems, 39–40
    Global Positioning Systems, 37–39
    remote sensing, 32–37
Dikes, 311
Dinosaurs, 322
Direct radiation, 71
Direct tide, 509
Dirt, 266
Disappearing stream, 404
Dissipation stage (thunderstorms), 180–182
Dissolved load, 422
Dissolving stage (midlatitude cyclones), 178
Distortion, on maps, 24, 25
Diurnal cycle, 54
Dolomite, 317

Dome dune, 491, 492
Donner Pass, 160
Doppler, Christian, 189
Doppler Effect, 189
Doppler radar, 189
Douglas County, Minnesota, 394
Downcutting, 423
Downdraft, 180
Downwelling current, 132, 133
Drainage basins, 411–413
Drainage density, 413–414
Drainage divide, 411, 412
Drainage patterns, 413
Drinking water contamination, 401
Drought, 164–166
    and desertification, 499–501
    SNWA categories of, 544, 547
    in southwestern U.S., 544, 546–547
Drought-evading plants, 250
Drought-resistant plants, 250
Drumlin, 463
Dry adiabatic lapse rate (DAR), 149–150
Dust Bowl, 301, 302, 500–502
Dust pneumonia, 501
Dust storms, 501

E

Earthflows, 385
Earthquakes, 348–360
    defined, 348
    as natural hazards, 353–360
    predicting, 356–357
    seismic processes, 348–353
Earthquake Lake, 384
Earth structure, 306–322
    asthenosphere, 306, 308
    crust, 308–309
    inner core, 306–307
    interactions within, *see* Tectonic processes
    liquid outer core, 306, 307
    lithosphere, 306, 308
    mantle, 306, 307
    rocks and minerals in crust, 310–322
Earth–Sun geometry, 42–57
    axial tilt of Earth, 48, 49
    and climate, 207
    and glaciations, 467
    human perceptions of, 53–56
    orbit of Earth around Sun, 45–47
    and Quaternary climate changes, 224–226
    rotation of Earth, 47–49
    and seasons, 50–53
    and shape of Earth, 44–45
    and the Universe, 44
East African Rift system, 340
Easterly wave, 190
East River (Colorado), 431
Economic development, 540
Ecosystems, 236–237. *See also specific topics, e.g.:*
    Plant geography
Ecotone, 243, 498
E horizon, 281
Electromagnetic energy:
    defined, 60
    and remote sensing, 35–36
Electromagnetic radiation:
    temperature and amount of, 61
    temperature and wavelength of, 60
Electromagnetic spectrum, 60–62
    defined, 60
    and photosynthesis, 239
El Niño, 133–134, 484, 533
Eluviation, 270, 273
Emissivity, 35
Emperor Seamount Chain, 366
Enchanted Rock (Texas), 376
End moraine, 462
Energy flows, 6, 70–76. *See also* Global energy system
    in atmosphere, 71–74
    heat transfer, 70–71

interaction of solar radiation and Earth's surface, 74–76
Energy resources:
    fossil fuels, 227–229, 319
    oil, 318
    solar power, 80–81
    unconventional oil sources, 560–564
    U.S. consumption of, 228
    wind power, 125–128
England, bluffs in, 515
Enhanced Fujita Scale, 185, 186
Enriquillo-Plantain Garden fault system, 353
Entisols, 284, 285, 296, 298
Entrenched meanders, 432–433
Environmental lapse rate, 86, 87, 149
Environmental management, GIS used in, 40
Environmental Protection Agency (EPA), 401
Eolian landforms, deposition and, 488–497
Eolian processes:
    and arid landscapes, 480–485
    defined, 485
    deposition, 488–497
    desertification, 497–502
    erosion and transport, 485–488
EPA (Environmental Protection Agency), 401
Ephemeral streams, 410
Epicenter, 349–350
Equator, 16, 17
    circumference of Earth through, 44
    day length at, 79
    defined, 15
    insolation at, 78, 79
Equatorial trough, 116
Equilibrium line, 450
Equinox:
    autumnal, 53
    and day length, 55
    defined, 51
    Sun position at, 55
    vernal, 51
Equivalent projections, 25–26
Eritrea, Africa, sand storm, 105
Erosion:
    coastal, 515–519, 530, 531
    eolian, 485–488
    as fluvial process, 420–422
    glacial, 457–461
    and isostatic adjustment, 308, 309
    soil, 277
Esker, 464
Eurasia:
    climate of, 209, 214, 215
    Mollisols in, 292
Europe:
    acid rain in, 379
    Alfisols in, 287
    arable land in, 549
    atmospheric pressure systems in, 110
    biomes in, 245, 246, 249
    climate of, 212, 214
    coastal dune fields in, 522
    forest in, 256
    human footprints in biomes, 256
European Alps:
    avalanche hazards in, 387
    vertical zonation in, 252–253
European Union, greenhouse emissions and, 231
Eustatic change, 507, 508
Evaporation, 140, 148–149
    defined, 74, 140
    heat removal through, 74
Evaporites, 317
Evapotranspiration, 148–149, 392
    actual, 203
    defined, 148
    potential, 203
    and soil-water loss, 267
Exfoliation, 375, 376
Explosive volcanoes, 361–364
Extreme weather events, global warming and, 231
Extrusive igneous rocks, 311–313

F

Fahrenheit, Gabriel Daniel, 90
Fahrenheit scale, 90, 91
Fall (autumnal) Equinox, 53
Farmland loss, 10
Fault(s):
    defined, 348
    types of, 350–353
Fault escarpment, 352
Federal Emergency Management Agency (FEMA), 194, 195
Feldspar, 313
Felsenmeer, 374
Felsic, 312, 313
FEMA (Federal Emergency Management Agency), 194, 195
Ferrel Cells, 119
Fertile Crescent, 550, 551
Fertility, soil, 299–301
Fetch (wind), 511
Field capacity, 268, 392
Fiery Furnace (Arches National Park), 378
Firn, 448–449
Fish Springs, 547
500-mb pressure surface, 174–176
Fjords, 508
Flash floods, 438–439
Flint, 317
Flooding, 417–419
    coastal, 532
    dams for protection from, 440–442
    in Mississippi Basin, 418–419
    from sea-level rise, 231
Floodplains:
    defined, 418, 431
    evolution of, 427–438
Flood stage, 418
Florida:
    beaches in, 531
    biomes in, 246
    sinkhole in, 404
Flows, 385–386
Fluid volcanoes, 364–365
Fluvial landforms, 420–438
    stream gradation, 422–427
    stream valleys and floodplains, 427–438
Fluvial processes:
    erosion and deposition, 420–422
    and landforms, 420–438
    stream gradation, 422–427
    stream valleys and floodplains, 427–438
Fluvial systems, 408–444
    channel flow, 416–419
    drainage basins, 411–413
    drainage density, 413–414
    drainage patterns, 413
    fluvial processes and landforms, 420–438
    human interactions with, 438–444
    hydraulic geometry, 416
    origin of streams, 410–411
    stream ordering, 414–415
Foehn winds, 129, 130
Fog, 153, 156
Foliated rocks, 320
Food chain, 237
Food production, 538
    and arable land supply, 548–549
    in Australia, 556
Foredune, 521
Foreshore, 519
Forest biomes, 242–247
    afforestation, 256
    changes in forest cover, 256, 257
    deforestation, 256–259
Fossils, and continental drift, 334
Fossil fuels:
    and carbon cycle, 227–229
    defined, 227
    sources of, 319
Fracking, 562–564
France, mistral wind in, 130

Free dunes, 491, 492
Freezing, 139–140
Freezing rain, 156
Freshwater, 141, 390, 548. *See also* Groundwater
Frictional forces, on airflow, 114–115
Fringing reefs, 527, 528
Fronts, 170–173
Frontal uplift, 157, 158, 171, 172
Frost heaving, 374, 475
Frost thrusting, 475
Frost wedging, 373–374
Fumeroles, 368

G

Gabbro, 313
Galaxies, 44
Gamma rays, 60
Ganges River Delta (India), 526, 527
Garlock Fault, 353
Gases:
    constant, 63
    variable, 63–69
Gass, Octavius, 543
Gavins Point Dam, 443
Gelisols, 284, 285, 291–293
General circulation models (GCMs), 230
Geographical awareness, 4
Geographical literacy, 10–11
Geographic grid, 14–22
    great circles, 14–15
    latitude, 15–17
    longitude, 17–19
    small circles, 15
    United States Public Land Survey, 19–22
    using, 19
Geographic Information Systems (GIS), 39–40
Geography, 4–5
    defined, 4
    subdisciplines of, 4, 5
Geologic time, 322–329
    in perspective, 324–325
    proposed Anthropocene Epoch, 328–329
    and rock cycle, 325–328
    "telling," 324
Geo Media Library, 10
Geomorphology:
    defined, 342
    desert, 480–485
Georgia, biomes in, 246
Geostationary Operational Environmental Satellites
    (GOES), 34–35, 64
Geostationary orbits, 34–35
Geostropic winds, 114
Geosynchronous orbits, 34
Geothermal activity, plate movement and, 335
Geothermal gradient, 341
Geysers, 367
Giant panda conservation, 565–570
GIS (Geographic Information Systems), 39–40
Glacials, 465–467
Glacial abrasion, 457
Glacial cycles:
    and axial tilt, 225–226
    reconstructing, 221–223
    and water level change, 507–508
Glacial drift, 461
Glacial erratics, 457
Glacial grooves, 457
Glacial Lake Missoula, 468, 469
Glacial landforms, 456–465
    from deposition of glacial drift, 461–465
    from glacial erosion, 457–461
Glacial mass budget, 449–450, 470
Glacial outwash, 463
Glacial plucking, 457
Glacial striations, 457
Glacial till, 461–462
Glacial trough, 460
Glaciers, 446–476
    continental, 455–456
    defined, 448

development of, 448–449
    glacial mass budget, 449–450
    history of glaciation on Earth, 465–470
    human impact on, 470–473
    landforms created by, 456–465
    land surface covered by, 446
    in mountainous regions, 452–455
    movement of, 450–451
    periglacial processes and landscapes, 473–476
Glacier National Park (Montana), 470, 471
Glen Canyon Dam, 442
Global energy system, 58–83
    and climate feedbacks, 230–231
    and composition of atmosphere, 62–70
    and electromagnetic spectrum, 60–62
    flow of solar radiation, 70–76
    global radiation budget, 76–80
    and solar constant, 61
    solar energy production, 80–81
Global Forest Resources Assessment, 256
Global Positioning Systems (GPS), 37–39, 336
Global radiation budget, 76–80
Global temperature patterns, 84–103
    annual range of surface temperature, 98–100
    and atmospheric structure, 86–89
    large-scale geographic factors for air temperature,
        90, 93, 94
    local factors influencing air temperature,
        93–98
    surface and air temperatures, 89–101
    temperature scales, 90, 91
    urban heat islands, 101–102
Global warming. *See also* Climate change
    and deforestation, 257
    and glacier fluctuations, 470
    in past climates, 218
    potential benefits of, 231
    predicting, 230–232
Global water balance, 142
Globes, 23
Gneiss, 320
Gobi Desert (Mongolia), 209, 250, 480
GOES (Geostationary Operational Environmental
    Satellites), 34–35, 64
Goosenecks of the San Juan River, 433
GPS (Global Positioning Systems), 37–39, 336
Graben, 352
Graded streams, 423, 425, 427, 434–436
Gradient, 416
Grand Canyon, 305, 325–327, 427, 428
Granite, 313
Granular soil structure, 278
Graphic scale, 27
Grassland biomes, 242, 247–249
Great Barrier Reef, Australia, 528, 555
Great circles, 14–15
Great Lakes:
    coastal sand dunes along, 523
    coastlines of, 507
    evolution of, 467, 470
Great Lakes region:
    coastal dunes in, 523
    deforestation in, 258
    pine forest, 258
    primary succession on coastal dunes, 254
    solar energy potential in, 80
    Spodosols in, 291
    Wisconsin glaciation in, 467
Great Plains:
    biome in, 249
    desertification in, 500–502
    drought in, 164–166
    Dust Bowl, 301, 302, 500–502
    Mollisols in, 292, 294
    native grassland in, 259
    Ogallala Aquifer in, 395–400
    prairie dogs on, 273
    riparian vegetation in, 254
    shale play in, 562
    wheat production n, 537
    wind farms in, 126

Great Rift system, 340
Great Smoky Mountains (North Carolina), 344
Great Unconformity, 326–327
Great Valley (California), 160
Greenhouse effect:
    and carbon dioxide, 64–65
    defined, 64
    and methane, 64
    and water vapor, 64
Greenhouse gases:
    and global warming, 230–232
    and prehistoric temperature, 229
Greenland:
    climate of, 215, 216
    on maps, 25, 26
    Viking colonies in, 223
Greenland ice cap, 9, 222
Greenland Ice Sheet, 456, 470, 472
Greensburg, Kansas, tornado, 188
Grinnell Glacier, 470, 471
Groins, 531–532
Groin fields, 532
Ground ice, 475–476
Ground moraine, 462, 463
Groundwater, 141, 390–402
    contamination of, 301, 400–401
    fracking and contamination of, 563
    human interactions with, 395–402
    and karst topography, 402–406
    movement and storage of, 392–394
    nitrate contamination of, 301
    and subsidence, 400
Groundwater mining/extraction, 398–400, 547
Guangxi province, China, 405
Guguan Island, 342
Guilin, China, 405
Gulf(s), 506–507
Gulf Coast:
    barrier islands along, 524
    climate of, 211, 212
    Hurricane Katrina, 193, 194
Gulf of Aden, 340
Gulf of Mexico, 506, 507
    air masses formed over, 170
    interaction of waves and shoreline, 514
    oil spill in, 530
Gulf Stream, 130, 132
Gullies, 421
Gust front (thunderstorms), 180
Gyres, 132

H

Hadley cells, 116–118
Hail, 156
Haiti, 2010 earthquake in, 348, 353
Half Dome (Yosemite National Park), 375, 376
Half-life, 324
Hanging valleys, 460
Hawaii:
    Andisols in, 300
    formation of, 366
Hawaiian hotspot, 365–366
Haystack karst, 405
Hayward Fault, 353
Headlands, 515, 516
Heat index, 90, 92–93
Heat islands, urban, 101–102
Heat transfer, 70–71
    latent, 74
    sensible, 74
Heliostats, 80
Heterotrophs, 236
Higgens, Patillo, 559
High clouds, 153
Highland (*H*) climates, 205, 206, 216–217
High latitudes, 16, 17
High Plains (Ogallala) Aquifer, 395–400
High-pressure ridge, 176
High-pressure systems, 108–110, 116
Hillslopes, fluvial erosion on, 421

Himalaya Mountains, 347
  climate of, 216
  creation of, 338
  formation of, 346, 347
Histosols, 284, 285, 291, 292
Hogback ridge, 345
Hoh Rainforest, 212
Holocene Climate Optimum, 223, 225
Holocene Epoch, 223, 323, 324, 329, 538
Homestead Act of 1862, 22, 500
Homesteading, 22, 500, 543
Hoodoo, 372
Hook echo, 189
Hoover Dam, 440, 442, 544
Horizontal lenses (ground ice), 475
Horn, 459
Horse Hollow Wind Energy Center (Texas), 126, 127
Horse latitudes, 118
Horst, 352
Hot and dry desert biome, 241, 242, 249, 250
Hot low-latitude desert climate (*BWh*), 208, 209
Hot low-latitude steppe climate (*BSh*), 208, 209
Hotspots, 365–368
Hubbard glacier (Alaska), 454
Hudson Bay region, climate of, 215
Hudson Canyon, 508
Hue (soil color), 275, 276
Human/environment interactions, 5, 10, 536–571
  acid rain, 379–380
  with coastlines, 529–534
  coping with destructive hurricanes, 192–197
  desertification, 497–502
  with fluvial processes, 438–444
  and future climate change, 226–232
  giant panda conservation, 565–570
  and glaciers, 470–473
  with groundwater, 395–402
  heat index and wind chill, 90
  perceptions of Earth–Sun geometry, 53–56
  and plant geography, 255–262
  and population growth, 538–541
  soils, 299–302
  soil salinization in arid and semi-arid lands, 548–558
  solar energy production, 80–81
  tornado monitoring, 188–189
  unconventional oil supplies development, 558–564
  urban heat islands, 101–102
  urban sprawl, 541
  water issues in American Southwest, 542–548
Human footprint, on biomes, 255–256
Humboldt Current, 132
  and climate, 210
  and El Niño, 133
  and surface temperatures, 99
Humid climates, soil formation in, 273
Humid continental hot-summer climates (*Dfa, Dwa*), 214, 215
Humid continental mild-summer climates (*Dfb, Dwb*), 214, 215
Humidity, 142–148
  dew-point temperature, 145–146
  maximum, 143
  relative, 143–145
  specific, 143, 144
Humid subtropical hot-summer climates (*Cfa, Cwa*), 211–213
Humus, 266, 267
Hurricane(s), 169, 190–197
  classification of, 191–192
  coping with, 192–197
  defined, 190
  forecasting, 197
  typical track of, 193
Hurricane Floyd, 193
Hurricane Katrina, 169, 193–195, 524
Hurricane Rita, 524
Hurricane Sandy, 10, 194–196, 532–533
Hydraulic fracturing (fracking), 562–564
Hydraulic geometry, 416
Hydroelectric generation, 440

Hydrogen bonding, 138
Hydrologic cycle, 141–142, 392
Hydrolysis, 377–378
Hydrosphere, 8, 138–140
Hygroscopic water, 392

I

Ice, 138–140. *See also* Glaciers
Icebergs, 455
Ice caps, 452, 453, 466
Ice-cap climate (*EF*), 216, 217
Ice core analysis, 221–223
Ice crystallization, 156, 157
Ice fields, 452, 453
Iceland, volcanic eruption in, 366–367
Ice sheets, 456, 466–468, 470, 472
Ice wedges, 475
Igneous rocks, 311–313
Illinoisan glaciation, 466, 467, 508
Illuviation, 270, 273
Inceptisols, 284, 285, 298–299
India:
  biomes in, 244, 247
  climate of, 207
  coastal dune fields in, 522
  economic development in, 540
  human footprints in biomes, 256
  industrialization in, 540
  monsoon in, 125
  oil demand in, 559
Indian Ocean, 506
  coral bleaching in, 533
  tsunami in, 356–359, 506
  typhoons or cyclones in, 190
Indirect radiation, 74
Indirect tide, 509
Indonesia, risk of coastal flooding in, 532
Industrialization, 538, 540, 558
Industrial Revolution, 538, 558
Infrared satellite images, 36
Inner core (Earth), 306–307
Inorganic materials, in soils, 266
Insect-borne diseases, global warming and, 231
Inselbergs, 481
Insolation:
  and air temperature, 93, 94
  defined, 70
  and seasons, 78–80
Interfluves, 412
Interglacials, 465
Intergovernmental Panel on Climate Change (IPCC), 230, 231, 540
Intermittent streams, 410
Internal deformation. in glaciers, 450, 451
International Date Line, 48
International System of Measurement (SI), 90, 91
Interpolation, 30
Intertropical Convergence Zone (ITCZ), 116–118, 123–125, 203, 498
Intrusive ice, 476
Intrusive igneous rocks, 311–313
IPCC, *see* Intergovernmental Panel on Climate Change
Irrigation:
  center-pivot, 398
  in San Joaquin Valley, 552–555
  and soil salinization, 550, 552–555
Island mountains, 481
Isobars, 30
Isohyets, 30
Isolines, 29–30
Isopachs, 30
Isostatic adjustment (isostacy), 308, 309
Isostatic depression, 456
Isotherms, 30
Itaipu Dam, 442
Italy, L'Aquila earthquake in, 356–357
ITCZ, *see* Intertropical Convergence Zone

J

Jackson Hole, Wyoming, 463

Japan:
  earthquake and tsunami in, 358–359
  and greenhouse emissions, 231
  and North Pacific garbage patch, 539
Jefferson, Thomas, 19, 20
Jetties, 532
Joints, 373
Joshua Tree National Park (California), 378
Jurassic Period, 217, 322

K

Kalahari Desert (Africa), 249, 480
Kame, 464
Kansan glaciation, 465, 466
Kansas:
  slumps in, 385
  water table in, 399
  wind energy in, 126
Kansas City, 501
Kansas River, 412, 429, 441–442
Karst topography, 402–406
  caves and caverns, 402–404
  defined, 402
  landforms at the surface, 404–405
Katabatic winds, 129
Kelvin scale, 90, 91
Kesterson National Wildlife Refuge, 539
Kesterson Reservoir, 553, 554
Kettle lake, 464
Kilt Rock, Scotland, 516
Kinabatangan River (Borneo, Malaysia), 235
Kinetic energy, 89
Kivas, 377
Knickpoints, 425
Köppen, Wladimir, 203, 204
Köppen Climate Classification System, 203–217
  arid and semi-arid (*B*) climates, 205, 206, 208–210
  and biomes, 242. *See also individual biomes*
  highland (*H*) climates, 205, 206, 216–217
  mesothermal (*C*) climates, 205, 206, 211–214
  microthermal (*D*) climates, 205, 206, 214–215
  polar (*E*) climates, 205, 206, 215–216
  tropical (*A*) climates, 205–208
Krumholz zone, 253
Kyoto Protocol, 231, 232
Kyrgyzstan hillslopes, 384

L

Laccoliths, 312
La Conchita debris flow (California), 386
Lagoons, 521, 525
Lahar, 364
Lake(s). *See also specific lakes*
  coastal dunes along, 523
  coastlines of, 507
  in Minnesota, 394
  natural, 394
  oxbow, 428, 429, 431
  water contained in, 141
Lake Albert (Uganda), 137
Lake Chad, 20
Lake Erie, 425, 426
Lake Itasca (Minnesota), 410, 411
Lake Mead, 440, 546
Lake Michigan:
  and Antrim Creek, 424
  coastal sand dunes along, 522, 523
  groins along, 531
  primary succession on coastal dunes, 254
  winds along, 512
Lake Missoula (Montana), 469
Lake Nasser, 442
Lake Pontchartrain, 193, 194
Lake Victoria (Africa), 340, 410, 411
Land breezes, 129
Land cover, 569
Landform(s):
  defined, 342
  depositional coastlines, 519–529

Landform(s) (cont.)
  desert, 481–485, 487–494
  erosional coastlines, 515–519
  fluvial, 420–438
  glacial, 456–465
  and ground ice, 475–476
  karst, 402–406
Landmass movement, climate change caused by, 224
Landsat system, 34
Landscape(s):
  arid, 478–485, 493–495, 497
  coastal, 530
  defined, 342
  and desertification, 499, 500
  karst, 402–406
  and periglacial processes, 473–476
Land-sea wind systems, 128–129
Land skin temperature, 95
Landslides, 384–385
Land surfaces, unequal heating of, 110–111
La Niña, 164, 165, 484, ñ64
L'Aquila, Italy, earthquake, 356
Large-scale maps, 27–29
Larsen B Ice Shelf, 470, 472–473
Last Glacial Maximum, 508
Las Vegas, Nevada, 542
  daily temperature cycle at, 97
  water issues in, 542–548
Latent heat, 74, 140
Latent heat of condensation, 140
Latent heat of freezing, 140
Latent heat of melting, 140
Latent heat of sublimation, 140
Latent heat of vaporization, 140
Lateralization, 286
Lateral moraine, 462
Latitude, 15–17
  and air temperature, 90
  and climate, 202, 203. See also individual
    climates
  defined, 15
  difference between longitude and, 19
  and relative humidity, 144
  and Sun angle, 44, 45
Laurentide Ice Sheet, 466, 467, 470
Lava, 311, 313
Lava dome, 361
Leader (lightning), 182
Leap Years, 45
Leeward side, 129, 159
Legend (maps), 22
Levees:
  artificial, 439–440
  natural, 431
Level of condensation, 151
Lewis and Clark Lake, 443
Life expectancy, 539
Lightning, 182–184
Limb, 342
Limestone, 316, 317, 403
Liquid outer core (Earth), 306–307
Lithification, 313
Lithosphere, 7–8, 306, 308
Lithospheric plates, 334–335
Little Ice Age, 223, 230, 470
Littoral drift, 513
Littoral processes, on coastlines, 512–514, 520
Livestock, overgrazing by, 260
Loamy sand, 277
Local relief, 381
Local wind systems, 128–130
Location, in physical geography, 5
Lodgement till, 461
Loess, 494–495, 497
Loma Prieta earthquake, 353
London, smog in, 88
Longitude, 17–19
Longitudinal dunes, 492, 493
Longitudinal profile (stream gradient), 424
Longshore current:
  and beach erosion, 532

and coastline deposition, 519–520
  defined, 513
Longshore drift, 513, 520, 525
Longwave radiation:
  counterradiation, 64
  defined, 60
  and radiation budget, 76–77
Los Angeles, California, 13
  Santa Ana winds in, 130
  smog in, 88
Louisiana:
  biomes in, 246
  Hurricane Katrina, 193, 194
Low clouds, 153
Lower mantle, 306, 307
Low latitudes, 16, 17
Low-pressure systems, 108, 109, 116
Low-pressure troughs, 176

M

Madison slide (Montana), 384
Magma, 311
Magnetic fields:
  of Earth, reversal of, 336, 337
  of the Sun, 46
Maldive Islands, 528
Mammatocumulus cloud, 185
Mammatus clouds, 185
Mammoth Care (Kentucky), 402
Manhattan Schist (New York City), 458
Manila, Philippines, 439
Mantle, 306–307
Maps, 22–31
  isolines, 29–30
  map projections, 23–27
  map scale, 27–29
Map projections, 23–27
Map scale, 22, 27–29
Marble, 320
Marcellus Shale, 562
Marianas Islands, 341, 342
Marianas Trench, 342
Marine terraces, 517
Marine west-coast climates (Cfb, Cfc), 212, 214
Maritime (term), 94
Maritime (moist) air masses, 170
Maritime locations, air temperature for, 94, 96–98
Maritime Polar air mass, 170
Maritime Tropical air mass, 170
Maritime vs. continental effect:
  and air temperature, 93, 94, 96–98
  and climate, 202, 203
  defined, 94
Maroon Bells area (Colorado), 30, 201
Mars:
  carbon dioxide in atmosphere of, 65
  sand dunes on, 496
Marshall Islands, 533
Mars rovers, 33
Mass wasting, 381–387
  avalanches, 386–387
  defined, 381
  flows, 385–386
  landslides, 384–385
  rockfall, 382–383
  soil creep, 383–384
Matterhorn (Switzerland), 460
Mature stage (thunderstorms), 180, 181
Mauna Loa (Hawaii), 366
  carbon dioxide from, 65
  lava flow at, 311, 313
Maximum humidity, 143
Meandering stream, 11, 428–431
  defined, 422
  entrenched meanders, 432–433
Mechanical (physical) weathering, 372–375
Medial moraine, 462
Medieval Warm Period, 223
Mediterranean dry-summer climates (Csa, Csb), 211,
  212
Mediterranean Sea, 506

Mediterranean woodland and shrub biome, 241, 242,
  245–246
Mercator, 26
Mercator, Gerhardus, 25
Mercator projections, 25
Meridian:
  defined, 17
  Prime, 17–19
  principal, 20
Meridional flow, 121
Mesas, 483
Mesa Verde (Colorado), 376–377
Mesocyclones, 185
Mesopause, 88
Mesosphere, 86, 88
Mesothermal (C) climates, 205, 206, 211–214
Mesozoic Era, 322–323
Metamorphic rocks, 318–320
Meteorites, 88
Meteorology, 202
Metes and bounds system, 19
Methane, in atmosphere, 64
Mexico City:
  coastal dune fields in, 522
  smog in, 87, 88
Michigan:
  climate of, 214
  drumlins in, 463, 465
  Entisols in, 298
  shale play in, 562
  stump prairie in, 259
  time zone for, 47
Microclimate, 252
Microorganisms, in soil, 273
Microthermal (D) climates, 205, 206, 214–215
Mid-Atlantic Ridge, 336, 339
Middle clouds, 153
Middle East:
  Aridisols in, 295
  cultural diversity in, 5
  Great Rift system, 340
  oil reserves in, 559, 560
Midlatitudes, 16–17
Midlatitude circulation, 119–122, 179
Midlatitude coniferous forest biome, 241, 242, 246
Midlatitude cyclones, 173–180
  cyclogenesis, 177–179
  defined, 173
  interaction of upper and surface airflows, 176–177
  upper airflow and 500-mb pressure surface,
    174–176
Midlatitude deciduous forest biome, 241, 242, 245
Midlatitude forest biomes, 243
Midlatitude grassland biome, 241, 242, 247–249
Mid-oceanic ridge, 339
Midway Atoll, Hawaii, 539
Midwestern United States:
  acid rain from, 379
  Alfisols in, 287
  arable land in, 549
  glaciers in, 465
  groundwater contamination in, 301
  Histosols in, 291
  Mississippi River basin floods, 418–419
  Mollisols in, 292
  natural lakes in, 394
  soil fertility in, 300
Milankovitch, Milutin, 224
Milankovitch theory, 224–226, 467, 470
Milky Way Galaxy, 44
Millennium mine (Canada), 562
Minerals:
  in crust, 310
  defined, 310
  silicate, 312, 313
Minnesota:
  air masses over, 170
  glacial striations in, 457
  lakes in, 394, 464
  Sioux Quartzite in, 457
Mississippian Period, 246, 322

Mississippi River:
    barge traffic on, 444
    meanders of, 430
    satellite image of, 444
    sediment in, 422
    source of, 410, 411
Mississippi River basin:
    and dams, 442
    floods in, 418–419, 439, 440
    as largest watershed, 412, 413
Mississippi River Delta, 436–437
Missoula, Montana, 469
Missouri River, 412, 442, 443
Missouri River Basin, 287, 442–443
Mistral wind, 130
Moderate Resolution Imaging Spectroradiometer
        (MODIS) satellite, 95
Mohorovicic discontinuity, 306, 308
Mojave Desert (United States), 208, 209, 249, 374, 543
Mollisols, 284, 285, 292–295
Monocline, 341, 342
Mono Lake, 547
Monsoon, 124–125
Montana:
    glacial trough in, 460
    wildfire in, 248
Mont Blanc, 20
Montreal Protocol, 69
Mont-Saint-Michel, France, 521
Monument Valley, Utah, 483
Moon:
    circle of illumination on, 47
    and tides, 508–511
Moorea, Tahiti, 528
Moraine, 462
Mountains:
    formation of, 342–347
    major chains, 346
Mountain breezes, 129
Mountainous regions, glaciers in, 452–455
Mount Capulin (New Mexico), 361
Mount Etna (Sicily), 106
Mount Everest, 108, 346, 347
Mount Fuji (Japan), 362
Mount Hood (Oregon), 3, 455
Mount McKinley (Alaska), 107
Mount Nyaragongo (Democratic Republic of the
        Congo), 364
Mount Rainier, 363, 364
Mount St. Helens, 363, 364
Mount Washington (New Hampshire), 122
Movement, in physical geography, 5
Muck, 291
Mud, 315
Mudflows, 385
Mudpots, 368
Multipath error, 39
Municipal planning, GIS used in, 40
Munsell Color Chart, 275
Murray-Darling River basin, 556, 557

N

Namib Desert (Africa), 491
Namibia, coastal dune fields in, 522
National Aeronautic and Space Administration (NASA),
        33, 34
National Hurricane Center, 197
National Oceanic and Atmospheric Administration
        (NOAA), 34–35
Natural gas, 562
    formation of, 318
    as power source, 319
Natural hazard(s), 10
    defined, 353
    earthquakes, 348–360
    hurricanes, 169, 190–197
    tornadoes, 184–189
    volcanoes, 360–368
Natural levees, 431
Natural Resources Conservation Service (NRCS), 301,
        302

Natural systems, 6, 7
Navajo Sandstone, 315
Neap tides, 510, 511
Nebraska:
    aquifer saturated thickness in, 396
    loess landscape in, 497
Nebraskan glaciation, 466
Nebraska Sand Hills, 491, 493
Nelson Lake, Illinois, 219
Netherlands:
    risk of coastal flooding in, 532
    windmills in, 126
Net radiation, 76–79
    and air temperature, 93, 94
    defined, 76
    and evapotranspiration, 148
Nevada, biomes in, 249, 250
Never-Summer Range (Colorado), 385
New England, acid lakes in, 379
New Jersey, Hurricane Sandy's destruction in, 10, 195,
        196, 532
New Madrid Seismic Zone, 353
New Mexico, Aridisols in, 295
New Orleans, Louisiana, 193–195, 440
New South Wales, Australia, 69
New York, drumlins in, 463, 465
New York City:
    Hurricane Sandy's destruction in, 195, 196
    on maps, 28
New Zealand:
    biomes in, 245
    climate of, 212
    fjords in, 508
    rias in, 508
Next Generation Weather Radar (NEXRAD), 188
Niagara Falls, 425–426
Night, 47, 53–54
Nile River Delta, 289, 410, 411, 436
Niles, Kansas, 441–442
Nimbostratus clouds, 153, 155
Nitrate contamination, 301
Nitrogen, in atmosphere, 63
NOAA (National Oceanic and Atmospheric
        Administration), 34–35
Nonfoliated rocks, 320
Normal fault, 350–352
North America:
    acid rain in, 379, 380
    air masses over, 170, 171
    Alfisols in, 287
    biomes in, 245–247, 249, 250, 259
    climate of, 209, 212, 214, 215
    coastline changes in, 508
    forest in, 256
    passive margin of, 338
    surface temperatures in, 98–100
    unconventional oil and gas deposits in,
        560–562
    unconventional oil supplies development in,
        558–564
North Atlantic, atmospheric pressure over, 108, 109
North Atlantic gyre, 132
North Carolina, risk of coastal flooding in, 532, 533
North Cascades, 459
North Dakota:
    Bakken Shale in, 562, 564
    Mollisols in, 294
Northern Hemisphere:
    air flow around pressure systems in, 108
    annual surface temperature range in, 98–100
    climates in, 211, 214
    Coriolis force in, 112, 113
    defined, 15
    factors of large-scale atmospheric circulation in,
        115
    insolation in, 78–80
    ITCZ in, 123
    ocean currents in, 130–132
    permafrost in, 474
    seasons in, 47, 50–55
    Summer Solstice in, 50

Sun slope in, 252
    surface temperatures in, 98–100
North Head, New Zealand, 522
North Pacific garbage patch, 539–540
North Pacific gyre, 539–540
North Pole:
    daily insolation at, 78–80
    day length at, 79
    insolation at, 78, 79
Northwest Ordinance of 1787, 20
Norway, fjords in, 508
NRCS (Natural Resources Conservation Service), 301,
        302
Nunataks, 452

O

Obama, Barack, 195
Obsidian, 313
Occluded front, 178
Occluded stage (midlatitude cyclones), 178
Ocean(s), 506–507
    defined, 506
    sea-surface temperatures, 96–97
    temperature effects with, 96
    water stored in, 140, 141
Ocean floor (seafloor):
    and magnetic reversals, 337
    and plate movement, 336
Oceanic circulation, 130–134
Oceanic conveyor belt, 132
Oceanic crust, 306, 308, 336
    collision of two plates of, 341, 342
    collision with continental crust, 340–341
Oceanic trench, 341
Offshore, 519
Offshore bar, 519
Ogallala (High Plains) Aquifer, 395–400
O horizon, 281
Oil, 318
    global prices of, 559
    peak oil production, 559–561
Oil shales, 561, 562
Oil spill (2010), 530
Oil supply:
    developing unconventional sources for, 560–564
    economic geography of, 559
    and industrialization, 558
    known oil reserves, 560
Oklahoma:
    dust storms in, 501
    oil fields in, 558
Old Royal Observatory (England), 17, 18
Olympic Mountains, 5
Olympic National Park (Washington), 212, 258, 259
OPEC (Organization of the Petroleum Exporting
        Countries), 559
Open stage (midlatitude cyclones), 178
Orbital eccentricity, 224–225
Orbital precession, 225, 226
Orbit of Earth, 45–47, 224–226
Oregon:
    biomes in, 246
    coastal dunes along, 522
    Pacific Coast in, 518
Organic matter, in soils, 266, 267
Organic sedimentary rocks, 314, 318, 319
Organisms, soil formation and, 272–273
Organization of the Petroleum Exporting Countries
        (OPEC), 559
Orogeny, 344
Orographic uplift, 157–161
Outer Banks, North Carolina, 512
Outlet (streams), 410
Outwash plain, 463–464
Overgrazing, 260
Overthrust fault, 351, 352
Overthrust fold, 342
Overturned (recumbent) fold, 342
Oxbow lakes, 428, 429, 431
Oxic horizon, 287
Oxidation, 378

Oxisols, 284–287
Oxygen:
    in atmosphere, 63
    in photosynthesis, 237–239
Oxygen isotope ratios, 221–223
Ozone, in atmosphere, 66–69
Ozone hole, 67–69
Ozone layer, 66–68, 88

P

Pacific Island region, 533
Pacific Northwest:
    air masses over, 170
    avalanche hazards in, 387
    clear cutting in, 259
    coastal features of, 518
    dams in, 442
    forest in, 256
    glaciers in, 454
Pacific Ocean, 506
    air masses formed over, 170
    and El Niño, 133–134
    hurricanes in, 190
    ocean currents in, 131, 132
    surface water flow in, 133
    trash in, 539–540
Pacific Ring of Fire, 362–363
Padre Island, Texas, 530
Paired terraces, 434
Pakistan, monsoon flooding in, 438
Paleocene Epoch, 322, 323
Paleoclimatologists, 218
Paleozoic Era, 322, 328
Palm Beach, Florida, 530, 531
Palsas, 476
Palynologists, 218
Pangaea, 334, 335, 337
Parabolic dunes, 492, 493
Paraguay, Itaipu Dam and, 442
Parallels, 17. See also Latitude
Parana River, 442
Parent material:
    defined, 271
    residual, 271
    and soil formation, 271–272
    transported, 272
Particulates, in atmosphere, 69–70
Passive margins, 338
Passive solar heating, 80, 81
Past climates:
    causes of change in, 224–226
    reconstructing, 217–224
Patagonia, climate of, 209
Patagonia ice field (Chile), 453
Patterned ground, 476
Peaks, 421
Peak-cluster depressions, 405
Peak-forest plains, 405
Peak oil production, 559–561
Peat, 291
Peds, 277
Pediment, 483
Pedogenic processes, 268–271
Pedoturbation, 289
Pennsylvania, first oil well in, 318
Pennsylvanian Period, 322, 337
Perched dunes, 523
Perched water table, 395
Perennial streams, 410
Periglacial processes, 473–476
    defined, 473
    ground ice, 475–476
    permafrost, 474
Perihelion, 47
Perito Moreno Glacier (Argentina), 470
Permafrost, 474
Permian Period, 322
Petroleum. See also Oil supply
    economic geography of, 559
    formation of, 318

pH:
    defined, 278
    soil, 278, 279
Phanerozoic Eon, 322
Philippines, 539
Phoenix, Arizona, 541
Photosynthesis, 237–241
Physical geography, 5–11
    cause-and-effect relationships in, 9–10
    defined, 6
    defining, 5–8
    and geographical literacy, 10–11
    and human/environment interactions, 10
    overlap of other areas and, 4
Physical weathering, 372
Piedmont glaciers, 453
Pingos, 476
Pinnacles, 483
Pixel, 34
Place, in physical geography, 5
Plane of the ecliptic, 45
Planets, 44
Plant ecologists, 239
Plant geography, 234–262
    ecosystems, 236–237
    factors influencing, 251–255
    and global biomes, 241–251
    human influence on, 255–262
    and photosynthesis process, 237–241
Plant succession, 254, 355
Plateaus, 483
Plate boundaries, volcanic arcs at, 361–364
Plate convergence, 340–347
Plate divergence, 339–340
Plate movement, 335–338
    continental drift, 336–337
    passive margins, 338
    plate convergence, 340–347
    plate divergence, 339–340
    transform plate margins, 338–339
Plate tectonics, 334–338. See also Tectonic processes
    defined, 334
    lithospheric plates, 334–335
    plate movement, 335–338
Platy soil structure, 278
Playas, 483, 484
Pleistocene Epoch, 323, 324
    eustatic changes during, 507, 508
    glacial periods during, 466, 467, 470
    sea-level change in, 507
Pliocene Epoch, 323
PLSS (Public Land Survey System), 21
Pluto, 44, 45
Plutons, 311, 312
Pocket beaches, 515, 516
Podzolization, 290
Point bars, 428
Polar cell, 122
Polar circulation, 122–123
Polar (E) climates, 205, 206, 215–216
Polar easterlies, 123
Polar fronts, 119, 120
Polar front jet stream, 120, 121, 177
Polar High, 122–123
Polar ice sheets, 140, 141
Pollen records, 218–219
Pools, 428
Population growth, 538–541
    in Australia, 556–558
    in developing countries, 540
    in Las Vegas metropolitan area, 542–545
Pore ice, 475
Port Huron moraine (Michigan), 462
Potential evapotranspiration (potential ET), 203
Potential natural vegetation, 241
Power plants:
    hydroelectric, 440
    natural gas, 319
    solar, 80
PRC (Pseudo-Random Code), 37
Precambrian, 322, 466

Precipitation, 156–164
    in Australia, 555
    average annual, in U.S., 213
    and biomes, 242. See also individual biomes
    in climate classification, 204
    in Dust Bowl, 501
    processes of, 156–164
    types of, 156, 157
Pre-Illinoisan glaciations, 467
Pressure gradient force, 111–113
Primary succession, 254
Prime Meridian, 17–19, 47
Principal meridian, 20
Process, 6
Progradation, 519
Projection theory, 23
Proterozoic Eon, 322
Proxy data, 218
Pseudo-Random Code (PRC), 37
Public lands, overgrazing in, 260
Public Land Survey System (PLSS), 21
Pueblo cliff dwellings, 376–377
Puerto Rico, tower karst in, 405
Puget Sound, Washington, 5
Pulaski county, Georgia, 302
Purcell Trench Lobe, 468
Pu'u 'O'o Volcano (Hawaii), 360
P waves (primary waves), 349–350
Pyroclastic material, 361

Q

Qinling Mountains (China), 569
Qionglai Mountains (China), 567
Quartz, 310, 313, 315
Quaternary Period, 323
    carbon dioxide and average global temperatures in, 229–230
    climate fluctuations during, 218, 224–226
    eustatic sea-level changes in, 508
    glacial cycles of, 221–223
Queensland, Australia, 514
Quelccaya ice cap (Peru), 448

R

Radar, 36, 37
Radial dikes, 328
Radial drainage pattern, 413, 414
Radiation. See also Solar radiation
    defined, 70
    to and from Earth, 61
    electromagnetic, 60
    heat transfer, 70–71
    net, 76–79
Radiation budget:
    defined, 76
    global, 76–80
Radiation fog, 153
Radioactive isotopes, 324
Radioactivity, 324
Radiometric dating, 324
Radio waves, 60
Rain, 156
Rainbows, 66
Raindrop coalescence, 156, 157
Rain shadow, 160
Rain shadow effect:
    in arid/semi-arid regions, 480
    from Rocky Mountains, 247, 296
    from Sierra Nevada, 250, 543
Rajasthan, India, 32
Rappahannock River (Virginia), 417
Rattlesnake Creek (Kansas), 400
Rayleigh scattering, 73–74
Recessional moraine, 462
Rectangular drainage pattern, 413, 414
Recumbent fold, 342
Red Sea, 340
Reflection:
    defined, 73
    of solar radiation, 74–75

Region, in physical geography, 5
Regolith, 271–272
Rejuvenated streams, 426
Relative humidity, 143–145, 148
Relief:
 defined, 274
 local, 381
 and soil formation, 274
Remote sensing, 32–37
 and biosphere, 261
 in climate analysis, 217
 defined, 32
 in Mars studies, 496
 plate movement measured by, 336
Representative fraction (map scale), 27, 28
Republican River, 412
Reservoirs:
 behind dams, 440–444
 carbon, 227
Residual parent material, 271, 272
Respiration, 238, 239
Retrogradation, 517
Return period, 418
Reverse fault, 351, 352
Reversing dunes, 492, 493
Revetments, 530
Rhine River (Europe), 415
Rhône glacier, 470
R horizon, 281
Rhyolite, 313
Rias, 508
Richter scale, 350
Ridge and Valley Province (Appalachian range),
 344
Rifting, 339, 340
Rills, 421
Rime ice, 122
Riparian zone, 254, 355, 418
Rip-rap, 530
Riss glaciation, 507, 508
Rivers, 410. See also specific rivers
 channel flow, 416
 confluence of, 412
 and evolution of human societies, 410
 geographical concentration of, 7
 sources of, 410, 411
River floodplain, 421
River mouth, 436
Robinson, Arthur H., 26
Robinson projection, 26, 27
Roche moutonnée, 457, 458
Rock(s), 310–322
 defined, 310
 igneous, 311–313
 metamorphic, 318–320
 radioactive isotopes in, 324
 rock cycle, 320–322
 sedimentary, 313–318
 and weathering, 372
Rock cycle, 320–322, 325–328
Rockfall, 382–383
Rock outcrop, 311
Rock sea, 374
Rock structure:
 defined, 341
 and desert landforms, 482–483, 487
Rocky Mountains, 334
 avalanche hazards in, 387
 climate of, 216
 and continental divide, 412
 ice caps covering, 465
 Mollisols by, 292
 rain shadow from, 247, 296
 relief in, 274
 shale play in, 562
Root wedging, 374
Rossby waves, 120, 121
Rotation of Earth, 47–49
Rub al-Khali (Arabian Peninsula), 119
Runoff, 410, 411
Runout (avalanches), 387

Russia, 256
 biomes in, 246, 249, 250
 climate of, 215
 loess deposits in, 495
 Mollisols in, 292

S

Saddles (landform), 421
Saffir–Simpson scale, 191, 192
Sahara Desert (Africa), 5, 249, 256
 Aridisols in, 295
 sand dunes in, 491, 493
 as sand sea, 491
 and STH pressure system, 480
Sahel (Africa), 498–500, 549
Salic horizon, 296
Salinity, 507
Salinization:
 and Aridisol formation, 295, 296
 defined, 295
 soil, in arid and semi-arid lands, 548–558
Saltation, 486, 487
Salt-crystal growth, 375
San Andreas Fault, 13, 338, 352–354
San Clemente Island, California, 517
Sand dunes, 488–493
 classification of, 491–493
 coastal, 521–523
 formation of, 488–491
 on Mars, 496
San Diego, California, 97
Sand seas, 491
Sand sheets, 491
Sandstone, 315
San Francisco, California:
 earthquakes in, 353–354
 pocket beach in, 516
 sea fog in, 153, 156
 temperatures in, 94
 tidal range in, 511
Sangre de Cristo Mountains, 327–328
San Joaquin River, 552
San Joaquin Valley, California:
 precipitation in, 552
 soil salinization in, 552–555
 subsidence in, 400
San Juan River (Utah), 433
San Luis Drain, 553–554
Santa Ana winds, 130, 131
Saõ Paulo, Brazil, 228
SAR (Synthetic Aperture Radar), 36
Satellites:
 GPS, 37–38
 remote sensing, 33–37
Saturated soil, 267
Saturated zone(s):
 defined, 393
 recharging, 395
Saturation, 143
Saturation curve, 143
Scale, map, 22
Scandinavia, climate of, 215
Scarp, 352
Scattering, 73–74
Schist, 320
Scientific method, 8
Scotland, bluffs in, 515, 516
Scree, 382
Sea(s), 506–507
 defined, 506
 water contained in, 141
Sea arches, 517
Sea breezes, 128
Sea caves, 517
Seafloor, see Ocean floor
Seafloor spreading, 339
Sea fog, 153
Sea-level change, 231
 coastline impacts of, 532–533
 eustatic, 507, 508
 and glacial periods, 507

Seamounts, 366
Seasons:
 and air temperature, 90, 93
 and Earth's orbit, 47
 and Earth–Sun geometry, 50–53
 and insolation, 78–80
 and migration of pressure systems, 123–125
 and net radiation patterns, 78, 79
 Sun position changes in, 54
Seasonality, climate and, 203. See also individual
 climates
Sea stack, 518
Sea-surface temperatures, 96–97
Seawalls, 530–531
Secondary succession, 254
Sections (USPLS), 20, 22
Sediment:
 and dust pneumonia, 501
 eolian transport of, 485–487
Sedimentary rocks, 313–318
Sediment load, in streams, 422–427
Sediment transport, on coastlines, 512–514
Segregated ice, 475, 476
SEGS (Solar Energy Generating Systems), 81
Seismic processes, 348–353
Seismic waves, 306
Semi-arid and cold desert biome, 241, 242, 250
Semi-arid climates, 205, 206, 208–210, 273
Semi-arid deserts, 480
Semi-arid lands, soil salinization in, 548–558
Semi-arid steppe (BS) climates, 480
 cold midlatitude steppe climate (BSk),
 208, 209
 hot low-latitude steppe climate (BSh), 208, 209
Sensible heat, 74
Severe thunderstorms, 182–184
Shale, 315, 316
Shale play, 562
Shape of Earth, 44–45
Sheeting structure, 375
Sheet runoff, 411
Shelf cloud, 185
Shell Oil, 559
Shenandoah National Park, 37
Shield volcanoes, 364–365
Shooting stars, 88
Shoreline protection, 530–531
Shortwave radiation:
 defined, 60
 and radiation budget, 76
Shuttle Radar Topography Mission (SRTM), 37
Sial, 308
Siberia, 256
 air masses over, 170
 climate of, 215
 Polar High over, 123
Sierra Nevada, 160, 161
 frost wedging in, 373
 landscape of, 352
 precipitation along, 160
 rain shadow from, 250, 543
Silicate minerals, 312, 313
Sill, 311
Siltstone, 315
Sima, 308
Simpson Desert (Australia), 480
Singapore, smog in, 67
Sinkholes, 404, 405
Sioux Quartzite, 457, 458
Skjaldbreiður (Iceland), 364
Sky, colors in, 73–74
Skykomish River (Washington), 418
Slab avalanche, 386
Slash and burn agriculture, 256–258
Slate, 320
Sleet, 156
Slickensides, 289
Sling psychrometer, 144
Slip face (sand dunes), 490
Slope:
 defined, 251

Slope (cont.)
  on sand dunes, 490
  and soil erosion, 277
  and soil formation, 274
  and vegetation distribution, 251, 252
Slump, 384, 385
Slump terraces, 385
Small circles, 14, 15
Small-scale maps, 27–29
Smoky Hill River, 412
Snow, 156, 448
Snowball Earth, 466
Snowy Range (Wyoming), 253
SNWA (Southern Nevada Water Authority), 544, 547
Soil(s), 264–303
  chemical composition of, 278–280
  classification of, 283–299
  defined, 266
  factors in forming, 271–275
  human interactions with, 299–302
  measurable characteristics of, 275–278
  pedogenic processes, 268–271
  and photosynthesis process, 238–239
  properties of, 266–268
  soil profiles, 280–283
  soil science, 283
  water in, 392–393
Soil additions, 268–270
Soil conservation, 501–502
Soil Conservation Service, 302
Soil creep, 383–384
Soil depletion, 268, 269
Soil Erosion Service, 302, 501
Soil fertility, 299–301
Soil-forming factors, 271–275
  climate, 272, 273
  defined, 271
  organisms, 272–273
  parent material, 271–272
  relief, 274
  time, 274
Soil-forming processes, 268–271
Soil horizons, 280–284. See also specific soil orders,
    e.g.: Alfisols
  calcic horizon, 293
  defined, 280
  evolution of, 282–283
  oxic horizon, 287
  salic horizon, 296
Soil order(s), 283–300
  Alfisols, 284, 285, 289, 290
  Andisols, 284, 285, 299, 300
  Aridisols, 284, 285, 295–296
  categories of, 283, 284
  defined, 283
  Entisols, 284, 285, 296, 298
  Gelisols, 284, 285, 291–293
  Histosols, 284, 285, 291, 292
  Inceptisols, 284, 285, 298–299
  Mollisols, 284, 285, 292–295
  Oxisols, 284–287
  Spodosols, 29, 284, 285, 291
  Ultisols, 284, 285, 287–288
  Vertisols, 284, 285, 288, 289
Soil profile(s), 280–283
  defined, 280
  through time, 282–283
Soil properties, 266–268
Soil salinization:
  in ancient Sumeria, 550–551
  in arid and semi-arid lands, 548–558
  in Australia, 555–558
  current global extent of, 551–552
  defined, 549
  in San Joaquin Valley, California, 552–555
Soil science, 283
Soil structure, 277–278
Soil taxonomy, 283
Soil transformations, 268–270
Soil translocations, 268–270
Soil-water belt, 392

Soil–water budget, 268
Solar cells, 80
Solar constant, 61
Solar energy, in photosynthesis, 239
Solar Energy Generating Systems (SEGS), 81
Solar flares, 46
Solar noon, 53
Solar power, 80–81
Solar power plants, 80
Solar radiation, 46. See also Global energy system
  atmospheric flow of, 70–76. See also Energy
    flows
  insolation, 70
  interaction of Earth's surface and, 74–76
  and Sun angle, 44, 45
Solar system, 44
Solifluction, 383, 384
Solstice(s):
  and axial tilt, 50
  and day length, 55
  defined, 51
  position of ITCA during, 124
  Summer, 50–53, 55
  Winter, 50, 53, 55
Solum, 281
Soma, Japan, 359
Sonoran Desert (United States), 249
South Africa, biomes in, 249
South America:
  Alfisols in, 287
  arable land in, 549
  Aridisols in, 269
  biomes in, 244, 247, 249
  climate of, 207
  deforestation in, 256
  and ITCZ migration, 124, 125
  Mollisols in, 292
  surface temperatures in, 99
South Dakota, 288
Southeast Asia:
  arable land in, 549
  biomes in, 244
  climate of, 207
  monsoon in, 124
South Equatorial Current, 132, 133
Southern Alps (New Zealand):
  avalanche hazards in, 387
  climate of, 216
  landscapes of, 333
Southern Hemisphere:
  air flow around pressure systems in, 108
  annual surface temperature range in, 98–100
  climates in, 211, 214
  Coriolis force in, 112, 113
  defined, 15
  factors of large-scale atmospheric circulation in,
    115
  maritime effect in, 100
  ocean currents in, 132
  seasons in, 50–54
  Summer Solstice in, 50
  surface temperatures in, 99, 100
Southern Nevada Water Authority (SNWA), 544,
    547
Southern Ocean, 506
South Pacific Ocean, 528–529
Southwestern United States:
  alluvial fans in, 484–485
  Aridisols in, 295
  arroyos in, 484
  canyons in, 483
  desert geomorphology in, 481–485
  desert pavement in, 494
  droughts in, 547
  salt-crystal growth in, 375
  water issues in, 542–548
Spanish Peaks (Colorado), 326–328
Spatial analysis, 5
Spatial resolution, 34
Spearman, Texas, 187
Specific humidity, 143, 144

Spelunkers, 403
Spheroidal weathering, 378
Spits, 520–521
Splash erosion, 421
Spodic horizon, 290
Spodosols, 29, 284, 285, 291
Spring (vernal) Equinox, 51
Spring tides, 510, 511
Spurs, 421
Squall line, 180, 181
Sri Lanka, 358
SRTM (Shuttle Radar Topography Mission), 37
Stable air, 162, 163
Stalactites, 404
Stalagmites, 404
Stars, 44
Star dunes, 492, 493
Starting zone (avalanches), 387
Staten Island, New York, 532
Stationary fronts, 171
Stefan–Boltzmann law, 61
STH Pressure System, see Subtropical High Pressure
    System
Stock, 311
Storms:
  midlatitude cyclones, 173–180
  thunderstorms, 180–189
  tropical cyclones, 190–197
Stratocumulus clouds, 155
Stratopause, 88
Stratosphere, 85, 86, 88
Stratovolcanoes, 361
Stratus clouds, 153–155
Stream(s). See also Fluvial systems
  braided, 422–423
  and desert geomorphology, 484–485
  disappearing, 404, 405
  ephemeral, 410
  graded, 423, 425, 427, 434–436
  human interactions with, 438–444
  intermittent, 410
  meandering, 11, 422, 428–433
  origin of, 410–411
  perennial, 410
  rejuvenated, 426
  sediment load of, 422–427
  trunk, 412
Stream competence, 434
Stream discharge, 416–419, 442
Streamer (lightning), 182
Stream gauge, 416
Stream-gauging stations (U.S.), 7, 416
Stream gradation, 422–427
Stream hydrograph, 417
Stream hydrology, 416
Stream ordering, 414–415, 435
Stream power, 434
Stream valleys, 427–438
Strike-slip faults, 351–353
Subarctic climates (Dfc, Dwc, Dwd), 215
Subduction, 341
Sublimation, 140
Subsidence, 400
Subsolar point:
  and axial tilt, 50
  defined, 45
  seasonal migration of, 52
  and seasons, 54
Subtropical High (STH) Pressure System, 118,
    119
  and arid/semi-arid regions, 480
  and climate, 203
  and desertification, 498
Suburbanization, 559
Succulents, 250
Sumatra, coral bleaching in, 533
Sumatra-Andaman earthquake, 354, 355
Sumatra-Andaman tsunami, 355, 357
Sumatra Fault, 354
Sumeria, ancient, soil salinization in, 550–551
Summer monsoon, 125

Summer Solstice, 50–53
  and axial tilt, 50
  Sun position at, 51, 55
Sun, 46
  energy flows from, 6
  in solar system, 44
  and tides, 508–511
Sun angle, 44, 45, 56
  angle of incidence, 74
  defined, 44
  in passive solar heating, 80
  at Summer Solstice, 51, 55
Sun Belt, 544
Sunspots, 46
Sun-synchronous orbits, 34
Supercell thunderstorms, 184–185
Supercontinent, 334
Surface temperature, 89, 96
  annual range of, 98–100
  and climate change, 230, 231
  and land skin temperature, 95
  in rural vs. urban landscapes, 101
Surface tension (water), 138
  defined, 267
  holding water to soil particles, 267, 268
Surface waves (seismic), 350
Surging, 451
Suspended load, 422
Suspension, 486, 487
Sustainability:
  in agriculture, 558
  and carrying capacity, 538
S waves (secondary waves), 349–350
Swiss Alps, 346
  climate of, 216
  formation of, 346
  glaciers in, 447, 470
  Matterhorn, 459, 460
Switzerland, limestone bed in, 379
Synclinal valley, 346
Syncline, 342
Synthetic Aperture Radar (SAR), 36
Systems theory, 6

T

Taconic Orogeny, 344
Taklamakan Desert (China), 95, 480
Taliks, 474
Talus, 382, 383
Talus apron, 383
Talus cones, 383
Tarns, 458, 459
Tasmania, Australia, 515
Technological development, population growth and, 538–539
Tectonic processes, 332–368
  earthquakes, 348–359
  passive margins, 338
  plate convergence, 340–347
  plate divergence, 339–340
  plate tectonics, 334–338
  transform plate margins, 338–339
  volcanoes, 360–368
Temperature. *See also* Air temperature; Global temperature patterns; Surface temperature
  and amount of electromagnetic radiation, 61
  and biomes, 242. *See also individual biomes*
  in climate classification, 204, 205. *See also individual climates*
  of Sun, 46
  and wavelength of electromagnetic radiation, 60
  and weathering, 374
Temperature inversions, 87–88
Temperature scales, 90, 91
Temporal lag, 93
Terminal moraine, 462
Tertiary Period, 323
Tescott, Kansas, 441–442
Texas:
  interaction of waves and shoreline, 514
  jetty on coast of, 532

oil fields in, 558, 559
  water table in, 399
Texture, soil, 276, 277
Thailand, risk of coastal flooding in, 532
Thalweg, 428
Thames River (London), 409
Thematic maps, 22
Thermohaline circulation, 132
Thermosphere, 86, 88–89
Thomson, William, Lord Kelvin, 90
Thornthwaite, C. Warren, 203
Thornthwaite climate system, 203
Three Gorges Dam (China), 442
Threshold point, 381
Thunder, 184
Thunderstorms, 180–189
  along cold front, 173
  clouds associated with, 185
  defined, 172
  evolution of, 180–182
  severe, 182–184
  tornadoes, 184–189
Tides, 508–511
Tidewater glaciers, 453
Tilt obliquity, 225
Time:
  geologic, 322–329
  and soil evolution, 282–283
  and soil formation, 274
Time of day:
  and air temperature, 93, 94
  and relative humidity, 145
Times Square, New York City, 228
Time zones, 47–49
Toe (slump), 385
Tokyo, Japan, 101
Tombolo, 521
Tools of geographers, 12–41
  digital technology, 32–40
  geographic grid, 14–22
  maps, 22–31
Topeka, Kansas:
  temperatures in, 94
  tornado in, 187
Topographic maps, 30
Topographic winds, 129–130
Topography:
  and climate, 202, 203
  defined, 30
Tornadoes, 184–189
  classification of, 186
  cyclones vs., 174
  distribution of, 187
  evolution of, 186
  monitoring of, 188–189
Tornado Alley, 187
Tornado outbreak, 188
Tornado warning, 188
Tornado watch, 188
Tower karst, 404
Tohoku earthquake, 357–368
Townships, 20
Track (avalanches), 387
Trade winds, 118
Transformations, soil, 268–270
Transform faults, 352
Transform plate margins, 338–339
Transgressive dune fields, 521–522
Translocations, soil, 268–270
Transpiration, 148
Transport, eolian, 485–488
Transported parent material, 272
Transverse ridges, 491, 492
Tree line, 253
Tree ring patterns, 218–221
Trellis drainage pattern, 413, 414
Triangulation, 38
Tributaries, 412
Triple junction, 340
Tri-State Tornado, 187
Tropical circulation, 116, 118

Tropical (*A*) climates, 205–208
Tropical cyclones, 190–197
Tropical deciduous forest and scrub biome, 241, 242, 244–245
Tropical depression, 190
Tropical easterlies, 118
Tropical forest biomes, 243, 257
Tropical monsoon climate (*Am*), 207
Tropical rainforest biome, 241–244, 257
Tropical rainforest climate (*Af*), 207
Tropical savanna biome, 241, 242, 247
Tropical savanna climate (*Aw*), 207, 208
Tropical storm, 190
Tropic of Cancer, 16, 53
Tropic of Capricorn, 16, 53
Tropopause, 85, 87
Troposphere, 86–88
Trunk stream, 412
Tsunamis, 355
  along Japanese coast, 357–359
  defined, 355
  in Indian Ocean, 356–359
  Sumatra-Andaman, 355, 357
Tulare Lake, 552
Tundra biomes, 241–242, 250–251
Tundra climate (*ET*), 215–216
Typhoons, 190. *See also* Hurricane(s)
Tyrannosaurus Rex, 322

U

Ubac slope, 252
Ultisols, 284, 285, 287–288
Uluru (Australia), 481
United Nations Environmental Programme (UNEP), 230
United States. *See also specific regions*
  acid rain in, 379–380
  annual precipitation in, 213
  Aridisols in, 295
  artificial levees in, 439
  biomes in, 246, 249
  climates of, 209
  coastal dune fields in, 522
  coastal population in, 529
  dams in, 442
  destructive hurricanes in, 192–193. *See also individual storms*
  drought in, 164–166
  energy consumption in, 228
  first oil well in, 318
  glaciers in, 454
  and greenhouse emissions, 231–232
  human footprints in biomes, 256
  industrialization in, 540
  karst landscapes in, 402
  loess deposits in, 495
  major earthquakes in, 353
  major watersheds in, 413
  Mollisols in, 295
  and North Pacific garbage patch, 539
  oil supply in, 559, 560
  risk of coastal flooding in, 532
  rivers in, 7
  solar energy potential in, 81
  stream-gauging stations in, 7
  surface temperatures in, 99
  temperatures in, 229
  time zones in, 47, 49
  tornadoes in, 5, 187–188
  unconventional oil resources in, 561, 562
  Utisols in, 287
  vegetation in, 255
  wind energy in, 126
U.S. Department of the Interior, 20
U.S. Drought Monitor, 165, 166
United States Army Corps of Engineers, 439, 442, 443
United States Geological Survey (USGS):
  Landsat system, 34
  and San Joaquin Valley groundwater, 554
  stream-gauging stations of, 7, 416
United States Public Land Survey (USPLS), 19–22

Universal Time Coordinated (UTC), 47–48
Universe, 44
Unloading, 375
Unpaired terraces, 434
Unsaturated zone, 392
Unstable air, 162, 163
Updrafts, 184
Upper mantle, 306–308
Upwelling current, 132, 133
Urban heat islands, 101–102
Urbanization, fluvial systems and, 438–439
Urban sprawl, 541
Ururashraju Glacier (Peru), 470
USGS, see United States Geological Survey
USPLS (United States Public Land Survey), 19–22
Utah:
    biomes in, 249, 250
    monocline in, 342, 343
UTC (Universal Time Coordinated), 47–48

V

Valle de Luna (Valley of the Moon), 20
Valleys:
    anticlinal, 345
    hanging, 460
    stream, 427–438
    synclinal, 346
Valley breezes, 129
Valley glaciers, 448, 453, 454
Vancouver, British Columbia, 98
Variable gases, 63–69
Vegetation:
    and climate, 241. See also Biome(s)
    climax, 254
    distribution of, see Plant geography
    potential natural, 241
    and soil salinization, 549–550
Venezuela, oil reserves in, 559
Venice, Italy, 400
Ventifacts, 487, 488
Venus:
    carbon dioxide in atmosphere of, 65
    mapping, 37
Vernal equinox, 51
Vertical zonation, 252–253
Vertisols, 284, 285, 288, 289
Vishnu Schist, 326
Visible light, 60
Volcanic activity, climate change caused by, 224
Volcanic arcs, 361–364
Volcanic glass, 313
Volcano(es), 360–368
    defined, 360
    explosive, 361–364
    fluid, 364–365
    hotspots, 365–368
Volga River, 417

W

Wales, sea arches in, 517
WAR (wet adiabatic lapse rate), 151
Warm fronts, 171–172
Warm sector (midlatitude cyclones), 178
Warren Dunes State Park (Lake Michigan), 523
Washington, D.C.:
    infrared satellite image of, 40
    latitude and longitude of, 19
Washington state, coastal dune fields in, 522

Water. See also specific topics, e.g.: Fluvial systems
    in American Southwest, 542–548
    and arid/semi-arid regions, 480
    in atmosphere, 136. See also Atmospheric
        moisture
    chemical composition of, 138
    groundwater, 390–402
    hydrogen bonding, 138
    hydrosphere, 138–140
    hygroscopic, 392
    oceans, 96–97, 140, 141, 506–507
    in photosynthesis, 237–238
    seas, 506–507
    in soils, 267–268
    thermal properties of, 138–140
Water depletion, 398–400
Water-level fluctuation, 507–508
Water rights, in American Southwest, 543–544
Watersheds, 411–413
    drainage density of, 414
    in the U.S., 413
Water table, 395
    perched, 395
    and soil salinization, 549
Water vapor, 138
    in atmosphere, 64
    atmospheric circulation and movement of, 125
    in photosynthesis, 239
    and urban vs. rural temperatures, 101
Water wars, 548
Waves, 511–512
Wave amplitude:
    electromagnetic waves, 60
    water waves, 511
Wave base, 511
Wave crest, 511
Wave-cut bluffs, 515–517
Wave-cut platforms, 517
Wave height, 511
Wavelength:
    electromagnetic waves, 60
    of water waves, 511
Wave refraction, 515
Wave trough, 511
Weather:
    defined, 202
    and thermal properties of water, 139
Weathering, 372–381
    and acid rain, 379–380
    chemical, 375–379
    defined, 372
    differential, 372
    mechanical, 372–375
Wegener, Alfred, 334, 335
Westerlies, 119–120
Westerly winds, 120
Western United States:
    arable land in, 548
    canyons in, 483
    overgrazing in, 260
    upright landforms in, 482
Wet adiabatic lapse rate (WAR), 151
White Mountains (California), 220, 252
Wicking, 296
Wilcox county, Georgia, 302
Wildfires, in California, 130, 131
Wilting point, 268, 392
Wind(s), 110. See also Eolian processes
    Chinook, 129

    and Coriolis force, 112–114
    directions of, 109
    erosion by, 485–487
    and evaporation, 148, 149
    fluid behavior of, 485
    foehn, 129, 130
    and frictional forces, 114–115
    geostropic, 114
    in hurricanes, 191
    in Intertropical Convergence Zone, 118
    katabatic, 129
    land-sea wind systems, 128–129
    and littoral processes, 512–513
    local wind systems, 128–130
    in midlatitudes, 119–120
    mistral, 130
    monsoonal, 124–125
    and pressure gradient force, 111–113
    Santa Ana, 130, 131
    topographic, 129–130
    trade, 118
    westerly, 120
Wind abrasion, 487
Windbreaks, 501
Wind chill, 90, 92
Wind energy, 125–128
Windmills, 126
Wind patterns, and heating of land surfaces, 110–111
Windward side, 129, 159
Winter monsoon, 125
Winter Solstice, 53
    and axial tilt, 50
    defined, 53
    Sun position at, 55
Wisconsin, drumlins in, 463
Wisconsin glaciation, 466, 467, 470, 507
WMO (World Meteorological Organization), 230
Wolong Giant Panda Reserve Center, 567
Wolong Nature Reserve (China), 566–570
World Meteorological Organization (WMO), 230
World Wildlife Fund (WWF), 569
Wrangell-St. Elias National Park (Alaska), 448
Written (verbal) scale, 27, 28
WWF (World Wildlife Fund), 569
Wyoming:
    groundwater contamination from fracking in, 563
    riparian vegetation in, 255

X

Xerophytic plants, 249, 250

Y

Yardangs, 487, 488
Yellowstone hotspot (Wyoming), 366–367
Yellowstone National Park, 366, 367
Yosemite Falls, 460
Yosemite National Park (California), 375, 376
Yuma, Arizona, 300

Z

Zonal flow, 121
Zone of ablation (or wastage), 449, 450
Zone of accumulation, 449